Social Ecology

Human-Environment Interactions

VOLUME 5

More information about this series at http://www.springer.com/series/8599

Helmut Haberl · Marina Fischer-Kowalski
Fridolin Krausmann · Verena Winiwarter
Editors

Social Ecology

Society-Nature Relations across Time and Space

Springer

Editors
Helmut Haberl
Institute of Social Ecology
Alpen-Adria University
Vienna
Austria

Fridolin Krausmann
Institute of Social Ecology
Alpen-Adria University
Vienna
Austria

Marina Fischer-Kowalski
Institute of Social Ecology
Alpen-Adria University
Vienna
Austria

Verena Winiwarter
Institute of Social Ecology
Alpen-Adria University
Vienna
Austria

ISBN 978-3-319-33324-3 ISBN 978-3-319-33326-7 (eBook)
DOI 10.1007/978-3-319-33326-7

Library of Congress Control Number: 2016941606

Printed on acid-free paper

This Springer imprint is published by Springer Nature
The registered company is Springer International Publishing AG Switzerland

Foreword I

It is widely acknowledged that global resources are becoming increasingly scarce due to growing human populations, their increasing wealth and changing environmental conditions. Not surprisingly, this challenge to the continued well-being of people on earth has attracted the interest of researchers from a number of disciplines. A growing number of research groups, institutes, and scientific journals have emerged to meet the need for knowledge about the complex interrelation between social and environmental factors of the earth system.

A common denominator of many of the statements emerging from these scientific communities—not least circles that address human–environment interactions—is a persistent claim that more thorough insights can be gained if different scientific disciplines join forces in their efforts to analyze the problems in question.

However, the challenge of bridging the Natural/Social Science divide is huge, as has been experienced in many research efforts targeted at, for example, sustainable natural resource management in human-dominated ecosystems. Even when defining basic labels such as 'interdisciplinary research,' 'multidisciplinary research,' and 'transdisciplinary research,' there seems to be a lack of a common agreement on the vocabulary. Almost all researchers seem to have their own individual definitions. It is clearly beyond the intention of this short preface to provide a thorough summary of this debate. Nevertheless, it seems to be understood that interdisciplinary research goes further than just working together on a given theme—it implies the existence of a common systemic or conceptual framework that enables the disciplines to learn from and contribute to one another (McNeill 1995, p. 1; Price 1990).

Central to many discussions of obstacles to interdisciplinary research are the epistemological and cultural differences between the Natural and Social Sciences, a divide exposed by the scientist Charles Percy Snow in his influential 1959 Reed Lecture presented at Cambridge. This has remained an important concern, and no definitive answers to the question of how best to overcome the difficulties have been provided. However, the increasing motivation to analyze human–environment relationships in a broader context has fostered a number of enlightening studies, many of which have been related to larger programmatic efforts.

These efforts share the paradigm that human social and natural systems interact, coevolve over time and have substantial impacts upon one another. The pathways of change of the coupled human–environment systems are formed by continuous, dynamic interactions among numerous changing factors, social as well as biophysical, which enable and constrain human choices of resource management strategies.

Early advancements were, for example, anchored in the UNESCO Man and the Biosphere (MAB) Programme, established in 1971 to promote interdisciplinary approaches to management, research and education in sustainable use of natural resources. Later, specifically in the 1990s, a large number of international and national research initiatives promoted interdisciplinary approaches as crucial to research on resource management and sustainability. For example, research programs launched by the International Geosphere-Biosphere Programme (IGBP) and the International Human Dimensions Programme on Global Environmental Change (IHDP) have helped develop conceptual frameworks in support of interdisciplinary mindsets. The scholarly recognition of 'sustainability science' in recent decades—in terms of scientific journals in their own right as well as pillars of recognized, well-established journals such as the *Proceedings of the National Academy of Sciences of the United States of America* (PNAS)—further underlines the growing recognition of interdisciplinary research.

It is widely acknowledged that it is much safer career-wise to work solely within the confines of some component problem in a single-disciplinary context than to deal with complex multidisciplinary linkages that require time and effort to overcome differences in vocabularies, problem-solving frameworks, and accepted standards of research.

Against this background, this book is unique. It presents a series of outcomes from a research environment supporting a group that has been remarkably persistent and successful in pursuing an interdisciplinary research agenda with a shared conceptual platform. The Vienna Social Ecology School was established in 1986 as a small group within what was then called the 'Interuniversity Research Institute for Distance Education' (IFF). Based on their common platform, the researchers from this internationally well-connected group deal with materials, energy, society, land use, and food production—together termed the metabolism of societies—and the environmental impacts of this metabolism. By sharing theoretical approaches, creating a consistent metric for material and energy flows for social systems and developing their large databases, the researchers in this group have contributed successfully to the advancement of science within the field of society–nature coevolution pertaining to history, to current development processes and to a future sustainability transition.

I am convinced that this book, which presents examples of the research group's collective outcomes, will be read with interest by a wide range of researchers with interest in human environmental interaction and pathways toward sustainability.

Copenhagen, Denmark, Anette Reenberg
September 2014

References

McNeill, D. (1995). *The challenge of interdisciplinary and policy-oriented research*. Working Paper 1995.8. Centre for Development and the Environment (SUM), University of Oslo.

Price, M.F. (1990). Humankind in the biosphere. The evolution of international interdisciplinary research. *Global Environmental Change, 1*(1), 3–13.

Foreword II

For more than a quarter century, the Institute of Social Ecology in Vienna has been a trail-blazer in research on the human–environment interface. One hears continual calls for greater interdisciplinary and transdisciplinary research on environmental issues, but such calls often lead to modest and sometimes merely rhetorical responses. In contrast, the scholars at the Institute of Social Ecology have been demonstrating how to conduct such research and the resulting benefits since its foundation in 1986. Established and long directed by Marina Fischer-Kowalski, originally trained as a sociologist, and now directed by (human) ecologist Helmut Haberl, the Institute has served as home for an extremely talented and diverse set of scholars from numerous disciplines in the Natural and Social Sciences. The resulting hotbed of interdisciplinary and transdisciplinary work has generated an enormous amount of important scholarship that provides both intellectual and practical insights into the relationships between human societies and their environments. This volume provides an excellent summary and synthesis of the wide range of work produced at the Institute of Social Ecology and a superb introduction to the 'Vienna School of Social Ecology.'

Vienna Social Ecology draws from but extends a range of ecological fields from several disciplines, including Cultural Ecology, Human Ecology, Environmental History, Industrial Ecology, and Ecological Economics. Social Ecology shares with these diverse fields a basic paradigm that recognizes that social and natural systems interact, coevolve over time, and impact one another in substantial ways that can be both beneficial and harmful. But it goes beyond these fields with a more sophisticated conceptualization of socioecological systems that captures the complexities of societies, their biophysical environments *and* the interactions between the two. Employing concepts such as social metabolism as the link between societies and their environments, colonization as the process by which societies modify their environments to meet their needs and socioecological transitions as crucial historical transformations in human use of the environment (termed sociometabolic regimes), the Vienna School provides a powerful set of lenses for viewing the relationships between societies and their environments.

Fischer-Kowalski, Haberl, and their many colleagues couple these insightful theoretical and conceptual innovations with path-breaking work that operationalizes and empirically examines the multifaceted phenomena involved in socioecological relations. Their work is nicely captured in this volume, which first lays out some of the key conceptual and theoretical premises of the Vienna School and then illustrates the value of its approach through numerous examples of empirical studies. These studies ably illustrate the value and flexibility of Social Ecology. The focus of these studies ranges from contemporary issues such as human health to long-term analyses of land and energy use and from local concerns such as the Vienna sewage system to the global economy.

The totality of this work provides rich insights into the complex origins and nature of what we conceive of as 'environmental problems' at various scales, and these insights further our understanding of 'sustainability' and the difficulties in achieving it. This work offers guidance for both social scientists and natural scientists involved in ecological research, especially in the domains of Coupled Human and Natural Systems, Long-Term Ecological Research (which the Vienna School suggests is better conceptualized as Long-Term *Socioecological* Research), and Sustainability Science. This volume, and the Institute of Social Ecology more generally, provides exemplars for these and related efforts to execute truly interdisciplinary and transdisciplinary research on ecological problems and sustainability.

In its relatively brief history, the Institute of Social Ecology has accomplished a great deal, helping to advance our understanding of ecological problems and sustainability with powerful conceptual and empirical work, as documented in this volume. In addition, the Fischer-Kowalski/Haberl team has trained a large number of highly talented researchers (well represented in this volume), several of whom have taken influential positions in a wide range of other institutions. This diffusion of socioecological competence, coupled with the continued vitality of the Institute of Social Ecology itself, ensures that the work of the Vienna School of Social Ecology will continue to gain influence. The seeds planted by Marina Fischer-Kowalski have not only taken root but are now yielding fruit and will do so for generations to come—to the benefit of those generations.

Stillwater, Oklahoma Riley E. Dunlap
October 2014

Contents

Foreword I v
Anette Reenberg

Foreword II ix
Riley E. Dunlap

Authors and Contributors xv

Abbreviations xxv

List of Figures xxxiii

List of Tables xxxix

Introduction xli
Helmut Haberl, Karl-Heinz Erb, Marina Fischer-Kowalski, Robert Groß, Fridolin Krausmann, Christoph Plutzar, Martin Schmid and Verena Winiwarter

Part I The Conceptual Repertoire

1 **The Archipelago of Social Ecology and the Island of the Vienna School** 3
Marina Fischer-Kowalski and Helga Weisz

2 **Core Concepts and Heuristics** 29
Marina Fischer-Kowalski and Karl-Heinz Erb

3 **Transitions in Sociometabolic Regimes Throughout Human History** 63
Fridolin Krausmann, Helga Weisz and Nina Eisenmenger

4 **Beyond Inputs and Outputs: Opening the Black-Box of Land-Use Intensity** 93
Karl-Heinz Erb, Tamara Fetzel, Helmut Haberl, Thomas Kastner, Christine Kroisleitner, Christian Lauk, Maria Niedertscheider and Christoph Plutzar

5 **'Society Can't Move So Much As a Chair!'—Systems, Structures and Actors in Social Ecology** 125
Daniel Hausknost, Veronika Gaube, Willi Haas, Barbara Smetschka, Juliana Lutz, Simron J. Singh and Martin Schmid

6 **Why Legacies Matter: Merits of a Long-Term Perspective** 149
Verena Winiwarter, Martin Schmid, Helmut Haberl and Simron J. Singh

7 **Toward a Socioecological Concept of Human Labor** 169
Marina Fischer-Kowalski and Willi Haas

Part II Empirical Approaches to Socioeconomic Metabolism

8 **Long-Term Trends in Global Material and Energy Use** 199
Fridolin Krausmann, Anke Schaffartzik, Andreas Mayer, Nina Eisenmenger, Simone Gingrich, Helmut Haberl and Marina Fischer-Kowalski

Method Précis: Energy Flow Analysis 212
Helmut Haberl

9 **More Than the Sum of Its Parts: Patterns in Global Material Flows** 217
Andreas Mayer, Anke Schaffartzik, Fridolin Krausmann and Nina Eisenmenger

Method Précis: Material Flow Analysis 234
Nina Eisenmenger

10 **Boundary Issues: Calculating National Material Use for a Globalized World** 239
Anke Schaffartzik, Nina Eisenmenger and Dominik Wiedenhofer

Method Précis: Life Cycle Assessment 253
Michaela C. Theurl and Anke Schaffartzik

Method Précis: Input-Output Analysis 257
Anke Schaffartzik

11 **How Circular Is the Global Economy? A Sociometabolic Analysis** 259
Willi Haas, Fridolin Krausmann, Dominik Wiedenhofer and Markus Heinz

12 Material Stocks and Sustainable Development 277
Dominik Wiedenhofer, Willi Haas, Michael Neundlinger and Nina Eisenmenger

Part III Empirical Approaches to Land Use and Colonization of Ecosystems

13 Livestock Grazing, the Neglected Land Use 295
Karl-Heinz Erb, Tamara Fetzel, Thomas Kastner, Christine Kroisleitner, Christian Lauk, Andreas Mayer and Maria Niedertscheider

Method Précis: Using Geographic Information Systems in Social Ecology 311
Christoph Plutzar, Christine Kroisleitner, Tamara Fetzel and Karl-Heinz Erb

14 Systemic Feedbacks in Global Land Use 315
Helmut Haberl, Karl-Heinz Erb, Thomas Kastner, Christian Lauk and Andreas Mayer

Method Précis: Human Appropriation of Net Primary Production (HANPP) 332
Helmut Haberl

15 A Burning Issue: Anthropogenic Vegetation Fires 335
Christian Lauk and Karl-Heinz Erb

16 How Far Does the European Union Reach? Analyzing Embodied HANPP 349
Helmut Haberl, Thomas Kastner, Anke Schaffartzik and Karl-Heinz Erb

17 Africa's Land System Trajectories 1980–2005 361
Maria Niedertscheider, Tamara Fetzel, Helmut Haberl, Fridolin Krausmann, Veronika Gaube, Simone Gingrich, Christian Lauk, Christoph Plutzar and Karl-Heinz Erb

18 Of Birds and Bees: Biodiversity and the Colonization of Ecosystems 375
Christoph Plutzar, Karl-Heinz Erb, Veronika Gaube, Helmut Haberl and Fridolin Krausmann

Part IV Empirical Approaches to Long-Term Socioecological Research

19 Long-Term Risks of Colonization: The Bavarian 'Donaumoos' 391
Martin Schmid

Method Précis: Working with Historical Material 411
Verena Winiwarter

20 A Forest Transition: Austrian Carbon Budgets 1830–2010 417
Simone Gingrich, Christian Lauk, Thomas Kastner, Fridolin Krausmann, Helmut Haberl and Karl-Heinz Erb

Method Précis: Carbon Accounting 428
Karl-Heinz Erb

21 From Energy Source to Sink: Transformations of Austrian Agriculture 433
Fridolin Krausmann

22 The Philippines 1910–2003: A Century of Transitions 447
Thomas Kastner, Karl-Heinz Erb and Fridolin Krausmann

23 How Tourism Transformed an Alpine Valley 459
Robert Groß

24 Cleaning a Metropolis: The History of Vienna's Sewage System 475
Sylvia Gierlinger and Michael Neundlinger

Part V Empirical Approaches to Working with Stakeholders

25 Planning, Residential Decisions and Energy Use in Vienna 489
Veronika Gaube, Alexander Remesch and Barbara Smetschka

Method Précis: Agent-Based Modeling 501
Alexander Remesch and Veronika Gaube

26 Time Use, Gender and Sustainable Agriculture in Austria 505
Barbara Smetschka, Veronika Gaube and Juliana Lutz

Method Précis: Functional Time Use Analysis 519
Lisa Ringhofer and Marina Fischer-Kowalski

27 Complex Disasters on the Nicobar Islands 523
Simron J. Singh and Willi Haas

Method Précis: A Methodological Guide to Local Studies 539
Simron J. Singh and Willi Haas

28 Island Sustainability: The Case of Samothraki 543
Panos Petridis and Marina Fischer-Kowalski

Method Précis: Transdisciplinary Research 555
Willi Haas

29 Health Through Socioecological Lenses—A Case for Sustainable Hospitals 559
Ulli Weisz and Willi Haas

Perspectives on Social Ecology: Learning for a Sustainable Future 577
Verena Winiwarter

Index 591

Authors and Contributors

About the Authors

Riley E. Dunlap is Regents Professor and Laurence L. and Georgia Ina Dresser Professor at the Department of Sociology at Oklahoma State University and a specialist in the field of Environmental Sociology. A Fellow of the American Association for the Advancement of Science and the American Psychological Association, he is also past President of the International Sociological Association's Research Committee on Environment and Society. Currently he chairs the American Sociological Association's Task Force on Sociology and Global Climate Change and has served on the Advisory Board of the Institute of Social Ecology since 2004.

Nina Eisenmenger (Master's in Human Ecology and Environmental Economics, PhD in Social Ecology) is Assistant Professor at the Institute of Social Ecology, Alpen-Adria University Klagenfurt Wien Graz (Austria) and coordinates the thematic area Social Metabolism at the Institute. Her research focus is on material flow accounting (MFA), the sustainable use of resources and resource efficiency, and international trade, structures of the world economy, and the link between economic and biophysical accounting systems. She is involved in various national and international research projects and has been the consultant on material flow analysis, sustainable use of resources and resource productivity for Eurostat; the European Commission; Statistics Austria; the Austrian Federal Ministry of Agriculture, Forestry, Environment and Water Management; and the OECD.

Karl-Heinz Erb is associate professor of Land Use and Global Change at Alpen-Adria University (Austria). His research focuses on interactions between humans and global environmental systems, with a particular focus on Global Land-System Science. He is analyzing changes in land use and land cover and their consequences for ecosystem structures and functioning, such as carbon stocks and flows from the national to the global scale. In 2010, he was awarded an ERC Starting Independent Researcher Grant by the European Research Council for the project 'Land Use Intensity from a Socio-Ecological Perspective'.

Tamara Fetzel holds a Master's in Environmental System Sciences with special focus on Geography and is a PhD candidate at the Institute of Social Ecology (SEC), Vienna, Alpen-Adria-University (Austria). In her research, she focuses on land system change on grasslands, with a particular focus on uncertainties related to global datasets.

Marina Fischer-Kowalski is professor at Alpen-Adria University and founded the Vienna-based Institute of Social Ecology. Her background is in Sociology (PhD at the University of Vienna). She has taught at Griffith (Australia), Roskilde (Denmark), Yale University (USA) and the Universidad Federal de Rio de Janeiro (Brazil). Currently, she is President of the International Society for Ecological Economics. As an expert member of UNEP's International Resource Panel, she became lead author of its publication 'Decoupling Resource Use and Environmental Impacts from Economic Growth' (2011). She is interested in social metabolism across history on global, national and local scales, and she investigates how social metabolism relates to social dynamics (quality of life, equity, division and quality of labor, time use) and to environmental change. She works on transitions in complex systems and engages in transdisciplinary sustainability research. She has published approximately 300 articles. Her most recent book (2007) was *Socioecological Transitions and Global Change: Trajectories of Social Metabolism and Land Use*.

Veronika Gaube (Master's in Ecology, PhD in Social Ecology) works at the Institute of Social Ecology, Vienna, Austria. Her research interests include Land-Use Science, substance (nitrogen) flows of urban and rural regions and integrated modeling of socioecological energy. She has experience with the development of sustainability indicators, participatory modeling, modeling techniques (such as agent-based modeling) and GIS in inter- and transdisciplinary research. She teaches competencies for sustainability indicator development, modeling techniques and agricultural studies.

Sylvia Gierlinger holds a Master's in Social and Human Ecology and is a PhD researcher at the Institute of Social Ecology (Alpen-Adria University) in Vienna (Austria). Her research focuses on urban metabolism, Long-Term Socioecological History and Environmental History. Currently, she is part of the URBWATER project team investigating the long-term dynamics of city-river relations during industrialization.

Simone Gingrich holds a PhD in Social Ecology and an MSc in Ecology. She is senior researcher and lecturer at the Institute of Social Ecology, Vienna, and member of the Center for Environmental History (ZUG), Vienna. She works in the field of Long-Term Socioecological Research (LTSER) and focuses on empirical analyses of environmental change during industrialization in Europe. Using historical sources, she generates long time series of various environmental indicators at local, regional and national scales. Her work is inspired by the idea that today's sustainability problems can only be adequately addressed if understood in their historical context.

Robert Groß holds a Master's in Interdisciplinary Environmental Science (University of Vienna) and is a doctoral candidate at the Institute of Social Ecology (Alpen-Adria University). He studies Alpine communities and their transformation into winter tourism destinations during the 20th century. His research focuses on the impact of large technical systems (LTSs) on local landscapes and how both were shaped according to rationalization efforts, changing snow conditions and economic decisions. In 2012, he published his first book, *Wie das 1950er Syndrom in die Täler kam: Umwelthistorische Überlegungen zur Konstruktion von Wintersportlandschaften am Beispiel Damüls in Vorarlberg.*

Willi Haas is a graduated engineer in Mechanical Engineering and Management and a senior researcher and lecturer at the Institute of Social Ecology, Vienna. His research focuses on society-nature interactions at the local level to better understand both how global (environmental) change affects the local and how local action contributes to global problems. He also investigates the potentials and limitations of the circular economy as an option to achieve a more sustainable future. He explores the interplay between energy transitions and human labor as future transitions will have far-reaching—yet, in many cases, underestimated—consequences for the future of labor. Furthermore, he studies the links between sustainability and health because human health is highly vulnerable to environmental impacts, whereas human reproduction significantly contributes to environmental pollution. His research focuses on inter- and transdisciplinarity as a research approach that promotes communication across boundaries of disciplines and beyond Science with significant consequences for Science itself.

Helmut Haberl has a PhD in Ecology (University of Vienna, 1995), habilitation in Human Ecology (University of Vienna, 2001). He is currently Director of the Institute of Social Ecology, Vienna, Alpen-Adria University (Austria). His research foci include Integrated Land-System Science, Long-Term Socioecological Research, sustainability indicators, theory of society-nature interaction and energy in environmental and socioeconomic systems. He was the IPCC-AR5 lead author for the chapter 'Agriculture, Forestry and Other Land Use' of Working Group III and lead author for two chapters in the Global Energy Assessment. Haberl is a past member of the Scientific Committee of the European Environment Agency and of the Scientific Steering Committee of the Global Land Project. He is the author of more than 100 papers in international peer-reviewed journals and numerous other publications.

Daniel Hausknost (PhD in Politics and International Relations and Master's in Politics and Philosophy) is a senior researcher and lecturer at the Institute of Social Ecology in Vienna. His research focus is on questions of political agency within structural constraints, particularly with regard to environmental politics and socio-ecological transitions. He is interested in the capacities and limits of democratic regimes to enact purposive environmental change and in the relationship between the fossil energy system and modern political institutions. He is involved in several international research projects. Prior to his work as a researcher, he worked for several international environmental NGOs.

Markus Heinz holds a Master's in Physics from the University of Vienna with special focus on Computational Physics. He is a PhD candidate at the Institute of Social Ecology (SEC), where he worked as a junior researcher for several years. His main research interests are in modeling social metabolism and land-use change.

Thomas Kastner holds a PhD in Environmental Science from the University of Groningen in the Netherlands (2012) and a Masters in Ecology from the University of Vienna (2007). He has been involved in various national and international research projects, mainly in the area of Land-Use Science. His research interests include the impact of dietary patterns on land demand, relations between international trade and land-use change and long-term changes in land-use systems.

Fridolin Krausmann is professor of sustainable resource use at the Institute of Social Ecology at Alpen-Adria University in Vienna (Austria). In his research, he focuses on socio-metabolic transition processes. He studies long-term changes in the use of energy, materials and land in local rural and urban systems, in national economies and at the global scale. His research has contributed to the development of socioecological methods (e.g., material flow analysis, human appropriation of net primary production) and their adaptation for application in Environmental History. In early 2016, he was appointed Vice Dean for research at the Faculty for Interdisciplinary Studies.

Christine Kroisleitner holds a Master's in Geography with a specialization in Climatology. She worked from 2012 to 2014 at the Institute for Social Ecology in Vienna and is now a Research Associate at the Institute of Geography and Regional Sciences at the University of Graz. Her work examines spatial patterns of environmental and climate change.

Christian Lauk holds a Master's in Biology from the University of Vienna (2005) and a PhD in Human and Social Ecology from the University of Klagenfurt (2013). As a researcher and lecturer at the Institute of Social Ecology (SEC), Vienna, his research addresses different aspects of biomass metabolism and its interrelation with land use, particularly greenhouse gas emissions from agriculture, socioeconomic carbon stocks, and anthropogenic vegetation fires.

Juliana Lutz (PhD in Sociology) works at the Institute of Social Ecology, Vienna, Austria. Her research interests address social drivers of land-use change, focusing on social changes within farming systems, sustainable food systems and alternative, local food supply. She teaches participatory methods within socioecological research.

Andreas Mayer has an academic background in Sociology and the interdisciplinary field of Social and Human Ecology. He works as a junior researcher and lecturer at the Institute of Social Ecology (SEC), Vienna, Alpen-Adria University (Austria). His main research focus is social metabolism, resource conflicts, food security, and Environmental Sociology.

Michael Neundlinger has an academic background in Social Anthropology, History and Social Ecology and is a PhD researcher at the Institute of Social Ecology (Alpen-Adria University) in Vienna (Austria). Recently, he has worked and published within the interdisciplinary field of Environmental History with a focus on Vienna from the eighteenth to the twentieth century. Currently, he is studying the transformation of Austrian agriculture during the process of industrialization. Using historical GIS and material and energy flow accounting, he explores issues of socio-ecological inequality in preindustrial agro-ecosystems.

Maria Niedertscheider holds a Master's in Ecology and is a PhD researcher at the Institute of Social Ecology (Alpen-Adria University) in Vienna (Austria). Her research focuses on land system changes at various spatial and temporal scales, with particular attention to land-use intensity.

Panos Petridis has a background in Biology and holds a Bachelor's and two Master's degrees from the Universities of Bristol, Edinburgh and Plymouth (UK). Currently, he is a researcher and doctoral candidate at the Institute of Social Ecology in Vienna, working on issues of island sustainability, protected area management and socioecological transitions.

Christoph Plutzar is a biologist and social ecologist. He holds academic courses on Geographic Information System (GIS). Since 2008, he has worked at the Institute of Social Ecology, mainly in the area of Land Use and Colonization of Ecosystems. The focus of his work is handling spatially explicit research questions with GIS, especially within the topics of land use, biodiversity and wilderness.

Anette Reenberg was, until her retirement in 2014, Professor of Landscape & Agricultural Geography at the Department of Geosciences and Resource Management (Section of Geography) at the University of Copenhagen (Denmark). She has a scientific background in human environmental aspects of the Geographical Sciences (with a doctoral thesis addressing human-environment interaction and sustainability in Sahelian land-use systems). More specifically, her research addresses issues related to natural resource management and land-use strategies, including adaptation to climate change. Her focus is on land-use and land cover systems viewed from the perspective of landscape ecological and interdisciplinary Land-Change Science, that is, relating land-use dynamics to their larger-scale driving forces, which are of a biophysical, cultural, socioeconomic, institutional or demographic nature. Global land uses, telecoupling and large land acquisitions are recent issues of interest. She has extensive experience as an international research coordinator.

Alexander Remesch holds a Master's in Sociology from the University of Vienna and is a PhD student and junior researcher at the Institute of Social Ecology (SEC), Vienna. His main research interests are in the modeling of socioecological systems and in societal time-use patterns within the context of sustainability.

Lisa Ringhofer (PhD in Sociology) is the co-founder of the development consultancy network Triple Minds (www.triple-minds.com), where she works as a senior consultant mainly in the area of 'environment for development'. She is also affiliated with the Institute of Social Ecology, with her research focusing on the social metabolism and time use of local rural systems and their impacts on development. She has performed extensive field research and has coordinated various development projects in Bolivia, Nicaragua, and Mozambique.

Anke Schaffartzik holds a Master's of Social and Economic Sciences in Human and Social Ecology from Alpen-Adria University, Vienna, and a Bachelor of Science in Environmental and Resource Management from the Brandenburg University of Technology, Cottbus. As a researcher and lecturer at the Institute of Social Ecology (SEC), Vienna, her work centers on questions of resource extraction, use and consumption at the national and global levels. She has contributed to the methodological development of material flow accounting (MFA), especially to applications of input-output analysis. She is currently focusing on assessing the upstream material requirements of traded goods.

Martin Schmid is associate professor at Alpen-Adria University and founding member of the Center for Environmental History in Vienna, Austria. He holds a Master's in History (2002) and a PhD in Environmental History (2007), both from the University of Vienna. In 2014, he achieved his habilitation in Environmental History at Alpen-Adria University. In his research, he focuses on agro-ecosystems in the early modern period and their transformation in the process of industrialization. He contributed to the development of concepts for interdisciplinary Environmental History and for Long-Term Socioecological Research (e.g., socio-natural sites, colonization of natural systems) and studies the environmental history of rivers, wars, and cities.

Simron J. Singh (PhD in Human Ecology) works on the theoretical, conceptual, and empirical aspects of society-nature interactions across time and space within the framework of sustainability and development discourses. He has conducted extensive field research in the Indian Himalayas and the Nicobar Islands. His main focus is on understanding the metabolism of local rural systems and their long-term dynamics using biophysical variables such as material and energy flows, land use and time use. Singh has published extensively and is the recipient of several awards and fellowships.

Barbara Smetschka holds a Master's in Social Anthropology as well as a postgraduate degree as 'Science Communicator'. She works at the Institute of Social Ecology, Vienna, Austria. Her research interests include gender and sustainability studies, time-use studies, integrated socioecological modeling, participatory research and inter- and transdisciplinary research. She has experience with participatory modeling in transdisciplinary research projects, where time use serves as an integrating concept. She teaches competencies for inter- and transdisciplinary sustainability research.

Michaela C. Theurl has a background in Organic Agriculture and holds a Masters in Ecology. She works at the Institute of Organic Agriculture (FiBL), Austria, and at the Institute of Social Ecology. Her research focuses on differences in agricultural food production systems and consumption, especially on life cycle analysis approaches for climate assessments.

Helga Weisz is head of the department Transdisciplinary Concepts & Methods at the Potsdam Institute for Climate Impact Research (PIK) and is professor of Industrial Ecology and Climate Change at Humboldt University Berlin. Her background is in microbiology (Mag.rer.nat. University of Vienna) and cultural studies (Dr. phil. Humboldt University Berlin). She was guest professor at several universities, including the University of St. Gallen and Yale University. Her research interests are social and urban metabolism, complex network analysis of trade networks, theory of society-nature interaction and the industrial metabolism-climate change nexus. Recent publications: the chapter Urban Energy Systems in the Global Energy Assessment (2012) and a special feature on Industrial Ecology in PNAS (2015). She has been a member of the IIASA council since 2014.

Ulli Weisz holds a PhD (Dr. phil.) in Human and Social Ecology and an M.Sc. (Mag.rer.nat.) in Ecology. She is a senior researcher and lecturer at the Institute of Social Ecology, Vienna, and at the Institute of Palliative Care and Organizational Ethics (both Alpen-Adria University). In her research, she addresses links between sustainable development and human health using inter- and transdisciplinary methods. She investigates the health co-benefits of sustainability and health promotion strategies at different scales. Both her practical experiences in her former profession as a qualified nurse and her academic background help her in 'doing transdisciplinarity', especially in hospitals.

Dominik Wiedenhofer holds a Master of Science in Human and Social Ecology from Alpen-Adria University, Vienna, and a Bachelor of Science in Environmental and Bio-Resource Management from the University of Life Sciences, Vienna. During an exchange semester at Flinders University and as a visiting researcher at the University of Sydney, both Australia and Nagoya University, Japan, he accumulated valuable expertise for his two main research areas: sustainable consumption (especially carbon, energy and material footprinting) and the modeling of dynamic material stocks and flows, with the goal of understanding long-term socioecological transitions of the social metabolism and potentials toward a circular economy.

Verena Winiwarter trained as a chemical engineer. She then earned her MA in Medieval History and her PhD in Environmental History at the University of Vienna under the direction of Karl Brunner. She is a full professor of Environmental History at the Faculty for Interdisciplinary Studies of Klagenfurt University.With Martin Knoll, she wrote the 2007 introductory volume *Umweltgeschichte: Eine Einführung* (Environmental History: An Introduction). With Hans-Rudolf Bork, she wrote the first popular book on Environmental History in German, which won the Austrian 'Book of the Year 2015' award: *Geschichte unserer Umwelt. 60 Reisen durch die Zeit*. She has directed several research projects funded by the Austrian Science Fund (FWF) and is a corresponding member of the Austrian Academy of Sciences. Since January 2010, she has served as the Dean of the Faculty for Interdisciplinary Studies.

Contributors

Riley E. Dunlap Department of Sociology, Oklahoma State University, Stillwater, Oklahoma, USA

Nina Eisenmenger Institute of Social Ecology (SEC), Alpen-Adria University, Vienna, Austria

Karl-Heinz Erb Institute of Social Ecology (SEC), Alpen-Adria University, Vienna, Austria

Tamara Fetzel Institute of Social Ecology (SEC), Alpen-Adria University, Vienna, Austria

Marina Fischer-Kowalski Institute of Social Ecology (SEC), Alpen-Adria University, Vienna, Austria

Veronika Gaube Institute of Social Ecology (SEC), Alpen-Adria University, Vienna, Austria

Sylvia Gierlinger Institute of Social Ecology (SEC), Alpen-Adria University, Vienna, Austria

Simone Gingrich Institute of Social Ecology (SEC), Alpen-Adria University, Vienna, Austria

Robert Groß Institute of Social Ecology (SEC), Alpen-Adria University, Vienna, Austria

Willi Haas Institute of Social Ecology (SEC), Alpen-Adria University, Vienna, Austria

Helmut Haberl Institute of Social Ecology (SEC), Alpen-Adria University, Vienna, Austria

Daniel Hausknost Institute of Social Ecology (SEC), Alpen-Adria University, Vienna, Austria

Markus Heinz Institute of Social Ecology (SEC), Alpen-Adria University, Vienna, Austria

Thomas Kastner Institute of Social Ecology (SEC), Alpen-Adria University, Vienna, Austria

Fridolin Krausmann Institute of Social Ecology (SEC), Alpen-Adria University, Vienna, Austria

Christine Kroisleitner Institute of Geography and Regional Sciences, University of Graz, Graz, Austria

Christian Lauk Institute of Social Ecology (SEC), Alpen-Adria University, Vienna, Austria

Juliana Lutz Institute of Social Ecology (SEC), Alpen-Adria University, Vienna, Austria

Andreas Mayer Institute of Social Ecology (SEC), Alpen-Adria University, Vienna, Austria

Michael Neundlinger Institute of Social Ecology (SEC), Alpen-Adria University, Vienna, Austria

Maria Niedertscheider Institute of Social Ecology (SEC), Alpen-Adria University, Vienna, Austria

Panos Petridis Institute of Social Ecology (SEC), Alpen-Adria University, Vienna, Austria

Christoph Plutzar Institute of Social Ecology (SEC), Alpen-Adria University, Vienna, Austria

Anette Reenberg Section of Geography, University of Copenhagen, Copenhagen, Denmark

Alexander Remesch Institute of Social Ecology (SEC), Alpen-Adria University, Vienna, Austria

Lisa Ringhofer Triple Minds Consultants, Vienna, Austria

Anke Schaffartzik Institute of Social Ecology (SEC), Alpen-Adria University, Vienna, Austria

Martin Schmid Institute of Social Ecology (SEC), Alpen-Adria University, Vienna, Austria

Simron J. Singh Faculty of Environment, University of Waterloo, Waterloo, Ontario, Canada

Barbara Smetschka Institute of Social Ecology (SEC), Alpen-Adria University, Vienna, Austria

Michaela C. Theurl Institute of Organic Agriculture (FiBL), Vienna, Austria; Institute of Social Ecology (SEC), Alpen-Adria University, Vienna, Austria

Helga Weisz Potsdam Institute for Climate Impact Research, Potsdam, Germany

Ulli Weisz Institute of Social Ecology (SEC), Alpen-Adria University, Vienna, Austria

Dominik Wiedenhofer Institute of Social Ecology (SEC), Alpen-Adria University, Vienna, Austria

Verena Winiwarter Institute of Social Ecology (SEC), Alpen-Adria University, Vienna, Austria

Abbreviations

a.s.l.	Above sea level
ABM	Agent-Based Model(ing)
AIC	Akaike Information Criterion
ANPATR	Andaman and Nicobar Protection of Aboriginal Tribes Regulation
APWD	Andaman Public Works Department
BC	Before Christ
bif	Burning of crop residues in the field
BioBaM	Biomass-Balance Model
$\text{BIODIVERSITY}_{pot}$	Potential biodiversity
BMLFUW	Bundesministerium für Land- und Forstwirtschaft, Umwelt und Wasserwirtschaft (= Austrian Federal Ministry of Agriculture, Forestry, Environment and Water Management)
BMR	Basic Metabolic Rate
BMVIT	Bundesministerium für Verkehr, Innovation und Technologie (= Austrian Ministry for Transport, Innovation and Technology)
C	Carbon
cal	Calorie (1 cal ≈ 4.2 J)
cap	Capita
CE	Common Era
CEDEFOP	European Centre for the Development of Vocational Training
CH_2	Methylene
CH_4	Methane
cm	Centimeter
CO_2	Carbon dioxide
COP	Conference of the Parties
COPD	Chronic Obstructive Pulmonary Disease
CSR	Corporate Social Responsibility
DCC	Domestic Carbon Consumption
DDT	Dichlorodiphenyltrichloroethane

DE	Domestic Extraction
DEC	Domestic Energy Consumption (= Direct Energy Input (DEI) – (energy) exports)
DEI	Direct Energy Input (= Domestic (energy) Extraction (DE) + (energy) imports)
dm	Dry matter
DMC	Domestic Material Consumption (= Direct Material Input (DMI) – (material) exports)
DMI	Direct Material Input (= Domestic (material) Extraction (DE) + (material) imports)
DPO	Domestic Processed Output
DPSIR model	D = Drivers, P = (environmental) Pressures, S = States (of the environment), I = (environmental) Impacts, and R = (policy) Responses
EEA	European Environmental Agency
EF	Ecological Footprint
EFA	Energy Flow Analysis/Accounts
EFQM	European Foundation for Quality Management
eHANPP	Embodied Human Appropriation of Net Primary Production
EJ	Exajoule (1 EJ = 10^{18} J)
Emergy	Embodied energy
EROEI	Energy Return On Energy Investment
EROI	Energy Return On Investment (= Energy Output ÷ Energy Input)
ERP	European Recovery Programme
EU	European Union
EU25	The 25 member states of the European Union before the year 2007
EU27	The 27 member states of the European Union before the year 2013
Eurostat	The statistical office of the European Union
EW-MFA	Economy-Wide Material Flow Accounting
FAO	Food and Agriculture Organization of the United Nations
FCA	Full Carbon Accounting
FSU	Former Soviet Union
FTU	Functional Time Use
FW	Fresh Weight
FWF	Fonds zur Förderung der wissenschaftlichen Forschung (= Austrian Science Fund)
g	Gram (1 g = 10^{-3} kg)
GAEZ	Global Agro-Ecological Zones
GDP	Gross Domestic Product
GEOSS	Global Earth Observing System of Systems
GHG	Greenhouse Gas

GIS	Geographic Information System
GJ	Gigajoule (1 GJ = 10^9 J = one billion joules)
GLC	Global Land Cover
GPP	Gross Primary Production
Gt	Gigaton (1 Gt = 10^9 t = 1 Pg = one billion tons)
GWP	Global Warming Potential
ha	Hectare (1 ha = 100 ares = 10^{-2} km^2 = 10^4 m^2)
HANPP	Human Appropriation of Net Primary Production
$HANPP_{harv}$	Human Appropriation of Net Primary Production through harvest
$HANPP_{luc}$	Human-induced changes in productivity due to land conversion and land use, including soil degradation
HCWH	Health Care Without Harm
HPH	Health Promoting Hospitals network of the WHO
HYDE	Hundred Year Database for Integrated Environmental Assessments
IAASTD	International Assessment of Agricultural Knowledge, Science and Technology for Development
IAM	Integrated Assessment Model
IBP	International Biological Programme
ICT	Information and Communication Technology
ICU	Intensive Care Unit
IEA	International Energy Agency
IFF	Fakultät für Interdisziplinäre Forschung und Fortbildung der Alpen-Adria-Universität (= Faculty for Interdisciplinary Research and Continuing Education at Alpen-Adria University; formerly known as 'Interuniversitäres Forschungsinstitut für Fernstudien' = Interuniversity Research Institute for Distance Education)
IGBP	International Geosphere-Biosphere Programme
IHDP	International Human Dimensions Programme on Global Environmental Change
IIASA	International Institute for Applied Systems Analysis
iLUC	Indirect Land-Use Change
IO	Input–Output
IOT	Input–Output Table
IPAT equation	$I = P \times A \times T$ (I = (environmental) Impact, P = Population, A = Affluence, T = Technology)
IPBES	Intergovernmental Platform on Biodiversity and Ecosystem Services
IPCC	Intergovernmental Panel on Climate Change
ISO	International Organization for Standardization
ISSC	International Social Science Council
IUCN	International Union for Conservation of Nature

IZT	Institut für Zukunftsstudien und Technologiebewertung (= Institute for Future Studies and Technology Assessment)
J	Joule
kcal	Kilocalorie (1 kcal = 10^3 cal)
kg	Kilogram (1 kg = 10^3 g)
kJ	Kilojoule (1 kJ = 10^3 J)
km	Kilometer (1 km = 10^3 m)
km^2	Square kilometer (1 km^2 = 1 km × 1 km)
KTQ	Kooperation für Transparenz und Qualität im Gesundheitswesen (= Cooperation for Transparency and Quality of Healthcare)
kW	Kilowatt
kWh	Kilowatt-hour (1 kWh = 10^3 watt-hours = 3.6 MJ)
l	Liter
LACA	Latin America and the Caribbean
LAU	Large Animal Unit
LCA	Life Cycle Analysis/Assessment
LCI	Life Cycle Inventory
LCIA	Life Cycle Impact Assessment
LPJ-DGVM	Lund-Potsdam-Jena Dynamic Global Vegetation Model
LT	Lifetime
LTER	Long-Term Ecological Research
LTSER	Long-Term Socioecological Research/Long-Term Socioeconomic and Ecosystem Research
LULUCF	Land Use, Land-Use Change and Forestry
m	Meter
m^2	Square meter (1 m^2 = 1 × 1 m)
m^3	Cubic meter (1 m^3 = 10^3 l)
MA18	Magistratsabteilung 18 der Stadt Wien—Stadtentwicklung und Stadtplanung (= Viennese Urban Planning Department)
MAB	Man and the Biosphere Programme (United Nations)
MagSW	Magistrat der Stadt Wien (= Vienna Municipal Administration)
MEA	Millennium Ecosystem Assessment
MEFA	Material and Energy Flow Accounting/Analysis
MENA	Middle East and North Africa
MF	Material Footprint
MFA	Material Flow Accounting/Analysis
Mha	Mega hectare (1 Mha = 10^6 ha = 10^4 km^2 = 10^{10} m^2)
MI	Material Intensity
MJ	Megajoule (1 MJ = 10^6 J = 1 million joules)
MLP	Multi-Level-Perspective
mm	Millimeter

MNP	Milieu en Natuur Planbureau (= Netherlands Environmental Assessment Agency; also known today as PBL—Planbureau voor de Leefomgeving)
MRIO	Multi-Regional Input–Output
Mt	Megaton (1 Mt = 10^6 t = one million tons)
MuSIASEM	Multi-Scale Integrated Analysis of Societal and Ecological Metabolism
N	Nitrogen
NGO	Non-Governmental Organization
NHS	National Health Service (United Kingdom)
NOx	generic term for the mono-nitrogen oxides NO and NO_2 (nitric oxide and nitrogen dioxide)
NPP	Net Primary Production
NPP_{act}	Actual Net Primary Production
NPP_{eco}	Net Primary Production/biomass remaining in the ecosystem after harvest (= NPP_t)
NPP_{pot}	Potential Net Primary Production
NPP_t	See NPP_{eco}
OECD	Organisation for Economic Co-operation and Development
ÖIR	Österreichisches Institut für Raumplanung (= Austrian Institute for Regional Studies and Spatial Planning)
PAME	Participatory Assessment, Monitoring and Evaluation
PCA	Partial Carbon Accounting
PEC	Primary Energy Consumption
Pg	Petagram (1 Pg = 10^{15} g = 1 Gt = one billion tons)
PJ	Petajoule (1 PJ = 10^{15} J)
PNAS	Proceedings of the National Academy of Sciences of the United States of America
PTB	Physical Trade Balance (= imports – exports)
Pu	Plutonium
QOL	Quality Of Life
quad	Quadrillion Btu (1 quad = 10^{15} Btu ≈ 1.1 EJ)
r^2	Coefficient of determination
RCU	Respiratory Care Unit
REDD	Reducing Emissions from Deforestation and forest Degradation; see also UN-REDD
REX	Raw material equivalents of Exports
RIM	Raw material equivalents of Imports
RMC	Raw Material Consumption (= Domestic Extraction (DE) + Raw material equivalents of Imports (RIM) – Raw material equivalents of Exports (REX))
RME	Raw Material Equivalent
RMI	Raw Material Input (= Domestic Extraction (DE) + Raw material equivalents of Imports (RIM))
RMU	Respiratory Management Unit

ROW	Rest Of the World
RTB	Raw material Trade Balance (= Raw material equivalents of Imports (RIM) – Raw material equivalents of Exports (REX))
s	Second
SEI	Stockholm Environment Institute
SES framework	Social-Ecological Systems framework
SET	Socioecological Transition
SFA	Substance Flow Analysis
SI	International System of Units (abbreviated from French: Le **S**ystème **I**nternational d'Unités)
S-LCA	Social-LCA
SMCE	Social Multi-Criteria Evaluation
SNS	Socio-Natural Site
SO_2	Sulfur dioxide
SRIO	Single-Region Input–Output
SSA	Sub-Saharan Africa
SUME	Sustainable Urban Metabolism for Europe (= EU project)
t	Ton
tce	tons of coal equivalent (1 tce $\approx$ 29.3 GJ)
TJ	Terajoule (1 TJ = 10^{12} J)
TMC	Total Material Consumption (= Total Material Requirement (TMR) – exports – unused extraction of exports)
TMR	Total Material Requirement (= Domestic Extraction (DE) + unused (domestic) extraction + imports + unused extraction in country of origin)
toe	tons of oil equivalent (1 toe $\approx$ 41.9 GJ)
TPES	Total Primary Energy Supply
TRAFO	'Transdisziplinäres Forschen' (= 'Transdisciplinary Research,' project financed by the Austrian Ministry of Science)
UK	United Kingdom
UNCTAD	United Nations Conference on Trade and Development
UNECE	United Nations Economic Commission for Europe
UNEP	United Nations Environment Programme
UNESCO	United Nations Educational, Scientific and Cultural Organization
UNFCCC	United Nations Framework Convention on Climate Change
UN-REDD	United Nations collaborative initiative on Reducing Emissions from Deforestation and forest Degradation; see also REDD
US	United States (of America)
USGS	United States Geological Survey
USSR	Former Union of Soviet Socialist Republics
UTE	Urban Time and Energy (= Austrian Science Fund project)

W	Watt (1 W = 1 J/s)
WBGU	Wissenschaftlicher Beirat der Bundesregierung Globale Umweltveränderungen (= German Advisory Council on Global Change)
WCED	World Commission on Environment and Development
WHO	World Health Organization
WSTP	Wiener Stadtphysikat (= Vienna Health Authority)
WWI	World War I (1914–1918)
WWII	World War II (1939–1945)
WWTP	Wastewater Treatment Plant
yr	Year

List of Figures

Figure 1.1 The material metabolism of a national economy 12
Figure 1.2 Interrelation between the Human Development Index and the ecological footprint (EF) 14
Figure 1.3 The conceptual model of society-nature interaction developed by the Vienna Social Ecology School 21
Figure 2.1 Social systems as hybrid systems 33
Figure 2.2 Stylized model of the product life cycle, from extraction to production, consumption and deposition, in physical and economic terms 45
Figure 2.3 Levels of potential societal colonizing interventions in natural systems 47
Figure 2.4 From a control spiral induced by colonizing interventions to a structural coupling of social and natural systems 51
Figure 3.1 The biophysical constraints of the agrarian regime 72
Figure 3.2 Historical development of domestic material consumption (DMC) and metabolic rates in selected industrial countries 81
Figure 4.1 The three dimensions of land-use intensity 105
Figure 4.2 Human appropriation of net primary production (HANPP) as an indicator of land-use intensity 112
Figure 4.3 Human alteration of carbon stocks in vegetation in Europe.... 113
Figure 4.4 A socioecological metabolism perspective on land-use change and land-use intensification 113
Figure 6.1 Average yearly flows of carbon (MtC year^{-1}) in Austria in the periods 1830–1880 and 1986–2000 154
Figure 6.2 Front page of a 17th-century treatise showing the town of Huancavelica with mines in the mountains and smelters in the foreground 158

Figure 7.1 Primary energy consumption (PEC) and working hours in the UK, 1870–2000, and Austria, 1950–2010 184
Figure 7.2 Annual energy consumption, working hours and energy intensity of working hours for Germany and Italy, 1870–1998 186
Figure 7.3 Annual working hours per employee and per inhabitant 189
Figure 7.4 Variation in the quality of work and its institutional form by sociometabolic regimes......................... 190
Figure 8.1 Global use of energy (DEC) and materials (DMC) in the period 1850–2005, share of biomass in DMC and DEC, global material exports by material groups, 1950–2010 202
Figure 8.2 Global metabolic rates (DMC and DEC per capita per year) and material intensity, 1870–2009 204
Figure 8.3 Share of world regions in global material use (DMC) and physical net trade (PTB, imports minus exports) by world region.................................... 207
Figure 8.4 Metabolic rates (material use, DMC/cap/yr) by world regions (1950–2010) and by material groups (2010) 208
Figure 8.5 Flow chart of an energy flow analysis consistent with the material flow analysis approach as used in socioeconomic metabolism studies 215
Figure 9.1 Average country-wide metabolic rates (DMC per capita per year) in 2010.................................... 219
Figure 9.2 Share of six major world regions and country groupings in the global extraction of biomass, fossil energy carriers and minerals in 1950 and 2010.......................... 223
Figure 9.3 Comparison of exports and domestic extraction in 2010 for five main material categories 225
Figure 9.4 Relation between population and DMC of biomass as well as total DMC in 2010 for 175 individual countries 228
Figure 9.5 Relation between GDP and fossil energy carriers as well as total DMC in 2010 for 175 individual countries 230
Figure 9.6 Relation between income (GDP/cap/yr) and metabolic rates (DMC/cap/yr) in 2010 for 175 individual countries 231
Figure 9.7 Material flow accounting scheme....................... 235
Figure 10.1 Schematic representation of material flows between an economy (Austria) and the rest of the world (ROW)....... 242
Figure 10.2 Austria's trade flows, balances material consumption and the raw material equivalents (RMEs) thereof in 2007..... 246
Figure 10.3 Development of Austria's resource consumption by component, 1995–2007 249
Figure 10.4 Opening the 'black box'—a schematic look at the production of traded goods 250

Figure 10.5 Life cycle concept of a product system 254
Figure 10.6 Example of a flow chart comprising the supply chain of tomatoes 254
Figure 11.1 Sociometabolic flow chart of the global economy in 2005 266
Figure 12.1 Economy-wide use of construction minerals, with estimated inputs and outputs from stocks in 2009 283
Figure 12.2 Modeled material flows and the quantitative relationship to the stock 284
Figure 12.3 Projections of construction mineral inputs and waste/recycling flows from stocks in 2020 for two recycling scenarios 285
Figure 13.1 National-scale comparison of grazing land according to Erb et al. (2007) with FAOstat data (FAO 2011) on permanent pasture 300
Figure 13.2 Pattern and extent of global grazing lands in percent-per-gridcell representation according to different sources 301
Figure 13.3 Comparison of four grazing land maps 303
Figure 13.4 Total feed supply of ruminants in the year 2000, breakdown by region 305
Figure 13.5 Global map of grazing intensity. **a** Biomass grazed per gridcell, **b** grazing intensity expressed as grazed biomass in percent of actual NPP on grazing land................. 306
Figure 14.1 Global calorific intake per capita for main food categories in 2000 and variants for 2050.......................... 319
Figure 14.2 Average crop yields in 2000 and 2050 for three different variants in a regional breakdown 320
Figure 14.3 Variants of global livestock feeding efficiencies in the year 2050 for ruminant meat, monogastric species as well as milk, butter and other dairy products............. 322
Figure 14.4 Box plots showing the area potentially available in the year 2050 for purposes other than food production on good-quality land.................................. 325
Figure 14.5 The concept of the human appropriation of net primary production (HANPP) 333
Figure 16.1 Concept of the calculation of the embodied human appropriation of net primary production (eHANPP) 353
Figure 16.2 Human appropriation of net primary production (HANPP) on national territory and embodied HANPP (eHANPP) related to consumption in the EU27..................... 356
Figure 16.3 National 'self-sufficiency' of European countries in the year 2007 in terms of biomass-based products, measured by the ratio of HANPP/eHANPP 357
Figure 17.1 HANPP per unit area in the year 2000, excluding unproductive areas 365

Figure 17.2 Productivity losses in Africa, total and the fraction resulting from human-induced dryland degradation 366
Figure 17.3 The trend in HANPP and its components as percentage of NPP_{pot} from 1980 to 2005. 367
Figure 17.4 Dynamics of indicators of land-use intensity in Africa compared to the world average. 369
Figure 18.1 Colonization and the relationship between society and biodiversity, based on the socioecological research paradigm. 379
Figure 18.2 Link between energy flow and species richness using the species-energy hypothesis. 382
Figure 18.3 Regression analysis between the NPP remaining after harvest and the breeding bird species richness in Austria 384
Figure 18.4 DPSIR (Driving forces, Pressures, States, Impacts and Responses) model related to biodiversity 385
Figure 20.1 The forest transition in Austria: land use and carbon stocks in terrestrial ecosystems, 1830–2010 421
Figure 20.2 Austria's socioeconomic carbon budget 1830–2010: domestic carbon consumption and socioeconomic carbon stocks 422
Figure 20.3 Carbon (C) flows and stocks in Austria's ecosystems and society in MtC/yr and Mt, ca. 1850, and ca. 1990. 424
Figure 20.4 Systematics of carbon flows and carbon accounting systems. 429
Figure 21.1 Nitrogen flows in the agricultural production system of Theyern, 1830 and 1999. 438
Figure 21.2 Agricultural energy inputs and outputs in Theyern in 1830 and 1999 439
Figure 22.1 Land use/cover in the Philippines 1910–2003, and indexed development of cereal production according to increases in area, cropping frequency and yields. 449
Figure 22.2 Energy use in the Philippines throughout the 20th century, with a focus on the nation's biomass metabolism 454
Figure 23.1 Author's depiction of land cover in Damüls in 1857 based on the Austrian Land Register 462
Figure 23.2 T-bar lift '1800', looking toward Faschina. Picture postcard of the '1800 T-bar Lift'. 466
Figure 23.3 Picture of snow groomer (ca. 1968) 468
Figure 23.4 Development of the piste areas and ski lift facilities in the Damüls ski region. Terrain map of the Damüls ski region. 470
Figure 24.1 Vienna and its waterscape in its current boundaries. 477

Figure 24.2 Population of Vienna 1800–2011 and its domestic energy consumption (DEC) 478
Figure 24.3 Constructing the main intercepting sewer: construction site on Marxergasse 481
Figure 25.1 Example of an implemented procedure for residential location decision-making using weights ranging from 1 to 3 494
Figure 25.2 Distribution of family households in 2001 and 2050 in two scenarios 496
Figure 25.3 Distribution of households' energy consumption for heating, electricity and transport. 497
Figure 26.1 Overview of the transdisciplinary research process 507
Figure 26.2 The 'sustainability triangle' 507
Figure 26.3 The 'sustainability triangle' for the farm agent 512
Figure 26.4 Model interface: sustainability scenario in the village of Hainfeld 513
Figure 26.5 Age/sex group segregation in time use, Campo Bello 2004 and 2006 521
Figure 27.1 Material flows on Trinket Island, 2000–2001. 528
Figure 27.2 Energy flows on Trinket Island, 2000–2001. 529
Figure 27.3 Comparing metabolic throughput of materials and energy—pre- and post-tsunami 535
Figure 28.1 Comprehensive model of the island's socioecological system. . . . 546
Figure 28.2 Scheme for a transdisciplinary research process that aims at integrating life-world and science perspectives in generating integrated knowledge 556
Figure 29.1 The sustainability triangle for hospitals 570

List of Tables

Table 2.1 Societal stocks and corresponding flows in empirical research . . . 37
Table 3.1 Metabolic profiles of sociometabolic regimes. . . . 67
Table 3.2 Examples of the variability of metabolic profiles within the agrarian metabolic regime . . . 74
Table 4.1 Systemic insights provided by the socioecological metabolism approach, as illustrated by prominent land-use strategies . . . 115
Table 6.1 Typology of legacies resulting from societal interventions in nature. . . . 164
Table 7.1 Daily working hours by sociometabolic regime for an average inhabitant and an average day of the year . . . 173
Table 7.2 Share of urban population in 1500 by world region (settlements with 2500 or more inhabitants). . . . 181
Table 8.1 Overview of some common energy units . . . 213
Table 9.1 Domestic extraction, domestic material consumption and physical trade balances for the years 1950 and 2010 for six country groupings . . . 222
Table 10.1 Indicators derived from economy-wide material flow accounting and raw material equivalent accounts. . . . 243
Table 10.2 Input-output table for a hypothetical three-product economy. . . . 257
Table 11.1 Comparison of material flow accounting and life cycle analysis as frameworks to analyze the circularity of economies . . . 265
Table 11.2 Potential and limits of the Circular Economy . . . 273
Table 15.1 Comparison by region of burned area/biomass with biomass extracted for food, feed, fibre, and bioenergy . . . 339
Table 21.1 The industrialization of Austrian agriculture in the 20th century: selected indicators. . . . 441

Table 22.1 Development of population, gross domestic product and indicators of agricultural intensification in the Philippines during the 20th century 451
Table 25.1 Total population, number of households and estimated energy use for Vienna in 2001 and for two scenarios in 2050. 497
Table 26.1 The four areas of time use for production and reproduction. 510
Table 26.2 Settings for scenarios. 515
Table 26.3 Daily labor time invested by age/sex groups in Campo Bello (Bolivia) as observed in 2004 and 2006 521
Table 29.1 'The Sustainable Hospital': Key features of the project series 564
Table 29.2 Planning of care provision for long-term ventilated patients: Comparison of the 2-step model with the proposed 3-step model. 571
Table 1 Three levels of learning/change. 582

Introduction

Helmut Haberl, Karl-Heinz Erb, Marina Fischer-Kowalski, Robert Groß, Fridolin Krausmann, Christoph Plutzar, Martin Schmid and Verena Winiwarter

Abstract

Social Ecology is inter- and transdisciplinary research field concerned with society–nature interactions. It integrates contributions from the Natural Sciences with those from the Social Sciences and Humanities to provide a scientific basis for sustainability. Socioecological research addresses spatial scales from the local to the global and is concerned with the past, present, and future. This chapter serves as an introduction to a volume that takes stock of 25 years of research at the Institute of Social Ecology in Vienna.

Keywords

Social ecology, Social metabolism, Colonization of natural processes, Socioecological transitions, Environmental history, Long-term socioecological research (LTSER)

Why Social Ecology?

Climate change, biodiversity loss, deforestation, the finiteness and the potential depletion of natural resources, poverty, hunger, and unequal access to necessities for livelihood, such as land, food, and energy, are among the plethora of challenges commonly included in the daunting agenda summarized by two magic words: 'sustainable development,' which some believe to be an oxymoron (Hall 2004). The concept of sustainable development can be traced to eighteenth century German forestry (Gottschlich and Friedrich 2014), where it denoted the imperative that the harvest of trees must not exceed their regrowth in the forest.

'Sustainable development' gained international prominence with the publication of the *World Conservation Strategy* (IUCN et al. 1980) and, in particular, with the report *Our Common Future*, issued by the World Commission on Environment and Development (WCED; United Nations 1987; see also Chap. 29). The concept was proposed in an effort to overcome the stalemate between those aiming to respect ecological 'limits to growth' (Meadows et al. 1972) and those seeing growth as imperative for developing countries to overcome poverty, hunger, and high infant mortality (World Bank 1978) or as the natural course of progress, as many Neoclassical economists see it (Rohlf 2008). The notion of sustainable development was successful. It promised the reconciliation of continued economic growth for an equitable and just development of the world's poorer regions, or intragenerational justice, with intergenerational justice, or the imperative to keep the global resources and ecosystems as well as the climate system in a condition suitable for future human existence and prosperity (Fischer-Kowalski and Haberl 1997; Haberl et al. 2004). This reconciliation was expected to be achieved by 'decoupling' social and economic development from the use of biophysical resources and, ultimately, from environmental pressures and degradation (United Nations 1987; World Bank 1992).

The turn from the 'limits to growth' deadlock to a debate on how to reconcile development with environmental protection was a challenge for the scientific community. Viable solutions require collaboration across scientific disciplines—in other words, interdisciplinarity. They even require collaboration between scientists and stakeholders from different social groups, such as policy-makers, administration, business and industry, civil society, and many more—usually denoted today as 'transdisciplinarity' (Dressel et al. 2014). The call for inter- and transdisciplinary programs of 'sustainability science' emerged (e.g., Clark and Dickson 2003; Kates et al. 2001), and new interdisciplinary research fields, such as Ecological Economics (Martinez-Alier 1987), Industrial Ecology (Ayres and Simonis 1994), and Environmental History (McNeill 2000; Sieferle 2001), gained significance.

As Marina Fischer-Kowalski and Helga Weisz explain in Chap. 1, Social Ecology, as understood in the 'Vienna School' at the Institute of Social Ecology in Vienna, is one of the islands in the archipelago of scientific approaches situated amid the continents of the Social Sciences, the Humanities and the Natural Sciences. Initially, a small group of researchers designed environmental reports and indicators to make visible and quantifiable the roles of human agency, socioeconomic contexts and institutions in altering the environment and placing ecosystems under pressure. The Institute of Social Ecology has since become a vibrant and larger but closely interacting group of scientists from the Social Sciences (in particular, Sociology and Anthropology), History and the Natural Sciences (in particular, Ecology), with the aim of forging an inter- and transdisciplinary research agenda.

Aims and Scope of this Volume

The scholarly literature on sustainability research is growing with the urgency of the issues. This book is the first comprehensive overview of Social Ecology that outlines its contributions to these scientific and societal challenges. It presents the current state of the art in Social Ecology, demonstrating the Vienna School's attempts to tackle global sustainability problems through innovative inter- and transdisciplinary scientific research. The core axioms of the Vienna School of Social Ecology are that social and natural systems interact, coevolve over time and have substantial impacts upon one another, with bidirectional causality. Social Ecology addresses energy and society, land use and food production, metabolism of societies, and the short- and long-term environmental impacts of human activities. It offers a conceptual approach to society–nature coevolution pertaining to history, to current development processes and to a future sustainability transition.

This book takes stock of Social Ecology and discusses its relation to neighboring fields and its epistemic and conceptual foundations. It introduces concepts and epistemological questions in a series of overview chapters. The strengths of socioecological research are highlighted by presenting inter- and transdisciplinary research on a variety of themes that touch upon all areas of the globe, covering spatial scales from urban and rural investigations to national and global analyses as well as temporal scales across past centuries and extending into the future. We introduce important socioecological methods in *Method Précis* attached to thematically related empirical chapters. This exemplification of current socioecological thinking is suitable for classroom use and extends the research frontier of Social Ecology in new and innovative directions. The book has been written by a diverse team of natural scientists, social scientists, and historians, building upon more than 25 years of shared experience.

The themes discussed include the following:

- Resource flows (materials and energy) and their measurement across time and space, the limits of these flows and their environmental and socioeconomic implications
- Relations between production and consumption as well as the role of trade in shaping the patterns and trajectories of resource use
- Land use and its relation to ecosystems and to socioeconomic material and energy flows
- The significance of urban development and city–hinterland relations and how they change over time
- Fundamental long-term changes in the relations between societies and their environment (‘socioecological transitions’) and the effects of these changes on land use and resource flows
- Connections among labor, gender relations, and energy and resource use
- Human pressures on biodiversity and ecosystems
- Climate change and drivers of greenhouse gas emissions in terms of resource flows and land use.

Socioecological approaches are interdisciplinary in nature. The applications are manifold, ranging from science to practice and from the local to the global. We discuss them throughout these chapters in their respective thematic contexts. Social Ecology aims to provide scientific foundations and to intervene, through stakeholder engagement, in various fields of discourse and policy to support sustainability. The book touches upon many pertinent issues, including the following:

- Spatial and temporal scales
- Long-term trajectories
- Historical legacies
- Future scenarios
- Energy transitions, both past and prospective
- Opportunities and limits of stakeholder involvement
- Complex systems thinking
- Sophisticated statistical data analysis across usually separate domains.

The process of creating this volume involved a stringent supervision by the editors and several workshops with all the authors—a process that lasted more than 3 years. Through this joint process, we were encouraged to develop socioecological theory further and to achieve common viewpoints while maintaining the richness of various scientific approaches and flavors. The book assembles a substantial amount of original material, and it differs from many edited volumes because of its high internal consistency. It builds upon and refers to many earlier efforts (such as Armitage et al. 2007; Becker and Jahn 2006; Berkes et al. 2003; Gunderson and Holling 2002). Recently, a number of books have appeared with a similar ambition to capture some of the essence of socioecological thinking (such as Allwood et al. 2012; Baccini and Brunner 2012; Giampietro et al. 2012; Glaser et al. 2012; González de Molina and Toledo 2014; Kander et al. 2013). We hope to match their achievements by the breadth of approaches offered, sustained by a large team disciplined in a long tradition of jointly developing and adhering to a shared theoretical paradigm and washed in the waters of many scientific traditions.

Box: Adventures in Social Ecology

Inter- and transdisciplinary research can result in challenging experiences due to the encounters with colleagues with different backgrounds and, hence, different fundamental concepts, languages, bodies of literature, and habits. Here, we present four experiences we found to be particularly enlightening—and entertaining. To maintain their flavor as personal narratives, we identify the respective authors and keep the text in the first-person singular.

Walk the Line—experienced by Martin Schmid

I was trained as a historian and archeologist and came in contact with Social Ecology as a young student in the mid-1990s. From the very beginning, my relationship with Social Ecology was highly ambivalent. Being a green—in every respect—student of History, I was fascinated by the theoretical explicitness of Social Ecology and by its emphasis on the materiality (not to say 'naturalness') of the world. Even more, I enjoyed the academic culture in the group that has now become the Institute of Social Ecology. I simply liked how teachers interacted with each other and with us, students. What a contrast to most of my experiences at the History Department of the University of Vienna, in intellectual as well as social terms. At the same time, I found social ecologists to be irritating, with their diagrams made up of circles and boxes, their databases filled with numbers, their talk of systems interacting with each other, and their strange Luhmannian idea of a society made up of communications. I had only been disciplined to view the world as a world of actors—humans, of course, driven by the motives that are highly variable over time and are culture dependent. But in the world of Social Ecology, there were no actors and no motives one could reveal using what historians call 'source critique.'

One of my first papers turned that irritation into a product. I discussed and criticized the concept of 'colonization of natural systems' by using it as an Environmental History interpretation of an agricultural handbook from the sixteenth century. The result was a rather clumsy culturalist critique of Social Ecology, and in that sense, quite typical for a freshman with a background like mine. Nevertheless, the Institute was generous in publishing this piece as a Social Ecology Working Paper.

This story points to one of the most important qualities of Viennese Social Ecology. The institute has been successful in walking the thin line between developing and caring for a common conceptual ground on the one side and dealing with diversity on the other side, including internal diversity and the diversity of research fields in its academic neighborhood. Viennese Social Ecology was and (as proven by this book) is an intellectual environment where one can see what is needed to realize interdisciplinarity. This requires both openness to and curiosity for other, sometimes irritating, perspectives. But it also requires a clear theoretical standpoint. Fortunately, Social Ecology has such a clear standpoint, with a shared paradigm according to which social and natural systems interact and coevolve over time. From our diverse disciplinary backgrounds, we all know about the limitations of that paradigm. We know in which sense and under which (disciplinary) pre-decisions it is false. This is what makes it a source of scientific rigor.

Almost 20 years have passed since I made first contact with Social Ecology. Now I am engaged in Environmental History at the institute. Clearly, the institute has changed since then. Yet some things have not changed. It is still an intellectual pleasure for me to criticize the shared paradigm of Social Ecology from a historian's perspective and to search for conceptual alternatives such as 'socio-natural sites' (SNSs; Chaps. 6 and 23 in this volume; Winiwarter and Schmid 2008), alternatives that follow the overall aim of integrating the Natural and Social Sciences with the Humanities but are better suited to involve the latter in such an endeavor. Social Ecology has become an important part of my academic identity that allows me to talk at the *Deutscher Historikertag* (German Historians' Day) about how ignorant most historians are of nature and to talk to an audience of limnologists about natural scientists' ignorance of society and culture at a conference about the Danube. Sometimes, it happens that I find myself arguing for and defending Social Ecology against culturalist and solely actor-centered approaches.

As a Socioeconomist, What is Your Take on That?—experienced by Helmut Haberl

Several times, even repeatedly with the same colleagues, I have had the following experience. After discussing a biophysical phenomenon such as biodiversity loss or climate change with a group of several natural scientists, such as ecologists or meteorologists, they turn to me and ask, 'As a socioeconomist, what is your take on that?'

The irony of the story is twofold. First, I was trained as a biologist, earth scientist, and mathematician and have studied biophysical phenomena such as energy use, biomass flows, and carbon cycles throughout my scientific career. Although I have been working with social scientists, historians, and economists for decades, I have no formal training in the Social Sciences and Humanities and would never regard myself as a social scientist, historian, or economist. Second, of course, 'socioeconomics' is not a scientific discipline at all. Clearly, there is a plethora of Social Science disciplines as well as specializations in the Humanities, such as Sociology, Anthropology, Macroeconomics, Microeconomics, History, Philosophy, Pedagogy, and many more. No scholar from any of these disciplines would regard him- or herself as a 'socioeconomist'; this seems to be a species of scientists recognized more by natural scientists than by their own close colleagues.

Perhaps the issue is deeper. I suspect that those (honorable and well-intended) natural scientists who are motivated to link up with social scientists, economists or scholars from the Humanities often find it difficult to communicate with those strange creatures from remote continents—to take up the metaphor stressed by Fischer-Kowalski and Weisz in Chap. 1 (and vice versa, of course!). Several times in such encounters, my impression was that these fine colleagues regard me as an ambassador of foreign and somehow uncanny territories who has the advantage of speaking their language and knowing their culture well enough to facilitate exchange. Thus, perhaps we Social Ecologists should take it as a sign of appreciation when we have similar experiences!

Have You Ever Read Hegel?—experienced by Robert Groß

Having successfully applied for a scholarship, I went to Aalborg University in spring 2013. A fellow Ph.D. student convinced me to present at a conference dealing with theory and empirical research in the Social Sciences and Humanities. All participants were cultural, intellectual or social historians, whereas I consider myself a human ecologist (with a Natural Science background) doing Environmental History. My paper addressed the concept of 'socio-natural sites' (SNSs), applied to the case study of Damüls/Vorarlberg (Chap. 23). The concept of SNSs is based on the assumption that social practices and material arrangements (e.g., landscapes) are structured by each other. If one or both changes, either by human hands or due to non-human influence (e.g., a shift in snowfall pattern), the SNS transforms (Chap. 6). The first question after I had delivered my paper left me dumbfounded for a moment: 'Had you read Hegel before writing your paper? The transformation of an SNS corresponds with Hegelian Dialectics.'

What is the point of this story? According to Pierre Bourdieu, social relations among established disciplines, which are considered theoretically pure and more empirically and application-based areas of scholarship (such as the parvenu 'interdisciplines' Social Ecology and Environmental History), are never neutral. Rather, representatives of higher-rated disciplines (e.g., Philosophy) claim interpretational sovereignty by referring to their prophets of historiography (Bourdieu 1984). One could argue that my colleague used Hegel as a metaphor for doing pure, intellectual history. My point is that he considered the study of materiality as historical force to be heretical. It is obviously a sensitive issue in historiography.

What is the potential of the heresy, of the concept of SNSs when communicating with historians? To me it became evident when other conference colleagues presented their respective papers, using 'practices' to analyze series of pictures taken by immigrants to depict motifs linked to their concepts of home. Most of the pictures displayed material artifacts, but the researchers avoided the nexus between practices and objects. Thus, no one asked the presenter for his or her relation to the masterminds of Philosophy (e.g., Hegel); rather, the discussion was one-sided, narrowed to the cultural background (gender, age, race and nationality) of the photographer. My impression of this discussion is that of the heretic. To look at practices this way means not to use the full analytical potential of praxeology and to sell old wine in new skins. The investigation of the transformation of practices and materiality—that is, the SNSs approach—bears the potential to structure narrations. Furthermore, the concept opens black boxes of communities, historiography and Social Ecology. From my point of view, the concept of SNSs is useful because it avoids dead-end discussions about what 'nature' and 'culture' mean, which have pervaded scholarship for much longer than Social Ecology and Environmental History have existed.

How Does This Fit?—experienced by Christoph Plutzar

Equipped with an extensive background in Natural Sciences (Biology, Geographic Information Systems), I began work at the Department of Vegetation Ecology at the University of Vienna in 1998. My job was to develop a map of avian species richness in Austria as part of a project within the Austrian 'Cultural Landscapes' research program and as a contribution to my Ph.D. project.

Soon after, the organizers of the 'Cultural Landscapes' program promoted a seminar entitled 'Ecological Orientations,' organized by the group known as the Viennese Institute of Social Ecology. My project leader encouraged me to participate in this course because of the Institute's reputation within cultural landscape research, and he thought it would be helpful in developing contacts. Because I needed to take courses for my Ph.D. study anyway, I decided to join—without any idea what to expect.

The seminar lasted three days and took place at a nice venue far out in the Lower Austrian countryside. Marina Fischer-Kowalski and Verena Winiwarter were teaching approximately 20 students with very different backgrounds, and I was one of the few in that crowd who had never been exposed to Social Ecology. Retrospectively, I do not think I recognized the life-changing elements of this seminar then, but I remember realizing the advantages of the concepts that were discussed. Trained as a natural scientist, I perceived the unique socioecological concepts of nature and its interactions with society almost as a sacrilege—but on second thought, I recognized how important and helpful that integrated interdisciplinary approach was. I was excited about these revelations without anticipating that I was ever to become part of that Institute and contribute to this book in the remote future.

Later, I was pleased to collaborate in joint projects that allowed me to dig deeper into Social Ecology. Infected with concepts such as the Human Appropriation of Net Primary Production (Chap. 14), I was soon labeled a 'socioeconomist' in my traditional surroundings. Several times, I realized that someone expected some 'socioeconomic' input or expertise from my side, which always resulted in disappointment on both sides. Questions like 'socio – what?' or 'how does this fit?' made me aware that an interdisciplinary approach was alien to most people, not to only researchers but also—and maybe especially—to non-scientists. I remember a couple of instances when someone, after having received an explanation of what I am doing, called out, 'Wow, that's really interesting!' But to be honest, my interlocutors often did not show so much enthusiasm. I do not know why, but many people turn reticent when it comes to interdisciplinary research. A strange animal such as Social Ecology can occasionally even intensify this reserve.

In conclusion, my 'socioecological adventure' may have driven me away from my roots, but in exchange, I had the possibility to participate in the challenges and excitements of interdisciplinary socioecological work—an experience that has become an important part of my life.

The Organization of this Book

In an attempt to display the strengths of Social Ecology in its continuous cross-fertilization of conceptual interdisciplinary thinking and empirical research, we have organized this book into five parts. The first part is entitled *The Conceptual Repertoire*, the second part *Empirical Approaches to Socioeconomic Metabolism*, the third part *Empirical Approaches to Long-Term Socioecological Research*, the fourth part *Empirical Approaches to Land Use and Colonization of Ecosystems*, and the fifth part *Empirical Approaches to Working with Stakeholders*. Interspersed in the second part are so-called *Method Précis*, short, non-technical explanations of important socioecological methods. These are attached to appropriate chapters. If a method is used in several chapters, the Precis is attached to the chapter where it first plays a substantial role. At the end, we supply an outlook section.

Part I—The Conceptual Repertoire

Chapter 1 opens the set of conceptual pieces by drawing a map of the ocean separating the three continents reigned over by the Natural Sciences, the Social Sciences and the Humanities. On that map, Marina Fischer-Kowalski and Helga Weisz identify the Social Ecology archipelago and, within that, the Viennese School of Social Ecology, among other interdisciplinary research fields. They portray Social Ecology as centered on a shared paradigm according to which human social and natural systems interact, coevolve over time and have substantial impacts upon one another, with causality pointing in both directions. Social Ecology addresses energy and society, land use and food production, the metabolism of societies and the environmental impacts of human activities. It offers a conceptual approach to society–nature coevolution pertaining to history, current development processes and a future sustainability transition.

Chapter 2 follows suit with an in-depth discussion of the conceptual and methodological repertoire of the Viennese Social Ecology. In this contribution, Marina Fischer-Kowalski and Karl-Heinz Erb outline the concepts of socioeconomic metabolism and the colonization of nature and discuss their epistemological foundations. Socioeconomic metabolism refers to material, substance and energy flows related to socioeconomic activities. The colonization of nature denotes purposive interventions into natural systems aimed at improving their utility for societal purposes. The chapter poses questions such as the following: can we conceive of social systems as 'hybrid systems,' a structural coupling between communication systems (in the tradition of sociological systems theory) and biophysical elements such as a human population, infrastructure and animal livestock? Is such a conceptualization consistent with notions of biological and cultural evolution and with complex systems theory?

The metabolism of human societies underwent major changes during the course of human history and is in a transition process now, as discussed in depth in Chap. 3 by Fridolin Krausmann and colleagues. Since the time of Paleolithic hunter-gatherers, the amount of materials extracted and used by humans and their impact on their environment has grown by several orders of magnitude. The chapter recalls major stages in human history through the lens of societal use of materials and energy. It introduces the notion of sociometabolic regimes and discusses the characteristics of resource use in hunter-gatherer societies, agrarian societies and the emergence of the industrial metabolic regime. It then discusses the variability within and among those regimes and the related environmental and sustainability problems.

Land use is a prime example of the human colonization of natural systems, as shown by Karl-Heinz Erb and colleagues in Chap. 4. Land use involves the colonization of ecosystems, organisms and, increasingly, the genomes of crop plants. This text focuses on land-use intensity, an important but far under-researched aspect of land use. It shows how the core concepts of Social Ecology—socioeconomic metabolism and the colonization of nature—open innovative avenues to research on land-use intensity. The authors argue that the strengths of the socioecological method inventory include the strict application of first principles, a sound and meaningful system boundary between society and nature and the accessibility for Social as well as Natural Science approaches. These features are seen as prerequisites for guiding data collation and organization, which allow researchers to investigate the feedback cycles between social and natural systems that constitute the trade-offs and synergies of the land system.

As apparent in several narratives in the box *Adventures in Social Ecology*, the tension between actor-centered and systems approaches represents an intellectual powerhouse of Social Ecology. In Chap. 5, Daniel Hausknost and colleagues tackle this conundrum head on. Their vantage point is that the emphasis on systems in many socioecological studies can profit from a complementary focus on actors in empirical socioecological research. Actors and their agency play an important role in transdisciplinary research, in local studies and in Environmental History. How do these actor-centered areas of research connect to the systems-centered theoretical framework of Social Ecology? How is agency accommodated in systems, and how can systems and their structures be influenced by actors? This chapter explores these questions both theoretically and in relation to concrete research examples. In doing so, it highlights some of the unresolved theoretical questions in Social Ecology and points toward possible ways they can be resolved.

In attempting to understand long-term changes, Social Ecology has employed a variety of different perspectives. One that has proven useful in many contexts is the concept of legacies, as Verena Winiwarter and colleagues discuss in Chap. 6. What one can learn from looking at things in a long-term comparative approach is exemplified by the 'fossil-energy-powered carbon sink' in Austria's agrarian–industrial transition and by the case of colonial mercury and silver mining in Central and South America. The importance of long-term legacies for the course of human history as well as our current predicament becomes visible, demonstrating the strength of an interdisciplinary socioecological approach to better understand current sustainability issues.

Labor is a central category in many Social Sciences, and it is central in Social Ecology because the process of colonization consists of physical interventions into Nature and requires human labor. In Chap. 7, Marina Fischer-Kowalski and Willi Haas outline a socioecological concept of labor. Throughout human history, socioecological transitions strongly affected human labor. The previous transition into the prevailing fossil-fuel-based sociometabolic regime fundamentally changed labor's characteristics in terms of physical power, knowledge and empathy. Lifetime spent on labor was reduced, and a new form of institutional organization became dominant: wage labor. Consequently, these authors ask how a transition away from the use of fossil fuels will shape labor's future.

Part II—Empirical Approaches to Socioeconomic Metabolism

The growing size and changing structure of socioeconomic metabolism is a major cause of the sustainability problems humanity is facing at the beginning of the twenty-first century (Ayres and Simonis 1994; Baccini and Brunner 2012; González de Molina and Toledo 2014). Socioeconomic metabolism is a core concept of Social Ecology. During the last two decades, the Vienna team has made important contributions to the advancement of the concept. It has developed related methods, and it has conducted a large number of empirical studies. This part of the book assembles contributions analyzing patterns and trends of socioeconomic metabolism and their drivers from global to national economies and across different temporal scales.

In the first contribution to this section, Fridolin Krausmann and colleagues (Chap. 8) investigate the evolution of global material and energy use since the take-off of global industrialization in the mid-nineteenth century. They show that global resource use has grown by more than an order of magnitude, much faster than population but slower than gross domestic product (GDP). They find that after a slow-down of global growth related to the stabilization of resource use at a high level in industrial countries, growth accelerated in the last decade, driven by emerging economies. This work reveals the full dimension of the metabolic transition and its impact on the extraction, trade and use of materials and energy at the planetary scale. Completing this historical transition globally may well be physically impossible due to resource constraints, but even if it were possible, it would wreak havoc with the earth's biotic and climatic systems.

In Chap. 9, Andreas Mayer and colleagues take a closer look at the variability of patterns of material consumption across individual countries. Their analysis of material flows in 180 countries shows that large differences in material flows can prevail even within groups of countries with a similar level of industrialization and economic development. The authors identify four major factors underlying the differences in global patterns of national material flows: resource availability, trade, population and GDP. The text highlights the potentials and limitations of indicators for direct material flows. These indicators provide important information about the material use patterns resulting from the domestic production structures, but they fail to adequately reflect the significance of final consumption.

Anke Schaffartzik and colleagues (Chap. 10) take up this point and address an issue that has recently received much attention in material flow studies: growing trade volumes and the deeper integration of all economies into global markets have posed a new challenge to material flow accounting (MFA). Addressing this challenge requires methods and approaches that expand MFA from a production-centered perspective to one that targets the resource flows related to consumption. The text critically discusses recent methodological developments in the field and, in its empirical part, shows how the shift to a consumption-based approach changes the material flow accounts for the Austrian economy.

The idea of a Circular Economy has become a prominent notion in the political discourse about sustainable resource use. Industry and governments in both industrial and emerging economies like to promote the term. In Chap. 11, Willi Haas and colleagues apply a sociometabolic approach to assess the degree of circularity of the global economy. Their analysis shows that only 7 % of all materials entering the global economy in 2005 were recycled. They identify the high share of materials used to generate thermal energy and the large and growing amount of materials accumulating in stocks as major obstacles to closing material loops, and they discuss the most effective steps toward a Circular Economy.

In Chap. 12, Dominik Wiedenhofer and colleagues investigate the role material stocks play in resource use patterns and sustainable development. Building and maintaining stocks of buildings and infrastructure is a major driver of resource use in both emerging and industrial economies, and large amounts of materials accumulate in growing stocks. The chapter presents an estimate of biophysical stocks of nonmetallic construction materials in residential buildings, roads and railways in the EU25, as well as related material input and output flows. Using a scenario approach, the authors assess stock-related material flows and recycling potentials for 2020. They identify proper management of existing transportation networks and residential buildings and a deceleration of the ongoing stock expansion as important steps toward more sustainable resource use.

Part III—Empirical Approaches to Land Use and Colonization of Ecosystems

Land use has transformed the face of the earth (Thomas et al. 1956; Turner et al. 1990) and is recognized as a pervasive driver of global environmental change (Foley et al. 2005). Since its inception, socioecological research has been concerned with land use, generating innovative concepts and avenues for empirical research as it progressed. This part of the book opens with a contribution by Karl-Heinz Erb and colleagues (Chap. 13), who discuss the role of livestock grazing, 'the neglected land use.' The observations are impressive: livestock consumes approximately 60 % of all the biomass used globally for human purposes (Krausmann et al. 2008), and livestock grazes approximately 36 % of the earth's lands, an area that is considerably larger than any other land-use category (Erb et al. 2007). The chapter discusses how socioecological methods can help tackle the conundrum of how to better understand the important process of livestock grazing in the earth system.

In Chap. 14, Helmut Haberl and colleagues discuss how socioecological methods—above all, the concept of 'human appropriation of net primary production,' or HANPP—can help researchers explore systemic feedbacks in global land use. The authors argue that a biophysical approach capable of linking land-use maps with biomass flows can help researchers understand how changes in one land use affect the entire system. Such changes create feedback that may cause havoc in unexpected places—for example, when the implementation of biofuels drives up food prices (Coelho et al. 2012) or when 'indirect land-use change' results in greenhouse gas emissions that may, in some cases, render biologically produced energy more destructive to the climate than fossil fuels (Chum et al. 2012).

Anthropogenic vegetation fires are another aspect of the global land system a socioecological approach can address. As Christian Lauk and Karl-Heinz Erb show in Chap. 15, human-induced vegetation fires play a central role in past and present society–nature interactions. There is evidence that tens of thousands of years ago, hunter-gatherers were already employing fires as a hunting technique. Today, vegetation fires continue to be an integral part of shifting cultivation and traditional pastoralism, and they are a crucial tool for the clearing of forests. At the same time, vegetation fires represent a risk that threatens infrastructures and contributes to climate change and air pollution. This text convincingly demonstrates how socioecological thinking can contribute to analyzing to what extent and under which circumstances human-induced vegetation fires are sustainable.

International trade plays an increasingly important role in supplying societies with biophysical resources and products, and land-based products such as food, feed, fiber and bioenergy are no exception. In Chap. 16, Helmut Haberl and colleagues show how extending the HANPP approach can lead to the consumption-based concept of 'embodied HANPP' to analyze teleconnections related to global biomass trade. Using the European Union as an example, they discuss how embodied HANPP can be estimated using bilateral trade matrices of biomass-based products. They show that the EU27 increasingly depends on lands outside its territory, and they discuss what that implies in terms of the land-related policies of the European Union.

Just as livestock grazing may be a neglected land use, Africa seems to be the neglected continent in terms of regional research efforts. Maria Niedertscheider and colleagues (Chap. 17) address that gap by using HANPP to analyze changes in Africa's land systems. They show that African land-use systems are unique compared to other world regions, with land-use intensity being lower than on any other continent. They analyze the current stage and main determinants of African land systems and their changes since 1980. Biomass harvest increased markedly in Africa, but this increase was driven mainly by cropland expansion rather than yield increases, which were more important almost everywhere else. Consequently, harvest growth was associated with considerable increases in the human domination of ecosystems in Africa.

Biodiversity is an important aspect of earth's ecosystems as it is related to the provision of goods and services that are essential to human society. In Chap. 18, Christoph Plutzar and colleagues discuss concepts and causes of biological diversity and—based on the socioecological interaction model—the relationship between human societies and biodiversity. Using empirical examples from Austria, they show how HANPP helps researchers study pressures on biodiversity related to changes in land use and its intensity. The results underline the potential of socioecological indicators to facilitate analysis of the interrelations between biodiversity and society.

Part IV—Empirical Approaches to Long-Term Socioecological Research

It has long been clear to ecologists that a systemic understanding of ecological systems requires long-term observations. Over the course of the twentieth century, Long-Term Ecological Research (LTER) at designated sites was established, albeit mostly in 'pristine' areas with as little human 'disturbance' as possible. Later, with the insight that most of the earth's land cover is influenced by humans and that the study of human-influenced land is badly needed, even urban LTER sites were set up. LTSER, Long-Term Socioecological Research, does more: it integrates the study of social systems and ecosystems at nested spatial levels over time to elucidate their interaction (Redman et al. 2004; Singh et al. 2013). The chapters in this section apply socioecological concepts and methods to historical case studies and illustrate their potential for LTSER and Environmental History. The chapters span a broad range in terms of both approaches and topics addressed, thereby showcasing the added value that can be gained from combining different scientific disciplines in long-term studies.

Martin Schmid, in Chap. 19, reconstructs the dramatic transformation of a peculiar landscape, the Donaumoos, a wetland along the left bank of the Upper Danube in Bavaria that was systematically drained from the 1770s onward. Drainage started during the first phase of the transition from an agrarian to an industrialized sociometabolic regime. The environmental and social consequences of this project were heavily contested already among contemporaries. With a long-term perspective covering more than 250 years, this Environmental History of the Donaumoos exemplifies how societies are trapped in a 'risk spiral,' meaning that solving one problem always results in new risks. Experts in the eighteenth century discussed major interventions into 'natural' systems with great passion. By revisiting a discourse of experts in the age of enlightenment, this chapter also contributes to a historical reflection of the term and the idea of 'colonization of nature.'

Chapter 20, on Austrian carbon budgets from 1830 to 2010, is written by an interdisciplinary team. Simone Gingrich and her co-authors offer an explanation of the 'forest transition,' the somewhat paradoxical fact that forests regrow with industrialization in many parts of the world. Empirical evidence on Austria's carbon budget in the period 1830–2010 shows that forests grew not only in area but also in wood density, resulting in the accumulation of ca. 240 MtC (megatons of carbon, 23 % of the initial stock) over the period. At the same time, society used more carbon, mostly in the form of fossil fuel. Austrian society amassed socioeconomic carbon stocks—in total, approximately 110 MtC—throughout the period, with construction wood being the main component. Annual carbon sequestration rates were well below fossil fuel emissions to the atmosphere. The authors conclude that the carbon sink in Austria's ecosystems and society was mainly a by-product of increased fossil-fuel use that accompanied industrialization. This has important consequences for environmental policy because it suggests that carbon sequestration cannot be expected to play a major role in transitions toward a low-carbon society.

Traditional low-input agriculture had to organize local land, labor and livestock resources in a way that maintained soil fertility and stable yields, albeit at a low level. Industrialization transformed the socioecological functioning of agriculture and its role in socioeconomic metabolism. In Chap. 21, Fridolin Krausmann discusses Austrian farming systems and offers a sociometabolic perspective on changes in the colonization of nature. Industrialization turned agriculture into a high input/high output system producing high yields but consuming more energy than it produces. By formalizing the functional interrelations of agricultural systems into a sociometabolic model, it becomes possible to reconstruct this process of transformation for the case of Austria.

Whereas Austria is a typical case for the industrialized part of the world, the Philippines serve as a case study of land-use transitions and sociometabolic regimes in the developing world. In Chap. 22, Thomas Kastner and colleagues reconstruct the rapid land-use transition of the archipelago during the twentieth century. Forest cover decreased from approximately 70 % in 1900 to less than 25 % in 2000, while cropland areas and grasslands expanded. Land-use change is linked to the transition from an agrarian to an industrial sociometabolic regime. During the twentieth century, population density rose more than tenfold. Fundamental changes in the nation's agricultural system ensued. During the first part of the century, food supply was maintained by expanding cultivated areas, while intensification became the dominant strategy thereafter. Fossil energy resources played a crucial role in this process. Massive changes in land-use intensity led to negative environmental impacts and increased dependency on fossil fuels and mineral resources. The Philippines are still in the midst of the transition toward an industrial society. The option space for future development is markedly smaller today than it was for those nations that pioneered the agrarian–industrial transition.

How tourism transformed an Alpine valley is the story Robert Groß tells in Chap. 23. Tourism moves global flows of capital, people and knowledge and thereby fundamentally transforms materiality, social relations, communities and life-worlds. This paper addresses the Environmental History of the alpine community of Damüls in Austria under the influence of tourism in the twentieth century. Environmental History seeks to understand historical society–nature relations and people's perception of nature in the past. Such a project poses a twofold challenge for environmental historians. They need to conceptualize nature as an independent historical factor without reducing it to a social construct, but at the same time, they need to address the social construction of 'beautiful landscapes' as an integral part of the tourism industry. These viewpoints can only be bridged dialectically, and this chapter uses the concept of socio-natural sites (SNSs, see Chap. 6 in this volume; Winiwarter and Schmid 2008) to tackle that conundrum. The example of Damüls is a telling example of the restless transformation, supported by the post-World War II Marshall Plan, of an Alpine sport arena built for the sake of skiers.

The agrarian–industrial transition plays out in specific ways in urban areas. In Chap. 24, Sylvia Gierlinger and Michael Neundlinger present an urban case elucidating the output side of social metabolism by discussing the cleaning of a metropolis, namely, nineteenth century Vienna and, in particular, its sewage system. The rapid transformation of the city posed a veritable challenge for the existing disposal infrastructure. City officials responded in particular ways to these challenges. Vienna's sewage system was built by incorporating many small creeks, thereby creating long-lasting legacies for river-city relations.

Part V—Empirical Approaches to Working with Stakeholders

Socioecological research aims to improve our understanding of transition processes and to support ongoing transition processes toward a more sustainable society. These chapters show how the approaches, concepts and methods discussed in the previous sections can bear fruit when applied together with stakeholders in practical settings, that is, in transdisciplinary projects.

Chapter 25, written by Veronika Gaube and colleagues, deals with the interaction of top-down and bottom-up decision-making determining the energy use in a city. Urban planning has to address a changing urban population and to provide the infrastructure required for the population's socioeconomic and environmental living conditions. Households play a major role: they are affected by urban planning decisions and are co-responsible for the environmental performance of a city. This chapter presents an agent-based decision model for the city of Vienna. The model assesses spatial patterns of energy use by household types. It shows that changes in preferences related to green areas in the vicinity of living quarters strongly influence the spatial distribution of households within the city area. The model also analyzes how the distribution of different households strongly affects the spatial patterns of energy use.

In Chap. 26, Barbara Smetschka and colleagues discuss how the analysis of time use can enrich socioecological research. They present two case studies of Austrian rural areas characterized by small-scale farming. Time use is crucial for decisions on the production strategies of farms; farmers avoid longer working hours as much as they try to maintain their income. The authors show that small-scale farming with a mixed production and cultivation of landscapes tends to increase the workload, particularly for women in a traditionally gendered working environment. Two alternatives are frequently chosen: farmers may either adopt less-sustainable methods of production or stop farming altogether. A better option may be to produce for the growing market for sustainable products with a new work organization that is attractive to young people and does not place a higher burden on female than male farmers. This presupposes (and can help finance) a local and regional infrastructure with care facilities. It also requires an innovative organization of labor and a fair distribution of the workload between sexes. Finally, it allows more leisure time for young farmers, men and women alike.

How does the 'local' relate to the 'global' and vice versa? Chap. 27, by Simron J. Singh and Willi Haas, illustrates this complex scale interaction, drawing on a case study from the remote Nicobar Islands in India. These islands, home to a rich tropical biodiversity and indigenous cultures, were subject to devastation by the 2004 tsunami. The tsunami took thousands of lives, devastated the villages, livestock and material culture and uprooted most of the coconut trees, a key base of subsistence. The overwhelming international aid that followed undercut the local self-determination of the islanders and resulted in a strong reliance on markets and cash flows. It led to surging material and energy flows dependent on aid and to fundamental changes in society–nature interactions. The new metabolic profile has high potential to drive large-scale land-use change and negatively affect local ecosystems. It remains to be seen how the Nicobarese, now without aid, will be able to recover not only from the first tsunami but also from the second tsunami of aid that washed away their modes of life that had been so well adjusted to their natural environment.

Chapter 28, on island sustainability, draws on several years of research and social involvement on the Greek island of Samothraki. Panos Petridis and Marina Fischer-Kowalski outline the challenges and the synergies of simultaneously conducting research 'on' a sustainability transition and research 'for' a sustainability transition. They explore the factors that cause island societies to prosper and sustain themselves and those that lead to collapse. In the past, a number of historical collapses, in the sense of breakdown of complexity and rapid population decline (Tainter 1990), have occurred on Samothraki. Currently, there is a fragile situation of slow decline of the population and rising ecological challenges, related mainly to excessive goat herding and tourism, that might lead to another 'tipping point' in the wake of the Greek economic and governance crisis and climate change. Meanwhile, the island community has decided to make an effort at turning the whole island into a Man and the Biosphere (MAB) Reserve according to UNESCO standards. Building upon a sociometabolic understanding of socioecological systems and using systems thinking and modeling efforts, the authors identify environmental and social tipping points for Samothraki intended to support that move. In line with the LTSER tradition, they argue that analyzing society-environment relations for different phases of the island's history and insights from past collapses can help to identify threats and possible ailments. Finally, this chapter reflects on both the process and the outcome of conducting transdisciplinary research.

Chapter 29, by Ulli Weisz and Willi Haas, reflects on the interrelations between two important societal concerns: sustainable development and health. This field of research underpins the key notion of 'health co-benefits.' The search for co-benefits aims to utilize synergies between both concerns, particularly between climate-change mitigation and health gains. It intends to inform policies in both arenas and to inspire collaboration. Although cross-cutting issues such as energy, agro-food systems and transport receive increasing international recognition, the health care system has rarely been addressed. The health care system takes responsibility for the reproduction of human health, usually through highly energy- and material-intensive forms of therapy. The health care system hence contributes to environmental problems and thereby contributes to health threats. In a transdisciplinary series of hospital projects involving scientists and health care practitioners, the authors ask how sustainability could be conceptualized for hospitals in line with both a socioecological understanding of sustainable development and with the 'hospitals' reality.' The approach aims to avoid unintended long-term side effects of health care—hospitals' core business—by expanding quality criteria for decision-making to include sustainability. The results of the testing phase convinced political actors in the health care system to make changes in their therapeutical settings. The authors demonstrate that 'health co-benefits' are a valuable argument within the sustainability debate.

Outlook and Conclusions

A recent compilation of global change indicators (Steffen et al. 2015) suggests that the Great Acceleration of human activities on earth—and the related pressures on its biophysical systems—is ongoing. There are no signs of humanity achieving any reductions in greenhouse gas emissions; on the contrary, they are continuing to grow, and faster than expected (Le Quéré et al. 2014). No comprehensive indicators exist to reliably monitor the global biodiversity trend (Jones et al. 2011), but there is little to suggest that biodiversity loss is slowing. Evidence is mounting that extinctions are already altering key processes that affect the productivity and functioning of ecosystems worldwide (Hooper et al. 2012). Recent research suggests that the world is running into constraints of resource extraction for several important resources simultaneously (Seppelt et al. 2014), thereby fueling a 'peak everything' debate. All of this is aggravated by financial crises in the world's industrial centers, particularly affecting the US, Europe and Japan. Moreover, there are mounting concerns over growing inequalities in the distribution of incomes and wealth (Piketty 2014).

Sustainable development may hence be expected to remain high on the agenda, and with it the challenges of inter- and transdisciplinary knowledge production, or the 'co-production of knowledge,' as it has recently been termed (Mauser et al. 2013). This book summarizes the contributions Social Ecology can offer to both Science and practice. It aspires to bridge the Social Sciences with the Natural Sciences and the Humanities to analyze society–nature interactions in a manner that is useful in designing interventions in Science and society:

- In relation to the Social Sciences, it aims to overcome the blindness of the Social Sciences to biophysical processes, enabling the Social Sciences to engage in sustainability discourses often dominated by the Natural Sciences and, in so doing, improving the ability to communicate with the Natural Sciences and the Humanities.
- In relation to the Natural Sciences, it aims to improve the ability to communicate with the Social Sciences and the Humanities, developing concepts for empirical research that link to social and cultural phenomena and overcoming arrogance and negligence vis-à-vis other realms of scholarship.
- In relation to the Humanities, it aims to improve the ability to communicate with the Natural and Social Sciences, to establish links with system approaches and to accept the role of natural phenomena in shaping the course of history.

This book shows both the conceptual depth and the empirical richness this agenda has fostered at the Institute of Social Ecology over the last few decades. It is based on the sustained effort of a group of researchers from various fields who have grappled with several problems, including the following: intricate conceptual issues; huge databases and data gaps; fundamentally different models, bodies of literature and academic cultures; and language disparities among the scientific 'continents.' This group has also striven to maintain productive and enjoyable working relations within a growing team. Finally, this group has addressed challenges in raising sufficient money to make all the above happen. We are enormously grateful for friendly support from colleagues, partners in practice and many other people and institutions. We are also grateful for the funding we have received from a great number of sponsors and funding agencies. There are too many supporters to list here, but most chapters thank them in their respective acknowledgement sections. We hope that you, the readers, will profit from this work. We would be truly grateful for your feedback.

One person truly deserves to be singled out here, however. Elise Harder accompanied us as editorial assistant throughout the whole book writing and production process. Without her help this book would not exist. Thank you so much, Elise!

References

Allwood, J.M., Cullen, J.M., Carruth, M.A., Cooper, D.R., McBrien, M., Milford, R.L., Moynihan, M.C. & Patel, A.C. (2012). *Sustainable materials: with both eyes open*. Cambridge: UIT Cambridge.

Armitage, D., Berkes, F. & Doubleday, N. (2007). *Adaptive co-management: Collaboration, learning, and multi-level governance*. Vancouver: UBC Press.

Ayres, R.U. & Simonis, U.E. (1994). *Industrial metabolism: Restructuring for sustainable development*. Tokyo, New York, Paris: United Nations University Press.

Baccini, P. & Brunner, P.H. (2012). Metabolism of the Anthroposphere: Analysis, Evaluation, Design, 2nd edition, Cambridge, Mass.

Becker, E. & Jahn, T. (2006). *Soziale Ökologie: Grundzüge einer Wissenschaft von den gesellschaftlichen Naturverhältnissen*. Campus, Frankfurt.

Berkes, F., Colding, J. & Folke, C. (2003). *Navigating social-ecological systems: Building resilience for complexity and change*. Cambridge, UK: Cambride University Press.

Bourdieu, P. (1984). *Homo academicus*. Paris: Ed. de Minuit.

Chum, H., Faaij, A., Moreira, J., Berndes, G., Dhamija, P., Gabrielle, B., Eng, A.G., Lucht, W., Makapo, M., Masera Cerruti, O., McIntyre, T., Minowa, T. & Pingoud, K. (2012). Bioenergy, In: Edenhofer, O., Pichs-Madruga, R., Sokona, Y., Seyboth, K., Matschoss, P., Kadner, S., Zwickel, T., Eickemeier, P., Hansen, G., Schlömer, S. & von Stechow, C. (Eds.), *IPCC Special Report on Renewable Energy Sources and Climate Change Mitigation* (pp. 209–332). Cambridge, UK: Cambridge University Press.

Clark, W.C. & Dickson, N.M. (2003). Sustainability science: The emerging research program. PNAS 100, 8059–8061.

Coelho, S., Agbenyega, O., Agostini, A., Erb, K.-H., Haberl, H., Hoogwijk, M., Lal, R., Lucon, O., Masera, O. & Moreira, J.R. (2012). Land and Water: Linkages to Bioenergy, in: Johansson, T., Patwardhan, A., Nakicenonivc, N., Gomez-Echeverri (Eds.), *Global Energy Assessment. International Institute of Applied Systems Analysis (IIASA).* (pp. 1459–1525). Cambridge, UK: Cambridge University Press.

Dressel, G., Berger, W., Heimerl, K. & Winiwarter, V. (2014). *Interdisziplinär und transdisziplinär forschen: Praktiken und Methoden*. Bielefeld: Transcript Verlag.

Erb, K.-H., Gaube, V., Krausmann, F., Plutzar, C., Bondeau, A. & Haberl, H. (2007). A comprehensive global 5 min resolution land-use data set for the year 2000 consistent with national census data. *Journal of Land Use Science*, 2, 191–224.

Fischer-Kowalski, M. & Haberl, H. (1997). Tons, Joules and Money: Modes of Production and their Sustainability Problems. *Society & Natural Resources*, 10, 61–85.

Foley, J.A., DeFries, R., Asner, G.P., Barford, C., Bonan, G., Carpenter, S.R., Chapin, F.S., Coe, M.T., Daily, G.C., Gibbs, H.K., Helkowski, J.H., Holloway, T., Howard, E.A., Kucharik, C.J., Monfreda, C., Patz, J.A., Prentice, I.C., Ramankutty, N. & Snyder, P.K. (2005). *Global Consequences of Land Use. Science*, 309, 570–574.

Giampietro, M., Mayumi, K. & Sorman, A.H. (2012). The Metabolic Pattern of Societies. Where economists fall short, Studies in ecological economics. London: Routledge.

Glaser, M., Krause, G., Ratter, B. & Welp, M. (2012). Human-Nature Interactions in the Anthropocene. Potentials of Social-Ecological Systems Analysis, Studies in Environment, Culture and Society (ed. M. Glaser & H. Egner). Routledge, London.

González de Molina, M. & Toledo, V.M. (2014). The Social Metabolism. A Socio-Ecological Theory of Historical Change. Springer, Environmental History 3, Dordrecht.

Gottschlich, D. & Friedrich, B. (2014). Das Erbe der Sylvicultura oeconomica. Eine kritische Reflexion des Nachhaltigkeitsbegriffs. (The Legacy of Sylvicultura oeconomica. A Critical Reflection on the Notion of Sustainability). GAIA - Ecological Perspectives for Science and Society 23, 23–29.

Gunderson, L. & Holling, C.S. (2002). Panarchy: understanding transformations in human and natural systems. New York: Island Press.

Haberl, H., Fischer-Kowalski, M., Krausmann, F., Weisz, H. & Winiwarter, V. (2004). Progress towards sustainability? What the conceptual framework of material and energy flow accounting (MEFA) can offer. Land Use Policy 21, 199–213.

Hall, C.A.S. (2004). The Myth of Sustainable Development: Personal Reflections on Energy, its Relation to Neoclassical Economics, and Stanley Jevons. J. Energy Resour. Technol. 126, 85–89.

Hooper, D.U., Adair, E.C., Cardinale, B.J., Byrnes, J.E.K., Hungate, B.A., Matulich, K.L., Gonzalez, A., Duffy, J.E., Gamfeldt, L. & O'Connor, M.I. (2012). A global synthesis reveals biodiversity loss as a major driver of ecosystem change. *Nature*, 486, 105–108.

IUCN, UNEP, WWF (1980). World Conservation Strategy - Living Resource Conservation for Sustainable Development. International Union for Conservation of Nature and Natural Resources (IUCN), Switzerland: Gland.

Jones, J.P.G., Collen, B., Atkinson, G., Baxter, P.W.J., Bubb, P., Illian, J.B., Katzner, T.E., Keane, A., Loh, J., Mcdonald-Madden, E., Nicholson, E., Pereira, H.M., Possingham, H.P., Pullin, A.S., Rodrigues, A.S.L., Ruiz-Gutierrez, V., Sommerville, M. & Milner-Gulland, E.J. (2011). The Why, What, and How of Global Biodiversity Indicators Beyond the 2010 Target. *Conservation Biology*, 25, 450–457.

Kander, A., Malanima, P. & Warde, P. (2013). Power to the People. Energy in Europe Over the Last Five Centuries. Princeton: Princeton University Press.

Kates, R.W., Clark, W.C., Corell, R., Hall, J.M., Jaeger, C.C., Lowe, I., McCarthy, J.J., Schellnhuber, H.J., Bolin, B., Dickson, N.M., Faucheux, S., Gallopin, G.C., Grübler, A., Huntley, B., Jäger, J., Jodha, N.S., Kasperson, R.E., Mabogunje, A., Matson, P., Mooney, H., Moore, B., O'Riordan, T. & Svedin, U. (2001). Sustainability science. *Science*, 292, 641–642.

Krausmann, F., Erb, K.-H., Gingrich, S., Lauk, C.& Haberl, H. (2008). Global patterns of socioeconomic biomass flows in the year 2000: A comprehensive assessment of supply, consumption and constraints. *Ecological Economics*, 65, 471–487.

Le Quéré, C., Peters, G.P., Andres, R.J., Andrew, R.M., Boden, T.A., Ciais, P., Friedlingstein, P., Houghton, R.A., Marland, G., Moriarty, R., Sitch, S., Tans, P., Arneth, A., Arvanitis, A., Bakker, D.C.E., Bopp, L., Canadell, J.G., Chini, L.P., Doney, S.C., Harper, A., Harris, I., House, J.I., Jain, A.K., Jones, S.D., Kato, E., Keeling, R.F., Klein Goldewijk, K., Körtzinger, A., Koven, C., Lefèvre, N., Maignan, F., Omar, A., Ono, T., Park, G.-H., Pfeil, B., Poulter, B., Raupach, M.R., Regnier, P., Rödenbeck, C., Saito, S., Schwinger, J., Segschneider, J., Stocker, B.D., Takahashi, T., Tilbrook, B., van Heuven, S., Viovy, N., Wanninkhof, R., Wiltshire, A. & Zaehle, S. (2014). Global carbon budget 2013. *Earth System Science Data*, 6, 235–263.

Martinez-Alier, J. (1987). *Ecological Economics*. Energy, Environment and Society. Oxford, UK: Blackwell.

Mauser, W., Klepper, G., Rice, M., Schmalzbauer, B.S., Hackmann, H., Leemans, R. & Moore, H. (2013). Transdisciplinary global change research: the co-creation of knowledge for sustainability. *Current Opinion in Environmental Sustainability*, Open issue 5, 420–431.

McNeill, J.R. (2000). *Something New Under the Sun: An Environmental History of the Twentieth-Century World*. New York: W.W. Norton & Company.

Meadows, D.H., Meadows, D.L., Randers, J. & Behrens, W.W. (1972). The Limits To Growth. A Report for the Club of Rome's Project on the Predicament of Mankind. New York: Universe Books.

Piketty, T. (2014). *Capital in the Twenty-First Century*. Cambridge: Harvard Press.

Redman, C.L., Grove, J.M. & Kuby, L.H. (2004). Integrating Social Science into the Long-Term Ecological Research (LTER) Network: Social Dimensions of Ecological Change and Ecological Dimensions of Social Change. *Ecosystems*, 7, 161–171

Rohlf, W.D.J. (2008). *Introduction to Economic Reasoning*, Seventh Edition. ed. Boston: Addison-Wesley.

Seppelt, R., Manceur, A.M., Liu, J., Fenichel, E.P. & Klotz, S. (2014). Synchronized peak-rate years of global resources use. *Ecology and Society*, 19.

Sieferle, R.-P. (2001). *The subterranean forest: Energy systems and the industrial revolution*. Cambridge: The White Horse Press.

Singh, S.J., Haberl, H., Chertow, M., Mirtl, M. & Schmid, M. (Eds.) (2013). Long-Term Socio-Ecological Research. Springer, Dordrecht, Heidelberg, London, New York.

Steffen, W., Broadgate, W., Deutsch, L., Gaffney, O. & Ludwig, C. (2015). The trajectory of the Anthropocene: The Great Acceleration. *The Anthropocene Review*, 2, 81–98.

Tainter, J.A. (1990). The Collapse of Complex Societies. Cambridge: Cambride University Press.

Thomas, W., Sauer, C.O., Bates, M. & Mumford, L. (1956). Man's Role in Changing the Face of the Earth. Chicago: The University of Chicago Press.

Turner, B.L.I., Clark, W.C., Kates, R.W., Richards, J.F., Mathews, J.T. & Meyer, W.B. (1990). The earth as transformed by human action: global and regional changes in the biosphere over the past 300 years. Cambridge: Cambridge Univeristy Press.

United Nations (1987). Our Common Future. Report of the World Commission on Environment and Development. UN Document. http://conspect.nl/pdf/Our_Common_Future-Brundtland_Report_1987.pdf.
Winiwarter, V. & Schmid, M. (2008). Umweltgeschichte als Untersuchung sozionaturaler Schauplätze? Ein Versuch, Johannes Colers "Oeconomia" umwelthistorisch zu interpretieren, In: Knopf, T. (Ed.), Umweltverhalten in Geschichte und Gegenwart: Vergleichende Ansätze. Narr Francke Attempto Verlag GmbH + Co. KG, Tübingen, pp. 158–173.
World Bank (1978). The World Development Report, 1978. The World Bank, Washington, D.C.
World Bank (1992). World Development Report 1992: Development and the Environment. The World Bank, Washington, D.C.

Part I
The Conceptual Repertoire

Chapter 1
The Archipelago of Social Ecology and the Island of the Vienna School

Marina Fischer-Kowalski and Helga Weisz

Abstract Social Ecology is an interdisciplinary research field rooted in the traditions of both the Social Sciences and Natural Sciences. The common denominator of this research field is not a shared label but a shared paradigm. Related labels that extend beyond Social Ecology include Human Ecology, Industrial Ecology, Ecological Economics and Socioecological Systems Analysis. The core axioms of the shared paradigm are that human social and natural systems interact, coevolve over time and have substantial impacts upon one another, with causality working in both directions. Social Ecology offers a conceptual approach to society-nature coevolution pertaining to history, to current development processes and to a future sustainability transition. This chapter reviews several academic traditions that have contributed to the emergence of this paradigm and then describes the research areas belonging to the field. One cluster deals with society's biophysical structures (such as energy and society, land use and food production and social metabolism, the field covered by Industrial Ecology and Ecological Economics). Other clusters identify the environmental impacts of human societies (such as the IPAT and footprint approaches), biohistory and society-nature coevolution. Another research area considers regulation, governance and sustainability transitions. In the last section, we describe the distinguishing characteristics of the Vienna Social Ecology School.

M. Fischer-Kowalski (✉)
Institute of Social Ecology (SEC), Alpen-Adria University,
Schottenfeldgasse 29, 1070 Vienna, Austria
e-mail: marina.fischer-kowalski@aau.at

H. Weisz
Potsdam Institute for Climate Impact Research, Telegrafenberg A56,
P.O. Box 601203, 14412 Potsdam, Germany
e-mail: helga.weisz@pik-potsdam.de

H. Haberl et al. (eds.), *Social Ecology*, Human-Environment Interactions 5,
DOI 10.1007/978-3-319-33326-7_1

Keywords Industrial ecology · Human ecology · Ecological economics · Social metabolism · Land use · Food systems · Sustainability transition · Society-nature coevolution · Energy and society · Natural resource use · Development · Environmental-impacts · Ecological footprint · Social-ecological systems

1.1 Introduction

We use the metaphor of an archipelago to describe Social Ecology as situated between the mainland of the Natural Sciences and Engineering on the one side and the mainland of the Social Sciences and the Humanities on the other side. Some islands are large and even have their own shipping lines (i.e., scientific journals), others are small and need to use foreign bypassing ships, and some are populated by isolated tribes. The populations on the archipelago are of mixed disciplinary origin and speak different scientific languages. They do not necessarily share a common name. They do share a few important features, however. They look at natural and social systems as systems in their own right that interact with one another, they believe causality between these systems works in both directions, and they search for less destructive and more sustainable ways in which the two systems can interact.

Social Ecology draws on traditions from several scientific disciplines (see Sect. 1.2) from the Social and Natural Sciences. Whatever the discipline of origin, the common motive for moving in the direction of socioecological research is a critical attitude toward the outcome of decades of differentiation and specialization among the academic disciplines (Latour 1991). The lack of intellectual cooperation, particularly from the 1970s onward, is considered detrimental to society's ability to properly understand and address its relation to the—increasingly strained—natural environment.

The common denominator of this research field is not so much a shared label—names extend beyond Social Ecology to Human Ecology, Industrial Ecology, Ecological Economics and Socioecological Systems Analysis—but a shared paradigm. The core axioms in this paradigm are that human social and natural systems interact, coevolve over time and have substantial impacts upon one another. What follows from this paradigm is a need to develop concepts and methods that allow us to address social and natural structures and processes on an equal epistemological footing. In various strands of the research, this challenge has been and is being resolved in different ways and at different levels of depth and consistency.

In the following essay, we will first reconstruct some of the earlier academic roots of social ecological thinking and then discuss several research traditions that address the biophysical features of human societies, such as energy, land use and social metabolism. Then, we will review approaches to identifying the environmental impacts of human activities. A third part is devoted to biohistory, and it reviews theoretical and empirical efforts to analyze the society-nature coevolution.

Finally, we will turn toward issues of regulation and governance, focusing on what to address as part of a sustainability transition. The last section characterizes the specifics of the Vienna Social Ecology approach.

1.2 Academic Traditions Contributing to the Emergence of Social Ecology

The academic roots of Social Ecology can be traced as far back as the 19th century, when the Natural and Social Sciences had not yet fallen into their respective epistemic boxes, which made later disciplinary crossovers so difficult. There are excellent reviews reconstructing such roots in the political economies of Adam Smith, David Ricardo, Karl Marx and Thomas Malthus (Fischer-Kowalski 1998; Martinez-Alier 1987; Sieferle 1990). These reviews illustrate the debates on the interrelations among population, land, food, technology and economic development. Whereas Smith, Ricardo and Malthus insisted on natural limitations for economic growth (in particular, land), Marx was the first to claim technological development (and thus human ingenuity) as the key driver of economic growth, thus overcoming natural limitations.

Another influential field was geography. George Perkins Marsh's book *Man and Nature: or, Physical Geography as Modified by Human Action* (1864) inspired at least two major efforts to comprehensively account for human-induced changes in the Earth system. One was the Princeton Conference on *Man's Role in Changing the Face of the Earth* (Thomas 1956). Another was the conference *The Earth as Transformed by Human Action*, held in 1987 at Clark University (Turner et al. 1990). In 1969, the German geographer Neef explicitly talked about the 'metabolism between society and nature' as a core problem of geography (Neef 1969). Since then, geographers have played a major role in Social Ecology.

Cultural Ecology, as brilliantly reviewed by Orlove (1980), is another important predecessor of later socioecological research. The beginnings of Cultural Anthropology (as in the works of Morgan 1877/1963) were, like Sociology, marked by evolutionism, that is, the idea of universal historical progress from more 'natural' and barbaric to more advanced and civilized social conditions. Then, Cultural Anthropology split into a more functionalist and a more culturalist tradition. The functionalist line retained a focus on the society-nature interface. Leslie White, one of the most prominent anthropologists of his generation and an early representative of the functionalist tradition, rekindled interest in energetics. White described the vast differences in the types of extant societies as social evolution, and the mechanisms propelling it were energy and technology (White 1943). Julian Steward's 'method of cultural ecology' considered the quality, quantity and distribution of resources within the environment. His approach can be illustrated by the early comparative study *Tappers and Trappers* (Murphy and Steward 1955), where two cases of cultural (and economic) change are presented in which tribes traditionally living from subsistence hunting and gathering (and some horticulture) completely change their ways of living because of their changing metabolism. The authors analyze this dynamic as an irreversible shift from a subsistence economy to dependence upon trade.

Despite some early calls for an 'ecology of man' (Adams 1935; Darling 1956; Sears 1953), Biological Ecology was reluctant to engage in Human Ecology before the environmental debate of the 1970s (Young 1974). Moreover, when the first influential texts by biological ecologists on Human Ecology finally appeared (e.g., Ehrlich and Ehrlich 1970; Ehrlich et al. 1973), they took a route that remained typical of most work by biological ecologists in this field until only recently: that of humans as agents of disturbance in ecosystems. This conceptualization of societies as one aggregated universal actor ignores the internal complexity unique to social systems and generates the misleading idea that society can be viewed as analogous to a single rational person. In addition, the exclusive focus on humans changing the environment prevented an understanding of mutual influences between society and nature. Together, these biases created severe barriers to interdisciplinary approaches toward the society-nature interaction. That ecologists tended to favor 'natural' ecosystems over 'human-dominated' ones as study objects may have contributed to these biases, but the most important factor was probably that many biological ecologists simply did not recognize the need to develop a more complex approach that would require conceptualizing socioeconomic systems as entities of a different kind than natural systems. Many ecologists may have been reluctant to engage in interdisciplinary cooperation. Even worse, neo-Malthusian concepts played an important role in the bioecological approaches toward Human Ecology (above all in the work of Paul Ehrlich), a point that hampered cooperation with social scientists. This changed substantially during the revived environmental debates of the 1970s, when major international research programs, such as *Man and the Biosphere* (MAB) by the United Nations Educational, Scientific and Cultural Organization (UNESCO) and the *International Human Dimensions Programme on Global Environmental Change* (IHDP), were launched, stimulating and supporting interdisciplinary work across the 'great divide' (Snow 1956) of the Social and Natural Sciences.

Meanwhile, the historical sciences, particularly the tradition of the Annales-School (Fernand Braudel), paved a path toward social ecological reasoning. For example, Braudel viewed the history of the Mediterranean as an outcome of interaction between social and natural processes. M. Godelier went further in formulating his core hypothesis in the introduction to *The Mental and the Material*: 'Human beings have a history because they transform nature. It is indeed this capacity which defines them as human. Of all the forces which set them in movement and prompt them to invent new forms of society, the most profound is their ability to transform their relations with nature by transforming nature itself' (Godelier 1984, p. 1). This way of looking at history is related to the Marxist tradition, but it transcends this tradition by moving in an ecological or coevolutionary direction. The classic reading of Marx leads to a discussion of changing 'modes of appropriation of nature' through the development of new means of production, that is, technology. Godelier's reading stresses the fact that human appropriation of nature modifies nature, and this modified nature in turn stimulates social change. Godelier thus deviates from typical Social Science by viewing nature as historically variable, not as static, and his core hypothesis attributes societies' historical dynamics to a feedback process from nature. The study of Environmental History since the 1970s is increasingly working along this basic idea of mutuality of nature-society relations (Winiwarter and Knoll 2007).

For Sociology, some claim the so-called Chicago School of Human Ecology (Park, Burgess, Duncan) as an entry point to the modern reading of Social Ecology. This school used analogues from biological ecology to analyze urban development (e.g., hierarchy, competition, succession). For them, however, the natural environment was reduced to spatial structure. For example, Duncan's *POET model* (population, organization, environment and technology) for describing social processes in no way referred to natural processes or conditions except space (Beus 1993). Some reviewers from the German Human Ecology tradition took a different view (Bruckmeier 2004). Catton and Dunlap (1978), for example, called for Sociology to move beyond Durkheim's dictum that 'a social fact can be explained only by another social fact' (as cited in Beus 1993, p. 94) and to abandon the 'human exceptionalism paradigm' in favor of an ecological paradigm in which the human species is one among other species on earth, sharing their susceptibility to nature. Although frequently cited, this appeal has not yet given rise to substantially new theoretical approaches, although there is a growing body of empirical research from environmental sociologists. A decade later, Beck (1986) started to publish on 'risk society', proposing that modern society in its latest stages should be characterized by its ways of creating and handling environmental risks and redistributing their consequences among its members rather than as a traditional industrial society occupied with emancipation from natural forces and efforts to legitimately handle social inequalities. The neo-Marxist tradition within Sociology tended to become narrowly focused on theories of capital, class and the state. Even the Marxist concept of 'control of the means of material production' was narrowed; private property and ownership signify a purely social or economic relationship, not a coupling between social actors and natural objects. There are, however, contemporary positions where this tradition is retained and explicitly linked to ecological concerns, such as the influential World Systems Theory (cf. Ciccantell and Bunker 1998; Goldfrank et al. 1999; Hornborg and Crumley 2007; Wallerstein 1999). Within Human Geography and Environmental Sociology, efforts have been made to link the economic requirement of capital accumulation to both economic growth and the continuing (over)exploitation of natural resources (Harvey 2014; Schnaiberg and Gould 1994). Another strand is exemplified by Foster (2000), who seeks to reconstruct and build upon Marx's materialist conception of history and his notion of 'metabolism of nature and society'. Jänicke (1988) and, later, Mol and Spaargaren (2000) opened a debate on 'ecological modernization' that claimed that the technological and organizational learning processes of modern societies increasingly led to the amelioration of environmental impacts. All these approaches, however, fall short of an epistemic turn that would allow for relations between the social and the natural that are more symmetrical.

With the German sociologist N. Luhmann, Sociology's influence on Social Ecology reached its peak and turning point. Luhmann's social theory builds on interdisciplinary systems theory, as exemplified in the works of H. v. Foerster, G. Bateson, H. Maturana and F. Varela, and on the formal epistemology of the mathematician G.S. Brown (Luhmann 1984/1995). The resulting general definition of systems as a self-referential operation (termed operationally closed) implies generally conceiving of systems as entities that reproduce their own boundaries toward the environment. Functionally, this means such systems create and use an internal mode of operation, e.g. communication in social systems, that distinguishes the system from its environment.

Combining this interdisciplinary background with a painstaking knowledge of the sociological and philosophical tradition, Luhmann arrives at the logical conclusion that a social system is not composed of humans but of communication between humans. It follows that social systems should be specified by defining how, what and to what effect humans communicate. The focal interest of the Social Sciences, then, would be to study the successes and failures of these communications in the short and long term. This implies distinguishing the operations of social systems from the consciousness of individual persons on the one hand and the socially organized physical condition of these persons as biological organisms on the other.

This theoretical architecture, enriched by a theory of communication and a theory of sociocultural evolution, enabled Luhmann to develop a social theory (Luhmann 1997/2012) that was unprecedented in its reach, complexity and sophistication. Socioecologically, Luhmann's theory thus marks the antipode to Ehrlich's understanding of the social as an aggregated and essentially undifferentiated human population.

Applying social systems theory to socioecological research, however, is an intricate and demanding task. Under the lens of this theory, the seemingly compelling idea of social and natural systems directly interacting with each other needs to be replaced by a concept of a complex network of structurally coupled systems in which physical embedding and societal self-regulation are attributed to three different system types: natural, human and communication (Sieferle 1997, 2011; Weisz 2002; Weisz and Clark 2011; Weisz et al. 2001).

In his book *Ecological Communication*, Luhmann applies his theory to investigate the question of why modern societies are facing so many difficulties in adequately reacting to the disturbances they create in their natural environment, even if these disturbances might turn out to be detrimental in the long run (Luhmann 1986/1989). In essence, Luhmann attributes the inability of modern societies to cope with global environmental change to exactly the same social structures (i.e., functional differentiation) that constituted the decisive evolutionary advantage of modern societies over traditional ones. Social Ecology noted that this was contingent on an unprecedented ability to utilize energy resources, which in turn created environmental change at a planetary scale (Weisz et al. 2001). Any contemporary observer of the international COP negotiations[1] to reach a binding climate mitigation treatment will find ample evidence for Luhmann's analysis from 1986. In conclusion, Luhmann's social theory allows for, but does not directly explore, the relational biophysical conditions of sociocultural evolution.

1.3 Society's Biophysical Structures

1.3.1 Energy and Society

The idea of energetic evolutionism, namely, that the control of energy matters for society and even determines the advancement of civilization, has a long tradition in social theory, prominently represented by Spencer. In his *First Principles* in 1862, the process of societal advance and the differences in stages of advancement among societies can be accounted for by energy: the more energy a society is able to consume, the more advanced it is. Societal progress is based on energy surplus. First, a surplus enables social growth and social differentiation. Second, it provides room for cultural activities beyond basic vital needs. Similarly, the beginnings of Cultural Anthropology were marked by energetic evolutionism (as in the works of Morgan 1877/1963). Along a less ideological vein, Cottrell (1955) offered a careful analysis of the relevance of the sources and amounts of societally available energy for social processes. The physicist V. Smil published periodic compendia, from *Energy in the Biosphere and Civilization* (1991) to *Energy in Nature and Society* (2008), that compiled encyclopedic knowledge on how energy matters socially and economically. Another physicist, R. Ayres, has presented convincing theoretical and empirical evidence that the expenditure of useful work (i.e., exergy) was and is key to economic growth (Ayres and Warr 2005). The historian Sieferle (1982/2001b) analyzed the rise of the United Kingdom (UK) in industrialization and political hegemony as an outcome of its 'subterranean forest', that is, its use of coal, which gave the UK access to many times more energy than if its entire territory had been covered with forest that was harvested and burned as an energy source. Current research is stimulated by the issue of reducing fossil fuel consumption both to address the impending 'peak oil' (and peak fossil fuels not too far away; Murphy 2012) and to avoid dangerous climate change and its potential consequences for economies and societies. What would be the potential consequences of changing society's energy base and of possibly reducing the energy intensity of social processes altogether?

[1]Since the mid-1960s, annual 'Conferences of the Parties' (COP) have been held within the United Nations Framework Convention on Climate Change (UNFCCC). The 'parties' are countries classified by the Convention into various groups with different obligations.

1.3.2 Land Use and Food Production

That societies extend over territories and restructure them for social purposes, with severe consequences for social as well as natural processes, is not a new issue in the Social Sciences; indeed, it has gained substantial momentum. Boserup (1965, 1981), in her continuation and in her critique of Neo-Malthusianism, was at odds with the development policies of her time by arguing for and empirically demonstrating social learning processes in the face of population growth and food scarcity. She was able to show that traditional agriculture found ways to accommodate feeding more people by intensifying land use (and not, as in the Malthusian paradigm, by extending agricultural area). Land-use intensity is an essential aspect of the human use of terrestrial ecosystems. In the course of history, the intensification of land use allowed humans to overcome Malthusian traps and to both support population growth and improve the supply of food and other products dependent on photosynthesis. It helped to achieve increases in agricultural production without requiring proportional increases in the area of agricultural land. However, thanks to intensification, most industrialized countries increased the volume of agricultural output despite shrinking agricultural areas in the last several decades, if not centuries. In the industrial part of the world, we find reforestation instead of the long-term deforestation of the past (although possibly at the expense of deforestation in developing countries). However, increasing land-use intensity has often been associated with detrimental effects on ecosystem functioning, such as soil degradation, groundwater and air pollution and biodiversity loss. Such processes have had negative effects on the ability of ecosystems to sustain vital ecosystem services, thereby potentially jeopardizing human well-being in the end. Under traditional agriculture (which prevailed worldwide until the 1960s), increased food output per unit area was achieved through increased investment of human labor (Boserup 1981; Netting 2010). However, this generated an incentive for high fertility to provide the necessary labor power, and this drove population growth. When fossil fuel use allowed for the industrialization of agriculture (mineral fertilization, pesticides, tractors), this mechanism changed. Agriculture turned from a supplier to a consumer of energy (Pimentel et al. 1973) and started to create toxicological hazards, such as those documented in R. Carson's famous book *Silent Spring* (1962), which examined the risk of poisoning along the whole food chain. Currently, the debate centers more on the risks of genetic engineering (see, for example, the Nature Special Feature 2013). The issue of land use and land cover change has mobilized a large research community, most recently the international program on *Future Earth* (http://www.icsu.org/future-earth), which addresses food security, diets, carbon emissions, biodiversity losses, climate and habitat change in broad interdisciplinary cooperation.

1.3.3 Social Metabolism/Ecological Economics/Industrial Ecology

As early as Marx, social metabolism with nature was at the core of human labor and society (Marx 1867/2010), leading to a philosophical/sociological debate on capitalism introducing a 'metabolic rift' (Foster 2000; Schmidt 1971) between humans and the environment. These considerations were (unknowingly) reintroduced from quite another angle by Ayres and Kneese. They claimed that the common failure of Economics results from viewing the production and consumption processes in a manner that is somewhat at variance with the fundamental law of the conservation of mass (Ayres and Kneese 1969, p. 283). They argue that there must be uncompensated externalities unless one of the following three conditions prevail. Condition one: all inputs of the production process are fully converted into outputs without unwanted residuals along the way. Condition two: all final outputs (commodities) are utterly destroyed or made to disappear in the process of consumption. Condition three: the property rights are so arranged that all relevant environmental attributes are in private ownership, and these rights are exchanged in competitive markets. They state that none of these conditions can be expected to hold; thus, environmental policies addressing wastes and emissions inevitably fall short of succeeding unless the full process of industrial metabolism (Ayres and Simonis 1994) is taken into account. Similarly, Georgescu-Roegen (1971), in arguing that mainstream economic theory and modern economies are at variance with thermodynamics and the law of conservation of mass, established a theoretical foundation for Ecological Economics.

In the 1990s, an operational picture of the full material metabolism of industrial societies emerged, its respective indicators were developed in an internationally comparative way and the World Resources Institute published two consecutive influential reports on the new model (Adriaanse et al. 1997; Matthews et al. 2000). The basic model places material flows within a wider picture of social metabolism (Fig. 1.1) that has become something like a paradigmatic mind model of the field.

By conceptually linking metabolic flows with biophysical stocks in this way, it became possible to define boundaries for social systems (both vis-à-vis their natural environment and vis-à-vis each other) and to create a consistent metric for material and energy flows for social systems on other scales (local communities, firms or cities). For nation states, material flow accounting (MFA) has become a regular part of public statistics in Japan, in the European Union (EU) and in several other countries (Fischer-Kowalski et al. 2011). This allows the provision of reliable annual accounts of material use in physical terms and their comparison across time and with economic accounts.

On the global scale, this is easier to do because it is only necessary to add up the extraction of raw materials that occurs during a year, and one can ignore the complex network of trade that distributes these resources to the world's countries. The International Resource Panel (UNEP 2011) saw strong public approval for publishing the eightfold increase of global resource extraction during the

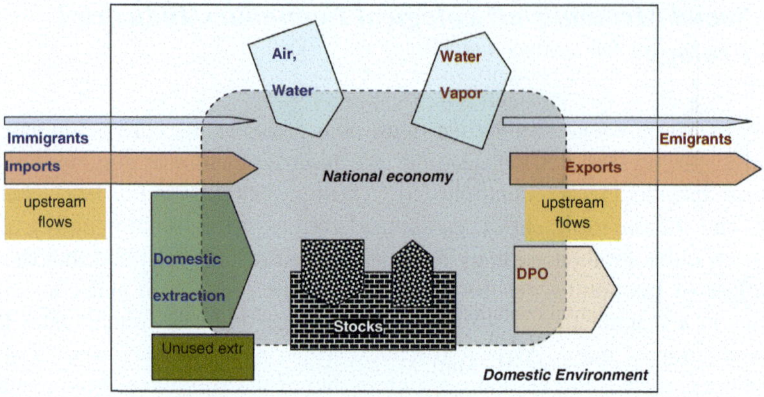

Fig. 1.1 The material metabolism of a national economy. (Source: Fischer-Kowalski et al. (2011), modified from Matthews et al. 2000). DE = Domestic extraction (amount of materials extracted from national territory for direct use). Imports = direct material input from trade (weight at border). Exports = material amounts exported (weight at border). DMI = direct material input = DE + imports. DMC = domestic material consumption = DE + imports–exports. Indirect (or embedded) material flows upstream of imports (and exports) can be expressed as raw material equivalents (RMEs). Total material requirement (TMR) = DE + unused (domestic) extraction + imports + unused extraction in country of origin. Total material consumption (TMC) = TMR—exports—unused extraction of exports. Domestic processed output (DPO) consists of wastes, emissions, dissipatively used materials and deliberate deposition (e.g., fertilizers). Balancing items: air and water contained in materials and that evaporate during production processes or that are drawn into commodities during production (e.g., oxygen in combustion)

20th century, with each of the following fractions increasing: biomass, construction minerals, fossil energy carriers and metals (incl. industrial minerals). Whereas societies at the beginning of the century had reproduced themselves mainly on biomass inputs (i.e., firewood and food for humans and animals), they increasingly turned to so-called nonrenewable resources, such as fossil fuels and ores (Krausmann et al. 2009). During this century, biophysical stocks also increased. The human population, for example, increased fivefold. In addition to the substantial population growth, metabolic rates—that is, resource use per person—increased, doubling from less than five metric tons per person per year to nearly ten metric tons. At the same time, the world gross domestic product (GDP; at constant prices) and average income per person increased 23-fold. Such a ‘decoupling’ of resource use and income is mainly due to technological progress, which allows the production of more value with less input but also feeds into further growth of resource consumption.

There is a rich body of literature comparing the resource requirements of nation states (e.g., Weisz et al. 2006 for the EU) along with their resource efficiencies (e.g., Schandl and West 2010 for Asia and the Pacific), their trade patterns (e.g., Dittrich and Bringezu 2010) and their growth in biophysical stocks (Müller 2006; Pauliuk et al. 2013). On the other end of the metabolic process, there is particular research interest in greenhouse gas emissions (which can be calculated from fossil fuel use, livestock numbers and steel and cement production) directly occurring within countries or indirectly caused by trading. These emissions are, of course, very relevant for climate policies.

Increasingly, the metabolism of cities also comes into view. City planning is an important means to reduce resource consumption while maintaining the same levels of welfare. Substantial amounts of energy for heat and transportation, construction materials and land can be saved through appropriate spatial structures (Kennedy et al. 2007; Weisz and Steinberger 2010).

Giampietro et al. (2012) choose a somewhat different metabolic approach. They undertake a 'Multi-Scale Integrated Analysis of Societal and Ecological Metabolism' (MuSIASEM) that systematically relates human labor, exosomatic energy use and economic output to describe the metabolic patterns of various types of social systems (from households to farms to national economies, stratified into sectors). This approach is seen as a continuation of Georgescu-Roegen's (1971) foundational work on 'Bioeconomics', influencing the emergence of Ecological Economics.

1.4 Identifying Environmental Impacts of Human Activities

In the 1970s, the so-called *IPAT equation* (Eq. 1.1) was developed from a debate about the relative importance of population growth, on the one hand, and growth in affluence, on the other, in determining human impacts on the environment (Chertow 2001; Ehrlich and Holdren 1971). IPAT is the lettering of the following formula:

$$\mathbf{I} = \mathbf{P} \times \mathbf{A} \times \mathbf{T}, \tag{1.1}$$

where **I** stands for (environmental) impact, **P** for population, **A** for affluence and **T** for technology. This formula has been repeatedly applied to estimate various environmental impacts such as land use, resource use, pollution, CO_2 emissions and the *ecological footprint* (see below). In more statistically elaborate applications (e.g., Dietz et al. 2007), regression analysis is used to determine the relative weight of the components and to calculate nonlinearities and interactions, respectively. Empirical results do not strongly confirm the original hope attached to this equation, namely, that improvements in technology would neutralize at least some of the detrimental effects of population growth and increasing affluence; in some cases, the contrary has even been found. Schandl and West (2012) found the increasing affluence of Asian and Pacific countries to be enhanced by technology changes in their impact on CO_2 emissions. This is highly plausible as economic growth in developing countries typically implies a shift toward using fossil fuels.

Another widely used approach to describing human environmental impact is the so-called ecological footprint (EF). In 1996, Wackernagel and Rees published the

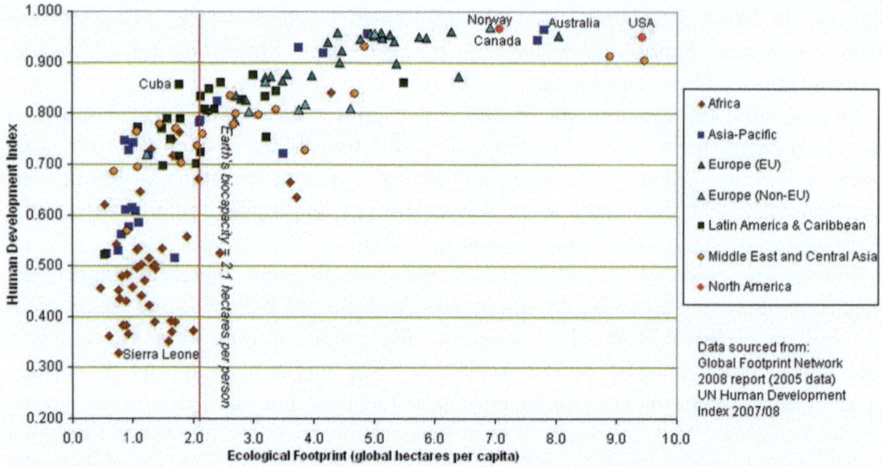

Fig. 1.2 Interrelation between the Human Development Index and the ecological footprint (EF). (Source: received from the Global Footprint Network by data request in August 2013, http://www.footprintnetwork.org)

book *Our Ecological Footprint: Reducing Human Impact on the Earth*. Ecological footprint analysis compares human demands on nature with the biosphere's ability to regenerate resources and provide services. It does this by assessing the biologically productive land and marine area—the 'global hectares'—required to produce the resources a population consumes and to absorb the corresponding waste using prevailing technology. Assessment of the per capita EF is a means of comparing consumption and lifestyles and checking them against nature's ability to provide for this consumption. Despite several modifications, there are still several methodological criticisms of this indicator, such as how different land productivities are taken into account (see Haberl et al. 2004) and how trade can be integrated into the picture (see Grazi et al. 2007). Nevertheless, it is doubtlessly one of the most powerful tools in public communication (Fig. 1.2).

The tool can inform policy by examining to what extent a nation uses more (or less) than is available within its territory or to what extent the nation's lifestyle would be replicable worldwide. The footprint can also be a useful tool to educate people about carrying capacity and overconsumption, with the aim of adjusting personal behavior. EFs may be used to argue that many current lifestyles are not sustainable. Such a global comparison also clearly shows the inequalities of resource use on this planet at the beginning of the 21st century.

From a Social Science perspective, these indicators and analyses of environmental impact fail to account for complexity on the social system side of the process, and they usually lack a coevolutionary perspective. While humans can have detrimental impacts upon nature, the storyline does not include how nature hits back, nor does it allow understanding the adjustments human societies make (or are forced to make).

Something of this type is attempted by the so-called *DPSIR model* (see Fig. 18.4) used by the Organisation for Economic Co-operation and Development (OECD) and the European Environmental Agency (EEA), where D = drivers, P = (environmental) pressures, S = states (of the environment), I = (environmental) impacts and R = (policy) responses. This is postulated to be a causal chain in which, for certain social reasons ('drivers'), pressures are exerted upon the environment that trigger changes there, and in the end, the loop is closed by society reacting to those changes with (presumably ameliorating) policies. Still, insofar as this model focuses on social processes, it does so only very narrowly (Stanners et al. 2007).

1.5 Biohistory and Society-Nature Coevolution

There is a long tradition in the Social and Historical Sciences of distinguishing qualitatively different modes of societal organization, of subsistence, of production and of stages of civilization. The distinctions drawn and the criteria upon which they are drawn vary, but they hardly account for society-environment relations or the environmental consequences of human activity.

It is the special achievement of Sieferle (1997) to regard the modes of societal organization not simply as socially or socioeconomically distinct but to systematize them so that they can be characterized as socioecological patterns, comprising social organization (in the widest sense of the word), concomitant modifications of the environment and intended or unintended environmental impacts. Key to the distinctions Sieferle draws is the source of energy and the dominant conversion technology of the energy a society uses. The charm of this classification is that it helps understand the differences in functional problems societies face when trying to establish and maintain themselves within their environment and the evolutionary advantages and drawbacks that occur, thereby providing some clue to the directionality of change. Sieferle distinguishes the hunting and gathering mode, the agrarian mode (with some subdivisions) and the industrial mode. The energy system of hunter-gatherers is 'passive solar energy utilization'. They live on the products of recent photosynthesis (plants and animals for food, firewood for heat). That they use fire to cook (rather grill) their food widens the spectrum of edibles; nevertheless, only a very small fraction of their environment qualifies as food. Its collection requires mobility, both on an everyday basis and seasonally, and allows only for very low population densities. In contrast, the agrarian mode—an offspring of the Neolithic revolution that occurred (although at different times) on all continents but Australia—is based on 'active solar energy utilization'. This means that certain areas are cleared of their natural vegetation and that solar energy in these areas is, as far as possible, monopolized for edible plants (Netting 2010). In effect, this leads to extensive deforestation of the Earth (and the enrichment of the atmosphere with the CO_2 that previously had been stored in trees and soils), to a sedentary way of life and to a large human labor burden (that even increases with progress in technologies to raise returns on land; Boserup 1965, 1981). The sedentary way of life (plus milk from livestock and ceramics to boil liquids) allows for much greater fertility, and the large labor burden motivates people to have children to share the labor. Thus, high population growth creates high population densities and an expansion of the agrarian mode across the world. Control of territory, tools, livestock and stored reserves is essential, and frequent territorial conflicts produce specialized classes of people to defend and attack territories, social hierarchies to control them and urban centers. In many parts of the world, these systems develop into major empires and civilizations that ever again collapse (Diamond 2005; Tainter 1988).

In the 17th century, a new energy regime emerged: a fossil fuel-based energy system that supplied society with an amount of energy never before accessible. In the UK, the use of coal instead of the increasingly scarce fuel wood allowed a process of urban growth and manufacture. Meanwhile, textile production for export became very profitable, and sheep gradually crowded out farmers growing food. The invention of the steam engine finally kicked off what is known as industrialization. This turn of history in Europe ('The European Special Course', Sieferle 1997, 2001a), as some argue, could also have happened in the East (Pomeranz 2000) or, perhaps, not at all. It caused large-scale ecological and social transformations and continues to spread from the industrial core countries (currently comprising approximately 20 % of the world's population) to the (much larger) rest of the world at an accelerating speed (Fischer-Kowalski and Haberl 2007). It remains an open question whether the ultimate exhaustion of fossil fuels, a detrimental transformation of the Earth's climate system, or politically guided change will bring this energy regime to a close; it will have sustained itself for a much shorter period than the previous regimes.

There is also an interesting new research area emerging from ecological research that addresses long-term processes and observes a global network of local and regional habitats across time, the sites of so-called Long-Term Ecological Research (LTER). Recently, this research has extended to the social processes and has become an LTSER network (Long-Term Socioecological Research, see Singh et al. 2013). A new term that emerged in this context is 'socio-natural sites' (SNSs), denoting places where a long history of human interventions in the environment has generated ever-changing structures in a coevolution of social and natural processes (Winiwarter et al. 2013).

1.6 Regulation, Governance and Sustainability Transitions

The good governance—or lack thereof—of the commons is a long-standing socioecological theme. Taking a point of departure from Hardin's (1968) *Tragedy of the Commons*, Elinor Ostrom's book *Governing the Commons* (1996) stimulated a rich strand of research (and won her a Nobel Prize in Economics). Her work was foundational for the new Institutional Economics. The focus of her research was on how humans interact with ecosystems to maintain long-term sustainable resource yields. She conducted field studies, for example, on the management of pastures and irrigation networks by locals and documented how societies have developed diverse institutional arrangements for managing natural resources and avoiding ecosystem collapse in many cases, even though some arrangements have failed to prevent resource exhaustion. Ostrom (1996) identified several 'design principles' of stable, local common pool resource management, such as the following:

- Clearly defined boundaries (effective exclusion of external un-entitled parties);
- Collective-choice arrangements that allow most resource appropriators to participate in the decision-making process;
- Effective monitoring by monitors who are part of or are accountable to the appropriators;
- A scale of graduated sanctions for resource appropriators who violate community rules and mechanisms of conflict resolution that are cheap and easily accessible.

In her later work, these principles were expanded to include several additional variables believed to affect the success of self-organized governance systems, including effective communication as well as internal trust and reciprocity.

Ostrom and her many co-researchers have developed a comprehensive 'Social-Ecological Systems (SES) framework', within which much of the still-evolving theory of common-pool resources and collective self-governance is now located (Ostrom 2009). A strong research community that utilizes these approaches is the so-called Resilience Alliance, a network of institutions and people sharing a paradigm of socioecological systems, which they define as 'a multi-scale pattern of resource use around which humans have organized themselves in a particular social structure (distribution of people, resource management, consumption patterns and associated norms and rules).' The aim of resilience management and governance is to keep the system within a particular configuration of states (system 'regime') that will continue to deliver the desired ecosystem goods and services. The system should not move into an undesirable regime from which it is either difficult or impossible to recover (see also Gunderson and Holling 2002). The Resilience Alliance network[2] publishes the influential open access journal *Ecology and Society*. A somewhat related approach has been advanced by R. Scholz and colleagues, who address 'human-environment systems' (Scholz 2011).

The core concept employed in the Frankfurt approach to Social Ecology is that of 'societal nature relations' (*gesellschaftliche Naturverhältnisse*). The focus of the Frankfurt approach is on the *relations* between society and nature in terms of the various societal *regulations* that define these relations. Operationally, this approach focuses on what they consider *basic* societal nature relations, which,

[2]http://www.resilience.org/index.php/key_concepts.

being related to basic human needs, are indispensable for individual and societal reproduction and development. The link to the concept of human needs turns societal nature relations into an irreducibly *normative* concept: the basic societal nature relations should be *regulated* in such a way that *all* humans are able to meet their basic needs (Becker et al. 2011, p. 79). The Frankfurt approach defines as its 'epistemic object' the 'crisis of societal nature relations' (Becker and Jahn 2006, p. 19). This definition is normative in that it presupposes the existence of a crisis, that is, a radical deviation of the 'is' state from an 'ought' state of societal nature relations. The purpose of Social Ecology is thus to generate the knowledge necessary to understand this crisis and to react to it in the sense of helping establish the 'ought' state of societal nature relations. The core research question of Social Ecology is thus, 'How can the crisis-ridden societal nature relations be perceived, understood and actively shaped?' (ibid., p. 12). In the 1990s, the German government established an interdisciplinary research program on 'social-ecological research' that enforced an orientation toward basic needs and demanded the strong involvement of stakeholders, thus strengthening the policy relevance of socioecological research in Germany over many years.

A somewhat related approach to managing coupled human-environment systems draws upon the Dutch societal transitions management school. In contrast to the resilience alliance tradition, the Dutch school focuses on technical and social systems rather than ecological systems. The core concern is the existence of 'persistent' and 'wicked' problems in social system functioning that can only be overcome by a systemic transition. Hence, a socioecological transition (SET) is a transition between two dynamic equilibria, that is, a shift from one more or less stable state to another. The typical model of a transition is the S-curve, which allows for the distinction of discrete phases of transition. There is a 'pre-development phase', in which some processes start to deviate from the dominant pattern; next is the 'take-off phase', where a departure from the original equilibrium can be observed; then there is an 'acceleration phase', where change accelerates in a non-incremental, disruptive and potentially chaotic manner; and finally, there is a 'stabilization phase', where the rate of change declines and a new dynamic equilibrium is reached (Rotmans et al. 2001). The nature of transitional dynamics is described in terms of a generic pattern that consists of a sequence of mechanisms that result in irreversible changes in the system. A key pattern is denoted by 'niches'—individual technologies, practices and actors outside or peripheral to the regime—as the loci for radical innovation (Geels 2005). Niches emerge and cluster, and by empowering a niche cluster, a niche regime unfolds. This niche regime becomes more powerful as the incumbent regime weakens. Finally, the niche regime becomes dominant and takes over the incumbent regime. The underlying mechanisms are variation and selection, adaptation, emergence, clustering, empowerment, transformation, decay and development. Transition management draws together a selective number of frontrunners in a protected environment: an arena. To effectively create a new regime, agents are needed at a certain distance from the incumbent regime. However, the continuous link with the regime is important. Therefore, regime agents are also needed, particularly change-inclined regime agents. Because of its methodological concept of transition management, this approach is frequently denoted the ‘Multi-Level-Perspective’ (MLP) and is linked to the term ‘adaptive management’. It was addressed by both the Global Energy Assessment (GEA 2012) and the United Nations Environment Programme (UNEP) to frame their new modeling approach (UNEP 2013).

Fischer-Kowalski and Rotmans published a comparative analysis of the Vienna and Dutch approaches to socioecological transitions (SETs). The Vienna approach to Social Ecology employs a clearly defined notion of SETs, which are conceived as shifts between ‘sociometabolic regimes’. They define a sociometabolic regime ‘as a dynamic equilibrium of a system of society-nature interaction’ (Fischer-Kowalski and Rotmans 2009). Binder et al. (2013) recently published a systematic comparison of frameworks for analyzing social-ecological systems (SES), identifying ten different frameworks. The broadness of this spectrum shows how strongly current research communities feel a need to systematically address society-nature interrelations.

Many of the Social Ecology approaches reviewed here are highly visible among the Natural Sciences as well, particularly those segments that address issues of sustainability. Although they may be less well received among social scientists who follow their traditional disciplinary pathways, an open mind toward paradigmatic change could involve the Social Sciences more intensively in an interdisciplinary discourse about humanity’s long-term future on Earth.

1.7 The Distinguishing Characteristics of the Vienna Social Ecology School

The beginnings of the Vienna Social Ecology School date back to the year 1986, when the then ‘Interuniversity Research Institute for Distance Education’ (IFF) employed Marina Fischer-Kowalski to start a program on society and environment. She arrived with a funded research project on ‘Social causation of burdens on the environment’. The name Social Ecology, later chosen in distinction to Human Ecology, was born out of the team’s conviction that it was not the human species that mattered but rather the social (and economic and technical) organization this species was evolving. The name may have been a bit misleading as there existed an older US American tradition of the same name. This movement—exemplified in the Institute for Social Ecology in Vermont—emerged from the idea of deep ecology (see Bookchin 1984) and continues to be centered on eco-activism and a new environmental ethic (see Lejano and Stokols 2013). The Viennese understanding of Social Ecology was fundamentally different. It insisted on interdisciplinarity across the ‘great divide’, a term coined by Snow (1956) to denote the rift between science and the Humanities, and it is basically functionalistic. This Social Ecology has much more in common with its sister fields of Human Ecology, Ecological Economics, Industrial Ecology, Ecological Anthropology,

Environmental Sociology and Environmental History, all of which address sustainability issues in a more or less interdisciplinary way.

The above-mentioned fields of research still bear the traces of their disciplinary roots and thus have similar but distinct views on sustainability. Traditional Human Ecology builds largely on ecological concepts (Young 1974) but entertains a rather simple understanding of society and the economy. Ecological Anthropology takes most of its empirical insights from studying non-Western cultures, and it receives much theoretical inspiration from biology (see Lutz 2001). Ecological Economics focuses on transforming or even replacing the body of theory known as Neoclassical Economics with a new understanding that seeks to integrate the physical aspects of the economic process into the center of economic theory (Ayres and van den Bergh 2005; Boulding 1966; Daly 1977; Georgescu-Roegen 1971; Gowdy and Erickson 2005; Martinez-Alier 1987). Industrial Ecology, partly born out of the realization that Ecological Economics has engaged too much in academic disputes, pays less attention to Macroeconomics and is mostly interested in technology transitions, material flows and practical applications (Ayres and Ayres 2002; Bourg and Erkman 2003) along with a strong, yet so far theoretically insufficiently developed, aspiration to incorporate the 'human dimension'. Finally, within Environmental History and Environmental Sociology, it is still a matter of debate whether the idea of addressing the relations between nature and history or nature and society in a biophysical sense should be a legitimate core question (Benton 1991; Winiwarter and Knoll 2007), although progress is being made within Sociology (Dunlap 2015).

It is easy to detect what these fields have in common: they address interactions between a 'social' and a 'natural' domain, a topic increasingly considered to be the core of 'sustainability sciences' (Kates et al. 2001). For sustainability science, understanding and transforming society-nature relations is simply 'the sustainability challenge'. In the same vein, the distinguishing element is also obvious: it is precisely how the 'social' is specified.

The ambition of Viennese Social Ecology is to conceptualize society

- comprehensively: not solely as economy, technology, culture, or Western industrialized societies;[3]
- as sufficiently complex: as an autopoietic system, not as an aggregate of humans or groups of rational actors;[4]
- as historically variable: implying that from a long-term and world historical perspective, different modes of subsistence (or sociometabolic regimes) are distinguished.[5]

---

[3]This might be considered a heritage from Ecological Anthropology.

[4]Maintaining the heritage from Maturana and Varela (1975) and Luhmann (1984/1995).

[5]This reflects the heritage from classical political economy and universal history.

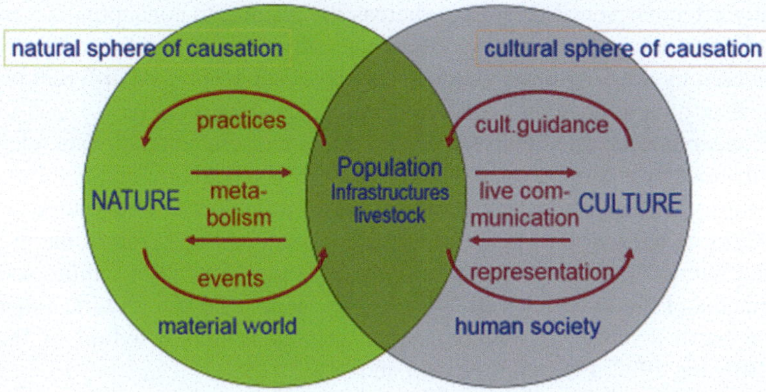

Fig. 1.3 The conceptual model of society-nature interaction developed by the Vienna Social Ecology School. (Elaborated after Fischer-Kowalski and Haberl 2007, p. 13; Fischer-Kowalski and Weisz 1999)

At the same time, it seeks to incorporate a sufficiently complex and realistic understanding of the material world, which means taking Natural Science concepts seriously and incorporating them at a conceptual and empirical level.

The metatheoretical position of Viennese Social Ecology is an epistemology based on distinctions that are not justified by arguing that something is ontologically given but are rather selected for their usefulness to inform insights, pose new research questions and foster interdisciplinary cooperation. In this context, the distinction between culture and nature is particularly relevant. What would be the justification for Social Ecology to start with the distinction between a natural and a cultural realm? This justification lies in the possible means of intervention. Intervention within society must refer to cultural meaning. Although physical interventions can be very effective—though sometimes not very targeted—in creating new communication, society can ultimately only be reached through communication (Luhmann 1984/1995).[6] This is a decisive observation as unsustainable development is a problem of society and not a problem of nature. Conversely, interventions in nature, or into the physical world, can only be effective by means of physical forces—nature is not susceptible to cultural or symbolic action. In effect, society must be conceived not as a communication system only but as having access to and being able to control physical forces, that is, to develop means of 'colonizing' nature via physical interventions (see Fig. 1.3).

One of the early insights of the Vienna team was that a sufficiently complex concept of society as a whole would be essential for a theoretically ambitious and

[6]Imagine 9/11 and no one talking about it!

practically effective Social Ecology, whereas the attempt to conceptualize 'nature' as a whole was not very promising. The complexity of the natural world can better be considered by drawing upon various meso-level concepts from different Natural Sciences than by aiming at an overall understanding of nature.[7]

The *conceptual model* shown in Fig. 1.3 contains some characteristic elements that are distinct from similar conceptual models. It is a heuristic that highlights the intersection of the Cartesian distinction between the material and symbolic (cultural) realms as mutually exclusive domains, on the one hand, and of the material world and human society, on the other hand, as comprising all of culture and specific elements of the material world. Therefore, the natural and cultural spheres of causation partly overlap in society; human society is thus a hybrid of the two realms (Boyden 1992; Fischer-Kowalski and Weisz 1999).[8]

Social metabolism is the key link between society and the natural environment. To reproduce its biophysical structures, society requires a continuous flow of energy and materials that need to be extracted from and eventually released to the environment (Ayres and Kneese 1969). In the same vein, communication is the key link between individual human consciousness (subsumed under *population* in Fig. 1.3) and *culture* (Luhmann 1984/1995).

Following the bended arrows in a recursive way, the conceptual model describes the society-nature coevolution as a self-referential dynamic with the selective forces being contingent on the internal selection pressures of the systems coevolving. Society intervenes in nature (through labor, technology and capital, summarized as *practices* in Fig. 1.3) to modify it according to its needs (e.g., agriculture, construction activities). Society's biophysical structures are susceptible to physical forces from nature, and through communication, these forces are represented culturally, interpreted as rewards for society's efforts (e.g., a large harvest), as catastrophes (e.g., a flood), or as potentially irrelevant. In the other direction, culture supplies guidance/programs for collective decisions and actions; certain culturally guided regulations lead to physical alterations in natural processes that, in turn, may or may not lead to new forces, intended or unintended, exerted from nature upon society. These changed forces might become culturally represented in one way or another (or even pass unregistered) and may or may not modify cultural guidance/programs for future action upon nature.

Conceptually, this heuristic allows us to draw upon Luhmann's fundamental distinction among communicative (termed *culture* in Fig. 1.3), conscious and physical modes of operation. Consequently, we can draw upon all important insights that are contingent on this theoretical architecture. At the same time, this heuristic conceptually allows for a direct embedding of society into the physical

[7]Natural scientists would not even consider engaging in something like finding an overall concept of nature—this has always been the realm of philosophy.

[8]Recent similar conceptualizations may be found in Liu et al. (2007), who discuss the complexity of coupled human and natural systems, and in Becker (2013), who emphasizes the importance of hybrid structures.

environment by including the human population, the livestock population and all physical artifacts (including infrastructures, buildings, technical equipment and all other kinds of products, usually summarized as *biophysical structures*) into the definition of society.

This conceptualization has benefits and costs. The most important benefit is that a consistent quantitative empirical program can be built upon it. The highly visible contributions to quantify the social metabolism, in terms of material, energy and land use, produced by the Vienna Social Ecology School have amply demonstrated this point (Fischer-Kowalski and Haberl 2007). These insights refer to very different historical circumstances, both contemporary and historical. Contemporary (or industrial, or modern) society-nature relations may be and have been analyzed, but equally well-founded insights into the social metabolism of hunter-gatherers or agrarian societies have been gained, together guiding the comparison across history. Even more so, an explicit theory about modes of society-nature coevolution and the stages in this process could be developed (see Chap. 3 in this volume). Such a quantitative approach strongly facilitates communication with natural scientists and appeals to them—yet another benefit.

One obvious cost is that a focus on quantifying societal material, energy and land use typically alienates social scientists and historians. This could, in principle, be balanced by a more direct exploration of the cross-cutting potential in the underlying heuristic. The overarching term 'communication', for example, comprises economic and monetary processes as well as legal processes or decision-making as different media of communication used by different subsystems. The core message of this conceptual model is, therefore, the insight that communication and physical forces operate in different systems and that humans and social systems, as receptive both to communication and to physical forces, are systems in their own right, located at the interface between the symbolic and the material worlds. Nonetheless, the cultural realm is still underexplored in the scholarly work that was guided by this model. One reason might be that a systems theory perspective, if taken seriously, demands a precision in specifying the reference system that does not allow easy conceptual access to other Social Science theories, especially those centered on social actors or values and attitudes. Overall, as will be documented throughout this book, the Viennese School of Social Ecology stands for empirical analysis in a liberal interpretation of methodological pluralism.

References

Adams, C. C. (1935). The relation of general ecology to human ecology. *Ecology, 16*, 316–335.

Adriaanse, A., Bringezu, S., Hammond, A., Moriguchi, Y., Rodenburg, E., Rogich, D., & Schütz, H. (1997). *Resource flows: The material basis of industrial economies*. Washington, D.C.: World Resources Institute.

Ayres, R. U., & Ayres, L. W. (2002). *Handbook of industrial ecology*. Cheltenham, UK, Lyme, US: Edward Elgar.

Ayres, R. U., & Kneese, A. V. (1969). Production, consumption and externalities. *American Economic Review, 59*(3), 282–297.
Ayres, R. U., & Simonis, U. E. (1994). *Industrial* METABOLISM: *Restructuring for sustainable development*. Tokyo, New York, Paris: United Nations University Press.
Ayres, R. U., & van den Bergh, J. C. J. M. (2005). A theory of economic growth with material/energy resources and dematerialization: Interaction of three growth mechanisms. *Ecological Economics, 55*(1), 96–118.
Ayres, R. U., & Warr, B. (2005). Accounting for growth: The role of physical work. *Structural Change and Economic Dynamics, 16*(2), 181–209.
Beck, U. (1986). *Risikogesellschaft. Auf dem Weg in eine andere Moderne*. Frankfurt a.M: Suhrkamp.
Becker, E. (2013). Social-ecological systems as epistemic objects. In M. Glaser, G. Krause, B. M. W. Ratter, & M. Welp (Eds.), *Human-nature interactions in the Anthropocene. Potentials of social-ecological systems analysis* (pp. 37–59). London: Routledge.
Becker, E., Hummel, D., & Jahn, T. (2011). Gesellschaftliche Naturverhältnisse als Rahmenkonzept. In M. Groß (Ed.), *Handbuch Umweltsoziologie* (pp. 75–96). Wiesbaden: VS Verlag für Sozialwissenschaften.
Becker, E., & Jahn, T. (2006). *Soziale Ökologie. Grundzüge einer Wissenschaft von den gesellschaftlichen Naturverhältnissen*. Frankfurt: Campus.
Benton, T. (1991). Biology and social science: Why the return of the repressed should be given a (cautious) welcome. *Sociology, 25*(1), 1–29.
Beus, C. E. (1993). Sociology, human ecology, and ecology. *Advances in Human Ecology, 2*, 93–132.
Binder, C. R., Hinkel, J., Bots, P. W. G., & Pahl-Wostl, C. (2013). Comparison of frameworks for analyzing social-ecological systems. *Ecology and Society, 18*(4), 26.
Bookchin, M. (1984). *The ecology of freedom. The emergence and dissolution of hierarchy*. Palo Alto, CA: Cheshire Books.
Boserup, E. (1965). *The conditions of agricultural growth. The economics of agrarian change under population pressure*. Chicago: Aldine/Earthscan.
Boserup, E. (1981). *Population and technological change—A study of long-term trends*. Chicago: The University of Chicago Press.
Boulding, K. E. (1966). The economics of the coming spaceship earth. In H. Jarrett (Ed.), *Environmental quality in a growing economy. Essays from the sixth RFF forum* (pp. 3–14). Baltimore: John Hopkins Press.
Bourg, D., & Erkman, S. (2003). *Perspectives on industrial ecology*. Sheffield: Greenleaf Publishing.
Boyden, S. V. (1992). *Biohistory: The interplay between human society and the biosphere—Past and present*. Paris, Casterton Hall, Park Ridge, New Jersey: UNESCO and Parthenon Publishing Group.
Bruckmeier, K. (2004). Die unbekannte Geschichte der Humanökologie. In W. Serbser (Ed.), *Humanökologie. Ursprünge - Trends - Zukünfte* (pp. 45–120). München: Ökom Verlag.
Carson, R. (1962). *Silent spring*. Boston: Houghton Mifflin Company.
Catton, W. R, Jr., & Dunlap, R. E. (1978). Environmental sociology: A new paradigm. *The American Sociologist, 13*(1), 41–49.
Chertow, M. R. (2001). The IPAT equation and its variants. Changing views of technology and environmental impact. *Journal of Industrial Ecology, 4*(4), 13–30.
Ciccantell, P. S., & Bunker, S. G. (1998). Space and transport in the world-system. Westport: Greenwood Publishing Group.
Cottrell, F. (1955). *Energy and society. The relation between energy, social change, and economic development*. New York, Toronto, London: McGraw-Hill Book Company. [Revised edition: 2009].
Daly, H. E. (1977). *Steady-state economics. The economics of biophysical equilibrium and moral growth*. San Francisco: W.H. Freeman and Company.

Darling, F. F. (1956). The ecology of man. *American Scholar, 25*(1), 38–46.
Diamond, J. (2005). *Collapse. How societies choose to fail or succeed.* New York: Viking
Dietz, T., Rosa, E. A., & York, R. (2007). Driving the human ecological footprint. *Frontiers in Ecology and the Environment, 5*(1), 13–18. See also www.stirpat.org (STIRPAT: "a research program in structural human ecology"). Accessed August 24, 2013.
Dittrich, M., & Bringezu, S. (2010). The physical dimension of international trade: Part 1: Direct global flows between 1962 and 2005. *Ecological Economics, 69*(9), 1838–1847.
Dunlap, R. (2015). Environmental sociology. In *Encyclopedia of the social and behavioural sciences*. New York: Elsevier (in print).
Ehrlich, P. R., & Ehrlich, A. H. (1970). *Population, resources, environment: Issues in human ecology*. San Francisco: Freeman.
Ehrlich, P. R., Ehrlich, A. H., & Holdren, J. H. (1973). *Human ecology, problems and solutions*. San Francisco: Freeman.
Ehrlich, P. R., & Holdren, J. P. (1971). Impact of population growth. *Science, 171*(3977), 1212–1217.
Fischer-Kowalski, M. (1998). Society's metabolism. The intellectual history of material flow analysis, part I: 1860-1970. *Journal of Industrial Ecology, 2*(1), 61–78.
Fischer-Kowalski, M., & Haberl, H. (2007). *Socioecological transitions and global change: Trajectories of social metabolism and land use*. Cheltenham, UK, Northhampton, USA: Edward Elgar.
Fischer-Kowalski, M., Krausmann, F., Giljum, S., Lutter, S., Mayer, A., Bringezu, S., et al. (2011). Methodology and indicators of economy wide material flow accounting. State of the art and reliablity across sources. *Journal of Industrial Ecology, 15*(6), 855–876.
Fischer-Kowalski, M., & Rotmans, J. (2009). Conceptualizing, observing and influencing social-ecological transitions. *Ecology and Society, 14*(2), 1–3. url:http://www.ecologyandsociety.org/vol14/iss2/art3
Fischer-Kowalski, M., & Weisz, H. (1999). Society as a hybrid between material and symbolic realms. Toward a theoretical framework of society-nature interaction. *Advances in Human Ecology, 8*, 215–251.
Foster, J. B. (2000). *Marx's ecology. Materialism and nature*. New York: Monthly Review Press.
GEA. (2012). *Global energy assessment—Toward a sustainable future*. Cambridge, UK, Laxenburg, Austria: Cambridge University Press and the International Institute for Applied Systems Analysis.
Geels, F. W. (2005). The dynamics of transitions in socio-technical systems: A multi-level analysis of the transition pathway from horse-drawn carriages to automobiles (1860–1930). *Technology Analysis & Strategic Management, 17*(4), 445–476.
Georgescu-Roegen, N. (1971). *The entropy law and the economic process*. Cambridge, MA: Harvard University Press.
Giampietro, M., Mayumi, K., & Sorman, A. H. (2012). *The metabolic pattern of societies. Where economists fall short*. New York: Routledge.
Godelier, M. (1984). *L'idéel et le matériel. Pensée, économies, societés*. Paris: Fayard.
Goldfrank, W. L., Goodman, D., & Szasz, A. (1999). *Ecology and the world-system*. Westport: Greenwood Publishing Group.
Gowdy, J., & Erickson, J. D. (2005). The approach of ecological economics. *Cambridge Journal of Economics, 29*(2), 207–222.
Grazi, F., van den Bergh, J. C. J. M., & Rietveld, P. (2007). Spatial welfare economics versus ecological footprint: Modeling agglomeration, externalities and trade. *Environmental & Resource Economics, 38*, 135–153.
Gunderson, L., & Holling, C. S. (Eds.), (2002). *Panarchy: Understanding transformations in human and natural systems*. Washington D.C.: Island Press
Haberl, H., Wackernagel, M., Krausmann, F., Erb, K.-H., & Monfreda, C. (2004). Ecological footprints and human appropriation of net primary production: A comparison. *Land Use Policy, 21*(3), 279–288.

Hardin, G. (1968, December 13). The tragedy of the commons. *Science, 162*, 1243–1248.
Harvey, D. (2014). *Seventeen contradictions and the end of capitalism*. Oxford: Oxford University Press.
Hornborg, A., & Crumley, C. L. (2007). *The world system and the earth system. Global socioenvironmental change and sustainability since the Neolithic*. Walnut Creek: Left Coast Press.
Jänicke, M. (1988). Ökologische Modernisierung: Optionen und Restriktionen präventiver Umweltpolitik. In U. E. Simonis (Ed.), *Präventive Umweltpolitik* (pp. 13–26). Frankfurt/New York: Campus.
Kates, R. W., Clark, W. C., Corell, R., Hall, J. M., Jaeger, C. C., Lowe, I., et al. (2001). Environment and development: Sustainability science. *Science, 292*(5517), 641–642.
Kennedy, C. A., Cuddihy, J., & Engel-Yan, J. (2007). The changing metabolism of cities. *Journal of Industrial Ecology, 11*(2), 1–17.
Krausmann, F., Gingrich, S., Eisenmenger, N., Erb, K.-H., Haberl, H., & Fischer-Kowalski, M. (2009). Growth in global materials use, GDP and population during the 20th century. *Ecological Economics, 68*(10), 2696–2705.
Latour, B. (1991). *Nous n'avons pas jamais été modernes. Essai d'anthropologie symétrique*. Paris: Editions la Découverte.
Lejano, R. P., & Stokols, D. (2013). Social ecology, sustainability, and economics. *Ecological Economics, 89*, 1–6.
Liu, J. G., Dietz, T., Carpenter, S. R., Alberti, M., Folke, C., Moran, E., et al. (2007). Complexity of coupled human and natural systems. *Science, 317*(5844), 1513–1516.
Luhmann, N. (1989). *Ökologische Kommunikation: Kann die moderne Gesellschaft sich auf ökologische Gefährdungen einstellen?* Opladen: Westdeutscher Verlag. (Original work published 1986).
Luhmann, N. (1995). *Soziale Systeme. Grundriß einer allgemeinen Theorie*. Suhrkamp, Frankfurt a.M. (Original work published 1984).
Luhmann, N. (2012). *Theory of society* (Vol. 1). California: Stanford University Press. (Original work published 1997).
Lutz, J. (2001). Ecological thought and the history of sociology. In J. Mucha, D. Kaesler, & W. Winclawski (Eds.), *Mirrors and windows. Essays in the history of sociology*. Torun: Nicholas Copernicus University Press.
Marsh, G. P. (1864). *Man and nature; or physical geography as modified by human action*. London, New York: Scribners & Sampson Low.
Martinez-Alier, J. (1987). *Ecological economics. Energy, environment and society*. Oxford: Blackwell.
Marx, K. (2010). *Capital. A critique of political economy* (Vol. I). E-Book: www.marxists.org/archive/marx/works/1867-c1/. (Original work published 1867).
Matthews, E., Amann, C., Fischer-Kowalski, M., Bringezu, S., Hüttler, W., Kleijn, R., et al. (2000). *The weight of nations: Material outflows from industrial economies*. Washington, D.C.: World Resources Institute.
Maturana, H. R., & Varela, F. G. (1975). *Autopoietic systems. A characterization of the living organization*. Urbana-Champaign, Illinois: University of Illinois Press.
Mol, A. P. J., & Spaargaren, G. (2000). Ecological modernisation theory in debate: A review. *Environmental Politics, 9*(1), 17–49.
Morgan, L. H. (1963). *Ancient society*. Cleveland: World Publications. (Original work published 1877).
Müller, D. B. (2006). Stock dynamics for forecasting material flows–Case study for housing in The Netherlands. *Ecological Economics, 59*(1), 142–156.
Murphy, D. J. (2012). Fossil fuels: Peak oil is affecting the economy already. *Nature, 483*(7391), 541.
Murphy, R., & Steward, J. H. (1955). Tappers and trappers: Parallel process in acculturation. *Economic Development and Cultural Change, 4*, 335–355.

Nature Special Feature. (2013). GM crops: Promise and reality. *Nature, 497* (7447), 5–152. http://www.nature.com/news/specials/gmcrops/index.html

Neef, E. (1969). Der Stoffwechsel zwischen Gesellschaft und Natur als geographisches Problem. *Geographische Rundschau, 21*, 453–459.

Netting, R. M. (2010). *Cultural ecology*. Long Grove: Waveland Press.

Orlove, B. S. (1980). Ecological anthropology. *Annual Review of Anthropology, 9*, 235–273.

Ostrom, E. (1996). *Governing the commons. The evolution of institutions for collective action*. Cambridge: Press Syndicate of the University of Cambridge.

Ostrom, E. (2009). A general framework for analyzing sustainability of social-ecological systems. *Science, 325*(5939), 419–422.

Pauliuk, S., Wang, T., & Müller, D. B. (2013). Steel all over the world: Estimating in-use stocks of iron for 200 countries. *Resources, Conservation and Recycling, 71*(2), 22–30.

Pimentel, D., Hurd, L. E., Bellotti, A. C., Forster, M. J., Oka, I. N., Sholes, O. D., & Whitman, R. J. (1973). Food production and the energy crisis. *Science, 182*(4111), 443–449.

Pomeranz, K. (2000). *The great divergence: China, Europe, and the making of the modern world economy*. Princeton: Princeton University Press.

Rotmans, J., Kemp, R., & van Asselt, M. (2001). More evolution than revolution: Transition management in public policy. *Foresight, 3*(1), 15–31.

Schandl, H., & West, J. (2010). Resource use and resource efficiency in the Asia-Pacific region. *Global Environmental Change, 20*(4), 636–647.

Schandl, H., & West, J. (2012). Material flows and material productivity in China, Australia, and Japan. *Journal of Industrial Ecology, 16*(3), 352–364.

Schmidt, A. (1971). *The concept of nature in Marx*. London: New Left Book.

Schnaiberg, A., & Gould, K. A. (1994). *Environment and society: The enduring conflict*. New York: St. Martin's.

Scholz, R. (2011). *Environmental literacy in science and society*. New York: Cambridge University Press.

Sears, P. B. (1953). Human ecology, a problem in synthesis. *Science, 120*, 959–963.

Sieferle, R. P. (1990). *Bevölkerungswachstum und Naturhaushalt. Studien zur Naturtheorie der klassischen Ökonomie.*. Frankfurt a.M.: Suhrkamp

Sieferle, R. P. (1997). *Rückblick auf die Natur: Eine Geschichte des Menschen und seiner Umwelt*. München: Luchterhand.

Sieferle, R. P. (2001a). *Europe's special course: Outline of a research program*. Stuttgart: Breuninger Stiftung.

Sieferle, R. P. (2001b). *The subterranean forest. Energy systems and the industrial revolution*. Cambridge: The White Horse Press. (Original work published in German 1982. Der unterirdische Wald. Energiekrise und Industrielle Revolution. C.H.Beck, München).

Sieferle, R. P. (2011). Cultural evolution and social metabolism. *Geografiska Annaler: Series B, Human Geography, 93*, 315–324.

Singh, S. J., Haberl, H., Chertow, M., Mirtl, M., & Schmid, M. (Eds.). (2013). *Long term socio-ecological research. Studies in society nature interactions across spatial and temporal scales*. Dordrecht, Heidelberg, New York, London: Springer.

Smil, V. (1991). *General energetics. Energy in the biosphere and civilization*. Manitoba, New York: Wiley.

Smil, V. (2008). *Energy in nature and society. General energetics of complex systems*. Cambridge, MA: MIT Press.

Snow, C. P. (1956, October 6). The two cultures. New Statesman (p. 413).

Spencer, H. (1862). *First principles*. London: Williams and Norgate.

Stanners, D., Bosch, P., Dom, A., Gabrielsen, P., Gee, D., Martin, J., et al. (2007). Frameworks for environmental assessment and indicators at the EEA. In T. Hak, B. Moldan & A. L. Dahl (Eds.), *Sustainability indicators. A scientific assessment. SCOPE 67* (pp. 127–144). Washington, Covelo, London: Island Press.

Tainter, J. A. (1988). *The collapse of complex societies*. Cambridge: Cambridge University Press.

Thomas, W. L, Jr. (1956). *Man's role in changing the face of the earth*. Chicago: The Chicago University Press.

Turner, B. L. I., Clark, W. C., Kates, R. W., Richards, J. F., Mathews, J. T., & Meyer, W. B. (1990). *The earth as transformed by human action: Global and regional changes in the biosphere over the past 300 Years*. Cambridge: Cambridge University Press.

UNEP. (2011). *Decoupling resource use and environmental impacts from economic growth*. Paris: United Nations Environmental Programme.

UNEP. (2013). GEO 5. Global environmental outlook. E-Book: http://www.unep.org/geo/GEO5_ebook/index.html

Wackernagel, M., & Rees, W. E. (1996). *Our ecological footprint: Reducing human impact on the Earth*. Gabriola Island, BC; Philadelphia, PA: New Society Publishers.

Wallerstein, I. (1999). Ecology and capitalist costs of production: No exit. In W. L. Goldfrank et al. (Eds.), *Ecology and the world-system* (pp. 3–11). Westport, Connecticut: Greenwood Press.

Weisz, H. (2002). *Gesellschaft-Natur Koevolution: Bedingungen der Möglichkeit nachhaltiger Entwicklung*. Berlin: Humboldt Universität.

Weisz, H., & Clark, E. (Eds.), (2011). Society-nature coevolution: An interdisciplinary concept for sustainability. *Geografiska Annaler, Series B, Human Geography, 4*, 281–287.

Weisz, H., Fischer-Kowalski, M., Grünbühel, C. M., Haberl, H., Krausmann, F., & Winiwarter, W. (2001). Global environmental change and historical transitions. *Innovation, 14*(2), 115–142.

Weisz, H., Krausmann, F., Amann, C., Eisenmenger, N., Erb, K.-H., Hubacek, K., & Fischer-Kowalski, M. (2006). The physical economy of the European Union: Cross-country comparison and determinants of material consumption. *Ecological Economics, 58*(4), 676–698.

Weisz, H., & Steinberger, J. K. (2010). Reducing energy and materials flows in cities. *Current Opinion in Environmental Sustainability, 2*(3), 185–192.

White, L. A. (1943). Energy and the evolution of culture. *American Anthropologist, 45*(3), 335–356.

Winiwarter, V., & Knoll, M. (2007). *Umweltgeschichte. Eine Einführung*. Köln: Böhlau

Winiwarter, V., Schmid, M., Hohensinner, S., & Haidvogl, G. (2013). The environmental history of the Danube river basin as an issue of long-term socio-ecological research. In S. J. Singh et al. (Eds.), *Long term socio-ecological research* (pp. 103–122).

Young, G. L. (1974). Human ecology as an interdisciplinary concept: A critical inquiry. In A. Macfayden (Ed.), *Advances in ecological research* (pp. 1–105). London, New York: Academic Press.

Chapter 2
Core Concepts and Heuristics

Marina Fischer-Kowalski and Karl-Heinz Erb

Abstract This chapter outlines the basics of our socioecological theory. It starts with the question of why entities such as 'culture' have been so successful that an evolving species like humankind could become the dominant power on the planet. It explains social systems as 'hybrids', a structural coupling between a (cultural) communication system and interconnected biophysical elements. In what sense are humans, domestic animals and artifacts hybrids? In what sense do these elements 'belong' to a certain cultural (communication) system? The constitutive operation is 'colonization'. Human beings are culturally 'colonized', as are their livestock and their artifacts. These hybrid elements and the metabolic flows required to maintain them determine the social system's impact upon the 'rest of nature'. This influence happens through the metabolic exchange of energy and materials (which in part occurs unintentionally, such as breathing or evaporation) and through 'labor', or culturally guided human action. The sociometabolic model is described in the following section as an interrelation of stocks (human population, territory, livestock and artifacts) and flows (energy and materials). It has systematic similarities with national accounting and is thus useful for addressing many research questions, such as the resource productivity of a national economy or its energy intensity. To some extent, it is the description of an economy, at any time in history, using biophysical instead monetary parameters.

Keywords Evolutionary success · Communication system · Colonizing interventions · Hybrids · Sociometabolic model

M. Fischer-Kowalski (✉) · K.-H. Erb
Institute of Social Ecology (SEC), Alpen-Adria University, Schottenfeldgasse 29, 1070 Vienna, Austria
e-mail: marina.fischer-kowalski@aau.at

K.-H. Erb
e-mail: karlheinz.erb@aau.at

H. Haberl et al. (eds.), *Social Ecology*, Human-Environment Interactions 5,
DOI 10.1007/978-3-319-33326-7_2

2.1 The Basic Socioecological Model Revisited

Sustainability science requires a non-reductionist macro-model of society-nature interactions. Such a model should describe these interactions in a way that complies with the categorial frameworks of natural as well as social systems. The theory's scope must be capable of providing an explanation for why entities such as 'culture' or 'society' have been so successful under the conditions of organic evolution that an evolving species like humankind has become the dominant power on the planet.[1] It should allow us to grasp the enormous differences among various modes of human subsistence, such as hunter-gatherer, agrarian, or industrial societies, and to explain the transitions between them. We believe that a theory of society-nature relations relevant for a sustainability transition must allow for, if not promote, recourse to long-term, historical and archaeological perspectives and must be useful across a range of spatial scales. The theory should be equally compatible with the Natural Sciences, particularly Biology and the Technical Sciences, as well as the Social Sciences and Economics.

In the following paragraphs, we explore whether the socioecological model we developed (see Chap. 1; in particular Fig. 1.3) lives up to such standards. We discuss its ability to provide an adequate epistemological basis as well as a research program by explaining in detail the individual components of the model.

The model's principal logic aligns with the Cartesian differentiation between *res extensa* (the natural sphere of causation) and *res cogitans* (the cultural sphere of causation). The next section will seek to explain the meaning of such a 'cultural sphere of causation', or cultural systems, within the context of theories of evolution. In contrast to the classical Cartesian dualism, however, our model introduces areas of reality belonging to both 'spheres'. It is exactly these areas that constitute decisive 'hybrid' parts of social systems.

In the sections thereafter, we will explain our core concepts: social metabolism and colonization. Both concepts focus on the 'hybrid' parts of social systems.

2.1.1 Why Talk About Culture? A Digression into Evolutionary Theory

To a substantial degree, we owe this model to a long-standing intellectual exchange with the environmental historian Rolf Peter Sieferle. One of his contributions consisted of an outline of the evolution of culture as a specific human evolutionary strategy (Sieferle 1997, 2011). Sieferle's point of departure is organic evolution, that is, the process by which changes in the genetic composition of

[1]Cf. the increasingly influential concept of 'the Anthropocene', which names an entire geological era after the dominant human species (Crutzen 2002).

populations of organisms occur in response to environmental changes. In the following paragraphs, we briefly reproduce his line of argument.

Each organism has a distinct identity. It lives in a temporal continuum such that its identity does not change during its individual history. An organism is also a morphological unit, a physical body with clear borders and the ability to defend these borders, to repel intruders and to heal injuries. In the functional center of an organism, there is an information complex, namely, its genome, which is fixed on a physical substrate (i.e., DNA or RNA) containing instructions that guide the synthesis of a larger physical complex, the organism's phenotype. The genetic information, however, does not solely lead to the development of the organism itself. It also leads to the synthesis of extrasomatic artifacts, such as spiders' nets or birds' nests—a feature Dawkins (1982) dubbed an 'extended phenotype'. From this perspective, modifications of and control over environmental conditions are not solely privileges of the human species. Along the same vein, there is learning, and thus information storage, in the nervous system of animals beyond the information encrypted in DNA. In many cases, however, individuals cannot communicate this information to other members of the species, and thus it is lost with the death of the individual. In many other cases in the zoological realm, individually acquired information is transmitted to other members of the same species without being fixed in the genome. Examples include the specific itineraries of migratory birds or the choice and use of tools by apes. In such cases, information does not flow only from parents to offspring but in any direction, avoiding the slow path of inheritance by variation and selection and permitting a much more rapid spread. Thus, culture is not restricted to the human species. However, its predominance as a strategy is specific to human populations. Sieferle asserts:

> Cultural tradition as such, that is non-genetic acquisition, storage, processing and transmission of information is not a specific feature of humans. A specific human feature, however, is to make this strategy dominant… Cultural evolution proper stabilizes the tradition of behavioral complexes by intergenerational transmission of information in human groups. This is the basic strategy of *Homo sapiens*, its evolutionary special path… When cultural evolution started, it continued organic evolution, and its emergence must have been awarded as a successful adaptation… Culture is not, however, merely an instrument of adaptation. Culture developed specific system properties that soon gave it characteristics that are not exclusively adaptive. (Sieferle 2011, p. 317)

Why was the emergence of culture awarded in evolution, and what renders it a successful strategy? Sieferle explores three possible explanations: culture might provide benefits for coping better technically with environmental conditions; culture might have been preferred in sexual selection; or culture may have benefitted intragroup cooperation. He opts for the third explanation: cultural evolution can be understood as a way to constitute stable cooperative groups of biologically unrelated persons; culturally defined groups can be more easily circumscribed by the use of symbols than genetically distinguished groups, with the chance of rapidly redefining boundaries; and cheaters and free riders can easily be identified and discriminated against. Thus, cultural evolution is primarily a social phenomenon and serves to help people address problems of complexity within groups, producing highly integrated and delineated 'pseudospecies'.

In organic evolution, when selection acts mainly in one direction—from the environment to the phenotype—a stable adaptation can be Sieferle argues that cultural evolution represents a positive feedback loop. Social expected. In contrast to this view, communication is recursive, and content is transmitted back and forth between members of a community so that patterns of plausibility rapidly emerge and move in any direction. Thus, a one-way adaptation to given circumstances is complemented or even replaced by open self-referential dynamics that allow for a high degree of freedom and velocity. 'In this sense, cultural evolution is a phenomenon *sui generis*, it is not an organism's (or a species' or a population's) method of survival. It has a high potential to depart from its original function, to generate peculiar autopoietic traits…' (Sieferle 2011, p. 311).

2.1.2 Society as a Hybrid System

This evolutionary reading of culture as an autopoietic system is one of the foundations of our socioecological model. It explains why we do not regard human societies as mere subsystems of the biosphere, and it complies with the sociological understanding of social systems as systems of recursive communication (Luhmann 1997/2012). Because the purpose of our theory is to understand society-nature interrelations, in contrast to Luhmann, we choose the term 'social system' for the structural coupling between a (cultural) communication system and interconnected 'hybrid' elements (see 'hybrid sphere' in Fig. 2.1) depending on the type of social system. The term 'social systems' can be applied to households, communities, cities, nation states, or organizations.[2] These social systems serve different functions. Each of them links a specific communication system with a human population, a set of artifacts and, eventually, animal livestock (see Fig. 3.1). We borrow the term 'hybrid' from Latour (1991), who uses it for material objects that are cultural artifacts and whose biophysical and cultural features are inseparably interwoven.

In what sense are humans, domestic animals and artifacts hybrids? The human population consists of organisms that must reproduce themselves biologically. Each individual possesses a consciousness that constitutes and reproduces itself through communication with others. Livestock consists of domesticated animals that provide services to humans, such as food, fiber, labor and the concentration of reactive nutrients. Of course, the reproduction of livestock is subject to biological processes, but it is also subject to long-term, culturally guided genetic selection by breeding (and possibly even by direct interventions in the genetic disposition of organisms by genetic engineering). Many behavioral aspects of livestock are culturally determined: where they live, what they eat, how they behave, if and with

[2]We use the term 'society' for social systems with the key function of sustaining a certain human population in a certain territory, such as local communities, cities and nation states.

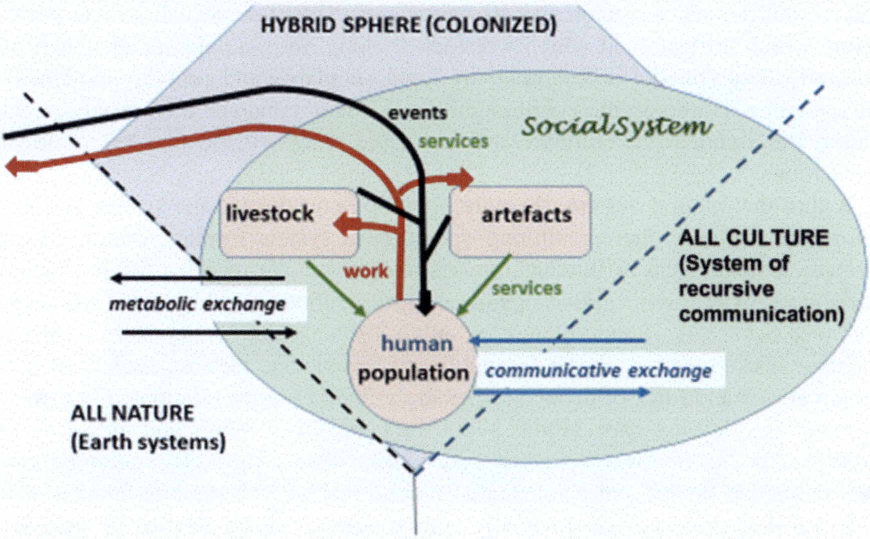

Fig. 2.1 Social systems as hybrid systems. (Detail from Fig. 1.3 in Chap. 1). *Legend*: The *green area* describes the social system; it extends beyond the hybrid sphere, crossing the dividing line into the cultural sphere. The human population, itself a hybrid phenomenon, is at the core of the social system. It connects to the cultural sphere by communication (*blue arrows*) and to the other hybrid and biophysical elements by experiencing events (*black arrows*), by performing work ('colonizing interventions', *red arrows*) and by receiving services (*green arrows*). All elements of the hybrid part of social systems engage in metabolic exchange with the natural environment

whom they mate and how long they live. Finally, all human artifacts, such as houses, roads, infrastructures, tools, machinery, vehicles, fixtures and fittings, are hybrids. A house is built according to a cultural program yet is subject to the laws of thermodynamics and to all sorts of natural impacts. The roof will develop leaks, the heating will break down, the façade will suffer weather damage—and if such natural processes take place without further intervention, the house will become uninhabitable and 'renaturalized'.[3] As long as the house is part of the social fabric, however, it is subject to a cultural program that determines what a proper house looks like, to whom it belongs and how it will be maintained.

In what sense do those elements 'belong' to a certain cultural (communication) system? On a generic level, the answer is simple and clear and follows from the basic assumptions about autopoietic (or autocatalytic) systems: it is the cultural

[3]Even an abandoned ruin is, of course, 'anthropogenic', but according to our understanding, it is no longer maintained by society and thus 'renaturalizes' and no longer belongs to the biophysical elements of society. It could also be 'reintegrated' once again, of course, and find use as a display piece.

system that defines and maintains its boundaries. Therefore, which human population, which artifacts and which livestock 'belong' to a certain social system is culturally determined (for example, by legal standards and property relations). This relation is represented within culture. Under concrete circumstances, of course, this boundary is complex and fuzzy, requiring specification by empirical conventions.

Within the cultural system, the core operation constituting the system is communication. What is the constituting operation of system formation among such diverse elements such as humans, houses and cows? We theorize the following. There does exist a constitutive operation, an operation we address as 'colonization' (see Sect. 2.3). Human beings are culturally 'colonized': they learn a certain language and a set of symbolic meanings and knowledge, they are socialized into certain norms and rules of behavior, they are guided by price relations in a certain currency and develop expectations about the behavior of others and they seek to mold their bodies in a way that suits a particular culture. This colonization permeates minds and bodies, but a substantial natural residue will remain: humans will always follow gravity even if socially embarrassing, will be plagued by diseases and will act out their desires in culturally inappropriate ways. It takes a continuous cultural effort to reproduce and modify human beings as suitable elements of the social system. The same type of operation is required to maintain and reproduce animal livestock as useful service providers; even more so, there is a tendency to 'renaturalize' and to require continuous colonizing interventions to feed, mate, give birth and die according to culturally prescribed modes. Nonliving elements such as artifacts that were produced according to cultural prescriptions renaturalize as well: they are subject to natural forces leading to an increase in entropy and demand continuous colonizing interventions to maintain their functionality for the social system.

We therefore think it is plausible to look upon the process of colonization as the core system-constituting operation between and among the elements of social systems, as specified in Fig. 2.1. Nevertheless, there is a substantial asymmetry between the human population and the other hybrid elements: only humans actively participate in the cultural discourse, and it is through their consciousness that cultural meaning is transformed into physically effective action.[4] The communication arrows between culture and the human population as the only one hybrid element illustrate this asymmetry (Fig. 2.1).

[4]On these grounds, Sieferle (2011) objects to including biophysical elements beyond the human population in the social system. We would maintain, however, that this distinction is somewhat fuzzy; it is not just animals that continue (at least for some time) to function the way they have been culturally conditioned (for example, laying many eggs or returning to a house). In the age of information and communication technology (ICT), many artifacts, as long as they are supplied with energy (which is also the precondition for humans), perform the tasks they have been designed for.

Why bother with such a complex and elaborate specification of social systems? As will become obvious in what follows and in many other chapters of this book, a sound understanding of system boundaries and of the necessary conditions of the self-reproduction of social systems is indispensable to conceptualize and quantify the metabolic exchange (see Fig. 2.1) between social systems and their natural environment as well as to understand conditions for sustainability.

The hybrid elements and the metabolic flows required to maintain them determine the social system's influence upon the 'rest of nature', with nature understood as the environment of the social system. This takes place, on the one hand, through the metabolic exchange of energy and materials (which, in part, occurs entirely unintentionally, such as breathing or evaporation) and, on the other hand, through 'labor'. Labor is understood here as intentional action with a given purpose (i.e., cultural meaning) that is physically effective. By means of work, interventions are made in natural conditions (both within and beyond the 'hybrid sphere'), which are thereby altered. However, purposeful changes, just like metabolism, also have unintended side effects, which the next level of intentional intervention will have to address. This involves societies in an ongoing control spiral (or risk spiral, see Müller-Herold and Sieferle 1998; Tainter 1988) and historical change.

These concepts—societal metabolism, colonizing interventions and their side effects—describe in large part what the Natural Sciences term 'human pressures upon the environment' (Smeets and Weterings 1999). The intended effects follow from the ways in which the social system functions: why and how does it pursue particular aims? The unintended effects, on the other hand, must be explained in terms of the functioning of the natural system. Society's pursuit of intended effects may ultimately be altered by this route.

In the natural 'environment', there are 'events' that affect the biophysical structures of society. These events can be ordinary processes such as rainfall, rust, or sun bleaching, many of which providing society with services, or more dramatic, rare events such as earthquakes, floods and epidemics. Events are either noticed or unnoticed by society and are (or are not) accorded certain meaningful interpretations. These meaningful interpretations take place in the consciousness of human beings, yet technical infrastructures can also provide cultural feedback (for example, measurement stations for air pollution). In any case, these meaningful interpretations must find a cultural representation in some form if they are to have effects on the cultural system. For example, the observation that bees are dying in huge numbers has to be communicated. Whether this communication culturally resonates as God's punishment (and so we must purge our sins) or as an insight that pesticide use in agriculture is poisoning bees is something over which nature has no influence—but communicating humans do wield such influence. Whichever interpretation prevails will provide cultural guidance to which responses to choose and will instruct human action.

Metaphorically, the relation between culture and the hybrid elements of social systems follows a hardware-software logic: the biophysical structures (including humans) are the hardware, and culture accounts for the software. If the hardware breaks down, the software loses its ability to work, but as long as it is functioning, the hardware behaves largely in accordance with the program. The program may also be altered (through communication). According to our approach, the actions (or behaviors) of human beings, cows, or power stations are seen not as culturally determined but as culturally guided by programs within particular spheres of action (which can also bring about failure, as the 2011 Fukushima nuclear disaster demonstrates). When these programs are put into practice, experiences are gained that may again find cultural representation, and through this, societal learning takes place.

2.2 Social Metabolism: Heuristic Definitions and Assumptions

The following section provides an overview of the key terminology in Social Ecology and an assessment of its consistency and usefulness. We discuss this in the following section in two parts: first, we focus on the terminology of societal metabolism; second, we look at the colonizing interventions of society in natural systems.

2.2.1 Sociometabolic Stocks and Flows and the Key Role of Population and Territory

The metabolism of social systems encompasses biophysical stocks, flows and—as we are gradually beginning to better understand, certainly in analogy to organic metabolism—the mechanisms regulating these flows. The starting point of sociometabolic analysis is the definition of the social system as a household, a business, a village, a nation state, or the global community. This definition will direct the research focus. It is important to note here that the social system defines its boundary; the boundary is not simply an arbitrary construct of the researcher. Delineating this boundary and deciding which elements are part of the system and which are not are decisive operations of the system itself. This presupposition is one of the lessons Social Ecology has taken from modern systems theory: a (complex) system constitutes itself through the marking and reproduction of its boundaries, and this fact must be recognized and respected from the research perspective. Social systems mark and reproduce their boundaries through the functions that they fulfill. Accordingly, they define their stocks and flows differently—along ownership boundaries, accepted entitlements, or governance responsibility, for example (see also Netting 1981, 1993).

In our past and current research, we predominantly focused on social systems whose function lies in the reproduction of a particular human population within

Table 2.1 Societal stocks and corresponding flows in empirical research

Stocks	Flows
Human population (structured according to age, gender, form of subsistence and other characteristics)	Demographic reproduction Migrations Lifespan/working time
All biophysical stocks (population, infrastructure, livestock, durable goods and tools)	Energetic input/output Material input/output
Territory (structured according to different characteristics)	Freshwater balance Net primary production (NPP) of plants Mineral resources

a particular territory. As explained above, we term this type of social system 'society' regardless of the scale at which this system operates. Both locally and nationally organized societies define 'who and what belongs' to them, or, more specifically, the biophysical stocks they wish to sustain and that they are operationally required to reproduce. This involves the biophysical stocks set out in Table 2.1, which require maintenance through corresponding flows.

Table 2.1 lists the societal stocks and their corresponding flows. A key component of a society is its human population; this is the actual point of reference regarding sociometabolic reproduction. Its size is significant in at least two respects: as the consumers of natural resources and as producers (labor power) who prepare natural resources for consumption and make colonizing interventions in natural systems. Their use of resources is determined by their way of life (as well as their demographic reproduction) and their level of consumption (in further chapters of this book, we discuss metabolic profiles and metabolic rates, see Chap. 3). Their role as producers determines the extent of intervention in natural systems through technologies and working time. A social system that cannot reproduce its population within certain varying ranges cannot maintain itself (see, for example, the fate of peripheral villages in industrial countries).

The size of the population engaged in exchange of this kind and its area density (i.e., the relationship between population and territory) was already viewed as highly relevant by one of Sociology's founding fathers, Durkheim, in his 1893 treatises (Durkheim 1893/2007) on the social division of labor. He regarded certain threshold values for population density as a prerequisite for the emergence of any kind of differentiated division of labor, and with this, specific forms of interdependence and 'solidarity'. The agricultural and development historian Ester Boserup (1965, 1981) has similarly analyzed the relationship between societal development and population density. In contrast to animal populations, for which a maximum sustainable population density—in ecology, termed 'carrying capacity'—exists, the interrelationship between territory and human population size is more complex and is determined by numerous factors, especially technology and, thus, technological change (Boserup 1981; Grübler 1998). Even in the case of hunter-gatherer or early agrarian societies, certain key resources (e.g., flints, salt) were exchanged over wide distances, and in modern societies, resource exchange via international trade is surging (Kastner et al. 2014a; Lambin and Meyfroidt 2011).

The territory as 'stock' is questionable. Is it a biophysical stock a society must reproduce? A territory is not only a natural space, a collection of given conditions within a topographical and natural area; it is also a political realm of power, within which particular interventions are legitimate. For this area, a society defines specific access and usage rights. Resources from the territory may flow into social metabolism without third parties thereby accruing rights to compensation. The human population present within this space is subject to society's operating rules. The functional boundary of the territory is maintained through operations of the social system, such as legal, military or economic activities, and the legitimacy of these operations is negotiated and needs to be recognized by other social systems, for example, neighbors or superordinate social systems. The spatial boundary is defined topographically. It is not a system boundary constituted through natural intersections, even if it may coincide with certain natural system boundaries, such as mountain ranges, watersheds, or water bodies. It soon becomes apparent that the area delineated by the topographical and the functional boundary need not be identical. In functional terms, for instance, the realm of power sometimes includes citizens who are located outside the territorial borders, whereas, conversely, some areas within the territory may have 'extraterritorial' status (e.g., the embassies of other countries). Water bodies often enjoy special status. The legal 'customs barrier' may, for example, be in the hinterland rather than at the coast, the topographical boundary.

Must societies 'reproduce' their territory? On the one hand, history is full of territorial conflicts; a social system that has its territory removed (or even only the resources offered by this territory, for example, through the diversion of a river) is doomed to fail. What role does the territory play in the metabolic reproduction of the social system's population? Of key significance is the fact that territory offers the population a legitimate physical 'common living space', that is, it serves as a 'repository for humans and their infrastructures' (Weichhart 1999). This 'repository function' provides the opportunity to participate in the consumption of the so-called 'free goods', the ecosystem services (Daily 1997a) within the territory (for example, clean air and water). In most cases, however, it means more than this. The state (or other political entities) is in some sense answerable to the 'common good' and thus to ensuring (at least minimal) conditions supporting the reproduction of its human subjects. In any case, at least within its territory, it must ensure that their metabolic reproduction is possible. The territory is therefore meaningful in containing natural resources which economic processes can appropriate. It also provides an outlet for the depositing of waste products from these processes, and it is a source of various non-provisioning ecosystem services.[5]

[5]However, under typical conditions, societies are not required to meet their own resource needs entirely from within their own territory but may regularly make additional use of the resource bases of other social systems (and, indirectly, of their territories) through exchange, trade or tribute obligations.

The relationship of a territory and its human population is not homogenous across space. Trade of resources allows for the emergence of differentiation among social systems. The earliest and most ubiquitous of these differentiations is that between urban centers and their 'hinterland'. A center-periphery structure regularly seems to appear when a territory acquires a certain population density (cf. Boserup 1981). However, we have been able to demonstrate elsewhere (Fischer-Kowalski et al. 2013, 2014a) that the size of urban centers and the share of their population relative to the population scattered across the territory and that provides their food is strictly limited under preindustrial conditions to a few percentage points.[6]

In particular, for agrarian societies, the energetic costs of transport and the human and animal labor power required to perform work are in direct competition with the products that require transportation as both depend on the same energetic basis, namely, the solar energy existing in the form of plant biomass. By this process, transport is translated into the energy required. In turn, this energy is translated into the area available for harvesting and into the labor power required to perform harvesting. This area is then translated into transportation distances, which must be overcome (Fischer-Kowalski et al. 2013; McNeill 2001; Sieferle et al. 2006). Model-based observations make it clear that the conditions of premodern agrarian societies place strict limitations on material flows between territories and that bulk raw materials (basic foodstuffs, sources of energy and construction materials) are used relatively locally.[7] There is an exception where territories are connected by advantageous waterways.

The availability of fossil fuel-based transport technology has lifted these constraints, resulting in a complex worldwide pattern of resource and commodity exchange across territories. Empirical research has shown that many mineral and fossil resources flow from low- to high-income industrialized countries and are directly related to or embedded in commodity flows (UNEP-IRP 2011). In contrast, international trade networks for biomass-based products are still dominated by directions of flow from sparsely to densely populated areas, almost independently of the economic performance of the individual nations (Erb et al. 2009b; Haberl et al. 2012; Kastner et al. 2014a, b).

Ester Boserup offers a different perspective on the interrelation between population and territory that is linked to the considerations detailed above

[6]However, researchers from the World Systems Theory have shown that under conditions of economic and social dominance, members of dominant societies can also draw on the resource bases of other territories under favorable military and transport conditions (Chew 2001; Ciccantell and Bunker 1998; Goldfrank et al. 1999).

[7]Under these general conclusions, of course, counter examples come to mind, such as the relatively wide geographic reach of marble for the opulent buildings of Antiquity or the reputation of Egypt and Spain as the 'bread basket' of ancient Rome. However, these examples only provide evidence for exceptional cases; one should be aware that these spectacular material flows constituted only a small share of the total metabolism of these societies.

(Fischer-Kowalski et al. 2014a). For her, a higher population density means, on the one hand, having access to a greater quantity of manpower and thus the opportunity to construct and maintain infrastructures (e.g., agricultural, transport) and to educate an intellectual elite capable of further developing these infrastructures. On the other hand, a higher population density means, as a matter of principle, lower resource density for the area (i.e., resource scarcity). Therefore, the advantages of higher density only emerge when technological or economic and military means outweigh resource scarcity. Nonetheless, the close connection between territory and the size of the population that can be sustained from its resources applies particularly to agrarian societies as their energetic basis consists almost entirely of solar energy stored in territorially dependent biomass. The same close link is not applicable to fossil fuel-based industrial societies, in which energetic supply does not depend on the size of the territory and in which machines can replace human labor. This replacement alters the relationship between requisite working power and demographic reproduction. Whereas under the conditions of the agrarian society, a vicious cycle among nutrition requirements, working power requirements and fertility rates prevails (which Boserup overlooks, cf. Fischer-Kowalski et al. 2014a), the conditions of industrial societies lead to a drastic reduction in fertility rates.

In contemporary industrial societies, the interlinkage between the sociometabolic requirements of the population and the territory has weakened and given way to globalized supplies. At the same time, the failure of the cultural systems of these societies to maintain a sufficiently integrated communicative base among their (increasingly culturally diverse) population segments within their territory is a major threat to system stability.

2.2.2 On the Relevance of Animal Livestock for Social Metabolism

Domesticated animals play a key role in the interaction between society and its natural environment. Livestock, comprising monogastric species such as pigs and poultry as well as ruminant species such as cattle, sheep and goats, can be regarded as 'live artifacts' providing essential services to humans. They deliver protein-rich food (meat, eggs and milk), fibers such as wool and hides, and building and handicraft materials such as bones and horn. The provision of technical power (draft power) that substitutes for human labor and reactive nutrients—contained in manure—that help prevent soil degradation in agriculture play a decisive role in a society's capacity for colonizing interventions in terrestrial ecosystems.

Preindustrial societies, in particular, vary markedly by the quantities of livestock they own. On the one hand, there is a difference between pastoral and cropland farming societies. Pastoral societies possess a large animal stock to make use of territories that are difficult to use in other ways because of unfavorable climatic conditions or topographic constraints. Cropland farming societies may possess varied stocks of animals depending on the demand for working and draft animals, the amount of marginal land that cannot easily be used for other purposes except grazing (e.g., high alpine pasture and forests) and the extent to which livestock provides fertilizer for cropland (FAO 2011; Krausmann et al. 2003).

Animal livestock is one of the key consumers determining a society's sociometabolic flows. According to thermodynamic laws, the provision of livestock products or services requires a substantially higher input-output ratio in quantitative terms. Large energetic losses are associated with the endosomatic metabolism of animals: as a rule of thumb, only one-fifth to one-tenth of their feed input is converted to growth (meat production), milk, or draft power, and the rest is lost to the maintenance of body functions. Furthermore, animals have to be kept alive in periods when their services are not required, such as during winter, when no mechanical energy is required on cropland fields.

Despite these inefficiencies, livestock plays a central role for many human populations. This becomes evident when abandoning the narrow perspective of efficiency in terms of biophysical input-output evaluation. From a broader perspective, many livestock systems appear particularly efficient. The services the animals provide, albeit associated with large feed requirements, come at little social cost. Livestock is, in principle, able to maintain itself and requires—depending on species and breed, of course—only a comparatively small labor input compared to cropping. Tasks such as herding and defense against predators are, to some extent, performed by livestock themselves quite naturally and efficiently (e.g., a cattle herd with bulls or a sheep herd with dogs) and thus require little input from humans. Livestock can also use areas that are far from the centers where their products are consumed due to their unique ability to transport the goods they provide, such as meat or muscle power. Beyond all this, the ability to digest biomass that is not digestible by humans, such as fiber-rich grasses or branches, makes livestock serve as highly valuable 'grazing machines'. Thus, livestock can indeed be regarded as a powerful means of expanding a society's resource base (Erb et al. 2012; FAO 2011; Herrero et al. 2013). Ruminant livestock can graze in almost all ecosystems, including forest understory, natural grasslands, shrublands and even semi-deserts—in short, land useless for cropping (see Chap. 13). This reduces competition for fertile land (see Chap. 14), and it broadens the source of nutrients for cropping. The German term 'Mistvieh', now a swear-word literally translated as 'dung-animal', can serve as a vivid illustration of this function of livestock: in Alpine preindustrial agriculture, old and otherwise useless domestic animals were not slaughtered but kept to feed from forests to extract nitrogen and nutrients, and their dung was collected for fertilizing cropland (Glatzel 1999).

Monogastric species (such as pigs) can also increase the efficiency of socioeconomic biomass use as they can be fed on biomass of lower nutritional quality and on food waste. Sharing a similar resource base, however, monogastric livestock can come into competition with human nutrition. From a materialistic perspective, the emergence of food taboos serves as a cultural regulation to prevent detrimental effects of this competition (e.g., Harris 1977/1991).

Thus, livestock represents a highly efficient means of colonization as a system with high input-output efficiency. In many preindustrial societies, livestock also plays a key role in safeguarding against environmental fluctuations. Its ability to maintain itself renders livestock a valuable living capital stock that can be liquidized (i.e., slaughtered) during harsh times but requires little maintenance during favorable times. This is a key aspect of livestock for food security that doubtless found entry in the many cultural regulations concerning the coexistence of humans and their livestock (FAO 2011).

In industrial society, with its virtually unlimited potential for increasing yields by removing the restrictions on area regarding transport and provision of fertilizer (McNeill 2001; Sieferle et al. 2006), livestock loses its central role as a working power and supplier of fertilizer. Animals are increasingly transformed into a supplier of food only, particularly protein, and the linkage between area and livestock numbers is loosened (Krausmann et al. 2008; Naylor et al. 2005). However, the global number of livestock has been increasing steadily, nearly doubling since 1961 (in overall numbers; in body mass, the increase is probably even greater). Today, cattle are by far the largest animal group on earth in terms of body mass, at twice the mass of human bodies (FAOSTAT 2014) and many times the mass of wild animals. This fact alone, particularly in light of their central role in human sustenance, makes it surprising that data on livestock and grazing is so scarce (see Chap. 13).

2.2.3 Artifacts, Infrastructure and Material Flows 'from Cradle to Grave'

In terms of quantities of metabolic flows, both the numbers of livestock and the stocks of built infrastructure have an impact. The latter are important because their construction and maintenance require space, flows of materials and energy, and working time. The overall functioning of the social system depends on infrastructure; society cannot function without supply structures for housing, water, gas, electricity, waste disposal structures, transport, power stations, river engineering, disaster protection, harbors and workplaces. Wiedenhofer et al. (2015), for instance, estimate that one-third of all nonmetallic minerals used in Europe are directed at maintaining existing stocks of residential buildings as well as road and railway infrastructure, illustrating their high maintenance costs as well as their long-term path dependencies and legacy effects (see Chaps. 19, 23 and 24). Failures to maintain infrastructures can put social systems at risk (see the debate on the collapse of the Roman Empire among Sieferle 2008, Fischer-Kowalski 2009 and Weisz 2009).

Defining practicable, meaningful and sound indicators and metrics that allow the complex nature of sociometabolic reproduction to be depicted has been a major challenge. For material flows of national economies, a methodology has been developed with intensive international cooperation (particularly with the World Resources Institute, the Wuppertal Institute and the Environment Agency of Japan, see Matthews et al. 2000). This methodology has become a part of standardized official statistics in Japan, the European Union and, increasingly, other parts of the world (see Fischer-Kowalski et al. 2011), and there are now annual reports on social metabolism at the level of nation states.[8] In our contribution to these international efforts, we have extended the accounting scheme from material flows to energy flows in a manner that complies with the above model of societal stocks. The now highly standardized methodological inventory of 'material and energy flow analysis' (MEFA) provides a biophysical accounting framework. It is rooted in the society-nature interaction model as described above and quantifies all flows in a social system that contribute to the reproduction of this system's biophysical stocks in mass (e.g., fresh weight, dry matter) or energetic units (e.g., joules, kilowatt-hours) and per unit of time, usually per year (see Haberl et al. 2004). The resulting indicators deviate from standard technical energy accounting by including the primary energy required for the endosomatic metabolism of livestock and humans. This conceptual bond of stocks and flows at the level of social systems has helped define system boundaries unambiguously and create a logical and clear accounting framework that is applicable at various scales and for very different historical periods.

It is no coincidence that the sociometabolic model has systematic similarities with national accounting—a connection used to address many research questions, such as the resource productivity of a national economy or its energy intensity. To some extent, the sociometabolic model is a description of the economy using biophysical instead of monetary parameters. Thus, the sociometabolic model accounts for the fact that natural systems react only to material interventions, as argued above.

Another important insight yielded by this model is that all input flows in a social system ultimately exit the system as output flows—the so-called 'material balance principle', which is ultimately grounded in the physical laws of thermodynamics (cf. the argumentation of Ayres and Kneese 1969). This provides an explanation of why many environmental measures to combat environmental pollution have proved inadequate. A dissipative system cannot be blocked from behind; to reduce emissions and waste, one must reduce the input into the system.[9]

---

[8]We have made great efforts (see also several related examples in this volume) to develop and apply an analogous methodology at other scales as well, such as local communities (see, e.g., Fischer-Kowalski et al. 2011; Grünbühel et al. 2003; Singh et al. 2001).

[9]Of course, it is also true that where such volumes are involved, certain qualitative parameters, such as the toxicity of material flows, fall by the wayside. Other measures, such as those from life cycle analysis (LCA; see also Method Précis Life Cycle Assessment), may also be consulted.

Finally, accounting flows on the system level yields the insight that the input flows of social systems are very different if measured in physical or economic values. For example, biomass in the form of food and feed constitutes a large share of the overall metabolism (approximately one-third in industrial societies and nearly 100 % in preindustrial economies), whereas it amounts to only a few percent in economic accounting. Similarly, the consumption of fossil fuels is far more important in biophysical terms than their monetary value would suggest.

There is yet another important insight, one that empirically emerged through analyses that compared the material flows of different economies, but is grounded in fundamentally different features of the economic and the biophysical life cycle. In the economic life cycle—from resource extraction to production, processing, trade and end use—monetary values exhibit a trend opposite that of material volumes. In economic terms, the value of a (future) product increases with each step in its preparation and processing. All the raw and auxiliary materials that will be incorporated in a product are, in sum, much less valuable than the product. The value of the product reaches its maximum at the point of sale to the end user. In the course of its utilization, the product progressively loses value, and when finally deposited as waste, it actually acquires a negative value (i.e., its disposal has to be paid for). As far as material volumes are concerned, the opposite is true: during resource extraction, large volumes must be set in motion with a corresponding amount of energy, and at every further step of processing, waste products and emissions are created, and the mass of the product as such becomes smaller. At the point of sale to the end user, the mass is minimal and finally also turns into waste or emissions (because, as we know, matter cannot cease to exist).

This insight is highly relevant when comparing countries with respect to their 'resource productivity' (an indicator measured as unit of gross domestic product, GDP, gained per one unit of material input). According to such a calculation, the national economy of Chile, for example, whose mining industry contributes half of the entire supply of copper to the world markets, has a high domestic material consumption (DMC/inhabitant) and a very low resource productivity (GDP/DMC). In contrast, a nation such as England, which imports almost all of its industrial products and specializes in banking and insurance services, has a very low level of domestic material consumption despite a much higher standard of living than Chile measured as GDP/inhabitant, and a much higher resource productivity. To grasp and interpret these differences, it is necessary to transcend the observation unit of one society by analyzing the network of societies interacting via trade. This can be done with the help of so-called hybrid multi-regional input-output models (MRIOs), both in monetary and physical units (Hertwich and Peters 2009; Lenzen et al. 2012; Peters and Hertwich 2004). By including the 'upstream flows' of imported (and exported) commodities, the so-called 'raw material equivalents' (RMEs) of traded products can be calculated (see Fig. 2.2 above and Schaffartzik et al. 2014), thus balancing the bias of a country's position further up or down the value chain. With the help of MRIOs, a new indicator for countries

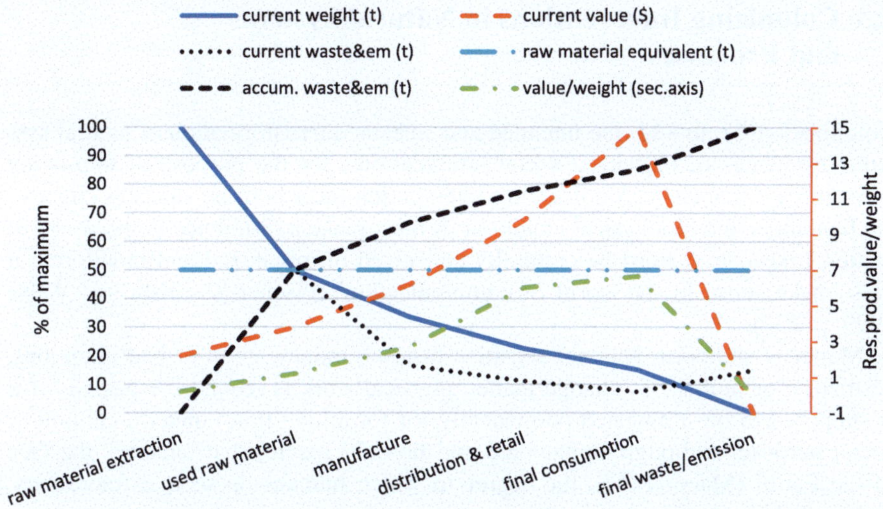

Fig. 2.2 Stylized model of the product life cycle, from extraction to production, consumption and deposition, in physical and economic terms. Model assumptions: From raw material extraction to used raw material, about 50 % of material is discarded (difference between 'total material requirement' and 'domestic extraction'); at each further stage, 33 % of the weight is lost to wastes/emissions. From one stage to the next, the value increases by a factor of 1.5 but drops to zero after consumption (stylized facts). The raw material equivalent (RME) corresponds to used raw material (domestic extraction, DE). See Haas et al. 2015 (For the meaning of the indicators and their abbreviations, see Fig. 1.1 in Chap. 1.). 'Value/weight' is typically accounted for as 'resource productivity'; the inverse is 'material intensity'

has been developed, the 'material footprint' (Tukker et al. 2014; Wiedmann et al. 2013). The material footprint (MF) portrays a country's share in the annual worldwide extraction of natural resources irrespective of whether the country extracts these resources directly from its domestic territory or draws on other territories by trade. Alternative complementary accounting systems that relate to land 'embodied' in the consumption of land-based products (i.e., biomass) are presented in Chap. 16 in this volume.

The rising monetary value of commodities along the stages of processing is, of course, due to the investment of labor and capital at each stage. Based on hybrid MRIO models, efforts have recently been made to quantify the amount and quality of labor input associated with traded commodities (Alsamawi et al. 2014; Simas et al. 2014a, b). This research allows the sociometabolic focus on single social systems, as depicted in Fig. 2.1, to be shifted toward the global network of human societies in biophysical and social terms. The monopoly of Economics in portraying the functioning of the world's societies is being challenged by the sociometabolic paradigm.

2.3 Colonizing Interventions in Natural Systems and Processes

Colonization[10] refers to 'the intended and sustained transformation of natural systems, by means of organized social interventions, for the purpose of improving their utility for society. A colonizing intervention must both be causally effective in changing some biophysical condition; it must make a difference in the world of matter. Likewise, it must be culturally conceived of, organized and monitored; it must make sense in the world of communication' (Fischer-Kowalski and Weisz 1999, p. 234).

Matter is stubborn. It is not sensitive to laws, money, good or bad intentions, aesthetics, or morals. To change matter, physical work is required, especially if it is supposed to be subjugated permanently (or for a long time) under a cultural or social purpose. Although humans are not the only species that 'changes the face of the Earth' (Marsh 1864), the degree to which humans do so is extraordinary. Although humans do change their natural environment by their metabolic processes, as explained in the previous section, colonization has a far greater impact. Colonization addresses natural systems across all hierarchical scales of organization (Fig. 2.3). It targets the level of atoms through exploiting energy flows caused by controlled (and uncontrolled) atomic decay, and it targets the level of macromolecules (particularly the genomes of many organisms) through domestication and genetic engineering. Furthermore, colonization alters the properties of molecules, cells and tissues, and it modifies organisms and their morphology and behavior. Some interventions are directed at entire ecosystems, such as the clearing of pristine forests for agricultural purposes.

2.3.1 System Theoretical Considerations

The concept of colonizing interventions builds on the theory of autocatalytic (Maturana and Varela 1987; Varela et al. 1974) systems. Many natural systems (certainly all live systems) are autocatalytic, that is, complex systems that cannot be 'controlled' or 'steered' from the outside. They can only be 'irritated'

[10]In any interdisciplinary field, it is advisable to use special terminology as sparingly as possible. It is important not to use a term stemming from a specific discipline in a markedly different way from its usual application, and it is important to avoid any choice of terminology that might foster the view that this particular discipline is in any way superior to others. In other words, we have to consider terminology not only as a tool serving the interests of research but also as touching on territorial and hegemonic issues between disciplines. With the term 'colonization', we refer to the Latin term *colonus*, which means farmer. In contrast, one may also associate the term with colony and colonialism, which refers to the subjugation and exploitation of a country by a dominant power. Both connections provide quite meaningful connotations.

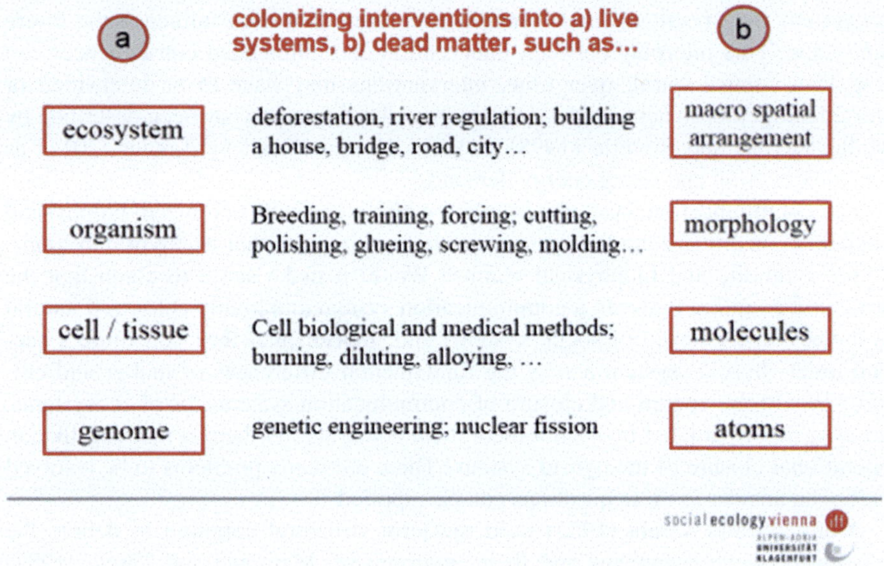

Fig. 2.3 Levels of potential societal colonizing interventions in natural systems

(to varying degrees, up to the point of destruction), always with some uncertainty in outcome and usually some unintended side effects. In organizational systems theory (e.g., Willke 1990), an outside effort at irritating a social system, with a particular goal or a certain intention in mind, is called an 'intervention'. We adopt this term to address a social system's effort to change the behavior of complex natural systems.[11] Usually, colonizing interventions seek to push a natural system out if its current state into a state beyond equilibrium, which is often a more fragile state. A pot is more fragile than a clump of clay; a river will continuously work on its dam to break it; a dog will forget how to obey; and a house will be washed away to the soil. Thus, there is a tendency for colonizing interventions to be self-denying, their objects falling back to their original equilibrium or some other state. There needs to be continuous monitoring and readjustment for the desired state (and the desired 'services' these states deliver) to be maintained. These interventions typically induce both intended and unintended changes in natural systems, sometimes at higher organizational levels. Intended changes in a grassland

[11]Let us consider, for example, constructing a chair from a piece of wood. The intention is clear: there exists a meaningful cultural program as to what features chairs should have. The intervention may still fail; the wood may not be homogenous and may break, or the glue may not stick. However, even if the end result is a nice chair, it is not a physical object under perfect social control. It still follows its natural destiny: the wood evaporates, worms may start eating their share or the surface may rot. I take continuous intervention to prohibit these natural fates—or else relinquish the chair altogether.

ecosystem may result in microclimatic alterations that then influence the entire landscape. This interplay between intervention and unintended consequences can lead to a control spiral: over time, interventions may have to be intensified or shifted altogether to achieve the original goal. This dynamic has been described by Müller-Herold and Sieferle (1998) as the 'risk spiral' and by Tainter (2011) as complexification and rising energy demand.

Do communication systems coupled with intensively colonized biophysical elements constitute complex systems in their own right, that is, 'hybrid systems' of the symbolic and biophysical realms? Would it make sense to claim that the structural coupling between a communication system and highly colonized natural systems is the essence of social systems and, indeed, that they constitute a second-order (hybrid) system across the fundamental differences of matter and culture? Would the operational closure of communication systems, or of ecosystems, necessarily be violated by such a theoretical construct? Is there something like an operational closure of the hybrid system? There are many problems to be resolved before an answer to these questions can be supplied.

Maturana and Varela (1975) used the term structural coupling to denote the relation between organisms and their environment. Maturana and Varela (1987) see the interactions between organisms and their environment as determined by the organism; the organism decides which components it extracts from the environment, and the organism's structures determine in which form they are being returned. Although numerous mutual causalities are conceivable, Maturana/Varela insist that changes in the environment may trigger changes in the organism, but they can never causally determine the metabolic processes of the organisms themselves. These arguments also very much reflect the processes we have addressed as social metabolism above (if we may, for the moment, equate the organism with the social system), and there we would fully agree with the Maturana/Varela point of view.

Our concept of colonization, however, draws an additional distinction between a colonized compartment and the rest of the natural environment. We would immediately agree that, biophysically speaking, the colonized compartment is no distinct entity—colonization is a matter of degree. However, the social system draws a distinction between colonized and uncolonized on the level of communication as well as operatively. It is defined by the communication system to which people, livestock and infrastructure 'belong'. These relations of belonging or property can be complex. For example, a factory may be on national territory and under its jurisdiction, but taxes may be due elsewhere, and the owner may not be a formal citizen of the country but a permanent resident. Moreover, these relations may not be well aligned among different subsystems. A relation of belonging or property does not imply complete control over something, but it does imply an interest in its reproduction and a certain liability, both of which are usually well represented within the communication system. This relation also matters biophysically. The metabolism of 'colonized' natural elements is typically taken care of socially. Thus, they become part of the social system's material stocks, and the flows required for their reproduction, by definition, become part of the social metabolism. Thus, in analogy to the Maturana/Varela perspective, the metabolic processes of social systems are determined by the functional requirements of those biophysical elements that are communicatively defined as 'belonging' to them and that are considered worth reproducing.

Boundaries, however, are fuzzy. They are fuzzy in the social definitions of 'belonging'. For example, whether a certain road 'belongs' to a municipality and needs to be maintained by it or whether the responsibility lies with the respective state may be ambiguously defined. Boundaries are also fuzzy in terms of causal control. For example, how much of the road's present state is controlled by colonizing interventions on the part of the social system and how much is controlled by natural processes of weathering and plant regrowth is a matter of degree. Relations may become more sophisticated, of course, as the literature on socio-technical systems illustrates[12] (see Appelbaum 1997).

For live elements, boundaries become even more complex. With livestock, there are usually clear property relations, and both the metabolism and the biological reproduction of the animals are largely under social control. In the historical background, there has been a long chain of colonizing interventions in animal reproduction that have led to changes on the genomic level (see Fig. 2.3), something that one day may be achieved much faster by a direct technical intervention in the genome. However, most metabolic requirements remain a matter of the nature of the organism (and in this sense, it remains operationally closed). If the social system uses livestock as working animals, it also needs to intervene in their behavioral characteristics (organismic level in Fig. 2.3) to train them for certain tasks and make them abstain from certain unwelcome habits. The same, of course, applies to pets. For plants, the main social property relation is the land they grow on. Nevertheless, among the plants growing on someone's land, large distinctions may exist concerning their degree of colonization. For example, there may be a carefully dressed flower bed where the plants have been genetically modified, have had additional colors injected into their roots (intervention in tissue, see Fig. 2.3), are carefully cut (intervention in the organism) and are sprayed with pesticides (intervention in ecological conditions, population dynamics, predator-prey relationships). However, an unwanted side effect of colonization may be that weeds take advantage of favorable conditions by, for example, freeriding on the watering. Traditionally, the plant's nutrient and water intake can be influenced, but its biological reproduction is not so easily influenced. Modern biotech firms are working hard on extending their control over plants' fertility, but sometimes they cannot achieve more than an intervention in the communication system, such as making the use of next-generation seed illegal, though it is still functionally possible.

Even if the situation remains relatively stationary, that is, if a continuous flow of colonizing interventions secures a continuous level of ecosystem services, a strong interdependency is created between the social system in demand of certain

---

[12] In this literature, an effort is made to define the social organization, its processes and rules and the technical equipment it uses for communication and production as one complex system.

resources or services and the natural systems (or elements) concerned. A steady flow of monitoring and labor is required to keep natural systems in a specific colonized state and to secure the flow of resources and services. To be able to provide these benefits, communication processes and social organization have to be set up in specific ways and have to be readjusted if there is any change in the natural systems.

In the real world where interdependencies are dynamic, the social system is continuously under threat of losing its resources and services. The natural 'partner system' has to be maintained in a certain state, which in itself is subject to fluctuations and variations. Society thus has to reorganize itself permanently to secure its ability to provide the continuous flow of work, energy and materials required to maintain the natural system in its colonized state. This reflexive mechanism ranges from individuals having to organize to be able to regularly take their pills (i.e., a medical colonization of the human body), to dog owners who need to find the time to walk their dogs regularly, to villagers having to take care of the uphill forest to protect them from avalanches, to communities having to organize their communal budget and institutional regulations for protection against hazards, to employers depending on the functioning of the public transport system for their workers to be in place, and so forth. In effect, one may see this relation as a structural coupling between the cultural or communication system(s) and particular highly colonized natural systems intimately linked to them in mutual functional interdependence. In that sense, every social is a hybrid entity resulting from the structural coupling (achieved by colonizing interventions) between a communication system and a set of natural elements. One can presume a coevolutionary dynamic, as suggested in Fig. 2.4: the social investment in natural systems with the intention of transforming them in a desired way would achieve at least part of the intended outcome, would change the natural system and would lead to certain returns in terms of resources or services. The supply of these resources/services would in turn change the social system, adjusting it to this inflow. In parallel, the required investment of time, energy, materials and attentive observation would transform the social system (e.g., population dynamics, time and labor organization, technology development, education and energy supply).

Here again, the work by Boserup (1965, 1981) provides illustrious examples. In the long run, a preindustrial society achieves higher agricultural yields per area at the price of higher labor investments, which in turn allows for or even stimulates population growth and again increases the need for higher yields (see also Fischer-Kowalski et al. 2014b). A similar coevolutionary dynamic can be illustrated for the case of oil. As a country such as the United States (US) gradually runs out of this resource (as occurred in the 1970s) and because it has built most of its economy around it, it needs to shift to imports. In so doing, it transfers a large amount of money to countries that can supply this resource but do not have a modern structure. Thus, resources flow into (from the enlightened 'modern' view) archaic structures that feel culturally threatened and in turn pose major security risks to the US (and other 'modern' societies). To better cope with this situation, the US shifts toward the new colonization strategy of fracking in its own territory,

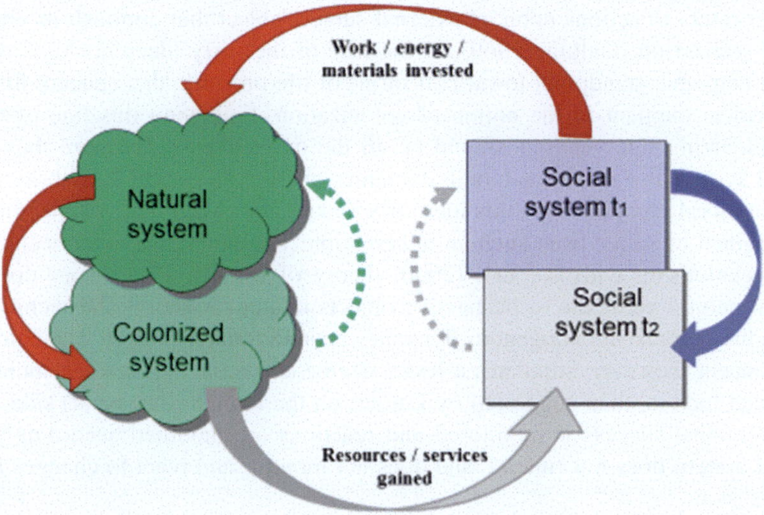

Fig. 2.4 From a control spiral induced by colonizing interventions to a structural coupling of social and natural systems

a well-stimulation technique in which rock is fractured by a hydraulically pressurized liquid. This technique requires a much higher labor and energy input and is associated with many detrimental effects on the land. The full coevolutionary cycle might end with both the need for higher security investments and higher investments in energy generation, and with it, more environmental damage.

On a more generic level, Tainter (1988, 2011) has described such processes of increasing mutual interdependency in his theory of societal collapse. He claims that societies build up complexity in their effort to resolve the problems that they encounter. An increase in complexity is always related to a need for a higher resource input, which results in a self-defeating process; each additional increase in complexity contributes less to problem-solving. However, although Tainter describes this process as purely social (with resources coming in as some passive element from the environment), we feel it could be better framed as a coevolutionary process: the changes in the environment due to societal problem-solving generate much of the problems that need to be solved.

2.3.2 How Can the Concept of Colonizing Interventions Be Made Operational?

The points of departure of our theoretical conception of metrics for colonizing interventions were simple. On the one hand, there was peace-researcher Johan Galtung's concept of 'structural violence' as an extension of direct violence. This concept refers to actions upon an external subject/object that diminish its chances of self-realization (Galtung 1969). In the face of the early 'deep ecology' debate and an empathic standpoint toward all forms of life on earth, this concept fits well as a critical element in the notion of colonization. Following this line of thinking, colonizing interventions should be all the more intense the more they tip a natural system away from its inbuilt dynamics or from its natural equilibrium. The implicit moral component of this idea only makes sense with live systems that can be imagined to suffer from such an intervention. The functional component bears some relationship with Latour's (2000) theory of actants. This theory does not assume natural elements to be passive objects of human action and technology; rather, they represent endogenous dynamics beyond human control. The colonization concept, however, stipulates a lesser degree of symmetry between social and nonsocial 'actors' than suggested by Latour: on the human (the social) side, there are intentions, success is monitored and reactions are modified accordingly; the natural system does not 'intend' and does not monitor and react to changes in the social system.

On the other hand, there was the idea to quantify the intensity of the intervention as proportional to the effort the social system puts into it. This effort could be expressed as the amount of human labor (measured, for example, in time units) on the one hand and the amount of technical energy invested on the other. In Land-System Science, such metrics are widespread, particularly in agronomic research. In combination with output metrics (e.g., agricultural yield), they have been used to describe the production function of land agriculture (and, less so, forestry), and, following the law of diminishing returns, they can be used to economically optimize land-based production (Erb et al. 2014). The biophysical indicator 'energy return on investment' (EROI; Hall and Klitgaard 2012; Hall et al. 2009) builds upon such metrics and has proven useful to describe the socioeconomic process underling land-use transitions (Krausmann et al. 2003).

However, the intensity of intervention is not necessarily a robust indicator of the effectiveness of colonization. A high amount of effort can be very ineffective at achieving a desired state of the natural system; vice versa, a small—and even declining—amount of input can be very effective in achieving a colonization goal. This discussion was punctuated by a publication by Schmid (*Magie in der Kolonie* 2006, or 'Magic as Colonizing Force') that experimentally suggested that magical rites be considered modes of colonization. Could we consider extensive rain-dance rituals performed with enormous effort to be colonizing interventions? This debate reinforced the idea that both the effort on the part of the social system and the effectiveness in changing natural conditions needed to be part of the concept. That effort mattered was well illustrated by a series of publications from our team that investigated the so-called Biosphere II experiment in Arizona (Haberl et al. 1998; Winiwarter 2000). This experiment at creating and running a quasi-artificial world in a materially closed glasshouse container was interpreted by the authors as an effort at 'total colonization'. The experiment failed; despite a tremendous effort in data management and an enormous continuous input of energy (28,000 gigajoules, GJ, per 'bionaut' per year—the European average is 200 GJ per inhabitant per year!), the bionauts lost weight and had to work much more than expected to manage the ecosystems and obtain a necessary minimum of food, oxygen and water.

A stringent empirical operationalization of the colonization concept on generic terms remains a challenge. The analytical framework 'human appropriation of net primary production' (HANPP), provides an example of such an operationalization for the broader field of 'colonization of terrestrial ecosystems' (see also Method Précis on Human Appropriation of Net Primary Production). Net primary production (NPP) denotes the annual biomass production of autotrophic organisms (mainly plants), that is, the balance between gross primary production and autotrophic respiration, and is a key process in the Earth system, providing the energetic basis of plant growth as well as for the trophic energy supply of all heterotrophic species, including humans. The HANPP concept had already been proposed in the late 1980s by Vitousek et al. (1986), motivated by questions about the scale of human activities within the context of global natural processes (Lieth 1973; Whittaker and Likens 1973, 1975). According to Vitousek et al., HANPP shows a huge range depending on the inclusiveness of definition, amounting to 3 % if only final biomass products are accounted for and up to 38 % if all NPP 'co-opted' as well as potential NPP losses are accounted for. The theoretical background of 'colonization theory' allowed the development of sound and meaningful definitions and the introduction of system boundaries that enhanced the analytical quality and the suitability of the HANPP indicator as an environmental pressure indicator (Erb et al. 2009a; Haberl et al. 2007, 2014). These further developments provided guiding principles for empirical research, much of which is documented in this volume (cf. Chaps. 14, 16, 17, 18 and 22 in this volume).

In this definition, HANPP indicates how much of the potentially available annual plant biomass production (NPP) is appropriated through human colonizing interventions, through deliberate changes of the land cover (for example, the conversion of pristine forest to grassland or arable land, or the sealing of soils by the construction of cities and roads) and/or through harvest. In other words, HANPP quantifies the difference between the NPP of the potential (i.e., hypothetically undisturbed) vegetation and the NPP that remains in the ecosystem after harvest. However, this indicator does not capture the social side of this process or the feedbacks on social structure. Two elements of these social features have also entered our empirical work: one is the labor burden in agriculture, which is dependent on colonizing interventions (see Chap. 26 in this volume), and the other is the issue of culturally molded diets. Diets may be considered an outcome of the coevolutionary process. A strongly meat-based diet results, as one possibility, from the existence of large but arid (or cold) lands that can support grazing animals but not farming, or it may result from abundant land in pioneer situations where there is still very low population density. However, the accustomed diet, reinforced economically, can lead to 'unadapted' colonization practices in regions that do not bear these features (e.g., the Netherlands), generating severe environmental side effects (that might trigger another coevolutionary dynamic). It can also lead to severe human health impacts by promoting overweight. For the world population, which is expected to become even larger, such diets will be unaffordable. Nevertheless, it seems difficult to leave such coevolutionary pathways.

Additional aspects of colonization entail an Environmental History perspective on society-nature interactions. In recent research, we analyze the coevolutionary pathways between a large river (the Danube) and the city of Vienna across five centuries (see Chap. 19 in this volume). This allows colonization to be localized spatially; the concept of socio-natural sites (*sozio-naturale Schauplätze*, SNSs) denotes a concrete place molded by a long sequence of colonizing interventions. A natural setting (such as a particular fertile region at a large river), as soon as it is occupied by people, is immediately modified by colonizing interventions. The particular interventions are part of the culturally predefined practices by which people at a certain time seek to meet their needs. They depend on their cultural repertoire, and they depend on the opportunity structure the environment offers. In all cases, they need to address food and shelter, securing the energetic and material metabolism of the group now and into the future. At these SNSs, material arrangements are created (by colonizing interventions) that are favorable for certain historical practices under the given opportunity structures. These socio-natural arrangements in turn facilitate certain practices and inhibit others: paths invite movement, whereas fences prohibit it. Natural arrangements have a similar selective effect: a large river makes crossing difficult, but it eases downward transportation; it threatens its surroundings with flooding, but it also fertilizes the land. If material arrangements are to favor certain practices, a continuous flow of labor and materials must be invested in their maintenance (colonization). The relationship between socio-natural arrangements and practices is coevolutionary. Arrangements (such as city walls) may become an obstacle to certain practices (in the case of population growth, for example), or they may provide an opportunity structure for unwanted practices (enemies may use the bridge across the river). When reconstructing the history of a river course and a city that lives at and from the river, these concepts relating to colonization have proved empirically useful.

Another approach, also with a historical note, seeks to quantify human labor time in relation to the colonizing practices employed. If it is correct to assume that colonizing interventions, within a sociometabolic regime, have a certain self-defeating tendency, then the requirement of human labor time should increase the longer the particular regime is in place. This time may come to a limit if human self-reproduction time (sleeping, resting, eating and hygiene) is squeezed too far to allow for a healthy life. Using labor time in this way to compare the colonization effort required between different social systems and framework conditions has proved to be a fruitful empirical approach (see Ringhofer et al. 2014 and Chap. 7 in this volume).

2.3.3 Which Intellectual Services Does the Colonization Concept Provide in Contrast to Other Conceptualizations?

The mainstream conceptualization of relating society to the material world is dependent on technology. Technology refers to both the specific tools (from bows and arrows to steam engines, washing machines and computers) and the social practices utilized to address certain problems, such as cooking food with fire and moving around by car. There is a long intellectual tradition of relating changing social practices and social structures to changes in technology, to the extent that many believe technology is the core driver of human history (Grübler 1998). There is much less systematic inquiry into the concomitant changes of the natural environment that happen as a consequence of certain technologies. Framing the society-nature interaction around technology creates a story of increasing human control over natural processes and of unlimited progress; nature largely plays a passive role.

The concept of colonization is more modest and much more symmetrical. It does not suggest that one side could control the other but suggests a chain of coevolutionary mutual impacts that may completely run outside of human control. It also takes a wider range of social responses to problems into account, such as changes in fertility, rising or shrinking inequality and even changing need patterns.

An increasingly popular discourse centers on ecosystem services.[13] Ecosystem services denote the sum of services ecosystems provide for society. These can be provisioning services (e.g., the provision of food, feed and fibers), regulating services (such as carbon sequestration and water and air purification), or cultural services (nonmaterial services, such as cultural, spiritual, or recreational functions). At the very heart of this concept, originally formulated in the late 1990s (Daily 1997b), is the acknowledgement that ecosystems provide goods and products beyond those that are marketed and have a price. Most of these other services remain unpriced and are thus not subject to the regulating forces (of any kind) of markets. Thus, the effects of human activities on these non-provisioning services remain unrecognized and are often detrimental (Daily 1995; Foley et al. 2005).

Whereas colonization draws attention to what social systems do to nature, ecosystem services draw attention to what social systems need and get from nature, including the warning not to destroy the source of these benefits. The discourse on ecosystem services is very closely related to the discourse on nature conservation—in a situation that is conceived as fairly desperate (considering galloping biodiversity and habitat loss), it is an effort to appeal to human self-interest by attaching economic value to those ecosystem services that may come under threat. In this approach, society is identified largely with its economy. The economy is

[13]The recently founded journal *Ecosystem Services*, in its first issue, 1/2012, gives an excellent overview of the various features of the ecosystem services approach.

seen as the main regulatory force, but as poorly organized to adequately address the maintenance of ecosystem services (Farley 2012).

One might consider ecosystem services to be quite a complementary concept to colonization. Ecosystem services (although, perhaps, in a wider sense) are what colonizing interventions attempt to secure and enhance. When we relate colonizing interventions to the economy, we talk about their cost, whereas the ecosystem services discourse talks about the benefits. Thus, a closer interlinkage between the two approaches might prove productive.

Chapter 4 in this volume elaborates how the colonization concept can help bridge these two conceptualizations within the context of land-use research. The colonization concept, empirically operationalized by the HANPP framework, allows technological aspects of land-use intensification to be addressed. It provides data on the output intensity of land management (where output intensity denotes the output per unit area of ecosystems, such as agricultural yields) that can be combined with data on input intensity (e.g., capital or labor inputs) to study the production functions of land-based production. Concomitantly, the HANPP framework systematically assesses changes in ecosystem states that are intimately associated with the input-output relation of land-based production, that is, the alteration of the availability of trophic energy in ecosystems. Many unintended consequences related to land use are closely linked to changes in NPP, such as human-induced degradation and changes in biodiversity or changes in biogeochemical cycles, such as altered carbon storage and sequestration. These interlinkages resulted in the conceptualization of land-use intensity as a multidimensional process (Erb 2012; Erb et al. 2013, 2014), thus bridging technological and conservationist perspectives of land use.

2.4 Conclusions

The socioecological theoretical approach we have attempted to describe in this chapter is ambitious, and it draws more on social systems theory than may be easily accessible to many readers. This approach has seen international successes empirically: the notion of social metabolism[14] guided the methodological develop-

[14]We do not claim priority for discussing social metabolism. In 1991, Baccini and Brunner published a book on *The Metabolism of the Anthroposphere* (Baccini and Brunner 1991), and the book by Ayres and Simonis on *Industrial Metabolism* was—with quite a delay after the preparatory conferences under the same name in Tokyo 1988 and Maastricht 1989—published in 1994 (Ayres and Simonis 1994). In Fischer-Kowalski and Haberl (1993), the concepts of metabolism and colonization first appeared jointly. Fischer-Kowalski, then, was the first to situate the concept of metabolism explicitly within the traditions of social theory. In the *International Handbook of Environmental Sociology*, edited by M. Redclift and G. R. Woodgate, she was as bold as to announce society's metabolism as a 'rising conceptual star' (Fischer-Kowalski 1997). Her later reviews of the intellectual history of society's metabolism were among the most cited articles in the *Journal of Industrial Ecology* (Fischer-Kowalski 1998; Fischer-Kowalski and Hüttler 1998).

ment of material flow accounting (MFA) that is now common statistics, and the particular empirical reading of colonization as HANPP has made its career in the literature. These successes are due to two great virtues: the concept allows social and natural systems to be addressed symmetrically, taking explicitly into account the complex and distinct system characteristics on both sides, and it is applicable across all of human history, without a modernist bias.

Among the Environmental Sciences, it has gained a reputation for interdisciplinarity and for relating to the Social Sciences. Among the Social Sciences, gaining a stand is more difficult. Describing the economy in biophysical terms (which is what MEFA does) is a core theme of Ecological Economics, but it has no place in mainstream Economics. Sociology and Political Science are not well integrated, either among themselves or with each other, and neither may feel comfortable with an analytic and systemic perspective that does not leave much room for deliberative action. Apart from sociological systems theory, there is currently little coherent and comprehensive social theory. Among historians, the combination of a systemic and general theoretic approach, with findings represented largely quantitatively, appears very uncommon. However, if society is to find a more sustainable pathway, the task of creating theory and methods that provide guidance needs to be pursued.

Acknowledgments Karl-Heinz Erb gratefully acknowledges funding from the European Research Council (Grant 263522 LUISE).

References

Alsamawi, A., Murray, J., & Lenzen, M. (2014). The employment footprints of nations. *Journal of Industrial Ecology, 18*(1), 59–70.

Appelbaum, S. H. (1997). Socio-technical systems theory: an intervention strategy for organizational development. *Management Decision, 35*(6), 452–463.

Ayres, R. U., & Kneese, A. V. (1969). Production, consumption and externalities. *American Economic Review, 59*(3), 282–297.

Ayres, R. U., & Simonis, U. E. (Eds.). (1994). *Industrial metabolism: restructuring for sustainable development*. Tokyo, New York, Paris: United Nations University Press.

Baccini, P., & Brunner, P. H. (1991). *The metabolism of the anthroposphere*. Berlin: Springer.

Boserup, E. (1965). *The conditions of agricultural growth. The economics of agrarian change under population pressure*. Chicago: Aldine/Earthscan.

Boserup, E. (1981). *Population and technological change: A study of long-term trends*. Chicago: University of Chicago Press.

Chew, S. C. (2001). World ecological degradation: accumulation, urbanization, and deforestation, 3000 BC-AD 2000. Walnut Creek, CA: Altamira Press.

Ciccantell, P. S., & Bunker, S. G. (1998). Space and transport in the world-system. Westport, CT: Greenwood Publishing Group

Crutzen, P. J. (2002). Geology of mankind. *Nature, 415*, 23.

Daily, G. C. (1995). Restoring value to the world's degraded lands. *Science, 269*(5222), 350–354.

Daily, G. C. (1997a). *Ecosystem services: benefits supplied to human societies by natural ecosystems*. Washington, D.C.: Ecological Society of America.

Daily, G. C. (1997b). *Nature's services: Societal dependence on natural ecosystems*. Washington, D.C.: Island Press.

Dawkins, R. (1982). *The extended phenotype: The gene as unit of selection*. Oxford: W.H. Freeman and Company.
Durkheim, É. (2007). *De la division du travail social*. Paris: Presses Univ. de France. (Original work published 1893).
Erb, K.-H. (2012). How a socio-ecological metabolism approach can help to advance our understanding of changes in land-use intensity. *Ecological Economics, 76*, 8–14.
Erb, K.-H., Krausmann, F., Gaube, V., Gingrich, S., Bondeau, A., Fischer-Kowalski, M., & Haberl, H. (2009a). Analyzing the global human appropriation of net primary production—processes, trajectories, implications. An introduction. *Ecological Economics, 69*, 250–259.
Erb, K.-H., Krausmann, F., Lucht, W., & Haberl, H. (2009b). Embodied HANPP: mapping the spatial disconnect between global biomass production and consumption. *Ecological Economics, 69*, 328–334.
Erb, K.-H., Mayer, A., Kastner, T., Sallet, K. E., & Haberl, H. (2012). The impact of industrial grain fed livestock production on food security: an extended literature review. Commissioned by Compassion in World Farming, The Tubney Charitable Trust and World Society for the Protection of Animals. Social Ecology Working Paper. Institute of Social Ecology, Vienna.
Erb, K.-H., Haberl, H., Jepsen, M. R., Kuemmerle, T., Lindner, M., Müller, D., et al. (2013). A conceptual framework for analysing and measuring land-use intensity. *Current Opinion in Environmental Sustainability, 5*, 464–470.
Erb, K.-H., Niedertscheider, M., Dietrich, J. P., Schmitz, C., Verburg, P. H., Rudbeck Jepsen, M., & Haberl, H. (2014). Conceptual and empirical approaches to mapping and quantifying land-use intensity. In M. Fischer-Kowalski, A. Reenberg, A. Mayer, & A. Schaffartzik (Eds.), *Ester Boserup's legacy on sustainability: Orientations for contemporary research* (pp. 61–86). Dordrecht: Springer.
FAO. (2011). World Livestock 2011. Livestock in food security. Food and Agriculture Organization of the United Nations.
FAOSTAT. (2014). FAOSTAT: Statistical Database of the United Nations Food and Agricultural Organization. Retrieved from http://faostat.fao.org.
Farley, J. (2012). Ecosystem services: the economics debate. *Ecosystem Services, 1*(1), 40–49.
Fischer-Kowalski, M. (1997). Society's metabolism: on the childhood and adolescence of a rising conceptual star. In M. Redclift & G. R. Woodgate (Eds.), *The international handbook of environmental sociology* (pp. 119–137). Cheltenham, Northhampton: Edward Elgar.
Fischer-Kowalski, M. (1998). Society's metabolism. the intellectual history of material flow analysis, Part I: 1860–1970. *Journal of Industrial Ecology, 2*(1), 61–78.
Fischer-Kowalski, M. (2009). Den Zusammenbruch von Zivilisationen konzeptualisieren? Plädoyer für eine sozialökologische Gesellschaftstheorie. *GAIA* 13–14.
Fischer-Kowalski, M., & Haberl, H. (1993). Metabolism and colonization. Modes of production and the physical exchange between societies and nature. *Innovation—The European Journal of Social Sciences, 6*(4), 415–442.
Fischer-Kowalski, M., & Hüttler, W. (1998). Society's metabolism. The intellectual history of material flow analysis, Part II: 1970–1998. *Journal of Industrial Ecology, 2*(4), 107–137.
Fischer-Kowalski, M., & Weisz, H. (1999). Society as a hybrid between material and symbolic realms. Toward a theoretical framework of society-nature interaction. *Advances in Human Ecology, 8*, 215–251.
Fischer-Kowalski, M., Singh, S. J., Lauk, C., Remesch, A., Ringhofer, L., & Grünbühel, C. M. (2011). Sociometabolic transitions in subsistence communities: Boserup revisited in four comparative case studies. *Human Ecology Review, 18*, 147.
Fischer-Kowalski, M., Krausmann, F., & Smetschka, B. (2013). Modelling transport as a key constraint to urbanisation in pre-industrial societies. In S. J. Singh, et al. (Eds.), *Long term socio-ecological research. Studies in society nature interactions across spatial and temporal scales* (pp. 77–101). Dordrecht, Heidelberg, New York, London: Springer.
Fischer-Kowalski, M., Reenberg, A., Schaffartzik, A., & Mayer, A. (Eds.). (2014a). *Ester Boserup's legacy on sustainability: orientations for contemporary research*. Dordrecht: Springer.

Fischer-Kowalski, M., Krausmann, F., & Pallua, I. (2014b). A sociometabolic reading of the Anthropocene: modes of subsistence, population size and human impact on Earth. *Anthropocene Review, 1*(1), 8–33.

Foley, J. A., DeFries, R., Asner, G. P., Barford, C., Bonan, G., Carpenter, S. R., et al. (2005). Global consequences of land use. *Science, 309*, 570–574.

Galtung, J. (1969). Violence, peace and peace research. *Journal of Peace Research, 6*, 167–191.

Glatzel, G. (1999). Historic forest use and its possible implications to recently accelerated tree growth in central Europe. In T. Karjaleinen, H. Spieker, & O. Laroussine (Eds.), *Causes and consequences of accelerated tree growth in Europe, EFI proceedings, 27* (pp. 65–74). Joensuu: EFI.

Goldfrank, W. L., Goodman, D., & Szasz, A. (1999). Ecology and the world-system. Westport, CT: Greenwood Publishing Group.

Grübler, A. (1998). *Technology and global change*. Cambridge, UK: Cambridge University Press.

Grünbühel, C. M., Haberl, H., Schandl, H., & Winiwarter, V. (2003). Socioeconomic metabolism and colonization of natural processes in Sangsaeng village: material and energy flows, land use, and cultural change in Northeast Thailand. *Human Ecology, 31*, 53–86.

Haas, W., Krausmann, F., Wiedenhofer, D., & Heinz, M. (2015). How circular is the global economy? An assessment of material flows, waste production and recycling in the EU and the world in 2005. *Journal of Industrial Ecology, 19*(5), 765–777. doi:10.1111/jiec.12244.

Haberl, H., Weisz, H., & Winiwarter, V. (1998). Kontrolle und Kolonisierung in der zweiten Biosphäre. In C. Wächter et al. (Eds.), Technik gestalten. Interdisziplinäre Beiträge zu Technikforschung und Technologiepolitik (pp. 239–251). München, Wien: Profil-Verlag.

Haberl, H., Fischer-Kowalski, M., Krausmann, F., Weisz, H., & Winiwarter, V. (2004). Progress towards sustainability? What the conceptual framework of material and energy flow accounting (MEFA) can offer. *Land Use Policy, 21*(3), 199–213.

Haberl, H., Erb, K.-H., Plutzar, C., Fischer-Kowalski, M., & Krausmann, F. (2007). Human appropriation of net primary production (HANPP) as indicator for pressures on biodiversity. In T. Hák, B. Moldan, & A. L. Dahl (Eds.), *Sustainability indicators: A scientific assessment* (pp. 271–288). Washington, D.C.: Island Press.

Haberl, H., Steinberger, J. K., Plutzar, C., Erb, K.-H., Gaube, V., Gingrich, S., & Krausmann, F. (2012). Natural and socioeconomic determinants of the embodied human appropriation of net primary production and its relation to other resource use indicators. *Ecological Indicators, 23*, 222–231.

Haberl, H., Erb, K.-H., & Krausmann, F. (2014). Human appropriation of net primary production: Patterns, trends, and planetary boundaries. *Annual Review of Environment and Resources, 39*, 363–391.

Hall, C. A. S., Klitgaard, K. A. (2012). Energy return on investment. In: Energy and the Wealth of Nations (pp. 309–320). New York: Springer.

Hall, C. A. S., Balogh, S., & Murphy, D. J. (2009). What is the minimum EROI that a sustainable society must have? *Energies, 2*, 25–47.

Harris, M. (1991). *Cannibals and kings: The Origins of Cultures*. New York, NY: Vintage Books. (Original work published 1977).

Herrero, M., Havlik, P., Valin, H., Notenbaert, A., Rufino, M. C., & Thornton, P. K., et al. (2013). Biomass use, production, feed efficiencies, and greenhouse gas emissions from global livestock systems. *Proceedings of the National Academy of Sciences, 110*(52), 20888–20893.

Hertwich, E. G., & Peters, G. P. (2009). Carbon footprint of nations: A global, trade-linked analysis. *Environmental Science and Technology, 43*(16), 6414–6420.

Kastner, T., Erb, K.-H., & Haberl, H. (2014a). Rapid growth in agricultural trade: effects on global area efficiency and the role of management. *Environmental Research Letters, 9*, 034015.

Kastner, T., Schaffartzik, A., Eisenmenger, N., Erb, K.-H., Haberl, H., & Krausmann, F. (2014b). Cropland area embodied in international trade: contradictory results from different approaches. *Ecological Economics, 104*, 140–144.

Krausmann, F., Haberl, H., Schulz, N. B., Erb, K.-H., Darge, E., & Gaube, V. (2003). Land-use change and socio-economic metabolism in Austria—Part I: driving forces of land-use change: 1950–1995. *Land Use Policy, 20*, 1–20.

Krausmann, F., Erb, K.-H., Gingrich, S., Lauk, C., & Haberl, H. (2008). Global patterns of socio-economic biomass flows in the year 2000: a comprehensive assessment of supply, consumption and constraints. *Ecological Economics, 65*, 471–487.

Lambin, E. F., & Meyfroidt, P. (2011). Global land use change, economic globalization, and the looming land scarcity. *Proceedings of the National Academy of Sciences, 108*, 3465–3472.

Latour, B. (1991). *Nous n'avons pas jamais été modernes*. Editions la Découverte, Paris: Essai d'anthropologie symétrique.

Latour, B. (2000). When things strike back: a possible contribution of "science studies" to the social sciences. *British Journal of Sociology, 51*(1), 107–123.

Lenzen, M., Kanemoto, K., Moran, D., & Geschke, A. (2012). Mapping the structure of the world economy. *Environmental Science and Technology, 46*(15), 8374–8381.

Lieth, H. (1973). Primary produciton: terrestrial ecosystems. *Human Ecology, 1*, 303–332.

Luhmann, N. (2012). Theory of society, Vol. 1. Stanford University Press. (Original work published 1997).

Marsh, G. P. (1864). Man and nature; or, physical geography as modified by human action. London, New York: Scribners & Sampson Low.

Matthews, E., Amann, C., Fischer-Kowalski, M., Bringezu, S., Hüttler, W., Kleijn, R., et al. (2000). *The weight of nations: material outflows from industrial economies*. Washington, D.C.: World Resources Institute.

Maturana, H. R., & Varela, F. J. (1975). Autopoietic systems. A characterization of the living organization. Urbana-Champaign, Illinois: University of Illinois Press.

Maturana, H. R., Varela, F. J. (1987). The tree of knowledge: the biological roots of human understanding. New Science Library/Shambhala Publications.

McNeill, J. R. (2001). Something new under the sun: an environmental history of the twentieth-century world. W.W. Norton & Company.

Müller-Herold, U., & Sieferle, R. P. (1998). Surplus and survival: risk, ruin and luxury in the evolution of early forms of subsistence. *Advances in Human Ecology., 6*, 201–220.

Naylor, R., Steinfeld, H., Falcon, W., Galloway, J., Smil, V., Bradford, E., et al. (2005). Losing the links between livestock and land. *Science, 310*, 1621–1622.

Netting, R. M. C. (1981). Balancing on an Alp: ecological change and continuity in a Swiss mountain community. CUP Archive.

Netting, R. M. C. (1993). Smallholders, householders: farm families and the ecology of intensive, sustainable agriculture. Stanford Univ Press.

Peters, G. P., Hertwich, E. G. (2004). Production factors and pollution embodied in trade: theoretical development. Working Paper 5/2004, Industrial Ecology Programme, Norwegian University of Science and Technology.

Ringhofer, E., Singh, S. J., & Fischer-Kowalski, M. (2014). Beyond Boserup: the role of working time in agricultural development. In M. Fischer-Kowalski, A. Reenberg, A. Mayer, & A. Schaffartzik (Eds.), *Ester Boserup's legacy on sustainability: Orientations for contemporary research* (pp. 117–138). Dordrecht: Springer.

Schaffartzik, A., Eisenmenger, N., Krausmann, F., & Weisz, H. (2014). Consumption-based material flow accounting: Austrian trade and consumption in raw material equivalents (RME) 1995–2007. *Journal of Industrial Ecology, 18*(1), 102–112.

Schmid, M. (2006). Herrschaft und Kolonisierung von Natur: Ein umwelthistorischer Versuch zur Integration von Materiellem und Symbolischem. *Mitteilungen der Österreichischen Geographischen Gesellschaft (MÖGG), 148*, 57–74.

Sieferle, R. P. (1997). Rückblick auf die Natur: eine Geschichte des Menschen und seiner Umwelt. Luchterhand Munich.

Sieferle, R. P. (2008). Zusammenbruch von Zivilisationen. Eine konzeptuelle Analyse am Beispiel des Imperium Romanum. *GAIA, 17*, 213–223.

Sieferle, R. P. (2011). Cultural evolution and social metabolism. *Geografiska Annaler: Series B, Human Geography, 93*, 315–324.

Sieferle, R. P., Krausmann, F., Schandl, H., & Winiwarter, V. (2006). *Das Ende der Fläche*. Böhlau: Zum Sozialen Metabolismus der Industrialisierung. Köln.

Simas, M., Wood, R., & Hertwich, E. (2014a). Labor embodied in trade. *Journal of Industrial Ecology,*. doi:10.1111/jiec.12187.

Simas, M., Golsteijn, L., Huijbregts, M. A. J., Wood, R., & Hertwich, E. (2014b). The "Bad Labor" footprint: quantifying the social impacts of globalization. *Sustainability, 6*(11), 7514–7540.

Singh, S. J., Grünbühel, C. M., Schandl, H., & Schulz, N. (2001). Social metabolism and labour in a local context: changing environmental relations on Trinket Island. *Population and Environment, 23*, 71–104.

Smeets, E., & Weterings, R. (1999). Environmental indicators: Typology and overview. European Environment Agency Copenhagen.

Tainter, J. A. (1988). *The collapse of complex societies*. Cambridge: Cambridge University Press.

Tainter, J. A. (2011). Energy, complexity, and sustainability: a historical perspective. *Environmental Innovation and Societal Transitions, 1*, 89–95.

Tukker, A., Bulavskaya, T., Giljum, S., deKoning, A., Lutter, S., & Simas, M. et al. (2014). The global resource footprint of nations. Carbon, water, land and materials embodied in trade and final consumption calculated with EXIOBASE 2.1. Leiden, Delft, Vienna, Trondheim.

UNEP-IRP (2011). Decoupling natural resource use and environmental impacts from economic growth. A Report of the Working Group on Decoupling to the International Resource Panel. Fischer-Kowalski, M., Swilling, M., von Weizsäcker, E.U., Ren, Y., Moriguchi, Y., Crane, W., Krausmann, F., Eisenmenger, N., Giljum, S., Hennicke, P., Romero Lankao, P., Siriban Manalang, A., Sewerin, S.United Nations Environmment Programme, Nairobi.

Varela, F. G., Maturana, H. R., & Uribe, R. (1974). Autopoiesis: the organization of living systems, its characterization and a model. *Biosystems, 5*, 187–196.

Vitousek, P. M., Ehrlich, P. R., Ehrlich, A. H., & Matson, P. A. (1986). Human appropriation of the products of photosynthesis. *BioScience, 36*, 368–373.

Weichhart, P. (1999). Die Räume zwischen den Welten und die Welt der Räume. Zur Konzeption eines Schlüsselbegriffs der Geographie, in: Meusburger, P. (Ed.), Handlungszentrierte Sozialgeographie. Benno Werlens Entwurf in kritischer Diskussion. Stuttgart, Franz Steiner Verlag, pp. 67–94.

Weisz, H. (2009). Wenn alles bleiben soll, wie es ist, muss sich alles ändern. *GAIA, 18*, 104–106.

Whittaker, R. H., & Likens, G. E. (1973). Primary production: the biosphere and man. *Human Ecology, 1*, 357–369.

Whittaker, R. H., & Likens, G. E. (1975). The Biosphere and Man. In H. Lieth & R. H. Whittaker (Eds.), *Primary productivity of the biosphere, ecological studies* (pp. 305–328). Berlin, Heidelberg: Springer.

Wiedenhofer, D., Steinberger, J. K., Eisenmenger, N., & Haas, W. (2015). Maintenance and expansion: modeling material stocks and flows for residential buildings and transportation networks in the EU25. Journal of Industrial Ecology, published online January 1, 2015. doi:10.1111/jiec.12216.

Wiedmann, T. O., Schandl, H., Lenzen, M., Moran, D., Suh, S., West, et al. (2013). The material footprint of nations. Proceedings of the National Academy of Sciences. Early edition, doi:10.1073/pnas.1220362110.

Willke, H. (1990). Beobachtung, reflexion, supervision. *Soziale Systeme, 4*(2), 105–118.

Winiwarter, V. (2000). Die Grenzen der Kolonisierung von Natur. Anmeldung von Zweifeln an den Möglichkeiten einer geokybernetischen Steuerung. In: Kreibich, R., Simonis, U.E. (Eds.) Global Change - Globaler Wandel. Ursachenkomplexe und Lösungsansätze - Causal structures and indicative solutions. Berlin: Berlin Verlag, pp. 159–169.

Chapter 3
Transitions in Sociometabolic Regimes Throughout Human History

Fridolin Krausmann, Helga Weisz and Nina Eisenmenger

Abstract The metabolism of human society is dynamic; it has undergone major changes during the course of human history, and it is currently in a transition process. Since the time of Paleolithic hunter-gatherers, the amount of materials extracted and used by humans and the associated impact on the environment have grown by several orders of magnitude, and the quality and quantity of the related sustainability problems have changed. This chapter recalls major stages in human history through the lens of the societal use of materials and energy. It introduces the notion of sociometabolic regimes and explores the characteristics of resource use in hunter-gatherer societies, agrarian societies and industrial societies. It discusses the emergence of these metabolic regimes, their variability and their environmental and sustainability problems.

Keywords Material use · Energy use · Metabolic profile · Hunter-gatherers · Agrarian society · Industrial society · Industrialization · Energy transition · Energy system · Economic growth · Decoupling

F. Krausmann (✉) · N. Eisenmenger
Institute of Social Ecology (SEC), Alpen-Adria University,
Schottenfeldgasse 29, 1070 Vienna, Austria
e-mail: fridolin.krausmann@aau.at

N. Eisenmenger
e-mail: nina.eisenmenger@aau.at

H. Weisz
Potsdam Institute for Climate Impact Research,
Telegrafenberg A56, P.O. Box 601203, 14412 Potsdam, Germany
e-mail: helga.weisz@pik-potsdam.de

H. Haberl et al. (eds.), *Social Ecology*, Human-Environment Interactions 5,
DOI 10.1007/978-3-319-33326-7_3

3.1 Sociometabolic Regimes and Transitions

The metabolism of human society is dynamic; it has undergone major changes during the course of human history, and it is currently in a transition process. Since the time of Paleolithic hunter-gatherers, the amount of materials extracted and used by humans has grown by several orders of magnitude. Quite obviously, a major driver of this growth is population growth, but resource use has also grown because per capita consumption has multiplied. With the sheer size of the global metabolism, both the composition of the materials used and the way in which they are used have been transformed. Throughout most of human history, the endosomatic metabolism of human organisms determined their material needs, and renewable biomass has been by far the most important material. In contrast, the metabolism of industrial societies is dominated by exosomatic metabolism and the use of mineral resources. With these changes, the environmental impacts of our metabolism have also changed, and many of the sustainability problems human society is facing today are a direct consequence of its metabolism. Maintaining the functioning of society's metabolism without destroying the resource base and without deteriorating the natural environment by exceeding its capacity to absorb the outflows of our metabolism is a basic requirement for sustainability. This struggle is by no means a modern phenomenon or limited to the industrial age; it has accompanied humans throughout their history. It is now one of the major challenges humans are facing at the beginning of the 21st century.

This chapter recalls major stages in human history through the lens of the societal use of materials and energy. There is one important lesson to be learned here. Unlike the endosomatic metabolism of the human organism, the extrasomatic metabolism of human societies varies by orders of magnitude. When analyzing the constituents and sources of these variations, we recognize that historical variations in the social metabolism did not occur arbitrarily. Instead, certain sociometabolic types (modes of production or subsistence of human societies) can be distinguished for historical and for contemporary societies that share, at whatever point in time and irrespective of biogeographical conditions, certain fundamental systemic characteristics derived from the way they utilize and thereby modify nature. This justifies speaking of sociometabolic regimes that are above all distinguished by their type of energy system and the energy density for which it allows. The energy system sets the boundary conditions for the material system because the main purpose of socioeconomic energy use is to reproduce and move material structures. Therefore, the use of energy and materials is closely intertwined. In general, sociometabolic regimes share a main source of energy and the main technologies of energy conversion. Accordingly, they share many other basic characteristics, such as patterns and levels of resource use (*metabolic profile*), demographic and settlement patterns, patterns of use of human time and labor, institutional characteristics and communication patterns (Sieferle 2001). According to Fischer-Kowalski et al. (2011a), a regime can be characterized by the sociometabolic profile of a given society and by the associated modifications in natural systems that occur either as an unintended consequence (pollution, soil erosion) or as an intentional change induced by the society (e.g., land-use change). When a sociometabolic system transcends its boundary conditions and key biophysical requirements and a new metabolic regime emerges, we speak of a *metabolic transition* (Fischer-Kowalski and Haberl 2007).

From a global history perspective, it is possible to identify major sociometabolic regimes in which significant transitions are generally referred to as 'revolutions'. The most important are the Neolithic revolution, which marked the transition from the hunter-gatherer regime to the agrarian regime, and the industrial revolution, which marked the transition from the agrarian to the industrial regime (cf. De Vries and Goudsblom 2002; Sieferle 1997; Simmons 2008).[1] The regime of the current industrial society is, however, problematic as neither its reliance on exhaustible resources nor its huge outflows, which exceed the sink capacity of the earth system, allow for long-term existence (Sieferle 1997), an issue that will be discussed later.

Although fundamental differences exist among these three major metabolic regimes, there is also significant variation within them. We use the term *metabolic profiles*, which comprise a set of key metabolic indicators, to characterize the three sociometabolic regimes, their subtypes and the differences among them.

3.2 Foraging Societies and the Regime of Hunter-Gatherers

3.2.1 The Uncontrolled Solar Energy System of Foraging Societies

Hunter-gatherers live off the land. They extract roots and fruits and hunt various types of animals, much like other large mammals. What distinguishes the hunter-gatherer metabolism from that of other animals is the use of fire and the corresponding use of considerable amounts of fuel wood (Goudsblom 1992). It is assumed that humans began to use fire for cooking, for the provision of heat and for hunting at least 800,000 years ago (Goren-Inbar et al. 2004). Hunter-gatherers rely solely on recent biomass as an energy source. Unlike agriculturalists, however, they do not systematically manage ecosystems to increase the availability of usable biomass. This basic metabolic system of hunter-gatherers (or foraging societies) has thus been termed the *uncontrolled solar energy system* (Sieferle 2001).

[1]Some authors, such as Boyden (1992), further differentiate between simple agrarian societies (early farming) and advanced agrarian societies (early urban phase) and thus distinguish four regimes. However, with respect to the basic characteristics of their energy system and the energy density for which it allows, there is no fundamental difference between these different agrarian societies.

The lifestyle of hunter-gatherers is often mobile. Variability in local resource availability, which may be induced by temporal variations in biogeographic conditions (climate) and by the local exploitation and depletion of edible biomass, causes humans to move and prevents them from establishing permanent settlements. However, the difference between mobile and sedentary lifestyles is continuous, and the frequency of movement in the past may have differed greatly among societies. Under conditions of high resource density, such as in estuaries or on lake shores, even sedentary lifestyles may have been common (Simmons 2008).

3.2.2 The Metabolic Profile of Hunter-Gatherers

Although little is known about the specific metabolic characteristics of foraging societies, plausible ranges of metabolic profiles can be estimated based on available technologies and the key characteristics of the energy supply system. The literature suggests that the metabolism of hunter-gatherers is two to four times greater than the endosomatic metabolism of human beings (Boyden 1992; Fischer-Kowalski and Haberl 1997; Sieferle 2001; Simmons 2008). To meet their needs in terms of food and fuel, hunter-gatherers extract approximately 0.5 and 1 t/cap/year (metric tons of biomass per capita per year; see Table 3.1). This biomass corresponds to an annual metabolic rate of 5–15 GJ/cap/year (gigajoules per capita per year; 1 GJ = 10^9 J). The amount of fire wood used can vary widely, from approximately 100 kg/cap/year to many times this value (Smil 2013). Food often constitutes a smaller share (approximately 200 kg/cap/year or 3 GJ/cap/year). Biomass used for clothing and shelter contributes a minor part of material use. Biomass is the dominant resource and accounts for far more than 99 % of all material input (Table 3.1). Almost all the extracted materials are used as sources for primary energy in the form of wood and food, the non-energy fraction of material use remains very low, and the energy metabolism and the material metabolism are practically identical. Textbooks classify Paleo- and Neolithic societies into stone, iron or bronze age societies. This classification is somewhat misleading as the mineral component of material use was negligibly small during these periods of human history. The nonrenewable materials used by hunter-gatherers include little more than flint used to light fire and stones used to manufacture artifacts such as arrowheads or blades. The per capita use of metals such as iron or copper did not exceed a long-term average of several grams per capita per year.[2] Taken together, all mineral materials constitute far less than 1 % of the total material turnover in hunter-gatherer societies.

[2]Based on historical data on global cumulative copper production (Hong et al. 1996) and estimates of prehistoric human population sizes (Biraben Jean-Noel 2003), copper production was likely approximately 4–5 g/cap/year (grams per capita per year) in prehistoric times.

Table 3.1 Metabolic profiles of sociometabolic regimes

		Hunter-gatherers	Agrarian[a]	Industrial[b]	Factor industrial to agrarian
Energy use (DEC) per capita	(GJ/cap/year)	10–20	40–70	150–400	3–5
Useful work per capita	(GJ/cap/year)	<1	<5	30–50	5–15
Material use (DMC) per capita	(t/cap/year)	0.5–1	3–6	15–25	3–5
Population density	(cap/km^2)	<0.1	<40	<400	3–10
Agricultural population	(%)	–	>80	<10	0.1
Energy use density (DEC per area)	(GJ/ha/year)	<0.01	<30	<600	10–30
Material use density (DMC per area)	(t/ha/year)	<0.001	<2	<50	10–30
Biomass (share of DEC)	(%)	>99	>95	10–30	0.1–0.3
Share of non-energy use of materials	(%)	<5	<20	>50	3–10
Material stocks	(t/cap)	<0.01	<10	100–1,000	10–100

(Sources: The data compiled in this table are derived from empirical studies on material and energy flows in agrarian and industrial societies, e.g., Krausmann and Haberl 2002; Schandl and Schulz 2002; Sieferle et al. 2006; Weisz et al. 2006, and literature data, e.g., Malanima 2002; Simmons 2008)

DMC domestic material consumption; *DEC* domestic energy consumption

[a]Typical values for an advanced European agrarian sociometabolic regime (18th century). In agrarian societies based on labor-intensive horticultural production with low significance of livestock, population density may be significantly higher, whereas per capita use of materials and energy would be lower (see Sect. 3.3.2)

[b]Typical values for fully industrialized economies. In countries with high population densities per capita, the DMC and DEC values tend to be in the lower range, whereas per area values are high. The reverse is true for countries with low population densities, in which case per area values can be very low

The mobile lifestyle and the limited availability of energy prohibited the accumulation of artifacts. Almost all materials were consumed shortly after extraction, and physical stocks maintained by humans did not amount to more than a few kg/cap. Although the direct use of materials and energy per capita appears to be very small, unused extraction[3] due to fires used to support hunting may have been considerable. Sieferle et al. (2006) estimate that in some cases, human-induced fires may have destroyed several hundred tons of biomass for a few kilograms of hunted game per capita.

Like all human societies, hunter-gatherers must match their energy demand density with the available energy supply density. Natural vegetation is the supply system for both endosomatic energy (food) and exosomatic energy (fire wood). The energy supply density of biomass, defined as the continuous supply of power over one year, ranges between 0.1 and 1 W/m^2 (watts per square meter), or between 3.2 and 31.6 MJ/m^2/year (megajoules per square meter per year; Smil 1991).[4] This is the potentially available energy. However, only a very small share of the naturally available biomass in an ecosystem is edible for humans.

Simmons (2008) assumes an average territorial requirement of at least 25 km^2 to feed one person, but depending on biogeographic conditions, the area demand may vary widely. Average population densities of foraging societies have been estimated to range between 0.02 and 0.2 cap/km^2 (Simmons 1989). Low per capita demand for material and energy paired with low population density results in very low extraction rates per unit of area. A per capita material demand of 1 t (equal to 15 GJ) combined with an average population density of 0.2 cap/km^2 results in an extraction rate of 2 kg or 30 MJ of biomass per ha of land per year. As Boyden (1992) has noted, this means that foragers use only a very small share of the available biomass of an ecosystem. On average, less than 0.01 % of the annual net primary production (NPP) is extracted by hunter-gatherers. This share, however, could increase considerably if biomass burned by human-induced fires for hunting (i.e., unused extraction in terms of material flow accounting) is included in the calculation.

If low and high estimates of the metabolic rate (7–15 GJ/cap/year) and low and high estimates of population densities (0.02–0.2 persons per km^2) are combined with the lower and upper limits of energy supply densities for biomass, foraging societies used between 0.005 and 1 % of the annually available biomass energy potential.

In the absence of technologies that would allow for increased density of usable (edible) biomass, hunter-gatherer societies were forced to adapt their energy demand density to the available energy supply density. This imposed effective

---

[3]Unused extraction is defined as the amount of materials extracted, moved or transformed without the intention of using them. Large-scale burning of natural vegetation for hunting causes a transformation of biomass into CO_2, water and ashes, which are then released into the atmosphere. Quantitatively, it is the most important component of unused extraction in hunter-gatherer regimes.

[4]The energy supply density of biomass as defined here is equivalent to the net primary production (NPP).

biophysical limits on the structure and complexity of these societies in terms of population size and density,[5] the accumulation of artifacts and the establishment of permanent settlements.

It has also been argued that despite the low availability of edible biomass, the uncontrolled solar energy system allows for high energy yields of human labor (Boyden 1992). Hunter-gatherers typically invest only a few hours per day to secure their daily requirements of biomass, and the energy invested by humans is small compared with the energy output in terms of food and fuel. This leaves them considerable time for activities other than securing food and wood, a condition Sahlins (1972) called *affluence without abundance* (see also Müller-Herold and Sieferle 1998).

3.2.3 Environmental Impacts and Sustainability

The absence of purposeful management of ecosystems and low rates of resource extraction does not mean that the impact of foragers on their natural environment is negligible. Large-scale fire management to support hunting, as has been practiced, for example, by the Australian aborigines, has led to the establishment of new vegetation types and alterations at the landscape level. There is even an ongoing debate as to what extent hunters have contributed to the extinction of large mammals (e.g., Martin and Klein 1989). Like other mammals, hunter-gatherers tap into a flow of renewable biomass. Thus, at a very general level, the mode of subsistence of hunter-gatherers can be considered sustainable. Sustainability was, however, not guaranteed and could be threatened by external factors as well as socioeconomic processes. Fluctuations in natural conditions that have the potential to lower the capacity of ecosystems to provide sufficient food for human populations on a supraregional scale imposed an external threat to sustainability. Population growth may also have led to unsustainable situations. In contrast to agriculturalists, hunter-gatherers were unable to deliberately improve the carrying capacity of the ecosystems they inhabited, and growth was constrained by the size of the territory and the availability of edible biomass per unit of land. Continued population growth inevitably resulted in the overexploitation of natural systems. In this context, it has been argued that foragers established effective social mechanisms to keep the population density low and that the population remained well under the potential capacity to buffer the effect of the negative impact of variable natural conditions (Boyden 1992; Simmons 2008). Nevertheless, global population slowly increased. It is estimated that at the beginning of the Neolithic period, some 10,000 years ago, the world population was roughly 4–5 million people (Cohen 1995).

---

[5]Both population size and density are decisive preconditions for technological change (Boserup 1965).

3.3 Agrarian Societies and the Agrarian Sociometabolic Regime

3.3.1 *The Controlled Solar Energy System of the Agrarian Regime*

Like hunter-gatherers, agrarian societies are fuelled by a solar-based energy system and ultimately rely on the solar energy stored by living plants through the process of photosynthesis in biomass. In contrast to hunter-gatherers, agriculturalists actively manage terrestrial (and, in some cases, aquatic) ecosystems. The transition from a hunter-gatherer to an agrarian sociometabolic regime is commonly referred to as the Neolithic revolution. Its distinguishing characteristic is the large-scale adaptation of agriculture 6000–10,000 years before present, constituting a major change in how humans used their natural environment. Although it is undisputed that agriculture marks a major transition in the history of humans, scholars are still undecided about the reasons for the Neolithic revolution (Weisdorf 2005). There is well-established evidence that the Neolithic revolution occurred in parallel in several world regions, of which the Fertile Crescent is probably the best known. Most theories agree that agriculture was 'neither discovered or invented' (Harlan 1992). That is, the success of agriculture was not the result of spontaneous and superior innovation and its subsequent diffusion across the globe; rather, it was the result of a long evolutionary process. The factors that drove people to take up agriculture in several regions around the globe in a comparatively narrow period of time and whether these adaptations were driven by resource shortage or excess remain issues of scientific debate (cf. Cohen 1977; Fagan 2001; Sieferle 2010). Nevertheless, it is very likely that environmental change was an important initiating factor. A prominent line of argument assumes that climatic change at the end of the Pleistocene resulted in a deterioration of living conditions and higher fluctuations, which favored the adaptation of agricultural techniques and, in turn, increased stability and ultimately allowed for population growth (Mithen 2003).

In metabolic terms, the agricultural techniques of breeding plants, farming and domesticating animals (colonization of natural systems, Fischer-Kowalski and Haberl 1998) supports a much higher level of useful energy per unit of territory compared with a hunter-gatherer regime. It is this major characteristic that distinguishes the two regimes and entails fundamental differences in their metabolic profiles and their interactions with their environment. Although the major source of energy is still biomass and therefore the energy supply system remains a land-based, low energy density system, the proportion of usable energy per unit of land increases by two or three orders of magnitude (depending on natural conditions and agricultural techniques, see Table 3.1). Agricultural techniques enable human societies to increase the output of edible or otherwise useful biomass (above all, animal power to perform work and fibers for clothes) and energy resources per area and to systematically modify the carrying capacity of the ecosystems they inhabit. The energy system of agrarian societies has therefore been termed the *controlled solar energy system* (Sieferle 2001). Biomass is still the single most important source of energy for socioeconomic metabolism and amounts to more than 95 % of the primary energy supply (Malanima 2002; Smil 2008). The only other energy sources are water and wind power.[6] Wind and water mills provide for specific types of work, and the kinetic energy of waterways (floats, riverboats) and wind (sailing ships) are essential for long-distance transport. Although wind and water power were of considerable socioeconomic importance on a regional scale, for instance, in 17th-century Netherlands or 18th-century England, they are quantitatively of little significance (Warde 2007). They usually account for no more than a few percent of the energy supply. The lack of technological options to convert one form of energy into another imposed a constraint for (historical) agrarian societies. Until the invention of the steam engine, heat could not be converted into work. The provision of certain types of energy was closely related to certain types of land use. Typically, the provision of heat relied on fire wood and woodlands, the provision of food, the energy basis of human work, cropland and the supply of animal draft power on grassland. Consequently, a certain energy mix required a corresponding land-use mix.

The energy system of agrarian societies is based on land use. Although agriculture allows for a much larger extraction of usable biomass per unit of area and thus supports much higher population densities compared with the uncontrolled solar energy system of hunter-gatherers, the energy supply is still limited by the availability of land and the area productivity (yield of useful biomass) that can be achieved in the managed ecosystems. Clearly, there is a considerable range of variation in how much biomass can be extracted per unit of land depending on the specific biogeographical conditions, the type of land-use system, the available technology and the role of human and animal labor (see Sect. 3.3.2), but the fundamental limitations of a solar-based energy system remain in place. A sociometabolic regime depending on the harvest of solar energy converted by plants sustains itself within narrow limits of available energy and consequently faces limits to biophysical and economic growth (Wrigley 1988, 2010). Therefore, the land-use system and its limited potential to supply certain types and amounts of primary energy constitute a major bottleneck for biophysical growth.

Figure 3.1 illustrates the basic relations among land, labor and energy that constrain development and growth within the agrarian regime based on a controlled solar energy system. At the core of this energy system is an agricultural population that invests labor to cultivate terrestrial ecosystems and to produce food, feed, fiber and fuel. Within the agrarian metabolic regime, the harvest of biomass

[6]In some agrarian societies, considerable amounts of fossil fuels have been used. This was the case, for instance, in the Netherlands, where peat was an important fuel in the 17th and 18th centuries (De Zeeuw 1978), or in England, where coal was used in considerable quantities already in the 16th century (Warde 2007; Wrigley 2010). These cases were, however, quite exceptional, and even then, the overall share of fossil materials in domestic material consumption (DMC) or domestic energy consumption (DEC) remained low.

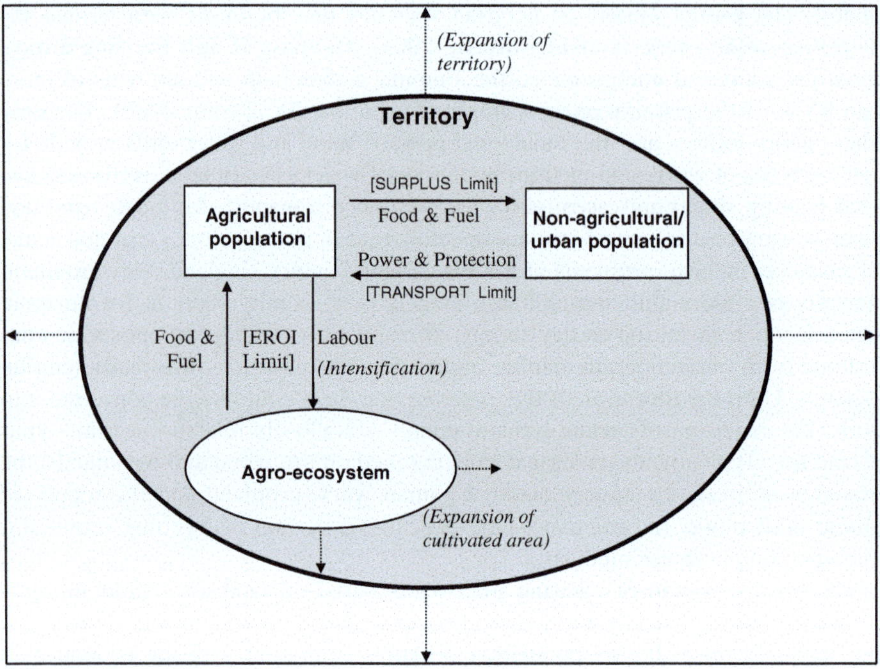

Fig. 3.1 The biophysical constraints of the agrarian regime. In a given territory, growth of the agricultural population can be based either on the *expansion of the cultivated area* or on *intensification*. Increasing biomass output per unit of land (intensification) is limited by diminishing marginal returns as the energetic return of agricultural production (EROI) declines with increasing investments of human labor (EROI limit). A growing urban population can be sustained by increasing the surplus rate, that is, by increasing the labor efficiency of the land-use system in the hinterland (SURPLUS limit) or by access to a larger rural hinterland (territorial expansion). Expanding the hinterland, however, increases transport distances and the (energetic) cost of transport (TRANSPORT limit) (see Sect. 3.3.1)

products from a given amount of land can be augmented within the limits imposed by natural constraints through intensification, that is, through increasing labor input (Boserup 1965). Under low-input conditions, agriculture[7] biomass output ultimately grows at a slower pace than labor investment, and marginal returns

[7]Low-input agriculture denotes agricultural production systems that operate in the absence of external energy subsidies or other off-farm resources. In a farming system relying solely on on-farm resources, the maintenance of soil fertility has to be based on a system of either area- or labor-intensive measures to optimize the utilization of locally available resources. These measures include the temporal and spatial rotation of different land-use types (e.g., shifting cultivation, three-field crop rotation with fallow), transfers of biomass and plant nutrients from extensively used woodlands or pastures to intensively used plots (e.g., by litter extraction or grazing in forests), recycling of residues and wastes, the best possible exhaustion of natural regeneration and renewability rates (biological fixation and deposition of nitrogen, soil processes), the establishment and maintenance of irrigation systems and measures to prevent or revert soil erosion (Kjaergaard 1994; Mazoyer et al. 2006; Netting 1993).

gradually diminish (Tainter 1988). At least on a larger scale, the energy output of land-use systems (in the form of the desired types of biomass for final use) has to exceed the amount of biomass-based energy that has been invested in the cultivation of the land. The working population has to produce enough food and raw materials to sustain their own living and that of the non-working fraction of the population (small children, old or sick people); otherwise, a stable population cannot be maintained. The ratio of energy output in the form of biomass for final use and human energy investment in biomass production is termed energy return on investment (EROI) and must be much larger than one.[8] Its minimum is given by basic demographic requirements, but an agricultural population may even produce a surplus that allows it to sustain a significant non-agricultural population and production (EROI limit). The maximum fraction of the non-agricultural population is ultimately limited by the physically attainable surplus rate, but in practical terms, it depends on social arrangements that determine how much the agricultural population is willing or how much it can be forced to give away (SURPLUS limit).

High transport energy costs are another important constraint for the metabolism of agrarian societies, and they impose severe limits for spatial differentiation and specialization on larger spatial scales (TRANSPORT limit). Under the conditions of the agrarian regime, only water transport allows for long-distance transport of bulk materials. The energy costs of overland transport increase to a prohibitive level after only a few kilometers (Bairoch 1993; Boserup 1981). This implies limitations for the spatial concentration and the exchange of staple food, feed and fuel. These are produced at low energy densities (with respect to both energy harvested per unit of land area and energy content per mass unit) and can only be transported over short distances if water transport is not an option. In addition to the surplus limit, high transport costs further constrain the sufficient supply of large urban populations and limit urbanization processes.

Under the conditions of the agrarian regime, the majority of the population lives on the land and off the land. The largest fraction of the population is engaged in agricultural production, with the share of non-agricultural population and urbanization typically lower than 20 %. The mobility of people is also limited, which hampers cultural exchange and thus the innovation and diffusion of technologies, but it supports local cultural diversity (Sieferle 1997).

3.3.2 Development Paths and Subtypes of the Agrarian Regime

Under the conditions of the agrarian regime, limitations to biophysical growth, whether caused by growth in the size of the population or by growing resource demand per capita, are set by biogeographical conditions, by elementary

[8]Cleveland assumes a minimum preindustrial agriculture EROI of five, but land-use systems with a higher EROI have also been observed (cf. Krausmann 2004; Leach 1976).

characteristics of biological processes and by ecosystem properties. For example, climate, photosynthesis, nutrient cycles and the decomposition of organic matter can constrain the theoretically attainable yield of useable biomass per unit of area. By deliberately managing the properties of ecosystems through cultivation, humans can influence and significantly increase (or decrease) the carrying capacity of the ecosystems they inhabit. This reduces their dependence on naturally given productivity and its fluctuations. Depending on climatic conditions, terrain, the history of population growth and the available technology, a variety of subtypes of the agrarian metabolic regime have evolved. These subtypes exhibit different development paths and considerable differences in the structure and level of biomass use and population density. Both the achievable output of useful biomass per unit of area (energy supply density) and the apparent consumption of biomass per capita (metabolic rate) are variable within the given ecological limitations. Together, the energy supply density and the metabolic rate determine the maximum population density that can be sustained in a given subtype of the agrarian regime. The examples of different subtypes of the agrarian regime shown in Table 3.2 demonstrate the variability of the metabolic characteristics.

Table 3.2 Examples of the variability of metabolic profiles within the agrarian metabolic regime

		Shifting 50y	Shifting 10y	Temperate mixed farming	Labor-intensive tropical cropland	Pastoralism
Cereal yield (real)	(t/ha sown area)	2	0.8	2	4	–
Energy extraction rate per unit area	(GJ/ha/year)	<2	<10	30	50	2.5
Metabolic rate (energy use/cap)	(GJ/cap/year)	10–20	10–20	45–75	10–20	250
Metabolic rate (material use/cap)	(t/cap/year)	0.7–1.5	0.7–1.5	3–6	1–2	18
Livestock/human	LAU/human	≪1	≪1	Roughly 1	0.1	≫1 (3.5)
Labor intensity		Low	Moderate	High	Very high	Very low
Population density	(cap/km^2)	<10	<30		Several 100	<2

Based on data from Coughenour et al. 1985 (pastoralism); Mazoyer et al. 2006; Netting 1993 (shifting cultivation); Sieferle et al. 2006 (temperate mixed farming). See also Smil (2013)
LAU Large animal unit
Note All biomass flows are given in energy units (gross calorific values) but can also be expressed in mass units: 1 GJ roughly equals 70 kg of biomass at a 15 % moisture content

3.3.2.1 Shifting Cultivation

In simple types of shifting cultivation or slash-and-burn agriculture, an exhaustion of soil fertility is counteracted by long fallow periods during which essential plant nutrients accumulate in the soil. After several decades of natural succession, the biomass regrowth is removed and burnt, and for one or two seasons, comparatively high crop yields per unit of sown area can be achieved. The overall land requirement for the whole cycle of shifting cultivation is much larger than the area sown in a particular year, so food output divided by the total land area in rotation remains small, and only low population densities can be supported. Labor input in long fallow systems is usually low, and the food output per unit of invested labor can be considerable (cf. Netting 1993). A simple hypothetical example derived from Mazoyer et al. (2006) illustrates the metabolic characteristics of a production system based on slash-and-burn agriculture. Energy use comprises fuel wood and food for humans, and it ranges from 10 to 20 GJ/cap/year. The used extraction of biomass per unit of area is very low (<1 GJ/ha/year), but several hundred GJ of biomass is burnt per capita per year for land clearing; that is, biomass destruction (i.e., unused extraction in terms of material flow accounting, MFA) is large compared with used extraction. This may be considered a rather wasteful use of the productive capacity of land. By investing more labor in land clearing, the fallow period can be reduced, for example, from 50 to 10 years. This reduces the regeneration period of the soils and leads to a reduction in yields per unit of sown area. However, because more area can be cropped per year, the total territorial yield can be multiplied. Consequently, population density and biomass extraction per unit of area can be increased while the amount of biomass burnt per capita decreases. The rate of used to unused extraction improves and area productivity grows, but at the expense of labor productivity. According to Smil (2013), such long fallow shifting cultivation supports population densities of between 20 and 30 cap/km^2, although regional variations can be very large.

3.3.2.2 Temperate Mixed Farming

Further reducing the rotation period to less than a year leads to land-use systems with permanent cultivation (two or three field rotations with fallow periods, crop rotation systems) and allows for further increases in population density at the cost of labor efficiency. Under temperate conditions, mixed farming systems that combine crop production with the multifunctional use of livestock have evolved. Livestock is used in plant nutrient management and to provide farm labor as well as milk, meat and other products. In addition to the land area used to grow crops, a significant fraction of the territory is used less intensively (with less labor input), such as woodlands or pastures, and serves as a nutrient reservoir for intensive crop fields. Long-term socioecological studies of mixed farming systems in Austria at the advent of the industrial revolution suggest that up to 30 GJ of primary biomass (food, feed, wood) was extracted at the large-scale average and in the long run

(Sieferle et al. 2006). As a result of the high demand for forage due to the high livestock-to-human ratio and the high fuel wood demand due to a cool climate, the average energy use per capita was high, typically ranging between 40 and 70 GJ/cap/year. In combination with the attainable area productivity, a population density of 45–75 cap/km^2 on large-scale averages was supported (Table 3.2).

3.3.2.3 Tropical Labor-Intensive Farming

Under tropical climate conditions (e.g., South and East Asia), a development path toward very labor-intensive cropping systems with irrigation and multiple harvests is common. Extensive land-use types are reduced, and a large fraction of the land is cultivated and used intensively. Most of the labor is supplied by humans. Animals are of less significance, and biomass extraction per unit of area can be similar to or greater than that in mixed farming systems. Less biomass is used per capita, and the average energy use can be as low as 10–20 GJ/cap/year. This means the carrying capacity can be much higher than in mixed farming systems. For the labor-intensive rice cultivation of smallholders in South and East Asia, local population densities of several hundred people have been recorded (Netting 1993). It is assumed that on larger scales, average population densities of 150–200 persons per km^2 have been supported by this subtype (cf. Boserup 1965; De Vries and Goudsblom 2002).

3.3.2.4 Pastoralism

Pastoralism is a subtype of the agrarian regime at the other extreme of the range. It prevails under arid or cold climate conditions, which are adverse for cultivation. Pastoralists make use of livestock to concentrate sparse biomass energy from large territories and to increase the output of useful biomass. Only very low population density can be sustained. The average population density of the Ngisonkyoka, a pastoralist community in Kenya, is only 1.3 cap/km^2 compared with a livestock (mostly cattle and sheep) of 4.5 large animal units (LAUs) per km^2. The high number of livestock compared with the population (3.5 LAU/cap) contributes to an extraordinarily high rate of biomass extraction per capita. In the case of the Ngisonkyoka, the energy use amounts to 260 GJ/cap/year, more than 84 % of which is biomass grazed by livestock. Wood fuel accounts for most of the reminder. Despite the high per capita rate, the biomass extraction per unit of area remains very low at 2.5 GJ/ha, or 7 % of the annual net primary productivity (Coughenour et al. 1985; Table 3.2).

These examples illustrate that from a long-term socioecological perspective on agrarian development, growth under the conditions of the agrarian regime is intrinsically related to population growth and leads to higher population densities. Population growth, in turn, usually drives a shift toward more labor-intensive land-use systems (e.g., from long fallow toward permanent crop rotation systems) (see also Fischer-Kowalski et al. 2011b). However, the land and biomass efficiency of land use also increases. Intensification means that more useful biomass per unit of land can be produced and that the ratio of used to unused biomass extraction is improved (Krausmann et al. 2012). Nevertheless, despite the considerable increases in the overall output of useful biomass, both the labor efficiency and the per capita throughput of biomass decline. Consequently, in the long run, continuous agrarian growth tends to result in declining per capita wealth: food and food quality (e.g., the share of meat in diets) decline with population growth, and per capita material and energy use begin to decline (Fischer-Kowalski et al. 2014).

3.3.3 Material Use in the Agrarian Regime

The inherent limitations of the biomass-based energy system, namely, the low energy density of the supply system, a lack of energy conversion technologies, a reliance on power delivered by humans and animals and the high energy costs of transport, shape the patterns of material use. Materials have to be extracted, moved and processed, all of which depend on the availability of power and useful work. Both are scarce in agrarian societies where humans and animals supply most of the useful work, and this constrains the movement of large amounts of bulk materials. Biomass continues to be the most important material and energy resource. Similar to hunter-gatherers, there is a large overlap between the material and the energy system. The amount of biomass used per capita largely depends on the significance of livestock,[9] on climatic conditions and on population density. Locally, it can also be influenced by the energy requirements of non-agricultural production, above all by energy-intensive mining and manufacturing. The amount of biomass used can

[9]One livestock unit requires an annual biomass intake of 3–5 t/year, compared with less than 0.5 t of food for human beings. The per capita level of biomass use in agrarian societies is therefore largely determined by the significance of livestock, that is, by the ratio of livestock to humans: the more livestock there is per capita, the higher the level of biomass use per capita and, consequently, overall energy use. Keeping livestock for the provision of draught power, nutrient management or milk production requires large amounts of feed and litter. In European land-use systems with a livestock-to-human ratio greater than one, far more than 80 % of all agricultural biomass is used in the livestock subsystem. The highest level of biomass use per capita, therefore, occurs where area under intensive herding is linked to a very low population density and very low rates of biomass extraction per unit of area. Figures from Coughenour et al. (1985) suggest that pastoralists in Kenya extract some 15 t of biomass per capita, with grazed biomass accounting for more than 95 % of their total material extraction. In contrast, biomass per capita is lowest in labor-intensive horticultural production systems with vegetable-based diets. Such farming systems typically entail high population density and high biomass harvest per unit of area (Hayami and Ruttan 1985; Krausmann et al. 2008a).

range between one and several metric tons per capita.[10] Other than biomass, small quantities of metal ores, mostly iron, gold, copper and tin, as well as salt, manufacturing (clay, quartz sand) and fertilizer minerals (marls) have been mined and used. The contribution of these nonrenewable raw materials to material consumption did not exceed a few percent (Table 3.2), although local variations may have been significant. Although the per capita apparent consumption of ores and other minerals was considerable in mining regions, averages at larger spatial scales were low and in the range of 0.01–0.1 t/cap.[11] The extraction and use of construction minerals such as clay for the production of bricks, dimension stone and sand and gravel was highly variable and typically amounted to less than 100 kg in simple agrarian communities and to several hundred kg in more advanced societies with a higher degree of urbanization and infrastructure networks (e.g., parts of the Roman Empire). However, even with considerable stocks of built structures (see below), the annual material flows for construction and maintenance were low because of the long lifespan of buildings and infrastructure. We estimate that at the eve of industrialization, less than 100 kg of gross ores and half a metric ton of sand, gravel and clay had been extracted in European countries. Reconstructions of the historical metabolism of advanced agrarian regimes such as those of Austria and the United Kingdom (UK) on the verge of industrialization (Schandl and Schulz 2002; Sieferle et al. 2006) indicate that the yearly consumption of all materials did not exceed 5 or 6 t/cap, of which biomass had a share of 80–90 % (Table 3.2).

In contrast to the mobility of hunter-gatherers, agriculturalists are characterized by a sedentary lifestyle. The emergence of permanent settlements was associated with the accumulation of physical stocks. In ancient civilizations, humans kept livestock and erected shelters and farm houses, urban centers and infrastructure networks emerged, and tools and other durable artifacts accumulated. Buildings and livestock accounted for the largest share of material stocks, whereas tools and other artifacts were of minor quantitative importance. It is difficult to estimate the size of physical stocks in agrarian societies. Stocks have varied between a few hundred kg/cap in simple agrarian societies with built infrastructures comprising little more than clay huts or wooden buildings to several metric tons in advanced agrarian civilizations with solid farm buildings, urban centers and a network of maintained roads. We assume that in general, physical stocks in agrarian societies rarely exceeded 10 t/cap on a large-scale average and were thus larger than in hunter-gatherer societies by several orders of magnitude (Table 3.1).

[10]The amount of biomass used per capita in agrarian societies is generally several times larger than in foraging societies, particularly when humans keep livestock. However, there is not necessarily a large difference in per capita consumption. As outlined above, the major difference between foraging and agrarian societies lies in the level of biomass extraction per unit of area, which is always several orders of magnitude above the level of hunter-gatherers.

[11]The use of copper in agrarian societies can illustrate the low level of mineral use in agrarian societies. The copper use in agrarian China or the Roman Empire was within 30–60 g/cap/year (own estimates based on Biraben 2003; Cohen 1995; Hong et al. 1996). Assuming an ore grade of 3.5 % (this equals average ore grades in US copper mines in 1880, see Ayres et al. 2004), the result is an annual gross copper flow of no more than 1.7 kg/cap.

3.3.4 Sustainability and the Agrarian Metabolic Regime

Although the metabolism of the agrarian regime is based on the exploitation of renewable resources and on harvesting flows rather than diminishing stocks, which is an important basis for long-term sustainability, it faces specific sustainability problems. The maintenance of successful exploitation of renewable flows is based on the management of soils that have to be considered nonrenewable on human time scales.[12] Their overuse or degradation immediately causes negative feedback at the local level: if yields decline, the ability to supply the local population with sufficient food and feed is at risk, and malnourishment is pending. Ecological sustainability, therefore, is a prerequisite for survival, and mismanagement is immediately punished. However, there is no guarantee against severe fluctuations and sustainability crises or even collapse. Growth can only be achieved within certain limits and is based on efficiency increases and the optimization of land use (Fig. 3.1). There tends to be positive feedback between population growth and biophysical growth, and agrarian societies show an overall tendency to increase land-use efficiency (output per unit of area) at the expense of labor efficiency (Boserup 1985; Fischer-Kowalski et al. 2014). This implies that in an agrarian regime, long-term per capita material wealth does not increase for the majority of the population over longer periods. To the contrary, in the long run, the material/energy output per capita reaches a limit or even starts to decline. Thus, in general, agrarian societies face sustainability problems related to the low density of the energy supply system, which is almost exclusively based on biomass, and related to the struggle over the long-term maintenance of soil fertility and the balance between food supply and population growth (cf. Grigg 1980; Kjaergaard 1994). Mismatches in this balance caused by population growth, soil degradation or fluctuations due to climatic conditions easily result in famine and demographic crises.[13] These crises are often triggered by epidemics or wars and lead to the destabilization of the labor-intensive land-use systems. Pollution problems occur only locally at mining sites or in urban agglomerations (Sieferle 2003b), but historically, the gradual expansion of agricultural land and related deforestation released large amounts of CO_2 (Houghton 2003). The global population increased slowly during the historical period of the agrarian regime, reaching some 700 million people at the beginning of the industrial revolution in the mid-18th century (Cohen 1995).

---

[12]Soils are built at average rates of 0.1 mm/year; thus, most soils are not renewable within time spans relevant for humans (Gerzabek et al. 2012).

[13]See the discussion on socioecological collapse and famines in Abel (1974), Diamond (2005) and Tainter (1988).

3.4 The Industrial Sociometabolic Regime

3.4.1 The Energy System of Industrial Societies

The energy system of the industrial regime differs radically from the solar-based energy systems of the other two regimes. Rather than tapping into renewable solar energy flows, industrial society relies on fossil resources to sustain its energy needs. The large-scale exploitation of accumulated and, at least at human time scales, nonrenewable stocks of energy carriers allowed for an increase in energy use per unit of area and per capita that was far above the limits of the agrarian regime.

One important reason for the decisive role of fossil fuels in the initial establishment and subsequent continuation of the new industrial metabolic regime is the energetic characteristics of fossil fuels, which are fundamentally different from those of biomass. First, the energy supply density of fossil energy carriers is many times higher than that of biomass (between 1000 and 10,000 W/m^2 for fossils fuels compared with 0.1–1 W/m^2 for biomass; Smil 1991). Second, fossil fuels initially appeared abundant and cheap, at least on time scales of several human generations. Third, with fossils fuels, large stocks are utilized, not small annual flows. Therefore, annual increases in energy use are not restricted by annual reproduction rates, at least until physical depletion is approached. In short, fossil fuels provide a cheap, high-density, abundant and stable energy source for several hundred years.

With the use of fossil fuels, the previous restrictions on the development of high-density energy demand systems (such as long-distance mass transport systems or mega cities, where energy demand densities can reach 1000 W/m^2) are eliminated. As Grübler (2004) has noted, the prevailing high-density energy demand characteristic of the industrial metabolic regime is in line with the fossil fuel supply, conversion and distribution system. The unprecedented energy supply density of fossil fuels in combination with a huge resource endowment allowed the multiplication of available useful energy and an unprecedented concentration and upscale of social demand systems. These are novel characteristics that constitute the core of the industrial sociometabolic regime. Two major processes contribute to a far-reaching decoupling of the energy system from the use of productive land. First, biomass as the primary energy source is substituted with area-independent[14] sources of energy in almost all applications. However, biomass remains an important raw material and the nutritional basis for humans and their livestock. In stark contrast to the declining share of biomass in the overall amount of energy and material consumption, the absolute amount of used biomass increases during industrialization, as shown in Fig. 3.2a–c, for major industrial economies. Both the

[14]Fossil energy carriers are, of course, not fully area-independent. Although the extraction of fossil energy carriers can effect considerable land areas at a regional scale, the overall extent of these areas is negligible compared with the land requirements of a biomass-based energy system (Smil 2003).

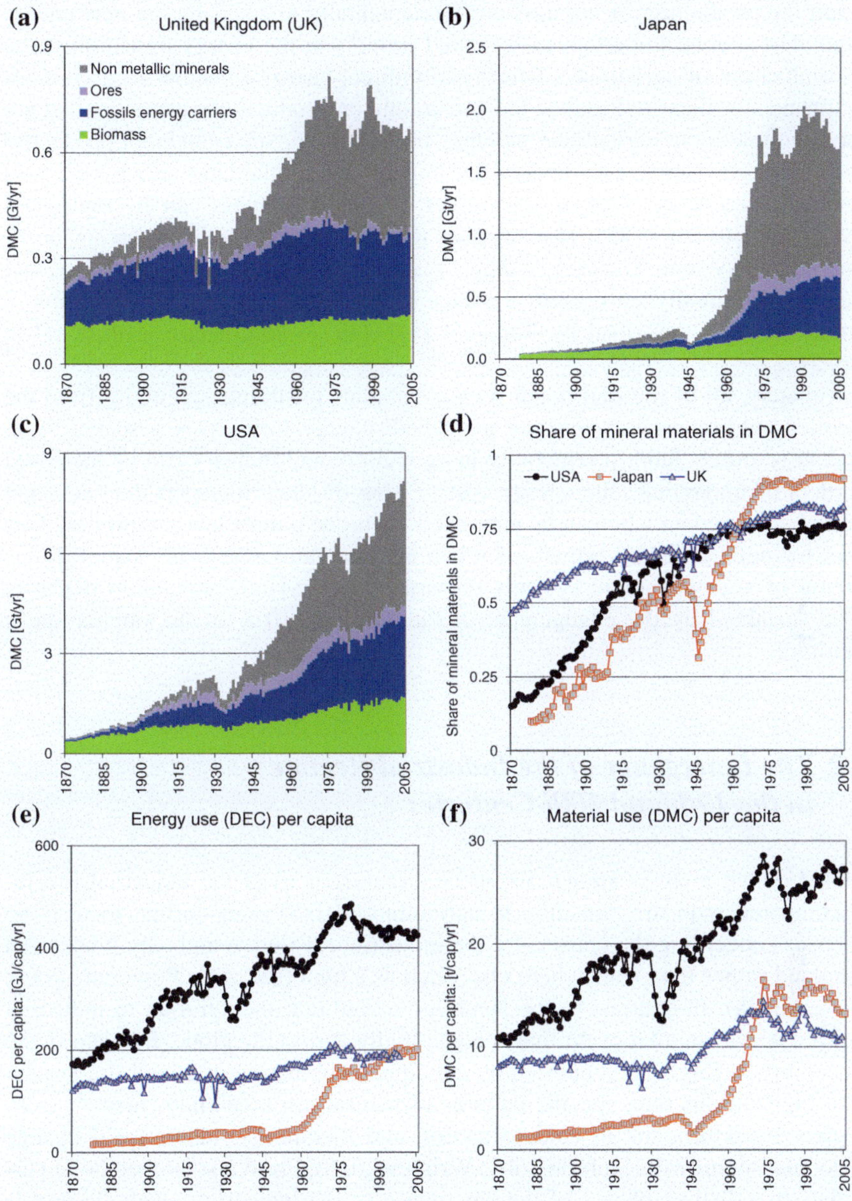

Fig. 3.2 Historical development of DMC and metabolic rates (DMC and DEC per capita per year) in selected industrial countries: Development of DMC by main material group in the period 1870–2005 in **a** the United Kingdom (UK), **b** the Japan and **c** USA. Development of the share of mineral materials in DMC in the three countries (**d**), of energy use per capita (**e**) and of material use per capita (**f**). (Sources: UK based on Eurostat 2007; Schandl and Schulz 2002; Japan based on Krausmann et al. 2011; USA based on Gierlinger and Krausmann 2012)

use and the production of biomass are fundamentally affected by the new energy system. The second important land-related process in the energy transition is the industrialization of agriculture. Direct and indirect energy subsidies (e.g., mechanizationand fertilization) reduce the constraints on productivity increases that are inherent in low-input agriculture, and they boost both the area and labor productivity of land-use systems. The demand for human and animal labor in agriculture is dramatically reduced, whereas the output of useful biomass can be multiplied (Hayami and Ruttan 1985). This reverses the role of agriculture in the socioeconomic energy system. The energy inputs result in declining energy efficiency, and agriculture eventually turns from a source of useful energy to an energy sink, a phenomenon first described by Pimentel et al. (1973) and discussed in detail in Chap. 21.

Ultimately, all of the traditional sociometabolic constraints stemming from the controlled solar energy system are abolished: energy turns from a scarce to an abundant resource, labor productivity in agriculture and industry can be increased by orders of magnitude, the energy cost of long-distance transport declines, and the number of people who can be nourished from one unit of land multiplies. This allows for unprecedented growth of urban agglomerations and the spatial differentiation of socioeconomic systems on larger scales. High population densities can be sustained by the engagement of far less than 10 % of the population in agriculture.

3.4.2 *The Emergence of the Industrial Regime in the 19th and 20th Centuries*

Although much more is known about the circumstances of the industrial revolution compared with the transition to agriculture 10,000 years earlier, there is no consensus concerning the reasons for the industrial revolution and why it occurred in England rather than somewhere else, such as China (see, e.g., Pomeranz 2000; Sieferle 2003a). In contrast to the Neolithic revolution, the agrarian to industrial transition was a singular event that spread rapidly across the globe. It began in the 18th century in England, supported by a unique combination of a land-use system with a high surplus rate, specific patterns of natural resource endowment,[15] technological breakthroughs in coal extraction and metallurgy, institutional change and population growth (Sieferle 2001; Wrigley 2010). From a sociometabolic perspective, two distinct phases of the emergence of the industrial metabolic profile can be distinguished (Krausmann and Fischer-Kowalski 2013). During the

[15]Among these factors were the combination of low availability of fuel wood with the occurrence of easily accessible deposits of both coal and iron in close vicinity. Finally, the 'river around Britain' allowed for long-distance transport among mining, manufacturing and consumption locations.

19th century, the energy transition in Europe, the United States and, later, Japan gained momentum through a positive feedback loop created by the emergence of the coal-steam engine and iron-ore railroad technology complex (Ayres 1990; Grübler 1998). Coal-based industrialization, while driving the expansion of the new industrial sociometabolic regime, was still characterized by a strong linkage between industrial production and a growing demand for human and animal labor. High population growth kept increases in per capita resource consumption low, and the rapidly growing population in urban industrial centers continued to rely on the delivery of nutrition from a largely preindustrial low-input agriculture.

This situation changed in the second phase of the metabolic transition, when petroleum (and, later, natural gas) replaced coal as the main source of energy. The large-scale use of petroleum began in the first half of the 20th century in the US, and in this phase, the emergence of a new technology cluster associated with the new key energy resource (Grübler 1998) was decisive: the internal combustion engine and the automobile as well as petrochemical industries and general electrification drove a rapid increase in energy and material use. The new technologies also facilitated a far-reaching decoupling of industrial production and human labor and the industrialization of agriculture and broke the link between industrialization and population growth. This period was characterized by a rapid multiplication of metabolic rates after World War II. Material and energy use per capita doubled in only two decades and then came to a halt with the oil price shocks of the 1970s. From then on, growth in per capita material and energy use in industrial countries stabilized at a high level, and overall growth slowed. For most industrialized countries, the 1970s saw the emergence of the general pattern of material and energy use that is currently characteristic of the industrial metabolic profile (Table 3.1).

3.4.3 The Metabolic Profiles of Industrial Societies

In industrial economies, material and energy use per capita exceed the level characteristic of advanced agrarian regimes by a factor of three to five (Table 3.1 and Fig. 3.2). Typically, more than 150 GJ of energy and 15 t of materials are used per inhabitant per year. A surge in agricultural output permits population densities up to 10 times larger than in most agrarian societies, whereas the share of agricultural population declines to only a few percent. Together, growth in metabolic rates and population lead to a multiplication of overall resource use. The material and energy use per unit of area can be 10–30 times above the level typical for the agrarian regime. Both the size of material flows and the composition fundamentally change. Fossil fuels substitute for biomass as the main energy carrier, and the use of mineral resources increases much faster than the use of biomass. The share of biomass in material and energy consumption is dwarfed and typically ranges between 10 and 30 %. During the early stages of industrialization, coal was by far the most important mineral resource, but other minerals, especially construction minerals, petroleum and natural gas, iron and copper and a wide variety of mining products, later gained in importance.[16] The large overlap between the energetic and material use of resources that prevailed in the agrarian regime diminished. In industrialized countries, approximately 50 % of all materials, or up to 10 t/cap/year, are used for non-energy purposes, mostly the accumulation of buildings and the built infrastructure. This also implies a shift from the dominance of throughput materials, which are used up within a brief period (biomass or fossil fuels used as energy carriers), toward the accumulation of materials that can be stored in socioeconomic stocks for several decades (ores, nonmetallic minerals). In contrast to throughput materials, which lose their functionality for society once they are used, accumulating materials can also be recycled (see Chap. 11).

Industrial societies build up large physical stocks of buildings, infrastructure networks and durable goods that have to be maintained. Existing estimates for Japan (Hashimoto et al. 2007; Tanikawa and Hashimoto 2009) and Switzerland (Rubli and Jungbluth 2005) suggest that the size of stocks of built infrastructure has tripled during the last 30 years; meanwhile, amounting to 200–400 t/cap in industrial societies. The enormous size socioeconomic stocks have reached can be illustrated when compared with biomass stocks in terrestrial ecosystems: these peak at less than 200 t/ha in mature forests compared with large-scale averages of 400–500 t/ha of physical stocks in industrial societies.

Total metal stocks stocks are significantly smaller but still enormous. The United States Geological Survey (USGS 2005) has calculated that the iron and steel in use in the US amounts to 15 t/cap, that of aluminum to 0.5 t and that of copper to 0.4 t. It is estimated that the accumulated stocks of iron in use in the US economy are roughly equal to the size of the remaining US iron stocks in identified ores. In addition to the stocks in use, large amounts of ores are accumulated in landfills (Gordon et al. 2006; Müller et al. 2006). In total, the amount of accumulated materials in the industrial regime is several times more than that of agrarian societies, and the maintenance of these stocks is a significant factor contributing to the high level of material and energy turnover in industrial societies. The accumulated stocks in infrastructure and buildings also reduce the flexibility to make a fast transition toward a structurally different metabolic regime.

Similar to the agrarian regime, the industrial regime displays a pronounced variety in the specific metabolic patterns, as indicated by the differences in material and energy use patterns in the UK, the US and Japan in Fig. 3.2 (see also the ranges given in Table 3.1). This should not be surprising because an accelerated global integration of markets in combination with prevailing spatial variety in environmental and socioeconomic conditions supported the formation of a global division of labor in terms of production and consumption. We therefore find that the robust structure and dynamic of the industrial regimes are superimposed by different metabolic varieties. In advanced industrial societies, differences in the composition of material and energy use and metabolic rates can be significant.

---

[16]Accordingly, the industrial regime has also been termed the ‘mineral economy’ as opposed to the organic economy of agrarian societies (Wrigley 1988).

Material and energy flow accounts show that in mature industrial societies, the per capita values of DMC and DEC can vary by a factor of two, as shown for the UK, the US and Japan in Fig. 3.2e, f. A host of biogeographic and socioeconomic factors are responsible for these differences. Densely populated countries with a high dependency on imports (such as Japan) typically have comparatively low rates of per capita material and energy use. On the other end of the scale are sparsely populated countries with high resource endowment per capita and high exports (such as the US). Their level of material and energy use can be more than twice as high (Krausmann et al. 2008b). Among other factors, this high use is related to the necessary size of the infrastructure networks (e.g., roads, electricity networks). For the same standard of supply of a population, these networks require more energy and materials if the country is sparsely populated. The climate zone, diet and resource endowment (the existence and accessibility of primary resources) have an equal impact (see Weisz et al. 2006).

3.4.4 What Drives Material and Energy Use in the Industrial Regime?

A number of highly interrelated structural attributes that mutually reinforce each other are responsible for the high level of material and energy consumption in the new regime. The new regime *facilitates* a high level of energy and material use of final users, and its operation *requires* a surge in natural resource use per se. This strong positive feedback between energy availability and the resulting possibilities to move and process materials and the material requirements to build and run these technologies has contributed to the unprecedented rise of material consumption observed during the historic period of industrialization, which is shown in Fig. 3.2 for the UK (a), the US (b) and Japan (c). The establishment and the area-wide diffusion of the new energy system and its metabolic regime require new infrastructures for extracting, distributing and processing energy carriers and material resources. Extensive infrastructures for transportation and communication, for grids and pipelines for supply, for the distribution and disposal of materials and for energy are essential features of the industrial regime and are reflected in the increasing consumption of mineral materials in the second half of the 20th century, particularly materials used for construction (Fig. 3.2a–c). Building, maintaining and operating industrial infrastructures requires huge amounts of energy as well as ores and industrial and construction minerals. During the history of industrialization, new technologies generally added new infrastructure requirements without actually replacing the old ones (e.g., rail, road and air transport infrastructures), and infrastructures and physical stocks continued to grow. Throughout the process of industrialization, infrastructures and the services they provide were responsible for a large fraction of socioeconomic energy and material use, and they contributed to unprecedented growth in material and energy use. In advanced industrial societies, infrastructures continue to have a significant influence on the high level of material and energy throughput, and they constitute physical legacies whose impact on future resource use can be considerable.

Another factor that characterizes the metabolic characteristics of the industrial regime is the high mobility of people, raw materials and goods. The new energy system triggered a new spatial organization of society based on a global division of labor and a large-scale spatial differentiation and concentration of resource extraction, industrial production, human dwellings and final consumption. Consequently, large quantities of raw materials and consumer goods are now shipped back and forth around the globe, and people travel ever greater distances for economic and leisure purposes. The physical volume of international trade grew from 0.8 Gt (gigatons) in 1950 to 10.6 Gt in 2010, and approximately 40 % of all extracted fossil fuels, ores and industrial minerals are shipped across international borders (Dittrich and Bringezu 2010; Schaffartzik et al. 2014). The intensity of overland road transport, for example, has reached several thousand metric tons and passenger kilometers per capita in industrialized economies, and it continues to grow. Throughout the process of industrialization, transport and mobility have been responsible for a considerable share of total energy and material throughput: during early periods of industrialization, rail transport consumed a large fraction of coal and iron supply, and since the second half of the 20th century, fuel demand for road and air transport has been the most important driver of energy use in industrial countries. Additionally, building up and maintaining a still-growing fleet of cars and aircraft consumes large quantities of energy and metal ores, and there is a strong positive feedback between mobility and infrastructure.

A third fundamental element of the metabolic pattern of modern industrial society is the emergence of a society of mass production and corresponding patterns of mass consumption. Starting with *Fordism* in the US in the 1930s, mass production and consumption became a general phenomenon in the industrial world in the 1950s and 1960s (Lutz 1989). The industrial production of huge amounts of consumer goods at affordable prices and the appearance of a large and increasingly wealthy middle class were major drivers in the increase in per capita resource use in the second half of the 20th century. These factors are still major drivers of the structure and size of the industrial metabolism, and they demand large amounts of energy and materials. The substitution of work derived from fossil fuels for human and animal labor facilitated the processing of much larger quantities of materials and was one of the preconditions for mass production and consumption. The increase in labor productivity not only resulted in a surge of industrial output but also freed large amounts of human time otherwise allocated to wage labor and household chores. With rising income and a higher living standard, this time was used for other, often material- and energy-intensive activities (e.g., tourism) and consumption. Households in industrialized countries are equipped with large stocks of energy and material-intensive consumer goods. The per capita use of living space grows linearly with income. The subsequent high energy demand for heating or cooling, light and the large number of electric devices for refrigeration, cooking, cleaning and communication contributes significantly to the high level of material and energy use in the industrial regime.

Finally, growth is a core element of industrial societies. This is in stark contrast to the agrarian regime, which was characterized by diminishing marginal returns and strong negative feedbacks from physical limitations, which prevented per capita economic growth for larger populations over longer periods. Economic growth is a basic feature of the industrial regime, and throughout the industrialization process, economic growth has been closely linked to physical growth, driven by both population growth and increases in per capita throughput. There has been a long-lasting trend of efficiency increases, that is, a decline in the amount of material and energy use per unit of GDP, in industrial countries. Together with a slowdown in population growth in most industrialized countries, this has contributed to reduced growth rates in material and energy demand, particularly during the last 30 years. In some cases, it has resulted in the stabilization of resource use at a high level. However, there is currently no evidence that material or energy use can decline over longer periods of time while the economy continues to grow (see, for example, Steger and Bleischwitz 2009).

3.4.5 Sustainability of the Industrial Regime

Whereas scarcity, poverty and an overexploitation of natural resources are always pending in the agrarian regime, the dominant trend within mature industrial regimes is that of abundance (however unevenly distributed). Due to its enormous material and energy use, the industrial regime currently faces output-related sustainability problems resulting from pressure upon the regional and global absorptive capacity of natural ecosystems for wastes and emissions. Some of these problems have been solved technologically (e.g., acid rain), but other local and global environmental problems of the industrial sociometabolic regime continue to emerge or worsen, including climate change and biodiversity loss. Environmental pressures have increased to the extent that it is assumed that human society has already left its safe operating space and is transgressing planetary sustainability boundaries (Rockström et al. 2009). The magnitude of the transformative effect of industrial society on its natural environment has even inspired prominent scholars to proclaim a new geological era, the Anthropocene (Steffen et al. 2011). However, it not only output-related environmental problems that are growing; the abundance experienced relative to the previous agrarian regime may reach limits as well. The current industrial sociometabolic regime is based on the use of exhaustible key resources. Rising prices for many key resources, from food to petroleum and key metals, indicate that the time of cheap and abundant resources may come to an end. Unless we assume unlimited substitutability of resources due to technological change, which is rather unlikely (Ayres 2007), the industrial sociometabolic regime, by definition, lacks the potential for sustainability. Sieferle (1997), therefore, considers it a transitory stage rather than a stable new regime. A transition to a sustainable industrial regime would imply that we need to find ways to drastically reduce the throughput of materials and energy in the industrial regime and to maintain the high energy density based on new and renewable energy resources.

3.4.6 *The Next Transition: The Metabolism of the 'Postindustrial' Society?*

During the last few decades, the growth of material and energy use in the industrial world has slowed down markedly and in some cases has even begun to decline, as, for example, in the UK (Fig. 3.2a) and Japan (Fig. 3.2c). Even if industrial economies have lost the strong momentum of biophysical growth, a high level of energy and material use is maintained. The trends of material and energy use in the last few decades give the impression that a mature industrial metabolic profile has emerged and stabilized. Any hopes that technological progress and the shift toward a postindustrial or service economy based on less material- and energy-intensive economic activities would lead to a significant decline in natural resource use in industrial economies remain unfulfilled. Despite the fast-growing service sectors and information and communication technologies, which contribute to economic growth at a lower material and energy intensity than the industrial sector, the postindustrial economy still builds on the industrial regime rather than replacing it. Moreover, instead of finding substitutes for its material- and energy-intensive metabolic profile, it adds to this profile. Infrastructures, mobility and mass production and consumption are not vanishing in the service economy.

In the industrial metabolic regime, economic development and social metabolism are as closely interlinked as they can be, and gains in efficiency, although sometimes enormous, have never led to a reduction in metabolism but have driven further growth (Ayres and Warr 2009). Although the historical perspective shows that technological solutions have often come into play in the past, it also reveals that these very solutions create new types of problems and that a spiral of risk is perpetuated. Finally, society will have to recognize that physical growth is limited and that it is more important to decouple the quality of human life from further material and energy use. This will not be achieved by means of technological solutions alone but will require far-reaching changes in society. Such changes will occur irrespective of whether the relevant political and economic actors wish them to. Those who advocate the concept of sustainable development believe it would be wiser to organize such a change proactively.

References

Abel, W. (1974). *Massenarmut und Hungerkrisen im vorindustriellen Europa*. Parey, Hamburg: Versuch einer Synthese.

Ayres, R. U. (1990). Technological transformations and long waves. Part I. *Technological Forecasting and Social Change, 37*, 1–37.

Ayres, R. U. (2007). On the practical limits to substitution. *Ecological Economics, 61*(1), 115–128.

Ayres, R. U., Ayres, L. W., & Warr, B. (2004). Is the U.S. economy dematerializing? Main indicators and drivers. In: J. C. J. M. van den Bergh & Janssen, A.M. (Eds.), *Economics of industrial ecology. Materials, structural change, and spatial scales* (pp. 57–93). MIT Press, Cambridge, Massachusetts, London, England.

Ayres, R. U., & Warr, B. (2009). *The economic growth engine: How energy and work drive material prosperity*. Cheltenham, UK and Northhampton MA, US: Edward Elgar.

Bairoch, P. (1993). *Economics and world history. Myths and paradoxes*. Harvester Wheatsheaf, New York.

Biraben, J.-N. (2003). The rising numbers of humankind. *Population and Societies, 394*, 1–4.

Boserup, E. (1965). *The conditions of agricultural growth*. Aldine/Earthscan, Chicago: The economics of agrarian change under population pressure.

Boserup, E. (1981). *Population and technological change—A study of long-term trends*. Chicago: The University of Chicago Press.

Boserup, E. (1985). The impact of scarcity and plenty on development. In R. I. Rotberg & T. K. Rabb (Eds.), *Hunger and history* (pp. 185–211). Cambridge, MA: Cambridge University Press.

Boyden, S. V. (1992). *Biohistory: The interplay between human society and the biosphere—Past and present*. Paris, Casterton Hall, Park Ridge, New Jersey: UNESCO and Parthenon Publishing Group.

Cohen, M. N. (1977). *The food crisis in prehistory. Overpopulation and the origins of agriculture*. New Haven, London: Yale University Press.

Cohen, J. E. (1995). *How many people can the earth support?*. New York, London: W. W. Norton & Company.

Coughenour, M. B., Ellis, J. E., Swift, D. M., Coppock, D. L., Galvin, K., McCabe, J. T., & Hart, T. C. (1985). Energy extraction and use in a nomadic pastoral ecosystem. *Science, 230*(4726), 619–625.

De Vries, B., & Goudsblom, J. (2002). *Mappae Mundi. Humans and their habitats in a long-term socio-ecological perspective*. Amsterdam: Amsterdam University Press.

De Zeeuw, J. W. (1978). Peat and the dutch golden age. A.A.G. *Bijdragen, 21*, 3–31.

Diamond, J. (2005). *Collapse*. Viking, New York: How societies choose to fail or succeed.

Dittrich, M., & Bringezu, S. (2010). The physical dimension of international trade: Part 1: Direct global flows between 1962 and 2005. *Ecological Economics, 69*(9), 1838–1847.

Eurostat. (2007). Economy wide material flow accounts and resource productivity. EU15 1970–2004. European Statistical Office, Luxembourg.

Fagan, B. M. (2001). *The little ice age: How climate made history 1300–1850*. New York: Basic Books.

Fischer-Kowalski, M., & Haberl, H. (1997). Stoffwechsel und Kolonisierung: Ein universalhistorischer Bogen. In M. Fischer-Kowalski, H. Haberl, W. Hüttler, H. Payer, H. Schandl, V. Winiwarter, & H. Zangerl-Weisz (Eds.), *Gesellschaftlicher Stoffwechsel und Kolonisierung von Natur* (pp. 25–36). Amsterdam: Gordon & Breach Fakultas.

Fischer-Kowalski, M., & Haberl, H. (1998). Sustainable development: Socio-economic metabolism and colonization of nature. *International Social Science Journal, 158*, 573–587.

Fischer-Kowalski, M., & Haberl, H. (2007). Conceptualizing, observing and comparing socioecological transitions. In M. Fischer-Kowalski & H. Haberl (Eds.), *Socioecological transitions and global change: Trajectories of social metabolism and land use* (pp. 1–30). Cheltenham, UK, Northampton, USA: Edward Elgar.

Fischer-Kowalski, M., Krausmann, F., Giljum, S., Lutter, S., Mayer, A., Bringezu, S., et al. (2011a). Methodology and indicators of economy wide material flow accounting. State of the art and reliablity across sources. *Journal of Industrial Ecology, 15*(6), 855–876.

Fischer-Kowalski, M., Krausmann, F., Mayer, A., & Schaffartzik, A. (2014). Boserup's theory on technological change as a point of departure for the theory of sociometabolic regime transitions. In M. Fischer-Kowalski, A. Reenberg, A. Mayer, & A. Schaffartzik (Eds.), *Ester Boserup's legacy on sustainability: Orientations for contemporary research* (pp. 23–42). Dordrecht: Springer.

Fischer-Kowalski, M., Singh, S. J., Ringhofer, L., Grünbühel, C., Lauk, C., & Remesch, A. (2011b). Socio-metabolic transitions in subsistence communities. *Human Ecology Review, 18*(2), 147–158.

Gerzabek, M. H., Winiwarter, V., Baumgarten, A., Blum, W. E. H., Butterbach-Bahl, K., Cushman, G., et al. (2012). Productive soils, global environmental cycles and energy crops. In Winiwarter, V., Gerzabek, M. H. (Eds.), *The challenge of sustaining soils: Natural and social ramifications of biomass production in a changing world, interdisciplinary perspectives* (pp. 48–71). Verlag der Österreichischen Akademie der Wissenschaften, Vienna: Austrian Academy of Sciences Press.

Gierlinger, S., & Krausmann, F. (2012). The physical economy of the United States of America: Extraction, trade and consumption of materials from 1870 to 2005. *Journal of Industrial Ecology, 16*(3), 365–377.

Gordon, R. B., Bertram, M., & Graedel, T. E. (2006). Metal stocks and sustainability. *Proceedings of the National Academy of Sciences of the United States of America, 103*(5), 1209–1214.

Goren-Inbar, N., Alperson, N., Kislev, M. E., Simchoni, O., Melamed, Y., Ben-Nun, A., & Werker, E. (2004). Evidence of Hominin control of fire at Gesher Benot Ya`aqov, Israel. *Science, 304*(5671), 725–727.

Goudsblom, J. (1992). *Fire and civilization*. London: Penguin Books.

Grigg, D. B. (1980). *Population growth and agrarian change. An historical perspective*. Cambridge, London, New York, New Rochelle, Melbourne, Sydney: Cambridge University Press.

Grübler, A. (1998). *Technology and global change*. Cambridge: Cambridge University Press.

Grübler, A. (2004). Transitions in energy use. In Cleveland, C. J. (Ed.), *Encyclopedia of energy* (pp. 163–177). London: Elsevier.

Harlan, J. R. (1992). *Crops and man*. Madison: American Society of Agronomy, Inc., Crop Science Society of America Inc.

Hashimoto, S., Tanikawa, H., & Moriguchi, Y. (2007). Where will large amounts of materials accumulated within the economy go?—A material flow analysis of construction minerals for Japan. *Waste Management, 27*(12), 1725–1738.

Hayami, Y., & Ruttan, V. W. (1985). *Agricultural Development. An international perspective*. Baltimore: John Hopkins University Press.

Hong, S., Candelone, J. P., Soutif, M., & Boutron, C. F. (1996). A reconstruction of changes in copper production and copper emissions to the atmosphere during the past 7000 years. *Science of the Total Environment, 188*(2–3), 183–193.

Houghton, R. A. (2003). Revised estimates of the annual net flux of carbon to the atmosphere from changes in land use and land management 1850–2000. *Tellus B, 55*(2), 378–390.

Kjaergaard, T. (1994). *The Danish revolution 1500–1800. An ecohistorical interpretation*. Cambridge: Cambridge University Press.

Krausmann, F. (2004). Milk, manure and muscular power. Livestock and the industrialization of agriculture. *Human Ecology, 32*(6), 735–773.

Krausmann, F., Erb, K.-H., Gingrich, S., Lauk, C., & Haberl, H. (2008a). Global patterns of socioeconomic biomass flows in the year 2000: A comprehensive assessment of supply, consumption and constraints. *Ecological Economics, 65*(3), 471–487.

Krausmann, F., & Fischer-Kowalski, M. (2013). Global socio-metabolic transitions. In S. J. Singh, H. Haberl, M. Chertow, M. Mirtl, & M. Schmid (Eds.), *Long term socio-ecological research. Studies in society—nature interactions across spatial and temporal scales* (pp. 339–365). Berlin: Springer.

Krausmann, F., Fischer-Kowalski, M., Schandl, H., & Eisenmenger, N. (2008b). The Global Sociometabolic Transition. *Journal of Industrial Ecology, 12*, 637–656.

Krausmann, F., Gingrich, S., Haberl, H., Erb, K.-H., Musel, A., Kastner, T., et al. (2012). Long-term trajectories of the human appropriation of net primary production: Lessons from six national case studies. *Ecolgical Economics, 77*, 129–138.

Krausmann, F., Gingrich, S., & Nourbakhch-Sabet, R. (2011). The metabolic transition in Japan: A material flow account for the period 1878 to 2005. *Journal of Industrial Ecology, 15*(6), 877–892.

Krausmann, F., & Haberl, H. (2002). The process of industrialization from the perspective of energetic metabolism. Socioeconomic energy flows in Austria 1830–1995. *Ecological Economics, 41*(2), 177–201.

Leach, G. (1976). *Energy and food production*. Guildford: IPC Science and Technology Press.

Lutz, B. (1989). Der kurze Traum immerwährender Prosperität. Eine Neuinterpretation der industriell-kapitalistischen Entwicklung im Europa des 20. Jahrhunderts. New York: Campus Verlag, Frankfurt am Main.

Malanima, P. (2002). *Energy systems in Agrarian societies: the European deviation*. Napoli: Consiglio Nazionale della Ricerche, Istituto di Studi sulle Società del Mediterraneo (CNR - ISSM).

Martin, P. S., & Klein, R. G. (1989). *Quaternary extinctions: A prehistoric revolution*. Tucson: University of Arizona Press.

Mazoyer, M., Roudart, L., & Membrez, J. H. (2006). *A history of world agriculture: From the neolithic age to the current crisis*. London: Earth Scan.

Mithen, S. (2003). *The little ice age: How climate made history 1300–1850*. London: Phoenix Paperback.

Müller, D. B., Wang, T., Duval, B., & Graedel, T. E. (2006). Exploring the engine of anthropogenic iron cycles. *Proceedings of the National Academy of Sciences of the United States of America, 103*(44), 16111–16116.

Müller-Herold, U., & Sieferle, R. P. (1998). Surplus and survival: Risk, ruin and luxury in the evolution of early forms of subsistence. *Advances in Human Ecology, 6*, 201–220.

Netting, R. M. C. (1993). Smallholders, householders. *Farm families and the ecology of intensive, sustainable agriculture*. Stanford: Stanford University Press.

Pimentel, D., Hurd, L. E., Bellotti, A. C., Forster, M. J., Oka, I. N., Sholes, O. D., & Whitman, R. J. (1973). Food production and the energy crisis. *Science, 182*(4111), 443–449.

Pomeranz, K. (2000). *The great divergence: China, Europe, and the making of the modern world economy*. Princeton: Princeton University Press.

Rockström, J., Steffen, W., Noone, K., Persson, A., Chapin III, F. S., Lambin, E. F., et al. (2009). Planetary boundaries: Exploring the safe operating space for humanity. *Ecology and Society, 14*(2), 32. (online) URL http://www.ecologyandsociety.org/vol14/iss2/art32/

Rubli, S., & Jungbluth, N. (2005). *Materialflussrechnung für die Schweiz*. Bundesamt für Statistik, Neuchatel: Machbarkeitsstudie.

Sahlins, M. (1972). *Stone age economics*. New York: Aldine de Gruyter.

Schaffartzik, A., Mayer, A., Gingrich, S., Eisenmenger, N., Loy, C., & Krausmann, F. (2014). The global metabolic transition: Regional patterns and trends of global material flows, 1950–2010. *Global Environmental Change, 26*, 87–97.

Schandl, H., & Schulz, N. B. (2002). Changes in United Kingdom's natural relations in terms of society's metabolism and land use from 1850 to the present day. *Ecological Economics, 41*(2), 203–221.

Sieferle, R. P. (1997). *Rückblick auf die Natur: Eine Geschichte des Menschen und seiner Umwelt*. München: Luchterhand.

Sieferle, R. P. (2001). *The subterranean forest. Energy Systems and the Industrial Revolution*. Cambridge: The White Horse Press.

Sieferle, R. P. (2003a). Der Europäische Sonderweg: Ursachen und Faktoren, 2. erweiterte Auflage. Schriftenreihe "Der Europäische Sonderweg", Breuninger-Stiftung, Band 1, Stuttgart.

Sieferle, R. P. (2003b). Sustainability in a world history perspective. In B. Benzing (Eds.), *Exploitation and Overexploitation in Societies Past and Present* (pp. 123–142). IUAES-Intercongress 2001 Goettingen. LIT Publishing House, Münster.

Sieferle, R. P. (2010). Lehren aus der Vergangenheit. Expertise für das WBGU-Hauptgutachten "Welt im Wandel: Gesellschaftsvertrag für eine Große Transformation". Berlin.

Sieferle, R. P., Krausmann, F., Schandl, H., & Winiwarter, V. (2006). *Das Ende der Fläche*. Böhlau, Köln: Zum gesellschaftlichen Stoffwechsel der Industrialisierung.

Simmons, I. G. (1989). *Changing the face of the earth*. Oxford, UK-Cambridge, MA: Blackwell.

Simmons, I. G. (2008). *Global environmental history 1000 BC to AD 2000*. Edinburgh: Edinburgh University Press.

Smil, V. (1991). *General energetics. Energy in the biosphere and civilization*. Manitoba, New York: Wiley.

Smil, V. (2003). *Energy at the crossroads. Global perspectives and uncertainties*. Cambridge Mass., London, England: The MIT Press.

Smil, V. (2008). *Energy in nature and society. General energetics of complex systems*. Cambridge, MA: MIT Press.

Smil, V. (2013). *Harvesting the biosphere: What we have taken from nature*. Yale: MIT Press.

Steffen, W., Persson, A., Deutsch, L., Zalasiewicz, J., Williams, M., Richardson, K., et al. (2011). The anthropocene: From global change to planetary stewardship. *AMBIO: A Journal of the Human Environment, 40*(7), 739–761.

Steger, G., & Bleischwitz, R. (2009). Decoupling GDP from resource use, resource productivity and competitiveness: A cross-country comparison. In R. Bleischwitz, P. J. J. Welfens, & Z. Zhang (Eds.), *Sustainable growth and resource productivity* (pp. 172–194). Greenleaf, Sheffield: Economic and global policy issues.

Tainter, J. A. (1988). *The collapse of complex societies*. Cambridge: Cambridge University Press.

Tanikawa, H., & Hashimoto, S. (2009). Urban stock over time: Spatial material stock analysis using 4d-GIS. *Building Research and Information, 37*(5–6), 483–502.

USGS. (2005). Metal stocks in use in the United States. Science for a changing world.

Warde, P. (2007). *Energy consumption in England and Wales, 1560–2000*. Naples: Consiglio Nazionale delle Ricerche, Instituto di Studi sulle Societa del Mediterrano.

Weisdorf, J. L. (2005). From foraging to farming: Explaining the neolithic revolution. *Journal of Economic Surveys, 19*(4), 561–586.

Weisz, H., Krausmann, F., Amann, C., Eisenmenger, N., Erb, K.-H., Hubacek, K., & Fischer-Kowalski, M. (2006). The physical economy of the European Union: Cross-country comparison and determinants of material consumption. *Ecological Economics, 58*(4), 676–698.

Wrigley, E. A. (1988). *Continuity, chance and change. The character of the industrial revolution in England*. Cambridge: Cambridge University Press.

Wrigley, E. A. (2010). *Energy and the English industrial revolution*. Cambridge: Cambridge University Press.

Chapter 4
Beyond Inputs and Outputs: Opening the Black-Box of Land-Use Intensity

Karl-Heinz Erb, Tamara Fetzel, Helmut Haberl, Thomas Kastner, Christine Kroisleitner, Christian Lauk, Maria Niedertscheider and Christoph Plutzar

Abstract Despite their central role in land-use transitions, changes in land-use intensity are only poorly understood, and databases for systematically analyzing change in land-use intensity are largely missing. This knowledge gap is critical because, due to the anticipated changes in global population numbers and food, fiber and energy demand, the development of strategies that aim to reap the benefits of land-use intensification (e.g., the reduced land demand for a certain level of production) while simultaneously avoiding detrimental social and ecological effects will become decisive in the near future. In this chapter, we first review existing approaches to analyzing land-use intensity and discuss existing barriers to

K.-H. Erb (✉) · T. Fetzel · H. Haberl · T. Kastner · C. Lauk · M. Niedertscheider · C. Plutzar
Institute of Social Ecology (SEC), Alpen-Adria University,
Schottenfeldgasse 29, 1070 Vienna, Austria
e-mail: karlheinz.erb@aau.at

T. Fetzel
e-mail: tamara.fetzel@aau.at

H. Haberl
e-mail: helmut.haberl@aau.at

T. Kastner
e-mail: thomas.kastner@aau.at

C. Lauk
e-mail: christian.lauk@aau.at

M. Niedertscheider
e-mail: maria.niedertscheider@aau.at

C. Plutzar
e-mail: christoph.plutzar@aau.at

C. Kroisleitner
Institute of Geography and Regional Sciences, University of Graz,
Heinrichstraße 36, 8010 Graz, Austria
e-mail: christine.kroisleitner@uni-graz.at

H. Haberl et al. (eds.), *Social Ecology*, Human-Environment Interactions 5,
DOI 10.1007/978-3-319-33326-7_4

land-use intensity research. We then elaborate on what the socioecological method inventory contributes to land system research. We argue that the concepts of socioeconomic metabolism and the colonization of nature are apt to significantly contribute to improvements in the analytical capabilities related to land-use intensity research. The strengths of the socioecological method inventory are its strict application of first principles, a sound and meaningful system boundary between society and nature and its applicability to Social and Natural Science approaches. These aspects are prerequisites for guiding the type of data collation and organization that allow investigation into the feedback cycles between social and natural systems that constitute the trade-offs and synergies of the land system.

Keywords Land-system science · Land use · Land-use intensity · Human appropriation of net primary production (HANPP) · Colonization of natural processes

4.1 Introduction

The awareness that land use represents a pervasive driver of environmental change has been rising in the last few decades (e.g., Foley et al. 2005, 2011; Turner et al. 2007). The ongoing global population growth, the rising per capita demand for land-based products such as food and timber and the prospects of a growing role of biomass as a renewable, supposedly environmentally benign source of energy suggest that humanity's pressures on land will continue to rise in the next decades (Coelho et al. 2012; Seto et al. 2012; Smith et al. 2013). For example, it is assumed that global agricultural output will grow by 70–100 % by 2050 (Alexandratos and Bruinsma 2012; Tilman et al. 2011). The area of the earth's fertile lands is limited, much of this area is already used more or less intensively for human purposes (see Chap. 14), and increasing the land area used for agriculture and forestry results in additional pressures on ecosystems and biodiversity (Lambin and Meyfroidt 2011; Pereira et al. 2012; Sala et al. 2000). Therefore, solutions are being sought that prevent a commensurate increase in area demand. As in the past (Krausmann et al. 2013), there is the expectation that growing land-use intensity (i.e., higher output per unit of land per year) and improved biomass conversion efficiency will help keep land demand at bay while allowing further increases in land-use outputs. At the same time, concerns are mounting that land use may be undermining its very foundations, such as by degrading the ability of ecosystems to continue supplying vital services (Millennium Ecosystem Assessment 2005), by exerting unacceptable levels of pressure on ecosystems or health (IAASTD 2009) or by jeopardizing soils (Lal 2012; Winiwarter and Gerzabek 2012).

Despite its importance, the intensity with which land is used is a currently under-researched aspect of land use (Erb et al. 2013a; Kuemmerle et al. 2013). In general terms, land-use intensity relates to the quantity, and sometimes the quality, of inputs and/or outputs per unit of land (see Sect. 4.5). Changes in land-use intensity within a certain land-cover type involve changes in inputs (e.g., labor, fertilizer or tractor hours), outputs (e.g., yields) or both. In contrast, changes in land cover refer to transitions from one land cover to another, such as from forest to cropland or from agricultural fields to urban settlements (see Chap. 14). The delineation of these two crucial categories, land-cover change and land-use intensity change, is not straightforward and strongly depends on the classification schemes and criteria used to define land-use or land-cover types. Nevertheless, there is common agreement that land-use intensity changes are more subtle than land-cover changes and are more difficult to observe on larger scales, such as via remote sensing technologies (Kuemmerle et al. 2013). This is probably the reason for the relative neglect of this aspect in the current mainstream of Land-Use Science. Another reason for the knowledge gaps regarding land-use intensity is that it is a complex, multidimensional concept in which different processes affect land systems at various levels. Due to this complexity, it is difficult to find suitable indicators and data sets able to integrate this multitude of aspects (Erb 2012; Erb et al. 2013a; Lambin et al. 2000; Shriar 2000).

Nevertheless, there is an increasing body of knowledge suggesting that changes in land-use intensity have far-reaching effects on the processes and patterns of the Earth system. Changes in the intensity of land use directly affect global carbon (Erb et al. 2013b), water (Portmann et al. 2010) and nutrient cycles (e.g., nitrogen, Gruber and Galloway 2008). These changes also affect biodiversity (Tscharntke et al. 2005) and change the biophysical properties of the land surface, including prarmeters such as albedo and sensible heat flux (Luyssaert et al. 2014).

Furthermore, the intensity of land use is a decisive factor in the amount of food, feed, fuel and fibers[1] extracted by society from land per unit area per year. In principle, increases in land-use intensity result in increases in production without proportional changes in the extent of area of a land-cover type, thus allowing for land sparing (i.e., production increases without proportional changes in land demand). In the course of history, it was land-use intensification that allowed humanity to overcome the so-called 'Malthusian trap' caused by limitations that hindered the increase of yields in agrarian societies (see Sect. 4.2 and Chap. 13). Land-use intensification also supported population growth and increases in biomass provision (Tilman 1999). Thanks to intensification, many industrialized countries have been able to dramatically increase agricultural output over the course of the last few decades, if not centuries. Since the early 1960s, intensification has brought approximately 2.7-fold increases in the global agricultural production of crucial resources (such as cereals), whereas harvested cropland area remained nearly stable in this period (FAOSTAT 2014). This 'communication' of land-use intensity with the extent of certain land-use types renders it a central aspect for past as well future sustainability challenges.

[1]Sometimes called the 'four Fs' of land-based resources.

In the future, land-use intensification processes will become even more decisive. The anticipated increases in global population numbers, the expected and partly necessary increases in nutritional demand and the increasing demand for renewable energy sources such as bioenergy will lead to considerable increases in the demand for biomass products from terrestrial ecosystems. However, these surges in biomass demand are restrained by planetary confines, most rigidly by the limit of 130 million km^2 of terrestrial surface that is available globally (see Chap. 14), of which only parts can be used for agriculture. Furthermore, the mandate to protect the shrinking pristine habitats of the Earth calls for strategies that allow for production increases without significant further expansions of land use. Putting this land to use would result in a drastic deterioration of biodiversity and other ecosystem services and in the release of considerable amounts of carbon to the atmosphere (Pan et al. 2011). Indeed, according to projections by the Food and Agricultural Organization of the United Nations (FAO; Alexandratos 1999) and the International Assessment of Agricultural Knowledge, Science and Technology for Development (IAASTD 2009), increases in agricultural intensity will contribute much more strongly to increasing production in the coming decades than the expansion of agricultural area.

However, land-use intensification poses a dilemma. It allows an increasing demand for biomass to be satisfied without proportional land-use expansions; however, depending on the appropriateness of technologies applied under the local soil and climate conditions, it may degrade many ecosystem services upon which humanity depends. In the past, increases in land-use intensity have often been associated with a plethora of unfavorable effects on ecosystem functioning, including soil degradation, groundwater and air pollution and biodiversity loss (Lindenmayer et al. 2012; Matson et al. 1997). The global increases in agricultural yield since 1960 have been associated with tremendous surges in agricultural inputs. For example, machinery employment (e.g., measured in the number of tractors) has increased 2.5-fold, and nitrogen fertilizer consumption has increased 7.3-fold (FAOSTAT 2014). Over the last few decades, the use of industrial inputs has increased to an alarming scale, and it increasingly affects many Earth system processes (Steffen et al. 2005).

Solving this dilemma is a major scientific and political challenge, motivating the call for 'sustainable intensification' (Cassman 1999; Tilman 1999; Tilman et al. 2011), or land-use strategies that allow the beneficial land-sparing effect of land-use intensification to be reaped while avoiding its detrimental social and ecological effects. The implementation of recent international strategies that aim to reduce greenhouse gas emissions from deforestation and forest degradation (REDD; www.un-redd.org), such as by conserving old-growth forests, also depends on area-saving technologies and therefore requires an improved understanding of the land system and its transitions.

However, there are considerable gaps in our knowledge of the underlying processes of intensification, and data that could be used to systematically analyze land-use intensification in adequate quality and/or quantity are still scarce (Kuemmerle et al. 2013). This situation hampers analytical capabilities.

Consequently, a causal understanding of the factors, mechanisms, determinants and constraints underlying land-use intensification is still largely unsatisfactory. Thus, improving the knowledge base is essential in light of the upcoming sustainability challenges related to land use. The prevailing limitations in understanding the complex dynamics of the global land systems can be illustrated by examples from the literature. For instance, empirical analyses of past land-use trajectories reveal that the combined effect of land-use intensification and land expansion on cropland is far from simple and straightforward. According to an analysis of decadal national time series of cropland expansion and yield increases, Rudel et al. (2009) show that increases in yields and declines in cropland have been infrequently paired and that between 1970 and 2005, agricultural intensification was generally not accompanied by decline or stasis in cropland area. Increases in production can create new demands, leading to an expansion of cultivated areas. This feedback loop has been denoted the 'rebound effect' or 'Jevons paradox' (Ceddia et al. 2013; Erb 2012; Lambin and Meyfroidt 2011). These issues cast doubt on straightforward interpretations or scenario-based extrapolations of the beneficial effects of land intensification strategies (see, e.g., Burney et al. 2010).

Pushing the frontiers of science related to land-system research presents a formidable challenge (Turner et al. 2007) as it requires innovative methods, novel datasets and perspectives that allow for the development of sound causal chains among the various factors, mechanisms, determinants and constraints underpinning land-use intensification processes.

In this chapter, we review approaches to analyzing land-use change as well as land-use intensity, and we summarize the current barriers to analyzing land-use intensity in mainstream Land-Use Science. Building on this body of knowledge, we discuss the potential contribution of the socioecological method inventory to the development of an analytical framework for conceptualizing and monitoring the complex, systemic interactions related to land-use intensification, including the feedback between production and consumption.

4.2 A Sketch of the History of Research on Land-Use Change

The analysis of land use, its drivers and its effects on ecosystems and landscapes as a topic of research dates back to the middle of the 19th century. One of the central 'founders' of Land-Use Science was George Perkins Marsh (1801–1882). Referring to traditions founded by F. de Saussure, A. von Humboldt, J.-B. Boussingault, C. Lyell and M. Somerville, George Perkins Marsh developed a comprehensive and critical perception of the human modification of the landscape in his book '*Man and Nature: or, Physical Geography as Modified by Human Action*', published in 1864 (see Goudie 2006; Simmons 1989), using examples from Europe, the United States and the Middle East.

Before that time, classical economists had already recognized the importance of the relationship among population growth, land-use intensity and the costs of food production. The British scholar T.R. Malthus (1766–1834) hypothesized about the interrelation between population growth and food production, concluding that (linear) increases in agricultural production would not be able to follow the (exponential) growth in population numbers. This would engender crises such as famine and war and would thus exert an endogenous population control, the so-called 'Malthusian trap' (Malthus 1798). Some years later, the classical political economist David Ricardo (1772–1823) applied the law of diminishing returns, which states that on a given plot of land, marginal returns decrease with each additional unit of input (i.e., along intensification), to develop his rent theory (Ricardo 1815). In his notion, the land rent land owners are able to charge on any parcel of land depends on the difference in fertility between that parcel and the least fertile plot in a region. Here, the least fertile plot serves as 'measure of reference'. It is characterized by an unfavorable input-output ratio and yields no rent for the land owner. Ricardo strongly criticized the Corn Laws, which were heavily debated at the time and eventually implemented in 1815. These laws were intended to prevent cereal imports by enacting high import tariffs. Ricardo argued that these tariffs would increase incentives to put less fertile domestic land under use, which would result in a deterioration of overall input-output ratios of land-based production and an increase in land rents for land owners. Because Ricardo regarded land owners as nonproductive forces of the national economy, he argued that this development would hamper economic development.

Ricardo's contemporary, the German Agronomist and Economist J.H. von Thünen (1783–1850), empirically studied and conceptualized optimal land use from an economic perspective. A land- and farm-owner himself, he established empirical relationships among market distance, agricultural production and land use. Inspired by his own farming operations, von Thünen elaborated on the linkages among different types of farming, such as the use of manure for the fertilization of cropland, as well as the costs of transport between markets located in towns and cities and their hinterlands. His most famous contribution was the spatial analysis of land use, the so-called 'von Thünen rings', which still represent standard knowledge in spatial economics and economic geography. This notion holds that for different cultivars, transport costs and (economic) yields differ. In all cases, however, transport costs increase with distance to markets, and locational rents (and land values) decrease. Certain forms of land use are thus restricted to 'rings' of profitability, competitiveness and land-use intensity around markets where the associated products can be sold (von Thünen 1826; see also Nelson 2002). Another seminal principle of von Thünen holds that agricultural intensification is only economically rational when the increases in the input costs are smaller than the additional profits from the increased production, an operationalization of the law of diminishing returns.

It took a century before research on land use and its drivers and impacts gained impetus again: in 1965, Ester Boserup challenged the Malthusian perspective, the prevailing view at that time, with an innovative conceptualization of the nature of agricultural innovation (Boserup 1965). According to Boserup's notion, the amount of agricultural production is not determined by available technology, but by population pressures, and new technologies are only adopted with population growth because they increase production at the expense of labor productivity. This insight is shared by the Russian agronomist Chayanov (1986; see also Turner and Ali 1996) and the American anthropologist Netting (1993). A central aspect of Boserup's notion is the observation that increasing land-use intensity allows agricultural production to increase in proportion with population growth. However, it results in a deterioration of labor productivity (i.e., working hours per person), a reduction in food quality and alterations of ecosystem properties, thus promoting further technological innovation. Boserup's ideas became foundational in studying the relations between agricultural expansion and population growth, and they inspired research on the nexus between land-use change and population growth as well as practical applications. (Brookfield 2001; Geertz 1963; Lambin et al. 2000, 2001; Turner and Doolittle 1978; Turner et al. 1977). Interdisciplinary research fields such as Cultural Ecology, Ecological Anthropology, Political Ecology, Ecological Economics and Land-System Science still draw from Boserup's perspectives on changes in population, technology and resource use (Erb et al. 2014; Turner and Fischer-Kowalski 2010).

A short time before, in 1955, W.L. Thomas—explicitly referencing Marsh—initiated a conference called *Man's Role in Changing the Face of the Earth* (Thomas 1956). This symposium was dedicated to the systematic analysis of the many ways people had affected the environment in a broad sense (see Simmons 1989). Whereas Marsh had mostly focused on deforestation, soil erosion and desertification, the symposium extended this approach and scrutinized the environmental effects of urbanization, waste disposal, industrialization and human impacts on the atmosphere, often adopting a historical perspective. However, despite the comprehensiveness of the associated publication, one omission is remarkable in retrospect: there is hardly any indication of the fact that humankind would indeed be capable of altering the biosphere's functioning as a whole.

This changed in the 1990s with the incipient availability of remote-sensing-derived land-cover data that illustrated the global extent of land-cover changes. In this period, the focus on intensity as a central aspect of land use lost momentum in land-use research. Analysis of the impacts of land use on natural processes gained importance in the 1970s, with a focus on the effects of land-surface processes on the global climate system, such as changes in albedo (Lambin et al. 2006). Later, this perception was underscored by the observation that land use considerably alters biogeochemical cycles, particularly stocks and flows of carbon in vegetation and soils (Houghton et al. 1983; Lambin et al. 2006). These new research directions, together with the novel datasets, brought land-cover changes into focus, put Natural Science approaches to the forefront of land-use research, and allowed Land-Use Science to move toward a central focus on Sustainability Science (Clark and Dickson 2003; Kates et al. 2001; Rindfuss et al. 2004).

Subsequently, Land-Use Science pursued an ample array of research topics on the drivers, determinants and impacts of land-use activities on terrestrial ecosystems across spatial and temporal scales (Lambin et al. 2006; Turner et al. 2007). Nevertheless, a Natural Science focus on the impacts of land use on the Earth system continued to dominate, reinforced by the methodological necessities associated with the handling and analysis of the powerful and increasingly available land-cover data from remote sensing. This diverted attention from intensity aspects of land use because most changes associated with intensification are unrelated to changes in land cover, and many aspects of land-use intensification are undetectable by remote sensing (Erb 2012; Erb et al. 2007; Verburg et al. 2011).

In recent years, however, intensification has moved again to the center of interest of a broader scientific community for three main reasons: first, the far-reaching, potentially detrimental ecological consequences of intensification (Matson et al. 1997); second, the emerging demand for land products such as bioenergy that will have to be covered within the confines of photosynthesis (Haberl et al. 2010; Smith et al. 2013); and third, the systemic interrelation between intensification and land expansion, which fuels the so-called land-sparing vs. land-sharing debate (Balmford et al. 2005; Grau et al. 2013).

In Sect. 4.3, drawing from debates outlined in Erb (2012), we briefly present intricacies related to the analysis of land-use intensity. We will discuss three decisive barriers that prevent the development of sound metrics for land-use intensity in mainstream Land-Use Science. This will represent the starting point for discussing the offerings of the socioecological method inventory for furthering analytical capacities as well improving the currently limited understanding of the processes, drivers, constraints and impacts of land-use intensity changes.

4.3 Intricacies of the Analysis of Land-Use Intensity

Three major characteristics of the current mainstream in land-use research contribute to the partial neglect of and ambiguities related to the analysis of land-use intensity. Overcoming these barriers, which are not isolated but mutually interdependent, is a central mandate of Land-System Science (Rounsevell et al. 2012):

1. **A prevalence of nominal scales**: Methodologically, the majority of land-use and land-cover analyses as well as datasets are based on classification systems that assign a discrete, homogenous land-use type (class) to each grid cell or polygon; they are based on nominal scales (Stevens 1946; see Method Précis on Geographic Information Systems). This methodological orientation is extremely powerful as it allows an intuitive and straightforward analysis of land changes, which are easily depicted as transitions from one land-cover class to another. Consequently, land changes can be quantified, mapped and traced through space and time. However, discrete classification systems suffer from difficulties of allocation, as in the case of fine-scale landscape mosaics, multiple uses (e.g., grazing on cropland, agroforestry) and gradients of human activity (e.g., grazing intensity). In mainstream Land-Use Science, such

problems are often addressed by increasing the spatial resolution of sensors and, hence, increasing the resolution of maps. But gains in resolution are cost intensive in terms of data collection, storage and handling, and they can result in a loss of spatial context and divert the focus from functional interdependencies among, for example, cropland, pastures and livestock (Erb et al. 2007; Verburg et al. 2009). Within the context of land-cover datasets, the disadvantages of discrete classification systems have been identified, leading to the development of continuous-field land-cover data such as 'tree cover' and 'herbaceous cover' maps (e.g., DeFries et al. 1995; Hansen et al. 2003). Likewise, for the analysis of functional processes such as the intensification of management intensity, rational scales, such as those that measure the magnitude of continuous quantities (e.g., flow of energy or materials, work input), would be much more suitable.

2. **A focus on land-cover change, particularly on transitions between cropland and forests**: The advances of satellite-based remote sensing resulted in the widespread availability of wall-to-wall fine-scale land-cover datasets. This moved the focus away from changes in intensity and management toward analyses of changes of land cover, namely, the biophysical properties of the Earth surface. Within the prevailing paradigm of analyzing impacts of land use on terrestrial ecosystems, particularly on the global carbon cycle within the context of climate change, the focus was directed to the analysis of croplands and forests. Other land uses for which the congruence between land cover and land use is less obvious and stringent experienced relative neglect (see Chap. 13). It should be noted at this stage of the discussion that, as outlined above, the technological advancement in land-cover data was instrumental in establishing Land-Use Science as a vibrant scientific discipline of its own. Furthermore, some aspects of the shortcoming of the almost exclusive focus on land-cover information have already been identified and addressed in Land-Use Science, and at least partial solutions have been proposed. For instance, the dominance of land-cover data is counterbalanced by efforts to reconcile land-cover data with socioeconomic information, such as agricultural census statistics. This led to the generation of reconciled land-use maps (Hurtt et al. 2011; Klein Goldewijk et al. 2007; Monfreda et al. 2008; Ramankutty et al. 2008). However, the restriction to agricultural land use is only seldom overcome in these approaches (see Erb et al. 2007).
3. **A predominance of Natural Science approaches**: Because the intensity with which land is used is greatly influenced by socioeconomic processes, options and capabilities, Natural Science-based approaches are insufficient for conceptualizing and quantifying land-use intensity and intensification processes. Although many Social Science-based studies on the processes and trajectories of land-use intensification exist, such data and analyses are scarce at the global scale (Liverman and Cuesta 2008). This scarcity, of course, is strongly linked to the abovementioned dominance of land-cover data. The need to counteract the underrepresentation of Social Sciences in land-use research was addressed by researchers aiming at 'socializing the pixel' (Geoghegan et al.

1998; Liverman et al. 1998). Another research strand aims to create typologies of human-environment systems, such as agricultural production systems or 'anthromes' (Ellis and Ramankutty 2008; Kruska et al. 2003), not just land cover. Although such approaches move the integrated nature of land systems into focus (Alessa and Chapin 2008; Verburg et al. 2009), they still suffer from problems resulting from the use of nominal scales, which cannot capture continuous gradients of land-use intensity.

Before discussing the ways in which the socioecological concept and method inventory can help to overcome these barriers, we explore the definition of land-use intensity in greater detail. In Sect. 4.4, we briefly summarize different approaches and indicators for measuring and analyzing land-use intensity. We refer the reader to other publications in which we explore in greater detail the dimensions, metrics and existing data of land-use intensity (Erb et al. 2013a; Kuemmerle et al. 2013).

4.4 Definitions of Land-Use Intensity

No generally accepted, comprehensive and systematic definition of land-use intensity or land intensification exists (Erb 2012; Erb et al. 2013a; Kuemmerle et al. 2013). In much of the scientific literature, a casual use of the term 'intensification' prevails, but precise definitions are scarce. Often, the term 'intensification' is used to denote an amalgam of complex changes related to agricultural industrialization processes (such as those during the 'Green Revolution') or to denote unspecific but detrimental socioecological effects of land use. Many definitions found in the literature relate to changes in intensity rather than to land-use intensity itself, such as intensification or disintensification.[2] Inspired by the classics in Economics, two major elements of intensity and intensification are commonly discerned in the literature (Lambin et al. 2000; Shriar 2000): output intensification and input intensification.

Output intensification is regarded by many authors as the most significant aspect of land-use intensification (Hunt 2000; Shriar 2000; Turner and Doolittle 1978). It denotes increases in (agricultural or forestry) production per unit area and time, such as tons of cereals per hectare per year. The central 'currency' of output intensity is thus the yield per unit area (Lambin et al. 2000). Although yield

[2]The term 'disintensification' was proposed by Brookfield (1972) to avoid the ambiguous intensification-extensification dichotomy. An alternative term for the reduction in land-use intensity is 'de-intensification', proposed by Boserup (1981). The meaning of extensification, however, is ambiguous, indicating both a reduction in intensity and the expansion of, e.g., cropland. The origin of this ambiguity may lie in the fact that an increase in agricultural production can be achieved via expansion of cropland, i.e., larger areas with the same or lower levels of yields per unit area, in contrast to intensification.

intuitively represents a straightforward indicator, there are many intricacies to its measurement. First, it is not straightforward to determine the unit in which output should be measured. Measurements in mass fresh weight, dry weight, energy content or nutritional value can give different results. Second, there are ambiguities related to the delineation of the unit of land. In some agricultural systems, a certain fraction of the land is left idle to recover fertility; however, the same unit of land can be cropped several times per year (multicropping). Fallow lands and multicropping complicate the calculation of output per unit area per cultivar as it is not possible to allocate fallow lands to individual cultivars (Erb et al. 2013a; Siebert et al. 2010). Third, because yields in agriculture and forestry are cultivar-specific and show large variation due to differences in climate, soil conditions, and management, the simple use of yield as an output intensity measure does not allow for comparative analyses across cultivars or even land-use types.

Input intensification relies upon the quantification of input variables per land unit per year, such as fertilizers, pesticides, employed labor and mechanical energy. This notion of input intensity is closely related to the definition of land-use intensification by Brookfield (2001), namely, the substitution of inputs of labor, skills and capital to land, most commonly in relation to increased agricultural output. Some authors go so far as to disregard the output component and relate land-use intensification to inputs alone (Brookfield and Hart 1971; Turner and Doolittle 1978). There are also many intricacies in the development of metrics for input intensity as most inputs are measured or measurable in different units, which hampers the practicability of input intensity measures. Consequently, authors often use single production factors as surrogate indicators for input intensification, such as the amount of nitrogen fertilizer or pesticides applied and their changes over time (see, e.g., Herzog et al. 2006; Temme and Verburg 2010). Although studies of this type capture central aspects of intensification, they cannot account for substitution effects, such as changes from mineral fertilizers to manure-based systems or changes to resource-sparing intensive high-tech applications such as precision farming.

Particular attention is paid to the relation of inputs and outputs in Economics, Agronomy, Systems Ecology and Anthropology. Although the focus of Economics is mostly on monetary relations, also biophysical aspects of this interrelation are studied, for example, in Agronomy. Here, the study of the relation between nutrient inputs and yields, specifically, outputs in terms of used biomass, is a central research topic. In Systems Ecology, the relation between energy inputs and outputs, denoted as energy return on investment (EROI; see below), has drawn attention. The relation between labor inputs and biomass outputs, an aspect of labor efficiency, is a common subject of research in Anthropology. However, in many Land-Use Science approaches, input and output intensification are often studied independently (Lambin et al. 2000; Netting 1993). Apparently, a tacit assumption (related to the abovementioned casual use of the term 'intensification') holds that increases in inputs would result in increased outputs. Empirical analyses, however, that would corroborate (or contradict) this assumption are rare (Netting 1993; Shriar 2000). The observation of 'agricultural involution' by Geertz (1963), i.e. the increase of inputs without corresponding increases in outputs to produce sufficient agricultural goods, suggests that more systematic analyses are highly desirable in this context. In this respect, Land-Use Science could benefit from the experiences of neighboring and underlying disciplines such as Anthropology.

A different way of operationalizing land-use intensity is to analyze system-level changes in ecosystems induced by land use. This approach is prevalent in Natural Science-oriented research disciplines that draw on the conception of land use as disturbance of ecological processes, closely related to the ecological concept of disturbance (Grime 1977; Pickett and White 1986; White 1979). Biophysical approaches are, for instance, methods that quantify the differences between the actual and potential (i.e., undisturbed, without human interference) state of the ecosystem, and they define the distance between the two states as a metric for intensity (Luyssaert et al. 2011). Global land-use datasets following this line of thought have been developed by scholars who use, for example, remote sensing to generate spatial inventories of the density of human artifacts on the Earth's surface (McCloskey and Spalding 1989; Sanderson et al. 2002). Another example is the 'inventory of human disturbance' by Hannah et al. (1994). This operationalization of land-use intensity does not aim to understand functional interrelations in the land system but generates inventories of the density of human artifacts, using them as a proxy for human activity.

This strand of thought gained recognition in Land-Use Science in the last few decades with the emergence of the concept of 'ecosystem services'. This concept, building upon notions of human-nature interactions formulated by Plato, gained importance in the second half of the 20th century in the course of attempts to estimate the economic value of 'natural capital' (for a review of the history of this concept, see Daily et al. 2009). In 2005, the Millennium Ecosystem Assessment (MEA; Millennium Ecosystem Assessment 2005) brought the concept to the attention of a broader audience and to the arena of political decision-making. The ecosystem service concept holds that ecosystems provide many essential services to society, many more than the provisioning services (e.g., the provision of biomass) that are part of the economic system and are marketed. These other services are usually grouped into regulating services (e.g., flood control, climate regulation, water purification) and cultural services (such as the aesthetic, spiritual and recreational functions ecosystems fulfill for society) and a group of services underlying provisioning, regulating and cultural services, the so-called supporting services. These supporting services encompass processes such as nutrient cycling, soil processes, primary production and biodiversity. Ecosystem service research is often aimed at valuing the importance of these services for society (see, e.g., Costanza et al. 1997, who estimated the economic value of ecosystem services at a minimum of approximately twice the global gross national product).

One central aspect of this concept that is directly related to land-use intensity research is that the maximization of the provision of certain ecosystem services (e.g., agricultural produce) has effects on many other services the ecosystem provides because they are connected via the supporting services. That human activities are decisive for the quantity and quality of ecosystem services delivered by

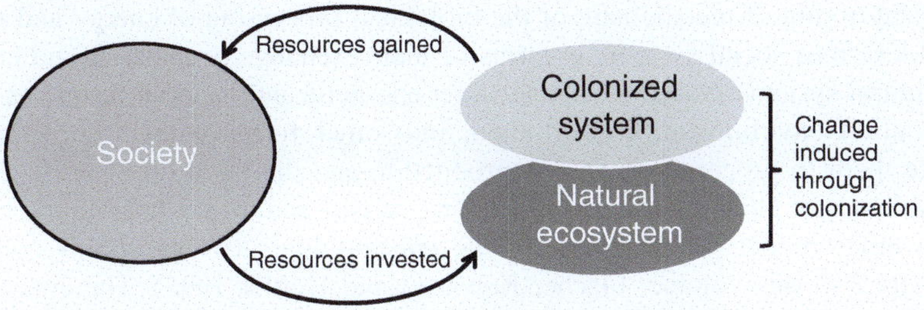

Fig. 4.1 The three dimensions of land-use intensity. Land-use intensity can be operationalized via inputs, outputs, or the system-level changes caused in ecosystems

ecosystems, obvious as it may seem from a socioecological perspective, has been recognized only very recently by the scientific community concerned with ecosystem services (e.g., Spangenberg et al. 2014).

All three dimensions of land-use intensity—inputs, outputs and system-level changes in ecosystems—need to be integrated to adequately address land-use intensity and its impact on society (Fig. 4.1; Erb et al. 2013a; Kuemmerle et al. 2013). Land-based production is the conversion of a combination of inputs into outputs dependent on the properties of the system, such as net primary production (NPP), standing crops and biodiversity, to name but a few. Ecosystem properties are altered in the course of land-based production, with far-reaching consequences for the dynamics of social systems and ecosystems.

4.5 The Contributions of Social Ecology to Land-Use Intensity Research

The theoretical and methodological approach of Social Ecology is useful to overcome some of the barriers to the analysis of land-use intensification, its underlying processes and its socioecological impacts. The material and energy flow analysis framework (MEFA framework; Haberl et al. 2004; Krausmann et al. 2004) denotes an interlinked set of methods to analyze society-nature interaction that extends the notion of socioeconomic metabolism (see Chaps. 3 and 8–12) by accounts that create an integrated picture of socioecological stocks and flows. In other words, it systematically combines socioeconomic material and energy flows with information on the associated changes in relevant patterns and processes in ecosystems and thus allows for the integration of Natural Science and Social Science perspectives. The MEFA framework has proven its utility for guiding the collection and analysis of biophysical information on socioeconomic activities (Haberl et al. 2001; Krausmann et al. 2004; see Chaps. 8–12 in this volume).

A particular contribution to the analysis of land-use intensity comes from the basic concept of socioeconomic metabolism itself. This concept—analogous to the biological notion of metabolism, or the biological processing of energy and materials—describes social systems in terms of their exchange of material and energy with natural systems. Social systems extract energetic and material resources from their natural environment, convert them into other forms within the economy, allocate them to processes such as reproduction and the build-up of artifacts and release them in altered form (as emissions, waste and waste heat) back to the natural environment (for a review of the epistemological roots of this concept, see Chap. 2 in this volume; Fischer-Kowalski and Hüttler 1998). The concept of socioeconomic metabolism and the need to focus on physical exchange processes between society and its environment follows from the conceptualization of society as a hybrid of the cultural system (recursive communication) and biophysical structures (such as the human population, artifacts and livestock). These biophysical structures serve as structural elements coupling natural and cultural processes. Under this notion, sustainability can be understood as a characteristic of the interactions between society and nature (Haberl et al. 2004). Based on this conception, material and energy exchanges between social and natural systems become a vital element of the observation and analysis of sustainable development.

The strength of the metabolism concept is that it introduces an unambiguous and meaningful system boundary between (the material realm of) a social system and the natural system with which it interacts in terms of energy and material flows (following thermodynamic principles, including the laws of conservation of mass and energy), and it can be consistently related to economic accounts. This delineation of social and natural systems via the biophysical component of society (see Chap. 2) allows for a systematic, consistent and comprehensive analysis of land-use intensity across different land uses. The focus of this approach is on the material structures and processes of society. When looking at processes related to land use, the biophysical structures of society include the human population, livestock and artifacts such as synthetic fertilizers or pesticides and machines used in agriculture. The focus on biophysical processes that occur at the interface between society and nature allows for the extension of so-called 'satellite accounts' to the economic system of national accounts. These 'satellite' accounts of environmental or social data provide additional information on gross domestic product (GDP) accounts. This focus thus allows the environmental dimension of socioeconomic activities to be explored. This is an important contribution of Social Ecology to Land-System Science as many dynamics and interactions in land systems cannot be captured with mere economic accounts. The socioecological metabolism approach allows the operationalization of biophysical constraints, minimum nutritional levels, overconsumption and their effects on human health in the land system in a manner consistent with and complementary to economic perspectives.

The biophysical perspective, however, is not a unique feature of the MEFA approach in Land-System Science. Two other prominent approaches that focus on sociometabolic processes and that are widespread in Land Science disciplines such as Agronomy and Ecological Economics should be discussed here: (a) the study of energy flows of farming systems and (b) life cycle analysis (LCA) of biomass products (see Method Précis on Life Cycle Assessment). These approaches study energy, material and substance flows and have the potential to complement and enrich each other. However, it is important to consider not only their similarities, namely, the study of energy and material flows, but also their differences, which are mainly related to a different system focus.

The introduction of a meaningful system boundary between the material components of social systems and ecosystems is a precondition for the systematic establishment of energy balances related to land-based production. Energy balances in agricultural systems have drawn attention since the 1970s, when researchers began to worry about the dependency of industrial agriculture on nonrenewable energy sources (Pimentel et al. 1973; Steinhart and Steinhart 1974). A large number of studies on the balance between energy inputs and outputs of systems related to the production of food have been published since then. Most of these studies focus on the crop production system (e.g., Bailey et al. 2003; Pimentel 2009) or the agricultural system of a specific region (e.g., Cleveland 1995) (see also Chap. 21 in this book, where the approach is applied at the level of regional land-use/production systems). In this type of accounting, energy inputs include all socioeconomic energy flows required to maintain the agricultural system, including the fossil fuels burned by agricultural machines and all energy needed to produce inputs such as fertilizers, pesticides and machines. Depending on the system boundary, it also includes the energy consumed by human metabolism during work. Some studies draw a wider system boundary, looking at the agrofood system of a certain region (e.g., Markussen and Østergård 2013; Pelletier et al. 2011). These studies include the energy inputs during both the agricultural production process and the processing, transport and storage of agricultural products. Other studies even include the biomass combustion in human-induced vegetation fires (see Chap. 15). When comparing different studies, it is important to bear in mind these differences concerning system boundaries.

In other words, the socioecological metabolism approach and its sound nature-society system boundary allow discerning different types of biophysical flows (e.g., used extraction, i.e., biomass that is harvested and used in socioeconomic processes, versus unused extraction, biomass that is not harvested but killed by land-use activities, such as felling losses). This socioeconomic differentiation of flows provides the basis for the development, quantification and assessment of land-use efficiency indicators. Tracing such indicators over time yields surprising insights into the processes underlying land-use change. For example, a comparison of energy input reveals that in contrast to the transition from a hunter-gatherer to an agrarian mode of subsistence (see Boserup 1965; Chayanov 1986), the transition from an agrarian to an industrial mode of subsistence is associated with drastic increases in labor productivity at the expense of energy efficiency. Although the amount of labor per unit of output decreases drastically with mechanization, the EROI of agriculture, that is, the amount of energy output divided by energy input, strongly declines during this transition, sometimes even below 1 (i.e., inputs larger than outputs; Krausmann et al. 2003). This would be impossible for an agrarian society which, with regard to its energy metabolism, depends almost exclusively on solar energy and thus on a positive energy balance. The industrial society, in contrast, has the option to energetically subsidize agriculture as the energy limitations are much less virulent (Fischer-Kowalski and Haberl 2007; McNeill 2001; Sieferle et al. 2006). It should be noted, however, that the choice of the system boundary is decisive for such interpretations. If, for example, the energy flow of human-induced vegetation fires is counted as socioeconomic energy flow, this conclusion is not valid, in which case preindustrial societies can also be characterized by negative EROI ratios (see Chaps. 15 and 22).

Analogous to energy balances, material as well as substance balances (e.g., focusing on nutrients such as nitrogen) are established at the farm level as a tool of operational management of agricultural enterprises and for the development of environmental performance indicators, the so-called 'farmgate balances'[3] (Van Beek et al. 2003). A particularity of these types of accounting should be noted: farm-gate balances usually account for the same flows as metabolism approaches, but the flows are listed in the opposite direction. Harvest, for instance, is an input from a metabolism perspective but an output in farm gate statistics; in contrast, fertilizers are socioeconomic outputs but are counted as inputs.

The focus of LCAs is typically on a certain product, such as lettuce (Hospido et al. 2009), apples (Blanke and Burdick 2005) or wheat (Meisterling et al. 2009), or the functions of this product (the so-called functional unit, see Method Précis on Life Cycle Assessment). LCAs include all inputs, such as energy and material, of an entire production chain, including production, transport, storage and disposal of the product. Many LCAs focus on impacts such as eutrophication or the climate effects of CO_2 produced during the life cycle of a certain product rather than only energy inputs and outputs. The main reason for this difference is the purpose of LCAs: they are intended to guide consumers in their choices between products with different environmental impacts. Overall, the sociometabolic approach has a closer relation to the energy balances discussed above than to LCAs due to its focus on a socioecological system instead of a specific product.

The methodological operationalization of the metabolism concept, the material flow analysis (MFA, see Method Précis on Material Flow Analysis; Adriaanse et al. 1997; Fischer-Kowalski et al. 2011), only focuses on socioeconomic resource flows. For example, domestic extraction includes only biomass that enters socioeconomic processing, that is, generally the primary products (e.g., cereal grains) and the fraction of the secondary products that are harvested (e.g., harvested straw used for bedding or as fodder). It does not include the fraction of biomass that is not harvested, such as the straw remaining on the fields (Krausmann et al. 2008). These flows, however, are of central importance for ecological processes, such as soil fertility (Blanco-Canqui and Lal 2009). Another indicator of the MFA framework, 'total material requirement', includes such 'hidden' flows, but such assessments within the MFA framework are regarded as much less rigid. Nevertheless, socioeconomic material or energy flows can be integrated consistently with ecological flows, as has been shown for biomass flows (Haberl et al. 2009).

---

[3]In German, 'Hoftor Bilanz'.

Land use not only includes withdrawals of biomass from ecosystems but also changes many ecosystem properties, such as its productivity. Thus, a sole focus on accounting flows of energy and material between society and nature is insufficient to capture the multidimensional nature of land use. The socioecological concept 'colonization of natural processes' takes this multidimensionality into account (see Chap. 2; Fischer-Kowalski and Haberl 2007; Fischer-Kowalski et al. 1997; Haberl et al. 2004) by studying system-property changes purposefully induced by human activities in natural systems, such as ecosystems. The sum of purposeful interventions that aim to bring natural systems into a desired state and to maintain that state once it is achieved are denoted as colonizing activities and include, for instance, agricultural activities such as weeding, mowing, planting and ploughing (see Chaps. 2 and 13–15). Thus, this concept is closely related to the system-level perspective of land-use intensity discussed above.

Land use is the example *par excellence* of the colonization of natural processes. The deliberate alteration of ecosystem properties to increase the usefulness of ecosystems is a key aspect of land-use and ecosystem management. Prominent examples of colonizing activities include the intentional selection of certain species and the exclusion of others (competitors or predators), the breeding of species or the purposeful change of biophysical constraints via irrigation or fertilization. The concept of colonization allows the establishment of systematic and consistent links between land-use-induced alterations in ecosystems and socioeconomic activities as well as the energy and material flows associated with these activities. This renders the socioecological method inventory highly suitable for developing 'pressure indicators', environmental indicators for developments in the release of substances, physical and biological agents or the use of resources. Pressure indicators are extremely valuable as they allow driving forces to be linked with alterations in the state of ecosystems (for which the lion's share of environmental indicators describing the quantity and quality of an ecosystem have been proposed) and, consequently, with the socioecological impacts of these alterations (Stanners et al. 2007).

A key aspect of colonizing interventions is that once the system properties have been altered in a (desired) direction, a continuous socioeconomic effort has to be organized to keep the natural systems in this state. Otherwise, the natural system would develop tendencies to reestablish the original states. An example is the climax state in ecosystems, where, for instance, without ploughing, weeding and harvesting, a cropland will run through successional states and will eventually reestablish the climax vegetation, such as a forest or a natural grassland, if no irreversible alterations of the ecological conditions occur.

In addition to deliberate, purposeful interventions in ecosystems, changes in land-use intensity are accompanied by unintended and sometimes undesired outcomes, as the ecosystem service concept vividly illustrates. Changes in species composition due to fertilizer application, nitrogen leaching, the toxic effects of pesticides, changes in the carbon content in soil and biota and the reduction of biodiversity (see Chap. 18) are prominent examples of such unintended outcomes.

Although these modifications of system properties are not the focus of the colonization concept sensu stricto, which focuses on deliberate alterations and societal investments to keep the system in a desired state, they can be included as long as information on socioeconomic activities can be causally linked to the alterations of ecological patterns and processes.

The establishment of sound indicators for colonizing interventions appears intricate. One approach could be to develop metrics for the socioeconomic efforts (labor, energy, technology) required to keep ecosystems in the desired state. Another approach—closely related to the concept of land-use-induced system-level changes discussed above—is to estimate the distance between actual ecosystem parameters and those one would expect to find in the absence of human activities, that is, those of the potential vegetation (Tüxen 1956). A prominent land-use indicator of this type is the human appropriation of net primary production, which will be presented in the following section (see Method Précis on Human Appropriation of Net Primary Production).

4.5.1 The Human Appropriation of Net Primary Production

The accounting framework 'human appropriation of net primary production', or HANPP, was proposed by Vitousek et al. in 1986 as a metric for the scale of human activities in relation to biosphere processes and has its roots in research on the relation between humans and primary productivity (Lieth 1973; Lieth and Whittaker 1975; Whittaker and Likens 1973). Primary productivity is the amount of energy fixed by autotrophic organisms, mainly plants, in a given period. Gross primary production denotes the total amount of energy fixed through photosynthesis. Net primary production (NPP) is the energy accessible to other organisms after the subtraction of the plant's own respiratory energy demands. NPP is the basis of heterotrophic food chains and carbon accumulation in the biosphere (i.e., build-up of carbon stocks in soil and biota).

Currently, the most widespread definition of HANPP (Erb et al. 2009a; Haberl et al. 2001, 2007, 2014; Krausmann et al. 2013) was inspired by the work of Wright (1990) on the relationship between NPP and biodiversity (see Chap. 18). In this definition, HANPP comprises biomass harvest and the productivity losses (or gains) resulting from land-use and land-cover change (see Method Précis on Human Appropriation of Net Primary Production).

HANPP comprehensively measures changes in trophic energy flows in ecosystems resulting from two processes: (a) human-induced changes in productivity due to land conversion, including soil degradation (Zika and Erb 2009), denoted $HANPP_{luc}$, and (b) the biomass withdrawn from ecosystems during harvest, denoted $HANPP_{harv}$. The latter measure also includes by-flows, such as the destruction of biomass during harvest or human-induced fires (Lauk and Erb 2009; see also Chap. 15 in this volume).

HANPP allows the monitoring of changes in the conditions of land systems from both socioeconomic and ecological perspectives (Krausmann et al. 2009). From a socioeconomic perspective, HANPP can express the NPP upstream requirements of biomass products, thus providing an indication of the HANPP intensity of biomass products processed and consumed. From an ecological perspective, HANPP denotes the difference between the potential energy flow in ecosystems and the energy flow that remains in the ecosystem after land-use-induced changes ($HANPP_{luc}$) and after harvest, that is, the amount of energy that is not monopolized by *homo sapiens* and is thus available to other biological species. These definitions are close to the ecological definitions. For example, $HANPP_{luc}$ is similar to the definition of stress, the alterations of the external (biophysical) constraints that determine productivity, and HANPP is close to the definition of disturbance, the withdrawal of biomass from an ecosystem (Grime 1977).

HANPP represents a valuable indicator of land-use intensity, integrating two dimensions. First, it operationalizes output intensity by assessing the amount of harvest ($HANPP_{harv}$). Second, it assesses changes at the system level, by comparing potential and actual energy flows in ecosystems. By using the unambiguous measure of reference NPP_{pot}, HANPP allows to overcome one above-discussed shortcoming of output metrics, the site-specific constraints on productivity, because NPP_{pot} only depends on natural determinants.

By comparing actual productivity levels with the levels given by a potential reference line (NPP_{pot}), HANPP fits well within the tradition of yield gap analysis, a prominent strand of Land-Use Science (Licker et al. 2010; Lobell et al. 2009; Neumann et al. 2010). Yield gap analyses aim to measure the distance between actual crop yields and potential reference yields to assess the capacity for future yield increases by means of management improvements. Usually, these studies define potential yields as yields that are achieved under optimum management conditions, such as optimum fertilization or irrigation (Mueller et al. 2012). The main difference between yield gap analysis and HANPP is the construction of the reference metric. The HANPP framework is related to NPP_{pot}, which is exclusively determined by natural parameters (e.g., precipitation, temperature, soil). Traditional yield gap approaches instead construct the reference line by integrating natural parameters with optimum management; this reference value is thus technology dependent. Thus, any technological advancement would imply increasing potential yields, which in turn automatically widen the yield gaps on non-adopting farms.

The combination of output intensity and system-level intensity renders HANPP a particularly well-suited analytical framework for land-system change (Fig. 4.2). It allows the calculation of integrated indicators of land-use efficiency, by combining outputs and system-level change metrics. One such integrated indicator is HANPP efficiency, the fraction of total HANPP that is associated with the production and consumption of biomass products (Erb et al. 2009b; Haberl et al. 2009; Niedertscheider et al. 2012), and its reciprocal value, the HANPP intensity

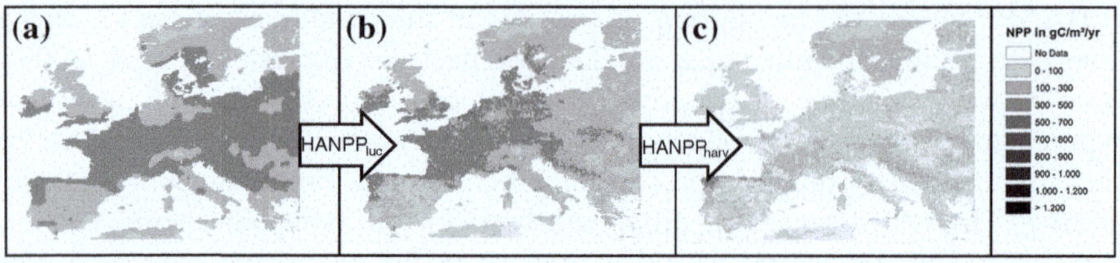

Fig. 4.2 HANPP as an indicator of land-use intensity: NPP flows in Europe 2000. **a** NPP_{pot}, **b** NPP_{act}, **c** NPP_{eco} and the respective 'distance metrics' between **a** and **b** $HANPP_{luc}$ (alteration of NPP) and between **b** and **c** $HANPP_{harv}$ (harvest of biomass), which in sum make up HANPP [Following the definition and using data described in Erb et al. (2009a), Haberl et al. (2007)]

of biomass products. Similar approaches are currently being developed for the analysis of global carbon flows, such as the carbon footprint concept (Hertwich and Peters 2009), which aims to quantify the upstream carbon emission related to the final consumption of products. Analogously, the water footprint concept aims to assess the upstream water demand related to agricultural goods, which yields insights into a central aspect of land use: its relation to global water cycles (Gerbens-Leenes et al. 2009; Hoekstra and Hung 2005).

4.5.2 *Beyond HANPP: Human-Induced Reduction of Carbon Stocks in Vegetation*

HANPP only monitors changes in ecological flows. It neither includes aspects of input intensification (e.g., use of nutrients, labor) nor provides information on ecosystem stocks, their changes over time and how they relate to socioeconomic processes. However, the HANPP framework can be expanded so that other aspects of land-use intensity can be depicted. For instance, the difference between actual and potential carbon stocks can be calculated (Erb 2004; Erb et al. 2008). Figure 4.3 shows an example of such an assessment for Europe. Following methodological approaches by West et al. (2010) and Gingrich et al. (2007), these maps assess the land-use-induced reductions in carbon stocks in Europe's vegetation for the year 2000. On average, land use resulted in a reduction of carbon stocks in vegetation by 60 % in both Eastern and Western Europe. In contrast, the actual $HANPP_{harv}$ is approximately 33 % of the potential NPP in Western Europe and 25 % in Eastern Europe (Haberl et al. 2007). HANPP is 40 and 52 % in the year 2000 in Western Europe and Eastern Europe, respectively. Thus, although output intensity is different in these two regions, the system-level changes are strikingly similar.

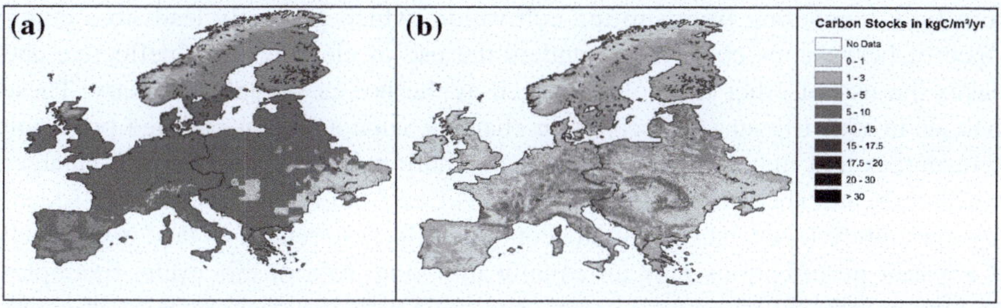

Fig. 4.3 Human alteration of carbon stocks in vegetation in Europe: **a** potential vegetation and **b** actual vegetation (Source: Own calculations based on biomass stock data from the International Biological Programme (IBP; Cannell 1982), the Intergovernmental Panel on Climate Change (IPCC; Eggleston et al. 2006), data on actual carbon stocks (FAO 2010) and the distribution of potential vegetation (Ramankutty and Foley 1999) and spatially explicit data on actual land-use patterns (Erb et al. 2007))

4.6 A Socioecological Approach to Land-Use Intensity

The socioecological method inventory, as well as its theoretical underpinnings, allows researchers to address the full cycle of land-use intensification and provides a basis for an integrated Land-System Science. It assesses inputs into ecosystems and the resulting changes in their functioning ('land-cover modification and change'), the outputs of products and the underlying cost-benefit relationships (Fig. 4.4). The unintended consequences of these interactions, associated with the system-level alterations discussed above, can also be assessed. Many unintended consequences appear on the side of the ecosystem, such as in the form of biodiversity loss or soil degradation (Lindenmayer et al. 2012; Verstraete et al. 2009).

An example from the seminal work by Boserup (1965) can illustrate these interlinkages. Shorter cropping cycles in swidden agriculture, driven by higher population numbers and increased food demand, prevent natural forest ecosystems

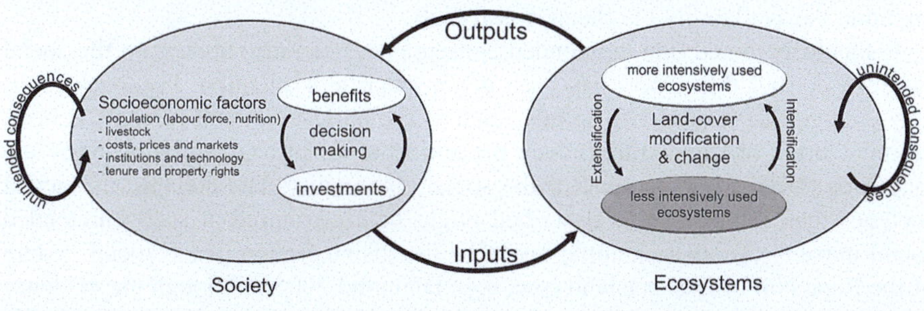

Fig. 4.4 A socioecological metabolism perspective on land-use change and land-use intensification

from fully recovering from shifting cultivation, which gradually leads to a dominance of herbaceous cover. This renders the use of clearing fires ineffective and makes the use of other technologies such as the hoe or plough necessary. These changes in turn trigger socioeconomic changes, such as the augmented provision of resources that may allow for a larger population, but they also trigger changes in social organization, necessitating the transition from hunter-gatherers to sedentary agriculture (see Chap. 3). Socioeconomic effects unrelated to land use as well as external perturbations play an equally important role in this cycle. Examples are human-induced increases in atmospheric CO_2 or the occurrence of droughts (Verstraete et al. 2009).

The strength of the metabolism concept for land-system research is that it allows the systematic analysis of the three dimensions of land-use intensity: (a) inputs, (b) outputs, (c) system-level ecological alterations, and (d) alterations in the social system as well as their dynamic interrelation. Table 4.1 provides examples of how a socioecological perspective allows the identification of system interrelations of higher orders (Erb 2012).

The 'forest transition' is a phenomenon that provides a prominent example of the power of the socioecological metabolism approach. It is the empirical observation that in many regions of the world, forests are currently regrowing after long periods of deforestation (Kauppi et al. 2006; Mather 1992; Meyfroidt and Lambin 2011; Meyfroidt et al. 2010; Rudel et al. 2005; see also Chap. 20 in this volume). Whereas a straightforward interpretation of this phenomenon would stress the 'improved' environmental performance at the regional scale, such as the increased carbon sink, the MEFA approach allows the assessment of the related feedback loops and underlying mechanisms, which include agricultural intensification (Erb et al. 2008) and the abandonment of practices such as forest grazing and litter raking (Erb et al. 2013b). It also allows the assessment of externalization effects due to trade (Kastner et al. 2011; Meyfroidt et al. 2010) and reduced domestic area demand, and it accounts for forest regrowth. The application of this socioecological, holistic approach reveals that the emerging carbon sink is not the result of an explicit land-use strategy but is part of a baseline development, intrinsically building upon the availability of (cheap) fossil fuels.

It should be noted that unintended consequences can also appear on the social system side, but this aspect has been given much less attention. Feedback loops may emerge that lead to risk spirals, such as the build-up of self-referential institutions, norms or rituals that reduce the ability to adapt to external changes and may even threaten the survival of the social community. The collapse of ancient societies that Tainter (1990) described might be an example of such unintended social impacts. Another example might be the disintegration of the social system of the Rapa Nui of Easter Island (see, e.g., Diamond 2005), although the evidence is inconclusive. The (further) development of analytical frameworks that are able to scientifically address the possibility of socioecological collapse remains high on the agenda of Sustainability Science (Costanza et al. 2007).

Table 4.1 Systemic insights provided by the socioecological metabolism approach, as illustrated by prominent land-use strategies

Land-use strategy	Intended benefit	Caveat introduced by a socioecological perspective
Land-use intensification	Provides more products (e.g., more food) and allows for land sparing, with overall beneficial effects for biodiversity, carbon sequestration, etc.	• Intensification can result in increased consumption due to increased resource availability (rebound effect), triggering further land-use intensification and expansion • Can be associated with reduced input-output ratios (decreased efficiency), heavily relies on fossil fuels, may result in a plethora of adverse ecological effects (soil compaction, nutrient leaching, toxic effects of pesticides, biodiversity loss, etc.)
Organic farming	Reduces resource use, particularly of nonrenewable resources, and reduces carbon emissions	• If not paired with reduced consumption, the increased area demand of organic farming can reverse the carbon-saving effect by triggering deforestation or reducing afforestation/regeneration and can increase the climate impact
Bioenergy supply from primary crops	Substitutes for fossil energy, reduces emissions	• Upper limits exist (confines of photosynthesis), even for technical potentials • Conflict with other land uses; land expansion/deforestation elsewhere • Increased harvest pressures reduce carbon stocks in perennial ecosystems—may result in large C emissions
Bioenergy production from crops or felling residues	Reduces environmental pressure per service unit, increases resource efficiency	• Using residues might have unintended consequences for, e.g., soil fertility or the carbon sink/storage capacities of ecosystems • Increasing the use-efficiency can reduce critical carbon flows in ecosystems
Reducing emissions from deforestation and forest degradation in developing countries (REDD)	Reduces carbon emissions, generates income in rural communities	• Land-use conflicts can result in considerable leakage and intensification/land expansion elsewhere • May decrease net income, self-sufficiency and food security in rural areas due to increased structural dependency on external markets

4.7 Conclusions

The formulation of land-use strategies aimed at harnessing beneficial aspects of land use for sustainability goals needs to be based on an improved systemic understanding of the underlying mechanisms and driving forces of land-use intensification processes (see Chap. 14). Land-use intensification may help reduce land-use competition through its ability to raise yields and hence reduce area demand. However, it may also create its own systemic feedbacks and adverse effects, including take-back or rebound effects (increased consumption) and adverse ecological effects of land-use intensification.

The need for interdisciplinary perspectives is growing, as are great sustainability challenges such as climate change, biodiversity conservation and food security. These challenges provide compelling cases for advancing our knowledge of land-use transitions and the related causal interlinkages between the drivers and the impacts of change. Land-use intensity is a vital but still under-researched process related to these challenges. It is directly linked to ecosystem change, agricultural and forestry production and human wellbeing.

The socioecological paradigm is well suited to contribute to the systemic advancement of our understanding of land use and its change over time, including land-use intensity. Its strength is that it is built upon first principles, such as the law of conservation of mass and energy, in compliance with stringent, meaningful system boundary definitions, and it is accessible to both Social and Natural Science approaches. It provides a theoretical basis to inform and guide data collection for the analysis of critical dimensions of land-use transitions and to advance systemic understanding of the many trade-offs related to land-use intensification and their feedback loops, which are at the very heart of ecosystem functioning and human well-being.

Acknowledgements Funding from the European Research Council for the Starting Independent Researcher Grant ERC-263522 'LUISE', from EU-FP7 265104 'VOLANTE' and from the Austrian Science Fund (FWF) project P20812-G11 is gratefully acknowledged. This book chapter contributes to the Global Land Project (http://www.globallandproject.org).

References

Adriaanse, A., Bringezu, S., Hammond, A., Moriguchi, Y., Rodenburg, E., Rogich, D., & Schütz, H. (1997). *Resource flows: The material basis of industrial economies*. Washington, D.C.: World Resources Institute.

Alessa, L., & Chapin, F. S, I. I. I. (2008). Anthropogenic biomes: A key contribution to earth-system science. *Trends in Ecology & Evolution, 23*, 529–531. doi:10.1016/j.tree.2008.07.002.

Alexandratos, N. (1999). World food and agriculture: Outlook for the medium and longer term. *Proceedings of the National Academy of Sciences, 96*, 5908–5914.

Alexandratos, N., Bruinsma, J. (2012). World agriculture: Towards 2030/2050. The 2012 revision. ESA Working paper No. 12–03. FAO, Rome.

Bailey, A. P., Basford, W. D., Penlington, N., Park, J. R., Keatinge, J. D. H., Rehman, T., et al. (2003). A comparison of energy use in conventional and integrated arable farming systems in the UK. *Agriculture, Ecosystems & Environment, 97*, 241–253.

Balmford, A., Green, R. E., & Scharlemann, J. P. W. (2005). Sparing land for nature: Exploring the potential impact of changes in agricultural yield on the area needed for crop production. *Global Change Biology, 11*, 1594–1605.

Blanco-Canqui, H., & Lal, R. (2009). Crop residue removal impacts on soil productivity and environmental quality. *Critical Reviews in Plant Sciences, 28*, 139–163.

Blanke, M., & Burdick, B. (2005). Food (miles) for thought—Energy balance for locally-grown versus imported apple fruit. *Environmental Science and Pollution Research, 12*, 125–127.

Boserup, E. (1965). *The conditions of agricultural growth: The economics of agrarian change under population pressure*. London: Earthscan.

Boserup, E. (1981). *Population and technological change: A study of long-term trends*. Chicago: University of Chicago Press.

Brookfield, H. C. (1972). Intensification and disintensification in Pacific agriculture: A theoretical approach. *Pacific Viewpoint, 13*, 30–48.

Brookfield, H. C. (2001). Intensification, and alternative approaches to agricultural change. *Asia Pacific Viewpoint, 42*, 181–192.

Brookfield, H. C., & Hart, D. (1971). *Melanesia: A geographical interpretation of an island world*. London: Methuen.

Burney, J. A., Davis, S. J., & Lobell, D. B. (2010). Greenhouse gas mitigation by agricultural intensification. *Proceedings of the National Academy of Sciences, 107*, 12052–12057.

Cannell, M. G. R. (1982). *World forest biomass and primary production data*. New York: Academic Press.

Cassman, K. G. (1999). Ecological intensification of cereal production systems: Yield potential, soil quality, and precision agriculture. *Proceedings of the National Academy of Sciences, 96*, 5952–5959.

Ceddia, M. G., Sedlacek, S., Bardsley, N. O., & Gomez-y-Paloma, S. (2013). Sustainable agricultural intensification or Jevons paradox? The role of public governance in tropical South America. *Global Environmental Change, 23*, 1052–1063.

Chayanov, A.V. (1986). The theory of peasant economy (1st edn). University of Wisconsin Press.

Clark, W. C., & Dickson, N. M. (2003). Sustainability science: The emerging research program. *Proceedings of the National Academy of Sciences, 100*, 8059–8061.

Cleveland, C. J. (1995). The direct and indirect use of fossil fuels and electricity in USA agriculture, 1910–1990. *Agriculture, Ecosystems & Environment, 55*, 111–121.

Coelho, S., Agbenyega, O., Agostini, A., Erb, K.-H., Haberl, H., Hoogwijk, M., et al. (2012). Land and water: Linkages to bioenergy. In T. Johannson, A. Patwardhan, N. Nakicenovic, & L. Gomez-Echeverri (Eds.), *Global energy assessment* (pp. 1459–1525). Cambridge: Cambridge University Press.

Costanza, R., d'Arge, R., De Groot, R., Farber, S., Grasso, M., Hannon, B., Limburg, K., Naeem, S., O'neill, R.V., Paruelo, J. (1997). The value of the world's ecosystem services and natural capital. *Nature 387*, 253–260.

Costanza, R., Graumlich, L., Steffen, W.W.L. (2007). Sustainability or collapse?: An integrated history and future of people on earth. Cambridge: The MIT Press.

Daily, G. C., Polasky, S., Goldstein, J., Kareiva, P. M., Mooney, H. A., Pejchar, L., et al. (2009). Ecosystem services in decision making: Time to deliver. *Frontiers in Ecology and the Environment, 7*, 21–28.

DeFries, R. S., Field, C. B., Fung, I., Justice, C. O., Los, S., Matson, P. A., et al. (1995). Mapping the land surface for global atmosphere-biosphere models: Toward continuous distributions of vegetation's functional properties. *Journal of Geophysical Research: Atmospheres, 100*, 20867–20882.

Diamond, J. (2005). Collapse: How societies choose to fail or succeed (revised edition). Harmondsworth: Penguin.

Eggleston, S., Buendia, L., Miwa, K., Ngara, T., Tanabe, K. (Eds.). (2006). 2006 IPCC guidelines for national greenhouse gas inventories. Volume 5: Waste. Hayama, Japan: Institute for Global Environmental Strategies (IGES).

Ellis, E. C., & Ramankutty, N. (2008). Putting people in the map: anthropogenic biomes of the world. *Frontiers in Ecology and the Environment, 6*, 439–447. doi:10.1890/070062.

Erb, K.-H. (2004). Land use–related changes in aboveground carbon stocks of Austria's terrestrial ecosystems. *Ecosystems, 7*, 563–572.

Erb, K.-H. (2012). How a socio-ecological metabolism approach can help to advance our understanding of changes in land-use intensity. *Ecological Economics, 76*, 8–14.

Erb, K.-H., Gaube, V., Krausmann, F., Plutzar, C., Bondeau, A., & Haberl, H. (2007). A comprehensive global 5 min resolution land-use data set for the year 2000 consistent with national census data. *Journal of Land Use Science, 2*, 191–224.

Erb, K.-H., Gingrich, S., Krausmann, F., & Haberl, H. (2008). Industrialization, fossil fuels, and the transformation of land use. *Journal of Industrial Ecology, 12*, 686–703.

Erb, K.-H., Haberl, H., Jepsen, M. R., Kuemmerle, T., Lindner, M., Müller, D., et al. (2013a). A conceptual framework for analysing and measuring land-use intensity. *Current Opinion in Environmental Sustainability, 5*, 464–470.

Erb, K.-H., Kastner, T., Luyssaert, S., Houghton, R. A., Kuemmerle, T., Olofsson, P., & Haberl, H. (2013b). Bias in the attribution of forest carbon sinks. *Nature Climate Change, 3*, 854–856.

Erb, K.-H., Krausmann, F., Gaube, V., Gingrich, S., Bondeau, A., Fischer-Kowalski, M., & Haberl, H. (2009a). Analyzing the global human appropriation of net primary production—Processes, trajectories, implications. An introduction. *Ecological Economics, 69*, 250–259.

Erb, K.-H., Krausmann, F., Lucht, W., & Haberl, H. (2009b). Embodied HANPP: Mapping the spatial disconnect between global biomass production and consumption. *Ecological Economics, 69*, 328–334.

Erb, K.-H., Niedertscheider, M., Dietrich, J. P., Schmitz, C., Verburg, P. H., Rudbeck Jepsen, M., & Haberl, H. (2014). Conceptual and empirical approaches to mapping and quantifying land-use intensity. In M. Fischer-Kowalski, A. Reenberg, A. Mayer, & A. Schaffartzik (Eds.), *Ester Boserup's legacy on sustainability: Orientations for contemporary research* (pp. 61–86). Dordrecht: Springer.

FAO. (2010). Forest resources assessment 2010. Main report. Rome: Food and Agriculture Organization of the United Nations.

FAOSTAT. (2014). Statistical Databases. http://faostat.fao.org [WWW Document]. URL http://faostat.fao.org.

Fischer-Kowalski, M., Haberl, H. (2007). Socioecological transitions and global change: Trajectories of social metabolism and land use. Cheltenham: Edward Elgar Publishing.

Fischer-Kowalski, M., Haberl, H., Hüttler, W., Payer, H., Schandl, H., Winiwarter, V., Zangerl-Weisz, H. (1997). Gesellschaftlicher Stoffwechsel und Kolonisierung von Natur: ein Versuch in Sozialer Ökologie. Fakultas.

Fischer-Kowalski, M., & Hüttler, W. (1998). Society's metabolism. *Journal of Industrial Ecology, 2*, 107–136.

Fischer-Kowalski, M., Krausmann, F., Giljum, S., Lutter, S., Mayer, A., Bringezu, S., et al. (2011). Methodology and indicators of economy-wide material flow accounting state of the art and reliability across sources. *Journal of Industrial Ecology, 15*, 855–876.

Foley, J. A., DeFries, R., Asner, G. P., Barford, C., Bonan, G., Carpenter, S. R., et al. (2005). Global consequences of land use. *Science, 309*, 570.

Foley, J.A., Ramankutty, N., Brauman, K.A., Cassidy, E.S., Gerber, J.S., Johnston, M., Mueller, N.D., O'Connell, C., Ray, D.K., West, P.C., Balzer, C., Bennett, E.M., Carpenter, S.R., Hill, J., Monfreda, C., Polasky, S., Rockstrom, J., Sheehan, J., Siebert, S., Tilman, D., Zaks, D.P.M. (2011). Solutions for a cultivated planet. *Nature 478*, 337–342.

Geertz, C. (1963). Agricultural involution: The process of ecological change in Indonesia. University of California Press.

Geoghegan, J., Pritchard, L. J., Ogneva-Himmelberger, Y., Chowdhury, R. R., Sanderson, S., & Turner, B. L. (1998). "Socializing the pixel" and "pixelizing the social" in land-use and land-cover change. In D. M. Liverman, E. Moran, R. R. Rindfuss, & P. Stern (Eds.), *People and pixels. Linking remote sensing and social science* (pp. 51–69). Washington, D.C: National Academy Press.

Gerbens-Leenes, W., Hoekstra, A. Y., & van der Meer, T. H. (2009). The water footprint of bioenergy. *Proceedings of the National Academy of Sciences, 106*, 10219–10223.

Gingrich, S., Erb, K.-H., Krausmann, F., Gaube, V., & Haberl, H. (2007). Long-term dynamics of terrestrial carbon stocks in Austria: A comprehensive assessment of the time period from 1830 to 2000. *Regional Environmental Change, 7*, 37–47.

Goudie, A. (2006). *The human impact on the natural environment* (6th ed.). Oxford, Cambridge: Blackwell.

Grau, R., Kuemmerle, T., & Macchi, L. (2013). Beyond "land sparing versus land sharing": Environmental heterogeneity, globalization and the balance between agricultural production and nature conservation. *Current Opinion in Environmental Sustainability, 5*, 477–483.

Grime, J. P. (1977). Evidence for the existence of three primary strategies in plants and its relevance to ecological and evolutionary theory. *American Naturalist, 111*, 1169–1194. doi:10.1086/283244.

Gruber, N., & Galloway, J. N. (2008). An earth-system perspective of the global nitrogen cycle. *Nature, 451*, 293–296.

Haberl, H., Beringer, T., Bhattacharya, S. C., Erb, K.-H., & Hoogwijk, M. (2010). The global technical potential of bio-energy in 2050 considering sustainability constraints. *Current Opinion in Environmental Sustainability, 2*, 394–403.

Haberl, H., Erb, K.-H., & Krausmann, F. (2014). Human appropriation of net primary production: patterns, trends, and planetary boundaries. *Annual Review of Environment and Resources, 39*, 363–391. doi:10.1146/annurev-environ-121912-094620.

Haberl, H., Erb, K.-H., Krausmann, F., Berecz, S., Ludwiczek, N., Martínez-Alier, J., et al. (2009). Using embodied HANPP to analyze teleconnections in the global land system: Conceptual considerations. *Geografisk Tidsskrift-Danish Journal of Geography, 109*, 119–130.

Haberl, H., Erb, K.-H., Krausmann, F., Gaube, V., Bondeau, A., Plutzar, C., et al. (2007). Quantifying and mapping the human appropriation of net primary production in earth's terrestrial ecosystems. *Proceedings of the National Academy of Sciences, 104*, 12942–12947.

Haberl, H., Erb, K.-H., Krausmann, F., Loibl, W., Schulz, N., & Weisz, H. (2001). Changes in ecosystem processes induced by land use: Human appropriation of aboveground NPP and its influence on standing crop in Austria. *Global Biogeochemical Cycles, 15*, 929–942.

Haberl, H., Fischer-Kowalski, M., Krausmann, F., Weisz, H., & Winiwarter, V. (2004). Progress towards sustainability? What the conceptual framework of material and energy flow accounting (MEFA) can offer. *Land Use Policy, 21*, 199–213.

Hannah, L., Lohse, D., Hutchinson, C., Carr, J. L., & Lankerani, A. (1994). A preliminary inventory of human disturbance of world ecosystems. *Ambio, 23*, 246–250.

Hansen, M., DeFries, R., Townshend, J., Carroll, M., Dimiceli, C., & Sohlberg, R. (2003). Global percent tree cover at a spatial resolution of 500 meters: First results of the MODIS vegetation continuous fields algorithm. *Earth Interactions, 7*, 1–15.

Hertwich, E. G., & Peters, G. P. (2009). Carbon footprint of nations: A global, trade-linked analysis. *Environmental Science and Technology, 43*, 6414–6420.

Herzog, F., Steiner, B., Bailey, D., Baudry, J., Billeter, R., Bukácek, R., et al. (2006). Assessing the intensity of temperate European agriculture at the landscape scale. *European Journal of Agronomy, 24*, 165–181.

Hoekstra, A. Y., & Hung, P. Q. (2005). Globalisation of water resources: International virtual water flows in relation to crop trade. *Global Environmental Change, 15*, 45–56.

Hospido, A., Milà i Canals, L., McLaren, S., Truninger, M., Edwards-Jones, G., Clift, R. (2009). The role of seasonality in lettuce consumption: A case study of environmental and social aspects. *The International Journal of Life Cycle Assessment 14*, 381–391.

Houghton, R. A., Hobbie, J. E., Melillo, J. M., Moore, B., Peterson, B. J., Shaver, G. R., & Woodwell, G. M. (1983). Changes in the carbon content of terrestrial biota and soils between 1860 and 1980: A net release of CO_2 to the atmosphere. *Ecological Monographs, 53*, 235.

Hunt, R. C. (2000). Labor productivity and agricultural development: Boserup revisited. *Human Ecology, 28*, 251–277.

Hurtt, G., Chini, L., Frolking, S., Betts, R., Feddema, J., Fischer, G., et al. (2011). Harmonization of land-use scenarios for the period 1500–2100: 600 years of global gridded annual land-use transitions, wood harvest, and resulting secondary lands. *Climatic Change, 109*, 117–161.

IAASTD. (2009). Agriculture at a crossroad. International assessment of agricultural knowledge, science and technology for development. Global report. Washington, D.C.: Island Press.

Kastner, T., Erb, K.-H., & Nonhebel, S. (2011). International wood trade and forest change: A global analysis. *Global Environmental Change, 21*, 947–956.

Kates, R. W., Clark, W. C., Corell, R., Hall, J. M., Jaeger, C. C., Lowe, I., et al. (2001). Sustainability science. *Science, 292*, 641–642.

Kauppi, P. E., Ausubel, J. H., Fang, J., Mather, A. S., Sedjo, R. A., & Waggoner, P. E. (2006). Returning forests analyzed with the forest identity. *Proceedings of the National Academy of Sciences, 103*, 17574–17579.

Klein Goldewijk, K., Van Drecht, G., & Bouwman, A. F. (2007). Mapping contemporary global cropland and grassland distributions on a 5 × 5 minute resolution. *Journal of Land Use Science, 2*, 167–190.

Krausmann, F., Erb, K.-H., Gingrich, S., Haberl, H., Bondeau, A., Gaube, V., et al. (2013). Global human appropriation of net primary production doubled in the 20th century. *Proceedings of the National Academy of Sciences of the United States of America, 110*, 10324–10329.

Krausmann, F., Erb, K.-H., Gingrich, S., Lauk, C., & Haberl, H. (2008). Global patterns of socioeconomic biomass flows in the year 2000: A comprehensive assessment of supply, consumption and constraints. *Ecological Economics, 65*, 471–487.

Krausmann, F., Haberl, H., Erb, K.-H., & Wackernagel, M. (2004). Resource flows and land use in Austria 1950–2000: Using the MEFA framework to monitor society-nature interaction for sustainability. *Land Use Policy, 21*, 215–230.

Krausmann, F., Haberl, H., Erb, K.-H., Wiesinger, M., Gaube, V., & Gingrich, S. (2009). What determines geographical patterns of the global human appropriation of net primary production? *Journal of Land Use Science, 4*, 15–33.

Krausmann, F., Haberl, H., Schulz, N. B., Erb, K.-H., Darge, E., & Gaube, V. (2003). Land-use change and socio-economic metabolism in Austria-Part I: Driving forces of land-use change: 1950–1995. *Land Use Policy, 20*, 1–20.

Kruska, R. L., Reid, R. S., Thornton, P. K., Henninger, N., & Kristjanson, P. M. (2003). Mapping livestock-oriented agricultural production systems for the developing world. *Agricultural Systems, 77*, 39–63. doi:10.1016/S0308-521X(02)00085-9.

Kuemmerle, T., Erb, K.-H., Meyfroidt, P., Müller, D., Verburg, P. H., Estel, S., et al. (2013). Challenges and opportunities in mapping land use intensity globally. *Current Opinion in Environmental Sustainability, 5*, 484–493.

Lal, R. (2012). Land degradation and pedological processes in a changing climate. *Pedologist* 315–325.

Lambin, E. F., Geist, H., & Rindfuss, R. R. (2006). Introduction: Local processes with global impacts. *Land-use and land-cover change* (pp. 1–8). Berlin: Springer.

Lambin, E. F., & Meyfroidt, P. (2011). Global land use change, economic globalization, and the looming land scarcity. *Proceedings of the National Academy of Sciences, 108*, 3465–3472.

Lambin, E. F., Rounsevell, M. D. A., & Geist, H. J. (2000). Are agricultural land-use models able to predict changes in land-use intensity? *Agriculture, Ecosystems & Environment, 82*, 321–331.

Lambin, E. F., Turner, B. L., Geist, H. J., Agbola, S. B., Angelsen, A., Bruce, J. W., et al. (2001). The causes of land-use and land-cover change: moving beyond the myths. *Global Environmental Change, 11*, 261–269.

Lauk, C., & Erb, K.-H. (2009). Biomass consumed in anthropogenic vegetation fires: Global patterns and processes. *Ecological Economics, 69*, 301–309.

Licker, R., Johnston, M., Foley, J. A., Barford, C., Kucharik, C. J., Monfreda, C., & Ramankutty, N. (2010). Mind the gap: how do climate and agricultural management explain the "yield gap" of croplands around the world? *Global Ecology and Biogeography, 19*, 769–782.

Lieth, H. (1973). Primary produciton: Terrestrial ecosystems. *Human Ecology, 1*, 303–332.

Lieth, H., Whittaker, R.H. (1975). Primary productivity of the biosphere. Berlin: Springer VI.

Lindenmayer, D., Cunningham, S., & Young, A. (2012). *Land use intensification: Effects on agriculture, biodiversity and ecological processes*. Melbourne: CSIRO Publishing.

Liverman, D. M., & Cuesta, R. M. R. (2008). Human interactions with the earth system: People and pixels revisited. *Earth Surface Processes and Landforms, 33*, 1458–1471. doi:10.1002/esp.1715.

Liverman, D., Moran, E.F., Rindfuss, R.R., Stern, P. (1998). People and pixels: Linking remote sensing and social science. Washington D.C.: National Academy Press.

Lobell, D. B., Cassman, K. G., & Field, C. B. (2009). Crop yield gaps: their importance, magnitudes, and causes. *Annual Review of Environment and Resources, 34*, 179.

Luyssaert, S., Hessenmöller, D., von Lüpke, N., Kaiser, S., & Schulze, E. D. (2011). Quantifying land-use and disturbance intensity in forestry, based on the self-thinning relationship. *Ecological Applications, 8*, 3272–3284.

Luyssaert, S., Jammet, M., Stoy, P. C., Estel, S., Pongratz, J., Ceschia, E., et al. (2014). Land management and land-cover change have impacts of similar magnitude on surface temperature. *Nature Climate Change, 4*, 389–393.

Malthus, T.R. (1798). An essay on the prinicple of population. Printed for J. Johnson, in St. Paul's Church-Yard, London. Retrieved from http://www.esp.org.

Markussen, M., & Østergård, H. (2013). Energy analysis of the Danish food production system: Food-EROI and fossil fuel dependency. *Energies, 6*, 4170–4186.

Mather, A. S. (1992). The forest transition. *Area, 24*, 367–379.

Matson, P. A., Parton, W. J., Power, A. G., & Swift, M. J. (1997). Agricultural intensification and ecosystem properties. *Science, 277*, 504–509.

McCloskey, J. M., & Spalding, H. (1989). A reconnaissance-level inventory of the amount of wilderness remaining in the world. *Ambio, 18*, 221–227.

McNeill, J.R. (2001). Something new under the sun: An environmental history of the twentieth-century world. WW Norton & Company.

Meisterling, K., Samaras, C., & Schweizer, V. (2009). Decisions to reduce greenhouse gases from agriculture and product transport: LCA case study of organic and conventional wheat. *Journal of Cleaner Production, 17*, 222–230.

Meyfroidt, P., & Lambin, E. F. (2011). Global forest transition: Prospects for an end to deforestation. *Annual Review of Environment and Resources, 36*, 343–371.

Meyfroidt, P., Rudel, T. K., & Lambin, E. F. (2010). Forest transitions, trade, and the global displacement of land use. *Proceedings of the National Academy of Sciences, 107*, 20917–20922.

Millennium Ecosystem Assessment. (2005). *Ecosystems and human well-being: Current state and trends* (Vol. 1). Washington, D.C.: Island Press.

Monfreda, C., Ramankutty, N., & Foley, J. A. (2008). Farming the planet: 2. Geographic distribution of crop areas, yields, physiological types, and net primary production in the year 2000. *Global Biogeochemical Cycles, 22*, 1–19.

Mueller, N. D., Gerber, J. S., Johnston, M., Ray, D. K., Ramankutty, N., & Foley, J. A. (2012). Closing yield gaps through nutrient and water management. *Nature, 490*, 254–257.

Nelson, G. C. (2002). Introduction to the special issue on spatial analysis for agricultural economists. *Agricultural Economics, 27*, 197–200.

Netting, R.M.C. (1993). Smallholders, householders: Farm families and the ecology of intensive, sustainable agriculture. Stanford: Stanford Univ. Press.

Neumann, K., Verburg, P. H., Stehfest, E., & Müller, C. (2010). The yield gap of global grain production: A spatial analysis. *Agricultural Systems, 103*, 316–326.

Niedertscheider, M., Gingrich, S., & Erb, K.-H. (2012). Changes in land use in South Africa between 1961 and 2006: An integrated socio-ecological analysis based on the human appropriation of net primary production framework. *Regional Environmental Change, 12*, 715–727.

Pan, Y., Birdsey, R. A., Fang, J., Houghton, R., Kauppi, P. E., Kurz, W. A., et al. (2011). A large and persistent carbon sink in the world's forests. *Science, 333*, 988–993.

Pelletier, N., Audsley, E., Brodt, S., Garnett, T., Henriksson, P., Kendall, A., et al. (2011). Energy intensity of agriculture and food systems. *Annual Review of Environment and Resources, 36*, 223–246.

Pereira, H. M., Navarro, L. M., & Martins, I. S. (2012). Global biodiversity change: The bad, the good, and the unknown. *Annual Review of Environment and Resources, 37*, 25–50.

Pickett, S.T.A., White, T.A. (1986). The ecology of natural disturbance and patch dynamics. New York: Academic Press.

Pimentel, D. (2009). Energy inputs in food crop production in developing and developed nations. *Energies, 2*, 1–24.

Pimentel, D., Hurd, L. E., Bellotti, A. C., Forster, M. J., Oka, I. N., Sholes, O. D., & Whitman, R. J. (1973). Food production and the energy crisis. *Science, 182*, 443–449.

Portmann, F.T., Siebert, S., Döll, P. (2010). MIRCA2000—global monthly irrigated and rainfed crop areas around the year 2000: A new high-resolution data set for agricultural and hydrological modeling. *Global Biogeochemical Cycles 24*, GB1011.

Ramankutty, N., Evan, A.T., Monfreda, C., Foley, J.A. (2008). Farming the planet: 1. Geographic distribution of global agricultural lands in the year 2000. *Global Biogeochemical Cycles 22*, GB1003.

Ramankutty, N., & Foley, J. A. (1999). Estimating historical changes in global land cover: Croplands from 1700 to 1992. *Global Biogeochemical Cycles, 13*, 997–1027.

Ricardo, D. (1815). *An essay on the influence of a low price of corn on the profits of stock*. London: John Murray.

Rindfuss, R. R., Walsh, S. J., Turner, B. L., Fox, J., & Mishra, V. (2004). Developing a science of land change: Challenges and methodological issues. *Proceedings of the National Academy of Sciences of the United States of America, 101*, 13976–13981.

Rounsevell, M. D. A., Pedroli, B., Erb, K.-H., Gramberger, M., Busck, A. G., Haberl, H., et al. (2012). Challenges for land system science. *Land Use Policy, 29*, 899–910.

Rudel, T. K., Coomes, O. T., Moran, E., Achard, F., Angelsen, A., Xu, J., & Lambin, E. (2005). Forest transitions: towards a global understanding of land use change. *Global Environmental Change Part A, 15*, 23–31.

Rudel, T. K., Schneider, L., Uriarte, M., Turner, B. L., DeFries, R., Lawrence, D., et al. (2009). Agricultural intensification and changes in cultivated areas, 1970–2005. *Proceedings of the National Academy of Sciences, 106*, 20675–20680.

Sala, O.E., Stuart Chapin, F., III, Armesto, J.J., Berlow, E., Bloomfield, J., Dirzo, R., Huber-Sanwald, E., Huenneke, L.F., Jackson, R.B., Kinzig, A., Leemans, R., Lodge, D.M., Mooney, H.A., Oesterheld, M., Poff, N.L., Sykes, M.T., Walker, B.H., Walker, M., Wall, D.H. (2000). Global Biodiversity Scenarios for the Year 2100. *Science 287*, 1770–1774.

Sanderson, E. W., Jaiteh, M., Levy, M. A., Redford, K. H., Wannebo, A. V., & Woolmer, G. (2002). The human footprint and the last of the wild. *BioScience, 52*, 891–904.

Seto, K., Bringezu, S., de Groot, D., Erb, K.-H., Graedel, T., Ramankutty, N., et al. (2012). Stocks, flows, and prospects of land. In T. Graedel & E. van der Voet (Eds.), *Linkages of sustainability* (pp. 71–96). Cambridge, MA: MIT Press.

Shriar, A. (2000). Agricultural intensity and its measurement in frontier regions. *Agroforestry Systems, 49*, 301–318.

Siebert, S., Portmann, F. T., & Döll, P. (2010). Global patterns of cropland use intensity. *Remote Sensing, 2*, 1625–1643.

Sieferle, R. P., Krausmann, F., Schandl, H., & Winiwarter, V. (2006). *Das Ende der Fläche. Zum Sozialen Metabolismus der Industrialisierung*. Köln: Böhlau.

Simmons, I. G. (1989). *Changing the face of the earth: Culture, environment, history*. Oxford: Blackwell.

Smith, P., Haberl, H., Popp, A., Erb, K.-H., Lauk, C., Harper, R., et al. (2013). How much land-based greenhouse gas mitigation can be achieved without compromising food security and environmental goals? *Global Change Biology, 19*, 2285–2302.

Spangenberg, J. H., Görg, C., Truong, D. T., Tekken, V., Bustamante, J. V., & Settele, J. (2014). Provision of ecosystem services is determined by human agency, not ecosystem functions. Four case studies. *International Journal of Biodiversity Science, Ecosystem Services & Management, 10*, 40–53.

Stanners, D., Bosch, P., Dom, A., Gabrielsen, P., Gee, D., Martin, J., Rickard, L., Weber, J.-L. (2007). Frameworks for environmental assessment and indicators at the EEA. In: Hak, T., Moldan, B., Dahl, A.L. (Eds.) Sustainability indicators. A scientific assessment. Scope 67 (pp. 127–144). Washington D.C.: Island Press.

Steffen, W. L., Sanderson, A., Tyson, P., Jäger, J., Matson, P. A., Moore, B, I. I. I., et al. (2005). *Global change and the earth system: A planet under pressure*. Berlin: Springer.

Steinhart, S. S., & Steinhart, C. E. (1974). Energy use in the US food system. *Science, 184*, 307–316.

Stevens, S. S. (1946). On the theory of scales of measurement. *Science, 103*, 677–680.

Tainter, J. (1990). The collapse of complex societies. Cambridge: Cambridge University Press.

Temme, A., & Verburg, P. (2010). Mapping and modelling of changes in agricultural intensity in Europe. *Agriculture, Ecosystems & Environment, 140*, 46–56.

Thomas, W. L. (1956). *Man's role in changing the face of the earth*. Chicago, Illinois: University of Chicago Press.

Tilman, D. (1999). Global environmental impacts of agricultural expansion: The need for sustainable and efficient practices. *Proceedings of the National Academy of Sciences, 96*, 5995.

Tilman, D., Balzer, C., Hill, J., & Befort, B. L. (2011). Global food demand and the sustainable intensification of agriculture. *Proceedings of the National Academy of Sciences, 108*, 20260–20264.

Tscharntke, T., Klein, A. M., Kruess, A., Steffan-Dewenter, I., & Thies, C. (2005). Landscape perspectives on agricultural intensification and biodiversity—Ecosystem service management. *Ecology Letters, 8*, 857–874.

Turner, B. L., & Ali, A. M. S. (1996). Induced intensification: Agricultural change in Bangladesh with implications for Malthus and Boserup. *Proceedings of the National Academy of Sciences, 93*, 14984–14991.

Turner, B. L., & Doolittle, W. (1978). The concept and measure of agricultural intensity. *The Professional Geographer, 30*, 297–301.

Turner, B. L., & Fischer-Kowalski, M. (2010). Ester Boserup: An interdisciplinary visionary relevant for sustainability. *Proceedings of the National Academy of Sciences, 107*, 21963–21965.

Turner, B. L., Hanham, R. Q., & Portararo, A. V. (1977). Population pressure and agricultural intensity. *Annals of the Association of American Geographers, 67*, 384–396.

Turner, B. L., Lambin, E. F., & Reenberg, A. (2007). The emergence of land change science for global environmental change and sustainability. *Proceedings of the National Academy of Sciences, 104*, 20666–20671.

Tüxen, R. (1956). Die heutige potentielle natürliche Vegetation als Gegenstand der Vegetationskartierung. *Angewandte Pflanzensoziologie, 13*, 5–42.
Van Beek, C. L., Brouwer, L., & Oenema, O. (2003). The use of farmgate balances and soil surface balances as estimator for nitrogen leaching to surface water. *Nutrient Cycling in Agroecosystems, 67*, 233–244.
Verburg, P. H., Neumann, K., & Nol, L. (2011). Challenges in using land use and land cover data for global change studies. *Global Change Biology, 17*, 974–989.
Verburg, P. H., van de Steeg, J., Veldkamp, A., & Willemen, L. (2009). From land cover change to land function dynamics: A major challenge to improve land characterization. *Journal of Environmental Management, 90*, 1327–1335.
Verstraete, M. M., Scholes, R. J., & Smith, M. S. (2009). Climate and desertification: Looking at an old problem through new lenses. *Frontiers in Ecology and the Environment, 7*, 421–428.
Vitousek, P. M., Ehrlich, P. R., Ehrlich, A. H., & Matson, P. A. (1986). Human appropriation of the products of photosynthesis. *BioScience, 36*, 368–373.
Von Thünen, J. H. (1826). Der isolierte Staat in Beziehung auf Landwirtschaft und Nationalokonomie oder Untersuchungen über den Einfluss, den die Getreidepreise, der Reichthum des Bodens und die Abgaben auf den Ackerbau ausüben. Friedrich Berthes, Hamburg. Retrieved from http://www.deutschestextarchiv.de/book/show/thuenen_staat_1826.
West, P. C., Gibbs, H. K., Monfreda, C., Wagner, J., Barford, C. C., Carpenter, S. R., & Foley, J. A. (2010). Trading carbon for food: Global comparison of carbon stocks vs. crop yields on agricultural land. *Proceedings of the National Academy of Sciences, 107*, 19645–19648.
White, P. S. (1979). Pattern, process, and natural disturbance in vegetation. *The Botanical Review, 45*, 229–299.
Whittaker, R. H., & Likens, G. E. (1973). Primary production: The biosphere and man. *Human Ecology, 1*, 357–369.
Winiwarter, V., Gerzabek, H. (Eds.). (2012). The challenge of sustaining soils: Natural and social ramifications of biomass production in a changing world, Interdisciplinary Perspectives. Verlag der Österreichischen Akademie der Wissenschaften.
Wright, D. H. (1990). Human impacts on energy flow through natural ecosystems, and implications for species endangerment. *Ambio, 19*, 189–194.
Zika, M., & Erb, K.-H. (2009). The global loss of net primary production resulting from human-induced soil degradation in drylands. *Ecological Economics, 69*, 310–318.

Chapter 5
'Society Can't Move So Much As a Chair!'—Systems, Structures and Actors in Social Ecology

Daniel Hausknost, Veronika Gaube, Willi Haas, Barbara Smetschka, Juliana Lutz, Simron J. Singh and Martin Schmid

Abstract From the socioecological perspective, society is conceived as a symbolic system that is coupled with biophysical elements. The biophysical and the symbolic components of society are considered to be coevolving. The expansion of the fossil energy regime, for example, was the result of changes in the symbolic systems of proto-industrial societies. At the same time, these systems were themselves transformed by the material dynamics the new energy regime released. Social Ecology has adopted complex systems theory as a metatheoretical framework to integrate the analysis of both symbolic and biophysical systems and their coevolution. This emphasis on systems in socioecological theory is balanced, to some extent, by a focus on actors in empirical socioecological research. The concept of actors and their agency plays an important role in transdisciplinary research, in local studies and in Environmental History. How are these actor-centered areas of research connected to the systems-centered theoretical framework

D. Hausknost (✉) · V. Gaube · W. Haas · B. Smetschka · J. Lutz · M. Schmid
Institute of Social Ecology (SEC), Alpen-Adria University, Schottenfeldgasse 29, 1070 Vienna, Austria
e-mail: daniel.hausknost@aau.at

V. Gaube
e-mail: veronika.gaube@aau.at

W. Haas
e-mail: willi.haas@aau.at

B. Smetschka
e-mail: barbara.smetschka@aau.at

J. Lutz
e-mail: julia.lutz@aau.at

M. Schmid
e-mail: martin.schmid@aau.at

S.J. Singh
Faculty of Environment, University of Waterloo, 200 University Avenue West, Waterloo, Ontario N2L 3G1, Canada
e-mail: simron.singh@uwaterloo.ca

H. Haberl et al. (eds.), *Social Ecology*, Human-Environment Interactions 5,
DOI 10.1007/978-3-319-33326-7_5

of Social Ecology? How is agency accommodated in systems, and to what extent can systems and their structures be influenced by actors? This chapter explores these questions both theoretically and in relation to concrete research examples. In doing so, it highlights some of the unresolved theoretical questions in Social Ecology and suggests possible ways they can be answered.

Keywords Complex systems theory · Agency · Structures · Coevolution · Structural coupling

5.1 Introduction

What defines Social Ecology as an interdisciplinary research program is its explicit focus on *society-nature interactions* (Fischer-Kowalski and Weisz 1999). This definition entails far-reaching ontological and epistemological commitments that frame the field of research and delineate its boundaries. First, to speak of society-nature interactions presupposes the existence of society and nature as distinct realms that are not reducible to each other. Thus, society cannot be conceived of as just another part of nature, nor can nature be understood as a mere social construction. This is what Fischer-Kowalski and Erb (2006) termed a 'realist' ontological framework.

If we regard nature and society as categorically distinct entities, the question arises of what kind of knowledge can be generated about their interaction. After all, natural scientists and social scientists have not always been on the best speaking terms. This 'great divide' between the epistemological realms of the Natural and Social Sciences may put the entire project of Social Ecology at risk unless it is effectively 'bridged' by a common metatheory that allows researchers from disciplines as far apart as Ecology and Sociology, Biology and Political Science or Engineering and Economics to construct a common 'object of knowledge', that is, to discuss society and nature in a common language. Because Social Ecology is an interdisciplinary endeavor by definition, it requires a practicable basis of cooperation between disciplines as different as those mentioned above. This basis is all the stronger the less content is 'lost in translation' from the Social to the Natural Sciences and vice versa. This was the rationale behind the decision in the 1990s by the Vienna group of social ecologists to adopt *general systems theory* (as elaborated by Humberto Maturana and Francisco Varela) as a metatheoretical framework under which the concepts of both nature and society could be operationalized and their interactions studied. Because both social and natural entities can be plausibly described in systems-theoretical terms, systems theory constitutes a promising common ground for interdisciplinary research on society-nature interactions and has therefore become the privileged epistemological vantage point in socioecological research.

However, socioecological research in practice transcends the systems-theoretical framework in several respects. 'Colonization' as a constitutive concept of the socioecological framework (see Chap. 2) has been defined as the '*intended* and sustained transformation of natural processes, by means of organized social interventions, for the purpose of improving their utility for society' (Fischer-Kowalski and Weisz 1999, p. 234; emphasis added). The emphasis on intentionality in the concept signals a form of agency that might exceed the explanatory confines of systems theory. This is a theoretical issue that has not been conclusively resolved within the theoretical framework of Social Ecology and that deserves further attention (see, for example, the systems-theoretical treatment of the possibilities and limits of purposive intervention in social systems in Willke 1994). On a more practical level, however, socioecological research frequently ventures into fields that abound with 'stakeholders' and 'agents' and that involve participatory processes in transdisciplinary research. These research endeavors pose important questions about the limits of the systems-theoretical framework and about interpretations of society that transcend it. Does it make sense to speak of 'stakeholders', let alone 'actors', in a systems-theoretical framework? What are the conditions and limits of 'participation' in an autopoietic system? What scope is there, for example, for 'political agency' to strive for radical change in the functional subsystem of politics? Can the economic system be deliberately transformed by intentional intervention from outside, or will it only change when the conditions for autopoietic self-reproduction are no longer given? Although some of these questions might be identified as simple problems of scale in that what looks like 'autonomous' agency on a micro level can be interpreted as recursive processes of autopoiesis from the bird's-eye view of systems theory, others address more fundamental questions about the limits of the systems paradigm on the one hand and the limits of agency on the other. These questions are no doubt stimulated by a kind of normative *unease*: if the social realm were all autopoietic with no scope for purposive intervention, what would be the point of engaging in problem-oriented research, as in Social Ecology? Social Ecology is driven by an intellectual and ethical urge not only to expose and define problems in the interaction between society and nature, but also to point toward ways of reorganizing these interactions in a more sustainable and less destructive way. This underlying hope for the possibility of purposive change informs the study of society-nature interaction and is based on the plausible assumption that social systems might not be as rigid as, for example, the laws of gravity and might be capable of answering to purposive intervention or 'agency'.

In this chapter, we address this tension between the systems-theoretical framework of Social Ecology and its agency-related components in an attempt to clarify to what extent this tension is plausible and can be made productive within the existing theoretical framework and to what extent it indicates the need for further theoretical elaboration and perhaps revision. In the next section, we retrace the arguments leading to the construction of the theoretical framework underpinning Social Ecology. We identify the neuralgic points in the framework where systems, structures and agency meet or where ambivalence between them occurs and leads to theoretical tensions. In doing so, we hope to prepare the ground for a further elaboration of the theoretical framework, leading to a more stringent and consistent paradigm. In section three, we present and discuss empirical research areas in Social Ecology that operate precisely at these neuralgic points and within the tensions among agency, structures and systems theory, including transdisciplinary studies, local studies, agent-based modeling (ABM) and Environmental History as well as Long-Term Socioecological Research (LTSER). These instances of empirical research help to contextualize and flesh out these theoretical tensions and might provide guidance for their resolution. Section four identifies some of the theoretical frontiers of Social Ecology that come into view when the role of agency, structures and systems in society-nature interactions is assessed both theoretically and empirically.

5.2 The Role of Systems Theory and Agency in Social Ecology

As Fischer-Kowalski and Erb note in Chap. 2 of this book, general systems theory as elaborated by Maturana and Varela (1975) and applied by Luhmann (1984) to social systems constitutes a cornerstone of the metatheoretical framework of Social Ecology. This framework has three major advantages. The first is the capacity of systems theory to serve as a 'bridge' between the Natural and Social Sciences in that it offers a rather formalized and uniform language to describe and explain phenomena in both realms. The second advantage is that systems theory can easily be operationalized in quantitative terms and is compatible with different modeling approaches—an advantage that is particularly striking in the realm of the Natural Sciences. The third advantage is that the concept of autopoiesis—that is, the process of recursive self-creation constitutive of operationally closed systems—allows for a conception of nature and culture as two operationally distinct entities that follow incommensurable logics of reproduction but are nevertheless structurally coupled in a way that could be termed 'coevolutionary'. Put differently, an autopoietic understanding of systems enables us to conceive of culture as a system of meaning that evolves according to its own rules of reproduction and therefore becomes operationally autonomous from nature while remaining deeply dependent on it through its biophysical metabolism with it. This, by implication, enables us to develop a more profound understanding of the environmental crisis of modern societies, whose metabolism with nature has exploded in quantitative terms and has started to create effects in nature that are now considered disruptive and problematic within the symbolic realm.

5.2.1 Society as a Hybrid

Although an autopoietic conception of systems is now commonplace in the Natural Sciences (especially in Biology and Ecology), it remains controversial to conceive of societies in these terms, especially outside the German-speaking world. In conceptualizing society, Social Ecology has drawn on the systems theory of Niklas Luhmann (1984, 1986), which is arguably the most elaborate and powerful social theory building on the concept of autopoiesis. For Luhmann, social systems consist of recursive communication and are thus essentially immaterial. Even human beings (and their mental systems) are not part of society but belong to its environment. Although this very radical notion of society allows Luhmann to explain the emergence of complexity and of functionally differentiated subsystems, it poses a critical problem to Social Ecology in that it does not offer an account of how social systems can influence natural systems (Fischer-Kowalski and Erb 2006). If social systems are strictly symbolic, some kind of interface is required that translates meaning (symbols) into physical action. The favored solution in the Vienna School of Social Ecology, as represented in this volume (see Chap. 2), is that the human being, as a 'hybrid' between the symbolic and material realms, constitutes this necessary interface. Consequently, Social Ecology distinguishes between 'society' and 'culture' such that the term 'culture' is reserved for Luhmann's system of recursive communication, whereas 'society' denotes a hybrid of culture and material components (the human body and its various material artifacts, including animal livestock) whose (re)production is (partly) controlled by culture. 'It is via these biophysical components of society that culture interacts with nature', as Fischer-Kowalski and Haberl (2007, p. 11) note. This theoretical move is, above all, of a pragmatic nature in that it allows for two distinct systemic realms to be analytically discerned (the symbolic realm and the material realm) while reserving the term 'society' for a rather underdetermined interaction-zone where those distinct systemic realms overlap. Thus, the disciplinary boundaries between the material and the symbolic are preserved, and interdisciplinary cooperation and communication are enabled in the new epistemic field that is constituted in defining 'society' as a hybrid.

An important consequence of this conceptual dislocation is that societies can no longer be conceived of as operationally closed systems in autopoietic terms. They 'consist rather of a 'structural coupling' of a cultural system with material elements' (Fischer-Kowalski and Haberl 2007; see also Fischer-Kowalski and Weisz 1999). This means that society as the central concept of Social Ecology is itself not a system but a hybrid formation that is suspended between and generated from different systems (the human body and its metabolic needs, some biophysical elements of nature, the nervous system of humans as the node of the symbolic world and culture as the 'objectified' system of communication). This is a rather open and conspicuously imprecise notion of society, and it begs the methodological question of how society as a non-system is possible in a world that is otherwise composed of systems (natural and cultural)—or to put it another way, what constitutes society as a conceptual unit if not its systematicity? How is this 'structural coupling' of the natural and the cultural to be understood if it is not the coupling of systems (as in Luhmann) or the coupling of a system and its environment (as in Maturana) but rather the coupling of a system with external 'elements'? However, this notion also offers a promising perspective on society as a hybrid realm that entails both systemic forces and instances of agency and 'free will'. The human as the interface between the symbolic world harboring the realm of meaning and the material world in which it lives and into which it physically interferes is the key to the socioecological understanding of society. It is never independent of those systemically ordered realms; meaning is not generated *ex nihilo* in a single nervous system but in the recursive operation of communications between several or myriad such systems. The human being is obviously dependent on the continuous reproduction of biological and ecological systems, but it can influence both realms intentionally in that it submits new communications to the cultural system (cf. Luhmann 1984, Chap. 4) and exerts physical work on the material world (Godelier 1986; Sieferle 1997). The human being translates symbolic meaning into work (for example, a worker who is told to lay a brick on a wall will usually know how to use their body to do exactly that), and it translates material realities back into the system of communication (a pilot reporting a storm to the tower or a farmer discussing soil conditions with their neighbor).

However, the human is not merely a mechanical interface or catalyst between two otherwise unconnected systems—it is also actively intervening into them and transforming them. It is embedded in and constituted by different systems, but it is also an agent modifying them both intentionally and inadvertently as side effects of its actions. Society is thus more than the sum of its elements in that it includes the (immaterial and material) systems that regulate the reproduction of these elements, but it remains operationally tied to the human being as the hybrid unit that carries out the operations that reproduce and modify these systems (operations of communication, work and, arguably, consumption). As we will see, however, this understanding of the human as the hybrid interface between culture and nature, or the symbolic and material realm, poses some methodological problems with regard to the material efficacy of 'society'. If the individual human being (and her machines as extended bodies) is the transmission belt between these realms, and if culture is the symbolic realm that drives these (human and artificial) bodies, what is the *differentia specifica* of society as an analytical unit? Is society conceptually required to influence biophysical objects, as Fischer-Kowalski and Haberl (2007, p. 10) suggest? In other words, does society require the agency 'to move so much as a chair' (ibid., p. 11), or does it suffice to assign this role to the human body? This question, esoteric as it may seem to non-theorists, will be defined as one of the most important theoretical frontiers of Social Ecology in the concluding section of this chapter as it has a decisive influence on understanding the relationship among systems, structures and agency in society-nature interactions. As such, it is a question relevant not only to theorists of Social Ecology but also to any interdisciplinary endeavor to understand society-nature interactions and the scope of purposive human intervention.

The social ontology thus presented (open questions notwithstanding) allows for a reconciliation of the otherwise disjunctive notions of systems, agents and structures in that society is made up of agents who are deeply embedded in (and constrained by) systems and their material and symbolic structures. To fully understand this conception of society, we need to address the role of these structures. The structure of a system, according to Maturana, refers to the relation of its elements at a certain point in time (Maturana and Varela 1990, p. 54; Riegas 1990, p. 336). The structure of a system thus denotes its concrete state and composition and not the underlying operations of reproduction that generated it. Although the recursive operations of a system remain the same, they allow for the emergence of diverse and complex (material and immaterial) structures that make up the world human beings inhabit. Thus, what humans as agents are confronted with are the structures of society (and of nature) and not the constitutive operations of the system (they are in the background, as it were, constantly reproducing and iterating the structures that surround us as meaning and matter). Take money as an example. Money as a universal equivalent emerged in the symbolic realm as a structure that became very powerful in the modern age. The underlying operation of buying and selling reaches back several thousand years in history (Graeber 2011; Simmel 1900/2011), but the historical convergence of certain societal structures (money, fossil energy, technology and science, for example) has ultimately led to the rapid development of new structures based on trade and commerce in what is called capitalism today. Hence, money has turned into an all-important symbolic structure of our world, although the binary operation on which it is based (buy/sell) has not changed. A single operation can thus have a large variety of effects in that it generates historically contingent structures that at the same time confine and enable the operation. Importantly, the effects of money are not confined to the symbolic realm but extend to the material world almost without limits in that money links its symbolic value to work and its material products and thus facilitates a social metabolism that has changed the face of the planet within just a few hundred years.

5.2.2 The Role of Structures

Symbolic structures, according to Luhmann, are 'expectations' and 'expectations of expectations'. This simply means that communication has to be recursive (and thus binding) to constitute meaning. Any utterance must connect to a preexisting structure of meaning or it will fail as communication and be perceived merely as 'noise' or nonsense. For a communication to be intelligible—and thus accepted, understood and perhaps responded to by the other(s)—the communicator must subject its own mind to a rigid selection process that is itself partly internalized 'conscious thought' and partly determined by the 'objective' (that is, independent of the single mind) structures of meaning within which the mind operates. Money is again a good example. The entire structure of meaning that money constitutes relies on the expectation that my ten-Euro bill will be accepted by anyone else for its universal exchange value. When I buy my lunch, the entire transaction is based on the expectation that the waiter at the restaurant will accept my bill and expects me to accept his price. My communication with the waiter will take the structure for granted: we might discuss the price of the dish but not the validity of money as such. It is very difficult to ignore this structure in modern societies or to 'do without' it. The universal exchange that money enables became the constitutive operation of a powerful subsystem (the economy) that moves billions of tons of matter around the globe and transforms the planet via physical work while being based on a simple symbolic operation.

Our world abounds with structures of meaning to which agents must adhere to be successful or even to survive, be they criminal laws, the rules of scientific peer review, the formal and informal rules of the labor market or the various structures of social distinction to which everyone in a society is subjected (Bourdieu 1984). However, the world also abounds with material structures, many of which are the result of physical work that is again the effect of symbolic structures. These material structures include physical infrastructures, such as roads, motorways and cities and their architecture, and the fields, pastures and pit mines that are the result of our ways of transforming the surface of the earth. The electric power plants that supply our computers are structures, as are our computers and the fiber cables connecting them to the Internet. They are concrete arrangements that emerged out of symbolic structures (think of the scientific knowledge and engineering skill required to build them!) and material structures understood as congealed physical work (including the matter it transformed). The term 'arrangement' is thus used within the socioecological strand of Environmental History and denotes precisely these congealed practices that mold matter into structures (see Chap. 6) and enable the continuation (and intensification) of practices.

In summary, Social Ecology works with a social ontology that involves the interplay of systems, structures and agents as constituent moments of society. Systems (both material and symbolic) are always temporalized in structures, and these structures confine the scope of agency within society. Agents are embedded in and 'thrown into' (to use a term of Heidegger 1927/2006) the structures that constitute their reality. Their agency takes place within and sometimes against these structures and includes intentional interventions as well as 'blind' routines. Structures emerge as the result of iterative operations in a system and as the result of conscious interventions that may dislocate or disrupt them. Structures are inert and durable as well as 'plastic' and variable (Maturana and Varela 1990, p. 182). There is scope for agency but not for change *ex nihilo*. The operations of a system remain immutable because they are constitutive of the system itself, but the structures of the system can be changed or the system (in extreme cases) eliminated. For example, it might be impossible to eliminate the buy/sell mechanism that is the constitutive operation of the economic system (without eliminating the economic system itself), but it might be possible to redefine its role within society by intervening in political and institutional structures. Such an intervention, however, would require specific 'conditions of possibility' within the preexisting structures of meaning. Ultimately, it would be nonsensical and impossible to eliminate 'communication' as the constitutive operation of the symbolic realm, but it is possible (to some extent) to intervene in the structures of meaning that result from communication and that offer scope for the construction of very different types of society.

Social Ecology deals with all three moments of society (its systems, structures and actors), but it has a methodological bias for the systemic level. This bias might be due to the fact that research on social metabolism commits scholars methodologically to systemic thinking, as in material and energy flow analysis (MEFA) and in several modeling approaches. The epistemological focus is on understanding what is going on in a system and not primarily on how to influence it. The concrete (and historically emerged) structures of social metabolism are analyzed, but there has been relatively little research on the role and scope of agency in changing these structures. The only prominent field of research in Social Ecology that explicitly addresses actors is 'colonization', a concept that relates to intentional and sustained activities of transforming nature for human purposes. However, although the systems-theoretical underpinnings of Social Ecology are quite elaborate, the role of actors is not yet well-defined. In the next section, we will assess and discuss several research fields within Social Ecology that explicitly or implicitly address actors (or 'agents') in an effort to arrive at a more coherent and systematic understanding of agency in Social Ecology. This understanding will be discussed toward the end of this chapter.

5.3 Systems and Actors as Cross-Cutting Issues in Social Ecology: Examples of Strands of Research

Social Ecology seeks sustainable solutions to societal problems. Based on the analysis of society-nature interactions, this search for solutions requires a fundamental restructuring of the way socioeconomic systems are organized (Haberl et al. 2011) and entails changes in both production and consumption while addressing societal activities such as nutrition, transport and mobility, housing and energy supply. If social metabolism, as one key concept of Social Ecology, means 'the whole of the materials and energy flows going through the industrial [and subsistence socioeconomic] system[s]' (Fischer-Kowalski and Hüttler 1999), then it is essential to recognize that a variety of actors in society make choices in production and consumption, and these choices determine environmental impacts.

To achieve fundamental changes and, ultimately, a qualitatively new state in a society, key system parameters need to be transformed, and actors' decisions and behaviors need to be adapted. Such an endeavor is quite challenging and requires a sound understanding of past and current transition dynamics to base interventions in ongoing change processes on such knowledge. In support of change, several methods, such as policy formulation, decision-making and monitoring to improve societal self-observation, can enhance a social system's potential for sustainable development. The methodological spectrum of Social Ecology includes, among others, systemic actor-oriented and organizational analyses and the use of historical sources as well as models and scenarios.

5.3.1 Transdisciplinarity

One methodological approach to fostering sustainability by including the perspective of actors is to arrange the production of knowledge in transdisciplinary settings. Thus, scientists and stakeholders collaborate in formulating problems so that they fit practical needs and scientific standards in integrating their different knowledge bases and producing results that are useful for the solution of problems and the advancement of science. This coproduction of knowledge serves to obtain better system, target and transformation knowledge and to find answers to '… three kinds of research questions: (a) questions about the genesis and possible development of a problem field, and about interpretations of the problems in the life-world; (b) questions related to determining and explaining practice-oriented goals; and (c) questions that concern the development of pragmatic means (technologies, institutions, laws, norms etc.) as well as the possibility of transforming existing conditions' (Pohl and Hirsch Hadorn 2007, p. 36), as our Swiss colleagues describe the principles of transdisciplinary sustainability research.

A wide range of methods are available that enable a variety of combinations. For example, scenario workshops can build on participatory modeling work using the decisions on key factors and values in the model as a baseline for envisioning different scenarios and as feedback on relations between factors and results. Another path for coproducing knowledge on transformative action is to develop decision support tools in and for participative research settings. Social multi-criteria evaluation (SMCE, Munda 2008) is a method to identify options for agency and political measures in stakeholder workshops. Assigning values and preferences and ranking alternatives is meant to be a joint endeavor that yields better knowledge for stakeholders and scientists alike (see also Method Précis on Transdisciplinary Research).

Transdisciplinary methods can empower those who make decisions on a daily basis. Everyday decisions of actors can, for example, be discussed using the heuristic of the sustainability triangle, which visualizes the interrelated ecological, economic and social factors of sustainability. Abandoning the perspective of sectoral/separated sustainability questions, the triangle takes on a systemic view of sustainability and elucidates the positive feedback loops among the increase in social well-being, wealth and resource use in industrialized economies and the specific dependencies among these factors. This allows for the discussion of possible interruptions of this spiral/helix. Actors can reflect upon the consequences of their actions and develop ideas on how to interrupt the vicious circle of unsustainable dynamics in a creative way. The sustainability triangle serves as a theoretical framework for asking whether and how actors, social systems, institutions and networks affect and are affected by resource use and by socioeconomic structures and dynamics and whether they foster or constrain a transition toward sustainability. With this knowledge, possible and potential interventions can be investigated. This framework enables such agents to develop alternative actions and political measures. It empowers actors with the possibilities of description and analysis, and it gives them a tool for envisioning and developing steps toward more sustainable pathways. There are several empirical examples of using the triangle conceptually and as a communication tool, some of which are described in Chaps. 26 and 29.

In transdisciplinary research, concepts and approaches that allow for constructive communication with and between actors are required, whether for modeling or other analytic purposes. For example, time and its use is both an analytical tool for investigating sustainability and a prevalent conception used in everyday discussions. Time-use data can be helpful in analyzing inequality and social dynamics. If we use time-use data as an indicator of changes in quality of life, we can introduce them via the triangle in the discussion and analysis of the development toward more sustainable solutions. An empirical example is given in Chap. 26 and the Method Précis on Functional Time-Use Analysis.

5.3.2 Formalized Models

The development of formalized models is becoming increasingly important in Social Ecology. The aim of these models is to help understand interrelations, to reconstruct past states of the system using incomplete datasets, to create forecasts or scenarios about future developments and to structure communication processes in a formative manner. The interactions of society and nature are based on human decision-making. Recent developments in computational science, however, have allowed for the application of numerical models for the systematic analysis and simulation of human decision-making and its direct and indirect effects. Modeling can be applied as a means for testing hypotheses about interrelations in complex human-environment systems (Van der Leeuw 2004), where reductionist approaches (e.g., limited to the analysis of social or biophysical parameters alone) are insufficient.

Model development has been found to provide a transdisciplinary platform that allows actors and experts to communicate on an equal footing throughout the research process. Participation of this kind is described as key to enabling social actors or social systems to learn from or be stimulated by the research process (Gaube et al. 2009; Hare and Pahl-Wostl 2002), and it represents a core methodology of Sustainability Science (Kates et al. 2001) and integrated Land Science (Turner et al. 2007). Participatory and transdisciplinary methods allow for mutual learning and for collaborative structuring of themes and aims, making sense for both practitioners and researchers. For example, participatory modeling as applied by Social Ecology not only helps us understand complex interrelations by reconstructing past states of a system and creating scenarios about future developments but also helps us structure communication processes and develop future scenarios and strategies together with stakeholders, empowering those directly affected. An empirical example is given in Chap. 26.

In terms of actors, the usefulness of models is not limited to their application as tools supporting transdisciplinary settings. Agent-based models (ABMs) are particularly useful for representing and understanding human decision-making. The strength of ABMs is their ability to simulate aggregate outcomes resulting from decisions made by many individual actors. The general application of ABMs has proven their utility in analyzing the dynamics of socioecological systems in which the decisions of actors influence biophysical dynamics, such as socioeconomic metabolism and land use, and vice versa. Using ABMs for the reconstruction of past decision-making and its impact on biophysical stocks and flows requires the reconstruction of the behavior of historical actors and depends upon consistent narratives (Van der Leeuw 2004). The challenge is to validate ABMs coupled with biophysical models by reconstructing past trajectories. Scale mismatch can occur when social and biophysical variables are combined (Fresco and Kroonenberg 1992). According to Gilbert (2007), different validation strategies are required by models that aim to formalize a theory (abstract model) or to describe a wide class of social phenomena (Boero and Squazzoni 2005). For an ABM, the two main components—agents and the examined environment—must be defined. For example, relevant agents for the food system in the long term are above all farms and consumers. Each farm and each consumer makes decisions concerning food production and diet dependent on a variety of framework conditions (e.g., availability of fossil fuels, chemical fertilizers, agricultural prizes and subsidies). In formalizing these decision-making processes and their linkages, an explicit actor-oriented understanding of society-nature interaction is required. Chapter 25 provides an empirical example of the use of ABMs to analyze urban residential decision-making.

5.3.3 Local Studies

In Social Ecology, society is conceived of as a cultural system that is structurally coupled with biophysical elements and that functions to reproduce a human population within a territory. This definition distinguishes 'society' from specific social systems (such as a firm or a friendship network), but it does not necessarily determine the location in a hierarchy of social units of a similar kind (for example, household, local community, state, federal state or the European Union). The scale is not such an important issue as long as it is understood that one (smaller) 'society' may be part of another (larger) 'society'. However, investigating socioecological systems (or social systems within a territory) across scales provides relevant information on society-nature interactions and their cumulative effects for sustainability analysis. Investigating 'local' systems is of great interest to some Social Ecology scholars because it furthers their understanding of the dynamic interplay among systems, agents, structures and decision-making processes and the way this interplay is related to sustainability (Singh et al. 2010; Chap. 27 in this volume). As such, local studies analyzing the decisions of (local) actors provide valuable insights into these processes: the interplay of systems, structures and agents, why certain decisions are made, and how these factors affect landscapes, ecosystems and society-nature interactions.

Here, 'local' refers to the sub-regional scale (such as a village or town, an island, an estuary or a valley), and local social systems are defined as systems that show some degree of social (cultural) integration, self-governance and systemic services and whose boundaries are socially (and not just geographically) defined. Direct empirical observation and primary data collection take place within these local social systems. Undertaking local studies entails methodological challenges in terms of primary data collection. It requires innovative and logical thinking in the field to generate the necessary data. The level of engagement and contact with the local community is far more intense than in studies that rely on secondary data. Often, the researcher encounters challenging situations due to the close proximity to local actors and stakeholders and becomes engaged in power struggles in some way.

There is a long tradition of studying local communities within Anthropology, Rural Sociology, Development Studies and Human Geography. 'Local studies' take a local view of a global problem by downscaling and investigating the way global processes affect the local and vice versa (e.g., subsidies, nature production, markets) and how this affects rural landscapes and society-nature interactions. For example, introducing transportation infrastructure into a village may bring the market closer to the people, which might in turn fuel the production of cash crops, which would require more imports (e.g., fertilizer, machinery, fossils, seeds) and the need for more capital. Therefore, a growing dependence on exports is likely to intensify food production through the use of agriculture inputs, which in turn will affect land use and increase the need for more labor. Increased population will lead to more pressure on the same territory or encroachment into new areas, or it may lead to migration. Social Ecology has emphasized the investigation of local 'rural' systems located in the transition economies of the global south (e.g., Fischer-Kowalski et al. 2011; Grünbühel et al. 2003; Ringhofer 2010; Singh et al. 2001). However, there is increasing interest in investigating the local and the rural in industrialized countries (e.g., Petridis et al. 2013; Chap. 28 in this volume).

5.3.4 Environmental History and LTSER

Environmental History and 'Long-Term Socioecological Research' (LTSER, Singh et al. 2013) are two varieties of long-term and historical approaches that bridge the gap between the Natural and Social Sciences and the Humanities (compare Chap. 6). With its ability to integrate the Natural Sciences and the Humanities, that is, to integrate research on the impact of human interventions in ecosystems with research on the socioeconomic and cultural reasons for such interventions, Environmental History aims to significantly extend the temporal scope of Sustainability Science.

To gain firm conceptual ground for interdisciplinary Environmental History, it is important to reflect the tensions and contradictions between system- and actor-centered approaches. Theoretical concepts are both epistemological and serve as tools for *interdisciplinary communication*. Thus, environmental historians must reflect, balance and bridge different approaches to be able to communicate with both the Natural Sciences and the Humanities. For most historians, systems theory is as strange and uncomfortable as Foucault's *discourse analysis* and Bourdieu's *praxeology* probably are for most system ecologists. The debate about nature's role in history has occupied many scholars in the field of Environmental History since its inception in the 1960s. One of the most pressing questions for some environmental historians is whether nature itself should be granted *agency* or should even be conceptualized as a historical *actor* (e.g., Steinberg 1991).

When Environmental History is based on a systems approach (similar to Social Ecology), it is particularly strong when it finds *explanations* for changes in patterns of society-nature interactions, for example, as a result of the transition from an agrarian to an industrial sociometabolic regime, and it is appropriate when it comes to quantifying changes in material and energy flows from and to 'nature' through social systems (e.g., Krausmann 2004; Sieferle et al. 2006). The system approach results in *explanations* for changes we observe from an etic perspective of an observer outside the system observed. Historians generally take an emic approach: they critically read and interpret their sources to *understand* why historical actors did what they did (Schmid 2006).

'However', wrote Theodore Schatzki, 'in challenging the long-standing Western theoretical practice of segregating society from nature, interactionist approaches [here, Schatzki notably refers to Viennese Social Ecology, authors' note] unwittingly uphold a key conceptual move that underlies such segregations: the separation of society from nature, the idea that theoretical work should begin from the presumption that society and nature are substantially, and not just analytically, distinct' (Schatzki 2003, p. 87). From a Humanities perspective, the distinction between nature and culture in Social Ecology can indeed become problematic. When one works with an analytical distinction, the risk of an ontological misconception is always at hand. 'There is no clear line between us and nature', wrote Richard White in his book on the Columbia River (White 1995, p. 109). Environmental History abounds in evidence that this observation is very much to the point. After environmental historians had explained—and, from time to time, defended—an interactionist approach among historians for years, some of them thought it was time to explore another *hybrid* concept. The result was the concept of 'socio-natural sites' (SNSs, Winiwarter and Schmid 2008; Winiwarter et al. 2013).

To bridge system- and actor-centered approaches, SNSs distinguish between human *practices* and material *arrangements* (instead of 'nature' and 'society'). Practices and arrangements are both understood as socio-natural hybrids. Arrangements are both the material precipitates and prerequisites of practices, and either can transform the other—if one changes, the other changes as well. The SNS is defined as the *nexus* of practices and arrangements. The concept is constructivist in the sense that it is a theory of second-order observation; by observing past observers through historical sources, it is the environmental historian who constitutes an SNS. Practices and arrangements are bound to each other, and the nexus between the two is *spatially* explicit, so it can be charted on a map. This is an important difference from the socioecological framework of nature-culture-interactions, where systems interact functionally (and not spatially) with each other.

5.4 Synthesis: The Interplay Among Actors, Structures and Systems and the Quest for Sustainability

Before we draw some preliminary conclusions and sketch some of the gaps in our understanding of the interplay among actors, structures and systems in Social Ecology, let us take stock of what we have learned so far in this essay.

First, we defined a social ontology according to which actors are constantly embedded in social systems and their symbolic and material structures. Social systems are based on self-referential symbolic operations and are therefore 'operationally closed', that is, essentially autonomous (but not autarkic) in their evolution. The result of the constant execution of these operations and of the interplay between different subsystems (which constitute each other's environments) is the emergence of concrete structures of meaning as well as material structures that result from the 'translation' of meaning into physical work (the plan of an architect is a symbolic structure resulting from a long evolution of meaning, but the building of the structure is its translation into physical labor and includes the transformation of large amounts of matter and energy). These structures in turn influence, regulate and sometimes even determine the generation of new meaning and the behavior of actors. They determine the 'possibilities' and 'impossibilities' of the social, so to speak. Certain forms of architecture, to continue the example, favor certain forms of communication and might even disable others. Laws 'enable' certain forms of behavior and 'disable' others, as do traditions or belief systems. Markets regulate possibilities in terms of prices, demands and supplies. Actors are immersed in all these structures, but they have learned (and are trained) to 'read' and understand the possibilities they have and to move within the structures of their world. In other words, actors are adapted to the systems and structures that constitute their world, and their 'actions' are usually adapted to the possibilities these structures provide. Acting against these possibilities, that is, acting to consciously change structures to generate other possibilities or to foreclose existing ones, is a relatively rare behavior in humans as the vast majority of human actions take the form of habitualized behavior, routines and internalized patterns of choice selection (Reckwitz 2003; Warde and Southerton 2012). However, the capacity in humans to act *against* the grain of structures does exist and is well documented (Tilly and Tarrow 2007). What it requires is (a) the ability to take the position of an 'observer' of the system, or, in systems-theoretical terms (Luhmann 1997), the ability to position oneself in the surrounding structures and to distinguish between these structures and the operations that constitute them, and (b) the ability to make a conscious attempt to intervene in them.

Both capacities, observation and intervention, must be distinguished analytically. Although the former is necessary for the latter to occur, the conditions of success for any attempt of structural intervention depends on many other factors that need to be discussed in detail. The mere potentiality of intervention by no means offers an indication of its chances of success. Both observation and intervention are represented in the methodological spectrum of Social Ecology. Material and energy flow accounting (MEFA), long-term socioecological research (LTSER) and land-use change research are fields of study that involve the observation of societal subsystems and their material dynamics without directly addressing the chances of actors intervening in these processes. They represent the systems perspective in Social Ecology, as it were. The more actor-focused research fields presented above, by contrast, address the question of intervention and its 'conditions of possibility' more directly. However, they, too, have a strong commitment to observation. For example, both local studies and Environmental History analyze the various ways in which actors are immersed in structures and arrangements and the extent to which willful agency is prescribed by them. In some cases, these studies reveal historical efforts to change structures or help to determine strategies for intervention at the local level. The same can be said for agent-based modeling (ABM) as a method that can be applied for both ends: it can help reveal the structural preferences of actors, that is, it can elucidate the underlying structures that shape the actions of a relevant population or group. If used in a participatory way, where first-order observations are fed back to the participating actors or where their own observations constitute the basis of the research framework, then ABM can indeed reveal opportunities for intervention and provide guidance in building appropriate strategies. The same is true for most transdisciplinary approaches within Social Ecology. Here, the idea is to strengthen the capacity of actors for societal self-observation and for identifying their own position within relevant structures and system dynamics to generate intervention strategies for specific problems. In transdisciplinary projects, a problem perceived by actors often presents the starting point for the research collaboration, which then proceeds to analyze the systemic and structural features underpinning the problem to produce the knowledge necessary to solve the problem by intervening in some of the structures identified.

Hence, actor-oriented approaches in Social Ecology rely on methods of societal self-observation to identify opportunities for intervention in the structures of social systems. However, these strands of research have so far mostly focused either on local scales (islands, cities, valleys, neighborhoods) or on concrete organizations such as hospitals and firms. A more general and substantial investigation of the conditions that allow purposive intervention in the hybrid structures that constitute the socioecological reality on our planet is lacking. For example, a structural analysis of entire political systems with regard to these conditions has not been conducted from a socioecological perspective (but see Fischer-Kowalski 2011 for a conception of framework conditions for sociometabolic transitions). The structural dependence of political institutions in modern democracies on the fossil energy regime (and its material metabolism) within which they emerged has only been presumed, not systematically investigated. Here, the famous 'systemic imperatives', or *Sachzwänge,* in democratic (as well as authoritarian) regimes need to be analytically linked to the material preconditions of regime stability. Conversely, the logics of political regime stability and legitimacy need to be better understood and their implications for the material reproduction of society studied. If it can be shown that the symbolic structures that constitute political systems depend on certain material structures for their stable reproduction, then the implications for climate change mitigation and a purposive sustainability transition would be massive: it would mean that for a successful intervention in the sociometabolic structures of modern societies, new and specific symbolic structures (institutions) are required and that the existing structures are not 'fit' to enable such societal intervention. Social Ecology could contribute crucial insights to the intricate relationships between the material and symbolic conditions of political and social system stability and thus help devise effective strategies for large-scale interventions in modern societies. This would require a methodological alignment of large-scale system observations in both the biophysical and symbolic realms in that the conditions of possibility for societal intervention are investigated not only on local scales and in individual organizational units but also on the national, supranational and global scales.

To do so, the structural relationship between actors as (potentially) willful units and the systems within which they operate and that constrain them need to be further investigated. In particular, the effectiveness of systemic-structural constraints and of the different modes of agency that try to break them needs to be better understood and properly theorized. First efforts in this direction have been undertaken in Hausknost (2014) and Hausknost and Haas (2013), where the 'transformative potential' of different modes of political agency are conceptually distinguished and assessed. The underlying observation is that political systems have developed strategies to relegate internal pressures for change into modes of agency that least destabilize the structure of the system. Consequently, pressures for radical change (such as pressures to decarbonize the economy or to move away from unsustainable modes of production) are relegated to modes of change that are systemically compatible but rather ineffective. At the same time, such modes of agency that would be most effective in transforming the sociometabolic regime but would destabilize existing symbolic structures are being suppressed by the system. This perspective allows for a conceptual integration of the system-actor divide in that it analyzes the ways in which symbolic systems tend to stabilize their structures by relegating purposive agency into modes of change that are least threatening to their internal integrity. To enhance the effectiveness of political agency, actors would need to focus on ways to overcome the system's power to suppress certain modes of agency instead of relying on the ineffective modes of agency the system offers. This analysis provides an actor-oriented perspective that does not deny the systemic constraints of agency but instead puts these very constraints at the center of its agenda. The crucial question then becomes what options for purposive agency there are (in the respective systems analyzed) to influence or reduce the systemic constraints that hold back purposive societal interventions.

To move this research agenda forward and to integrate it with the existing actor-oriented approaches in Social Ecology, the theoretical assumptions underpinning the socioecological conception of society as a hybrid between a symbolic system and material components need to be refined and revisited. One of the issues driving socioecological theory production has been the problem of the 'impotence' of the systems-theoretical conception of society with regard to the material world, or, as Fischer-Kowalski and Haberl (2007, p. 10) put it, 'How can a purely symbolic system make a difference in terms of influencing biophysical objects?' The programmatic demand implicit in this question is that 'society must not be so exclusively self-referential that it cannot move so much as a chair' (ibid., p. 11). The solution Social Ecology offers, as noted above, is to conceive of society as the *structural coupling* of a cultural system (recursive communication) with material elements (see also Chap. 2). This definition, however, denies society the status of a system as 'a unit that is able to reproduce a difference between itself and its environment' (Fischer-Kowalski and Weisz 1999, p. 244) because the material components (human bodies and their artifacts, including livestock, infrastructures and tools) do not belong to the system of communication nor do they constitute a separate system of their own. They belong to either ecological systems (human and livestock bodies) or the inanimate material world (buildings, roads, machines, computers, etc.). What ties them to the symbolic system of culture, however, is that they are all animated or created and reproduced by symbolic programs. The human body is the most immediate interface between the symbolic and the material realm. The artifacts that were created through the human body (and with the help of other artifacts that ultimately—in a historical perspective—were created by human bodies) are material representations of symbolic structures and thus carry the symbolic realm within them (the most conspicuous example perhaps being the computer as a 'hybrid' of material hardware and symbolic software that reproduces and multiplies communication at an ever-increasing rate; Miebach 2011).

Although this conception of society is intuitively compelling, it involves a major methodological challenge in that the term 'structural coupling' in complex systems theory has so far been reserved for systems alone and is difficult (or perhaps impossible) to apply to the relation of systems on the one hand and selected 'elements' on the other. In Luhmann (1997, p. 114), society as a symbolic system is structurally coupled *exclusively* to the cognitive systems of the individuals who constitute the population. These cognitive systems and the social system are cogenerative in that the existence of one is the precondition of the other's autopoiesis (Lippuner 2011, p. 312). Social systems depend on the performance of cognitive systems, which generate new communications and actualize the latent existence of society. The cognitive systems, conversely, depend on society as a system of meaning that enables them to communicate in the first place. The two systems are structurally coupled because each uses the other as a means of selection (and thus complexity reduction) in the common medium of language. Thus, society is coupled *exclusively* to individual cognitive systems, and, as Luhmann adamantly stresses, there is no direct coupling of society to any physical, chemical or biological entity (Luhmann 1997, p. 114). The cognitive systems, in turn, are coupled to the neurological systems of human bodies via the medium of perception (Lippuner 2011, p. 318). Thus, any material influence on society has to go through individual perception (and thus neurological systems) and then through individual consciousness (and thus cognitive systems) to be filtered into society (and thus systems of communication) (ibid., p. 331).

Hence, Luhmann seems to offer an answer to the question of the material efficacy of society. He would perhaps argue that it is never 'society' itself that moves a chair but a human body, which is structurally coupled to a cognitive system, which is structurally coupled to society. Hence, society 'acts' through the human body by way of a three-stage structural coupling: communication—consciousness—perception—body (Lippuner 2011, p. 331). Luhmann, in that sense, would endorse the socioecological concept of the human as a 'hybrid', but he would insist that 'society' can only 'act' in the material world through the individual and that the narrow zone of communication that can be consciously influenced by individual cognitive systems is the only channel through which society can 'help itself' (Luhmann 2002, p. 124). He would probably agree that every freight vessel crossing the ocean and moving thousands of tons of commodities is ultimately steered by human bodies that are directed by cognitive systems that are coupled to the system of communication that constitutes their world of meaning. Without the latter, however, there would be no ship, no commodities, no navigation and no trade. For this complex causal chain to work, the structural coupling of the systems involved must be very rigid and tight, and the scope for conscious intervention in the complex system of meaning that constitutes society must be all the more limited the more functionally differentiated and globally dispersed this tightly knit web of meaning is.

This dual perspective of society as a symbolic macro-structure and the individual as both the material 'agent' and the possible source of conscious intervention (which, of course, is only successful if it is 'selected' by society in that only the individual consciousness can 'think' and only society can 'communicate'; Luhmann 1997, p. 105) has not yet been theoretically pursued to the necessary depth in Social Ecology. An important fact is that there is no communication between individual minds that is not already 'social', and there is no 'communication' between the individual and society (because all communication is always already *in* society). *Nevertheless*, all meaning is generated in individual minds. Understanding this fact might yield interesting new perspectives in transdisciplinary and empirical approaches that try to identify opportunities for purposive intervention in society (cf. Fuchs 2003). After all, the structural coupling of individual minds with society presupposes that all new meaning is somehow generated in individual mind systems and that society might have an influence on the selection of meaning by devising institutions that create an enhanced openness to novelty and provide new selection mechanisms that differ from those 'naturally' provided by the functional differentiation of industrial societies. Are there ways in which society can 'help itself' by increasing the chances of selection of such communications that result in effective sociometabolic interventions? What, for example, might be appropriate institutions that allow for a greater openness of societal structures to integrate new patterns of meaning, new prescriptions and new attributions? Luhmann himself was skeptical in this regard (Luhmann 1986) because, in his view, the very independence of society from any 'direct' material influence has been the very precondition of its successful autopoiesis. Social Ecology has not pursued these questions far enough to come to any substantial conclusions. Instead, it decided at an early stage to separate the 'cultural' from the 'social' (see above) in that the former is merely symbolic and the latter involves the 'structural coupling' of the symbolic sphere with material 'elements'. It thus left the core question—how is the symbolic realm transmitted into the material realm and vice versa?—undertheorized and stopped at a promising but underdeveloped notion of hybridity that only says that there must be *some* kind of transmission ('structural coupling').

This question of the transmission mechanisms between the symbolic and the material and of the role 'conscious' meaning generation can play in influencing it is a major theoretical frontier of Social Ecology that needs to be further explored. The paradigm as such is compelling, but the very logic of social 'hybridity' must be filled with a deeper theoretical explanation. This can be done by either following Luhmann much further than the theorists of Social Ecology have done so far (as indicated above) or by following alternative paths of Social Theory that have proven successful in conceptualizing hybridity (for example, the Actor Network Theory of Latour (1993, 2004); see also Miebach 2011) and that might be in a better position to explain the 'material agency' of technology and other artifacts that are congealed symbolic structures that codetermine the generation of new meaning and the selection of possible actions.

Should Social Ecology insist that society (and not individual bodies coupled to it) must have the potency to 'move a chair', then it might indeed have to drop systems theory altogether and look for other social theories that provide a more encompassing notion of agency. Social Ecology's more recent interest in 'hybrid structures', that is, 'structures moulded both physically and culturally, in which the rules of the two realms are somehow superimposed upon one another' (Fischer-Kowalski and Steinberger 2011, p. 643), suggests an openness toward technosociological notions such as 'material agency' and actor networks. The compatibility of such notions with the systems-theoretical metatheory of Social Ecology needs to be examined in more detail, however. An integration of different sociotheoretical approaches under the umbrella of Social Ecology might be viable (as the theoretical developments in Environmental History suggest—see above). However, the price that will have to be paid in the long run might be increased theoretical complexity and thus the danger of undermining the basis for interdisciplinary communication and cooperation in Social Ecology. We are confident, however, that this frontier can be explored and 'conquered' by elaborating theoretical solutions to these problems that are at the same time complex enough to live up to the complexity of the problem and simple enough to enhance the interdisciplinary cooperation that is the main strength of socioecological research. In any case, Social Ecology is facing the exciting challenge of refining and further developing its theoretical paradigm to accommodate a more nuanced understanding of the ways in which systems, structures and actors interact in a world that is made up of symbolic meaning and physical matter.

References

Boero, R. & Squazzoni, F. (2005). Does empirical embeddedness matter? Methodological issues on agent based models for analytical social science. *Journal of Artificial Societies and Social Simulation, 7*(2).

Bourdieu, P. (1984). *Distinction: A social critique of the judgement of taste*. Cambridge, Mass.: Harvard University Press.

Fischer-Kowalski, M. (2011). Analyzing sustainability transitions as a shift between socio-metabolic regimes. *Environmental Innovation and Societal Transitions, 1*, 152–159.

Fischer-Kowalski, M., & Erb, K.-H. (2006). Epistemologische und konzeptuelle Grundlagen der Sozialen Ökologie. *Mitteilungen der Österreichischen Geographischen Gesellschaft, 148*, 33–56.

Fischer-Kowalski, M., & Haberl, H. (2007). Conceptualizing, observing and comparing socio-ecological transitions. In Marina Fischer-Kowalski & Helmut Haberl (Eds.), *Socioecological transitions and global change: Trajectories of social metabolism and land use* (pp. 1–30). Cheltenham: Edward Elgar.

Fischer-Kowalski, M. & Hüttler, W. (1999). Society's metabolism. The intellectual history of material flow analysis, Part II: 1970–1998. *Journal of Industrial Ecology, 2*(4), 107–137.

Fischer-Kowalski, M., Singh, S. J., Ringhofer, L., Grünbühel, C. M., Lauk, C., & Remesch, A. (2011). Socio-metabolic transitions in subsistence communities: Boserup revisted. *Human Ecology Review, 18*(2), 147–158.

Fischer-Kowalski, M., & Steinberger, J. (2011). Social metabolism and hybrid structures. *Journal of Industrial Ecology, 15*(5), 642–643.

Fischer-Kowalski, M., & Weisz, H. (1999). Society as hybrid between material and symbolic realms: Toward a theoretical framework of society-nature interaction. *Advances in Human Ecology, 8*, 215–251.

Fresco, L. O., & Kroonenberg, S. B. (1992). Time and spatial scales in ecological sustainability. *Land Use Policy, 9*(3), 155–168.

Fuchs, P. (2003). Der Eigen-Sinn des Bewusstseins. Die Person, die Psyche, die Signatur. Bielefeld: transcript.

Gaube, V., Kaiser, C., Wildenberg, M., Adensam, H., Fleissner, P., Kobler, J., et al. (2009). Combining agent-based and stock-flow modelling approaches in a participative analysis of the integrated land system in Reichraming, Austria. *Landscape Ecology, 24*, 1149–1165.

Gilbert, N. (2007). *Agent-based models*. Los Angeles: SAGE.

Godelier, M. (1986). *The mental and the material: Thought, economy and society*. London: Blackwell Verso.

Graeber, D. (2011). *Debt: The first 5000 years*. New York: Melville House.

Grünbühel, C. M., Haberl, H., Schandl, H., & Winiwarter, V. (2003). Socio-economic metabolism and colonisation of natural systems in a northeast Thai village: Material and energy flows, land use, and cultural change in SangSaeng. *Human Ecology, 31*(1), 53–86.

Haberl, H., Fischer-Kowalski, M., Krausmann, F., Martinez-Alier, J., & Winiwarter, V. (2011). A socio-metabolic transition towards sustainability? Challenges for another great transformation. *Sustainable Development, 19*, 1–14.

Hare, M., & Pahl-Wostl, C. (2002). Stakeholder categorisation in participatory integrated assessment processes. *Integrated Assessment, 3*(1), 50–62.

Hausknost, D. (2014). Decision, choice, solution: 'Agentic deadlock' in environmental politics. *Environmental Politics*, 23(3). doi: 10.1080/09644016.2013.874138.

Hausknost, D. & Haas, W. (2013). The role of innovation in a socio-ecological transition of the European Union. NEUJOBS Working Paper no. D 1.4, April 2013. Available at http://www.neujobs.eu/sites/default/files/NEUJOBS_Del1.4.pdf.

Heidegger, M. (2006). Sein und Zeit. Tübingen: Max Niemeyer Verlag (Original work published 1927).

Kates, R. W., Clark, W. C., Corell, R., Hall, J. M., Jaeger, C. C., Lowe, I., et al. (2001). Environment and development: Sustainability science. *Science, 292*(5517), 641–642.

Krausmann, F. (2004). Milk, manure, and muscle power. Livestock and the transformation of preindustrial agriculture in Central Europe. *Human Ecology, 32*(6), 735–772.

Latour, B. (1993). *We have never been modern*. Cambridge, Mass.: Harvard University Press.

Latour, B. (2004). *Politics of nature: How to bring the sciences into democracy*. Cambridge, Mass.: Harvard University Press.

Lippuner, R. (2011). Gesellschaft, Umwelt und Technik: Zur Problemstellung einer "Ökologie Sozialer Systeme". *Soziale Systeme, 17*(2), 308–335.

Luhmann, N. (1984). *Soziale Systeme. Grundriss einer allgemeinen Theorie*. Frankfurt/Main: Suhrkamp.

Luhmann, N. (1986). *Ökologische Kommunikation. Kann die moderne Gesellschaft sich auf ökologische Gefährdungen einstellen?*. Opladen: Westdeutscher Verlag.

Luhmann, N. (1997). Die Gesellschaft der Gesellschaft. (Vol. 2). Frankfurt am Main: Suhrkamp.

Luhmann, N. (2002). Einführung in die Systemtheorie. In D. Baecker (Ed.), Heidelberg: Carl-Auer Verlag.

Maturana, H. R., & Varela, F. J. (1975). *Autopoietic systems: A characterization of the living organization*. Urbana-Champaign, Ill.: University of Illinois Press.

Maturana, H. R., & Varela, F. J. (1990). *Der Baum der Erkenntnis: Die biologischen Wurzeln des menschlichen Erkennens*. München: Goldmann.

Miebach, B. (2011). Computer und soziale Systeme: Strukturelle Kopplung oder Material Agency? *Soziale Systeme, 17*(1), 97–119.

Munda, G. (2008). *Social multi-criteria evaluation for a sustainable economy*. Berlin, Heidelberg: Springer-Verlag.

Petridis, P., Hickisch, R., Klimek, M., Fischer, R., Fuchs, N., Kostakiotis, G., Wendland, M., Zipperer, Z., Fischer-Kowalski, M. et al. (2013). Exploring local opportunities and barriers for a sustainability transition on a Greek island. Social Ecology Working Paper 142. Vienna: Institute of Social Ecology.

Pohl, C., & Hirsch Hadorn, G. (2007). *Principles for designing transdisciplinary research*. Proposed by the Swiss Academies of Arts and Sciences: Oekom Verlag, München.

Reckwitz, A. (2003). Grundelemente einer Theorie sozialer Praktiken. *Zeitschrift für Soziologie, 32*, 282–301.

Riegas, V. (1990). Glossar. In Volker Riegas & Christian Vetter (Eds.), *Zur Biologie der Kognition: Ein Gespräch mit Humberto R. Maturana und Beiträge zur Diskussion seines Werkes* (pp. 329–337). Frankfurt a.m.: Suhrkamp.

Ringhofer, L. (2010). *Fishing, foraging and farming in the Bolivian Amazon: On a local society in transition.* New York: Springer.

Schatzki, T. R. (2003). Nature and technology in history. *History and Theory, 42*, 82–93.

Schmid, M. (2006). Herrschaft und Kolonisierung von Natur: Ein umwelthistorischer Versuch zur Integration von Materiellem und Symbolischem. *Mitteilungen der Österreichischen Geographischen Gesellschaft, 148*, 57–74.

Sieferle, R. P. (1997). *Rückblick auf die Natur: Eine Geschichte des Menschen und seiner Umwelt.* München: Luchterhand.

Sieferle, P., Krausmann, K., Schandl, H. & Winiwarter, V. (2006). Das Ende der Fläche: Zum gesellschaftlichen Stoffwechsel der Industrialisierung. Umwelthistorische Forschungen 2. Köln, Weimar & Wien: Böhlau.

Simmel, G. (2011). The philosophy of money. London: Routledge (Original work published 1900).

Singh, S. J., Grünbühel, C. M., Schandl, H., & Schulz, N. (2001). Social metabolism and labour in a local context: Changing environmental relations on Trinket Island. *Population and Environment, 23*(1), 71–104.

Singh, S. J., Haberl, H., Chertow, M., Schmid, M. & Mirtl, M. (Eds.) (2013). *Long-term socio-ecological research: Studies in society-nature interactions across spatial and temporal scales* (=Human-Environment Interactions, series edited by Emilio Moran, Vol. 2). Dordrecht: Springer.

Singh, S. J., Ringhofer, L., Haas, W., Krausmann, F. & Fischer-Kowalski, M. (2010). Local studies manual: A researcher's guide for investigating the social metabolism of rural systems. Social Ecology Working Paper 120. Vienna: Institute of Social Ecology.

Steinberg, T. (1991). *Nature incorporated: Industrialization and the waters of New England*. Amherst: University of Massachusetts Press.

Tilly, C., & Tarrow, S. (2007). *Contentious politics*. Boulder, CO: Paradigm Publishers.

Turner, B. L. I., Lambin, E. F., & Reenberg, A. (2007). The emergence of land change science for global environmental change and sustainability. *Proceedings of the National Academy of Sciences of the United States of America, 104*(52), 20666–20671.

Van der Leeuw, S. E. (2004). Why model? *Cybernetics and Systems, 35*(2–3), 117–128.

Warde, A. & Southerton, D. (Eds.) (2012). The habits of consumption (Vol. 12). COLLeGIUM: Studies across disciplines in the humanities and social sciences.

White, R. (1995). *The organic machine: The remaking of the Columbia river*. New York: Hill and Wang.

Willke, H. (1994). *Systemtheorie II: Interventionstheorie*. Stuttgart: Gustav Fischer Verlag.

Winiwarter, V., & Schmid, M. (2008). Umweltgeschichte als Untersuchung sozionaturaler Schauplätze? Ein Versuch, Johannes Colers „Oeconomia" umwelthistorisch zu interpretieren. In Thomas Knopf (Ed.), *Umweltverhalten in Geschichte und Gegenwart: Vergleichende Ansätze* (pp. 158–173). Tübingen: ATTEMPTO.

Winiwarter, V., Schmid, M., Hohensinner, S. & Haidvogl, G. (2013). The environmental history of the danube river basin as an issue of long-term socio-ecological research. In S. J. Singh, H. Haberl, M. Chertow, M. Schmid & M. Mirtl (Eds.), *Long-term socio-ecological research: studies in society-nature interactions across spatial and temporal scales* (pp. 103–122). Dordrecht: Springer.

Chapter 6
Why Legacies Matter: Merits of a Long-Term Perspective

Verena Winiwarter, Martin Schmid, Helmut Haberl and Simron J. Singh

Abstract Long-term approaches are an important concern of Social Ecology and Environmental History. This chapter outlines approaches developed within the Vienna School of Social Ecology to analyze society-nature interactions over long periods in a manner that combines the Humanities, the Social Sciences and the Natural Sciences. It discusses the interrelations between Environmental History and the novel research strand of Long-Term Socioecological Research (LTSER) that expands the Natural Science-based Long-Term Ecological Research (LTER) approach. The chapter also relates sociometabolic approaches to the concept of socio-natural sites (SNSs) as a specific configuration of 'practices' and 'arrangements', both conceived of as socionatural hybrids. Using the 'fossil-energy-driven carbon sink' in Austria's agrarian-industrial transition and colonial mining in Central and South America as research examples, it analyses the importance of long-term legacies for the course of human history as well as our current predicament, thereby demonstrating the strength of a truly interdisciplinary socioecological approach to contribute to a better understanding of current sustainability issues.

Keywords Carbon sink · Fossil energy · Mercury mining · Silver mining · Mexico · Huancavelica/Peru · Deforestation · Organic machine · Eternity costs · Environmental legacies · Socio-natural sites (SNSs)

V. Winiwarter (✉) · M. Schmid · H. Haberl
Institute of Social Ecology (SEC), Alpen-Adria University,
Schottenfeldgasse 29, 1070 Vienna, Austria
e-mail: verena.winiwarter@aau.at

M. Schmid
e-mail: martin.schmid@aau.at

H. Haberl
e-mail: helmut.haberl@aau.at

S.J. Singh
Faculty of Environment, University of Waterloo,
200 University Avenue West, Waterloo, Ontario N2L 3G1, Canada
e-mail: simron.singh@uwaterloo.ca

H. Haberl et al. (eds.), *Social Ecology*, Human-Environment Interactions 5,
DOI 10.1007/978-3-319-33326-7_6

6.1 Introduction: Long-Term and Historical Approaches to Social Ecology

The interdisciplinary field of Social Ecology studies the relationship between society and nature. The investigation of this relationship and its transformation over longer periods, incorporating contributions from the Historical Sciences, is a distinguishing feature of Viennese Social Ecology. From the beginning, long-term and historical approaches have been and continue to be a constitutive element of Viennese Social Ecology. Important terms such as 'socioecological regimes' and 'energetic transition' (see Chaps. 2 and 3) were formulated in (universal) historical works (Sieferle 1997). They acquire meaning through a perspective that incorporates the common history of society and nature over centuries and millennia.

This long-term approach takes its inspiration not least from the conviction that the particularities of contemporary relationships between society and nature can only be properly understood if contrasted with the conditions of the past. The US environmental historian McNeill (2000) employed numerous, primarily quantitative parameters to show that in the 20th century, humanity altered ecosystems worldwide to a more profound and wide-ranging degree than ever before. Research fields such as 'Global and Climate Change' base their argumentation on a historical approach. They formulate statements regarding contemporary situations and characterize and evaluate these situations by means of comparison with situations from the past. The debate in Environmental Science around the definition of the 'Anthropocene' (Crutzen and Stoermer 2000) as a distinct epoch in Earth's history and its appropriate chronological boundaries is a case in point for a long-term approach. However, the historical approach in these research fields usually remains implicit. Thoroughly researched contributions from Historical Sciences are rare. 'Long-term' and 'historical' are not, after all, identical. Whereas the former refers to a consideration of longer time frames (independent of any concrete epistemological interest), historical research aims at the source-based reconstruction of human-influenced events, actions and structures in the past. Environmental History regards these actions, events and structures as the expression of a constantly changing relationship between society and nature.

Social Ecology as an interdisciplinary research field combines approaches from Natural and Social Sciences and the Humanities. By creating terms such as 'colonization of natural systems' and further developing others, such as 'social metabolism' (see Chaps. 1–3), Social Ecology has succeeded in conceptualizing the relationship between society and nature in a way that is not hegemonic and, in so doing, has avoided determinism in all its forms (such as biologism or culturalism).

These conceptual foundations of Social Ecology emerged 20 years ago through critical analysis of other disciplines and their 'blind spots'. Sociology, to give but one example, continues to adhere to Durkheim's dictum that social facts can only be explained in terms of other social facts (see Chap. 2). It thereby constrains its own capacity to contribute to current environmental debate. Regardless of whether one sees this as a good thing, this debate is fundamentally characterized by the

Natural Sciences and rests on their results. The interdisciplinary research field of Social Ecology is an irritation for such older and better-established research fields. Its interdisciplinary approach calls into question decisions constitutive of other sciences. Explaining social facts by means of social facts is, after all, only one option among others. There are similar issues regarding the relationship between historical research and interdisciplinary Environmental History.

Establishing the relationship between society and nature as a subject of scholarly research relies on the critical analysis of different disciplines, but it equally requires involving these other disciplines in this debate and motivating them to make their own contributions to it—and this is where the communicative challenge of interdisciplinary research lies. Communicative connections need to be encouraged, and appropriate terms and concepts need to be introduced to develop Social Ecology's own connectivity to other disciplines. Socioecological research over long periods must consider two interfaces in particular: the Historical Sciences (which today see themselves primarily as part of the Humanities) and the Ecological Sciences. Viennese Social Ecology established two different approaches: Environmental History and Long-Term Socioecological Research (LTSER). The two are similar in terms of their scientific interest in the relationship between society and nature and its transformation over longer periods. They also share the conviction that concentrating solely on either sociocultural or natural aspects is unhelpful and results in an inadequately narrow explanatory approach.

Just as Environmental History and LTSER communicate with different scientific communities, they also necessarily differ in terms of their conceptual basis. Environmental History's sphere of intervention is the field of history. Its task is to communicate to other historians the plausibility of a scientific perspective that sees 'Nature' as an influencing force within history. Its message to the Historical Sciences is that 'Nature' is more than stage scenery and is not merely the relatively static background to historical events. It is also more than the '*longue durée*' (Braudel 1977, 2001) and more than the very gradually evolving spatial setting for human existence. 'Nature' is the coauthor of history; as already mentioned, however, an environmental program can only be put into practice with the cooperation of the Natural Sciences. 'Socio-natural sites' (SNSs) provide an appropriate conceptual basis (Winiwarter and Schmid 2008; Winiwarter et al. 2013; see also Chap. 23). Social Ecology's analytical differentiation between society and nature is abandoned in favor of a hybrid (Fischer-Kowalski and Weisz 1999) and, indeed, socionatural approach. The analytical and systemic differentiation between society and nature has proven to be less easily applicable to the actor-centered approaches of the Social Sciences and Cultural Studies (see Chap. 5).

LTSER explores the blind spot of another scientific field, that of Ecology. This research aims to show that society has exerted and continues to exert an influence on nature (and by no means only in the form of) 'anthropogenic disturbance'. Ultimately, LTSER aims to convince ecologists that we can hardly continue to engage with 'natural ecosystems' but should instead be discussing 'socio-ecological', 'coupled socio-ecological' or 'human-environmental systems' (regarding these terms, see Singh et al. 2013a). Whereas a systemic approach of this kind has proven to be a barrier to communication between the Environmental and Historical Sciences, it has proven useful in the communication between the LTSER community and Ecology. Consequently, the conceptual basis for LTSER is systems theory. Interactions between society and nature are addressed as a question of socioecological metabolism (Haberl et al. 2013a) and land-use change.

This chapter illustrates both variants of long-term and historical research in Social Ecology by means of two examples. At first sight, the examples appear to address very different phenomena in the common history of nature and society: on the one hand, we discuss carbon flows and stocks as a focus of LTSER, and on the other hand, we provide a colonial environmental history of the long-term consequences of historical mining activities. However, both examples show that the conceptual 'lenses' through which we view the past have tangible political consequences. Human relationships with nature, even those of several centuries back, continue to have an impact today. The conditions of possibility for our relationship with nature were already determined in the past. We will discuss this insight in detail in the conclusions.

6.2 Including the Social Dimension in Ecology: From LTER to LTSER

In recent decades, vast resources have been invested in the worldwide establishment of institutions for the monitoring and documentation of and research into changes in ecosystems over longer periods. The often-used acronym 'LTER' stands for 'Long-Term Ecological Research' (http://www.ilternet.edu/). LTER sites are equipped for long-term monitoring with the instruments, data storage capacity and comparative spatiotemporal measurement protocols for a range of ecosystem parameters. LTER undertakes the documentation and analysis of changes in patterns and processes in ecosystems over longer periods and thereby contributes to a better understanding of global change. The key goals of LTER include identifying the effects of global change as early as possible and understanding its consequences (Redman et al. 2004).

At the inception of LTER, natural or near-natural ecosystems were the focus—in line with the aim of identifying possible effects of global change, which also depends upon the ability to distinguish these effects from changes resulting from direct human intervention. This is easier to achieve in the case of little-utilized (ecologists often use the term 'undisturbed') ecosystems because fewer causes of change need to be identified. The disadvantages of this approach, however, lie in the fact that natural or near-natural ecosystems are becoming progressively less important while the significance of ecosystems that are more or less intensively utilized by humans continues to increase. Thus, a focus on ecosystems that are utilized little, if at all, by humans cannot provide much relevant insight into how ecosystems might be sustainably managed within the context of global change (Haberl et al. 2006).

Efforts have been underway for some time now to expand the original concept of LTER to include economic and institutional dimensions (Collins et al. 2011; Redman et al. 2004; Singh et al. 2010). However, this entails a significant increase in the complexity of research designs, moving from an empirical/analytical/model-oriented approach to the inter- and transdisciplinary concept of LTSER, 'Long-Term Socioecological Research' (Collins et al. 2011; Haberl et al. 2006; Singh et al. 2013b). This requires not only an expansion of the inventory of empirical methods to include historical and socioeconomic data collection procedures but also far-reaching changes to research processes through the expansion of expertise to include the Social and Economic Sciences, the Humanities and Cultural Studies and, ultimately, the inclusion of stakeholders. These changes entail the need to develop interdisciplinary terminology, concepts, models and syntheses.

The effort, however, results in a significant benefit because this expansion of the research perspective allows results to be obtained that are directly relevant in a social and therefore political and economic context. Only through the widening of the research approach does it become possible to arrive at conclusions with direct relevance for sustainability. LTSER allows the recognition of the feedback loops incurred in social and economic processes and thereby provides crucial insights into phenomena important for global change.

6.2.1 The Fossil Fuel-Driven Carbon Sink

In the following section, we aim to illustrate LTSER's possibilities to elucidate global change through the example of changes in socioecological carbon flows and stocks in Austria from 1830 to the present. Carbon is important for a number of reasons. First, a large proportion of the human impact on the global climate can be ascribed to changes in the global stocks and flows of carbon: the extraction of coal, oil and gas as well as the energetic utilization of these material resources leads to the release of carbon dioxide (CO_2) into the Earth's atmosphere. CO_2 is one of the most important greenhouse gases (GHGs) that cause global warming. Humanity also impacts carbon stocks and flows through changes made to ecosystems, such as the clearing of forests and the cultivation of pastures and agricultural land. The absorption of CO_2 by terrestrial ecosystems, the so-called 'terrestrial carbon sink', is a key process because of its capacity to decrease the velocity of climate change. It is also accounted for (albeit not comprehensively) in climate change agreements such as the *Kyoto Protocol*.

With a purely ecological approach such as LTER, an analysis of ecological carbon flows and stocks in Austria would lead to the following conclusion. The quantity of biomass produced annually by ecosystems through photosynthesis—essentially, what is termed 'net primary production', or NPP—increased in the last 170 years by approximately one-third. Despite a massive (+70 %!) increase in human harvesting of biomass, the biomass stocks in ecosystems increased as well. The stocks of

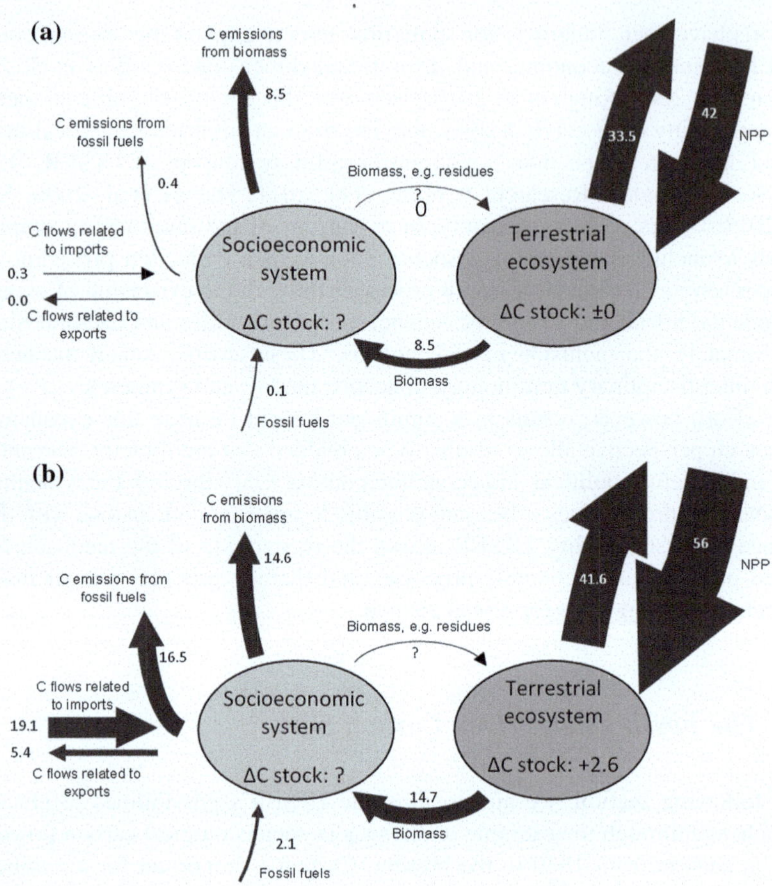

Fig. 6.1 Average yearly flows of carbon (MtC year^{-1}) in Austria in the periods **a** 1830–1880 and **b** 1986–2000. *Arrow sizes* are drawn proportionally to the magnitude of the carbon flow (except those that are unknown, marked with '?'). Imports and exports include carbon in fossil fuel products and biomass products. Biomass transfers from the socioeconomic system to the terrestrial ecosystems are unknown, as are carbon stock changes in the socioeconomic system. (*NPP* Net primary production). (Reprinted with permission from Haberl et al. 2013b)

carbon in soils and vegetation grew from approximately 1.04 GtC[1] in 1880 to ca. 1.23 GtC in 2000, i.e., by approximately 18 % (Erb et al. 2008; Gingrich et al. 2007). This means that the ecosystems in Austria store significant quantities of carbon each year—approximately 2.6 MtC/year (megatons of carbon per annum[2]) by the end of the 20th century (Fig. 6.1b). This corresponds to approximately 16 % of

[1] 1 GtC = 10^9 tC = 1 billion tons of carbon; one ton of carbon is approximately equivalent to 2 tons of dry-matter biomass.

[2] 1 MtC = 10^6 tC = 1 million tons of carbon.

the CO_2 emissions produced through the combustion of fossil fuels in Austria, indeed a significant amount.

What are the reasons for this development? It was caused primarily by the expansion of forested area as well as by increased planting density and, therefore, the quantity of carbon stored in soils, trees and other vegetation per unit area. The increase in carbon storage can be attributed to a reduction in overuse of woodland, partly because of decreasing demand for fuel wood but also because of greatly reduced nutrient removal via forest grazing or litter use[3] (Erb et al. 2008; Gingrich et al. 2007; Haberl et al. 2013a, b). In a purely ecological approach, one may attempt to explain these phenomena through processes such as increased nutrient input (e.g., in the form of nitrogen from combustion gases) or the so-called CO_2-fertilization effect (increased plant growth as a result of higher atmospheric concentrations of CO_2).

Only a comprehensive socioecological analysis can show further potential causes of the phenomena observed (Fig. 6.1). In the last 170 years, a socio-ecological transition has taken place in Austria from an agrarian to an industrial society. The energy system has fundamentally changed from a biomass-based agrarian energy system to the fossil fuel-based energy system of industrial societies. This led not only to a decrease in the demand for fuel wood, which was largely replaced by coal, oil and gas, but also to profound changes in agriculture. Yields per area unit per year for major cereal crops increased by factors between four and seven. Thus, despite a clear reduction in agricultural land area, significantly more agricultural biomass could be produced. Additionally, draught animals were replaced by tractors, which also reduced the requirement for agricultural land (Krausmann and Haberl 2007). Imports and exports grew exponentially but remained largely balanced. The spectacular increase in the supply of food and other agricultural raw materials, including fibers and bioenergy, became possible through increases in domestic productivity, accompanied by ever-greater integration into the global market.

On a systemic level, we can therefore speak of a 'fossil-energy-driven carbon sink'. Only fossil energy has made it possible to increase agricultural yields at such a high rate by means of synthetic fertilizers and tractors. The replacement of wood as a fuel source by fossil energy carriers must also be considered (Haberl et al. 2013b). What proportion of the increase in NPP is attributable to direct interventions such as the increased use of fertilizers and irrigation or the return of forested land ('forest transition', see Chap. 20), and what role is played by climatic and biogeochemical feedbacks (such as changes in precipitation and temperature or higher CO_2 concentrations in the atmosphere)? At present, such questions can only be addressed through modeling. A recent study (Erb et al. 2013) presents strong indirect evidence that historical legacy effects and subtle changes in management that are not captured by widely used land-use/carbon models play a

[3]The use of foliage and young branches as bedding for livestock.

substantial role. These findings question the validity of conventional approaches to attribute changes in the carbon balance of biota and soils to either human activities or global change feedbacks—a hugely important task required to gauge humanity's impact on the global carbon cycle (Le Quéré et al. 2012). The example shows that linking socioeconomic with ecological approaches is also very fruitful on the level of basic research. Only through an LTSER approach can direct (land use) and indirect (climate change) drivers of important biophysical processes, such as carbon emissions and carbon sequestration, be recognized and interpreted correctly. Solid LTSER will be required to adequately address these highly complex but hugely important questions.

LTSER, we have attempted to show, offers a link between ecological processes and the societal drivers that are not included in merely ecological research. LTSER thus embraces socioecological approaches. We have also seen the value of a long-term perspective. Historians can contribute to this broader agenda of LTSER by offering long-term data on society. In turn, historical research could benefit from including ecological perspectives.

6.3 How Does 'Nature' Feature in History? From History to Environmental History

Nature plays an important role in history, whether 'nature' refers to the inner nature of people or to the outer nature surrounding them. Without accounting for biophysical, material phenomena, historical developments cannot be explained; consequently, neither can the present, which is always the result of earlier structures and events.

Because power depends upon (among other things) access to resources, even political history cannot be told immaterially. However, the greater part of the Historical Sciences continues to focus on the immaterial. Great men are no longer the mainstay of study, the importance of economic, social and political structures have long been recognized, but the natural conditions of these structures continue to be underestimated. The long-term consequences of interventions in nature are rarely examined.

6.3.1 Colonial Mining in South and Central America

Like the above description of LTSER, this section at least partly addresses forests and wood. We shall show that the current situation in some regions of Mexico cannot be understood without recourse to history. As even the standard narrative of the Historical Sciences would hold, the economic and political power of Spain from the 16th century onward relied on the exploitation of its colonies (see, e.g., Waszkis 1993). North and South America established their hegemony on the global silver market from approximately 1570 until the 20th century (Nriagu 1994). People and nature paid the price for this hegemony. Rich silver mines were located in 'Virreinato de Nueva España', the 'Viceroyalty of New Spain', which existed from 1535 until 1822 in what is now Mexico. In the 'Virreinato del Perú' and the 'Virreinato de Nueva Castillia' (from 1542), an essential agent for the extraction of silver could be found, the highly toxic liquid metal mercury.

Both types of mines ruined the landscape. Pollutants such as sulfur dioxide (SO_2), mercury and salt entered the biosphere in large quantities. Previously undisturbed areas were settled and deforested, and the lungs of those working in the mines were filled with dust, which caused silicosis. Mining accidents regularly caused the deaths of hundreds of mine workers. The exploitation of nature went hand in hand with the exploitation of indigenous peoples. The mercury mine in the Peruvian town of Huancavelica was fittingly called a 'public slaughterhouse', or *mina de la muerte*, a mine of death. All those who worked in the service of the Spanish throne's insatiable desire for precious metals were more or less chronically poisoned with mercury.

Silver was extracted from the middle of the 16th century onward by means of amalgamation. In shallow-walled, open enclosures known as *patios*, finely ground silver ore was mixed with water, salt, mercury, iron and other substances and then regularly stirred over a period of weeks or months depending on the temperature. Either donkeys or workers who treaded barefoot in the knee-deep toxic sludge performed the mixing. Amalgamation reduced the amount of fuel wood and improved the yield of precious metals so that mercury became an essential processing material, particularly once the richest ores had been removed and the remainder was extracted from less-concentrated material. There were only three mercury mines worldwide: Almadén in Spain, Idrija in today's Slovenia and Huancavelica in Peru.

The Santa Bárbara mercury mine was situated above the town on the eastern slopes of the Western Cordillera at an altitude of ca. 4000 m a.s.l. The Spaniards had begun their activities there at the end of the 16th century, initially with opencast mining. During the winter, the mine often flooded, and miners had to work while standing knee-deep in ice-cold water. For those involved in smelting, lung infections brought about by the temperature change between the hot mercury kilns and the icy conditions were a common cause of death (Brown 2001). One of the earliest sources informing about the horrible working conditions was written by Felipe Waman Puma de Ayal (*1534 or ca. 1550, †1615) at the beginning of the 17th century. His major work *Nueva Corónica y buen gobierno* ('New Chronicle and Good Government') is a depiction of the life of the Incas and a profound critique of the Spanish conquerors. It includes a view of Huancavelica (known then as *La Villa Rica de Oropesa*). As can be seen in Fig. 6.2, the town (referred to as *uilla*) is in the center. Above left, the image probably shows the mercury mine (*socabón*); above right, the silver mines of Potosí (*Guayna Potocí, minas de plata*); below in the foreground, smelting kilns, possibly the enclosures known as '*patios*' and other arrangements required for smelting the metals. The title makes clear reference to the working conditions: *LA VILLA RICA DE OROpesa*

Fig. 6.2 Front page of a 17th-century treatise showing the town of Huancavelica with mines in the mountains and smelters in the foreground. (Reprinted from Poma de Ayala 1615/1616)

de Guancabilca, minas de azogue. En la dicha mina se acaua los yndios—'... Huancavelica, mercury mines. In this mine the Indios perish'.

When the walls of the open-cast mine threatened to collapse, the mine's operators resorted to tunnel construction. Conditions during the first half of the 17th century were particularly grim. The *mita*, the colonial system of forced labor carried out by the indigenous population, supplied the mines with cheap labor. Labor was so cheap that neither machinery to transport the ore above ground nor any form of safety system were put in place. The indigenous workers knew the dangers and sometimes went to desperate measures to avoid forced labor. The Spanish Viceroy reported to the King in 1630 that native Indian mothers were mutilating their sons to save them from a painful death as a result of mercury poisoning (Brown 2001).

Chronic mercury poisoning leads to tremors, pallor, discoloration and inflammation of the gums, loosening of teeth, abnormal salivation, anemia, difficulties in speaking, loss of appetite and a loss of muscular control. Even at low doses, poisoning causes personality changes. The medical literature records irritability, impatience, hypercritical or unsociable behavior, depression, anxiety, memory loss, obsessive-compulsive behavior and concentration problems. The toxic metal transgresses the placenta, causing miscarriage and stillbirth or defects in the embryo. Such effects are commonly recorded for miners and their wives and children, who often accompanied them at work, whether in the mercury mine or during the amalgamation process at the silver smelting works (Robins 2011).

Whereas the *mita* labor period at the silver mines in Potosí (today part of Bolivia) lasted one year, it was limited to two months at the mercury mine because the Spanish colonial masters knew that the symptoms of mercury poisoning improved over time. By limiting exposure to two months, the workers could be brought back periodically. In any case, a report from 1604 suggests that the graves of deceased mining and smelting works laborers contained puddles of mercury where the bodies had decayed (Robins 2011).

In 1680, the 13 mercury kilns operating in Huancavelica and producing a relatively high output of 594 t of mercury (not including the smuggled portion, which was a continuous factor) emitted 69 t of mercury. The entire population of the town, not merely the laborers from the mines and kilns, was exposed to 30–100 times the current maximum safe exposure levels, and near the smelting kilns, the air pollution likely exceeded 1000 times the current safe level for inhalation (Robins et al. 2012).

If one assumes that during mercury extraction roughly 25 % of this highly volatile metal vaporizes, then ca. 17,000 t of mercury would have entered the atmosphere here between 1564 and 1810. Worldwide, more than 236,000 t of mercury vapor were released into the atmosphere between 1550 and 1930 as a result of silver and gold extraction. The lower the concentrations of silver were in the ore, the more mercury was required. Whereas roughly 1.5 kg of mercury was needed per kilogram of silver in 1740, by the end of the colonial regime between 1790 and 1810, 2.4–2.9 kg of mercury was used per kilogram of silver (Nriagu 1994; Robins and Hagan 2012).

The Huancavelica mine closed in 1974 because yields had dwindled. Soil studies published in 2012 show that the current inhabitants of Huancavelica (population 40,000) are exposed to mercury concentrations that are among the highest worldwide, posing a significant risk to their health. The soil in the locality contains up to 35,000 times as much mercury as unpolluted areas (Robins and Hagan 2012). Even today, mercury is used for amalgamation processes. Estimates suggest that each year, 350 t of mercury enter the atmosphere as a result.

The sites of the mercury and silver mines were linked by colonial rule. Mercury also poisoned the workers in the silver mines and was an agent of environmental change in silver mining regions. Silver mining history and the associated environmental change has been researched by Studnicki-Gizbert and Schecter (2010). From the beginning of the 16th century to the beginning of the 19th century, more than 50,000 t of silver and nearly 800 t of gold were shipped to Spain from Mexico alone.

To produce a single kilogram of silver, the Mexican mines needed the wood resources of more than 6000 m^2 of forest. In total, more than 315,000 km^2 of forest was destroyed, an area nearly four times the size of Austria. However, this figure does not even include the deforestation resulting from the miners' independent silver production. They received a part of their payment in the form of ore. The miners did not have mercury for their small and inefficient kilns—it was reserved for the large 'Haciendas de Beneficio'—so their use of wood for silver production was higher. Roughly 76,000 km^2 of deforestation can be attributed to their activities. Between 1558 and 1804, 20 % of the land area of Mexico was deforested in the pursuit of silver production (Studnicki-Gizbert and Schecter 2010)—more than the entire land area of modern Germany.

As Studnicki-Gizbert and Schecter show, the valley in which the silver mine of San Luis Potosí, a city at an altitude of ca. 1850 m a.s.l. in central Mexico, is located is a good example of the rapidity with which mining-related deforestation took place. In 1614, only 20 years after mining activities began, charcoal had to be transported to the mines over distances as great as 120 km. By the mid-17th century, travellers to the region encountered a treeless, denuded landscape in which only an occasional yucca palm had survived. The forests had fallen victim to charcoal production. The demise of a leguminous species, the Mesquite tree (from the *nahuatl* name *mizquitl*) that was one source of charcoal, proved particularly detrimental. In the subtropical heat, the shade provided by trees is a decisive ecological factor for forests. However, the Mesquite provides more than this: only leguminous plants can use atmospheric nitrogen directly as a nutrient source. This benefits other plants as well because they grow on the enriched soils. Numerous animals depend upon the protein-rich bean pods of the Mesquite tree as a food source. As soon as the tree stock on the mountains around the silver mines had been cleared, desertification processes began. Even today, reforestation of the area remains impossible.

Deforestation also proved to be an effective weapon against the indigenous population. The Guachichites lived as hunter-gatherers, cultivating only a little maize. As excellent archers, they had managed to prevent the expansion of the

Spaniards to the north for a long period. Their diet consisted to a significant degree of the bean pods from the Mesquite tree, their main source of flour. The destruction of the forest removed their hunting grounds and, with them, their means of survival. The clearance of the forests left the Guachichites facing starvation and broke their resistance. This allowed the Spanish colonial leaders to open negotiations with them and to offer peace in return for food. '*Paz por compra*', peace through payment, was the term given by the Spanish to this tactic for the pacification, settlement and Christianization of the indigenous population. This new way of life entailed the cultivation of cropland and pasture—a further factor preventing reforestation.

Deforestation as a result of silver mining was followed by overgrazing, first by beef cattle and then by sheep. The once-forested land became ever more desiccated, and the colonial exploitation led to a transformation of the landscape that continues today (Studnicki-Gizbert and Schecter 2010).

At the beginning of this chapter, we noted that the long-term consequences of interventions in nature are rarely a center of focus. The above case illustrates how greed for gold and silver denuded the Mexican landscape, profoundly changed its ecology and caused deaths and disease. Colonial exploitation had long-term effects. Spanish history would have been different without colonial exploitation. The devastated landscapes and diseases continue to affect the present political and economic situation of former colonies. Identifying the distant reasons for the dire postcolonial developments in many former colonies certainly puts them in perspective.

6.4 Long-Term Legacies of Human Interventions in Natural Systems

Environmental History aims to establish the power nature wields over the course of history (within the context of the Historical Sciences), to include LTSER's Social Science perspectives within Ecology and to ensure that a sufficiently complex concept of society is employed in this context. These dialogues within science are necessary to address environmental policy issues by offering the required basic scientific knowledge. The high levels of mercury pollution resulting from the mining activities begun 400 years ago in Huancavelica still have an impact today. The mountains of northern Mexico are barren because the forest ecosystem that once existed there was cleared. The soils quickly eroded, and because soil formation is a very slow process (depending on the parent material, topography, climate and biota), the soils might not be restored before the end of the next ice age.

To interpret the present situation, it is useful to look back at history. Relevant examples can easily be found, ranging from the drainage of bogs in the Netherlands around the year 1000 (see also Chap. 19), which continues to have an impact today, to the effect of Agent Orange on the forests of Vietnam, which have been replaced by the rampant growth of bamboo thickets.

Looking back in time also helps us gain a clearer picture of the future. How will people who discover nuclear waste dumps in 5000 CE (i.e., 3000 years from now) react? It is possible that they will—as absurd as it may sound—interpret them to be the sacred sites of a lost civilization and build shrines in their honor, where they will be subject to radioactive contamination. In principle, we cannot make assumptions about societies 3000 years into the future with any certainty because we simply cannot know what explanations they will find plausible. The isotopes of plutonium (Pu), currently used to produce weapons and to fuel power stations, have a far longer life than 3000 years. Their half-life is approximately 24,100 years, and even after approximately 250,000 years, toxic amounts will remain. Because it seems hardly possible to secure a final disposal site for nuclear waste and mark its location with warning signs that will be correctly understood 250,000 years or more from now, the opposite strategy has been discussed: it might be safer to make the disposal sites so completely invisible that no one will ever discover them. At the Finnish Onkalo Nuclear Storage Site, this is indeed the concept (Upson 2009). Of course, this remains nothing but a vague hope.

The legacies resulting from the utilization of radioactive materials are a particularly extreme example. However, the principle applies that we are burdening the future with the waste we are producing today, and not only radioactive waste. According to a 2006 expert report commissioned by the German Federal Ministry of Economics from the accounting firm KPMG, which is seen as taking a conservative approach by mining experts, the eternity costs of German hard coal mining amount to at least 12.5–13.1 billion € (Preuße and Sroka 2007).

Coal mining is another example of practices with a long-term impact. However, not all are as dramatic and cost intensive. Where future environmental policy-making is concerned, it seems indispensable to focus attention upon long-term impacts and to identify those that entail a particularly high social commitment as a consequence. Technologies that result in such long-term impacts require increased attention or, as a precaution, stricter licensing control procedures as an obligation to future generations given that intergenerational equity features prominently in sustainability discourse. If one examines the past with such a research interest, it becomes apparent that particularly SNSs that were or are used for the production of energy for societies are worth examining in more detail.

Watermills are among the oldest forms of energetic utilization of watercourses. Apart from the millwheel, the arrangement consists of a channel to direct water flow toward the watermill and, in some cases, the raising of the water table in the immediate vicinity. Although usually only a small part of the water volume of a watercourse is diverted into a millstream, it constitutes a long-term intervention. Most of the energy invested goes into constructing the arrangement, although the wooden millwheel must be regularly repaired. If one compares the remains of a mill with those of a modern hydropower plant, then those of the mill appear to be 'benign'. In the worst case, what remains is a walled millstream through which some of the water flow continues to be diverted. Sometimes, interventions at the site to create water storage remain visible after millennia, but they have no transformative impact on the ecology of the watercourse. By contrast, the long-term effects of a large hydropower plant are far more problematic. Indeed, differences begin at the point of construction: power plants are often constructed on dry land, after which a new river bed will be excavated so that the river is 'built into' the power plant.

White (1995) called such hybrids 'organic machines'. If one wishes to maintain a hydropower plant of this kind, a significant investment of energy is required, not only for the power plant itself but also to stabilize the river that is dammed as part of the process. The bedload held back by the power plant needs to be removed upstream of its dam and returned to the river downstream where it is needed to stabilize the streambed. Hydropower plants therefore have a local, regional and sectoral impact upon human practices that extends beyond energy use itself. The power plant's reservoirs are often several kilometers long, and water levels of more than 10 m at the head of the reservoir are by no means rare. Water temperature and river morphology change over time, and with this comes changes to fish fauna. Migratory fish disappear and with them the fishermen who have used the aquatic ecosystem for their particular form of resource extraction. River vessels, by contrast, encounter improved conditions for transportation, mainly because the construction of hydropower plants requires the riverbanks to be reinforced and the depth of navigable waterways regulated.

If a run-of-river hydropower plant is dismantled, it cannot be assumed that the new ecosystem will resemble the ecosystem that existed before the power plant was constructed because profound changes have taken place (Doyle et al. 2005). If it is not dismantled, the costs of creating and maintaining the site, which persist over a long period, must be taken into account.

Coal power plants create a dangerous legacy upstream of the value-added chain in terms of the eternity costs of the coal mining industry. Nuclear power stations, with their energetically intensive form of power generation, have an even greater long-term legacy. If one takes into account the costs of the final deposition of radioactive wastes, including their long-term safeguarding, it becomes clear that the majority of the costs pertaining to this technology relate to their maintenance and stabilization. By comparison, the construction of such facilities is cheap. The requirements in terms of monitoring long-term impacts are high in technical, economic and social terms, and these impacts last several hundred thousand years. Interim and final repositories would probably be identifiable long after the next ice age as a human legacy, unless the greenhouse gas (GHG) emissions produced by humans delay the coming of the next ice age even beyond this time (Berger and Loutre 2002).

We therefore suggest that interventions in nature can be categorized according to whether legacies are associated with them, and if so, which types of legacies. We distinguish among benign, problematic and wicked legacies, which we categorize according to their longevity, the costs they entail and their potential to transform society. Wicked legacies are primarily those associated with arrangements where the expected energy yield is high. In most cases, the societal costs of such energy sources arise mainly after their active utilization has ended. This categorization entails setting the (comparably short-lived) energy yield against the

Table 6.1 Typology of legacies resulting from societal interventions in nature. The characteristics of socio-natural sites (SNSs) can be categorized according to their longevity, the costs they entail and their potential for transforming human practices in the future

Characteristics	Arrangements with short-lived legacies	Arrangements with stable, long-lived legacies	Arrangements with transformative legacies
Type of legacy	Benign	Problematic	Wicked
Longevity of legacy	Short	Middle	Long
Maintenance requirements	Low	Middle	High
Energy expenditure centered on	Production	Production and maintenance	Maintenance
Energy harvest density	Low	Middle	High
Transformative potential (impact on practices)	Local, sectoral	Local, regional, sectoral (one or several)	Societal, global

(generally long-lived) costs and expenditures. Table 6.1 provides an overview of the characteristics of different legacies.

The last row of this table focuses on the long-term impacts on human society, or the impact such arrangements have on future possible or necessary societal practices. We differentiate benign long-term impacts from problematic and wicked ones by their scale. Benign impacts remain limited to the local level and to individual economic sectors. Arrangements with problematic long-term impacts will have an effect on practices at both local and regional levels and in one or more economic sectors.

In contrast, the transformative potential of arrangements with wicked long-term impacts is global and affects society as a whole. The far-reaching impacts of nuclear accidents such as those at Chernobyl and Fukushima provide a foretaste of the long-term impacts of nuclear technology: large tracts of land could become uninhabitable because of radioactive pollution, and armies of engineers and safety personnel would be required to render the radioactive materials secure in both a technical and social sense. Their stabilization efforts are a battle against natural and social dynamics. ‘Hazardous waste management’ will likely become an increasingly large economic sector that will have to address industrial-age legacies for future societies.

6.5 Conclusions

Every intervention in nature and every creation of an SNS has both desired (or at least, foreseen) consequences and unforeseen ‘side effects’. Indeed, unforeseen side effects have characterized human history. Since at least the Neolithic revolution, keeping stocks of food to provide stable amounts of seed and supplies despite variable yields has been a key factor of success. In addition to the desired impact—avoiding famine during times when harvests fail—such food stores have attracted all types of undesired 'dinner guests'; the existence of food stores has led to storage pests or vermin around the world. Following the argument proposed by Wilkinson (1973), humans have proven to be particularly innovative when they encounter problems. However, their solutions often lead to the creation of new problems. Tainter (1988) employed a number of historical examples to show that societies become more complex (e.g., develop new bureaucracies and structures) when they attempt to solve problems. However, this increased complexity also increases resource and energy requirements as marginal returns on investment decline, thus creating new problems, which may ultimately lead to a 'collapse'. In Environmental History, this combination is referred to as the 'risk spiral'. This term is intended to draw attention to the fact that instead of reducing risk, human responses tend either to displace risk elsewhere or to foster entirely new risks (Müller-Herold and Sieferle 1997). Long-term observation of the relationship between society and nature is a prerequisite to understand this development over time.

The conditions that enable societies to develop, as we initially proposed, are profoundly influenced by the long-term impacts of earlier interventions in nature, which, in many instances, happened in the distant past. This is a significant insight for the sustainability debate. It is not entirely new; the philosophy underpinning environmental sustainability certification and technology impact assessment is built upon such insights. What is new is the extent to which we need to take account of nature and its dynamics and how long ago significant changes happened. There are interventions with long-term impacts that predate the technology of modernity.

However, technologies, and to some extent environmental policies, are not commonly developed with long-term impacts in mind. Rather, they constitute short-term solutions to immediately pressing problems. If one takes seriously Heinz von Förster's variation of Kant's imperative, 'I always try to act so as to increase the number of choices', then one must give priority to long-term impacts when making decisions, such as those regarding new technologies (von Foerster 1973). This priority would be associated with a significantly different approach to evaluating technologies. The long-term perspective of Social Ecology, whether in its environmental history or its long-term socioecological manifestation, can contribute to the development of a 'sustainability assessment' that places particular emphasis on the problematic and dangerous long-term legacies of SNSs.

Acknowledgements Enric Tello Aragay (Universitat de Barcelona) assisted with the translation of the text depicted in Fig. 6.2.

Sources and References

Archival Material

Poma de Ayala, & Felipe Guaman (1615/1616). El primer nueva corónica y buen gobierno (1615/1616) (København, Det Kongelige Bibliotek, GKS 2232 4°). Autograph manuscript facsimile, annotated transcription, documents, and other digital resources. Retrieved from http://www.kb.dk/permalink/2006/poma/info/en/frontpage.htm. Last accessed March 20, 2013.

References

Berger, A., & Loutre, M. F. (2002). An exceptionally long interglacial ahead? *Science, 297*, 1287–1288.

Braudel, F. (1977). Geschichte und Sozialwissenschaften. Die longue durée. In M. Bloch, F. Braudel, L. Febvre & C. Honegger (Eds.), *Schrift und Materie der Geschichte. Vorschläge zu einer systematischen Aneignung historischer Prozesse* (edition suhrkamp 814, pp. 47–85). Frankfurt am Main: Suhrkamp.

Braudel, F. (2001). *Das Mittelmeer und die mediterrane Welt in der Epoche Philipps II. 3 Bände*. Frankfurt am Main: Suhrkamp.

Brown, K. W. (2001). Workers' health and colonial mercury mining at huancavelica, Peru. *The Americas, 57*(4), 467–496.

Collins, S. L., Carpenter, S. R., Swinton, S. M., Orenstein, D. E., Childers, D. L., Gragson, T., et al. (2011). An integrated conceptual framework for long-term social–ecological research. *Frontiers in Ecology and the Environment, 9*, 351–357.

Crutzen, P. J., & Stoermer, E. F. (2000). The 'Anthropocene'. *Global Change Newsletter, 41*, 17–18.

Doyle, M. W., Stanley, E. H., Orr, C. H., Selle, A. R., Sethi, S. A., & Harbor, J. M. (2005). Stream ecosystem response to small dam removal: Lessons from the Heartland. *Geomorphology, 71*, 227–244.

Erb, K.-H., Gingrich, S., Krausmann, F., & Haberl, H. (2008). Industrialization, fossil fuels and the transformation of land use: An integrated analysis of carbon flows in Austria 1830–2000. *Journal of Industrial Ecology, 12*(5–6), 686–703.

Erb, K.-H., Kastner, T., Luyssaert, S., Houghton, R. A., Kuemmerle, T., Olofsson, P., & Haberl, H. (2013). Bias in the attribution of forest carbon sinks. *Nature Climate Change, 3*, 854–856.

Fischer-Kowalski, M., & Weisz, H. (1999). Society as a hybrid between material and symbolic realms. Toward a theoretical framework of society-nature interaction. *Advances in Human Ecology, 8*, 215–251.

Gingrich, S., Erb, K.-H., Krausmann, F., Gaube, V., & Haberl, H. (2007). Long-term dynamics of terrestrial carbon stocks in Austria. A comprehensive assessment of the time period from 1830 to 2000. *Regional Environmental Change, 7*(1), 37–47.

Haberl, H., Erb, K.-H., Gaube, V., Gingrich, S., & Singh, S. J. (2013a). Tracing changes in socio-ecological stocks and flows of materials and energy across space and time—An overview of socio-metabolic LTSER approaches. In S. J. Singh, H. Haberl, M. Chertow, M. Mirtl, & M. Schmid (Eds.), *Long-term socio-ecological research. Studies in society-nature interactions across spatial and temporal scales* (pp. 29–52). Dordrecht, Heidelberg, New York, London: Springer.

Haberl, H., Erb, K.-H., Gingrich, S., Kastner, T., & Krausmann, F. (2013b). Human appropriation of net primary production, stocks and flows of carbon, and biodiversity. In R. Lal, K. Lorenz, R. F. Hüttl, B. U. Schneider, & J. von Braun (Eds.), *Ecosystem Services and carbon sequestration in the biosphere* (pp. 313–331). Berlin: Springer.

Haberl, H., Winiwarter, V., Andersson, K., Ayres, R., Boone, C., Castillo, A., et al. (2006). From LTER to LTSER: Conceptualizing the socio-economic dimension of long-term socio-ecological research. *Ecology and Society, 11*(2), 13. http://www.ecologyandsociety.org/vol11/iss2/art13/

Krausmann, F., & Haberl, H. (2007). Land-use change and socioeconomic metabolism: A macro view of Austria 1830–2000. In M. Fischer-Kowalski & H. Haberl (Eds.), *Socioecological transitions and global change* (pp. 31–59). Cheltenham: Edward Elgar.

Le Quéré, C., Andres, R. J., Boden, T., Conway, T., Houghton, R. A., House, J. I., et al. (2012). The global carbon budget 1959–2011. *Earth System Science Data Discussions, 5*, 1107–1157.

McNeill, J. R. (2000). *Something new under the sun. An environmental history of the twentieth century*. London: Allen Lane.

Müller-Herold, U., & Sieferle, R. P. (1997). Surplus and survival: Risk, ruin, and luxury in the evolution of early forms of subsistence. *Advances in Human Ecology, 6*, 201–220.

Nriagu, J. O. (1994). Mercury pollution from the past mining of gold and silver in the Americas. *The Science of the Total Environment, 149*, 167–181.

Preuße, A., & Sroka, A. (2007). Technische Abschätzung von Folgelasten des Steinkohlenbergbaus. *bergbau 12*, 540–546.

Redman, C. L., Grove, J. M., & Kuby, L. H. (2004). Integrating social science into the long-term ecological research (LTER) network: social dimensions of ecological change and ecological dimensions of social change. *Ecosystems, 7*, 161–171.

Robins, N. A. (2011). *Mercury, mining, and empire: The human and ecological cost of colonial silver mining in the Andes*. Bloomington: Indiana University Press.

Robins, N. A., & Hagan, N. A. (2012). Mercury production and use in colonial andean silver production: Emissions and health implications. *Environmental Health Perspectives, 120*, 627–631.

Robins, N. A., Hagan, N. A., Halabi, S., Hsu-Kim, H., Gonzales, R. D., Morris, M., et al. (2012). Estimations of historical atmospheric mercury concentrations from mercury refining and present-day soil concentrations of total mercury in Huancavelica, Peru. *The Science of the Total Environment, 426*, 146–154.

Sieferle, R. P. (1997). *Rückblick auf die Natur*. München: Luchterhand.

Singh, S. J., Haberl, H., Chertow, M., Mirtl, M., & Schmid, M. (2013a). Introduction. In S. J. Singh, H. Haberl, M. Chertow, M. Mirtl, & M. Schmid (Eds.), *Long-term socio-ecological research. Studies in society-nature interactions across spatial and temporal scales* (pp. 1–26). Dordrecht, Heidelberg, New York, London: Springer.

Singh, S. J., Haberl, H., Chertow, M., Mirtl, M., & Schmid, M. (Eds.). (2013b). *Long-term socio-ecological research. Studies in society-nature interactions across spatial and temporal scales*. Dordrecht, Heidelberg, New York, London: Springer.

Singh, S. J., Haberl, H., Gaube, V., Grünbühel, C. M., Lisievici, P., Lutz, J., et al. (2010). Conceptualising long-term socio-ecological research (LTSER): Integrating the social dimension. In F. Müller, C. Baessler, H. Schubert, & S. Klotz (Eds.), *Long-Term ecological research, between theory and application* (pp. 377–398). Dordrecht, Heidelberg, London, New York: Springer.

Studnicki-Gizbert, D., & Schecter, D. (2010). The environmental dynamics of a colonial fuel-rush: silver mining and deforestation in New Spain, 1522 to 1810. *Environmental History, 15*(1), 94–119.

Tainter, J. A. (1988). *The collapse of complex societies (new studies in archaeology)*. Cambridge: Cambridge University Press.

Upson, S. (2009). Finland's nuclear waste solution. Scandinavians are leading the world in the disposal of spent nuclear fuel. Retrieved from http://spectrum.ieee.org/energy/nuclear/finlands-nuclear-waste-solution/0. Last accessed November 10, 2014.

von Foerster, H. (1973). On constructing a reality. In F. E. Preiser (Ed.), *Environmental design research* (vol. 2, pp. 35–46). Stroudberg: Dowden, Hutchinson & Ross.

Waszkis, H. (1993). *Mining in the Americas: Stories and history*. Cambridge: Woodhead Publishing Limited.

White, R. (1995). *The organic machine: The remaking of the Columbia river*. New York: Hill and Wang.

Wilkinson, R. G. (1973). Poverty and progress: An ecological model of economic development. London: Methuen.

Winiwarter, V., & Schmid, M. (2008). Umweltgeschichte als Untersuchung sozionaturaler Schauplätze? Ein Versuch, Johannes Colers „Oeconomia" umwelthistorisch zu interpretieren. In T. Knopf (Ed.), *Umweltverhalten in Geschichte und Gegenwart: Vergleichende Ansätze* (pp. 158–173). Tübingen.

Winiwarter, V., Schmid, M., Hohensinner, S., & Haidvogl, G. (2013). The environmental history of the Danube River basin as an issue of long-term socio-ecological research. In S. J. Singh, H. Haberl, M. Chertow, M. Mirtl, & M. Schmid (Eds.), *Long-term socio-ecological research. Studies in society-nature interactions across spatial and temporal scales* (pp. 103–122). Dordrecht, Heidelberg, New York, London: Springer.

Chapter 7
Toward a Socioecological Concept of Human Labor

Marina Fischer-Kowalski and Willi Haas

Abstract Within the socioecological framework, labor figures in two contexts. One context is social metabolism: extracting and transforming materials from nature for social consumption. This is not fundamentally different from what any other animal does to secure its food and the food for its offspring, and it is, as far as we know from hunter-gatherers, not very much work. The second context is what we term colonizing activities. Colonizing activities are deliberate interventions in natural systems to modify their functioning and truly demand labor from humans; the energy transition of the Neolithic revolution is the starting point for man as a laborer. We assume that human society is currently in another energy transition in which we are moving away from the use of fossil fuels. This transition will have as many implications for human labor as the transition toward the fossil fuel-based industrial society had. In the first section, we characterize quantitative, qualitative and institutional features of human labor from a socioecological perspective. We then focus on the interrelation between sociometabolic regimes and the amount of human life time spent on labor, the respective critical capacities of human labor power (physical power, intelligence/knowledge, empathy) and the institutional forms in which labor is employed. The third section speculates about the future: what might labor look like after the ongoing socioecological transition is completed? In light of the major changes in work and life induced by the fossil-fuel-based socioecological transition, what changes might we expect from a major societal transition away from fossil fuels?

M. Fischer-Kowalski (✉) · W. Haas
Institute of Social Ecology (SEC), Alpen-Adria University,
Schottenfeldgasse 29, 1070 Vienna, Austria
e-mail: marina.fischer-kowalski@aau.at

W. Haas
e-mail: willi.haas@aau.at

H. Haberl et al. (eds.), *Social Ecology*, Human-Environment Interactions 5,
DOI 10.1007/978-3-319-33326-7_7

Keywords Energy transitions · Labor burden · Colonizing activities · Employment · Wage labor · Physical work · Empathy · Sociometabolic regimes

7.1 Introduction

What is the role of human labor in the socioecological paradigm? In the work of Marx, labor constitutes a fundamental relation of man to nature and is a basic social relation, a relation within society.[1] So far, the socioecological approach corresponds to Marx. Moreover, for Marx, following in the footsteps of Adam Smith, labor is the source of all wealth, the (only) creator of value.[2] In our approach, we do not really discuss the origin of use value as we have been staying within a biophysical framework as far as labor is concerned. Within this framework, labor figures in two contexts. One context, which is basically the labor Marx addresses, is social metabolism: capturing, extracting and preparing materials from nature for social consumption. This is not fundamentally different from what any other animal does to secure its food and the food for its offspring. As we know from studies on hunter-gatherers, this need not be very much work. The second context, which is explicitly addressed by our model, is what we term colonizing activities. Colonizing activities, as explained in Chap. 1, are deliberate interventions in natural systems to modify their functioning. The classical case is agriculture, and we agree with Bauer (2013) that the Neolithic revolution is the starting point for man as laborer. In both contexts, human labor directly and intentionally interferes with natural systems, but with unintended side effects. Whereas in many chapters of this book we analyze the impacts of human labor on natural systems (e.g., on land cover change or emissions from human energy use), in this chapter we focus almost exclusively on the social and cultural aspects.

The point of departure for this chapter is the assumption that human society—willingly or unwillingly and slowly or rapidly—is in a transition away from the use of fossil fuels. We expect that this transition will have as many and equally far-reaching implications for human labor as did the transition toward a fossil-fuel-based industrial society. To better understand this situation, we make an effort to characterize historical linkages between energy transitions and human labor and their quantitative, qualitative and institutional features. To our knowledge, this effort has not yet been undertaken; such a broad venture would necessitate the format of a book to make a legitimate claim for scientific dignity. What we

[1] 'So far therefore as labour is the creator of use value, is useful labour, it is a necessary condition, independent of all forms of society, for the existence of the human race; it is an eternal nature-imposed necessity, without which there can be no material exchanges between man and Nature, and therefore no life' (Marx 1867, p. 30).

[2] 'We see, then, that labour is not the only source of material wealth, of use values produced by labour… labour is its father and the earth its mother' (Marx 1867, p. 30).

may hope for, however, is to be able to draw some useful distinctions, show interesting empirical findings derived from socioecological analyses and sketch our perspective.

In the first section of this chapter, we attempt to characterize quantitative, qualitative and institutional features of human labor from a socioecological perspective.

The second section uses these distinctions and focuses on the interrelation between sociometabolic regimes (see Chap. 3) and the amount of human lifetime spent on labor, the respective critical capacities of human labor power and the institutional forms in which labor is employed.

The third section speculates about the future: what might labor look like after the ongoing socioecological transition (SET) is completed or has come to the next stage? Most analyses of 'green jobs'[3] address a fairly close future and mainly consider the future of gainful employment.[4] We open the time horizon (which also means keeping it somewhat unspecific) and ask ourselves the following question: in light of the major changes in work and life induced by the fossil-fuel-based SET, what changes might we expect from a major societal transition away from fossil fuels?

7.2 Some General Distinctions to Characterize Human Labor Quantitatively, Qualitatively and According to Its Institutional Form

7.2.1 How Can Human Labor Be Characterized Quantitatively Across Different Sociometabolic Regimes?

The starting point is the question of how to distinguish 'labor' from other human activities. A socioecological perspective helps to circumvent some of the long-standing debates around this issue and suggests viewing human labor as an element of human time use within a social (distributional) context. As humans reproduce their lives in social groups, they cooperate in various forms of a 'division of labor'. Thus, it makes sense to regard all those activities that are subject to such a division and thus constitute social interdependencies as 'labor' and to only label those elements of time use that cannot be socially transferred to anyone else as 'non-labor'.[5]

[3]For example, CEDEFOP (2009), Eurofound (2012), OECD (2010).

[4]There are some exceptions, such as the project by Hans Böckler Stiftung (2001), 'Arbeit und Ökologie'. This project looked at a broad range of conceptions of 'labor', including household work and paid civil work.

[5]Note that this definition is far from any physical definition of work. From a sociometabolic perspective, a physical definition of labor would make it very difficult to line up with social definitions; physically speaking, digesting food is as much 'work' as collecting and cooking it.

Whereas traditional time use research (Gershuny 2000) addresses time use on a descriptive and individual level only, a socioecological analysis places the time-use and activity categories in a functional context of system reproduction.[6] Depending on functional linkages, labor (time) can be distinguished from non-labor (time). We follow Ringhofer (2010) in distinguishing the following system references for time use:[7]

The reproduction of the self (such as sleeping, eating, resting, learning, having fun): This class of activities cannot be subject to a social division of labor and thus cannot be considered 'labor'.

The reproduction of household and family: These activities also address personal reproduction but in an intersubjective mode (child bearing and rearing, food preparation, daily chores…). This clearly needs to be considered 'labor'.[8]

The reproduction of the community: participation in 'public affairs' on various scales beyond the family, such as collective decision-making, voting, participation in religious and public ceremonies, military service or shared infrastructure work. This clearly has labor-like features.

The reproduction of the economy at large: On a systemic level, this is time used for the production of goods and services for anonymous consumers; individually, it is the time used for income generation on a market (and is thus close to conventional economic definitions of labor).

This approach is in line with Marx's understanding in that he defines the amount of labor hours (spent with the average skill, effort and technical effectiveness available in Marx's day; see Marx 1867, p. 31) as a standard measure of human labor. However, it clearly deviates from Marx in including household and family reproduction in the definition of labor; his focus on 'productive' and paid work prevents him from even considering this.

What share of a population's time is devoted to each function varies considerably from one socioecological regime to another. It is also related to demographic structures because they determine the share of people fit for work as well as the dependency ratio. How activities are distributed among subgroups of the population by gender, age and status is highly variable. The higher the status of a group is, the more time for self-reproduction (type 1) it will typically be entitled to.

For the individual, the question is how much labor time it needs to survive and to reproduce. This demands a minimum of activities of second and fourth types presented above. Under favorable environmental and social conditions, the time required may be very little (see Ringhofer 2010; Sahlins 1972). Under unfavorable environmental or social conditions, the time required may be more than the individual can afford over a protracted period, so it will not reproduce and will not be

[6] M. Giampietro and K. Mayumi have a long research tradition of placing human time use within a framework of system reproduction, and they see it as a key link among demography, energy use and economic output (e.g., Giampietro et al. 2012). We follow a similar perspective.

[7] For more details, see the method section in Chap. 26.

[8] In her narrative on labor in classical Greek philosophy, Hannah Arendt shows that Aristotle considered this class of labor particularly unworthy of free citizens and destined to be performed by slaves (Arendt 1958, Chap. 2).

Table 7.1 Daily working hours by sociometabolic regime for an average inhabitant and an average day of the year

		Cases Numbers in (*decimal*) *hours* per day per person	Work in the economy	Household and family work
Local cases		Trinket (hunter-gatherers), India	0.8	2.1
		Campo Bello (swidden agriculture), Bolivia	2.5	2.1
		Nalang (permanent farming, traditional), Laos	3.5	2.1
Country cases	Early industrialization	Japan 1870 (traditional agriculture, beginning of industrialization)	4.4	n.d.
		Germany 1870 (traditional agriculture, cities industrialized)	3.3	n.d.
	Industrialized	France 1998/1999	2.1	2.7
		Netherlands 2000	2.1	2.4
		Germany 2001/2002	2.2	3.0
		Japan 2001	3.3	1.6
		UK 2000/2001	2.4	2.6

Sources: Trinket, Campo Bello and Nalang: Fischer-Kowalski et al. (2011); Japan 1870: Maddison (2001), p. 383; France 1998/1999, Netherlands 2000, UK 2000/2001: European Commission (2003), Eurostat database (2013); Germany 2001/2002: Eurostat database (2013), Statistisches Bundesamt (2006); Japan 2001: Statistics bureau (2011, 2013)

able to survive.[9] For the social community, demand for labor can best be seen as an economic relation, the relation between the benefit the community has (i.e., the marginal return upon additional labor) and the cost of this additional labor. In family and community relations, an additional potential laborer (an additional child, for example, or a second wife) may be sustained, although the benefit of the labor it can deliver is lower than the costs.[10] In strictly economic relations, additional labor will not be employed if the benefits do not outweigh the costs. Beyond social relations, what determines both variables, the benefits and the costs of additional labor, strongly depends on a society's energy regime (see Table 7.1).

[9]Young adult slaves working in the mines of ancient Rome had an average life expectancy of no more than two to three years (Scheidel and Krausmann 2011). Forced laborers in concentration camps in Hitler's Germany fared no better.

[10]Under tight ecological conditions, children or sick people may be sent away (such as children aged 10–12 from alpine villages in Europe in the past few centuries) or seek a ritual death (e.g., traditional Japan).

As long as labor only serves the collection and hunting of food and its preparation for consumption, it does not require much time. Moreover, an increase in labor time under such conditions may well be self-defeating: an increase in hunting and gathering in the same area will tend to deplete the sources of food and force the community to migrate. As documented in research from Cultural Anthropology (Gowdy 1998; Sahlins 1972), the hunter-gatherer regime requires the least amount of human labor from its members.[11] With the transition to the agrarian regime, the amount of labor increases, and it increases even more with the intensification of traditional agriculture (Table 7.1; also Boserup 1981; Ringhofer 2010). At first, the transition to the industrial regime may have increased labor time even further (Voth 2000), but later provided relief. This storyline is reflected in the data compiled in Table 7.1, using the working hours per day per inhabitant as an indicator. The reference to 'inhabitant' (in contrast to the more common 'inhabitant of working age') is fair insofar as the societies compared have a very different age structure and very different standards for appropriate working age. Work in the economy is lowest for the hunter-gatherer regime (0.8 h/day), rises with (traditional) agricultural intensification (with a maximum value of 4.4 h in Japan in 1870) and then drops to slightly more than two hours (Japan: 3.3) under contemporary industrial conditions.

The reason the maximum economic labor time is found under agrarian conditions is the increasing amount of colonizing activities required to feed increasing numbers of people on the same land (Boserup 1965). With the industrialization of agriculture, this situation has been greatly relieved by fossil-fuel-based technologies.

Notwithstanding doubts about limited data quality and comparability, the amount of daily time spent on household and family chores seems to be fairly invariable across sociometabolic regimes (see examples in Table 7.1).

Another approach to quantifying necessary labor is to determine the amount of time the population of a social system spends on food production. This is not so easy to estimate. The most encompassing indicator would be the proportion of the population's lifetime (in a particular year, for example) spent on food production and/or agriculture. In Japan in 1870, for example, if we assume that all the labor time given for 'work' in Table 7.1 above was for food production only, it would have been a lifetime average of 4.4 h per day. Another two to three hours per day would have to be spent on household and family reproduction. This amounts to a lifetime average of approximately seven working hours every day. Considering that in these societies children up to the age of ten (or younger for most working tasks) may make up 30–40 % of the population, the work demand on adults would

[11]Here, we need to address the ambiguity of type 3 activities (community reproduction). As repeatedly documented, hunter-gatherers often have very time-consuming social rituals. In time-use studies or statistics on working time, such rituals are normally not counted as 'labor time'. Considering that this may be the time required to maintain the cohesion of the community, it could count as labor time. However, the activities that belong to that class often bear a much more deliberate and entertaining character than household chores or agricultural work.

typically amount to 10–12 h per day. This seriously encroaches upon the minimal 50 % of lifetime required for basic personal maintenance and leaves no space for anything else.

Yet another way to approach this issue is to look at the proportion of the population that is freed from the necessity of making its living from food production. For several influential social theories, any excess lifetime that does not have to be spent on food production (or, in other terms, the socially available energy not consumed in basic provisioning of the population) is considered a measure of the 'civilization' or 'advancement' of a society (Morgan 1877; Spencer 1862; White 1943). This population share is mainly determined by two mechanisms: one is the productivity of agricultural labor, which depends on natural and technological conditions; the other is the incentive for or pressure upon farmers to produce a food surplus beyond their subsistence requirements. In preindustrial societies, the second mechanism could be estimated by adding up the proportion of agricultural produce that can be taken away from the producers by tithes and taxes. From this, one can conclude how many people can live from the surplus time freed from food-producing labor. If each farming family pays 10 % tithes and taxes, we can assume that at least one non-farm family can sustain itself on ten farm families—or, if we assume a more luxurious life for them, maybe only one non-farm family per 20 farming families. Such information on tithes and taxes exists for many agricultural systems and has been analyzed by historians (e.g., Kulke and Rothermund 2008 for India across history). Unfortunately, we do not know of any systematic compilation of such data across regions and time.

Another approach to this question is to determine the proportion of the urban (as opposed to rural) population. Although estimating urban populations is far from trivial (what size of settlement may be considered urban? How are the boundaries of urban settlements drawn?), a number of historians and modelers have generated historical (i.e., preindustrial) estimates of the size of urban populations. We will draw on these estimates in Sect. 7.3.

7.2.2 How Can Human Labor Power Be Characterized Qualitatively?

For a characterization of human labor power across sociometabolic regimes, we are looking for very abstract dimensions that bear a certain relation to energy because different sources and uses of energy characterize different sociometabolic regimes. One should be able to argue that these capacities are rooted in human nature in the sense that they are part of the natural equipment of every human being, and they should render themselves useful for social enhancement (or suppression) and training (or lack thereof). From the wide range of possible distinctions, we have selected the following three basic capacities for our analysis: physical power (as a capacity of the body), rationality/knowledge (as intellectual capacity) and empathy (as emotional/social capacity).

Physical power: Physical power is the capacity to alter physical objects through force. This capacity is related to the notion of exergy, which is the ability to perform work in a physical sense (see Ayres and Warr 2005). It is also related to the concept of energy efficiency: you can look at the human body as a kind of machine that requires a certain amount of energy input (=food) to perform a certain amount of work (exergy output). Physically speaking, the human body is not a very efficient machine because it needs much energy input just for living (its basic metabolism), and it can transform only a small amount of energy input into useful work and only for a limited fraction of its lifetime. The basic metabolic rate (BMR) depends upon age and body mass. For example, it amounts to 6.2 MJ/day (megajoules per day) for a 50 kg woman and to 9.6 MJ/day for an 80 kg man. This corresponds to fluxes of approximately 70–110 W (watts). However, heavy farm work or sports such as day-long cross-country skiing may double the daily energy demand, with typical kinetic efficiencies of approximately 20 %. Even eight hours of heavy physical work is equivalent to an output of only 2 MJ—corresponding to burning 1 kg of coal in a 10 % efficient engine (Smil 2008, p. 138). Another limitation is the relatively small 'installed power' of the human body; even well-trained young adults cannot sustain a power output flux beyond 150–170 W longer than a few minutes. The peak power delivery of trained humans is 8–12 kW (kilowatts) for several seconds (ibid., p. 134). 'Human effort, even at its best, is a most unimpressive source of mechanical energy', is Smil's summary (ibid., p. 138).

Rationality/knowledge: Rationality/knowledge represents the intellectual capacity to correctly anticipate the effects one's actions will have and to plan these actions deliberately. This capacity is related to information processing and learning from experience as well as from communication with others. Although the human brain in adults is responsible for approximately one-sixth of the BMR, brain work is, energetically speaking, light work—even intensive intellectual activity only marginally raises the brain's metabolic demand (Smil 2008, p. 128). Even more than with physical power, however, the perspective on the individual is too narrow. Rationality and knowledge should be looked upon as social properties, as being developed and maintained collectively, with individuals having only a certain share in this collective propensity. Of course, developing and maintaining a stock of knowledge and information processing generates a certain energy (and labor) demand at the social system level.

Empathy: Empathy is the capacity to emotionally anticipate and mirror the feelings of other living beings. Although modern brain research has demonstrated empathy to be an innate capacity of primates to sense the feelings of others and 'understand' (mirror) the intentions guiding the activities of others (Rizzolatti et al. 2006), this natural capacity should be expected to be strongly influenced by cultural features on the social system level. It should be looked upon not as an intellectual but as an emotional capacity, one that is crucial for human labor, which involves and functionally relies upon communication and caring for the needs of people or other living beings. Empathy as an emotional capacity rooted in a certain neuronal equipment must not be equated with a value orientation of altruism. The ability to mirror the feelings of others may just as well be used to manipulate or harm them more skillfully.

These qualitative features of human labor power have three types of interlinkages to sociometabolic regimes: they can be *functionally* (economically, technologically) more or less relevant for work performance; they may be *socially* (culturally) more or less valued and enhanced or suppressed (investment in education); and, finally, they may be *technologically* more or less supported and enhanced, or they may be more or less substituted by technology. We will follow these different pathways for our subsequent historical analysis.

7.2.3 How Can the Institutional Form of Labor Be Characterized?

In some highly stratified societies, it may be considered unworthy for the 'free man' to work for his subsistence altogether. This view is well represented, for example, by Aristotle, Hesiod and Xenophon for ancient Greece and by Cicero for ancient Rome. These thinkers highly valued a life of leisure and service to the polis for the free and self-determined citizen, but work under someone else's command was considered incommensurable with personal dignity. It is not the physical effort as such but the subjection to personal neediness or the will of others that is despicable. For Cicero, only work in the *artes liberales*, such as architecture, medicine and science, was acceptable; work as a craftsman, day laborer or merchant was unacceptable (cf. Bauer 2013, pp. 312 ff.).[12] In contrast, the Jewish-Christian tradition is marked by dual codes: man's mission is to subjugate (and redesign) the earth by his labor, and Adam was condemned to sustain his life by hard toil. Paul issued the decree that one who did not work should not eat. Later, the rule of the Benedictines saw work as a means to tame intemperance and bodily desires. With Protestantism, labor finally became a sacred duty for everyone whom God rewards with earthly wealth. Obedience is not debasing to a man but makes him agreeable to God (Bauer 2013, pp. 144 ff.).[13]

Upon reviewing the historical and anthropological literature, we identified the following broad classifications for the institutional form in which labor may be organized:

- As **family work** within personally interdependent **household systems** and a mutuality of obligations. Examples are subsistence agriculture, hunting and gathering and household work in most sociometabolic regimes.
- As **slavery**, where a master owns the laborer and has to take care of his/her reproduction or, if cheaper, buy a new one.

[12]In effect, working as a peasant was acceptable both in ancient Greece and Rome—however hard his toil, he was under his own command.

[13]This, of course, is a Western storyline of the cultural framing of labor. There is surely also a storyline for the East, one that we are unfortunately unaware of.

- As other kinds of collective, often **compulsory services**, such as the military, prisoner camps, cloisters and sometimes voluntary work, where individuals invest surplus time for the sake of a community, such as hospice services.
- As serfs within **manorial systems**, where the family receives land from the lord of the manor and owes a share of its produce as taxes and/or compulsory labor in return.
- As **self-employed** in one's own firm/enterprise. Such businesses are often household based, but they sell products and services in markets.
- As **wage labor**, where one is personally free to sell a certain quantity of time in a labor market. This form of labor has undergone an enormous differentiation process regarding professional and hierarchical specializations.

Whereas the Social Science discourses about the institutional form of labor tend to focus on self-determination, hierarchy and exploitation from the point of view of their moral and political legitimacy, a socioecological reflection needs to focus on economic and ecological functionality. Functionality as understood here refers to the effectiveness with which the natural resource base can be utilized for people's benefit at lowest environmental cost. The institutional form of labor is at least as relevant for this functionality as the technologies used.

Ultimately, in every pre-fossil-fuel social system, the amount of available labor depends on land (Sieferle et al. 2006). Because every laborer must use most of the resources he can generate by his labor power just to sustain himself and his family, any effort to have more labor power (and more riches) under control will ultimately lead to efforts to extend the territory. This, in turn, requires military power, which is again based on human labor and animal traction and thus may easily become self-defeating. Moreover, destroying the territorial competitor at least partially destroys his resource base, resulting in an enormous overall waste of resources through military conflicts.

If labor is mainly organized as family work within household systems, there can be only a small degree of division of labor and a low level of communication and learning. If there is a high labor burden, such as in agricultural systems, there is always the incentive to acquire additional wives and children to share the labor and thereby outgrow one's resource base. This organization of labor is also very vulnerable to attacks and raids and therefore may only be able to persist where there is a low population density.

Slavery, in contrast, presupposes territorial conquests that allow the social system to capture or purchase slaves for labor. This allows the system to save on a very costly aspect, namely, the family's investment in bearing children and raising them to adulthood. Thus, if a system is based on slave labor, it can afford, at the same land/food level, up to two-fifths more labor power than when it uses farmers or free laborers.[14] However, slavery is also costly because it

[14]This is estimated on the assumption of a life expectancy of 40 years; each age bracket of eight years is supposed to require the same amount of food (the first age bracket in this calculation also includes the extra food for pregnant and breastfeeding mothers and the food required for children who do not survive to the next age bracket). If slaves are sold and bought at age 16, the buyer saves on two-fifths of the lifetime investment while only losing a small part of the lifetime labor power.

requires a continuously high level of supervision and control, often made more difficult by a lack of shared language.[15] Moreover, losing young people to slavery is extremely costly for the conquered region; it may destroy the resource base there and create the worst enemies one may have. In the long run, it is probably not a very sustainable strategy. It requires continuous military expansion and supervision, and it undercuts the social learning processes to be gained from labor experiences.

Serfdom within manorial systems is a way to organize labor on a household and family basis while at the same time providing some military protection. It creates a tight coupling between the resource base (land) and labor. This retains the features of free family systems of mainly lateral differentiation (low degree of division of labor), low communication and learning and a tendency to outgrow the resource base through demographic expansion. It is better adapted to stable territorial relations, low supervision and peace than slavery systems and has thus proven to be a more efficient way to organize labor (McEvedy and Jones 1978).

Wage labor does not presuppose a coupling between a resource base and the laborer. Like slavery, it is fully flexible with regard to the material and purpose of the work. The burden of supervision is greatly relieved; because the laborer must sell his/her labor power in a market, it is in his/her interest to perform the work properly to retain market value. At the same time, the laborer must be willing to learn and will be confronted with various experiences and communication contexts that promote learning. This learning also yields benefits at the system level. It is critical, however, that there be a market with sufficient demand for labor and ways to maintain subsistence through non-working times. Wage labor as such is very well adapted to avoiding resource waste. Moreover, it does not create incentives for fertility. Children do not relieve one from labor; on the contrary, they consume time and create additional costs.

7.3 Human Labor in Different Sociometabolic Regimes[16]

As stated, although hunter-gatherers do work according to the standards of our time-use distinctions, their work is very close to what other social animals need to do to sustain themselves. With the transition to agriculture, labor becomes a much more pronounced feature of specifically human existence—both qualitatively and quantitatively.

[15]McEvedy and Jones (1978) consider this a main reason for population stagnation after a long period of growth in the ancient empires of the West and East, before the advent of feudalism.

[16]For a more elaborate explanation of sociometabolic regimes, see Krausmann and Fischer-Kowalski (2013).

7.3.1 Labor in the Agrarian Regime

7.3.1.1 Quantitative Features

Quantitatively, a critical question is how many people can be sustained from a certain piece of land and how many additional people not working the land (e.g., landlords, urban citizens, soldiers) can be subsidized. As Boserup (1981) has shown, there is a tendency to develop techniques that allow more people to live from a piece of land by intensifying land use at the expense of investing additional labor. The increased labor burden creates an incentive to have more children to share the workload. This triggers population growth, lowering labor productivity even more. If population pressure on the land is reduced by, for example, labor opportunities in urban centers, agricultural labor productivity may rise again and allow for increasing surplus production that then allows a larger urban population to be fed.[17] Nevertheless, working hours in mature agrarian systems tend to be very high (Clark and Haswell 1967, p. 3; Fischer-Kowalski et al. 2011; Fischer-Kowalski 2011). In the agrarian regime, the overwhelming majority of the population (including children and elderly) is occupied with food production most of their available lifetime. This is related to the relatively low energy return on investment (EROI)[18] of agriculture and the focus on humans and animals as the main sources of mechanical power (depending on land productivity, the EROI lies somewhere between 10:1 and 2:1; for example, maize has a 4.1:1 EROI in a draft animal agricultural system; see Pimentel et al. 1999; see also Chaps. 4 and 21 in this volume). The proportion of the population that can be sustained from the surplus of agricultural labor, even in advanced agrarian systems, ranges between 5 and 15 %. This (low) proportion is reconfirmed by typical rates of taxation and rates of urban populations across the preindustrial history of countries (see Table 7.2).

In Table 7.2, we present a few examples for the time around 1500, as this is a period in which we can expect urban settlements to sustain themselves exclusively on contributions from traditional agriculture. As becomes apparent from these numbers, population shares in settlements of 2500 or more inhabitants vary between 2 and 10 % across world regions. This means that on average for the world, it took 25 peasants to feed one urban citizen. Even in Renaissance Western Europe, 11 farmers had to contribute.

7.3.1.2 Qualitative Features

Qualitatively, the agrarian regime relies mainly on the **physical power** (and physical endurance) aspect of human labor power. This applies to the rural population

[17]This is at the core of the Nobel Laureate William Arthur Lewis' influential 'dual sector model' (Lewis 1955).

[18]More precisely, one ought to be talking about 'energy return upon energy investment' (EROEI) which measures the net energy gained by the effort.

Table 7.2 Share of urban population in 1500 by world region (settlements with 2500 or more inhabitants)

World region	Share of urban population (%)
Northern Africa	2.04
Japan	2.90
India	4.36
China	6.45
Western Europe	10.63
World	4.14

Source: Klein Goldewijk et al. (2010)

(constituting the very large majority) and, for example, to slave labor in mines and infrastructure building (if we think of the Roman Empire). There is little societal effort to improve the skills and knowledge base of those 90–95 % of the population in agricultural labor—as long as they feed themselves and deliver their tithes and taxes, they are left to themselves to organize their work. With the fellachs in ancient Egypt, with the lower castes in India, with the feudal serfs in Europe or with the African slaves in the Southern states of America, no effort is made to spread literacy or practical knowledge concerning work in agriculture.[19] Education, in the sense of societal investment in the skills and the knowledge capacity of physical labor, is largely absent. This investment remains reserved to a small minority of privileged, usually male urban elites liberated from the need to work for their subsistence, and is largely disconnected from what may be considered 'productive work' (Sohn-Rethel 1970). Religious castes and organizations contribute to the education of ideological elites[20] and to the religious indoctrination of children (e.g., Sunday schools, Qur'an schools) but do not convey functionally useful knowledge to those whose labor sustains society. Whatever great civilizational gains were achieved are rarely connected to the mass of human labor. The social tensions created by this highly unequal distribution of knowledge (even without any relation to practical application) in Europe were reflected in the widespread religious conflict over the Bible and the right of everyone to read it for himself in Europe from the 14th to the 16th century and in the efforts from above to maintain knowledge monopolies, as reflected in the burning of Giordano Bruno in 1600.

Technological enhancement and the replacement of human physical power are mainly sought in animal power: buffaloes, oxen and later horses, and elephants and camels in other regions are used for traction of ploughs, water pumps and

[19]Some exceptions exist, such as the Roman Empire's writers on agricultural technology, transmitted at least to an intermediate stratum of administrators of large estates.

[20]This certainly holds for the universities founded in Europe from the late 11th century onward. However, monasteries and religious communities in Europe (and probably similarly in the rest of the world) sometimes played an important role in systematic improvements of agriculture through experimentation and learning. Nevertheless, this does not mean that they transferred the control of such knowledge to their inferiors, let alone spread it among regional peasant populations.

carriages, whereas donkeys are used for carrying burdens. This technological solution draws on the human labor capacity for **empathy** because educating and guiding working animals requires a certain understanding and concern for their needs and feelings. Animals may be useful or even indispensable for having a higher power density than humans (details in Smil 2008, p. 174), but they do not quantitatively replace humans' physical labor (which is still required for working with and feeding those animals). In general, agrarian systems are marked by a substantial degree of public cruelty and by a cultural emphasis on heroism and the use of force. Thus, it should not be expected that the evolution of empathy will receive much social enhancement.

7.3.1.3 Institutional Form of Labor

Labor in agriculture (>90 %): In peripheral or unproductive regions (such as mountainous areas, marshes and sparsely populated arid regions), agricultural or pastoralist labor is usually organized as household-based family labor in accordance with family power structures and is largely subsistence oriented.

In more productive regions, labor is organized in some form of manorial (or, more precisely, feudal) systems with bonded serfs or slaves who owe a defined proportion of their produce in the form of naturalia, labor or money to the landlord.[21]

Labor outside agriculture (<10 %) can be organized as slave and compulsory labor (for example, in mining or construction) or as household-based self-employment (for example, in crafts, trades or transportation). The 'atomization of production was the rule…' (Christ 1984, p. 2).

Another common and important form of 'labor' is **military service**. It may be considered labor in the sense of providing food, animals, slaves and treasures by looting and by protecting one's own population from looting. However, it cannot be considered labor in the sense of producing resources; it does so only by redistributing resources between enemies and friends.

7.3.2 Labor in the Coal-Based Industrial Regime

7.3.2.1 Quantitative Features

Quantitatively, the unfolding of the coal-based industrial regime multiplies the demand for labor. Industrial labor is mainly the exertion of physical power, and much additional physical power is brought into the economy from coal-driven steam engines. Nevertheless, the demand for human labor is increasing so much

---

[21]It is interesting to note that an author such as Adam Smith comments on the relation between this form of labor organization and low labor productivity in agriculture (Smith 1776, Chap. 2).

that even the rapid population growth ('first demographic transition') can be absorbed. For the industrial workforce, only humanitarian legal efforts gradually achieve a reduction of daily working hours and a ban on child labor. Industrial labor is cheap, and the profits to be gained from it far surpass the surplus to be gained from land ownership. In this phase, there is clearly a positive relation between energy input into the economy and the number of labor hours: more energy does not replace but rather facilitates the use of additional human labor. In the UK, the total number of labor hours in the economy rose in line with energy use until the First World War (WWI) and in Austria, even until the end of World War II (WWII). There was a reduction of working hours when energy use stagnated, a pattern that corresponded to the period when coal dominated the energy regime. In contrast, after WWII (when the use of oil became dominant), energy use in both countries increased and labor decreased, reflecting a substitution of labor with technical energy. For the rest of the 20th century, there is no clearly discernible relation between energy use and labor hours (see Fig. 7.1).

Historians of time use (Voth 2000) have even documented that in the early phase of industrialization in the UK (18th century), the weekly working hours of urban laborers rose above the level of the previous agrarian conditions.

7.3.2.2 Qualitative Features

The **coal-based industrial regime** at the onset mainly added **physical power** through steam engines driving water pumps (in mines) and weaving looms in manufacture.[22] This additional physical power did not so much replace human labor as increase its demand because production processes could be realized at a lower cost and at a larger scale. For agriculture, there was only an indirect impact, which came through allowing (and, in part, forcing; see Wrigley 1988, 2010 for the UK) the rural population to migrate to urban centers and to **make their living on wage labor in manufacture**. Later, railways facilitated the transport of coal and food over large distances into these urban centers, thus allowing them to grow and at the same time stimulating rural surplus production. In their impact on manufacture (so-called proto-industry) and, later, industrial labor, steam engines did not improve the skill component in human labor but rather stupefied labor (see Marx and Engels 1848). The key capacity of human labor in agriculture as well as in urban wage labor remains **physical power and endurance**. Small skill segments, however, evolve further—in urban craftsmanship and engineering, in trade and finance, in the military and among civil servants.

During this regime, most countries started to introduce **publicly financed compulsory schooling for children** (including the lower classes in urban centers). Often under the supervision of the clergy, children were expected to learn

[22]The first and (for a long time in the UK) most important impact of coal was its use as a substitute for wood in heating and cooking in emerging urban centers that could not otherwise have grown in size (Krausmann and Schandl 2006).

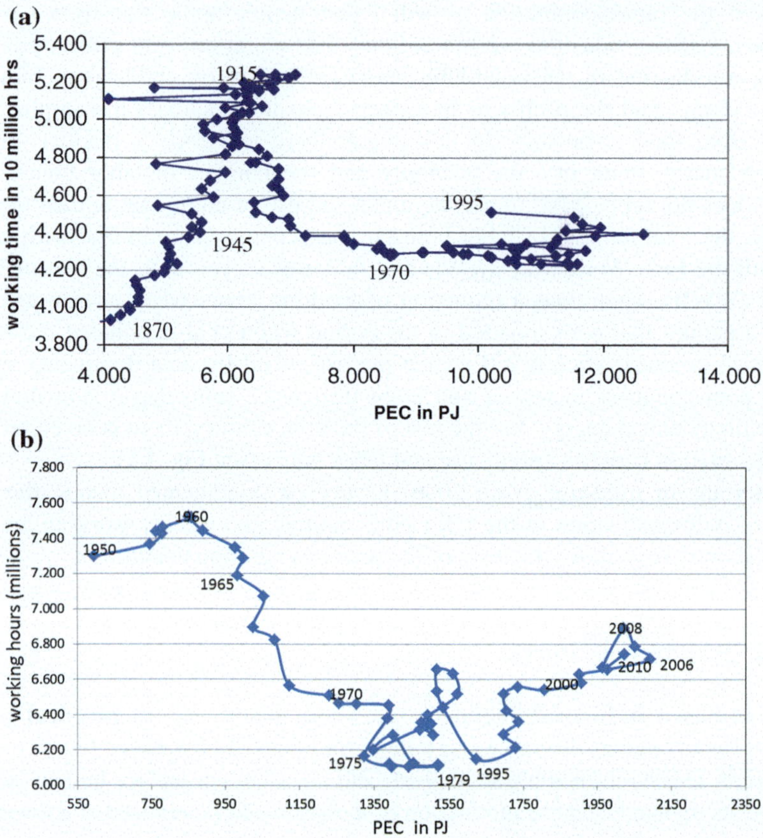

Fig. 7.1 **a** Primary energy consumption (PEC) and working hours in the UK, 1870–2000. (Source: after Krausmann et al. 2003; Schandl and Schulz 2002). **b** PEC and working hours in Austria, 1950–2010. (Source: PEC: Krausmann et al. 2003 (updated version including PEC data to 2010); working hours: TED (The Conference Board Total Economy Database™, January 2013))

reading, writing and simple forms of algebra in addition to religious beliefs. As Gellner (1988) plausibly argues, this mainly related to the functionalities of the modern nation state and the requirements of its military and had very little meaning for 'qualifying labor'. However, it created a need for teachers as perhaps the first laborers who were mainly qualified by formal education.

With **empathy**, one can observe an increasing cultural differentiation by gender: whereas men, in their work and beyond, are supposed to be tough and contain their emotions, women are supposed to be sympathetic and emotional. Empathy, one might say, becomes a female virtue, but a virtue after all (Badinter 1980; Elias 1939).

7.3.2.3 Institutional Form of Labor

The most spectacular change during this regime is the rise of free wage labor. Free wage labor, a very minoritarian form at first, increases to become the most dominant institutional form. Gradually, often by revolutions, serfdom and slavery are abolished.

In contrast to the landed aristocracy of the agrarian regime, industrialists see themselves as hard-working, as responsible for the labor process and as drivers of technical innovation. Capitalists do not see themselves as a leisure class but feel obliged to frugality and work ethics (Weber 1920).

During this phase, the separation of a sphere of production and gainful employment from the sphere of reproduction as a cozy and secluded home wisely governed by a housewife (who is not seen as 'working' but as exercising love and care) becomes an urban middle-class model that gradually spreads to other social strata (Bolognese-Leuchtenmüller and Mitterauer 1993).

7.3.3 Labor During the Rise of the Oil-Based Industrial Regime (Europe: Late 1940s to Early 1970s)

7.3.3.1 Quantitative Features

Primary energy consumption (PEC) in the economy rises, but overall labor hours decline; energy input per labor hour is no longer stagnant but rises rapidly. This novel 'substitution' effect of mechanical energy for human labor can be clearly seen in Fig. 7.1 for the UK after the World Economic Crisis in the 1920s and for Austria in the post-WWII period. From then on, the further increase in energy input is associated with a decline in labor hours: mechanical energy substitutes for labor. Until the early 1970s, there is a steady increase in energy input into the economy, and with the increase in energy, working hours decline. This is the 'golden age' of building up the welfare state, boosting private consumption, steadily increasing wage levels and reducing working time. It is also the 'golden age' of the consequences of expanding the education system becoming statistically visible in the rapid increase of 'white collar labor'[23] over 'blue collar labor' and the near disappearance of agricultural labor.

The same message Fig. 7.1 conveys for the UK and Austria, Fig. 7.2 conveys for Germany and Italy. With the implementation of the 'oil regime' after WWII, human labor hours in the economy completely dissociate from energy input. Whereas labor hours show a slight decline, energy use soars, as does the energy

[23]'White collar labor' versus 'blue collar labor' characterizes this distinction better than the more common distinction between industrial production and services, although these distinctions, of course, overlap. See also Peter Drucker, who coined the notions 'The employee society' (Drucker 1953) and 'Wissensgesellschaft' (Drucker 1969).

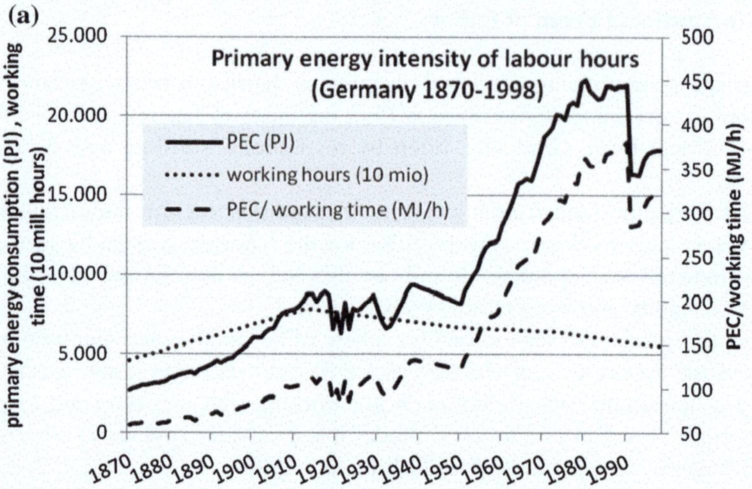

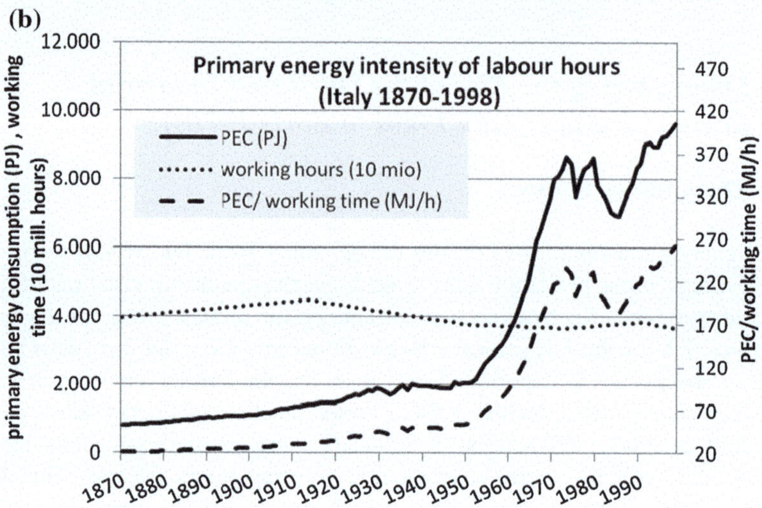

Fig. 7.2 Annual energy consumption, working hours and energy intensity of working hours for (**a**) Germany and (**b**) Italy, 1870–1998. (Sources: Cleveland (2011), Maddison (2001, 2008); PEC calculated based on background data from Pallua (2013))

intensity per labor hour. This same pattern can be found for all European countries. The decline in employment in agriculture, where the working hours per employee had been particularly high relative to all other economic sectors, plays an important role in the decline in labor hours.

Somewhat similar changes occur in the households: electric equipment (e.g., washing machines, vacuum cleaners, mixers) substitutes for physical effort from the housekeepers, and it raises the intellectual demand to handle those machines. As has been demonstrated in a number of studies, however, the overall impact is not to reduce household work because purchasing and servicing this equipment, in combination with larger homes and higher standards of order and cleanliness, costs considerable time. In combination with the gradual disappearance of servants, the household burden upon middle-class women tends to increase.

7.3.3.2 Qualitative Aspects of Human Labor

Liquid fossil fuels and electricity allow for the **substitution of the physical power dimension of human labor by decentralized energy services**. Key technologies are the internal combustion engine used for cars and multipurpose electro-motors linked to electricity grids or powered by batteries. Liquid fossil fuels used for tractors and in chemical conversion for mineral fertilizers and pesticides also substitute for a large part of physical human and animal labor in agriculture. In effect, physical strength and prowess lose much of their economic and, consequently, cultural value.

Instead, the **knowledge dimension of human labor** becomes much more important. There is unprecedented growth in public education and knowledge production. This is the 'golden age' of expanding the public education system, propagating equal opportunities and building up a skilled workforce with capacities in information and knowledge management rather than physical power and endurance. Knowledge production, information processing and communication become major economic activities. For the first time in history, knowledge production and learning cease to be class privileges and ideological bastions; they become secular, rational and functionally related to roles in the labor market.[24] In 1973, Daniel Bell published *The Coming of Post-Industrial Society*, in which he outlined a vision of a knowledge-based service society that would overcome both the farm's and the factory's hardships.[25]

Regarding **empathy**, the gendered picture predominates: toughness and rationality for men, empathy and emotionality for women. Women as loving housewives taking care of husbands and children becomes the majority model of middle-class life.

7.3.3.3 Institutional Form of Labor

In this phase, wage labor becomes the most dominant form of labor by far. Self-employment both in agriculture and in other sectors declines, whereas employed

[24]It would be a promising exercise to document this in the OECD reports on education from the early 1960s onward. This trend was often criticized by more traditional, humanistic educational professionals. Interesting, however, is that the previous tension between religious/denominational and public/secular education that had blocked educational reforms in so many countries for such a long time gradually faded away.

[25]In politics, the term knowledge society boomed much later (and is now, for example, part of the EU's future perspectives).

labor rises. The overall participation rates in gainful employment remain largely constant. Within wage labor, there is a shift from 'blue collar' to 'white collar' labor. From the end of WWII to the early 1970s, unemployment rates remain very low.

7.3.4 Labor in the Transition Phase from the Early 1970s Onward

7.3.4.1 Qualitative Features of Labor

One might draw the following analogy: just as technological development plus increasing fossil fuel use had substituted for much of human physical work, so information and communication technologies are now **substituting for knowledge work**. Substituting for knowledge work is inherently less energy-intensive than substituting for physical work, even if it is not optimized in this direction. Nevertheless, knowledge production and knowledge handling remain key features of human work.

Coinciding with the first world oil crisis in 1973, structural change in the relation between energy and labor becomes apparent: the trend of steeply increasing primary energy input in high-income countries is over and gives way, after some sharp fluctuations, to a more stationary energy consumption, both overall and per working hour (see Figs. 7.1 and 7.2). There is no longer a discernible correlation between energy use and working time.[26]

The reduction of physical work in Europe was, of course, also greatly enhanced by the externalization of industrial production to the world's periphery, where emerging economies with very low labor costs were prepared to produce the steadily increasing amount of industrial products that Europe and other rich regions of the world wished to consume. Studies of carbon emissions from trade (Hertwich and Peters 2009) have shown, for example, that the apparent domestic growth reduction in fossil-fuel-based energy was—at least to a certain degree—compensated for by rising fossil fuel combustion elsewhere.

Intellectual educational standards in the labor force continue rising, as does school and university enrollment. Qualified white collar work increases, whereas industrial blue collar work continues to decline.

There are indications that—connected to the rising importance of marketing, services and communication processes—the capacity for **empathy** is gradually losing its exclusive female label and becoming a more important qualification for work generally.

[26]For the same period, Ayres observed a loosening of the very tight ties between exergy and economic output (Ayres and Warr 2005). He interprets this as an effect of the shift toward information and communication technologies (ICT).

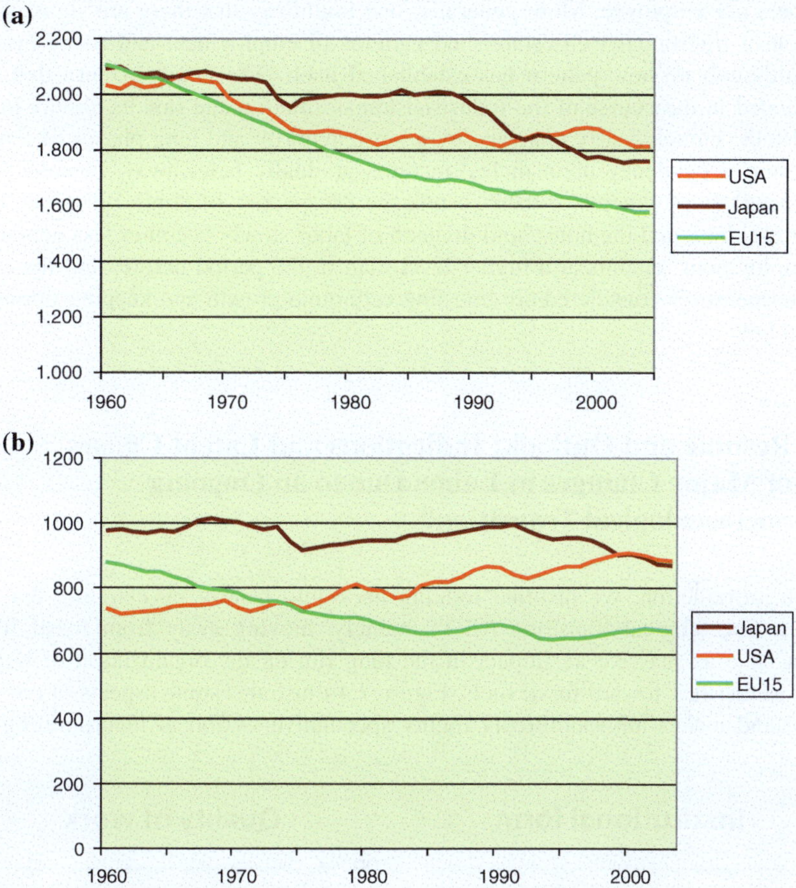

Fig. 7.3 **a** Annual working hours per employee. (Source: Maddison 2001, 2008; OECD 2000). **b** Annual working hours per inhabitant. (Source: Maddison 2008; OECD Stat Extracts, http://stats.oecd.org/Index.aspx?DatasetCode=ANHRS; own calculations)

7.3.4.2 Quantitative Features and Institutional Form

In Europe, the average annual working hours per inhabitant decline very little in the early 1970s, much less than before (Fig. 7.3b), but the working time per employee continues to decline (Fig. 7.3a). This is a symptom of increasing part-time work (particularly by females), unemployment and rising flexibility in the use of labor power.[27] Whereas Japan shows trends of declining working time similar to Europe, the US shows increasing working time per inhabitant, with stagnating

[27]In the sense of setting paid labor time on or off according to demand (in retail sale, for example, interrupting working time during the day when there is less demand or reducing cleaning services in offices during holiday periods).

numbers per employee. More generally, one might say that there are signs of the erosion of traditional well-established patterns of employment and rising insecurity, although no new pattern has established itself. The family pattern that was introduced in the course of the industrial transformation and saw its climax in the late 1960s, namely, early marriage by a large majority and long phases of female economic dependency upon males' income, gradually fades away. Females seek (and need) employment for their sustenance irrespective of family ties, they bear fewer children, and the household division of labor slowly becomes less gendered. Unemployment remains at a higher level than in the period before, and the main countermeasures considered are boosting economic growth and keeping immigration at bay.

7.4 Resume and Outlook: Indications and Latent Causes of Major Changes in Labor Due to an Ongoing Socioecological Transition?

In the introduction, we justified looking back into history by claiming that the next socioecological transition (SET), namely, moving away from fossil fuels, might have as massive an impact in the long run on the organization of human labor as the SET toward fossil fuels. Figure 7.4 illustrates some aspects of our storyline and makes an—admittedly highly speculative—effort at incorporating the

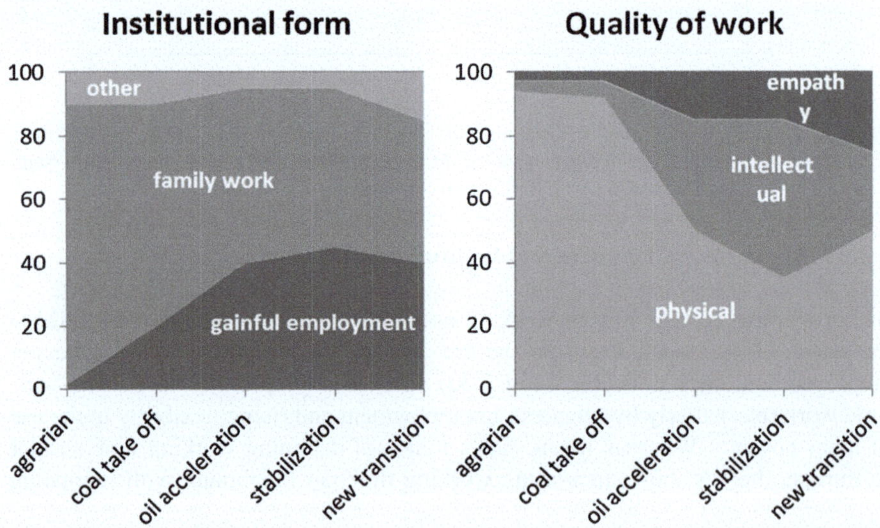

Fig. 7.4 Variation in the quality of work and its institutional form by sociometabolic regimes (work including market-oriented employment and nonmarket subsistence work, incl. household & nonmarket community work; Europe only)

structural changes we anticipate from a new sociometabolic transition and the effects these elements will have on the future of work.

The structural changes we envision in the course of the new transition in Europe would encompass the following:

- an energy shift away from fossil fuel use toward renewable energy
- a production and consumption shift away from energy—and materials-intensive products toward services enhancing human resources and capacities (supported by demographic change)
- an institutional shift toward low-maintenance infrastructures that have a lower risk of climate change impacts
- decreasing energy consumption (efficiency increases, savings)
- decreasing use of (virgin) raw materials (shift to nonmaterial energy sources, efficiency increases, recycling, reduced consumption)
- decline in outsourcing production from Europe (because of slowly decreasing wage differentials)

Why can we assume that these structural changes might lead to the changes in work we picture in Fig. 7.4?

In the right part of Fig. 7.4, we project[28] that the proportion of physical work, after a long period of decline, will rise again. This follows from the assumption that rising energy generation expenses and declining EROI will make energy more costly and less abundant. This is already observed for fossil fuels, where 'conventional' resources are becoming depleted and new, 'unconventional' resources that require much higher energy investments, such as tar sands, are increasingly used (Murphy and Hall 2010), although this trend has not yet had a major impact on energy prices. Some argue that a declining EROI can also be expected for renewable energy. In our reasoning, a decline in continuously available low-price energy could lead to a reduced substitution of human labor by mechanical energy and to an increased use of very intelligent but mechanical tools and devices. In urban areas, walking and cycling might substitute for motor-driven vehicles, in part because additional exercise benefits health.

The existing 'green jobs' reports (such as UNEP 2008) and the 'European Strategy Agenda 2020' do not elaborate on the of quality of 'green' labor in terms of physical work, intellectual capacity or empathy demanded. Recent studies for the US (Mattera et al. 2009) and for Austria (Leitner et al. 2012) identify forestry and agriculture,[29] the construction industry, waste management and trade and transport

[28]It should be noted that the numerical values in Fig. 7.3 are only illustrative. The reference frame of 100 % refers to the total of human working hours outlined in the time budget approach explained in Sect. 7.1. For these working hours, no reasonable statistics exist that would allow for a quantitative historical comparison of the quality and institutional form of labor as we attempt in Fig. 7.4.

[29]There are also other arguments for why the decline in agricultural labor in Europe may be reversed in the future. This reversal may occur due to, for example, a health-oriented increase in organic farming, decentralized energy generation or higher costs of fossil fuel-based (labor-saving) supplies.

as the main sectors of new ‘green jobs’. Physical labor is clearly in demand for these sectors. Another line of reasoning sees the increasing frequency of extreme climate events as a source of additional physical labor, be it in the form of gainful employment, of nonmarket civil services (‘other’) or of family labor in coping with such events. This is reflected on the left side of Fig. 7.4 as a possible increase in nonmarket forms of labor (family and ‘other’) at the expense of gainful employment.

The above figures also indicate a continuing process of substitution of (particularly medium-qualified) intellectual or knowledge work by ICT and, eventually, its global outsourcing to lower-income countries, also facilitated by the use of ICT. The only services that are very difficult to substitute by ICT and nearly impossible to outsource to other countries are those that involve face-to-face contact with the resident population, namely, various forms of caretaking. In view of an aging population that is increasingly culturally heterogeneous and demanding, we assume an increase in the type of work that is based on empathy (at the expense of medium-qualified intellectual work) in the institutional form of collective services, family work and gainful employment.

There are new framework conditions that may have a long-lasting structural impact on work beyond the features described above. After many decades of decline, there is now (since approximately the year 2000) a sharp rise in the prices of all raw materials (commodities). Although some believe this to be a transitional phenomenon due to lagging investments, we see many indications of approaching scarcity or of rising efforts in the extraction of material and energy (Mudd 2010). If this should be the case, it might have two substantial impacts. First, the share of jobs to supply society with material and energy would rise[30] due to both lower energy returns on energy investments and declining ore grades. Second, if commodity prices (including energy) continue to be high or even rise further, this could substantially alter business strategies. There could be a shift in the dominant mode of cost reduction from labor to resources. In this case, it is not the increase in labor productivity that would be the key measure but the saving of resources, possibly at the expense of more labor (see, for example, Dobbs et al. 2012). Macroeconomically, this would mean that there is a shift in relative prices between material goods and human labor and, consequently, a decline in demand for material goods and increasing demand for human labor. Macroeconomic growth, as far as it depends on rising labor productivity, would be impaired. Increasing the share of work in caretaking, as assumed above, would have an impact in the same direction, as labor productivity cannot be enhanced much my caretaking. Furthermore, resource-saving jobs, such as renovation, repair, remaking and reusing, might gain momentum. In effect, if the purchasing power of workers is reduced, distributional conflicts over wages should become more frequent.[31]

[30]Whereas Japan in 1870 had to spend approximately 37 % of the available work time to supply their society with energy (see calculations from Table 7.1), this has declined to approximately 8 % in Europe today. Thus, an increase of several percent does not seem unrealistic.

[31]This vision strongly resembles the projections of Randers (2012). For the OECD countries, he projects labor productivity increases will lose their dynamics, consumption will stagnate or even decline because of rising shares of investment (required in adaptation to climate change, for example) and social conflicts will increase.

Finally, our societies might become less energy intensive. If the world seeks to avoid dangerous climate change (i.e., a rise in average temperature beyond two degrees), most simulations assume a global decline in primary energy use of 1 % annually (see, for example, GEA 2012; WBGU 2011). If this assumption were to be realized globally, the required decline in primary energy use for Europe would need to be much steeper. Part of this decline can be realized by avoiding losses,[32] but more expensive energy might lead to lower use. Could our societies gradually slow down again?

Acknowledgements This work originated from our contribution to the EU FP7 project 'NEUJOBS' (www.neujobs.eu) in April 2012. Our task in this project was to provide theoretical guidelines for socioecological transitions in a way that is meaningful to our partner research institutions specialized in Labor Economics or educational and demographic analyses about the future of the European labor market. The ensuing discussions and the seriousness with which our assumptions were met encouraged us to proceed on this pathway.

References

Arendt, H. (1958). *The human condition*. Chicago: Chicago University Press.

Ayres, R. U., & Warr, B. (2005). Accounting for growth: The role of physical work. *Structural Change and Economic Dynamics, 16*(2), 181–209.

Badinter, E. (1980). *L'Amour en plus: Histoire de L'amour maternel XViie - XXe siècle*. Paris: Flammavarion.

Bauer, J. (2013). *Arbeit. Warum unser Glück von ihr abhängt und wie sie uns krank macht*. München: Blessing Verlag.

Bolognese-Leuchtenmüller, B., & Mitterauer, M. (1993). *Frauen - Arbeitswelten. Zur historischen Genese gegenwärtiger Probleme*. Wien: Verlag für Gesellschaftskritik.

Boserup, E. (1965). *The conditions of agricultural growth. The economics of agrarian change under population pressure*. Chicago: Aldine/Earthscan.

Boserup, E. (1981). *Population and technological change—A study of long-term trends*. Chicago: The University of Chicago Press.

CEDEFOP (European Centre for the Development of Vocational Training). (2009). *Future skill needs for the green economy*. Report. Luxembourg: Publications Office of the European Union.

Christ, K. (1984). *Die Römer. Eine Einführung in ihre Geschichte und Zivilisation*. München: C. H. Beck.

Clark, C., & Haswell, M. (1967). *The economics of subsistence agriculture*. London: McMillan.

Cleveland, C. J. (2011). *Podobnik energy datasets*. 2013. Available from http://www.digitaluniverse.net/energytransitions/view/exercise/51cbf0fa7896bb431f6a3cc8/

Dobbs, R., Oppenheim, J., & Thompson, F. (2012). *Mobilizing for a resource revolution*. In: McKinsey Quarterly January.

Drucker, P. F. (1953). The employee society. *American Journal of Sociology, 58*(4), 358–363.

Drucker, P. F. (1969). *The age of discontinuity: Guidelines to our changing society*. New York: Harper & Row.

[32]With coal-powered electricity generation, for example, a high proportion of the primary energy contained in coal is lost as waste heat. In the case of photovoltaic or wind power, this kind of loss does not occur. Another major case can be made for improving the insulation of buildings.

Elias, N. (1939). *Über den Prozeß der Zivilisation. Soziogenetische und psychogenetische Untersuchungen*. Basel: Haus zum Falken.
Eurofound/Gaušas, S., Caturianas D., Jonušauskaitė, G., Mackonytė, G., Muravjova, O., Nakrovas, V., Paliokait, D. A., Repeokait, D. D., Pocius, D., Stanusait, D. V., & Šndtanu, T. (2012). *Greening of industries in the EU: Anticipating and managing the effects on quantity and quality of jobs*. Report. Lithuania: Public Policy Management Institute.
European Comission. (2003). *Time use at different stages of life—Results from 13 European countries*. Working Papers and Studies, Luxembourg, 2003.
Eurostat database. (2013). *Population on 1 January*. Extracted via source dataset August 2013 from http://appsso.eurostat.ec.europa.eu/nui/submitViewTableAction.do
Fischer-Kowalski, M. (2011). Analyzing sustainability transitions as a shift between socio-metabolic regimes. *Environmental Innovation and Societal Transitions, 1*(1), 152–159.
Fischer-Kowalski, M., Singh, S. J., Ringhofer, L., Grünbühel, C., Lauk, C., & Remesch, A. (2011). Socio-metabolic transitions in subsistence communities. *Human Ecology Review, 18*(2), 147–158.
GEA. (2012). *Global energy assessment—Toward a sustainable future*. Cambridge University Press, Cambridge UK and New York, NY, USA and the International Institute for Applied Systems Analysis, Laxenburg, Austria.
Gellner, E. (1988). *Plough, sword and book*. London: Collins Harvill.
Gershuny, J. (2000). *Changing times: Work and leisure in post-industrial societies*. Oxford: Oxford University Press.
Giampietro, M., Mayumi, K., & Sorman Alevgül, H. (2012). *The metabolic pattern of societies. Where economies fall short*. New York: Routledge.
Gowdy, J. (1998). *Limited wants, unlimited means: A reader on Hunter-Gatherer economics and the environment*. Washington D.C.
Hans Böckler Stiftung. (2001). *Pathways to a sustainable future, results from the work & environment interdisciplinary project*. Düsseldorf: Hans Böckler Stiftung.
Hertwich, E. G., & Peters, G. P. (2009). Carbon footprint of nations: A global, trade-linked analysis. *Environmental Science and Technology, 43*(16), 6414–6420.
Klein Goldewijk, K., Beusen, A., & Jansen, P. (2010). Long-term dynamic modeling of global population and built-up area in a spatially explicit way: HYDE 3. *The Holocene, 2010*(20), 565–573. doi:10.1177/0959683609356587.
Krausmann, F., & Fischer-Kowalski, M. (2013). Global socio-metabolic transitions. In S. J. Singh, H. Haberl, M. Chertow, M. Mirtl, & M. Schmid (Eds.), *Long term socio-ecological research. Studies in society—Nature interactions across spatial and temporal scales* (pp. 339–365). Berlin: Springer.
Krausmann, F., & Schandl, H. (2006). Der soziale Metabolismus der Industrialisierung. Die Überwindung der energetischen Schranken des agrarischen Wirtschaftens. *GAIA, 15*(4), 285–293.
Krausmann, F., Schandl, H., & Schulz, N. B. (2003). *Long term industrial transformation: A comparative study on the development of social metabolism and land use in Austria and the United Kingdom 1830–2000*. Social Ecology Working Paper 70. Vienna: Social Ecology.
Kulke, H., & Rothermund, D. (2008). *Geschichte Indiens von der Induskultur bis heute*. München: Beck Verlag.
Leitner, A., Wroblewski, A., & Littig, B. (2012). Green Jobs. *Diskussion von Arbeitsbedingungen und Beschäftigungspotentialen*. Vienna.
Lewis, W. A. (1955). *The theory of economic growth*. London: Allen & Unwin.
Maddison, A. (2001). *The world economy. A millenial perspective*. Paris: OECD.
Maddison, A. (2008). *Historical statistics for the world economy: 1-2006 AD.*, online data base. http://www.ggdc.net/maddison/
Marx, K. (1867). *Capital*. London: Penguin. (Volume 1).
Marx, K., & Engels, F. (1848). *Manifest der kommunistischen Partei*. Berlin: Dietz Verlag.
Mattera, P., Dubro, A., Gradel, T., Tompson, R., Gordon, K., & Foshay, E. (2009). *High road or low road? Job quality in the new green economy. Report*. Washington: Good jobs first.

McEvedy, C., & Jones, R. (1978). *Atlas of world population history, facts on file*. Harmondsworth and New York: Penguin.

Morgan, L. H. (1877). *Ancient society*. New York: H. Holt and Co.

Mudd, G. M. (2010). The environmental sustainability of mining in Australia: Key mega-trends and looming constraints. *Resources Policy, 35*(2), 98–115.

Murphy, D., & Hall, C. (2010). Year in review-EROI or energy return on (energy) invested. *Annals of the New York Academy of Sciences, 1185*, 102–118.

OECD/Evans, J. M., Lippoldt, D. C., Marianna, P. (2000). *Trends in working hours in OECD countries*. Paris: OECD. http://www.olis.oecd.org/OLIS/2000DOC.NSF/LINKTO/DEELSA-ELSA-WD(2000)6

OECD/Martinez-Fernandez, C., Hinojosa, C., Miranda, G. (2010). *Green jobs and skills: The local labour market implications of addressing climate change*, Working document, CFE/LEED, OECD, www.oecd.org/dataoecd/54/43/44683169.pdf?contentId=44683170

Pallua, I. (2013). *Historische Energietransitionen im Ländervergleich. Energienutzung, Bevölkerung, Wirtschaftliche Entwicklung*. Social Ecology Working Paper 148. Vienna: Social Ecology.

Pimentel, D., Pimentel, M., Karpenstein-Machan, M., Srzednicky, G. S., Driscoll, R. H. (1999). *Energy use in agriculture: An overview*. CIGR E-Journal, *1* (International commission of Agricultural Engineering).

Randers, J. (2012). *2052: A global forecast for the next fourty years*. White River Junction (Vermont): Chelsea Green Publishing.

Ringhofer, L. (2010). *Fishing, foraging and farming in the Bolivian Amazon. On a local society in transition*. New York: Springer.

Rizzolatti, G., Fogassi, L., & Gallese, V. (2006). Mirrors in the mind. *Scientific American, 295*, 54–61.

Sahlins, M. (1972). *Stone age economics*. New York: Aldine de Gruyter.

Schandl, H., & Schulz, N. B. (2002). Changes in United Kingdom's natural relations in terms of society's metabolism and land use from 1850 to the present day. *Ecological Economics, 41*(2), 203–221.

Scheidel, A., & Krausmann, F. (2011). Diet, trade and land use: A socio-ecological analysis of the transformation of the olive oil system. *Land Use Policy, 28*(1), 47–56.

Sieferle, R. P., Krausmann, F., Schandl, H., & Winiwarter, V. (2006). *Das Ende der Fläche. Zum gesellschaftlichen Stoffwechsel der Industrialisierung*. Köln: Böhlau.

Smil, V. (2008). *Energy in nature and society. General energetics of complex systems*. Cambridge: MIT Press.

Smith, A. (1776). *An inquiry into the nature and causes of the wealth of nations*. Whitestone, Dublin.

Sohn-Rethel, A. (1970). *Geistige und körperliche Arbeit*. Frankfurt am Main: Suhrkamp.

Spencer, H. (1862). *First principles*. London: Williams and Norgate.

Statistics bureau. (2011). *Population by age (1920–2011). Japan statistical yearbook, Chapter 2 population and households*. Ministry of Internal Affairs and Communications.

Statistics bureau. (2013). *2001 survey on time use and leisure activities—Results on time use*. Ministry of Internal Affairs and Communications. Extracted August 2013 from: http://www.stat.go.jp/english/data/shakai/2001/jikan/zenkoku/zenkoku.htm

Statistisches Bundesamt. (2006). Zeitbudgets – Tabellenband I – Zeitbudgeterhebung 2001/2002, Statistisches Bundesamt, Wiesbaden, 2006.

TED: The Conference Board Total Economy Database™, accessed January 2013. Available from https://www.conference-board.org/data/economydatabase/

UNEP. (2008). *International panel for sustainable resource management (resource panel)*. 2008. Available from http://www.unep.fr/pc/sustain/initiatives/resourcepanel/index.htm

Voth, H.-J. (2000). *Time and work in England 1750–1830*. Oxford: Clarendon Press.

WBGU. (2011). *World in transition—A social contract for sustainability flagship report*. Berlin: WBGU [German Advisory Council on Global Change].

Weber, M. (1920). Die protestantische Ethik und der Geist des Kapitalismus. In M. Weber (Ed.), *Gesammelte Aufsätze zur Religionssoziologie*. Bd I, pp. 17–206. Tübingen: J.C.B.Mohr.
White, L. A. (1943). Energy and the evolution of culture. *American Anthropologist, 45*(3), 335–356.
Wrigley, E. A. (1988). *Continuity, chance and change. The character of the industrial revolution in England*. Cambridge: Cambridge University Press.
Wrigley, E. A. (2010). *Energy and the english industrial revolution*. Cambridge: Cambridge University Press.

Part II
Empirical Approaches to Socioeconomic Metabolism

Chapter 8
Long-Term Trends in Global Material and Energy Use

Fridolin Krausmann, Anke Schaffartzik, Andreas Mayer, Nina Eisenmenger, Simone Gingrich, Helmut Haberl and Marina Fischer-Kowalski

Abstract In the 20th century, the human population grew fourfold and the global economy grew 20-fold. This chapter explores how social metabolism has changed with these megatrends. It shows that material and energy use have grown faster than the population but less than the GDP, implying a growth in metabolic rates and some decoupling of resource use from economic growth. Since the beginning of the 21st century, global resource use has again accelerated, and much of the remaining world is transitioning from an agrarian to an industrial metabolic profile.

Keywords Metabolic transition · Metabolic profile · Metabolic regime · Global material and energy use · Trade · Industrialization · Population · Economic growth · Decoupling · Material and energy flow accounting (MEFA) · Material stocks

---

F. Krausmann (✉) · A. Schaffartzik · A. Mayer · N. Eisenmenger · S. Gingrich · H. Haberl · M. Fischer-Kowalski
Institute of Social Ecology (SEC), Alpen-Adria University, Schottenfeldgasse 29, 1070 Vienna, Austria
e-mail: fridolin.krausmann@aau.at

A. Schaffartzik
e-mail: anke.schaffartzik@aau.at

A. Mayer
e-mail: andreas.mayer@aau.at

N. Eisenmenger
e-mail: nina.eisenmenger@aau.at

S. Gingrich
e-mail: simone.gingrich@aau.at

H. Haberl
e-mail: helmut.haberl@aau.at

M. Fischer-Kowalski
e-mail: marina.fischer-kowalski@aau.at

H. Haberl et al. (eds.), *Social Ecology*, Human-Environment Interactions 5,
DOI 10.1007/978-3-319-33326-7_8

8.1 Introduction

The global human use of materials and energy has grown tremendously over the last few centuries. As argued in Chap. 3 of this book, resource use has been rising not only due to the multiplication of the world population but also due to economic growth and changes in the modes of subsistence. On average, one person in an industrialized country uses three to five times more materials and energy per year than an average member of an agrarian society. Beyond population growth, the growing socioeconomic metabolism is a major driver of global environmental change and a major challenge for global sustainability. The growing socioeconomic metabolism is associated with sustainability problems both on the input side (damage caused by extraction, resource scarcity and conflicts over scarce resources) and on the output side (wastes and emissions). Industrial metabolism, or the metabolic rates of mature industrial countries, has a bandwagon effect for the rest of the world. Effective strategies to stabilize global resource use and to reduce material and energy consumption require a downsizing of metabolic rates in the industrial world (UNEP 2011). In this chapter, we explore how the global human extraction and use of materials and energy has evolved during the process of industrialization and how the historical transition process from agrarian to industrial metabolic regimes as discussed in Chap. 3 of this volume is reflected in the development of global resource flows since the mid-19th century, a period extending from the early stages of coal-based industrialization in the industrial centers in Europe and North America to recent globalization processes that involve rapid industrialization in emerging economies such as China and Brazil. Such a long-term perspective reveals the full dimension of the metabolic transition and its impact on the extraction, trade and use of materials and energy at the planetary scale.

8.2 Methods and Data

This article discusses long-term time series data on global energy and material use. These data have been compiled using standard methods of material and energy flow accounting (MEFA) (Fischer-Kowalski et al. 2011; Haberl 2001). The highly aggregated data on material and energy flows we present fall within four main material groups (biomass, fossil energy carriers, ores and nonmetallic minerals for industrial use and construction minerals) and four main energy carriers or types (coal, oil and natural gas, hydropower and nuclear heat). These accounts rely largely on statistical data from international sources such as the Food and Agriculture Organization of the United Nations (FAO) (biomass), the International Energy Agency (IEA), the United Nations (fossil energy carriers) and the United States Geological Survey (USGS) (metals and nonmetallic minerals). Large material flows, however, are not covered in statistical sources and have been modeled from available data based on established procedures. Flows that had to be modeled include biomass grazed by livestock, sand and gravel used in construction, and gross ore production. These flows are modeled based on physical data (livestock numbers, bitumen and cement consumption) using standard estimation procedures (e.g., Eurostat 2012). On national levels, we calculate indicators such as domestic material consumption (DMC, material use) and domestic energy consumption (DEC, energy use). Details about data sources and methods can be found in the original publications presenting the datasets (Krausmann et al. 2009a; Schaffartzik et al. 2014). For the 1950–2010 period, we also show global material flows by six world regions: industrial countries (IND encompasses the OECD members of 2005 and European non-member states), the former Soviet Union (FSU), the Middle East and North Africa (MENA), Sub-Saharan Africa (SSA), Latin America and the Caribbean (LACA) and Asia (including all Asian countries except those included in other regions) The long time series data are available for download from the Institute of Social Ecology's material flow database at http://www.uni-klu.ac.at/socec/inhalt/1088.htm.

8.3 Long-Term Global Trends in Material and Energy Use

Growth in Metabolic Scale: Human-induced flows of materials and energy have greatly increased throughout human history. We estimate that during the period of hunter-gatherers, material use (DMC) generally grew more in parallel with population, and it reached some 7 million metric tons of biomass per year (0.7 Gt/yr; gigatons per year; 1 Gt = 10^9 t) by the advent of the Neolithic revolution. During the approximately 10,000 years from the emergence of agriculture to the onset of the industrial revolution in the 18th century, the size of the global metabolism further multiplied and climbed to roughly 2 Gt/yr in the mid-17th century, with biomass still accounting for more than 90 % of all materials used. Two factors contributed to the growth of material use in that period: the adoption of agriculture facilitated unprecedented population growth, and the mode of subsistence of agriculturalists was much more resource intensive than that of foraging societies, particularly because of livestock husbandry and a growing stock of built infrastructures and manufacturing. By 1850, when most European countries had joined the British path of coal-based industrialization, global material use had doubled to approximately 4 Gt/yr. With industrialization, the socioeconomic monitoring of resource use improved. Based on early statistical records, we are able to compile annual data series of global resource extraction.[1] Figure 8.1 shows the development of global energy use (DEC) and global material use (DMC) since the mid-19th century. The period from 1850 to the present covers a large part of

[1]At the global level, resource extraction and apparent consumption (resource use, DMC or DEC) of the resource are identical.

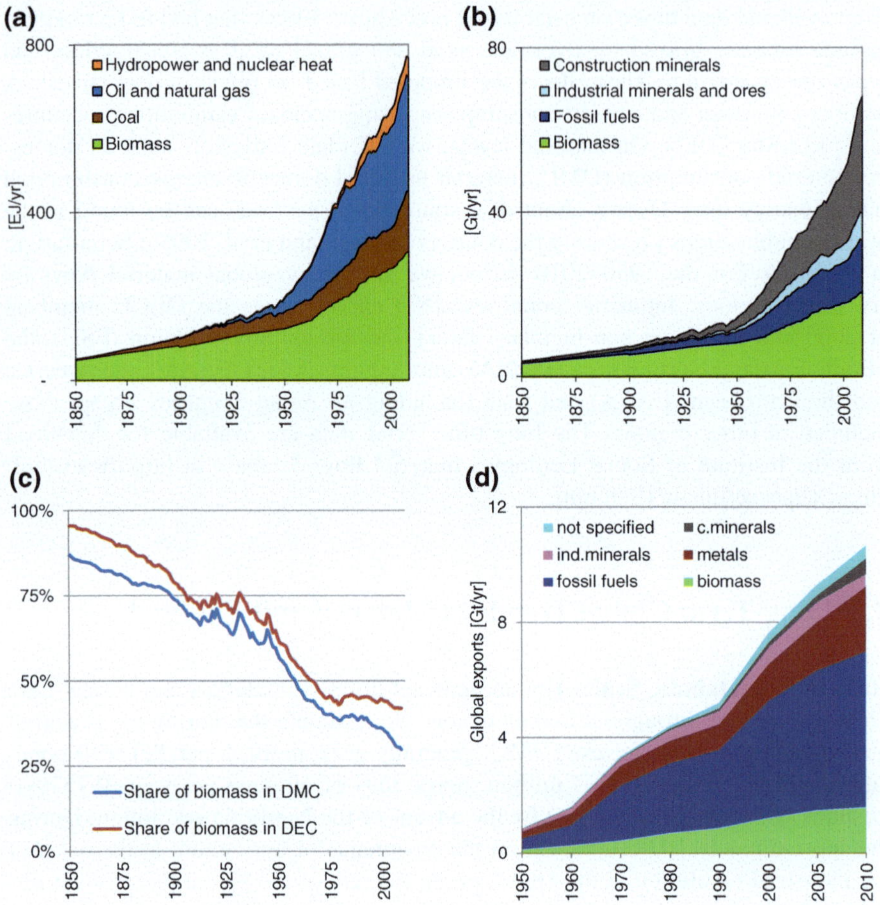

Fig. 8.1 Global use of energy (DEC) in EJ/yr (**a**) and materials (DMC) in Gt/yr (**b**) in the period 1850–2005. Share of biomass in DMC and DEC (**c**). Global material exports by material groups in Gt/yr, 1950–2010 (**d**). (Source: Krausmann et al. 2009a, 2013; Schaffartzik et al. 2014; reprinted with permission)

the history of Western industrialization and the subsequent spread of the industrial metabolic pattern across the globe.

The most striking feature is the overall growth of socioeconomic material and energy use during this 160 year period. Energy use grew by a factor of 14 and material use by a factor of 19. In 2009, roughly 730 EJ (exajoules; 1 EJ = 10^{18} J) of primary energy and 68 Gt of materials were extracted by human societies. During this 160 year period, global resource extraction grew continuously, with no prolonged periods of reduction in global resource use. Only major political and economic disruptions such as the two World Wars, the World Economic Crisis in the 1930s and the oil price shocks of the 1970s caused reductions in growth rates.

For a few years, they may have even induced a drop in resource use, but these events did not have a long-lasting impact on global physical growth. The latest upswing in growth occurred in the first decade of the 21st century, when growth rates were faster than ever before. It is not just resource extraction that has grown, however. Population growth and urbanization have exhausted local and regional resources, leading to a surge in trade flows (Fig. 8.1d). Records of international trade in mass units have only been available since the 1950s. These data show that trade was still low after World War II (WWII) but has continuously gained significance. In the past 60 years, the global physical export volume grew by one order of magnitude, from 0.9 Gt/yr to 10.6 Gt/yr. Thus, trade has increased much faster than extraction. Between 1950 and 2010, the share of exported materials in global extraction more than doubled, from only 7 to 16 %. Approximately 40 % of all extracted fossil energy carriers, metals and industrial minerals were shipped across international borders in 2010. Trade with biomass also quickly gained importance during this period, growing from 3 to 8 % of globally used biomass extraction.

Change in the Composition of Resource Use: The global economy in the mid-19th century was still largely based on biomass, reflecting the persistent dominance of the controlled solar energy system of the agrarian metabolic regime (Chap. 3). By 1850, biomass still accounted for 80 % of all material and more than 95 % of all energy inputs (Fig. 8.1). With the rapidly progressing industrialization of the economies of Europe and the US as well as the emergence of urban-industrial centers around the globe in the second half of the 19th century, the use of coal, metals and construction minerals surged, and the significance of biomass as a key resource gradually fell to 75 %. In the 50 years following WWII, the pace of this development further accelerated, and the share of biomass has now declined to roughly 30 %. With a share of 40 %, construction minerals became the largest material group, fossil fuels made up another 20 %, and ores and industrial minerals accounted for 10 %. Another aspect of the changes in the composition of material use is a shift in the relation between material and energy. In 1850, material and energy metabolism were largely identical. The largest part of all materials was still used for energy provision, mostly biomass in the form of food, feed and wood fuel. Non-energy use of materials was limited to some timber, clay, stone and sand for construction and a small amount of ores and minerals, totaling no more than 20 % of all materials. In the 160 years since, the consumption of metals and minerals has increased at a much faster pace compared to that of the sum of energy carriers, and the share of materials used as energy carriers has declined to 40 %.

Physical Stocks: Whereas the materials primarily used for energy provision (food, feed, fuel wood, coal, oil, gas) are typically used within a short period after extraction, most other materials accumulate in durable goods, buildings and infrastructures with a much greater lifespan. The rapid increase in accumulating minerals indicates that large physical stocks of artifacts have been built up at a global level. No solid global estimates of the amount of physical stocks are available, but the expansive networks of roads, railroads, pipelines and dams as well as the number of buildings and the stock of cars and machinery suggest large numbers. A recent estimate of the global growth of carbon stocks in timber, plastics and

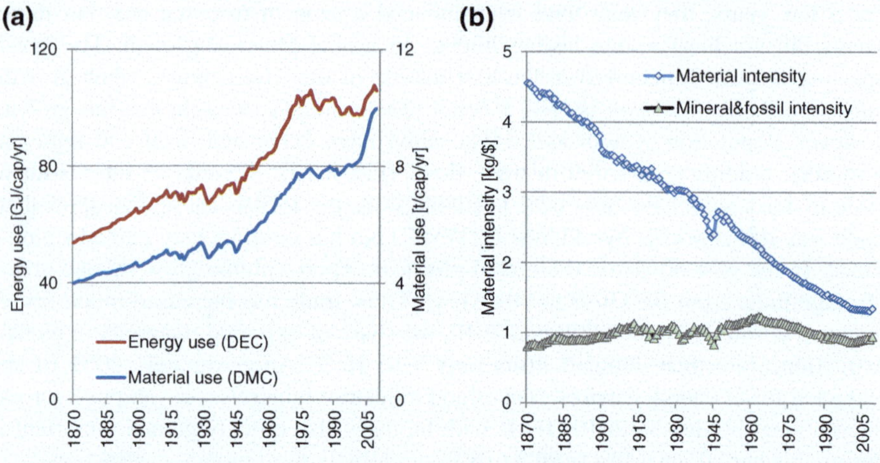

Fig. 8.2 **a** Global metabolic rates (DMC and DEC per capita per year) and **b** material intensity (DMC and mineral/fossil DMC per unit GDP, constant 1990 international $), 1870–2009. (Source: Krausmann et al. 2013; reprinted with permission)

asphalt indicates that global socioeconomic carbon stocks were 2.3 GtC (approximately 4.6 Gt dry matter biomass) in 1900 and increased to 11.5 GtC in 2008. A large fraction of that carbon is contained in long-lived wood products such as buildings and furniture, but bitumen and plastics have recently become a significant fraction (38 % in 2008) (Lauk et al. 2012). Stocks are of crucial importance for sustainable development. On the one hand, material stocks constitute metabolic legacies as they have a persistent influence on future resource demand. On the other hand, they can be viewed as 'mining sites' for future material demand when they have reached their maximum lifetime—provided they are designed in a way that permits reuse or recycling (Baccini and Brunner 2012). As long as stocks increase, that is, as long as flows into stocks are larger than outflows from the dismantling of abolished buildings and infrastructures, recycling cannot achieve an absolute reduction in the use of virgin materials (see Chap. 11).

Metabolic Rates: The third important aspect of the metabolic transition, in addition to the observed multiplication of the overall amount of resources used and the shift from biomass to fossil and mineral resources, is the growth in metabolic rates (resource use per capita). Material and energy use have grown much faster than the population, and the global average per capita consumption doubled and tripled in the 1870–2010 period (Fig. 8.2a). Energy use has grown from 47 to 110 GJ/cap/yr (gigajoules per capita per year; 1 GJ $= 10^9$ J) and material use from 3.4 to 10 t/cap/yr. These trends differ for different resources. The extraction of biomass generally grew in parallel with the population and remained fairly constant over time at approximately 3 t/cap/yr. This is quite remarkable given the far-reaching changes in the use of biomass. Several partly opposing factors were at work. With the energy transition from biomass to coal and oil, biomass was partly replaced as an energy carrier as wood fuel, and draft animals lost significance in the energy system. Equally important were efficiency gains. The ratio of primary biomass extraction to consumable biomass products improved due to technological changes, such as increases in corn-to-straw ratios and improvements in livestock conversion (Krausmann et al. 2009b; Smil 2000). Together, substitution and efficiency gains balanced increases in biomass demand due to changes in diets toward biomass-intensive animal products and toward growing paper and timber consumption. In contrast to biomass, the amount of fossil and mineral materials used grew much faster than the population, with the level of per capita consumption increasing from 0.4 to 7 t/cap/yr, driving the overall growth of metabolic rates.

Energy and material use per capita have not grown continuously; several phases can be identified during which growth accelerated. Metabolic rates grew steadily from 1870 to 1913 during the 'coal-based' period of industrialization but at a comparatively low average rate of 0.6 % per year. After a stretch of ups and downs in the first half of the 20th century caused by the two World Wars and the World Economic Crisis, growth accelerated during the 'oil phase' of the metabolic transition with the emergence of mass production and consumption. Between 1945 and 1973, the consumption of oil, metals and construction minerals multiplied, and average growth rates were high at 1.6 % per year. After the first oil price shock in 1973, growth abruptly came to a halt. A 26 year period of stagnation followed that lasted until the end of the 20th century, when resource use began to rise again and at higher rates than ever. In the first decade of the new century, material use has increased at an annual rate of 2.2 %.

Resource Use and Economic Growth: Population growth is one major driver of material and energy use as more people require more resources (Steinberger et al. 2010). However, we have also seen resource use grow much faster than the population, and metabolic rates have been multiplying. This increase in metabolic rates is closely related to economic development. In this section, we explore how the relation between economic growth and material use developed. Indicators of resource intensity measure the amount of materials or energy used per unit of economic output (GDP). Figure 8.2b shows that the global economy experienced a continuous decline in material intensity between 1870 and 2009. In 1870, more than 4.5 kg of materials were used per $ of economic output (GDP in constant 1990 international $). At the beginning of the 21st century, only 1.3 kg of resources were needed per $ output, less than one-third of the value observed in 1870. Despite these remarkable efficiency gains, longer periods of 'absolute decoupling' of economic growth and resource use (i.e., an absolute reduction of resource use while economic growth continues) never occurred at the global scale. As mentioned above, the only periods of a decline in material and energy use occurred during major economic depressions and lasted only a few years.

Moreover, if the biomass fraction is removed from the global DMC and we consider only mineral and fossil materials, the key resources of industrial development, the trend changes completely. Figure 8.2b shows that no gains at all have been achieved in the global productivity of minerals and fossil energy carriers (denoted as 'mineral productivity' in Fig. 8.2b) since the 19th century. In both 1870 and 2010, roughly 1 kg of fossil fuels and mineral materials were used to produce 1 $ of economic output. Mineral material intensity even increased in the 19th century and in the post-WWII growth period. Only in the late 1960s did mineral material intensity begin to decline, but that improvement has recently ground to a halt as growth in resource use has accelerated in the last decade. This dynamic demonstrates that global economic development is still tightly linked to the use of nonrenewable resources. In contrast, biomass is a resource mainly coupled with population. This could change if, for example, biomass regains significance as an energy carrier, as proposed in some climate change mitigation strategies. This would most likely establish a link between biomass and economic growth that has not existed in the last one-and-a-half centuries.

Regional Trends: Global trends in resource use obscure considerable regional differences. At the beginning of the 21st century, only 1.5 billion people, less than one quarter of the global population, lived in countries with a metabolic profile typical for the industrial regime. On average, one inhabitant in these countries uses approximately 17 t of materials and 300 GJ of primary energy per year (Chap. 3). Metabolic rates in the industrial countries have stabilized at this high level of resource use and have even begun a slow decline in some countries. During most of the time since the onset of fossil fuel use, economic development in this comparatively small part of the world has driven the observed global rise of metabolic rates and much of the overall growth in material and energy use. In the year 2000, industrialized countries still consumed more than 40 % of all globally extracted raw materials and primary energy sources and an even higher share of key natural resources such as copper (67 %), aluminum (72 %) and electricity (60 %). The dominant role of industrial countries in the global metabolic transition is even more obvious from a cumulative perspective on global resource use. According to data from Marland et al. (2007), four fifths of all CO_2 emissions since 1700 (and roughly an equal share of all fossil fuels that have ever been combusted on this planet) have been emitted in the currently mature industrialized countries, including the planned economies of Eastern Europe and the USSR. The overwhelming share of the globally extracted ores, industrial minerals and fossil energy carriers has been consumed or accumulated during the comparatively short history of industrialization by the currently industrialized economies. This dominance, however, is about to change.

Approximately 11 % of the global population lives in countries with a very low level of resource use (mostly in Sub-Saharan Africa). In these countries, an average inhabitant uses only 37 GJ of energy and 3 t of materials per year, the share of biomass in DEC is still 93 %, and mineral materials account for only one-quarter of the DMC (Krausmann et al. 2008). These patterns still closely resemble the metabolic profile typical for the agrarian regime (Chap. 3). The overwhelming majority of the global population lives in countries with an average metabolic profile somewhere between the extremes. Some of these countries are rapidly industrializing, such as China and Brazil, and are gradually adopting resource use

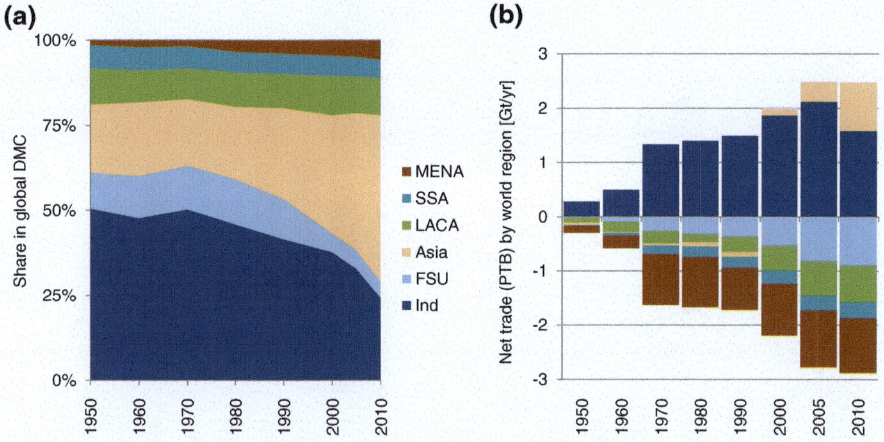

Fig. 8.3 **a** Share of world regions in global material use (DMC) and **b** physical net trade (PTB, imports minus exports) by world region. Due to inconsistencies in global trade data, global net imports are slightly larger than global net exports. MENA (Middle East and Northern Africa), SSA (Sub-Saharan Africa), LACA (Latin America and the Caribbean), FSU (former Soviet Union), Ind (Industrial countries. OECD members of 2005 and other European non-member countries), Asia (excl. countries included in FSU, IND and MENA)

patterns typical for industrial countries. Development in these countries increasingly drives global trends in material and energy use (e.g., Russi et al. 2008; Schandl and West 2010). This contributes to a remarkable regional shift in metabolic patterns.

Figure 8.3 shows that the Western industrial countries and the former Soviet Union (FSU) region, which together dominated global material use until the 1990s, rapidly give gave way to Asia, which now uses almost half of all materials extracted globally and continues to grow at a high rate. With the rising global significance of the Asia region, the share of the industrial region fell to less than one-third. The only region with even higher growth was the small but dynamic Middle East and North Africa (MENA) region, where oil-exporting countries experienced a 24-fold rise in resource use between 1950 and 2010. In contrast, the share of Latin America (LACA) and Sub-Saharan Africa (SSA) in global material use remained largely unchanged, with growth below the global average. This shift in the dominance of global resource use patterns is reflected in trade flows (Fig. 8.3b). For most of the observed period, the industrial region was the only net-importing region. In the last decade, however, Asia changed from a net-exporting region into the second-largest net-importing region, driven by high economic growth and a rapidly growing demand for energy carriers and raw materials. All other regions are providers of materials, with the net exports of the MENA region being largest followed by the resource-rich FSU and the LACA region.

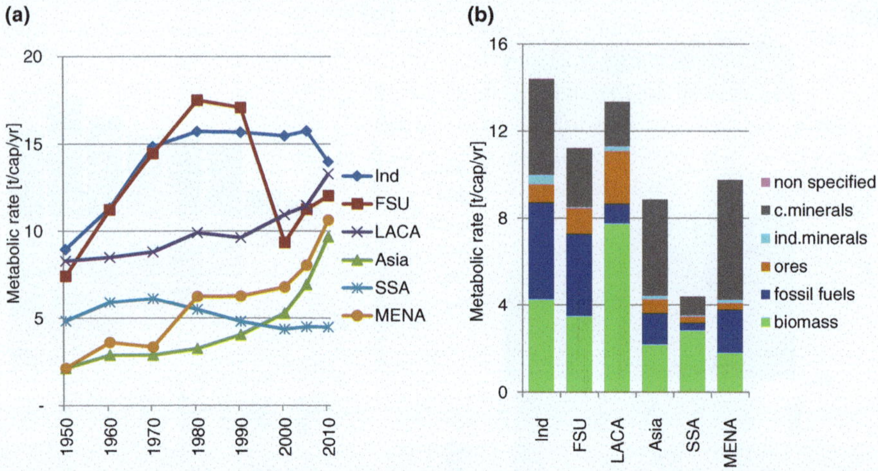

Fig. 8.4 Metabolic rates (material use, DMC/cap/yr) **a** by world regions (1950–2010) and **b** by material groups (2010). For Abbreviations of world regions see Fig. 8.3, *c.minerals* construction minerals; *ind.minerals* industrial minerals. (Source: Krausmann et al. 2013; reprinted with permission. Updated to 2010 using data from Schaffartzik et al. 2014)

Figure 8.4a shows the development of per capita material use since 1950. Whereas the industrial and FSU regions were the driving force of the 50 % increase in metabolic rates during the 1950s and 1960s, the surge in the last decade, when per capita DMC increased by 24 % after two decades of stagnation, was driven by growth in Asia (+67 %), MENA (+43 %) and FSU (+26 %). Growth in metabolic rates also gained momentum in Latin America, with a 16 % increase in this decade. In contrast, in SSA, the region with the lowest metabolic rates of the six world regions, material use per capita stagnated in the last decade after a period of decline.

Despite its rapid growth, the average metabolic rate of the Asia region is still much lower than that of the industrial region and is even below the global average (Fig. 8.4b). With respect to the composition of material use, however, Asia is already approaching the metabolic profile characteristic for industrialized countries, with a very high share of construction minerals and a rapidly growing fraction of fossil energy carriers. Biomass, in contrast, accounted for only one-quarter of material use in 2010. The average material use per capita in the resource-rich and raw-material-exporting MENA and FSU regions is high, and both regions have a material use composition similar to that of the industrial region. In contrast, LACA and SSA are notable; in both regions, biomass is still the most important material with a share above 50 % in DMC, whereas the DMC of fossil energy carriers and construction minerals is very low. The SSA region also has by far the lowest level of material use per capita. LACA and SSA are still quite far from the typical industrial metabolic profile, but both regions are important suppliers of materials for industrialization in Asia and the industrial region.

8.4 Conclusions

The global metabolic transition is an ongoing process. The aggregate global development of material and energy use reflects many of the characteristics of the transition from the agrarian regime to the industrial metabolic regime as outlined in Chap. 3 of this volume. During the last 150 years, humans have managed to continuously expand the carrying capacity of the planet by investing labor and fossil-fuel-based energy to colonize ecosystems and to gain access to and make use of new resources. The global population grew by a factor of six, from 8 to 48 persons/km^2. Biomass, the key resource of the metabolism of humanity for most of its existence, rapidly lost significance, and the per capita use of mineral and non-energy-use materials grew at higher rates than ever before. Global material and energy use grew faster than the population but at slower rates than the economic output. In the last decade, however, we have observed a recoupling, with resource use growing at the same pace as the economy, largely driven by surging demand in emerging economies, particularly the giant economy of China. Between 2000 and 2010, material use increased by 41 % and even outpaced global economic output as measured by GDP (The World Bank Group 2010).

With its enormous resource demand, human society is increasingly dominating natural systems (Vitousek et al. 1997), with the size of human-induced flows of materials and energy already on the same order of magnitude as natural flows. For example, the global annual terrestrial net primary production (NPP), the key energy source of all living beings, is estimated to be 120 Gt of biomass per year compared to 68 Gt of materials extracted by humans each year. The consequences of this large and rapidly growing resource consumption are dramatic. Humans already use or affect three-quarters of the earth's terrestrial surface. The remaining quarter is mostly too cold (arctic tundra) or too dry (deserts) to be used, and pristine forests are only a relatively small fraction of the remaining global wilderness areas (Erb et al. 2007). The rapid loss of biodiversity and other vital ecosystem services is one of the consequences of this excessive human domination (Millennium Ecosystem Assessment 2005). Greenhouse gas emissions, a direct consequence of the above-discussed patterns of resource use, have reached a level that is rapidly changing the global climate system (IPCC 2007). With the enormous growth in social metabolism since the beginning of industrialization, human influence on the biosphere has reached such a magnitude that it now seems justified to speak of a new geological era, the Anthropocene (Steffen et al. 2011). A global convergence in metabolic rates at the current metabolic profile of industrial countries paired with the expected growth in global population would lead to a doubling or tripling of global resource use. These additional resources are unlikely to be available, and we cannot expect the global ecosystem to have the capacity to absorb the corresponding wastes and emissions. It is thus obvious that the scale of socioeconomic metabolism cannot continue to grow as in the past, and humanity must find ways to return to a safe operating space within planetary boundaries (Rockström et al. 2009). This will require a significant reduction of resource use in the industrial world and ways to obtain wealth and quality of life in the emerging economies without emulating the metabolic profile of the old industrial core. The socioecological research assembled in this volume suggests that this will not be possible through gradual changes or technological fixes. A sustainability transition may well require similarly fundamental changes in society as the historical agrarian-industrial transition. Completing this historical transition on a global scale may be physically impossible due to resource constraints; however, in all cases, failing to make this transition would wreak havoc with the earth's biotic and climatic systems.

Acknowledgements This chapter is based on a presentation given at the Spring Meeting of the Materials Research Society in San Francisco (March 2013) and contains revised and updated text and figures from Krausmann et al. (2013). Global trends and patterns in material use. MRS Online Proceedings Library, 1545, (http://journals.cambridge.org/action/displayAbstract?fromPage=online&aid=9023020&fileId=S1946427413010750), reproduced with permission.

References

Baccini, P., & Brunner, P. H. (2012). *The metabolism of the anthroposphere: Analysis, evaluation, design*. Cambridge: MIT Press.

Erb, K.-H., Gaube, V., Krausmann, F., Plutzar, C., Bondeau, A., & Haberl, H. (2007). A comprehensive global 5 min resolution land-use dataset for the year 2000 consistent with national census data. *Journal of Land Use Science, 2*(3), 191–224.

Eurostat, (2012). Economy-wide material flow accounts (EW-MFA). Compilation Guide 2012.

Fischer-Kowalski, M., Krausmann, F., Giljum, S., Lutter, S., Mayer, A., Bringezu, S., et al. (2011). Methodology and indicators of economy wide material flow accounting. State of the art and reliability across sources. *Journal of Industrial Ecology, 15*(6), 855–876.

Haberl, H. (2001a). The energetic metabolism of societies, Part I: Accounting concepts. *Journal of Industrial Ecology, 5*(1), 11–33.

IPCC, (2007). Climate Change 2007. Synthesis Report. Contribution of Working Groups I, II and III to the Fourth Assessment Report of the Intergovernmental Panel on Climate Change. Cambridge University Press, Cambridge, UK and New York, USA.

Krausmann, F., Fischer-Kowalski, M., Schandl, H., & Eisenmenger, N. (2008). The global sociometabolic transition: Past and present metabolic profiles and their future trajectories. *Journal of Industrial Ecology, 12*(5/6), 637–656.

Krausmann, F., Gingrich, S., Eisenmenger, N., Erb, K.-H., Haberl, H. & Fischer-Kowalski, M. (2009a). Growth in global materials use, GDP and population during the 20th century. *Ecological Economics*, *68*(10), 2696–2705.

Krausmann, F., Haberl, H., Erb, K.-H., Wiesinger, M., Gaube, V. & Gingrich, S. (2009b). What determines geographical patterns of the global human appropriation of net primary production? *Journal of Land Use Science*, *4*(1), 15–34.

Krausmann, F., Schaffartzik, A., Mayer, A., Gingrich, S. & Eisenmenger, N. (2013). Global trends and patterns in material use. MRS Online Proceedings Library, 1545. (http://dx.doi.org/10.1557/opl.2013.1075).

Lauk, C., Haberl, H., Erb, K.-H., Gingrich, S. & Krausmann, F. (2012). Global socioeconomic carbon stocks and carbon sequestration in long-lived products 1900–2008. *Environmental Research Letters*, *7*(034023).

Marland, G., Boden, T. A. & Andres, R. J. (2007). Global, regional, and national CO2 emissions. In Anonymous (Eds.), *Trends: A Compendium of data on global change*. Oak Ridge,

Tenn., U.S.A: Carbon Dioxide Information Analysis Center (CDIAC), Oak Ridge National Laboratory, U.S. Department of Energy.
Millennium Ecosystem Assessment. (2005). *Ecosystems and human well-being: Synthesis*. Washington, DC: Island Press.
Rockström, J., Steffen, W., Noone, K., Persson, A., Chapin III, F. S., Lambin, E. F., Lenton, T. M., Scheffer, M., Folke, C., Schellnhuber, H. -J., Nykvist, B., de Wit, C. A., Hughes, T., van der Leeuw, S. E., Rodhe, H., Sörlin, S., Snyder, P. K., Costanza, R., Svedin, U., Falkenmark, M., Karlberg, L., Corell, R. W., Fabry, V. J., Hansen, J., Walker, B., Liverman, D., Richardson, K., Crutzen, P., Foley, J. A. et al. (2009). Planetary boundaries: Exploring the safe operating space for humanity. *Ecology and Society*, *14*(2), 32. [online] URL: http://www.ecologyandsociety.org/vol14/iss2/art32/.
Russi, D., Gonzalez-Martinez, A. C., Silva-Macher, J. C., Giljum, S., Martinez-Alier, J., & Vallejo, M. C. (2008). Material flows in Latin America: A comparative analysis of Chile, Ecuador, Mexico and Peru (1980–2000). *Journal of Industrial Ecology, 12*(5–6), 704–720.
Schaffartzik, A., Mayer, A., Gingrich, S., Eisenmenger, N. & Krausmann, F. (2014). The global metabolic transition: Regional patterns and trends of global material use, 1950–2010. *Global Environmental Change 26*, 87–97.
Schandl, H., & West, J. (2010). Resource use and resource efficiency in the Asia-Pacific region. *Global Environmental Change, 20*(4), 636–647.
Smil, V. (2000). *Feeding the world: A challenge for the twenty-first century*. Cambridge: MIT Press.
Steffen, W., Persson, Å., Deutsch, L., Zalasiewicz, J., Williams, M., Richardson, K., et al. (2011). The anthropocene: From global change to planetary stewardship. *AMBIO: A Journal of the Human Environment, 40*(7), 739–761.
Steinberger, J. K., Krausmann, F., & Eisenmenger, N. (2010). Global patterns of material use: A socioeconomic and geophysical analysis. *Ecological Economics, 69*(5), 1148–1158.
The World Bank Group. World Development Indicators (online version). Retrieved December 9, 2010, from http://data.worldbank.org/data-catalog.2010.
UNEP (2011). A Report of the Working Group on Decoupling to the International Resource Panel. In M. Fischer-Kowalski, M. Swilling, E. U. von Weizsäcker, Y. Ren, Y. Moriguchi, W. Crane, F. Krausmann, N. Eisenmenger, S. Giljum, P. Hennicke, P. Romero Lankao, A. Siriban Manalang, S. Sewerin (Eds.), *Decoupling natural resource use and environmental impacts from economic growth.* Nairobi: United Nations Environmment Programme.
Vitousek, P. M., Mooney, H. A., Lubchenco, J., & Melillo, J. M. (1997). Human domination of earth`s ecosystems. *Science, 277*(5325), 494–499.

Method Précis: Energy Flow Analysis

Helmut Haberl

The analysis of energy flows related to socioeconomic activities is a key method in Social Ecology. Energy flow analysis (EFA) is based on the physical notion of energy as the ability to do work. The abstract notion of energy encompasses mechanical (potential and kinetic) energy, radiant energy, chemical energy, heat, electric energy and nuclear energy. The functioning and sustenance of any biophysical system—including the biophysical structures of society—requires a continuous flow of energy (Smil 1991, 2008).

Energy flows between ecosystems and socioeconomic systems are an important part of socioeconomic metabolism. It is the purpose of this Method Précis to outline methods to account for and analyze such energy flows. Most sustainability problems are directly related to the quantity and quality of energy used by a society. Energy is an indispensable economic resource; almost all technologies used in production, transport, distribution and consumption activities require energy. In the Social Sciences, there is a long-standing, controversial discussion on the extent to which energy can be used to explain social phenomena. Answers range from the outright denial of any explanatory power of an influence of energy on society to the assumption that energy is perhaps the most important driver of socioeconomic change (Cottrell 1955; Martinez-Alier 1987). Although conventional energy statistics and balances are an important data source for socioecological studies of energy, many applications require a broader concept of EFA, explained below after a tour-de-force through important fundamentals of energy analysis.

Different types of energy can be measured with the same units. Unfortunately, different world regions and expert communities traditionally use different units. The energy unit according to the International System of Units (SI) is the joule (J). One joule is a very small amount of energy. In almost all practical contexts, one must deal with multiples constructed using prefixes, such as kilojoule (1 kJ = 10^3 J), megajoule (1 MJ = 10^6 J), gigajoule (1 GJ = 10^9 J), terajoule (1 TJ = 10^{12} J), petajoule (1 PJ = 10^{15} J) and exajoule (1 EJ = 10^{18} J). Table 8.1 provides factors to convert some often-used units into joules.

The laws of thermodynamics are indispensable when trying to understand energy. According to the First Law of Thermodynamics, the amount of energy is constant in a closed system. In other words, energy cannot be destroyed (hence, the term 'energy consumption' is incorrect; energy conversion or energy demand are more accurate formulations). The Second Law, or Entropy Law, states that in a closed system, energy conversions can only take place if entropy increases. Entropy is the foundation of measures of energy quality, such as exergy. For example, mechanical and electrical energy are of high quality (high exergy) because they can, in theory, be converted to other kinds of energy without loss. Other kinds of energy, such as heat, have a lower quality (lower exergy); their conversion to high-quality energy (e.g., mechanical energy) entails substantial losses even under idealized optimal conditions.

Table 8.1 Overview of some common energy units

Abbreviation	Unit	Conversion to Joules
kWh	Kilowatt-hour	1 kWh = 3.6 MJ
Cal	Calorie*	1 cal = 4.1868 J
Btu	British thermal unit	1 Btu = 1.055 kJ
Quad	Quadrillion Btu	1 quad = 10^{15} Btu = 1.055 EJ
Toe	Tons of oil equivalent	1 toe = 41.9 GJ**
Tce	Tons of coal equivalent	1 tce = 29.3 GJ**

* Note that 'food calories' always refer to kilocalories. 1 kcal = 1000 cal (10^3 cal)
** Indicative value: different conventions exist. If possible, always check for the conversion factor assumed

Another important distinction is that between energy and power. Energy refers to a defined amount of work performed regardless of the time required to perform the work. The amount of energy flowing or being converted per unit of time is defined as 'power', measured in the unit watt (W): 1 W = 1 J/s.

Due to the economic importance of energy, national statistical offices as well as international bodies (e.g., the International Energy Agency and the United Nations) regularly collect and publish energy statistics. Energy statistics report energy used in artifacts, such as in technical structures such as motors, power plants, furnaces and boilers, but they exclude biological energy flows such as food, feed and the physical work performed by draft animals and humans (IFIAS 1974). Although any collection of data on technical energy use in different units and related to different economic units, such as economic sectors, is referred to as 'energy statistics', an 'energy balance' provides a consistent view of the energy used in a national economy in one common unit (but, unfortunately, not always in joules). Important notions used in energy balances are the following:

- Primary energy, i.e., energy in the form in which it is extracted from the biosphere. Examples are wood, coal, crude oil, natural gas, water and wind power, solar energy and heat from nuclear fission and geothermal sources. Primary energy use is usually approximated with the indicator 'Total Primary Energy Supply', abbreviated TPES, which is the sum of the domestic extraction of primary energy plus the net import (= import minus export) of energy carriers.
- Conversion processes from primary to final energy, such as electricity and heat generation or the conversion of crude oil to products such as heating oil, gasoline or diesel fuel.
- Final energy is the energy sold to final consumers. This excludes the sale of energy to sectors that convert energy into other forms, e.g., electricity generation. Examples of final energy carriers are wood chips, coke, gasoline, gas, electricity and heat (if it is sold, but not if it is self-generated in a heating system).
- Useful energy is the energy actually flowing when a certain energy service is provided. Examples are the mechanical energy needed to shape, move or lift objects in stationary motors or vehicles; the heat required for heating or

processing purposes; chemical energy needed in the chemical industry; light (radiant) energy; and the data processing services of computers.
- Energy services are those services for which energy is used (Lovins 1979). Examples include transport; the establishment of suitable conditions within a building in terms of temperature, humidity or illumination; the cleaning of clothes; the shaping of objects; and chemical conversions. In contrast to primary, final and useful energy, energy services cannot be measured in energy units (or in any other common unit) and, hence, cannot be aggregated.

Two concepts are important when interpreting data in energy statistics and balances:

- There are two conventions for the energy equivalent of combustible materials (coal, oil, gas and biomass). The gross calorific value (higher heating value) measures the total amount of energy released in combustion measured in a bomb calorimeter. It includes the energy released in the condensation of water vapor in the flue gas. Because the energy released by condensation is often unusable, it is excluded in the second convention, the net calorific value (or lower heating value). The difference between the net and gross calorific value is generally 5–15 %.
- Different conventions exist to account for the primary energy equivalent of non-combustible energy technologies, such as hydropower, wind power, geothermal and nuclear energy. These conventions may differ by factors of up to three to four. Thus, understanding the conventions followed in a particular data source is mandatory to avoid misinterpretation or flawed comparisons.

Because the food, feed and muscle power of humans and draft animals are of fundamental importance in a socioecological context, even if food or work flows are relatively small compared to fossil fuel flows in industrial society, EFA methods have been developed to establish energy balances that include biomass and work flows. In principle, EFA follows the above-outlined logic of energy balances, that is, it traces the flow of energy through the socioeconomic system from primary energy supply to useful energy (Fig. 8.5), but it adds biomass and work to the technical energy flows considered in the conventional energy balance (Haberl 2001). Whereas conventional energy balances account for biomass only if it is used in artifacts (e.g., power plants, heating systems, as engine fuel), EFA also includes the biomass converted by endosomatic processes.

EFA considers all energy crossing the nature-society boundary based on its physical energy content, or the gross calorific value. The nutrition of humans and livestock is considered an energy conversion process within society. The processing of primary biomass to food and feed is regarded as a conversion from primary to final energy. Food and feed are defined as final energy and human as well as animal labor as useful energy (Haberl 2001). EFA can generate indicators such as the following:

- Direct energy input (DEI): all energy entering a society (the same concept as 'direct material input' used in material flow analysis, MFA).
- Domestic energy consumption (DEC): direct energy input minus export (the same concept as 'domestic material consumption' used in MFA).

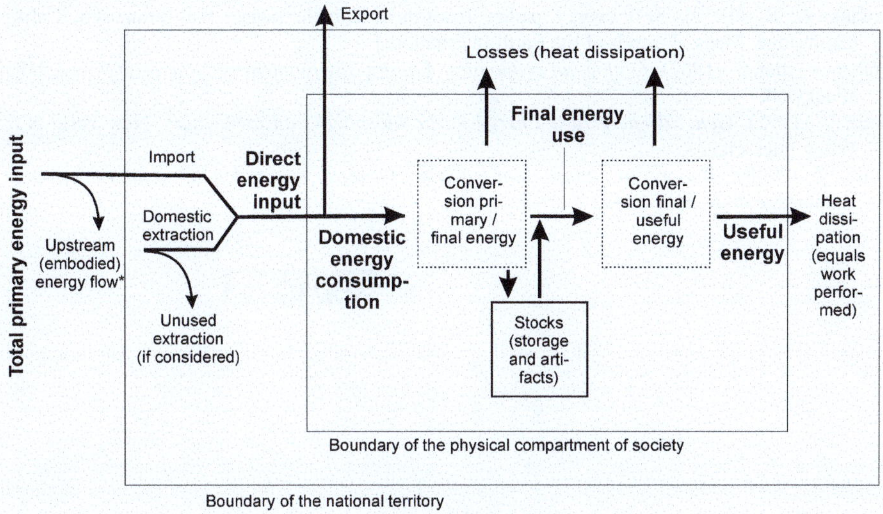

Fig. 8.5 Flow chart of an energy flow analysis (EFA) consistent with the material flow analysis (MFA) approach as used in socioeconomic metabolism studies. (Redrawn after Krausmann and Haberl 2002, Fig. 1). * Upstream flows related to import may (or may not) include foreign unused extraction—needs to be made explicit in each case

- Indicators of final and useful energy as they are used in conventional energy balances.
- 'Upstream flows': includes energy expended in providing imported energy carriers ('embodied' energy). Energy mobilized during energy extraction but not actually entering socioeconomic metabolism (e.g., crop residues) is termed 'unused extraction' and may (or may not) be considered appropriate for the respective research question.

EFA complements MFA methods to form the combined 'material and energy flow analysis' (MEFA) framework, which has been shown to be useful for studying sustainability and socioecological transitions (see also Chaps. 3, 6, 20, 21 and 22).

References

Cottrell, F. (1955). *Energy and society: The relation between energy, social change, and economic development*. New York: McGraw-Hill.

Haberl, H. (2001b). The energetic metabolism of societies Part I: Accounting concepts. *Journal of Industrial Ecology, 5*, 11–33.

IFIAS (1974). *Energy analysis workshop on methodology and conventions*. Stockholm: International Federation of Institutes for Advanced Study (IFIAS).

Krausmann, F., & Haberl, H. (2002). The process of industrialization from the perspective of energetic metabolism: Socioeconomic energy flows in Austria 1830–1995. *Ecological Economics, 41*, 177–201.

Lovins, A. B. (1979). Soft energy paths: Towards a durable peace. San Francisco, Calif., Cambridge, Mass.: Friends of the Earth International.

Martinez-Alier, J. (1987). *Ecological economics: Energy, environment and society*. Oxford, UK: Blackwell.

Smil, V. (1991). *General energetics: Energy in the biosphere and civilization*. New York, NY: Wiley Interscience.

Smil, V. (2008). *Energy in nature and society: General energetics of complex systems*. Cambridge, MA: MIT Press.

Chapter 9
More Than the Sum of Its Parts: Patterns in Global Material Flows

Andreas Mayer, Anke Schaffartzik, Fridolin Krausmann and Nina Eisenmenger

Abstract The basic characteristics of the size and composition of material flows depend on the respective stage of countries' metabolic transition from an agrarian to an industrial society. On a global level, resource use grew between 1950 and 2010 by a factor of 3.7 to 71 Gt (gigatons) per year. Moreover, the spectrum of resources used by industrializing societies broadened to include significant amounts of metals, nonmetallic minerals and fossil energy carriers in addition to biomass resources. However, there are large differences in material flow patterns within groups of countries with similar levels of industrialization and economic development. Of the multitude of possible approaches to interpreting and understanding differences in country-wide patterns of resource extraction and use, in this chapter, we will focus on the impacts of population, resource endowment, trade and economic wealth. The world's poorest and least developed countries typically have a very low metabolic rate, and biomass dominates their domestic material consumption (DMC). The richer a country is and the more industrialized it becomes, the higher its per capita DMC and the higher its share of mineral materials. This is particularly the case for fossil energy carriers, of which significant shares are traded internationally. Linking these drivers to patterns of material use is a worthwhile effort in understanding current developments of resource use and, hence, future directions toward sustainability.

A. Mayer (✉) · A. Schaffartzik · F. Krausmann · N. Eisenmenger
Institute of Social Ecology (SEC), Alpen-Adria University,
Schottenfeldgasse 29, 1070 Vienna, Austria
e-mail: andreas.mayer@aau.at

A. Schaffartzik
e-mail: anke.schaffartzik@aau.at

F. Krausmann
e-mail: fridolin.krausmann@aau.at

N. Eisenmenger
e-mail: nina.eisenmenger@aau.at

H. Haberl et al. (eds.), *Social Ecology*, Human-Environment Interactions 5,
DOI 10.1007/978-3-319-33326-7_9

Keywords Metabolic profiles · Drivers of social metabolism · Industrial metabolism · Material flow accounting (MFA)

9.1 Introduction

Beginning in the mid-18th century, the industrial revolution fundamentally changed society-nature relationships both quantitatively and qualitatively. The spectrum of resources used by industrializing societies broadened to include significant amounts of metals, nonmetallic minerals and fossil energy carriers in addition to biomass resources, which had dominated the metabolism of agricultural societies for thousands of years. In satisfying the metabolic needs of the newly industrialized societies, biomass was not 'replaced' by other resources, but overall resource use increased steeply, particularly in the past century (Schandl and Krausmann 2007). Industrialization has since developed from a Western European into a global phenomenon. Coal and petroleum resources enabled the movement of large mass flows of materials, whereas telecommunication became increasingly instantaneous in bridging spatiotemporal disconnects among regions, cities, countries and even continents. Between 1900 and 2009, the global material extraction increased by a factor of eight, from under ten gigatons per year (Gt/yr; $1 \text{ Gt} = 10^9 \text{ t} = 10^{12} \text{ kg}$) to nearly 70 Gt/yr. Although global extraction was still dominated by biomass (three-quarters of total material use) at the beginning of the 20th century, the share of renewable resources continuously declined to only approximately one-third in 2009 (Krausmann et al. 2009).[1] Although this is a common development around the globe, it cannot explain all differences in patterns of material flows across countries. To a considerable degree, basic characteristics of the size and composition of material flows depend on the respective stage of the transition from an agrarian to an industrial metabolism of countries (see Chap. 3). Even within groups of countries with similar levels of industrialization and economic development, however, large differences in material flow patterns can be observed (UNEP 2011; Weisz et al. 2006, see also Fig. 9.1). The method of material flow accounting (MFA) has undergone international harmonization and standardization (also see Method Précis on Material Flow Analysis), and high-quality global datasets are now available that provide information on direct material flows, including extraction, trade and use worldwide (Fischer-Kowalski et al. 2011). This makes it possible to scrutinize global patterns of material flows at the country-wide scale.

This chapter discusses differences in material flows across countries and among main material groups. Existing material flow accounts show that resource extraction and use differ greatly both in absolute and per capita terms and in composition from one country to the next. In the scientific literature, a multitude of

[1]For a more in-depth discussion of this global development, please see Chap. 8 in this volume.

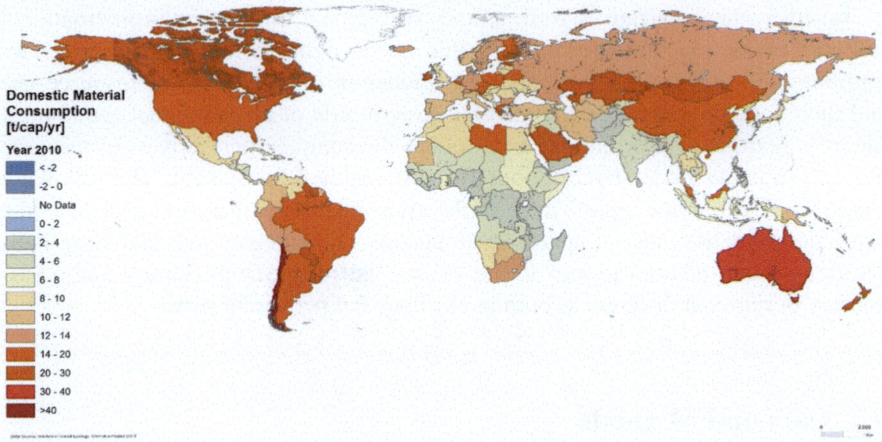

Fig. 9.1 Average country-wide metabolic rates (DMC in metric tons per capita per year) in 2010. (Source: derived from the global MFA database presented in Schaffartzik et al. 2014)

inter-linked factors are discussed that influence the size of annual economy-wide material flows (typically measured per capita or year or per unit of gross domestic product, GDP) of individual countries and lead to the observed differences. Among these are economic development, the structure of the economy, the role of an economy in the international division of labor and trade, population density, resource endowment and climate (Dittrich et al. 2012; Krausmann et al. 2008a, 2009; Schandl and Eisenmenger 2006; Steger and Bleischwitz 2011; Steinberger et al. 2010; Weisz et al. 2006). There are a multitude of approaches or narratives that can be used to interpret and understand differences in patterns of resource extraction and use. In this chapter, we will focus on tracing four narratives. **(1)** Resources are extracted where they occur and where it is technically and economically feasible to extract them. However, even when resources are available in a certain region, their extraction might be hindered by the lack of appropriate technology or financial means, the prevalence of conflicts, the specific property rights situation or other legislation. Hence, the availability of resources has an influence on material flow patterns, but other factors matter as well. **(2)** The places where resources are extracted increasingly differ from the places of processing and final consumption. Trade allows certain regions to specialize within the global division of labor and to become global producers or consumers of different resources. Trade thus translates between extraction and consumption and has an important impact on a country's material use patterns. **(3)** More people require more resources. Although population growth is an important driver of rising resource use, however, the relation between population and material use is more complex, and population density influences metabolic rates for different reasons. **(4)** Finally, in addition to population, economic development is an important driving force for material consumption and partly explains differences in metabolic rates across countries.

Together, these narratives form a story of regional differences in international resource extraction and use, a story that sheds light on some underlying factors but cannot provide an exhaustive explanation of all drivers of resource use and their specific impact on the material flow profile of all individual economies. Identifying these patterns allows us to reduce the complexity of drivers of resource use and to better understand options for sustainable development. The following chapter focuses on the significance of the abovementioned narratives for resource extraction and use. We compare developments of aggregate material flow indicators by main material groups in the second half of the 20th century and cross-country or regional differences both in absolute and per capita terms.

9.2 Data and Methods

We use data from a global database on direct material flows (extraction, trade and use) that provides country-specific data for roughly 60 different groups of materials for the 1950–2010 period.[2] We focus on four main material groups: biomass, fossil energy carriers, metals and industrial minerals and construction minerals. Although the methods used to compile these data follow the principles and standards of economy-wide material flow accounting (EW-MFA), the extraction of construction materials, the largest of the four main material groups, is systematically underestimated. Following standard estimation procedures, only sand and gravel for asphalt and concrete are included, whereas considerable amounts of these materials used as filling material or foundations are not included. This approach yields numbers 30–40 % lower than more-comprehensive estimates, but patterns across countries are well reflected. We use the indicators domestic extraction (DE), domestic material consumption (DMC; synonymously used with the term 'material use') and the physical trade balance (PTB; see Method Précis on Material Flow Analysis for details). It is important to note that in this chapter, we focus on direct material flows, that is, upstream requirements and resources used to provide the final products to end consumers are accounted for in the economy where the product was manufactured, not where it was consumed. Methods to calculate raw material equivalents are currently being developed, and the few existing data are still subject to considerable uncertainties (Wiedmann et al. 2013). This means that the data on DMC presented here reflect the production structures of an economy and not the consumption of the final consumers. However, the

[2]The database of this article covers 177 individual countries, which we have grouped into six major country groupings and world regions according to geographic and politico-economic criteria: the Western Industrial (W-Ind) grouping, the former Soviet Union and Allies (FSU-A), Asia (excluding Japan and Asian FSU countries), the Middle East and Northern Africa (MENA), Sub-Saharan Africa (SSA), as well as Latin America and the Caribbean (LACA). For a detailed description of regions, see Schaffartzik et al. (2014).

implications of trade for material flow indicators are discussed in more detail in the section on trade.

9.3 Global Material Flows

An initial global estimate of material use was published by Schandl and Eisenmenger (2006), who estimated global resource extraction at 48 Gt/yr for the year 2000. Since then, the methodology and standards of MFA for country-wide and global applications have been advanced and refined (Behrens et al. 2007; Dittrich et al. 2012; Krausmann et al. 2008b; Steinberger et al. 2010). The most recent global material flow estimate (Schaffartzik et al. 2014) places global material extraction and consumption at 70.9 Gt/yr in 2010, and growth in material use accelerated in the first decade of the 21st century. Between 2000 and 2010, the DMC grew more than twice as fast as the population (3.7 % annual growth for DMC compared to 1.3 % for population). In terms of the composition of material extraction and use, roughly two-thirds of total extraction in 2010 was nonrenewable materials (fossil and mineral materials). In fully industrialized countries, nonrenewable materials usually make up more than 67 % of the DMC. Among the nonrenewable resources, one-third are fossil fuels and two-thirds are all other mineral resources, including construction materials, metal ores and industrial minerals. In contrast, the least developed and the emerging economies are usually characterized by a lower share of nonrenewable resources in their metabolic profiles and a much higher share of biomass in the total DMC. Only a few large countries dominate global material flows. Nearly half of the globally extracted materials (47 %) are consumed by the three largest resource consumers: China (22.8 Gt/yr), the United States (5.4 Gt/yr) and India (5.4 Gt/yr).

The amount of materials used per capita has been denoted the metabolic rate (Fischer-Kowalski et al. 2011). On average, 10.3 t of materials were used per capita globally in the year 2010. Figure 9.1 shows that the average metabolic rates vary greatly across countries. The highest metabolic rates of above 30 t/cap/yr are found in the industrial countries of Northern America and Australia, but a few countries in the global south also show high rates of material use (e.g., Chile, Saudi Arabia, Ukraine). On average, the metabolic rate in industrial countries was significantly above the global average at 14.2 t/cap/yr. The region with the lowest metabolic rates is Sub-Saharan Africa where the per capita material use was only 4.5 t/yr. South and South Eastern Asia also stands out in terms of low levels of material use. Time series data show that global metabolic rates (DMC per capita) grew by an average annual rate of 2.8 % between 1950 and 2010, whereas the population grew only 1.7 % and the economy 3.9 % annually (see Chap. 8). Table 9.1 provides an overview of various MFA indicators across the six country groupings in 1950 and 2010, including annual growth rates for DE and DMC.

Table 9.1 Domestic extraction (DE), domestic material consumption (DMC) and physical trade balances (PTBs) for the years 1950 and 2010 for six country groupings. Data are given in million metric tons per year for total volumes, metric tons per year for per capita values and in percent for average annual growth rates. Negative PTB values indicate more exports than imports (i.e., net exports). (Source: derived from the global MFA database presented in Schaffartzik et al. 2014)

	DE						DMC						PTB			
Year	1950	2010	1950–2010	1950	2010	1950–2010	1950	2010	1950–2010	1950	2010	1950–2010	1950	2010	1950	2010
Unit	Mio t/yr	Mio t/yr	%	t/cap/yr	t/cap/yr	%	Mio t/yr	Mio t/yr	%	t/cap/yr	t/cap/yr	%	Mio t/yr	Mio t/yr	t/cap/yr	t/cap/yr
FSU-A	1,846	5,600	1.87	6.87	14.19	1.22	1,838	4,818	1.62	6.84	12.21	−0.97	−7	−781	−0.03	−1.98
MENA	304	4,903	4.74	3.75	12.98	2.09	1,149	3,903	2.06	2.15	10.33	2.65	−129	−1,001	−1.59	−2.65
SSA	865	4,146	2.65	4.89	4.84	−0.01	855	3,835	2.53	4.83	4.48	−0.13	−9	−311	−0.05	−0.36
LACA	1,374	7,371	2.84	9.92	15.51	0.75	1,260	6,698	2.82	9.09	14.10	0.73	−115	−672	−0.83	−1.42
Asia	2,632	34,565	4.39	2.16	9.40	2.48	2,585	34,977	4.44	2.13	9.64	2.55	−39	893	−0.03	0.25
W-Ind	5,706	14,864	1.69	8.95	13.00	0.62	5,999	16,351	1.69	9.41	14.30	0.70	293	1,487	0.46	1.30

9.4 Resource Availability

It is trivial but nevertheless important to note that resources can only be extracted where they occur and that none of the four material groups is distributed equally across the globe. Biomass and some minerals such as sand and gravel are more ubiquitous than point resources such as ores, many nonmetallic minerals and fossil energy carriers. Some of the point resources are highly concentrated and only occur in a few places on earth, such as sources of phosphate rock and some critical metals. Although data on known occurrences of a certain resource inform us of where extraction might occur, occurrence is not yet a reliable indicator of actual extraction activity; resource extraction is constrained by the sociopolitical, economic and technical feasibility of extraction, which is highly variable among resource deposits.

During the 60 years under analysis here, an important shift in terms of the locations of resource extraction can be observed. We first consider fossil and mineral resources. This shift is driven most noticeably by the gradual depletion of easily accessible deposits of nonrenewable resources (fossil energy carriers, ores, industrial minerals) in many industrial countries with a comparatively long history of resource exploitation, where ore grades are in decline and deposits of conventional fossil energy carriers are largely exploited (Giurco et al. 2010). In 1950, nearly three-quarters of the total extraction of fossil energy carriers still occurred in the Western Industrial grouping. By 2010, this share had decreased to slightly more than one-quarter (see Fig. 9.2) due to a smaller increase of extraction in these countries compared to the overall surge in global extraction. We find a similar decline in the relative importance of extraction of ores, industrial minerals and construction minerals (i.e. minerals) in the Western Industrial grouping, from nearly three-fifths to a little more than one-fifth. The Asian countries, in contrast,

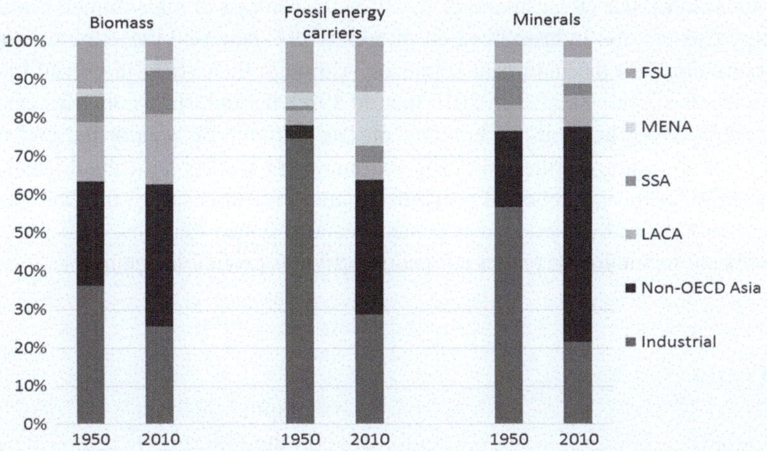

Fig. 9.2 Share of six major world regions and country groupings in the global extraction of biomass, fossil energy carriers and minerals in 1950 and 2010

have become significant consumers of resources in the last few decades. By 2010, they dominated world resource extraction and thus had essentially taken over the role held by the industrial region in the first half of the 20th century.

The countries of the Latin American region (LACA) are frequently referred to as role models of 'extractive economies' and of the perils of overly focusing the economy on resource extraction (Bunker 1985). Surprisingly, aggregate material flow data do not suggest that LACA's contribution to global resource supply and use increased in the last few decades. However, more disaggregated material flow datasets reveal that the region has indeed increased in significance as a global resource supplier of a number of specific material groups. The shares in global extraction of metal ores, namely, iron and copper ores (2.3 to 20 % and 18.8 to 47.3 %, respectively), as well as the shares of a few crops (e.g., oil-bearing crops from 6.9 to 20.5 %) grew constantly throughout the observed period. Nevertheless, the overall contribution of the region to global resource extraction has remained constant since 1950. However, as mentioned, resource extraction does not depend solely on resource endowment. Africa, a region well-endowed with resources, shows a small and even declining share, from 9 % of global extraction in 1950 to only 3 % in 2010. The prevalence of conflicts and a lack of infrastructure and financial capital are factors that inhibit higher resource extraction in this world region (Collier and Gunning 1999; Stürmer 2010).

Fossil and mineral resources are point resources, biomass is a nearly ubiquitous resource. With the exception of very arid ecosystems (deserts) and the arctic ice shelf, biomass is available in all terrestrial (and aquatic) ecosystems. Although all countries harvest biomass, extraction is less correlated with natural fertility (i.e., the biological productivity in a certain region, for example, measured in terms of net primary production), as one might expect (Erb et al. 2012). In contrast, technology plays a decisive role. The industrialization of agriculture (fertilization, irrigation, high-yielding cultivars) has drastically increased biomass yields per unit of area, and global biomass extraction grew more strongly between 1950 and 2010 than did the land area under agricultural use (Krausmann et al. 2013). In contrast to many minerals and fossil energy carriers, the industrialization of agriculture increased the extraction of biomass, particularly in the industrial countries. Although their share in overall biomass extraction was slightly smaller in 2010 than in 1950, the industrial countries extracted significantly higher amounts of biomass on less agricultural area at the end of our period of observation. Furthermore, for nonrenewable resources, technological innovations allow for the extraction of previously unused resources (e.g., oil shale) and the re-exploitation of old, abandoned deposits. For abandoned deposits, high prices and more efficient technologies render extraction activities profitable again.

9.5 Trade

Trade with raw materials, semi-manufactured and manufactured products is growing faster than the economy and has increased from 0.8 Gt/yr in 1950 to 10.6 Gt/yr in 2010, when roughly 15 % of all extracted materials were exported. From a needs perspective, one reason why significant amounts of resources are not consumed where they are extracted is the simple fact that populations and resources are not equally distributed across the globe. Due to urbanization, resource consumption is increasingly concentrated in densely populated areas that are often far from where the resources occur and are extracted. Trade allows populations to live and consume independently of local resource constraints. Trade also allows economies to develop irrespective of their domestic resource endowment. These drivers of trade are superimposed by economic drivers. Exports have high significance for the economic wealth of a country and for economic growth. From a world systems perspective (Wallerstein 1974), the global distribution of labor assigns countries different roles depending on the level of economic development, political power, endowment with natural resources or the size of the population. Production structures are increasingly organized in a global division of labor, and trade is required to exchange products along the production chain and to deliver goods and materials to the intermediate or final demand. As a result, production and consumption are organized in translocal markets.

Not all materials are traded equally. Heavy and bulk materials with low economic values (prices) are traded to a lesser extent than materials that have a higher value per weight. Trade in larger quantities was only possible after fossil-energy-based technologies (e.g., rail and ship transport) that made it cost effective to transport such quantities over large distances. However, abundant resources of low economic value per kilogram (such as most construction minerals) are still traded much less than scarce materials with a high price. A closer look at the resources that are traded (Fig. 9.3) reveals that in terms of physical trade volumes, fossil fuels are by far the most important traded resource: 5.4 Gt/yr, or 40 % of all extracted fossil materials were exported in 2010. The dependence of industrialized countries on fossil fuels and the fact that oil and gas (and, to a lesser extent,

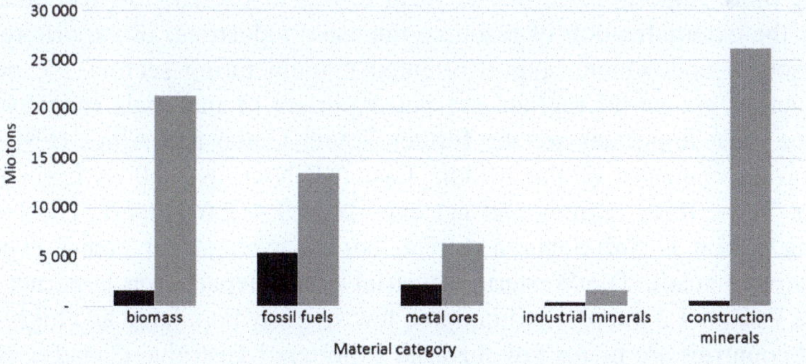

Share of exports in DE [%]	8	40	35	27	2

Fig. 9.3 Comparison of exports and domestic extraction (DE) in 2010 for five main material categories. The *right bar* in *gray* indicates DE in million metric tons, and the *left bar* in *black* shows exports in million metric tons. The numbers below the graphs (%) indicate the share of exports in DE

coal) are point resources that are highly concentrated in a few parts of the world are prime reasons why they are heavily traded. Fossil fuel exports are particularly important for the countries of the Middle East, Russia and a few other countries, such as Nigeria, Venezuela and Australia (in terms of coal). The countries that show the highest import dependence on fossil fuels are the highly industrialized countries in the Organisation for Economic Co-operation and Development (OECD) and China (oil and gas) as well as other countries that have little fossil fuel resources on their own territory. The same holds true for industrial minerals and ores, which are resources of strategic importance in the global economy. In total, 2.2 Gt/yr of metal ores (35 % of total extraction) were traded in 2010. The trade importance of industrial minerals is slightly lower, but it is still more than one-quarter of the extracted resources exported in 2010. Construction minerals, mostly sand and gravel, are ubiquitous resources and are hardly traded at all.

Biomass traditionally played a minor role in global trade. This is partly because a large share of biomass extraction, namely, biomass grazed by livestock, cannot be traded by definition. It is also because all countries engage in agriculture and forestry, and a high degree of self-sufficiency in staples has been an important policy target in many countries, thus contributing to a low share of exports in DE. However, this is changing in a globalized economy, where prices for agricultural products are increasing. Between 1950 and 2010, the share of biomass exports in the global extraction of biomass increased from 2 to 8 % (1.6 Gt/yr in 2010). For key products such as cereals and timber, the share is already higher, at 12 and 20 %, respectively.[3] Major biomass exporters are economies of low population density and high availability of land per capita, such as the United States, Australia, Brazil and Argentina; these countries are exporting an increasing share of their biomass extraction.

The physical trade balance (PTB = imports − exports) is an MFA-derived indicator that measures the form of trade dependence of an economy, that is, whether there are more imports (e.g., due to local resource scarcity, low resource availability or the externalization of resource-intensive industries) or exports (e.g., rich local resource endowment, export-oriented manufacturing sector). At the global scale, four of six world regions are net exporters of materials (negative PTB). The three most important are the former Soviet Union and Allies (FSU-A), the oil exporting countries of the Middle East (MENA), as well as Latin America (LACA). These three regions had net exports of 1.4–2.6 t/cap/yr. The fourth net exporting region is Sub-Saharan Africa, but its average per capita exports are comparatively small. The Western Industrial group depends on large net imports (positive PTB of 1.3 t/cap/yr), although a few sparsely populated and highly industrialized countries are major raw material exporters (Australia, Canada). Asia was a net exporting region through most of the observed period, but it became a net

[3]If upstream requirements are considered, the trade figures would be considerably higher. Erb et al. (2009) show for the year 2000 that biomass trade in mass units was only approximately 3 %, but in terms of embodied HANPP (eHANPP), approximately one-fifth was traded.

importing region in the last few decades, and it increasingly depends on imported resources.

Thus, trade follows economic strength rather than population. Across countries, material imports are significantly correlated with GDP, as are exports, to a lesser extent. Among the four material categories, the trade of biomass, ores and industrial minerals is most highly correlated with GDP. These resources are largely imported by regions with a long history of colonization that have developed mature, wealthy economies but have exploited their local resource base (at least to some extent). Fossil fuel exports (in contrast to imports) have no correlation with GDP. With few exceptions, rich countries import rather than export fossil fuels, and many large fossil fuel exporters have not yet invested their revenues in the diversification and industrialization of their economies (Steinberger et al. 2010).

Because DMC measures direct flows and not the raw materials used in the production process of imported and exported goods, it reflects material use in the production process rather than total material requirements associated with final consumption. Therefore, metabolic rates (DMC/cap/yr) and their high diversity are strongly shaped by extraction and production rather than patterns of final consumption; hence, they are shaped by the role a country plays within the global division of labor. Waste flows that occur during extraction and production processes are typically high. For countries with high exports of raw materials or manufactured goods, the waste flows occurring through extraction and production are accounted for under their DMC, not that of the importing country. These waste flows can be very large compared to the exported product (e.g., copper ore vs. exported concentrate or copper wire; forage vs. exported meat or cheese); thus, large exporters typically have a high DMC per capita, whereas importers have a comparably lower DMC per capita. This is e.g. the reason for Chile's extraordinary high DMC of 52 t/cap/yr: Chile is the world's largest exporter of copper, and the wastes of extracting and refining copper for export contributes 85 % to Chile's DMC. In contrast, Japan, which imports most of its raw materials, had a very low DMC of only 9.5 t/cap/yr in 2010.

9.6 Population

The following two sections focus on drivers of material consumption. In general, we find that the larger the country, the higher its DMC. However, the relation between the size of a country and its metabolism is less straightforward than it may seem at first glance because the size of a country can be measured in different ways.[4] Quite

[4]Comparing the sizes of countries using these units yields quite different rankings, and the range between the smallest and the largest differs by an order of magnitude. Across the 175 countries in our database, the size range is largest when measured in territory and GDP (range differs by a factor of approximately 50,000) and smallest when measured in terms of population (factor of 8000).

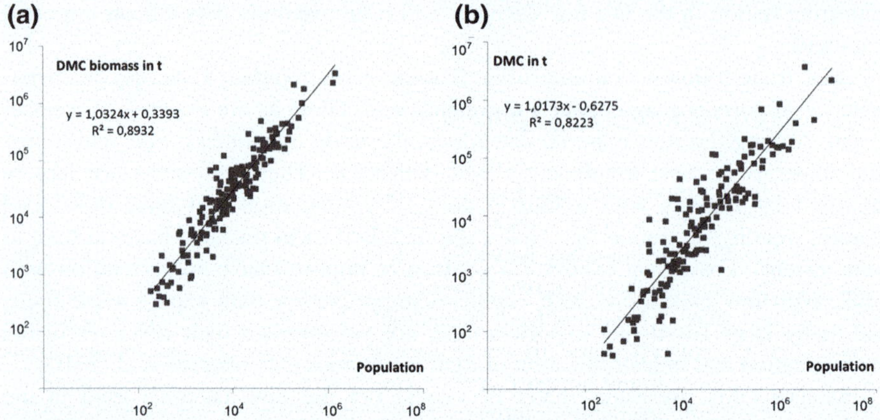

Fig. 9.4 Relation between population and (**a**) DMC of biomass and (**b**) total DMC in 2010 for 175 individual countries. Linear correlations between log-transformed variables

surprisingly, population (as a proxy for the size of the social system) proves to be a better predictor of total DMC than GDP (the size of its economy) and a much better predictor than territorial expansion (a rough proxy for a country's endowment with natural resources). Population explains 82 % of the variation of material use between individual countries in the year 2010 and is a main driver of the size of two dominant material groups: biomass and construction minerals. In contrast, it hardly explains differences in the DMC of fossil fuels, ores and industrial minerals. The link between material consumption and population is especially pronounced for biomass (see Fig. 9.4). Biomass is largely used to feed humans (either directly as food or indirectly as fodder for animals that are providing food to humans); that is, it is used to sustain the endosomatic metabolism of a population. Consequently, per capita biomass consumption does not vary as much as the consumption of those materials required for the sustenance of the exosomatic metabolism (fossil fuels, metallic and nonmetallic minerals).

Population does not just drive the overall size of social metabolism; there is also a relationship (albeit inverse and much weaker) between metabolic rates and population density. International MFA data show that many sparsely populated countries have high metabolic rates, whereas high-density countries often have below-average metabolic rates. Population density has an effect on the size of per capita resource extraction. Sparsely populated countries, by trend, are characterized by comparatively high rates of resource extraction because, on a per capita basis, their resource endowment is often statistically higher than that of countries with a very high population density. Sparsely populated countries therefore typically extract and export large amounts of materials (e.g., Australia, Russia or Brazil). In contrast, countries with a very high population density tend to have a lower relative endowment of natural resources, and they frequently have a longer history of resource exploitation than do low population density countries. Some have also exploited many of the best fossil fuel and mineral deposits, as can be seen for many central European countries (Krausmann et al. 2008a). Many high density countries appear to be net importers of materials—a factor that contributes to comparatively lower metabolic rates, as compared to net exporting countries (see Sect. 9.5).

Other density-related factors also influence the size of metabolic rates. Sparsely populated countries typically require higher inputs of energy and materials than densely populated countries to provide the same magnitude of services to their population (this applies to infrastructure and transport in particular, e.g., public transport systems in cities). Thus, cities and high population density can have a positive effect on reducing the material, energy and land input required to sustain a given population. All these arguments indicate that population density has an influence on the size and composition of metabolic rates and can help to explain material flow patterns, particularly for countries with very high or very low metabolic rates, although there are many exceptions.

9.7 Economic Development

In addition to population, economic development, measured in terms of GDP, is an important driver of material use. Income (GDP per capita) has frequently been used to investigate the relation between affluence and material use (Canas et al. 2003; Steinberger et al. 2013; UNEP 2011). As mentioned above, the transition from an agrarian to an industrial society results in growth in material use per capita and a shift from biomass toward fossil and mineral materials (Krausmann et al. 2009; Schaffartzik et al. 2014). Hence, material use usually grows with economic development, and it has been shown that differences in income can explain differences in metabolic rates (Bringezu et al. 2004). Nevertheless, similar to population, the link between GDP and material use is less straightforward than it may seem, and factors such as technology, population density and trade affect the relation (Behrens et al. 2007; Steinberger et al. 2010).

Figure 9.5 shows that total DMC is reasonably well correlated with GDP ($r^2 = 0.72$). However, the correlation between GDP and total material use is not linear, and it varies significantly among material groups. Fossil fuels show the highest correlation with GDP ($r^2 = 0.83$), indicating the tight coupling of industrial development and fossil fuel use. This also applies to construction minerals. The construction sector is a key sector in industrializing economies and an important driver of economic growth, with urbanization and the development of large infrastructures as core process of industrialization. The correlation with economic strength, however, is slightly weaker than the correlation with fossil fuels (Steinberger et al. 2010). The DMC of metals, in contrast, shows a surprisingly poor correlation with GDP, mainly because metal DMC is not a good indicator of actual metal use as it aggregates gross ore extraction with the trade of processed metals (see Sect. 9.5). However, studies investigating pure metal flows have shown high correlations between the use of all major metals and GDP (Graedel and Cao 2010). The weakest correlation is found between biomass and GDP, but although

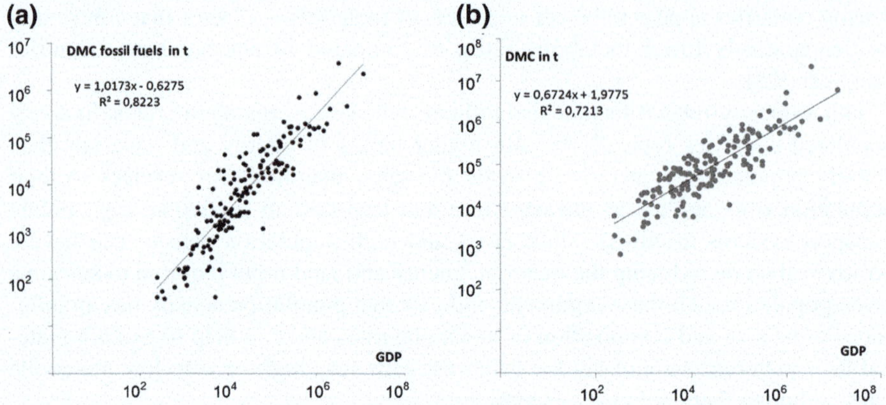

Fig. 9.5 Relation between GDP (constant 2000 US dollars) and (**a**) fossil energy carriers and (**b**) total DMC in 2010 for 175 individual countries. Linear correlations between log-transformed variables

the share of biomass in the total DMC decreases with income, rich countries still tend to consume more biomass than poor countries (Steinberger et al. 2010).

Metabolic rates increase with economic development. In the year 2010, the average metabolic rates of high-income countries and of the least-developed countries varied by a factor of nearly four (3.8 vs. 14.9 t/cap/yr). Income has been shown to be an important driver of metabolic rates, and it explains roughly 60 % of the differences in metabolic rates among countries (UNEP 2011, Fig. 2.7, p. 17). Over time, material use and metabolic rates grow with GDP, but the type of coupling changes with economic development. During the early stages of industrialization, material use and economic wealth both grow at similar rates. During the process of economic development at higher income levels, this strong coupling is disrupted, and the growth of material use slows compared to GDP (Steinberger et al. 2013). In most highly industrialized countries, metabolic rates have stabilized at a high level during the last few decades, whereas income has continued to grow. This at least partly indicates a certain saturation of material wealth in these economies, a phenomenon called decoupling of economic growth and material use (UNEP 2011). However, the observed decoupling trends of DMC and GDP have been questioned because of the distorting effects of trade: offshoring material-intensive industries and importing products can keep the DMC in high-income countries low despite material-intensive consumption patterns (Wiedmann et al. 2013) (Fig. 9.6).

As expected from the above discussion, the amount of materials required to generate one unit of GDP (material intensity; MI; measured as DMC per GDP) declines with economic development. On average, in 2010, the MI amounted to 1.4 kg/\$, which was 80 % less than in 1950 (2.5 kg/\$). Consequently, the countries with the lowest income have the highest MIs, and the richest countries require less materials per unit of GDP. Across the six world regions, the MI ranges from 3.2 kg/\$ in Sub-Saharan Africa to 0.7 kg/\$ in the Western Industrial

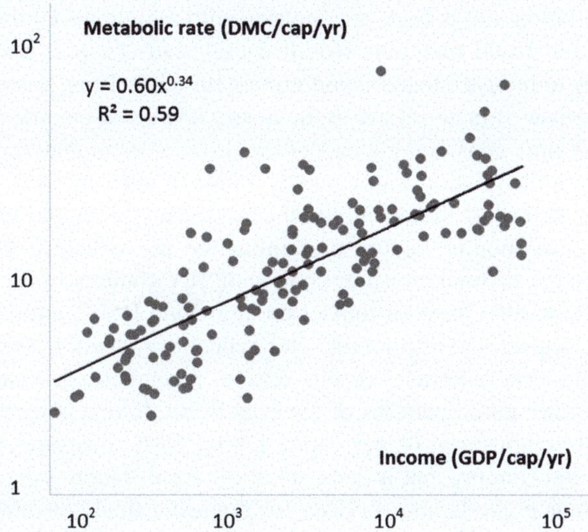

Fig. 9.6 Relation between income (GDP/cap/yr in constant 2000 US dollars) and metabolic rates (DMC/cap/yr) in 2010 for 175 individual countries. Correlations between log-transformed variables (power function)

grouping. Although MI decreases continuously in all other world regions, the two resource-exporting regions, the Middle East and Northern Africa and Latin America and the Caribbean, stand out in that as their MI stagnated or even increased during the last three decades. In these regions, material use, much of it related to the production of exports (see Sect. 9.5), grows with GDP, and there is no sign of decoupling.

9.8 Conclusions

On a very general level, all economies are part of the global metabolic transition from an agrarian to an industrialized mode of production and the corresponding material use trajectory. However, individual countries are at different stages of this transition, the speed of the transition varies across countries, and the emerging patterns of material flows differ depending on different underlying factors. The world's poorest and least-developed countries typically have a very low metabolic rate, and biomass dominates their DMC. The metabolic profiles of these least-developed countries show characteristics typical of the agrarian rather than the industrial metabolic regime. The richer and more industrialized a country, the higher its per capita DMC and the higher its share of mineral materials, particularly fossil energy carriers. The main material groups have different material characteristics and fulfill different functions in socioeconomic systems, and the size of their flows in an economy depends on different factors. Whereas biomass use grows with population and is high in countries with a high availability of land per capita, mineral and fossil materials (fossil energy carriers, ores and nonmetallic minerals) are key to industrialization and grow with GDP. Thus, population growth is an important driver of material use at the macro level, and income and economic development explain a considerable part of the differences in material flows across countries, although the relation is affected by a host of other factors.

Energy, the structure of the economy, the land-use system and historical legacies are the socioeconomic factors that influence the structure of a country's metabolism. Patterns of final consumption, such as variations in diet, mobility or heating and cooling, also have an impact on aggregate DMC, although less so as DMC measures apparent consumption and reflects production structures rather than final consumption patterns. For this reason, resource endowment and trade, which determine the spatial patterns of material flows, have a considerable impact on the level and composition of per capita DMC. DMC measures the aggregate resource use of an economy, but it does not allow for differentiating among materials used to produce goods and services for domestic final consumption and for exports, nor does it reflect resources used in the production of imported goods. However, DMC does provide important information about the material use patterns resulting from the domestic production structures. DMC thus informs about those material flow patterns for which the countries are immediately responsible, which they can influence directly by policies and technology, and that these resources can eventually become domestic waste flows.

In general, our analysis shows that material flow patterns are highly diverse among regions and countries both in terms of total material use and in terms of single material categories. Linking drivers to patterns of material use is a worthwhile effort in understanding current developments of resource use and, hence, future directions toward sustainability.

References

Behrens, A., Giljum, S., Kovanda, J., & Niza, S. (2007). The material basis of the global economy: Worldwide patterns of natural resource extraction and their implications for sustainable resource use policies. *Ecological Economics, 64*, 444–453.

Bringezu, S., Schutz, H., Steger, S., & Baudisch, J. (2004). International comparison of resource use and its relation to economic growth: The development of total material requirement, direct material inputs and hidden flows and the structure of TMR. *Ecological Economics, 51*, 97–124.

Bunker, S. (1985). *Underdeveloping the Amazon*. Chicago: University Press.

Canas, A., Ferrao, P., & Conceicao, P. (2003). A new environmental Kuznets curve? Relationships between direct material input and income per capita: evidence from industrial countries. *Ecological Economics, 46*, 217–229.

Collier, P., & Gunning, J. W. (1999). Why has Africa grown slowly? *Journal of Economic Perspectives, 13*, 3–22.

Dittrich, M., Giljum, S., Lutter, S., & Polzin, C. (2012). *Green economies around the world? Implications of resource use for development and the environment*. Vienna, Austria: SERI.

Erb, K.-H., Krausmann, F., Lucht, W., & Haberl, H. (2009). Embodied HANPP: Mapping the spatial disconnect between global biomass production and consumption. *Ecological Economics, 69*(2), 328–334.

Erb, K.-H., Haberl, H., DeFries, R. S., Ellis, E. C., Krausmann, F., & Verburg, P. H. (2012). Pushing the planetary boundaries. *Science, 338*, 1419–1420.

Fischer-Kowalski, M., Krausmann, F., Giljum, S., Lutter, S., Mayer, A., Bringezu, S., et al. (2011). Methodology and indicators of economy wide material flow accounting. State of the art and reliability across sources. *Journal of Industrial Ecology, 15*(6), 855–876.

Giurco, D., Prior, T., Mudd, G., Mason, L., & Behrisch, J. (2010). *Resourcing the future: Foresight to secure long-term national benefit from Australia's minerals.*

Graedel, T. E., & Cao, J. (2010). Metal spectra as indicators of development. *Proc. Natl. Acad. Sci. U. S. A., 107*, 20905–20910.

Krausmann, F., Erb, K.-H., Gingrich, S., Lauk, C., & Haberl, H. (2008a). Global patterns of socioeconomic biomass flows in the year 2000: A comprehensive assessment of supply, consumption and constraints. *Ecological Economics, 65*, 471–487.

Krausmann, F., Fischer-Kowalski, M., Schandl, H., & Eisenmenger, N. (2008b). The global sociometabolic transition. *Journal of Industrial Ecology, 12*, 637–656.

Krausmann, F., Gingrich, S., Eisenmenger, N., Erb, K.-H., Haberl, H., & Fischer-Kowalski, M. (2009). Growth in global materials use, GDP and population during the 20th century. *Ecological Economics, 68*, 2696–2705.

Krausmann, F., Erb, K.-H., Gingrich, S., Haberl, H., Bondeau, A., Gaube, V., Lauk, C., Plutzar, C., Searchinger, T. D. (2013). Global human appropriation of net primary production doubled in the 20th century. *PNAS* 201211349. doi:10.1073/pnas.1211349110

Schaffartzik, A., Mayer, A., Gingrich, S., Eisenmenger, N., Loy, C., & Krausmann, F. (2014). The global metabolic transition: Regional patterns and trends of global material flows, 1950–2010. *Global Environmental Change, 26*, 87–97.

Schandl, H., & Eisenmenger, N. (2006). Regional patterns in global resource extraction. *Journal of Industrial Ecology, 10*, 133–147.

Schandl, H., & Krausmann, F. (2007). The great transformation: A socio-metabolic reading of the industrialization of the United Kingdom. In M. Fischer-Kowalski, & H. Haberl (Eds.), *Socioecological transitions and global change: Trajectories of social metabolism and land use* (pp. 83–115). Cheltenham and Northampton: Edward Elgar.

Steger, S., & Bleischwitz, R. (2011). Drivers for the use of materials across countries. *Journal of Cleaner Production, 19*, 816–826.

Steinberger, J. K., Krausmann, F., & Eisenmenger, N. (2010). Global patterns of materials use: A socioeconomic and geophysical analysis. *Ecological Economics, 69*, 1148–1158.

Steinberger, J. K., Krausmann, F., Getzner, M., Schandl, H., West, J. (2013). Development and dematerialization: An international study. *PLoS One*, *8*. doi: 10.1371/journal.pone.0070385

Stürmer, M. (2010). *Let the good times roll? Raising tax revenues from the extractive sector in sub-Saharan Africa during the commodity price boom.* Deutsches Institut für Entwicklungspolitik, Bonn.

UNEP. (2011). Decoupling natural resource use and environmental impacts from economic growth. In M. Fischer-Kowalski, M. Swilling, E. U. von Weizsäcker, Y. Ren, Y. Moriguchi, W. Crane, F. Krausmann, N. Eisenmenger, S. Giljum, P. Hennicke, P. Romero Lankao, A. Siriban Manalang, S. Sewerin (Eds.), *A report of the working group on decoupling to the International resource panel.* Nairobi: United Nations Environmment Programme.

Wallerstein, I. (1974). *The modern world system.* San Diego, CA: Academic Press.

Weisz, H., Krausmann, F., Amann, C., Eisenmenger, N., Erb, K.-H., Hubacek, K., & Fischer-Kowalski, M. (2006). The physical economy of the European Union: Cross-country comparison and determinants of material consumption. *Ecological Economics, 58*, 676–698.

Wiedmann, T. O., Schandl, H., Lenzen, M., Moran, D., Suh, S., & West, J. (2013). The material footprint of nations. In *Proceedings of the National Academy of Sciences* 201220362.

Method Précis: Material Flow Analysis

Nina Eisenmenger

Material flow accounting (MFA) aims to provide a biophysical representation of society-nature interactions that complements purely monetary economic accounting systems. It quantifies all material flows into and out of a socioeconomic system. MFA accounts for solid, gaseous and liquid materials excluding water and air and is measured in physical units (mass, usually metric tons) (Eurostat 2001; OECD 2008).

MFA represents a broad family of different accounts, from national to local levels or from aggregate materials to single substances (OECD 2008). Similar physical accounts are used to measure stocks and flows of energy, carbon and other relevant substances (see Method Précis on Energy Flow Analysis). Within this broader family, economy-wide material flow accounting (EW-MFA) is most widely applied and is part of standard statistical reporting in the EU (2011). EW-MFA considers all material inputs and outputs to the system, but it treats the socioeconomic system itself and any processes therein as a 'black box'. Methodological harmonization and standardization has been intensely promoted in past few years, resulting in high consistency among available datasets (Fischer-Kowalski et al. 2011).

EW-MFA measures all material flows that are required for the establishment, operation and maintenance of socioeconomic biophysical stocks. By convention, these biophysical stocks include humans, manmade artifacts (infrastructure, buildings, vehicles, machinery, durable goods, etc.) and productive livestock (animal husbandry and aquaculture). It follows that two system boundaries need to be defined: one between the socioeconomic system and its natural environment and another separating it from other socioeconomic systems. Concerning the first boundary, material inputs flows are raw materials extracted from the domestic natural environment (domestic extraction, DE), and outputs are wastes and emissions released to the natural environment (domestic processed outputs, DPOs). Flows crossing the second boundary are imports from and exports to other national economies. Following the laws of thermodynamics, particularly the law of conservation of mass, material inputs equal material outputs corrected by stock changes (for more details, see Eurostat 2001). Stock changes are material flows to socioeconomic stocks with a lifetime exceeding one year, or materials released from physical stocks and transformed to wastes and emissions (Fig. 9.7).

Domestic extraction (DE) is defined as the raw material extracted from nature. It includes agricultural harvest and forestry as well as raw material extraction from mining and quarrying. Material flows are usually grouped along four main material categories:

1. *Biomass* extraction includes all harvested plant-based biomass, that is, all crops, timber, grazed biomass and crop residues further used in socioeconomic processing. In addition, animal biomass from fishing and hunting is included

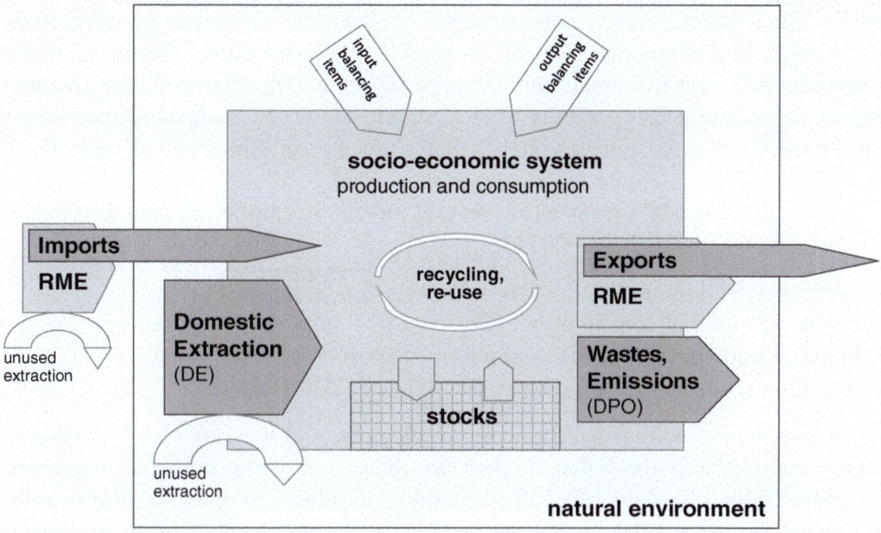

Fig. 9.7 Material flow accounting scheme. Legend: RME = Raw material equivalent. Input and output balancing items: to have a closed material balance, so-called balancing items have to be introduced into the MFA framework on both the input and output sides (water vapor, air as entry into combustion processes, etc.) (Eurostat 2001)

if it is extracted from populations of wild-living animals. Biomass production from domestic livestock is not counted as DE.

2. *Metallic minerals* comprise all ores accounted for as the mass leaving the mine, including the waste rock ('run-of-mine' approach).
3. *Nonmetallic minerals* comprise all minerals used in industrial processing and in construction, such as sand, clays, phosphate, salt and diamonds.
4. *Fossil energy carriers* include all coal, crude oil, natural gas and unconventional energy sources (such as gas hydrate, shale gas).

Material extraction data can be compiled from official statistics (for details, see Eurostat 2012). Some quantities are reported in units other than metric tons per year and must be converted into mass units. Some flows are only poorly or not at all reported in official statistics. In these cases, missing data have to be estimated using standardized MFA estimation procedures (Eurostat 2012; Krausmann et al. 2014). This applies to crop residues, grazed biomass and nonmetallic minerals used for construction purposes. In addition, gross ores often have to be estimated based on metal concentrates reported in statistics and the corresponding ore grades. Estimation procedures have been developed thoroughly over the years, and they aim to use biophysical data as a basis for the applied estimation procedure (Eurostat 2012).

In addition to materials extracted domestically, EW-MFA accounts for imports and exports (derived from foreign trade statistics, which report physical next to monetary units). Trade flows comprise products at very different stages of processing, namely, primary goods (copper ores, wheat), secondary products (copper wires, wheat flour) and final goods (mobile phones, cakes). This makes trade flows very different from material extraction, and the allocation of traded goods to one of the material categories is often difficult. However, correspondence tables have been developed (Eurostat 2012) that allow for an allocation of trade flows according to the main material content of the traded good.

Within the EW-MFA framework, several indicators can be calculated (Eurostat 2001; Fischer-Kowalski et al. 2011):

- Direct material input, DMI = DE + imports.
- Domestic material consumption, DMC = DE + imports − exports.
- Physical trade balance, PTB = imports − exports.
- Resource efficiency or resource productivity = GDP/DMC

MFA indicators are considered pressure indicators, and they are closely linked to socioeconomic activities—that is, they are indicative of the potential magnitude of environmental burdens related to economic activities. In order to discuss environmental impacts, MFA indicators need to be complemented by other indicators such as those from Life Cycle Assessments (see Method Précis on Life Cycle Assessment). The most prominent EW-MFA indicator is DMC, which is used as a key indicator for material use and, in relation to the GDP, as a proxy for resource efficiency in general (European Commission 2011). DMC represents the materials used within a socioeconomic system in the production process and/or in final consumption that are transformed into wastes and emissions.

Two other flows related to the EW-MFA framework should be mentioned.

In the course of extraction from nature, materials are moved or extracted without the intention of using them in socioeconomic processing or attributing economic value to them. These flows are commonly termed ‘unused extraction’ and include unused by-products in agriculture (straw and roots left on the fields), by-catch in fishery, overburden in mining and soil and rock excavated during the construction of infrastructure (Dittrich et al. 2012; Eurostat 2001). In most cases, no statistical data are available to account for this unused extraction, and the mass of these flows has to be estimated (Bringezu and Bleischwitz 2009). Obviously, the provision of data of sufficient quality, particularly for comparison across countries and time, is quite a challenge.

In recent years, the demand for indicators addressing the total raw material use of a country’s final consumption has been growing. In such an indicator, all raw materials used in the production process of traded goods need to be considered. MFA summarizes these raw material inputs as raw material equivalents (RMEs; Eurostat 2001; Schaffartzik et al. 2014; see also Chap. 10 in this volume) of traded goods. Adding (subtracting) the RME of imports and exports to (or from) the DMC yields the indicator raw material consumption (RMC; Schaffartzik et al. 2014; or Chap. 10 in this volume) or the material footprint (Wiedmann et al. 2015). RMC measures the amount of raw materials extracted and used at the global level to manufacture the products consumed in one particular country. Recent policy documents (European Commission 2011) consider indicators such as RMC to address issues such as outsourcing and its relevance for resource productivity in forging strategies for a more sustainable use of biophysical resources.

References

Bringezu, S., & Bleischwitz, R. (2009). Sustainable resource management. In *Global trends, visions and policies*. Sheffield: Greenleaf Publishing Limited.

Dittrich, M., Bringezu, S., & Schütz, H. (2012b). The physical dimension of international trade, part 2: Indirect global resource flows between 1962 and 2005. *Ecological Economics, 79*, 32–43.

EU. (2011). *Regulation on European environmental economic accounts*. Regulation (EU) No 691/2011 of the European Parliament and of the Council. Official Journal of the European Union.

European Commission. (2011). *Roadmap to a resource efficient Europe*. COM(2011) 571 final. Brussels: European Commission.

Eurostat. (2001). *Economy-wide material flow accounts and derived indicators. A methodological guide* (pp. 1–92). Luxembourg: Eurostat, European Commission, Office for Official Publications of the European Communities.

Eurostat. (2012). *Economy-wide material flow accounts (EW-MFA)*. Compilation Guide 2012. Luxembourg: European Statistical Office.

Fischer-Kowalski, M., Krausmann, F., Giljum, S., Lutter, S., Mayer, A., Bringezu, S., et al. (2011b). Methodology and indicators of economy wide material flow accounting. State of the art and reliability across sources. *Journal of Industrial Ecology, 15*(6), 855–876.

Krausmann, F., Weisz, H., Eisenmenger, N., Schütz, H., Haas, W., & Schaffartzik, A. (2014). Economy-wide material flow accounting. In *Introduction and Compilation Guide*. Social Ecology Working Paper No. 151. Vienna: IFF Social Ecology.

OECD. (2008). *Measuring material flows and resource productivity*. Volume I. The OECD Guide. OECD, Paris; MA, USA.

Schaffartzik, A., Eisenmenger, N., Krausmann, F., & Weisz, H. (2014b). Consumption-based material flow accounting Austrian trade and consumption in raw material equivalents 1995–2007. *Journal of Industrial Ecology, 18*(1), 102–112.

Wiedmann, T., Schandl, H., Lenzen, M., Moran, D., Suh, S., West, J., & Kanemoto, K. (2015). The material footprint of nations. *PNAS*,. doi:10.1073/pnas.1220362110.

Chapter 10
Boundary Issues: Calculating National Material Use for a Globalized World

Anke Schaffartzik, Nina Eisenmenger and Dominik Wiedenhofer

Abstract Due to the global fragmentation of supply and use chains, final consumption and the production of goods and services are often spatially disconnected. A country in which a large share of material and energy use is dedicated to the production of exports may seem to consume more material than a country that imports material-intensive products. Material flow accounting (MFA) is a well-established tool within environmental accounting, and the indicators it provides are increasingly used to inform policy-making on sustainability issues. Growing trade volumes and the deeper integration of all economies into global markets have posed a new challenge to MFA: how can we expand the scope of the accounts from a production-based perspective to one that includes consumption? In this chapter, we discuss the recent additions to the MFA method that seek to allocate material use to those economies where final consumption occurs rather than to those economies producing for export. These approaches are illustrated with a case study of the Austrian economy. This case study compares material use in Austria under production- and consumption-based approaches.

Keywords Material flow accounting (MFA) · Input-output analysis · Raw material equivalents (RMEs) · Upstream flows · Trade

A. Schaffartzik (✉) · N. Eisenmenger · D. Wiedenhofer ·
Institute of Social Ecology (SEC), Alpen-Adria University,
Schottenfeldgasse 29, 1070 Vienna, Austria
e-mail: anke.schaffartzik@aau.at

N. Eisenmenger
e-mail: nina.eisenmenger@aau.at

D. Wiedenhofer
e-mail: dominik.wiedenhofer@aau.at

H. Haberl et al. (eds.), *Social Ecology*, Human-Environment Interactions 5,
DOI 10.1007/978-3-319-33326-7_10

10.1 Growth, Globalization and Conceptual Challenges

The post-World War II period is often referred to as the 'golden age' of capitalism. The rapid economic growth that occurred in the process of repairing the damages of the war saw the regional depletion of domestic reserves of materials key to the industrial economies. With coal, crude oil and strategically important metals such as iron, aluminum and copper becoming scarce and expensive to extract within the industrialized countries, production in these economies began to depend more strongly on raw material imports. Industrialized countries increasingly specialized in the production of highly processed goods, exporting them to the growing number of economies integrated into the global markets. These processes resulted in increased import and export volumes, structural changes to the national economies participating in trade and an increasing regional differentiation of global production and consumption activities (WTO 2008). At the end of the 'golden age' in the 1970s, many industrialized countries were left with an acute awareness of their strategic dependence on conflict-ridden regions of the world as resource suppliers and their vulnerability to international price fluctuation. In the 1980s and 1990s, a second side effect of the increasing global trade volumes gained both academic and public attention: globalization had entailed global environmental problems. On the one hand, the rising imports of primary and secondary products into industrialized countries meant that, from the perspective of these economies, some of the most material- and energy-intensive and thus environmentally harmful industries were being offshored, and environmental problems were being outsourced to other countries. On the other hand, an understanding of many environmental problems as global rather than local in their consequences began to develop. The 1987 *Montreal Protocol on Substances that Deplete the Ozone Layer* and the 2001 *Third Assessment Report of the Intergovernmental Panel on Climate Change* (IPCC) *on Anthropogenic Global Warming* were milestones in this regard.

Globalization also made the study of society-nature relations increasingly complex. The concept of social metabolism postulates that economies require energy and material inputs as well as the discharge of wastes. This take on society-nature relations was developed at a time when those who criticized the dominant paradigm of boundless growth were first taking the stage (Ayres and Kneese 1969; Daly 1973; Meadows et al. 1972). Alongside the concept of social metabolism, a corresponding tool was developed that, instead of regarding an economy purely in monetary terms, accounts for society's biophysical inputs and outputs. Economy-wide material flow accounting (EW-MFA) documents material use in mass units, most commonly metric tons (t). The considered material flows include all materials an economy extracts from its domestic environment, imports from and exports to other economies and discharges to the environment (Eurostat 2007; see also Method Précis on Material Flow Analysis). From its beginnings in academia, this approach has continuously gained wider acceptance together with the notion that growth in a physically finite world must be limited. As a result, economies have implemented the monitoring of resource use and emissions as part of standard national accounting (see, for example, Eurostat 2009) and have developed policies aimed at improving their resource efficiency (e.g., European Union, Japan, China).

National energy, material and emission inventories, most prominently the EW-MFA, follow a production-based perspective: the resources used and emissions generated during the production of goods for export are attributed to the producing country and not to the country of final consumption. Only the weight of the traded product itself is included in material flow accounts. However, the globalization of environmental problems also took its toll on the dominant perspective within the social metabolism concept: for open economies, production-based inventories increasingly fail to adequately address the environmental pressures along the whole international supply chain associated with domestic consumption. In recognition of the increasing importance of global resource use mediated by international trade for environmental accounting and policy, several complementary accounting approaches have been developed, such as greenhouse gas emissions accounting (Peters et al. 2011). Although the idea of accounting for global resource use in material flow accounting (MFA) is not new (Eurostat 2001), it has only recently been put into practice.

The approaches that have been developed to account for global resource use open up old 'black boxes' and simultaneously draw new 'gray areas' to our attention. In this chapter, we illustrate how the effects of globalization have necessitated rethinking system boundaries for MFA, and we discuss the extent to which this brain twister remains unsolved.

10.2 Accounting for Consumption

Trade includes goods at very different stages of the production process, including basic commodities such as wheat or metal concentrates, intermediate goods such as flour or copper wires and final products such as bread or automobiles. In the course of the production process, a large amount of material is 'consumed', that is, transformed into wastes and emissions, and not physically included in the final product itself (Fischer-Kowalski and Amann 2001). If processed goods are traded, the wastes and emissions associated with their production occur somewhere other than the point of final consumption. For a country that extracts raw materials and further processes these for export, this means that a significant amount of waste and emissions related to final consumption elsewhere stay within its boundaries. Economies that specialize in high-end production tend to import goods that have already undergone some processing steps in the exporting economies, making them much 'lighter' in terms of their actual weight upon crossing the border but 'heavier' in terms of the weight of the materials used in their production. The lion's share of the total amount of raw materials required is thereby 'left behind' as wastes and emissions in other countries. Importing these goods rather than producing them domestically from raw materials results in a comparatively lower level of material use under the current EW-MFA framework. These flows are accounted for

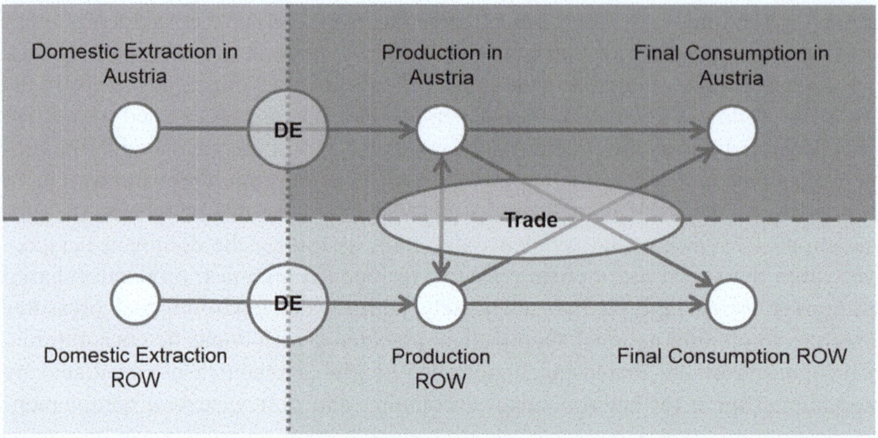

Fig. 10.1 Schematic representation of material flows between an economy (Austria) and the rest of the world (ROW). Production-based environmental accounting captures those material flows that cross the border between a socioeconomic system and its environment as domestic extraction (DE) or that cross the border between two socioeconomic systems as trade flows

as part of the domestic material consumption (DMC)[1] of the producing (and exporting) economy, with only the weight of the traded goods themselves attributed to the importing (and consuming) economy (see Fig. 10.1).

In assessing resource efficiency and dematerialization, the usefulness of standard MFA indicators is limited by the fact that a reduction in the indicator value can be due to either real improvements in the economy's technical efficiency or a relocation of material-intensive production stages to other countries. Therefore, it becomes necessary to account for the amount of material mobilized globally to satisfy domestic final demand (e.g., for final consumption in the Austrian economy as illustrated in Fig. 10.1). Methodologically, we require a calculation that allows us to estimate how much domestic extraction (DE) in the rest of the world (ROW) is ultimately used in producing goods for final consumption in a specific country, such as Austria in our example.

Attempts to develop consumption-based approaches for MFA date back to the 1990s. One of these approaches is the calculation of so-called raw material equivalents (RMEs) for imported and exported goods. The RMEs of traded goods consist of all the material inputs required to provide these goods, including the mass of the good itself. They are calculated for imports and exports, the assumption being that the RMEs of imports (RIM) must be included in an economy's consumption, whereas the RMEs of exports (REX) must be deducted to arrive at a complete balance of material inputs and outputs (Eurostat 2001). RMEs give rise to a new set of indicators within the MFA framework: the raw material trade balance (RTB), raw material consumption (RMC) and raw material input (RMI) (see Table 10.1).

[1]DMC = Domestic extraction (DE) + imports − exports.

Table 10.1 Indicators derived from economy-wide material flow accounting (EW-MFA) and raw material equivalent (RME) accounts

EW-MFA indicators	RME-based indicators
Direct Material Input (**DMI**) = Domestic Extraction (DE) + Imports	Raw Material Input (**RMI**) = DE + RME of Imports (RIM)
Physical Trade Balance (**PTB**) = Imports − Exports	Raw Material Trade Balance (**RTB**) = RIM − RME of Exports (REX)
Domestic Material Consumption (**DMC**) = DE + Imports − Exports	Raw Material Consumption (**RMC**) = DE + RIM – REX

The development of methods to account for RMEs can profit from the methodological achievements in the fields of energy and carbon footprints, and empirical applications of the RME concept and comprehensive RME accounts are currently being published in growing numbers. To estimate RME, it is necessary to trace material flows through the production system and into domestic or foreign final consumption. These links, represented by simple arrows in Fig. 10.1, are highly complex. In the past few years, several different approaches to gathering the required data and managing their complexity have been developed in empirical RME studies. We will describe the three main groups of approaches here, using them to illustrate the system boundary challenge for RME calculation.

10.2.1 Input-Output (IO) Approaches

The link between production and final consumption visible on the right-hand side of Fig. 10.1 has always been somewhat of a ‘black box’ in MFA because these flows cross neither the boundary between a socioeconomic system and its environment nor the boundary between two socioeconomic systems. As a tool for looking into this black box, IO models have become a mixed blessing for environmental accounting. IO data systematically cover the whole economy and the interrelations among sectors therein and are part of the system of national accounts. However, IO data are typically available in monetary units only, forcing those who use these data within environmental accounts to assume some degree of proportionality between physical and monetary flows. Although this assumed proportionality may be very reasonable for materials that are captured well by monetary accounts (e.g., fossil energy carriers that are almost exclusively traded via markets), it will constitute a potential source of error for those materials that have no direct monetary value. Biomass grazed by livestock, fuel wood extracted from forests, and subsistence farming are highly relevant in terms of the associated material use but are not covered in standard IO tables (see also Method Précis on Input-Output Analysis). In the following, we will keep the potential caveat of this approach in mind, touching upon it again later.

To allocate material consumption to final demand, two new system boundaries must be introduced into the MFA framework: one that separates production from consumption and one that separates production for domestic final consumption from production for foreign final consumption. There are two groups of IO approaches to RME calculation that differ in how they deal with the system boundary issue:

1. A single-region IO (SRIO) model applies only one IO table (that of the importing economy) using information on domestic inter-industry relations and technical coefficients to calculate the RMEs of exports and imports. The 'domestic technology assumption' is made whereby the system boundaries that can be identified for the domestic economy also hold true for the ROW, that is, the inputs required for the production of outputs to either exports or domestic final consumption are also applied to imported goods.
2. A multi-regional IO (MRIO) model introduces the boundary between production and consumption and foreign and domestic final consumption for multiple countries or regions, integrating the input-output data of different national economies into one IO table for the whole world. Here, the specific (monetary) inter-industry relations in the economies producing the traded goods are applied.

10.2.2 Coefficient or LCA-Based Approaches

To avoid having to make a proportionality assumption for monetary and physical flows, coefficient-based approaches use product- or resource-specific coefficients based on physical data to assess upstream material inputs. Although the IO approaches seek to reach economy-wide coverage, the coefficient approaches tend to focus on certain products or product groups in a bottom-up manner. The coefficients used are usually derived from life cycle analysis (LCA; see also Method Précis on Life Cycle Analysis). These inventories cover a large number of products and processes in great detail and, in contrast to IO data, are built from information on physical (and not monetary) flows. The main challenge here lies in ensuring full coverage while avoiding double-counting.[2]

[2]Full coverage may be hindered by truncation decisions, which have to be made in the tracing of any life cycle. Allocation decisions can potentially lead to double-counting and must be made whenever the same production process produces more than one good. Allocation can be based on different aspects of the products (economic value, mass/energy, material input avoided by coupled (instead of single-product) production). The results of these different allocation approaches can be completely opposite, so the decision of which procedure to use has a major impact on the results.

10.2.3 Hybrid Approaches

The term ‘hybrid approach’ is used here to refer to the combination of IO and LCA as developed by the LCA community to solve system boundary issues (Suh et al. 2003). In RME applications, an SRIO approach is extended by an LCA module for non-competitive imports, that is, for those products/activities not represented or not sufficiently represented by the domestic IO structures. This approach is also based on the proportionality assumption between physical and monetary flows. In the following, we will discuss the results and implications of such an approach using Austria as an example.

10.3 Austria’s Global Resource Use

Austria serves as a typical example of the growing significance of trade flows in the metabolism of industrial economies. From 1960 to 2008, the share of imports in direct material input grew from 16 to 34 %, and only two-thirds of all material input was of domestic origin. Of all the materials processed in Austria, 23 % leave the country as exports (Statistik Austria 2011). Biomass and nonmetallic minerals in particular are still mainly extracted domestically. Fossil fuels and many important metals are not (or no longer) available within Austria, so DE has either ceased or is decreasing. However, the strategic importance of these minerals leads to continuous demand and a growing dependence on imports. To assess Austria’s global resource use, we calculated the RME of Austria’s traded goods for the years 1995–2007 by the following material categories: biomass, metal ores, nonmetallic minerals, fossil energy carriers and other products.[3]

We use our case study of the Austrian economy (Schaffartzik et al. 2014) to test two conflicting hypotheses:

1. Austria is a net-importer of material goods, many of which are highly processed, and it has exhausted its own resource base for a number of key materials. Moreover, a large share of its imports are highly processed. Therefore, it is also a net-consumer of global resources in RMEs.
2. In the Austrian economy, a significant share of the resources extracted domestically and of the imported goods are used to produce exports for final consumption elsewhere, making Austria a net-supplier of global resources in RMEs.

[3]The category of other products had to be introduced because some of the product groups in the IO data are composed of more than one type of the four basic materials, and a dominant material cannot be unambiguously identified.

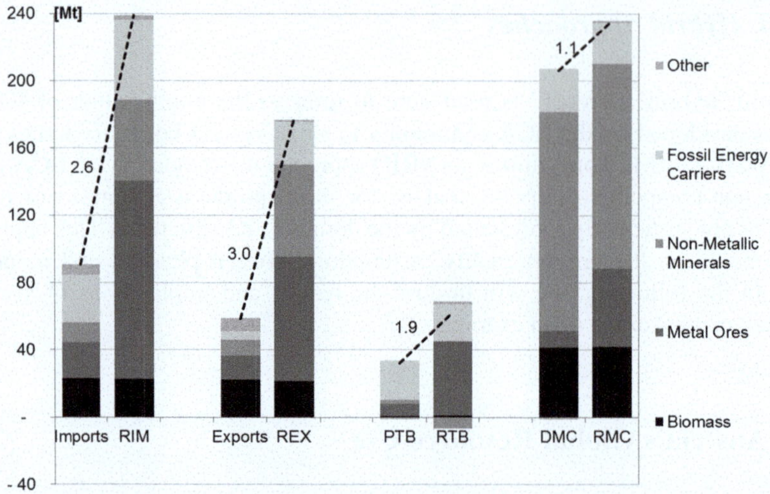

Fig. 10.2 Austria's trade flows, balances material consumption and the raw material equivalents (RMEs) thereof, 2007, in Mt (million tons)

10.3.1 Imports

In the year 2007, Austria imported approximately 91 Mt (megatons; 1 Mt = 10^6 t) of material, mainly from its European neighbors (Germany, Italy, Switzerland, the Czech Republic and Hungary) and from the large exporting economies of China and the USA. On average, approximately 11 t was imported per person. Austria has dwindling domestic reserves of fossil energy carriers, so the economy is especially dependent on imports of crude oil, natural gas and hard coal. The same holds for most metals, especially iron, copper and aluminum, for which imports have to meet demand both for final consumption and for intermediate use in the export-oriented metal manufacturing industries. However, it is not just domestically scarce resources that are imported; Austria has one of the highest shares of woodlands in Central Europe, and consequently, it has a large wood-processing industry. Nevertheless, in recent decades, this export-oriented industry has increasingly depended on imported roundwood. Wood was the largest fraction of biomass imports in 2007 (see Fig. 10.2).

Each ton of imports required upstream inputs of materials and energy in its extraction and production. Metals (mainly steel) and construction minerals were needed to build the infrastructure for oil and gas extraction, and energy carriers were required to run the mining processes and to fuel transport between production processes. In the mining of metals, large amounts of waste rock are extracted and must be accounted for. Even the felling of trees requires machinery and energy inputs and leads to some biomass losses. In 2007, Austria's 91 Mt of imports was associated with approximately 148 Mt of upstream material inputs, bringing the RIM to 239 Mt or, on average, 29 t/cap (metric tons per capita). RIM surpassed imports by a factor of 2.6 (see Fig. 10.2). Metal ores made up the largest share (49 %) of RIM, followed by fossil energy carriers and nonmetallic minerals (20 % each). Metals are not traded as ores but as either metal concentrates or products. From extraction[4] to concentration and processing in final products, a steady reduction in the material mass occurs. The incorporation of large amounts of metal (especially steel) in production technologies and infrastructures further contributes to the disproportional significance of metal ores in RIM. Fossil energy carriers are important (energy) inputs in all other production processes. Therefore, their RIM are also significantly higher (factor 1.7) than the imports. Austria's nonmetallic mineral RIM were 4.1 times larger than the imports because they are required in large amounts in production and transportation infrastructure.

10.3.2 Exports

Austria exported 33 Mt less than it imported in 2007, when total exports amounted to 58 Mt or 7 t/cap. Untypically for an industrialized economy, biomass made up the largest share of Austria's exports and consisted mainly of wood and wood- or pulp-based paper products. Austria's steel and metal manufacturing industry exported large quantities of highly processed metal products from iron, copper and aluminum. Crude oil and coal were exported in negligible amounts, and 2 Mt of natural gas represented the only notable export of fossil energy carriers. Exports play a very important role for the Austrian economy, not only in terms of provision of raw materials and products but also financially; in 2007, nearly 60 % of the gross domestic product (GDP) stemmed from exports (Statistik Austria 2012).

A large share of the material used in the Austrian economy is dedicated to the production of goods for export. Typical inputs into industrial production processes are metals and nonmetallic (especially construction) minerals for infrastructure and fossil energy carriers. For metals and fossil energy carriers, as well as for biomass (wood and pulp), this means some of the imported materials, including their upstream inputs, are used in the production of exported goods. Overall, approximately 118 Mt of materials were used in Austria in 2007 to produce 58 Mt of exports, so REX amounted to 177 Mt or 21 t/cap. REX surpassed exports by a factor of three (see Fig. 10.2). The ratio of REX to exports is thus slightly higher than that for RIM to imports (factor of 2.6), indicating that the Austrian economy's exports are more highly processed and incorporate more upstream material inputs than its imports. Metal ores are the largest component of Austria's REX (42 %), followed by nonmetallic minerals (31 %). Because fossil energy inputs are required in all production processes, this category represents the third-largest fraction (14 %).

[4]Following MFA conventions, we accounted for metal DE as the gross ore (see Eurostat 2001).

10.3.3 Trade Balance

Austria is a net-importer, meaning its physical trade balance (PTB = imports − exports) is positive. In 2007, Austria's PTB was slightly above 32 Mt (4 t/cap), of which fossil energy carriers contributed the major share (23 Mt, or 73 %). The second-largest material group in the PTB was metal ores (22 %). Fossil energy carriers and metals are resources with limited domestic availability in Austria. For biomass, both import and export flows are considerable but are nearly balanced, resulting in a PTB of just 0.9 Mt. In 2007, Austria's raw material trade balance (RTB = RIM − REX) was nearly twice as large as its PTB (factor of 1.9). The large amount of metal ores required as inputs for the production of imported goods accounts for the major part of this difference. The RIM exceeded the REX of metals by 44 Mt. The next-largest fraction in the RTB is fossil energy carriers (22 Mt). Nonmetallic minerals are a category of net imports in terms of their PTB (2 Mt) but are a category of net exports in terms of their RTB (−6.6 Mt). Because this category is dominated by construction minerals, this shift indicates that a dominant share of infrastructure within Austria is dedicated to the production of exports. Biomass remains a category of (fairly small) net imports, and the RIM exceed the REX by approximately 1.1 Mt (see Fig. 10.2).

10.3.4 Material Consumption

In 2007, Austria's domestic material consumption (DMC = DE + imports − exports, see also Method Précis on Material Flow Analysis) amounted to 206 Mt. This means that, on average, each Austrian consumed approximately 25 t/cap and approximately five times as much as the global average of 4.7 t/cap. Nonmetallic minerals made up the largest share at 63 % of DMC, or 16 t/cap. The dominant fraction within this material category is construction minerals required for housing, transportation and production infrastructure. Although these minerals are typically used in large quantities in all industrialized economies, they are low in price and ubiquitous and are therefore only traded in very small amounts. The second-largest category in DMC is biomass at 20 %, or 6 t/cap, followed by fossil energy carriers at 13 %, or 3 t/cap. Austria's global resource consumption, or raw material consumption (RMC = DE + RIM − REX), is slightly larger than its DMC, and in 2007, it amounted to approximately 236 Mt, or 28 t/cap. This means that if Austrian resource use is examined from a consumption-based perspective, we find that 3.6 t/cap more resources were used than if we follow a production-based perspective. Although we were able to find some evidence for both of our hypotheses, the impact of Austria's net-imports clearly outweighs the material mobilized within the economy for highly processed exports. Austria is a global consumer of materials in terms of RME.

Because Austria's exports and imports as well as RIM and REX nearly doubled between 1995 and 2007, its DMC and RMC stagnated (Fig. 10.3). As a consumer

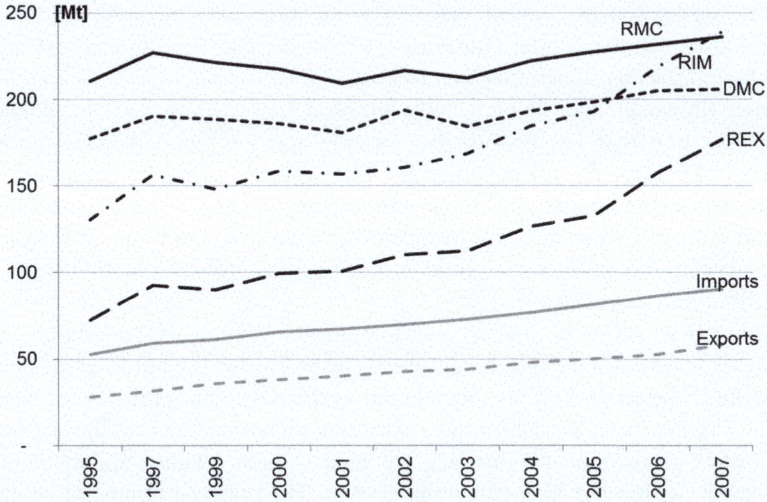

Fig. 10.3 Development of Austria's resource consumption by component, 1995–2007

of final products and as a supplier of finished and semi-finished goods, Austria is increasingly integrated into international and especially European supply chains, using additional resources in other economies to satisfy domestic final demand. Although this outsourcing of material requirements protects the domestic environment, it contributes to higher environmental impacts at the global scale. At the same time, both imported and domestically extracted resources are integrated into products Austria exports and are designated for final consumption elsewhere. The trade-off in environmental impacts involved largely depends on production differences. Although it can be assumed that Austrian environmental standards are comparable to those of its European trading partners, its high share of renewable energy (especially hydro-electricity) could potentially lead to Austria benefiting from trade to protect its domestic environment.

10.4 National Resource Use in a Global Perspective

The concept of social metabolism and its MFA tools were successful in demonstrating that economic activities are not without environmental consequences and that the biophysical size of a national economy must be considered alongside its monetary dimensions. As we have shown in our case study, Austria serves as an example of an open economy that, as can be said to varying degrees of all other countries, engages in economic activities and interactions outside its own territory. Trade plays an increasingly important role in meeting the national economy's resource demand and in maintaining its economic dynamics. A society's metabolism is no more confined by borders than financial transactions are, giving rise to an increasing spatial disconnect between production and consumption. This has major implications for standard measures of resource use, forcing us to rethink the system boundaries by which they are defined.

Fischer-Kowalski and Weisz (1999) propose that society can 'be conceived of [...] as a unit that has historically variable and imprecise material delimitations' (p. 244). They discuss whether any choice made as to the delimitations or boundaries in conceptualizing society will necessarily impact what is included in or excluded from its metabolism. In much the same way, the boundaries we draw when analyzing the materials required to satisfy a society's metabolic demand can follow different parameters.

Traditionally, MFA has been sensitive to two types of boundaries: those between a socioeconomic system and its environment for the purposes of defining DE and those between two socioeconomic systems for the purposes of defining trade (cf. Figure 10.1). To determine a society's global resource requirements and the according environmental impacts, we need to gain a better understanding of what goes on within the socioeconomic system. The material inputs into a system (DE and imports) are usually further processed. They may then end up as waste or emissions, be integrated into stocks, be consumed domestically or form part of foreign final consumption as exports (Fig. 10.4). IO data such as we have used for our Austrian case study allow us to trace monetary flows through the socioeconomic system. The IO tables provide us with information on how many and which types of inputs are required to produce one unit of output for final consumption, whether domestic or foreign. They thus open up the 'black box' of the socioeconomic system. At the same time, the use of IO data introduces a new system boundary as only those flows are accounted for that have a price.

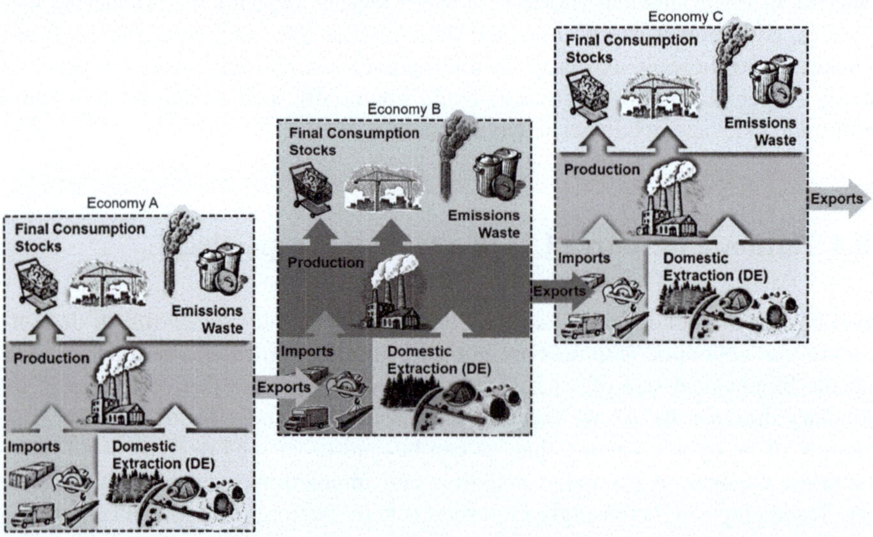

Fig. 10.4 Opening the 'black box'—a schematic look at the production of traded goods

Returning to our Austrian case study, we remember that on the one hand, the high relevance of trade flows is quite typical of an industrialized economy, whereas on the other hand, the large biomass export flows might be considered atypical. In the latter circumstance, the system boundary implied by the use of monetary input-output data could constitute a source of error. We anticipate, for example, that the biomass grazed by livestock in Austria's traditional agricultural systems in mountainous areas is insufficiently reflected by monetary data.

The effects of globalization, especially the increasing importance of trade, have made it necessary to re-think system boundaries in MFA. Although the IO approach has proven useful in tracing material flows through (and not just into and out of) the socioeconomic system, the exclusion of non-priced flows it introduces must be considered a shortcoming. The MFA community will continue to seek ways and means to offer not just production-based indicators but also consumption-based indicators of material use. As we have argued in this chapter, the consumption approach helps us better understand the global implications associated with an economy's resource use (as corresponding to its final demand). Notwithstanding, the production approach is better suited to analyze local pressures on the domestic environment because it accounts for the amount of material that will eventually be transformed into wastes and emissions in a country's territory. The consumption-based perspective offers an additional point of focus in the study of social metabolism rather than a rejection of the production-based perspective.

References

Ayres, R. U., & Kneese, A. V. (1969). Production, consumption and externalities. *American Economic Review, 59*(3), 282–297.

Daly, H. E. (1973). The steady-state economy: Toward a political economy of biophysical equilibrium and moral growth. In H. E. Daly (Ed.), *Toward a steady-state economy* (pp. 149–174). San Francisco: W.H.Freeman.

Eurostat. (2001). *Economy-wide material flow accounts and derived indicators. A methodological guide*. Luxembourg: Eurostat, European Commission, Office for Official Publications of the European Communities.

Eurostat. (2007). *Economy-wide material flow accounting. A Compilation Guide*. Luxembourg: European Statistical Office.

Eurostat. (2009). *Economy wide material flow accounts. Compilation guidelines for reporting to the 2009 Eurostat questionnaire* (Version 01 – June 2009). Luxembourg: European Statistical Office.

Fischer-Kowalski, M., & Amann, C. (2001). Beyond IPAT and Kuznets curves: Globalization as a vital factor in analysing the environmental impact of socio-economic metabolism. *Population and Environment, 23*(1), 7–47.

Fischer-Kowalski, M., & Weisz, H. (1999). Society as a hybrid between material and symbolic realms. Toward a theoretical framework of society-nature interaction. *Advances in Human Ecology, 8*, 215–251.

Meadows, D. L., Meadows, D. H., & Randers, J. (1972). *The limits to growth*. New York: Universe Books.

Peters, G. P., Minx, J. C., Weber, C. L., & Edenhofer, O. (2011). Growth in emission transfers via international trade from 1990 to 2008. *Proceedings of the National Academy of Sciences, 108*(21), 8903–8908.

Schaffartzik, A., Eisenmenger, N., Krausmann, F., & Weisz, H. (2014). Consumption-based material flow accounting: Austrian trade and consumption in raw material equivalents (RME) 1995–2007. *Journal of Industrial Ecology, 18*(1), 102–112.

Statistik Austria. (2011). *Materialflussrechnung* (*Material Flow Accounting*). Statistik Austria. http://www.statistik.at/web_de/statistiken/energie_und_umwelt/umwelt/materialflussrechnung/index.html. Accessed 09/2013.

Statistik Austria. (2012). *STATcube Statistical Database*. Statistik Austria. http://www.statistik.at/web_de/services/datenbank_superstar/index.html. Accessed 09/2013.

Suh, S., Lenzen, M., Treloar, G. J., Hondo, H., Horvath, A., Huppes, G., et al. (2003). System boundary selection in life-cycle inventories using hybrid approaches. *Environmental Science and Technology, 38*(3), 657–664.

WTO. (2008). *World trade report 2008: Trade in a globalizing world*. Geneva: World Trade Organization (WTO).

Method Précis: Life Cycle Assessment

Michaela C. Theurl and Anke Schaffartzik

Although the accounting systems used within Social Ecology often have an economy-wide scope, specific research questions may require a focus on the life cycle of one particular product. When the same good for final consumption is produced in different ways, each with its own technology adapted to a particular geographic location, it may be of interest to know which of these production paths causes the least environmental impact (see, for example Theurl et al. 2014). Life cycle assessment (LCA) is an accounting tool that follows a defined product[5] from 'cradle to grave'—from production to use and, finally, to disposal (or from 'cradle to cradle' if the product is recycled rather than disposed of) (Fig. 10.5). LCA considers both inputs from the environment (resource and land use) and outputs to the environment (wastes and emissions) that occur during the production, use and end-of-life phases.

The application of the life cycle concept in an LCA for a particular product is a challenging procedure that follows a four-step framework (cf. Baumann and Tillmann 2004): (1) goal and scope definition, (2) life cycle inventory (LCI), (3) life cycle impact assessment (LCIA) and (4) interpretation. LCA is an iterative approach, and once a step is finished, evaluation checks are carried out and findings are related to the initial model to improve the quality of the analysis. The LCA method has been internationally harmonized and, as a whole and in its individual steps, is governed by regulations of the International Organization for Standardization (ISO; Finkbeiner et al. 2006).

A distinctive feature of LCA compared with other methods used in Social Ecology is that it does not have readily defined system boundaries. Instead, in defining the goal and scope of an LCA study, the researcher must determine what is to be included in the assessment. This always entails a truncation decision as it is not possible (nor beneficial) to include all processes directly and indirectly related to the product in question. Depending on the research question, one can distinguish two different types of LCAs: attributional and consequential. Whereas an attributional LCA describes the environmental impacts of important physical flows from a product, a consequential LCA estimates possible environmental impacts arising from marginal changes in the output of a product and considers the consequences of changes that would be outside the system boundary in attributional LCAs (system expansion; Finnveden et al. 2009). Based on these specifications, a systems model is constructed that allows for the inclusion of the environmentally relevant flows in the life cycle of a given product (Fig. 10.6).

At this point, the amounts of inputs and outputs associated with the modeled system are calculated and summarized in the LCI. The LCIA corresponds to the

[5]For the sake of clarity, we will mainly describe LCA in terms of a product. However, LCA may also be applied to well-defined processes, services or even organizations.

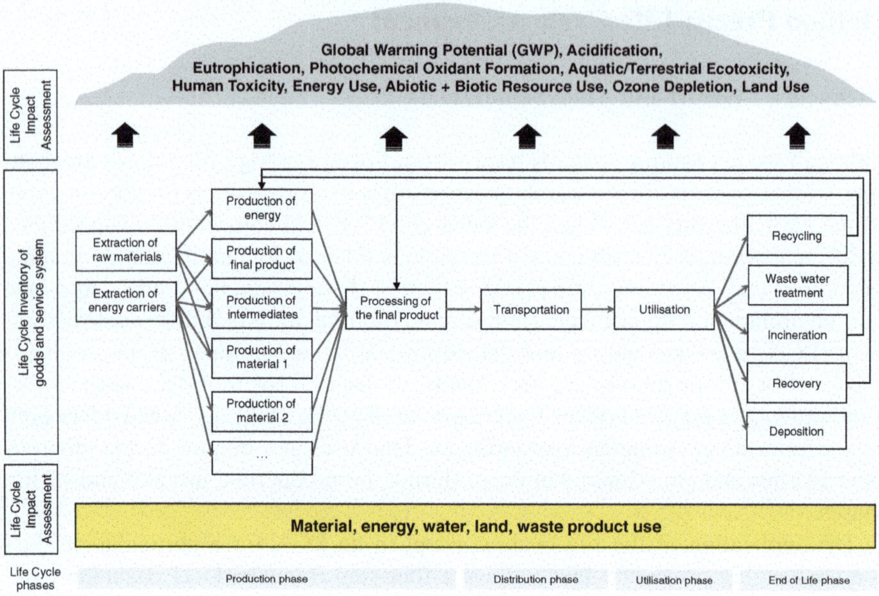

Fig. 10.5 Life cycle concept of a product system. (Modified after JRC 2013)

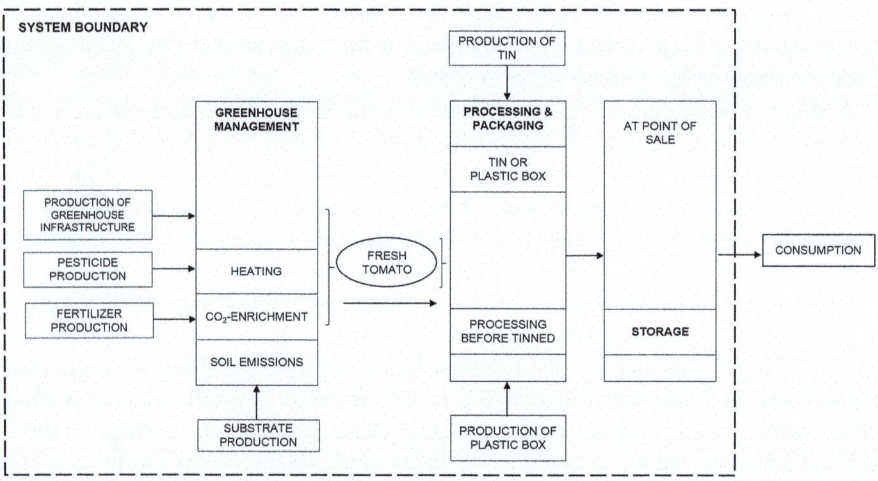

Fig. 10.6 Example of a flow chart comprising the supply chain of tomatoes. (Reproduced from Theurl et al. 2014; reprinted with permission)

translation of these flows into their impact on the environment. This step makes the LCA results more meaningful to consumers, policy makers and other concerned parties. Inputs and outputs are associated with specific impact categories; for example, the input of coal for combustion could be associated with global warming potential (GWP), whereas the application of fertilizers in agricultural production systems could be translated into both GWP and eutrophication. Again, the impacts included in the LCIA depend strongly on the initial definition of the goal and scope, which may prescribe a focus on human health, the natural environment or resource scarcity. The commonly used impact categories in LCA (Pelletier et al. 2007, also see Fig. 10.5) are the following:

- Global Warming Potential (GWP)
- Acidification
- Eutrophication
- Photochemical Oxidant Formation
- Aquatic/Terrestrial Ecotoxicity
- Human Toxicity
- Energy Use
- Abiotic Resource Use
- Biotic Resource Use
- Ozone Depletion
- Land Use

In LCA, environmental impacts are always related to a so-called functional unit representing the use of the product. For example, if tomatoes are the product under investigation, the functional unit might be 1 kg of tomatoes, and the global warming potential per kilogram of tomatoes would be calculated as an impact.

Although the list of impacts covers ecological and human health as well as resource depletion, social welfare—an important impact category in its own right—is entirely missing. Often referred to as social-LCA (S-LCA), this branch of analysis is currently under development. Impact categories in S-LCA include human rights, working conditions and governance. S-LCA provides information on social and socioeconomic aspects and, together with LCA, helps complete the analysis of socioeconomic activities.

LCA has proven to be a useful and informative tool, but some methodological caveats remain that require a cautious interpretation of LCA results. As Roy and colleagues (2009) note, multiple results may be obtained from LCAs for the same product. This has to do with the impact of truncation choices and how the environmental impacts are allocated to different products generated simultaneously. Despite the methodological standardization, the assumptions made in an LCA can have a strong impact on the results. If these assumptions are made transparent and their impact on the results is taken into account, the application of LCAs opens up our analysis of processes and flows both within and among economies and thus adds to our understanding of socioecological systems.

References

Baumann, H., & Tillmann, A. M. (2004). *The Hitch Hiker's guide to LCA. An orientation in life cycle assessment methodology and application*. Lund: Studentlitteratur.

Finkbeiner, M., Inaba, A., Tan, R. B. H., Christiansen, K., & Klüppel, H.-J. (2006). The new International standards for life cycle assessment: ISO 14040 and ISO 14044. *International Journal of Life Cycle Assessment, 11*, 80–85.

Finnveden, G., Hauschild, M. Z., Ekvall, T., Guinée, J., Heijungs, R., Hellweg, S., et al. (2009). Recent developments in life cycle assessment. *Journal of Environmental Management, 91*, 1–21.

JRC. (2013). *Introduction to LCA*, Figure: Scheme of a product system's life cycle with data collection of product and waste flows (blue lines) and resources (green) and emissions (grey arrows) followed by the impact assessment of the emissions and resource consumption. Available from http://lca.jrc.ec.europa.eu/lcainfohub/introduction.vm. Accessed October 10, 2013.

Pelletier, N. L., Ayer, N. W., Tyedmers, P. H., Kruse, S. A., Flysjo, A., Robillard, G., et al. (2007). Impact categories for life cycle assessment research of seafood production systems: Review and prospectus. *International Journal of Life Cycle Assessment, 12*, 414–421.

Roy, P., Nei, D., Orikasa, T., Xu, Q., & Okadome, H. (2009). A review of life cycle assessment (LCA) on some food products. *Journal of Food Engineering, 90*, 1–10.

Theurl, M. C., Haberl, H., Erb, K.-H., & Lindenthal, T. (2014). Contrasted greenhouse gas emissions from local versus long-range tomato production. *Agronomy for Sustainable Development, 34*, 593–602. doi:10.1007/s13593-013-0171-8.

Method Précis: Input-Output Analysis

Anke Schaffartzik

Monetary input-output tables (IOTs) are compiled by national statistical offices as part of the system of national accounts and are compatible with economy-wide material flow accounting (EW-MFA; see Method Précis on Material Flow Analysis) in that they systemically cover the whole economy in a top-down manner. In IOTs, inter-industry flows and final demand are reported annually. Table 10.2 shows a very simple IOT for a hypothetical economy that produces only three goods: wood, paper and books.

The values in the columns show which inputs are required for the production of all the wood, paper and books in this economy. For example, 90 units of wood, ten of paper and five of books are needed to produce the total amount of paper. The top-left 3 × 3 section of the table depicts the inter-industry flows, and the fourth column shows the final demand. The total output of each of the products corresponds to the sum of inter-industry use and final demand and is represented in the column on the far right. In monetary terms, value is added to a product during the production process. This happens because labor, taxes, interest, rent and profit have to be paid for. The value added is shown in the fourth row. Only as many units of a product can be consumed within the economy as are generated. Therefore, the totals by column and by row must match.

For applications in Social Ecology, information on environmentally relevant flows such as emissions, waste or material flows is introduced into the IOT to yield so-called *environmentally extended input-output tables*. Such calculations are based on a generalized form of the open static IO model elaborated by the economist Wassily Leontief (Leontief 1986). In the following example of an IOT extended by material flow data, we will speak of matrices and describe their size as n × n, meaning the matrix has n rows and n columns. We will also refer to n × 1 vectors as a special kind of matrix that has n rows and 1 column.

We will call the n × 1 vector of the total output of the economy **x** (the column on the very right in Table 10.2), and we will use **A** to denote the n × n matrix of inter-industry requirements. **A** may also be called the technology matrix because it informs us of the input required by each activity from other activities to produce one unit of output. For our example in Table 10.2, the **A** matrix would inform us that the production of one unit of paper requires 0.29 (=90/310) units of

Table 10.2 Input-output table for a hypothetical three-product economy

	Wod	Paper	Books	Final demand	Total output
Wood	85	90	0	20	195
Paper	10	10	90	200	310
Books	5	5	20	500	530
Value added	95	205	420	350	1070
Total	195	310	530	1070	2105

input from the wood-producing activity. We will denote the $n \times 1$ vector of final demand (second column from the right in Table 10.2) as $\mathbf{y}$. Then, Eq. 10.1 states in standard matrix notation that the total output of the economy is the sum of all intermediate and final consumption:

$$\mathbf{x} = \mathbf{A}\mathbf{x} + \mathbf{y} \tag{10.1}$$

Solving for total output $\mathbf{x}$ yields:

$$\mathbf{x} = (\mathbf{I} - \mathbf{A})^{-1}\mathbf{y} \tag{10.2}$$

where $\mathbf{I}$ is the $n \times n$ identity matrix, i.e., the matrix-equivalent to the number 1 – a matrix that is made up of ones along the diagonal and zeros in all other places. $(\mathbf{I} - \mathbf{A})^{-1}$ is the $n \times n$ matrix of total requirements, or the Leontief inverse. Each element of the Leontief inverse $\{l_{ij}\}$ shows the total, i.e., the direct and indirect, intermediate requirements of activity i in producing one unit of final demand of activity j. Direct requirements are directly delivered from one activity to another. Indirect requirements are the additional input required by third sectors in enabling the direct provisions between two sectors.

This model can be extended to include any kind of factor inputs required to produce a given final demand (Lenzen 2001). In most applications in Social Ecology, we consider material factor inputs. Then, $\mathbf{F}$ would be the $k \times n$ matrix of material use intensities, where k is the number of materials considered. Each element $\{f_{\alpha i}\}$ shows the direct input of material α into activity i per unit of output of that activity. The total (i.e., direct and indirect) material requirements $\mathbf{E}$ (k x n matrix with elements $\{e_{\alpha i}\}$) embodied in the final demand is given by Eq. 10.3:

$$\mathbf{E} = \mathbf{F}(\mathbf{I} - \mathbf{A})^{-1}\mathbf{y} \tag{10.3}$$

The IO methodology thus provides a way to explore the link between final consumption and the inputs required in production. By using environmentally extended IOT, Peters and Hertwich (2008) were able to show that in 2001, 8.3 % of the production-related CO_2 emissions in the United States occurred in the production of exports and that this share was 18.9 % on average for the Annex B countries of the *Kyoto Protocol* and 25.3 % for the Non-Annex B countries. An example of an extension by material flows is presented in this chapter.

References

Lenzen, M. (2001). A generalized input-output multiplier calculus for Australia. *Economic Systems Research, 13*(1), 65–92.

Leontief, W. (1986). *Input-output economics*. New York, Oxford: Oxford University Press.

Peters, G., & Hertwich, E. (2008). CO_2 embodied in International trade with implications for global climate policy. *Environmental Science & Technology, 42*(5), 1401–1407.

Chapter 11
How Circular Is the Global Economy? A Sociometabolic Analysis

Willi Haas, Fridolin Krausmann, Dominik Wiedenhofer and Markus Heinz

Abstract The Circular Economy is an appealing strategy for sustainable development that is being promoted by industries and governments in several industrial and emerging economies, such as the European Union, Japan and China. This chapter uses a sociometabolic approach to assess how circular or linear global material flows are at the turn of the 21st century. Analysis of the global material flows shows that 58 Gt (gigatons) of materials are extracted, 28 Gt are for energetic use, 26 Gt are additions to stocks, 4 Gt are consumed within a year and 4 Gt are waste rock. Of these flows, 4 Gt are recycled, so together with the 58 Gt of extracted materials, the global economy processed 62 Gt of materials. Thus, for 7 % of the global economy's inputs, the material loop is closed. An exploration of the potentials and limitations of the Circular Economy reveals that strategies targeting the output side (end-of-pipe) are limited given the present proportions of flows, whereas a shift toward renewable energy, a significant reduction of societal stock growth and decisive eco-design are required to advance toward a Circular Economy.

Keywords Circular economy · Recycling · Resource efficiency · Material flow accounting (MFA) · Life cycle assessment (LCA) · Energy transition · Stocks

---

W. Haas (✉) · F. Krausmann · D. Wiedenhofer · M. Heinz
Institute of Social Ecology (SEC), Alpen-Adria University,
Schottenfeldgasse 29, 1070 Vienna, Austria
e-mail: willi.haas@aau.at

F. Krausmann
e-mail: fridolin.krausmann@aau.at

D. Wiedenhofer
e-mail: dominik.wiedenhofer@aau.at

M. Heinz
e-mail: markus.heinz@aau.at

H. Haberl et al. (eds.), *Social Ecology*, Human-Environment Interactions 5,
DOI 10.1007/978-3-319-33326-7_11

11.1 Introduction

It is increasingly evident that humanity's resource use must be reduced. The material and energy resources to expand the current metabolic pattern of the industrial countries to the rest of the world are not available on this planet, nor are the capacities of the global ecosystems sufficient to absorb the outflows of this industrial metabolism. In this context, the notion of a Circular Economy and the concept of recycling as a key strategy to increase the circularity of resource flows in an economy have gained momentum (see definition box). They appear to be attractive possibilities to reduce the wasteful handling of resources by closing the loop of resource flows.

The principles of what we now discuss under the idea of a Circular Economy are not new; they have a long history. Preindustrial history and the early phases of industrialization provide numerous examples of how society was concerned with reusing and recycling materials and turning them into valuable resources. Waste flows of preindustrial societies were low, and until industrialization, entire societies were organized on the basis of largely closed loops. As late as 1900, the global resource flows of approximately 7 Gt (gigatons) still consisted of roughly 80 % biomass, most of it used as fuel, human food and feed for livestock (Krausmann et al. 2009). In the strict sense of the Circular Economy, this biomass cannot be reprocessed into the original or comparable products within the socioeconomic system.[1] However, by returning night soil, manure and ashes to the soil, wastes from biomass use were used to maintain soil fertility and provided the basis for the renewable production of biomass products. In preindustrial societies, all by-products and residues of biomass processing were used, and closing the loop between socioeconomic biomass flows and local ecosystems was essential.[2] In rural areas, reuse and recycling were organized mainly at the household level. In urban centers, effective collection methods existed for biomass wastes and wastes from products from metals and nonmetallic minerals, which were still used in low quantities prior to industrialization. In the city of Frankfurt, for example, prior to the First World War (WWI), an institution called 'Altwerk' specialized in repairing old

---

[1]In the public debate and, often, in scientific discourse, the notion of recycling is applied when biomass is composted and returned to fields as fertilizer and to improve soils. Here, the basic goal is to maintain a closed loop for essential plant nutrients such as nitrogen and phosphorus. In a strict sense, however, this manure and compost do not qualify as recycling. First, a significant part of the circle is not within the socioeconomic system—a precondition for recycling—but the biomass wastes are returned to ecosystems. Second, it is not the main material component that is recycled. The original product (e.g., food) is converted into an auxiliary or fertilizing product that supports soil fertility and plant growth, i.e., the production of new primary biomass.

[2]This was even the case in China in the 1970s. According to Rodale (1973), 'About 80 % of all fertilizer used is organic—either animal or human wastes, mulches or green manure crops' (p. 290).

products or, if this was no longer feasible, in remaking them into other products. Other examples include a comprehensive commercial selection system for scrap iron, which ensured high recycling rates of iron in Europe in the Middle Ages, or ragpickers, who effectively collected tatters from defined wards for paper production (Reith 2001). Until the onset of the metabolic transition, waste flows were generally very small, and rates of reuse and recycling (metals, glass and building materials) were high. Most end-of-life products were not considered waste but had considerable economic value. Only the more recent history of industrialization exhibits societal developments in the direction of the so-called throw-away societies or linear consumption patterns (Ellen MacArthur Foundation 2013; Hislop and Hill 2011).

Box: Definitions

Waste

'Waste' means any material or object the holder discards or intends to discard.

Circular Economy

The Circular Economy is a popular concept with a range of different definitions in the literature, as in (Allwood et al. 2010; Chen and Graedel 2012; Ellen MacArthur Foundation 2013; Hislop and Hill 2011; Mathews and Tan 2011; Moriguchi 2007; Preston 2012). Although these authors do not contradict each other, most of them fall short of providing a definition clear enough for analytical investigations. Some of them extend the concept to many environmental and social dimensions, thus blurring the differences from concepts such as green or sustainable economies. In this chapter, for the sake of analytical insight, we use a narrow definition that focuses on the core idea of the concept. This definition is as follows:

The Circular Economy strives to improve resource efficiency mainly by closing the resource loop and by stopping the wasteful use of resources. In general, these goals can be achieved by following certain priorities. The first priority is reduced material use for the same service, more intensive use of existing products (e.g., sharing or selling the service instead of individual ownership), longer life, more repair, more reuse and improved material efficiency in the production process. The second priority is end-of-life recycling (of the material contained in waste). To foster priorities 1 and 2, product design needs to be optimized to facilitate reuse, repair and recycling. For waste that cannot be recycled, options for cascading uses or downcycling must be identified and put to use. An example is biodegradable products made from plant-based materials that can be composted and used as fertilizer at the end of their service life.

Extending these principles of the Circular Economy across the economy requires far-reaching changes in the basic structures of industrial systems.

The Circular Economy aims to redesign an economy at the system level because this enables efficiency improvements that far outweigh potential savings from improving the efficiency of specific industrial processes or individual products and services.

In summary, the Circular Economy, as understood here, strives to minimize the relation between global material extraction and the inputs of virgin materials to the final consumption of products and services. Thus, the Circular Economy aims at an optimum level of closed loops—a complete circle without an unlimited supply of carbon-free energy and a large stockpile of inactive materials is practically unachievable (Ayres and Ayres 1999).

Recycling

'Recycling' is one of the measures a Circular Economy employs. Recycling means any recovery operation by which post-consumer end-of-life waste is reprocessed into products, materials or substances that can serve the original or comparable purposes. In practice, recycling is the main measure applied today, and it is already practiced at a global scale. 'Downcycling' refers to reprocessing, where the new product from these recycled materials has a lower material quality than the original product (e.g., plastic bottles become street boundary posts). Neither recycling nor downcycling include energy recovery or reprocessing into materials that are to be used as fuels or for backfilling operations (e.g., waste incineration for energy production).

Upcycling has gained attention recently. Upcycling refers to the process of converting waste materials into new products of better quality. In terms of mass flows, it is insignificant. The reprocessing of organic material within the economy (e.g., food waste into organic manure) is considered cascading use rather than recycling. Recycling in the context of a Circular Economy should not interfere with activities of the first priority, such as using less material for the same service (e.g., longer life, reuse and more intensive use).

Although the emergence of mass production and consumption superseded earlier preoccupations with material reuse and recycling, during the 1970s, the concept of improving the physical circularity of the economy by recycling gained attention again and was increasingly brought into debate (Wittl 1996). Numerous books, papers and readers about recycling were published, and recycling was promoted throughout Europe and the Americas. The need for recycling was argued from an environmental perspective, with an emphasis on the need to limit landfill capacities and to save resources. This new debate was initially driven by a fundamental criticism of wasteful handling of natural resources in the industrial economy (e.g., Meadows et al. 1972). Over the subsequent decades, private and public initiatives became bogged down by the practical challenges of recycling quantitatively minor but publicly quite visible material flows (e.g., packaging waste, batteries). Today, recycling is omnipresent in everyday life, and the younger generations in industrialized as well as many emerging economies are growing up with separate collection systems for glass, metals, paper and biowaste (Graedel and Reck 2012).

In the recent debate about pathways toward a more sustainable industrial metabolism, the idea of the Circular Economy has been revived and expanded to fit the environmental challenges of today (Allwood et al. 2010; Chen and Graedel 2012; Ellen MacArthur Foundation 2013; Hislop and Hill 2011; Mathews and Tan 2011; Moriguchi 2007; Preston 2012). Compared to the 1970s, there is a fundamental change in arguments. First, signs of resource depletion and sharp increases in both prices and related volatilities of raw material supply make the Circular Economy indispensable to maintain resource security. Second, a Circular Economy is instrumental for society to remain within 'planetary boundaries' and especially to mitigate greenhouse gas (GHG) emissions. This change in arguments reflects the fact that over the last few decades, two global framework conditions have decisively changed: the increasing acceptance of climate change and rising resource prices.

A critical examination of the literature reveals a lack of precise criteria with which specific measures can be evaluated for their contribution toward a Circular Economy. Hence, a diverse picture of measures and examples can be found in the literature. Some examples are encouraging. An industrial equipment provider claims, for example, that he already offers remanufacturing for a range of products for 30 years and that his end-of-life parts have a return rate of more than 90 % (Preston 2012). Other examples suggested as advancing the circularity of the economy are highly questionable, such as a television (TV) manufacturer who claims that for the same screen size, the product weight was reduced by nearly a factor of five over a period of ten years. However, because consumers own more TV sets and replace them at a much faster rate than 15 years ago (NPD DisplaySearch 2012) and because the average screen size has increased, it is likely that the net effects of the mentioned efficiency gains on overall resource input are small. Although these examples for implementing the Circular Economy show that resource efficiency per unit product or service may be improved considerably, they neglect system-wide issues such as rebound effects or problem shifts, which may drive further increases in material or energy consumption[3] (Alcott 2008; Van den Bergh 2012; Van den Bergh et al. 2011).

[3]The term rebound effects, or the Jevons paradox, refers to the observed phenomenon that greater efficiency usually leads to cost savings for the same amount of services. However, these savings are then spent either on more of the same service or on other products and services, leading to overall lower savings than anticipated. This rebound usually ranges from 10 to 50 %. A related term is carbon leakage effects, which refers to the externalization of emissions, such as to other regions. There is also the green paradox, which highlights that energy savings or the substitution of fossil fuels by renewable energies in important consumer markets might lead to globally reduced prices for fossil fuels, which may boost energy consumption in other regions.

Recycling is still the most prominent strategy toward a Circular Economy, and for some materials, it is already very advanced; end-of-life recycling rates for 18 metals such as iron, aluminum, copper and silver have increased over the decades and are above 50 % today (Graedel et al. 2011). Paper, glass and plastic, as well as construction and demolition waste, are other materials for which considerable efforts have been made to increase recycling rates. Considering all these improvements, one might assume that a Circular Economy is within reach and is only a matter of appropriate incentives and technologies. On what grounds can such a claim be made? How do we assess specific improvements and their overall contributions to closing material loops? The examples show that a proper discussion needs to be based on an appropriate analytical framework.

The next section discusses the explanatory power of different approaches for assessing the circularity of an economy and the specific measures for increasing it. In light of global sink and source constraints, we argue that a system-based approach is required to capture the overall effects of measures such as recycling, whereas a product-based approach can help assess resource efficiency gains over the life cycle of alternative products. Sociometabolic analysis, a system-based approach, is applied to assess the current state of the global economy's degree of circularity. In the final section, we discuss the current situation, including its dynamics and possible strategies for improvements in the material streams of material flow accounting (MFA). We show that for the latter life cycle assessment (LCA), approaches are required to guide the development toward a global Circular Economy with the potential to remain within planetary boundaries.

11.2 Choosing an Analytical Framework

Two different sociometabolic approaches come under consideration when assessing the global economy's circularity and its challenges. One is MFA, which accounts for all material flows necessary to reproduce a socioeconomic system (for more details, see Method Précis on Material Flow Analysis in this volume). The other is LCA (see Method Précis on Life Cycle Assessment in this volume), a cradle-to-grave analysis that accounts for all energy/material flows resulting from a single unit of product or service (Haberl et al. 2013; Rebitzer et al. 2004; see also Table 11.1). For both approaches, a consistent definition of system boundaries is essential because only those flows that cross the boundary between the socioeconomic system and its environment are accounted for. At the global level, in principle, both approaches pursue the same concept of defining the boundaries between society and environment.

Table 11.1 Comparison of material flow accounting (MFA) and life cycle analysis (LCA) as frameworks to analyze the circularity of economies

	MFA material flow accounting	LCA life cycle analysis
Characteristics		
Scheme	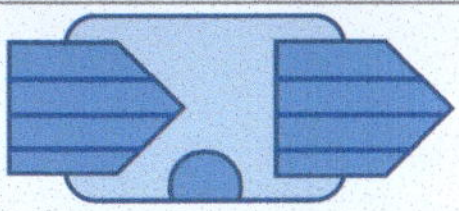	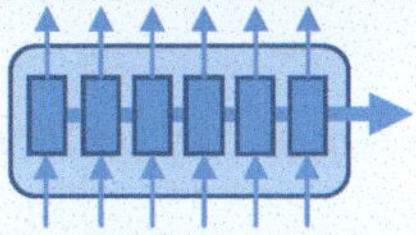
Focus	Socioeconomic system (e.g., local system, national or global economy)	Functional unit delivered by a product system
Reference	Accounting for a system's annual material flows that cross the system boundary to nature or other social systems	Assessing environmental pressures/ impacts associated with all stages of a product's life, from cradle to grave
Explanatory power	To discuss whether a system's material and energetic exchanges can be maintained in the long run, which is a minimum condition for ecological sustainability	To discuss which alternative product system delivers the same functional unit with lower environmental pressures/impacts
	Flows of local and/or national systems can be added up to higher aggregate levels without double counting	To a certain extent, pressures/ impacts associated with a functional unit can be summed up (e.g., across households), but with higher aggregations, this is prone to double counting
	MFA is compatible with different dynamic modeling approaches; combining MFA with LCA can reveal insights into potential resource savings for a social system	The potential to reduce environmental impacts by introducing alternative product systems with lower impacts can be calculated
Limitations	Socioeconomic system itself is a black box (products and processes are not visible) MFA accounts for input and output flows; environmental pressures/ impacts are not part of the MFA; needs to be combined with other approaches (e.g., Van der Voet et al. 2009) No standardized approach for estimating environmental impacts associated with material flows	Allocation problem: no single method provides a general solution to how environmental burdens should be shared in multi-functional processes among its functions or products Truncation error: the incompleteness resulting from the omission of environmental loads in the higher upstream layer of the product system; in some cases, this can be highly significant (see Reap et al. 2008)

As discussed in Table 11.1, the two approaches offer quite different insights and have different limitations. To assess the circularity of the global economy, we combine the two sociometabolic approaches.

11.3 Applying a Sociometabolic Analysis

To assess the global economy's circularity, we use an MFA framework. On the input side, the framework provides quantitative information on the extraction of materials by type. Data on global material extraction (which is equal to material use at the global scale) are available for 40 different material groups in metric tons per year for the year 2005 (Schaffartzik et al. 2014). At an aggregate level, we distinguish four main material groups: biomass, metals, minerals and fossil energy carriers. The simplified sociometabolic flow chart in Fig. 11.1 shows the aggregated flows of materials through the global economy, from extraction to processing to use or temporary accumulation in material stocks to recycling or final treatment before it leaves the socioeconomic system. The first phase is material extraction such as mining or harvesting, followed by processing steps such as refining fossil fuels or manufacturing furniture. For the use phase, two distinctions need to be made that are of high relevance for the Circular Economy.

The first distinction is whether a specific material is used as an energy carrier for energy production or as raw material for other processes. Energy-rich materials such as coal, oil and fuel wood are usually converted into technically useful

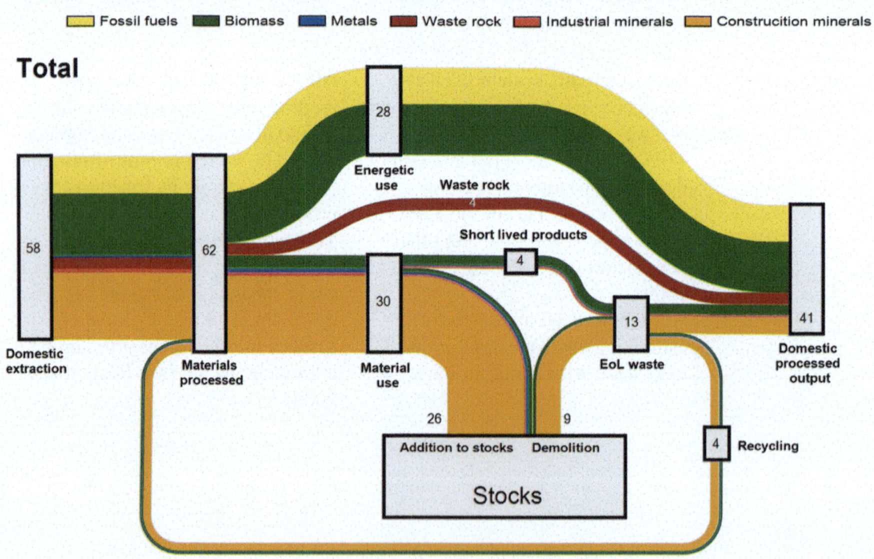

Fig. 11.1 Sociometabolic flow chart of the global economy in 2005

energy by combustion. This applies to the largest fraction of all fossil materials (only 2 % of all fossil energy carriers are used in material applications such as plastics) and part of the biomass (e.g., wood fuel). However, agricultural biomass used to feed humans and livestock also must be considered an energy carrier because it is converted into metabolic energy by catabolic processes. All these fossil and biomass materials used as energy carriers are converted into gaseous emissions (mainly CO_2) and other residues (combustion residues, excrements). None of these residues can be recycled in the sense that they can be used again for the original purpose. To a limited degree, cascade utilization or downcycling is possible, for example, when excrements are used as fertilizer or when ash is used in chemical processes, as backfilling or as grit in winter. However, closing the loop for these materials is not possible. Thus, by definition, the large fraction of materials that are used as energy carriers cannot be recycled and, thus, cannot participate in closing the loop within the global economy.

The second distinction applies only to the non-energy fraction of material inputs and concerns the useful service lifetime. We need to distinguish between materials that are used and discarded within one year and materials that remain in the socioeconomic system for a longer period and thus are added to the stock of artifacts such as buildings or machines. Products consumed within a year become end-of-life waste and are potentially directly available for recycling again. Typically, these are consumer goods such as packing, hygiene products, newspapers and batteries. In contrast to these consumables, a large amount of durable goods remains in the socioeconomic system for more than a year. These durable goods range from books, household appliances, furniture, machinery and cars to buildings and infrastructures. This fraction is not immediately available for recycling, and much of it remains in the economy for several decades. Although economies still increase their physical stocks, a considerable amount of stocks reach their end-of-life every year and are discarded or demolished. These material flows are another fraction that needs to be considered in light of a Circular Economy.

For all materials coming to their end-of-life within an economy that is striving for circularity, there are essentially three possible pathways to meet this goal: they might be reused or recycled to close the loop of specific flows, they might be used for a different purpose (cascade utilization, downcycling), or they might leave the economy as processed output (e.g., emissions).

To discuss these options, we combine an MFA-derived analysis on global material use with LCA results for resource savings of specific strategies and measures. In its practical application, LCA can provide quantitative information on potential resource savings if recycled materials are used instead of primary raw materials in specific processes for certain products. With the obtained factors (metric tons of savings per metric ton of material), the savings in input and output flows (e.g., for increased metal recycling rates) can be calculated and deducted from the overall inputs and outputs in the sociometabolic flow chart (for more details on methods and data sources, see Haas et al. 2015).

11.4 Current State of the Global Economy's Circularity

Applying the sociometabolic analysis to the global situation provides us with a rough picture of the degree of circularity of the global economy at the turn of the 21st century. Figure 11.1 shows the size of material flows for which loops have been closed by recycling and reuse as well as those flows that were throughputs. In the year 2005, 58 Gt of raw materials[4] were extracted and entered the global economy. Together with 4 Gt of recycled materials, the global economy processed 62 Gt of material (100 %). Approximately 45 %, or 28 Gt, of energy carriers were converted by technical or physiological combustion processes into gaseous emissions or solid wastes leaving the socioeconomic system. Another 42 % of the input, or 26 Gt, are additions to stocks, of which the largest part by far are nonmetallic minerals used in buildings and infrastructures. According to our estimations, at the same time, approximately 9 Gt of stocks reached their end-of-life and were demolished and were thereby potentially available for recycling. Only approximately 4 Gt, or 6.5 % of the global material input, were actually used for material purposes within this year. This means that in 2005, only 13 Gt, or one-fifth of all material inputs, became so-called end-of-life waste. According to our estimates, approximately 30 % of this end-of-life waste was actually recycled in 2005, whereas the other part was treated in waste plants and thus left the socioeconomic system as gaseous, liquid or solid outputs.

Our results indicate that the loops were closed only for approximately 7 % of the material inputs (4 Gt). Seventy percent of the inputs, or 41 Gt, are throughputs that leave the socioeconomic system within a year, of which 28 Gt are energy carriers, which by definition cannot be recycled. Almost one-third of all material inputs (30 % or 17 Gt) are net additions to socioeconomic stocks, which become available for recycling only after longer periods, often decades.

11.5 Challenges for a Global Circular Economy

For each of the four material categories, we discuss the present situation of circularity, the role recycling played to close loops in 2005 and the role recycling could play in the future. Finally, based on the analysis of the most important flows, we draw conclusions concerning the greatest challenges and the most effective ways of progressing toward a Circular Economy.

---

[4] Gt or gigatons equals 10^9 metric tons.

11.5.1 Fossil Energy Carriers

Of the 58 Gt of global material extraction per year, approximately 12 Gt are fossil energy carriers, of which approximately 98 % are used to produce energy (BIOIS 2011). With present-day technologies, the release of a fuel's energy is irreversible. With the exception of plastics and a few other material applications, recycling is not possible for this material category. However, due to source and sink problems, a transition toward a new energy system will eventually be necessary. There are different options for making this transition, some of which might conserve the present linearity of the system and others that have the potential to significantly improve the level of circularity. Carbon capture and storage is one option that conserves the economy's linearity. This technology increases the material and energy input required by fossil-fuel-powered plants per unit of energy output and therefore reduces the efficiency of energy production (Herzog 2011) and reinforces the economy's linearity. In contrast, a rising share of energy generated by solar wind geothermal and tidal power plants in the total energy mix could improve the circularity of the economy. Compared to fossil-fuel-powered plants, these technologies reduce material inputs and outputs: inputs because a constant supply of fuel is no longer needed and outputs because the production of wastes and emissions is much lower. As soon as electricity is produced with a low carbon intensity, a shift in transport technologies away from combustion engines and toward electric motors can further increase the overall material efficiency of the global economy. Recalculating the global material flows based on substituting half the currently used fossil energy carriers with wind power[5] would decrease the societal outputs by approximately 15 %. Because no recycling is possible for energetically used fossil energy carriers, its fraction is a stumbling block to the Circular Economy.

Approximately 2 % of all fossil energy carriers are used as material, mainly in the production of plastic, bitumen and lubricants. In the EU, the plastic recycling rates are between 21 and 33 %. However, in most cases, plastic is only downcycled to replace products of lower quality (e.g., food packaging to plastic bags or flower pots) (BIOIS 2011). For asphalt (a mixture of gravel and bitumen), in situ recycling in industrialized countries is already quite high, but quantitative assessments at the global level or for world regions are lacking.

11.5.2 Metals

Ores, the raw material from which metals are extracted, account for approximately 4.5 Gt (8 %) of global extraction. The actual metal content of these ores is only approximately 0.8 Gt; the remainder is tailings and processing slags of

[5]Jacobson and Delucchi (2011) believe it is feasible for wind power to cover 50 % of the global energy demand by 2030.

little further use. In principle, metals are infinitely recyclable. However, there are significant challenges to metal recycling. One general challenge holds that ‘after millennia of products made almost entirely of a handful of metals, modern technology is today using almost every possible metal, but often only once’ (Graedel and Reck 2012, p. 694). Furthermore, there is a tendency for metal products to become less recyclable over time. This is due to several factors, including the increasing significance of complex alloys and composite materials as well as the dissipative use of metals, nanomaterial-technologies and microelectronics. Although production processes make use of highly elaborate technologies, very crude technologies (shredding, crushing, magnetic sorting) are used in recycling mainly to keep costs low. At present, end-of-life recycling rates for many ‘base metals’ (e.g., copper, zinc) are slightly above 50 %. For two other metals, the recycling rates are significantly higher: iron, with a recycling rate of approximately 90 % (Graedel et al. 2011; UNEP 2011), and lead. Lead is an exception because the largest share of lead is used for just one product group, vehicle batteries, of which approximately 90–95 % is collected and recycled. There is also a wide range of metals and metalloids with recycling rates below 1 % (e.g., lithium, thallium). These metals are often used in very small quantities in individual products, which makes recycling difficult and costly. In general, metal recycling provides a positive net balance for both material and energy. This is because the processing steps from ore extraction to pure metal require moving and processing huge quantities of material (examples for ore grades: iron 43 %, copper 1 %, lead 12 %, aluminum 19 %; see Eurostat 2012), which requires large amounts of energy that could be saved when metals are recycled. Among the different end-of-life waste streams, metals constitute approximately 2 % of mass. If consequent eco-design is applied and if economic incentives are in place, the material loop of this stream could be closed to a very high degree. Additionally, this could substantially reduce the share of global industrial carbon emissions due to steel production, estimated to be 25 % in 2006 (Allwood et al. 2011). For each additional ton of recycled steel instead of steel made from virgin ore material, more than 70 % of emissions and more than 70 % of material extraction could be reduced.

11.5.3 Nonmetallic Minerals

Construction minerals represent the largest fraction of the nonmetallic material group. They make up nearly 40 % of global domestic extraction and 63 % of all end-of-life waste. Construction minerals comprise gravel, sand, limestone, gypsum, chalk and clays. Globally, this fraction is growing at very high rates. According to experts, nearly all types of construction materials can be recycled (Chong and Hermreck 2010), but recycling also has negative side effects arising from transport and processing. Insights into the potentials and limitations of construction mineral recycling can be gained from case studies using LCAs that systematically consider all inputs and outputs related to recycling. A US study

(Chong and Hermreck 2010) notes that even in cases reported as sustainable construction practices, there are significant inefficiencies that reduce the benefits from recycling. The saturation of local markets for recycled materials can become a critical factor because an increase in the distance between project sites and recycling facilities might compensate for the benefits of recycling. The study concludes that further increases in recycling activities depend on the existence of a market for recycled materials, regional recycling capacities, total energy used to recycle and awareness among workers and designers that they have options to use recycled materials in construction projects. An Italian case study (Blengini and Garbarino 2010) investigated specific recycling chains and proved ecological benefits for 13 of 14 environmental indicators over the use of natural aggregates. That study also emphasized the significance of distance and transport. They estimated that an increase of transportation distance by a factor of two to three for recycled aggregates increases the induced impacts such that they outweigh the avoided impacts. However, these are best practice examples. In general practice, the recycling rates for construction minerals are still low, and the construction sector remains skeptical about the use of recycled material (Blengini et al. 2012). Therefore, construction and demolition waste from buildings and roads is mainly used for backfilling to merely replace locally extracted sand and gravel with short production chains. In European countries, the recycling rates of construction materials are, on average, approximately 46 %. To further close loops for construction minerals, it is necessary that the eco-designs of today's buildings significantly increase lifetimes, improve recyclability and enhance regional flow management to keep transportation routes short.

There is a long tradition of recycling for glass, which is a relatively small fraction of nonmetallic minerals. The material inputs for the production of glass account for less than 0.5 %[6] of global extraction, and glass accounts for approximately 6 % of end-of-life waste. Recycling rates in industrialized countries range from 40 to 70 %. Glass can be re-melted and used in new glass products without loss of physical properties or quality.[7] However, according to the priorities of the Circular Economy, reuse would be far more favorable than recycling.

11.5.4 Biomass

Biomass is the second largest material flow at the global scale, and it accounts for roughly one-quarter of global material extraction. Of the 19 Gt of biomass, approximately 80 % is used energetically as food, fodder or fuel (which has the

[6]According to the *World Silica Sand Market* report (The Freedonia group 2012), extraction will grow to 278 Mt (million metric tons) in 2016, from approximately 175 Mt in 2005.

[7]Colored glass cannot be turned into clear glass products, but it can be recycled into other colored glass products.

smallest share). Although this biomass does not qualify for recycling, the efficiency of its use can be increased. One example is food waste: due to inefficiencies in the chain from harvest to consumption, approximately 20–30 % of total available food is wasted. Prevention of food waste has first priority. To date, however, there is little experience that could shed light on how to improve resource efficiency in this case. Cascade utilization of by-products, residues and excrements also has great potential to improve the overall resource efficiency, mainly by composting to substitute for industrial fertilizer. Another pathway to improve resource efficiency of biomass is reducing the size of the livestock system. Livestock is an inefficient biomass converter, and nearly 60 % of all biomass is used as fodder.

Approximately 12 % of biomass (approximately 4 % of global extraction) is wood used in construction, furniture and paper production. In the construction sector, the reuse of wood has increased. In Europe in 2005, approximately 13 % of the consumed round wood was recovered; of this, 2 % was reused and 34 % was used for energy generation (Merl et al. 2007). Paper is another fraction with a long recycling tradition. Currently, approximately 40–50 % of all paper is recycled. However, there is also enormous potential for prevention as not all paper products are actually usefully consumed (e.g., free newspapers, unsolicited bulk mail). The literature is inconclusive concerning the net balance of paper recycling and under which circumstances incineration is better than recycling. This is because LCA studies have been performed under different decision-making situations and have used different system boundaries (Ekvall 1999). A recent LCA-based comparison of alternative treatments of waste paper offered environmentally favorable or neutral profiles for bioethanol production when compared with recycling or incineration (Wang et al. 2012). The results for paper recycling are still varied.[8] Nevertheless, paper recycling will not play a significant role in closing the loop due to the small quantity of flows.

11.6 Conclusions

The sociometabolic approach shows that only 7 % of all materials entering the global economy are currently recycled and therefore used in closed loops. To conclude, we discuss the potential of certain strategic options by altering key variables, one by one, in our sociometabolic analysis for a better understanding of how this would change the global economy's circularity (see Table 11.2).

First, 50–100 % of the fossil energy carriers are replaced by energy from renewable sources. This reduces inputs by 12–21 % and outputs by 15–30 %, and it means an additional loop closing of 0.6–1.5 %. Second, the aggregate

[8] Paper recycling reduces the length of fibers. Therefore, the 'recycling' is always a downcycling from one quality level to the next (from high-quality printing paper to corrugated board).

Table 11.2 Potential and limits of the Circular Economy—compared to the 2005 sociometabolic flow chart, with 7 % circularity, the list strategy can achieve different input and output reductions and additional loop closing

	Strategy	Input reduction (%)	Output reduction (%)	Additional loop closing (%)
Fossil energy carriers	Energy transition to solar, wind and geothermal	12–21	15–30	+0.6–1.5
Metals	Eco-design, improved recycling technologies	0.6–1.9	0.7–2.5	+0.5–0.8
Nonmetallic minerals	Eco-design, local flow management, halting stock growth	13–28	0	+14–28
Biomass	Prevention of food waste, improved cascading use	3–8	6–12	+0.6–0.9
Sum		22–47	13–27	+14–27

recycling rate for metals is increased to 80 or 100 % from the current 71 %. Only small reductions in inputs and outputs are achieved in this way, and an increase in the circularity of less than 1 % can be expected. Third, infrastructure and building stocks are stabilized. This means net additions to stocks are reduced by 50 % compared with no net additions to stocks at all. This reduces inputs by 13–28 %, outputs remain the same. And loop closing increases by 14–28 %. Fourth, we reduce food waste by 50–100 %. Inputs would decrease by 3–8 %, outputs would decrease by 6–12 %, and loop closing would increase moderately. Combining all four strategies could achieve a reduction of inputs by up to 47 % and a decrease of outputs by up to 27 %. In such a future scenario, one-third of all materials used could be kept in the societal material loop.

From this, several lessons can be learned for policies aiming at the implementation of a Circular Economy.

- The energy transition, which is urgently required to stay within the planetary boundary for climate change, is mutually interlinked with the Circular Economy. No energy transition means a tight limit for circularity, and no Circular Economy means a stumbling block for climate mitigation.
- One of the most crucial issues is the handling of large and still-growing physical stocks. If the renewal of buildings and infrastructure can be managed with reduced net additions to stocks and increased recycling of demolition waste, it would have a considerable effect on the circularity of the global economy. How to achieve a positive material and energy net balance remains a challenge because the results are highly sensitive to transport distances. The eco-design of buildings and transport systems plays a crucial role in improving recycling opportunities.

- Against the background of an average growth rate in global material consumption of approximately 2.6 % per year (1950–2005), which even accelerated in the last decade, the discussed strategies are far from a realistic future scenario. Achieving a reversal of the global growth trend in resource consumption into a no-growth or even shrinking dynamic remains the greatest challenge.

References

Alcott, B. (2008). The sufficiency strategy: Would rich-world frugality lower environmental impact? *Ecological Economics, 64*, 770–786.

Allwood, J. M., Ashby, M., Gutowski, T. G., & Worrell, E. (2011). Material efficiency: A white paper. *Resources, Conservation and Recycling, 55*, 362–381.

Allwood, J. M., Cullen, J. M., & Milford, R. L. (2010). Options for achieving a 50 % cut in industrial carbon emissions by 2050. *Environmental Science and Technology, 44*(6), 1888–1894.

Ayres, R. U., & Ayres, L. W. (1999). *Accounting for resources, 2: The life cycle of materials, Cheltenham, UK & Northampton*. Massachusetts: Edward Elgar Publishing. ISBN 1-85898-923-X, retrieved 20 January 2015.

BIOIS. (2011). Analysis of the key contributions to resource efficiency.

Blengini, G. A., & Garbarino, E. (2010). Resources and waste management in Turin (Italy): The role of recycled aggregates in the sustainable supply mix. *Journal of Cleaner Production, 18*, 1021–1030.

Blengini, G. A., Garbarino, E., Solar, S., Shields, D. J., Hámor, T., Vinai, R., et al. (2012). Life cycle assessment guidelines for the sustainable production and recycling of aggregates: The sustainable aggregates resource management project (SARMa). *Journal of Cleaner Production, 27*, 177–181.

Chen, W.-Q., & Graedel, T. E. (2012). Anthropogenic cycles of the elements: A critical review. *Environmental Science and Technology, 46*(16), 8574–8586.

Chong, W. K., & Hermreck, C. (2010). Understanding transportation energy and technical metabolism of construction waste recycling. *Resources, Conservation and Recycling, 54*(9), 579–590.

Ekvall, T. (1999). Key methodological issues for life cycle inventory analysis of paper recycling. *Journal of Cleaner Production, 7*(4), 281–294.

Ellen MacArthur Foundation (Ed.). (2013). Towards the circular economy. Economic and business rationale for an accelerated transition.

Eurostat. (2012). Economy-wide material flow accounts (EW-MFA). Compilation Guide 2012.

Graedel, T. E., Allwood, J. M., Birat, J.-P., Buchert, M., Hagelüken, C., Reck, B. K., et al. (2011). What do we know about metal recycling rates? *Journal of Industrial Ecology, 15*(3), 355–366.

Graedel, T. E., & Reck, B. K. (2012). Challenges in metal recycling. *Science, 337*, 690–695.

Haas, W., Krausmann, F., Wiedenhofer, D., Heinz, M. (2015). How circular is the global economy? An assessment of material flows, waste production and recycling in the EU and the world in 2005. *Journal of Industrial Ecology*, doi:10.1111/jiec.12244

Haberl, H., Erb, K.-H., Gaube, V., Gingrich, S., & Singh, S. J. (2013). Socioeconomic metabolism and the human appropriation of net primary production: What promise do they hold for LTSER? In S. J. Singh., et al. (Eds.): *Long term socio-ecological research. Studies in society: Nature interactions across spatial and temporal scales* (pp. 29–52). Springer Netherlands: Human-Environment Interactions, Bd. 2.

Herzog, H. (2011). Scaling up carbon dioxide capture and storage: From megatons to gigatons. *Energy Economics, 33*(4), 597–604.

Hislop, H., & Hill, J. (2011). *Reinventing the wheel: a circular economy for resource security*. London.
Jacobson, M. Z., & Delucchi, M. A. (2011). Providing all global energy with wind, water, and solar power, Part I: Technologies, energy resources, quantities and area of infrastructure, and materials. *Energy Policy, 39*, 1154–1169.
Krausmann, F., Gingrich, S., Eisenmenger, N., Erb, K.-H., Haberl, H., & Fischer-Kowalski, M. (2009). Growth in global materials use, GDP and population during the 20th century. *Ecological Economics, 68*, 2696–2705.
Mathews, J. A., & Tan, H. (2011). Progress toward a circular economy in China. The drivers (and inhibitors) of eco-industrial initiative. *Journal of Industrial Ecology 15*(3), 435–457.
Meadows, D. L., Meadows, D. H., & Randers, J. (1972). *The limits to growth*. New York: Universe Books.
Merl, A. D., Humar, M., Okstad, T., Picardo, V., Ribeiro, A., & Steirer, F. (2007). Amounts of recovered wood in cost E31 countries and Europe. In C. Gallis (Ed.), *Management of recovered wood: Reaching a higher technical, economic and environmental standard in Europe Thessaloniki*. Thessaloniki: University Studio Press.
Moriguchi, Y. (2007). Material flow indicators to measure progress toward a sound material-cycle society. *Journal of Material Cycles and Waste Management, 9*(2), 112–120.
NPD DisplaySearch. (2012): Global TV replacement study. Retrieved from: http://www.displaysearch.com/cps/rde/xchg/displaysearch/hs.xsl/global_tv_replacement_study.asp. Retrieved January, 20 2015
Preston, F. (2012). *A global redesign? Shaping the circular economy*. London.
Reap, J., Felipe, R., Scott, D., & Bert, B. (2008). A survey of unresolved problems in life cycle assessment. *The International Journal of Life Cycle Assessment 13*(4), 290–300. doi:10.1007/s11367-008-0008-x.
Rebitzer, G., Ekvallb, T., Frischknechtc, R., Hunkelerd, D., Norrise, G., Rydbergf, T., et al. (2004). Life cycle assessment. Part 1: Framework, goal and scope definition, inventory analysis, and applications. Review. *Environment International 20*, 701–720
Reith, R. (2001). Recycling - Stoffströme in der Geschichte. In S. Hahn & R. Reith (Eds.), *Querschnitte 8: Umweltgeschichte*. Wien: Verlag für Geschichte und Politik.
Rodale, R. (1973). Notebook-energetic Chinese. *Ecologist, 3*(8), 289–290.
Schaffartzik, A., Eisenmenger, N., Gingrich, S., Mayer, A., & Krausmann, F. (2014). The global metabolic transition: Regional patterns and trends of global material flows, 1950–2010. *Global Environmental Change, 26*, 87–97.
The Freedonia group. (2012). World Silica sand market—Report of industry market research for business leaders, strategists, decision makers, Cleveland.
UNEP. (2011). Recycling rates of metals—A Status Report.
Van den Bergh, J. C. J. M. (2012). Effective climate-energy solutions, escape routes and peak oil. *Energy Policy, 46*, 530–536.
Van den Bergh, J. C. J. M., Truffer, B., & Kallis, G. (2011). Environmental innovation and societal transitions: Introduction and overview. *Environmental Innovation and Societal Transitions, 1*(1), 1–23.
Van der Voet, E., Van Oers, L., De Bruyn, S., De Jong, F., & Tukker, A. (2009). Environmental impact of the use of natural resources and products. CML report 184—Department Industrial Ecology, Leiden.
Wang, L., Templer, R., & Murphy, R. J. (2012). A life cycle assessment (LCA) comparison of three management options for waste papers: Bioethanol production, recycling and incineration with energy. *Bioresource technology, 120*, 89–98.
Wittl, H. (1996). *Recycling: Vom neuen Umgang mit Dingen*. Regensburg: Roderer.

Chapter 12
Material Stocks and Sustainable Development

Dominik Wiedenhofer, Willi Haas, Michael Neundlinger and Nina Eisenmenger

Abstract Material stocks are an important part of the metabolism of society. Due to their long service-lifetimes, these stocks induce long-term dynamics of resource use for their regular reproduction, triggering resource flows during construction, use, maintenance, refurbishment and at the end of their useful lifetime in the form of waste. This chapter explores the material stocks of residential buildings and transportation infrastructure in the EU25 and the way these stocks are related to the overall material consumption of construction minerals. Special focus lies on flows required for maintenance and reproduction versus expansion of the stock. The dynamics of stocks and flows are assessed from a systems perspective on inputs, end-of-life waste and recycling flows in 2009, and a trend scenario for 2020. Thus, we explore the potential impacts of the *European Waste Frame-work Directive*, which strives for a significant increase in recycling. We find that in the EU25, a large share of material inputs are directed at maintaining and refurbishing existing stocks. Proper management of existing transportation networks and residential buildings is therefore crucial for the size of future material flows. Halting, or at least decelerating, ongoing stock expansion is another promising avenue toward more-sustainable resource use.

D. Wiedenhofer (✉) · W. Haas · M. Neundlinger · N. Eisenmenger
Institute of Social Ecology (SEC), Alpen-Adria University,
Schottenfeldgasse 29, 1070 Vienna, Austria
e-mail: dominik.wiedenhofer@aau.at

W. Haas
e-mail: willi.haas@aau.at

M. Neundlinger
e-mail: michael.neundlinger@aau.at

N. Eisenmenger
e-mail: nina.eisenmenger@aau.at

H. Haberl et al. (eds.), *Social Ecology*, Human-Environment Interactions 5,
DOI 10.1007/978-3-319-33326-7_12

Keywords Dynamic material stocks and flows · Social metabolism · Recycling · Construction and demolition waste

12.1 Introduction

Ongoing efforts in the European Union to improve its environmental performance and move toward a sustainable development pathway led to the formulation of the Waste Framework Directive, which was put into national law by 2010 (European Parliament, Council 2008). In addition to pushes toward improved waste prevention and comprehensive national waste management plans, the directive mandates quantitative goals for increased recycling as an important step toward a 'recycling society'. Construction and demolition waste, which, after carbon emissions, is the second-largest waste stream of the European economies, has been targeted with a compulsory recycling rate of at least 70 % of weight by 2020 (ibid.).

These waste streams predominantly stem from the refurbishment and demolition of material stocks, which were usually built decades ago. In Europe, large stocks of buildings and infrastructure already exist, although systematic data are scarce. For example, residential buildings in Europe are currently a mixture of different ages and building types: 30–40 % were built in the 1945–1970 period and another 20–40 % in the 1971–1990 period (Nemry et al. 2010). Annual demolitions ranged from only 0.05 to 0.2 % of national dwelling stocks in the 1980–2005 period (Thomsen and Van der Flier 2009), resulting in a rapid accumulation of buildings. Studies of specific European countries have shown that the existing material stocks in buildings are quite large and will continue growing slowly, with saturation to be expected in the mid-21st century (Bergsdal et al. 2007; Müller 2006; Tanikawa and Hashimoto 2009). The necessary maintenance and refurbishment of these buildings to meet current energy and other standards have large implications financially, for material and energy use and for emissions (Nemry et al. 2010; Pauliuk et al. 2013).

Additionally, transportation networks such as roads and railways are a major consumer of construction minerals, especially when stricter building standards due to frequent natural disasters intensify material demand, as is the case in Japan (Schiller 2007; Tanikawa and Hashimoto 2009). For the European road and rail network, no detailed age/type data exist due to the much shorter service-lifetimes of this infrastructure compared to buildings, although the material turnover can be expected to be much higher than for buildings (Schiller 2007).

Although the focus of this chapter is on stocks and flows of construction minerals,[1] which have a relatively low direct environmental impact on a per ton basis, these

[1]Throughout this chapter, we use the term construction minerals to denote the group of materials we are actually investigating in this work. In standard material flow accounting, these minerals are part of the category 'non-metallic minerals', which includes salt, fertilizers and other non-classifiable nonmetallic materials. Construction minerals typically account for approximately 96 % of the nonmetallic minerals group on a weight basis.

materials constitute one-third of total annual resource use in industrialized countries, and their use is interlinked with fossil fuels and metals (Steinberger et al. 2010). For example, concrete, on a global level and during the whole production process, translates into 7 % of fossil-fuel-related greenhouse gas (GHG) emissions (Allwood et al. 2012). Concrete is often reinforced with steel, which, over its life cycle and for total global production, is responsible for another 9 % of global fossil-fuel-related GHG emissions (ibid.). Second, large-scale infrastructure and sprawling cities impact land-use patterns, causing habitat fragmentation and problems in water management. Third, the sheer size of construction and demolition waste flows and limited landfill capacities can become problematic. Additionally, these waste flows are frequently contaminated with toxic substances, such as asbestos or heavy metals, making specific treatment procedures necessary. This means that in addition to the stock and flow dynamics explored later, various other aspects of high environmental importance must be considered when thinking about material stocks.

12.2 A Socioecological Perspective on Material Stocks

From a socioecological viewpoint, the major criterion for sustainability is the capacity of society to functionally reproduce itself via its material and energetic metabolism in the long run (Haberl et al. 2004). But what is actually being reproduced? Based on the concept of social metabolism (Fischer-Kowalski and Haberl 2007), societies are conceived of as systems that include biophysical stocks (humans, livestock and artifacts such as residential buildings, roads and railways) and that interact with the natural environment (via biophysical flows of material and energy) to reproduce, operate, maintain and expand these stocks (Fig. 1.3). Artifacts are objects made by humans and maintained in their desired state through recurring labor inputs. Artifacts include buildings, tools, machinery, roads and so forth (Haberl and Zangerl-Weisz 1997, p. 63). These artifacts fulfill certain functions, such as facilitating mobility or providing warm and dry living space. During the service-lifetime of stocks, repeated inputs of materials, energy and labor are needed to cope with their physical deterioration, to maintain their functions and to adapt them to changing needs. This creates a long-term legacy of resource use patterns that are partly determined by the quality and quantity of material stocks and the socially desired functions provided by stocks. At some point, however, maintenance activities are discontinued because either the functions provided by these stocks are no longer needed or they become too costly. At that point, the stocks have arrived at the end-of-life and turn into output flows, with some share being recycled and the rest either landfilled or remaining where it had been used (i.e., it becomes a 'hibernating' or 'passive' stock). This means that the '[…] central attribute of a stock is its temporal durability. Seen thus, stocks are suitable for depicting the influences a system's history has on its present—and hence for analyzing temporal developments' (Faber et al. 2005, p. 155).

For this work, we are interested in the interrelations of existing material stocks and flows, and we draw on recently completed research on residential buildings and transportation networks in the 25 countries of the European Union (EU25) (Wiedenhofer et al. 2015). Specifically, the following aspects are of interest: the magnitudes of material inputs to and outputs from stocks; the recycling potential of end-of-life stocks; the relationship between stocks and economy-wide material use; and the distinction between materials used for maintaining existing stocks and those used for the net expansion of stocks. Such an investigation can help us critically evaluate the role of existing stocks in promoting or hindering more sustainable patterns of society-nature interactions, and it contributes to a better understanding where additional socioecological theory building might be fruitful.

12.2.1 Defining and Operationalizing 'Maintenance' and 'Expansion'

From the above conceptual reflections, it becomes clear that the main distinction of interest in this work is the difference between maintaining the existing stock and expanding it. Maintenance flows are defined as material inputs used to keep the stock extent constant (e.g., the number of buildings and the length of roads/railways) so that these stocks continue to provide a steady level of service to society. Maintenance therefore includes two material inputs. The first is ongoing renovations to, for example, roofs, tiles, non-load-carrying walls and worn-down layers of a road. Second, construction to replace stocks that were demolished because they were at the end of their service-lives is also included.

The built environment and transportation networks are also being expanded by way of, for example, net additional housing and road construction, thereby driving material stock growth. Our estimates here are based on empirical net growth rates for roads, railways and residential buildings as well as material intensities of up-to-date construction standards.

Empirically, only a minority of stock types in the EU25 exhibit an actual decline, such as certain state roads and single track railway lines in specific countries in specific years. Overall, decreases in stock extent cause outputs from stocks, but no new inputs, again in the absence of stock hibernation.

12.3 Current Research Approaches to Material Stocks and Flows

The major constraint of material stock accounting, especially on larger scales, is the lack of data on stock quantity, quality, characteristics, material composition and age distributions. Data are also lacking on demolition and maintenance activities. Moreover, retrospective studies face the problem of scale; although it is possible to exactly identify the material content of a single house, doing so for each building in a whole country is impossible. Different approaches exist to address this problem. Examples include representative building typologies; material intensities per floor space and building period; coupling of spatial databases, age distributions and building standards; linking consumption data for certain materials with service-lifetimes and waste factors; and combining input-output approaches with data capital investments, floor space, building regulations and construction activity.

Studies interested in the dynamics of stocks and flows over time often use dynamic lifetime distributions, where material inputs turn into stocks, remain in use for their lifetime and then turn into output flows again. These approaches require information on the lifetimes of each cohort, age distributions in standing stock and relatively long time series. In a simpler 'leaching model', the reciprocal of a lifetime (1 / LT) is interpreted as a constant fraction of the stock turning into outputs. This second approach is less data intensive and yields rough approximations of output flows if inputs are relatively steady, lifetimes are short and the period of observation is as long as the lifetime (Van der Voet et al. 2002). Which model is better depends on the topic under investigation, the level of accuracy required and data availability.

For this chapter on the material stocks and flows in residential buildings and transportation infrastructure in the EU25, we drew upon recently published research, to which we refer for detailed documentation (Wiedenhofer et al. 2015). In this research, bottom-up stock estimates were coupled with lifetimes, refurbishment rates and data on annual construction and demolition activity. Specifically, a typology of residential houses was coupled with data on annually completed and demolished dwellings and total dwelling stocks. These data were compiled from Eurostat, European Housing Statistics reports and other sources. Data on the extent of road and railway infrastructure were compiled from the statistical office of the European Union (Eurostat), the United Nations Economic Commission for Europe (UNECE) and the European Road Foundation. Data on material intensities for four road and two railway types were taken from the literature. Not covered are commercial and public buildings, such as factories, offices, hospitals and government buildings; bridges, tunnels and supporting infrastructures; and other public works, such as dams, sewers and underground structures.

For the business-as-usual trend scenario until 2020, the average national growth and demolition rates from 2003 to 2009 as well as the specific service-lifetimes of stock components that are subject to renovation were used. The slow-down of economic activity due to the banking and subsequent public debt crisis of 2007/08 is also taken into account, where for all countries that experienced dwelling stock growth of more than 2 % between 2003 and 2009 (Spain, France, Ireland) the average of the remaining EU25 countries, 0.78 %, is used for the scenario calculations from 2010 to 2020.

The construction materials covered in this study include concrete, asphalt and all other non-metallic minerals, such as bricks, stones, tiles, aggregates, sand and gravel. This covers 96 % of the so-called domestic material consumption of non-metallic minerals in the EU25, which is an important indicator in the material flow accounting system (Eurostat 2012).

The recycling rates for construction and demolition waste were sourced from a recent meta-study for the European commission (Monier et al. 2011). At this stage of the research, no explicit recycling loops or cascading uses are included in the model; rather, a comparison of the magnitudes of material flows is discussed. As a next step, one would need to consider first the temporal and spatial distribution of stocks and flows in the face of the severe economic limitations on transporting and storing large quantities of recycled construction and demolition flows. Second, the specific material qualities and quantities would need to be included to gain an understanding of the actual possibility of recycling and usability for certain purposes. This goes far beyond this modeling exercise as well as current data availability and must be left for further research.

12.4 Comparing Modeled Bottom-up Stock and Flow Results to Economy-Wide Material Consumption of the EU25

The inputs into stocks of residential buildings, roads and railways are modeled at 1907 Mt (1 Mt $= 10^6$ t $=$ one million metric tons), of which annual maintenance of the roads network makes up most, with 49 %, or 943 Mt per year. Maintenance inputs into buildings amount to 207 Mt, or 11 % of the total estimated inputs in 2009 (Fig. 12.1). Building replacement construction only takes up a small share of those inputs, with 23 Mt of concrete and 13 Mt of other construction minerals. Expansion of the housing stock used 27 %, or 519 Mt, of inputs, whereas the remaining 13 %, or 238 Mt, is related to the expansion of the road and railway network.

In the EU25, 3,137 Mt of construction minerals were used in 2009, the majority of which were sand and gravel (64 %), limestone (which is required for concrete) and gypsum (18 %) as well as various types of building stones (6 %). The 1,907 Mt of input flows discussed above would account for 61 % of the overall domestic material consumption (DMC) of construction minerals in the EU25 (Fig. 12.1). However, not all of those inputs are virgin materials; they are, to some extent, already replaced by recycling flows. This means that if all the estimated recycled materials are used to replace inputs into stocks and for other purposes not covered in this study, 1322 Mt of virgin materials, or 42 % of the DMC, are still required for stock maintenance and expansion.

The modeled outputs from stocks amount to 1269 Mt, of which construction and demolition waste from roads makes up the largest fraction, at 45 % or 573 Mt (Fig. 12.1). Outputs from railway stocks make up only 30 Mt. Buildings demolition waste is estimated at 112 Mt, or 9 % of total output from stocks. Overall

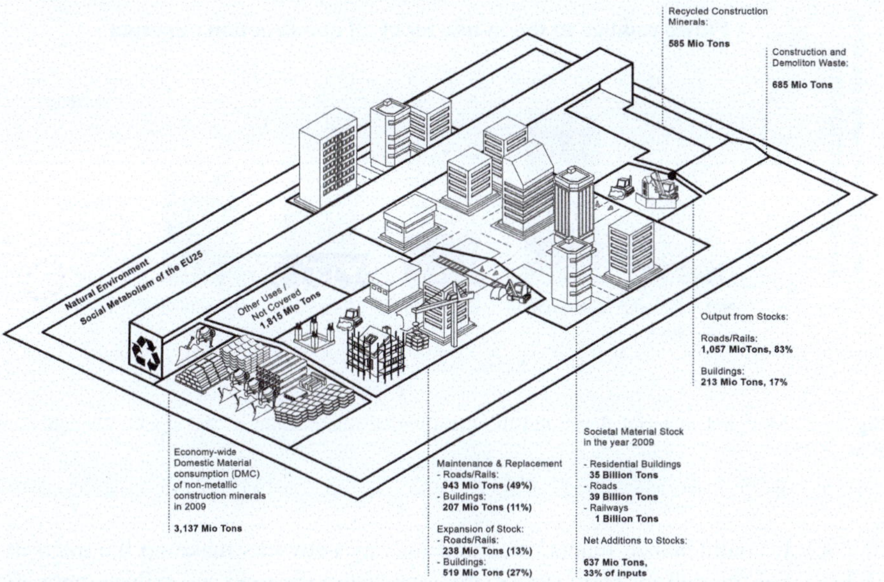

Fig. 12.1 Economy-wide use of construction minerals, with estimated inputs and outputs from stocks, in 2009 (DMC without 'other products, salt and fertilizer' and sourced from Eurostat MFA accounts; Wiedenhofer et al. 2015)

recycled construction minerals amount to approximately 584 Mt, of which 81 %, on average, results from maintenance and demolition of the roads and railway network, whereas the remainder stems from buildings stocks.

Estimates for the stocked construction minerals in roads, railways and residential buildings in 2009 are 39 Gt (1 Gt = 10^9 t = 1,000 Mt), 1 Gt and 35 Gt, respectively. From a per capita perspective, these are construction mineral stocks of 128 t in roads, 72 t in residential buildings and 3 t in railways in 2009 for the EU25, on average.

These results raise the question of where the 'remaining' 39–56 % of the DMC of construction minerals used annually are destined. Uncertainties in the model parameters, such as lifetimes, material intensities and demolition and recycling rates, definitely play a role. Additionally, the estimates presented herein do not cover all societal material stocks. For example, bridges, ports, airports, tunnels, public works, underground networks and commercial and public buildings are not included. Each of these stocks has a different service-lifetime and function for society and, therefore, different temporal implications for resource use and recycling opportunities. This is an issue for further research.

Interestingly, the masses of the estimated material flows into and from stocks are all in the range of only 0.8–1.6 % of stocks, a reflection of the long-term relationships between these stocks and material flows (Fig. 12.2). The major share of net additions to stocks is due to the expansion of the residential building stock

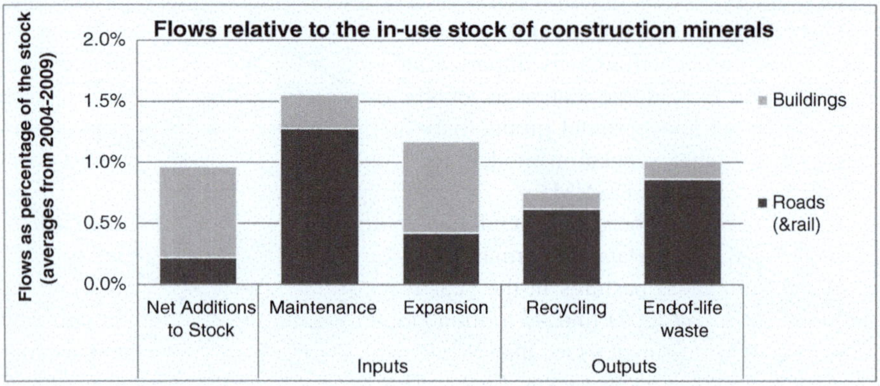

Fig. 12.2 Modeled material flows and the quantitative relationship to the stock (Wiedenhofer et al. 2015)

(0.7 %). In maintenance inputs, road and railway networks make up the majority (1.3 %). For the outputs from stocks, materials from the road and railway network make up most of the estimated recycling and waste streams.

12.4.1 A Business-As-Usual Outlook for 2020: Some Quantitative Effects of Increased Recycling Due to the 'European Waste Framework Directive'

The *European Waste Framework Directive 2008/98/EC* states in Article 11 that 'by 2020, the preparing for re-use, recycling and other material recovery, including backfilling operations using waste to substitute other materials, of non-hazardous construction and demolition waste [...] shall be increased to a minimum of 70 % by weight' from the European average of 46 % in 2004–2009 (Monier et al. 2011; Mudgal et al. 2011). Using the data and model described in Wiedenhofer et al. (2015), a business-as-usual scenario can be modeled in which only recycling rates are increased according to the EU Waste Framework Directive targets, whereas all other factors, such as demolition, growth rates[2] and service-lifetimes, are held constant (a so-called ceteris paribus assumption).

In such a business-as-usual scenario, annual inputs of construction minerals into residential buildings and the road/railway network in 2020 are estimated at 1829 Mt, which is a decrease of −4 % compared to 2009 (Fig. 12.1). This

[2]Housing growth rates for Spain, Ireland and France have been reduced to the remaining EU average for 2010 onward to take the economic recession into account (Wiedenhofer et al. 2015).

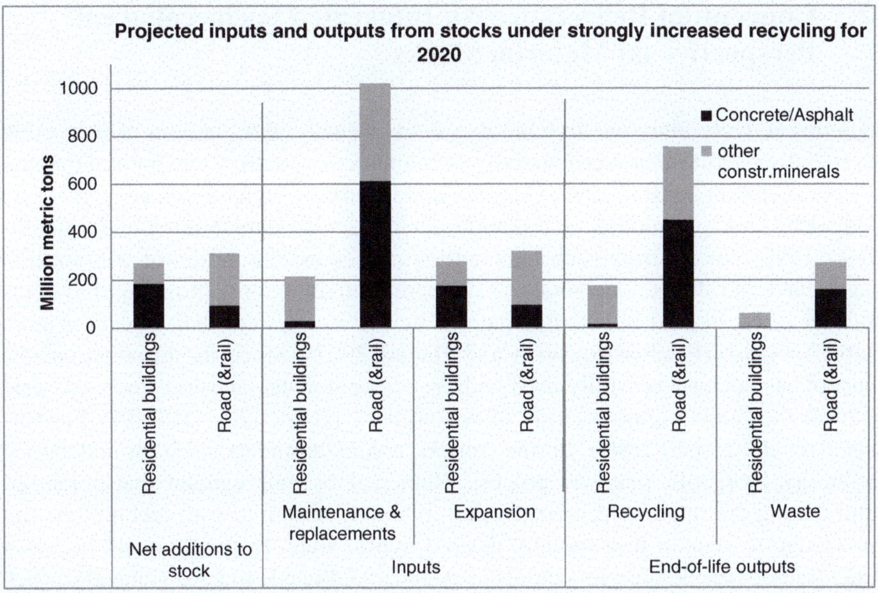

Fig. 12.3 Projections of construction mineral inputs and waste/recycling flows from stocks in 2020 for two recycling scenarios (Wiedenhofer et al. 2015)

decrease stems from reductions in expansion-related inputs (−22 % in 2020 compared to 2009), the majority of which is due to reduced housing expansion in the scenario (−46 % reduction of housing expansion material inputs). Maintenance inputs increase in the scenario (8 % from 2009) and make up the majority of the modeled inputs in 2020 (55 %), where all the concrete and 8 % of other construction minerals go into replacement construction. Modeled outputs from stocks amount to 1264 Mt, an increase of 5 % compared to 2009. Increased recycling rates in 2020 result in 932 Mt of recycled materials, an increase of 70 % over 2009. This means that if the Waste Framework Directive is fully implemented by 2020, 51 % of virgin inputs into the stocks of residential buildings, roads and railways could be sourced from recycled materials given the otherwise business-as-usual scenario of ongoing stock expansion (Fig. 12.3). If all recycled materials are used for maintenance alone, 75 % of these flows could be covered, material quality considerations aside. This suggests that increased recycling will probably not suffice to achieve significant reductions of overall virgin material use. The ongoing stock expansion, paired with currently very low demolition rates, indicates that these important steps toward improved recycling will only suffice to stabilize virgin material use.

12.5 Conceptual Reflections: An Integrated Socioecological Perspective on Material Stocks

Societies actively intervene in natural systems, usually with the aim of increasing the socially useful natural productivity of biomass and to moderate natural fluctuations or maintain constant and/or desired conditions for societal use, such as a road facilitating ease of mobility or a dam protecting settlements (Haberl and Zangerl-Weisz 1997). To describe such intervention processes, the notion of colonization was coined and defined as 'society's deliberate interventions into natural systems in order to create and maintain a state of the natural system that renders it more useful socially' (Fischer-Kowalski and Haberl 2007). Since its introduction, the concept has been successfully used within socioecological land-use studies to shed light on the human appropriation of net primary productivity (HANPP; Fischer-Kowalski and Haberl 2007). In this context, colonization has not been understood as merely a one-off, one-way process. Rather, it became evident that colonized natural systems require recurring inputs of labor, resource use, technology and knowledge to remain in a socially desired hybrid state. In turn, society becomes structurally dependent on its colonized systems and needs to be organized in such a way that these ongoing activities are feasible.

Although colonization has been elaborated in detail for biomass appropriation, it has not received similar attention when other 'socially useful purposes' are targeted when intervening in natural systems. Examples include building stocks of infrastructure for the provision of shelter and mobility of individuals and goods or energy supply, which in turn shape the specific ways and environmental implications of using the services provided by the stock (Pauliuk and Müller 2014). Future elaborations for these functions promise to be rewarding because they might enhance our understanding of how social organization is also interdependent on the quality, quantity and arrangement of its infrastructure and building stocks. The relationship among mobility-related energy use, consumption patterns, 'situated lifestyles' and urban form or more generally settlement patterns, is one more prominently discussed aspect (Heinonen et al. 2013; Minx et al. 2013). Stocks also play a crucial role in a Circular Economy because they strongly influence the amounts of materials available for recycling, the input requirements for maintenance and expansion (Chap. 11) and the resource requirements of utilizing the current stock. Theoretical work to further develop colonization for societal stocks might provide a better understanding of many of the empirical findings and implications of recent research on settlement patterns and stocks and my inspire new insights. Thus, understanding might be enhanced regarding a) how far these processes determine societal activities and efforts and b) how strongly they are actually interlinked, especially over time.

12.6 Practical Reflections: Policy Implications

The majority of residential buildings in the EU25 were constructed between 1945 and 1990, and their service-lives are expected to end in the mid-21st century. This will require a substantial increase in replacement construction, recycling and waste treatment (Bergsdal et al. 2008; Müller 2006). Based on past housing survival, these projections assume 60–120 years of service-life. However, empirical demolition rates for buildings are generally very low, with an EU25 average of 0.15 % (Thomsen and Van der Flier 2009; Wiedenhofer et al. 2015). Implicitly, this would be a lifetime for the entire stock of 667 years (inverse of the demolition rate), which is not a realistic figure but shows that this projected strong increase of demolitions is not yet occurring. Because demolitions are usually driven by factors other than actual technical end-of-life (Thomsen and Van der Flier 2011), lifetime extension and an increased focus on renovations and refurbishments are important policy options toward more sustainable and more efficient resource use. In this way, the material input as well as the waste and recycling side could be addressed while making the building stock more energy efficient (Pauliuk et al. 2013).

Roads (and other infrastructure types) must be included in considerations of more sustainable resource use as they constitute a large part of the material stock and induce large material flows because of their maintenance intensity (Hashimoto et al. 2009; Schiller 2007; Steger 2012). As a simple example, if rates of expansion, maintenance and demolition remained the same as between 2004 and 2009, it would take approximately 300 years[3] for the entire building stock to be fully maintained/replaced, whereas for the road network, it would only take 86 years. Because these recycling and waste flows are dominated by materials from roads (which are mostly aggregates and only some concrete and asphalt), their usefulness for further use after reprocessing is limited, and mainly downcycling does occur.

Furthermore, an inverse relationship between buildings density and material stocked in roads and other infrastructure has been reported for a case study in Germany (Schiller 2007), indicating the importance of spatial planning and reducing urban sprawl. Careful densification of settlements is also advocated for climate mitigation strategies to reduce the energy requirements of transportation. In combination, this could lead to reduced traffic loads, prolonging the lifetimes of roads and decreasing their large maintenance requirements and the monetary burden on state budgets. Thus, substantial co-benefits toward more sustainable material use and climate strategies might exist that must be understood more clearly.

Interestingly, residential building stocks are already starting to shrink regionally driven by rapid decreases in demand for housing due to population decline, even overtaking trends for more living space per capita and smaller household

[3]This ballpark number is calculated as the inverse of the relative size of the maintenance-related material flows to the stock, from Fig. 12.3.

sizes (Deilmann et al. 2009). In such regions, input requirements for maintenance could substantially decrease because demolished buildings would not need to be replaced. However, some of the recycled materials could potentially be difficult to use because long-distance transportation is neither economically nor environmentally efficient for such bulk materials. Furthermore, the provision of infrastructures such as roads/railways, sewers and energy grids could also be decreased. Overall, such regionally stagnating or even declining biophysical stocks pose very different challenges for future sustainable material use than expanding regions or cities do.

Trade-offs with the upstream production requirements of material use must be carefully considered. The existing building stocks that are highly inefficient with regard to their use phase (Nemry et al. 2010; Salat 2009) must be maintained/retrofitted or even demolished. However, replacements and maintenance as well as the expansion of the building stocks require careful material and design choices, with significant GHG emission benefits if passive solar and ventilation guidelines are applied and materials are chosen according to low life cycle energy impacts (González and García Navarro 2006). Generally, the dominant use of reinforced concrete over other construction materials (Bergsdal et al. 2007) is problematic when bricks, processed stones and wood have comparable structural and thermal properties but with significantly lower associated environmental burdens (Gustavsson and Sathre 2011; Sathre and O'Connor 2010), although biomass is not automatically carbon neutral, as the debate on biofuels and land-use impacts shows (Haberl et al. 2010). Furthermore, the large-scale availability of such quantities of timber is highly questionable given existing substantial societal impacts on ecosystems (Krausmann et al. 2013) and the envisaged demand for second-generation biofuels from forests, carbon sequestration potentials and other ecosystem services (Schulze et al. 2012). Having 'both eyes open' with regard to materials, energy and emissions efficiency, especially at the systems and product design level, seems to be a prerequisite for serious steps toward a more sustainable social metabolism (Allwood et al. 2012).

12.7 Conclusions

Interestingly, the maintenance-related material inputs of construction minerals into the stocks covered in this study amount to 34–58 % of the DMC of nonmetallic minerals in the EU25 in 2009, depending on how recycling is handled. The road network, with its relatively brief material service-lifetimes and therefore high maintenance requirements, plays a major role as a driver of material flows. Overall, the results of this work show a significant commitment of annual resource use for maintaining the existing extent of stock.

Strongly increased recycling in an otherwise business-as-usual scenario in 2020 is estimated to only cover 51 % of material input flows into residential buildings and roads/rails. This scenario is based on an increase in the recycling of construction and demolition waste from the European average of 47 % in 2004–2009 to the Waste Framework Directive's goal of a minimum of 70 % by 2020. Long-term dynamic material stock studies suggest a strong increase in building demolition and, therefore, waste flows and recycling over the next few decades. Currently, however, the recycling of construction and demolition waste very often actually means downcycling (e.g., using crushed concrete and bricks as a replacement for virgin sand and gravel), thereby only replacing lower-quality materials. These quality aspects and implications for recycling strategies in light of stock aging are an important issue for further research.

Based on the results presented in this paper and in line with other studies (Hashimoto et al. 2007; Müller 2006; Shi et al. 2012), the following insights emerge. The size of stocks and their service-lifetimes are the two most important factors driving material use necessary for renewal and maintenance. This means that a reduction of material use would be most easily achieved via a stabilization of existing stock extent ('steady stocks') and an effort to prolong the lifetimes of standing infrastructure and buildings. Solely focusing on the efficiency of new buildings cannot be expected to yield substantial short-term impacts because of the low turnover of materials in overall stocks, but it should have positive mid-term effects. On a strategic level, infrastructure and buildings that provide high-quality services with low maintenance and operation requirements should become a priority.

Finally, serious efforts toward more sustainable patterns of society-nature interactions require more systematic considerations on how to use the large material stocks accumulated in industrialized countries more efficiently and over a much longer period. Because most of these stocks were constructed during the rapid acceleration of the fossil-fuel-based system of high resource throughput, moving toward resource-saving strategies such as longer lifetimes, improved maintenance and renovation and an overall stabilization (or sometimes even shrinking) of material stocks amounts to a critical paradigm shift.

References

Allwood, J. M., Cullen, J. M., Carruth, M. A., & University of Cambridge. Engineering Dept. (2012). *Sustainable materials with both eyes open*. England: UIT Cambridge Ltd., Cambridge.

Bergsdal, H., Bohne, R. A., & Brattebo, H. (2008). Projection of construction and demolition waste in Norway. *Journal of Industrial Ecology, 11*, 27–39.

Bergsdal, H., Brattebo, H., Bohne, R. A., & Mueller, D. B. (2007). Dynamic material flow analysis for Norway's dwelling stock. *Building Research and Information, 35*, 557–570.

Deilmann, C., Effenberger, K.-H., & Banse, J. (2009). Housing stock shrinkage: Vacancy and demolition trends in Germany. *Building Research & Information, 37*, 660–668. doi:10.1080/09613210903166739.

European Parliament, Council. (2008). Directive 2008/98/EC of the European Parliament and of the Council of 19 November 2008 on waste and repealing certain Directives, Directive 2008/98/EC.

Eurostat. (2012). Economy-wide material flow accounts (EW-MFA). Compilation Guide 2012. Eurostat.

Faber, M., Frank, K., Klauer, B., Manstetten, R., Schiller, J., & Wissel, C. (2005). On the foundation of general theory of stocks. *Ecological Economics, 55*, 155–172.

Fischer-Kowalski, M., & Haberl, H. (2007). Socioecological transitions and global change: trajectories of social metabolism and land use. Elgar, Northampton, MA [u.a.].

González, M. J., & García Navarro, J. (2006). Assessment of the decrease of CO_2 emissions in the construction field through the selection of materials: Practical case study of three houses of low environmental impact. *Building and Environment, 41*, 902–909. doi:10.1016/j.buildenv.2005.04.006.

Gustavsson, L., & Sathre, R. (2011). Energy and CO_2 analysis of wood substitution in construction. *Climatic Change, 105*, 129–153. doi:10.1007/s10584-010-9876-8.

Haberl, H., Beringer, T., Bhattacharya, S. C., Erb, K.-H., & Hoogwijk, M. (2010). The global technical potential of bio-energy in 2050 considering sustainability constraints. *Current Opinion in Environmental Sustainability, 2*, 394–403. doi:10.1016/j.cosust.2010.10.007.

Haberl, H., Fischer-Kowalski, M., Krausmann, F., Weisz, H., & Winiwarter, V. (2004). Progress towards sustainability? What the conceptual framework of material and energy flow accounting (MEFA) can offer. *Land Use Policy, 21*, 199–213. doi:10.1016/j.landusepol.2003.10.013.

Haberl, H., & Zangerl-Weisz, H. (1997). Kolonisierende Eingriffe: Systematik und Wirkungsweise, in: Gesellschaftlicher Stoffwechsel und Kolonisierung von Natur: ein Versuch in Sozialer Ökologie. Verlag Fakultas, Amsterdam.

Hashimoto, S., Tanikawa, H., & Moriguchi, Y. (2007). Where will large amounts of materials accumulated within the economy go?—A material flow analysis of construction minerals for Japan. *Waste Management, 27*, 1725–1738.

Hashimoto, S., Tanikawa, H., & Moriguchi, Y. (2009). Framework for estimating potential wastes and secondary resources accumulated within an economy—A case study of construction minerals in Japan. *Waste Management, 29*, 2859–2866. doi:10.1016/j.wasman.2009.06.011.

Heinonen, J., Jalas, M., Juntunen, J. K., Ala-Mantila, S., & Junnila, S., (2013). Situated lifestyles: I. How lifestyles change along with the level of urbanization and what the greenhouse gas implications are—A study of Finland. *Environmental Research Letters 8*(2). doi:10.1088/1748-9326/8/2/025003

Krausmann, F., Erb, K.-H., Gingrich, S., Haberl, H., Bondeau, A., Gaube, V., et al. (2013). Global human appropriation of net primary production doubled in the 20th century. *Proceedings of the National Academy of Sciences, 110*(25), 10324–10329. doi:10.1073/pnas.1211349110.

Minx, J., Baiocchi, G., Wiedmann, T., Barrett, J., Creutzig, F., & Feng, K., (2013). Carbon footprints of cities and other human settlements in the UK. *Environmental Research Letters 8*(3). doi:10.1088/1748-9326/8/3/035039

Monier, V., Hestin, M., Trarieux, M., Mimid, S., Domrose, L., van Acoleyen, M., et al. (2011). Study on the management of construction and demolition waste in the EU., Contract 07.0307/2009/540863/SER/G2, Final report for the European Commission (DG Environment). Bio Intelligence Service S.A.S, Paris, France.

Mudgal, S., Tan, A., Carreno, A. M., Trigo, A. de P., Dias, D., et al. (2011). Analysis of the key contributions to resource efficiency (Final Report Annexes). BIO Intelligence Service, Paris.

Müller, D. (2006). Stock dynamics for forecasting material flows—Case study for housing in The Netherlands. *Ecological Economics, 59*, 142–156. doi:10.1016/j.ecolecon.2005.09.025.

Nemry, F., Uihlein, A., Colodel, C. M., Wetzel, C., Braune, A., Wittstock, B., et al. (2010). Options to reduce the environmental impacts of residential buildings in the European Union—Potential and costs. *Energy and Buildings, 42*, 976–984. doi:10.1016/j.enbuild.2010.01.009.

Pauliuk, S., & Müller, D. B. (2014). The role of in-use stocks in the social metabolism and in climate change mitigation. *Global Environmental Change 24*(0) (January 2014): 132–142. doi:10.1016/j.gloenvcha.2013.11.006

Pauliuk, S., Sjöstrand, K., & Müller, D. B. (2013). Transforming the Norwegian dwelling stock to reach the 2 degrees celsius climate target. *Journal of Industrial Ecology, 17*, 542–554. doi:10.1111/j.1530-9290.2012.00571.x.

Salat, S. (2009). Energy loads, CO_2 emissions and building stocks: Morphologies, typologies, energy systems and behaviour. *Building Research & Information, 37*, 598–609. doi:10.1080/09613210903162126.

Sathre, R., & O'Connor, J. (2010). Meta-analysis of greenhouse gas displacement factors of wood product substitution. *Environmental Science & Policy, 13*, 104–114. doi:10.1016/j.envsci.2009.12.005.

Schiller, G. (2007). Urban infrastructure: Challenges for resource efficiency in the building stock. *Building Research & Information, 35*, 399–411. doi:10.1080/09613210701217171.

Schulze, E.-D., Körner, C., Law, B. E., Haberl, H., & Luyssaert, S. (2012). Large-scale bioenergy from additional harvest of forest biomass is neither sustainable nor greenhouse gas neutral. *Global Change Biol. Bioenergy, 4*, 611–616. doi:10.1111/j.1757-1707.2012.01169.x.

Shi, F., Huang, T., Tanikawa, H., Han, J., Hashimoto, S., & Moriguchi, Y. (2012). Toward a low carbon-dematerialization society. *Journal of Industrial Ecology, 16*, 493–505. doi:10.1111/j.1530-9290.2012.00523.x.

Steger, S. (2012). Material stock and material flows of road infrastructres in Germany. Some results from the MaRess-Project. Presented at the Conference on Socio-Economic Metabolism Industrial Ecology, Darmstadt.

Steinberger, J. K., Krausmann, F., & Eisenmenger, N. (2010). Global patterns of materials use: A socioeconomic and geophysical analysis. *Ecological Economics, 69*, 1148–1158. doi:10.1016/j.ecolecon.2009.12.009.

Tanikawa, H., & Hashimoto, S. (2009). Urban stock over time: Spatial material stock analysis using 4d-GIS. *Building Research & Information, 37*, 483–502. doi:10.1080/09613210903169394.

Thomsen, A., & Van der Flier, K. (2009). Replacement or renovation of dwellings: The relevance of a more sustainable approach. *Building Research & Information, 37*, 649–659. doi:10.1080/09613210903189335.

Thomsen, A., & Van der Flier, K. (2011). Understanding obsolescence: A conceptual model for buildings. *Building Research & Information, 39*, 352–362. doi:10.1080/09613218.2011.576328.

Van der Voet, E., Kleijn, R., Huele, R., Ishikawa, M., & Verkuijlen, E. (2002). Predisting future emissions based on characteristics of stocks. *Ecological Economics, 41*, 223–234.

Wiedenhofer, D., Steinberger, J. K., Eisenmenger, N., & Haas, W. (2015). Maintenance and expansion: modeling material stocks and flows for residential buildings and transportation networks in the EU25. *Journal of Industrial Ecology*, published online January 1, 2015. doi: 10.1111/jiec.12216

Part III
Empirical Approaches to Land Use and Colonization of Ecosystems

Chapter 13
Livestock Grazing, the Neglected Land Use

Karl-Heinz Erb, Tamara Fetzel, Thomas Kastner, Christine Kroisleitner, Christian Lauk, Andreas Mayer and Maria Niedertscheider

Abstract Livestock production plays a key role in national and regional economies, for food security and poverty alleviation, the climate system, biodiversity, and the biogeochemical cycles of carbon, nitrogen and water. In the near future, the importance of livestock is bound to increase due to anticipated economic developments and associated changes in human diets in many parts of the world as well as the need to feed a growing, increasingly urbanized world population. Ruminant livestock production occurs on grazing lands that extend over approximately one-third of the terrestrial ice-free surface. In addition, a significant proportion of cropland produce is used as feedstock for animals, whether monogastric or ruminant species. Despite this central role of livestock and grazing in the Earth system, datasets that would allow for systematic, comprehensive analyses are extremely scarce, if not nonexistent, and prone to extreme

K.-H. Erb (✉) · T. Fetzel · T. Kastner · C. Lauk · A. Mayer · M. Niedertscheider
Institute of Social Ecology (SEC), Alpen-Adria University,
Schottenfeldgasse 29, 1070 Vienna, Austria
e-mail: karlheinz.erb@aau.at

T. Fetzel
e-mail: tamara.fetzel@aau.at

T. Kastner
e-mail: thomas.kastner@aau.at

C. Lauk
e-mail: christian.lauk@aau.at

A. Mayer
e-mail: andreas.mayer@aau.at

M. Niedertscheider
e-mail: maria.niedertscheider@aau.at

C. Kroisleitner
Institute of Geography and Regional Sciences, University of Graz,
Heinrichstraße 36, 8010 Graz, Austria
e-mail: christine.kroisleitner@uni-graz.at

H. Haberl et al. (eds.), *Social Ecology*, Human-Environment Interactions 5,
DOI 10.1007/978-3-319-33326-7_13

uncertainties. The largest data gaps relate to the extent and intensity of grazing activities. Furthermore, no statistics are available that allow an assessment of the amount of biomass that is harvested each year on grazing lands. In this contribution, we provide examples of how the socioecological method inventory can improve the knowledge base for grazing by, for example, allowing researchers to make a robust approximation of biomass flows due to grazing and to estimate the extent and intensity of grazing land use. Based on empirical analyses, we elaborate on the current state of research on global grazing, identify important uncertainties and outline possible contributions of socioecological research to this neglected but central aspect of global land use.

Keywords Grazing land · Global land-use patterns · Livestock grazing · Permanent and non-permanent pastures · Land-use accounting · Uncertainty

13.1 Introduction

Livestock, domesticated animals raised for agricultural purposes, plays an important role in society-nature interactions. In preindustrial societies, livestock fulfills a plethora of functions. In addition to providing proteins and energy for nutrition in the form of meat and milk, livestock was (and still is) essential as a provider of mechanical energy (muscle power) and (not so obvious but nevertheless vital) nutrients for other agricultural activities. One important aspect of livestock, particularly ruminant species (e.g., cattle, sheep and goats, that is, animals with a four-compartment stomach that are able to digest plant material with a high cellulosic content), is their ability to feed on land that is not directly usable for crop production due to environmental constraints. Furthermore, managing cattle is a relatively labor-extensive form of land use. Consequently, livestock has a high potential to broaden the resource base of society by allowing the use of otherwise inaccessible land. The German term 'Mistvieh' (in English, literally 'dung-animal'), a widely used swearword, provides an illustrious example of this function. In old times, this word denoted old and otherwise useless domestic animals that were kept to feed on land outside agricultural production (often forests) to collect nitrogen and nutrients that were then applied to arable fields in the form of manure (Glatzel 1999). Keeping livestock is also an important risk reduction strategy for vulnerable communities as animals represent a capital and nutrient stock and act as insurance when required (Herrero et al. 2009). Consequently, livestock herding is among the most widespread land-use activities today, particularly in regions with harsh environmental conditions (Kruska et al. 2003; Seré et al. 1996; Thornton et al. 2002).

The role of livestock changes fundamentally with the transition from an agrarian to an industrial mode of subsistence. In the course of this transition, the role of livestock as capital stock and for the provision of mechanical energy is increasingly replaced by other assets of the fossil-fuel-based economy. Nevertheless, livestock plays an essential role in the provision of food in the form of meat and milk. Today, livestock production accounts for 40 % of global agricultural gross domestic product (GDP), and at least 600 million of the world's poor depend on income from livestock (Steinfeld et al. 2006). Livestock products supply one-third of humanity's protein intake, causing obesity for some while remedying undernourishment for others (ibid.). Hence, livestock products are key within the context of global biomass production and consumption systems (Wirsenius 2003).

Livestock production is forecast to gain even greater importance (Bouwman et al. 2005; Erb et al. 2012). With increasing income and growing population numbers, the consumption of animal products, particularly meat, is expected to increase significantly. This will affect grasslands and croplands alike as a significant fraction of feedstuff is expected to be produced on cropland, and intensifying cropland-based livestock production is commonly regarded as a resource-efficient means of increasing food production (Erb et al. 2012).

These developments can aggravate the considerable detriments related to livestock production. Livestock production is recognized as a major contributor to greenhouse gas (GHG) emissions, particularly methane, a highly effective GHG (Steinfeld et al. 2006). In recent decades, livestock have emitted one-sixth to one-fifth of the global annual non-CO_2 GHG emissions (Herrero et al. 2009). An additional 17 % of emissions are attributed to land-use changes related to agriculture and deforestation for grazing (IPCC 2007). Expansion of livestock production is often considered a major driver of deforestation, especially in Latin America, with strong impacts on biodiversity and the global climate system (Szott et al. 2000), although causal interrelations underlying the livestock-deforestation link are debated (Kaimowitz and Angelsen 2008). Livestock herding is also an important driver of human-induced soil degradation. It is estimated that 20 % of the world's pastures and grasslands are affected by overgrazing, compaction and erosion created by livestock action (Steinfeld et al. 2006).

Grasslands represent the land cover type typically associated with livestock, particularly ruminants. Grasslands are the single largest land-cover type, estimated to extend over 50 million km^2, or approximately 40 % of ice-free land (Souttie et al. 2005). Despite this impressive role of livestock in society and the biosphere, there is surprisingly little concise and comprehensive quantitative information on global grazing. For instance, large uncertainties remain in the extent of livestock grazing (Ramankutty et al. 2008), and there is very little information on the amount of biomass consumed by livestock (Bouwman et al. 2005; Krausmann et al. 2008; Wirsenius 2000). Whereas the production of livestock products is relatively well documented (e.g., by the Food and Agriculture Organization of the United Nations, FAO), only fragmentary information is available for feedstuffs that are not traded or marketed, such as roughage, grazed biomass and fodder crops.

Due to these uncertainties and information gaps, assessments that aim to delineate the effect of livestock on global ecological, atmospheric or hydrological processes are bound to remain incomplete and associated with large uncertainties. The relative lack of global-scale studies and the prevalence of site-specific perspectives on ecosystem responses to managed grazing has led to a fragmented understanding of grazing's role in the Earth system (Asner et al. 2004). Tellingly, difficulties in producing grazing land inventories often lead to the decision to omit this land-use type in global assessments of land use and land cover change (e.g., the rapid land cover change assessments in the Millennium Ecosystem Assessment 2005), despite its importance in land-use research.

In this chapter, we discuss the contribution of socioecological approaches to livestock research. We present empirical, first-principle-based approaches to estimate the global extent of grazing land and the amount of biomass grazed on these areas. Consistent assessment of the extent of grazing lands and the amount of grazed biomass is essential to assess the land-use intensity of grazing (see Chap. 4). Surprisingly, estimation approaches build on the same modeling 'philosophy' and address knowledge gaps by subtractive approaches. In the case of area, all known land uses as well as the areas known to be devoid of land use are subtracted from the entire terrestrial surface, resulting in an area that is most likely used for grazing. Analogously, from an extrapolated total feed demand of the livestock, all known feed categories, mainly market feed, are subtracted. The remainder is defined as the amount of grazing or roughage required for livestock sustenance.

13.2 Estimating Global Grazing Areas

No official, robust data on the extent and distribution of grazing land currently exist. Grazing represents a land use, not a land cover, and it is thus not a specific crop or plant association but an activity that occurs on a wide range of ecosystems, from intensively managed artificial pastures to savannahs and semi-deserts. However, it also occurs on cropland, in cities and—particularly relevant for many biogeochemical cycles—in forests (Erb et al. 2013). With regard to grazing, there is only a very loose relation between land use and land cover, in contrast to, for example, forestry, which cannot occur in the absence of woody biomass, or cropland agriculture, which is bound to fields and agricultural infrastructure such as farms and roads. This limits the applicability of remote sensing techniques—the most commonly used technology in global land-use research for the establishment of wall-to-wall maps—in mapping and quantifying the extent of grazing land.

Available census data are limited in scope and suffer from large discrepancies in the definition of grazing land. The most important and only international dataset with global coverage is provided by the FAO (2011). This dataset, which is available at the spatial scale of nations, provides data on the extent of permanent pastures, or pastures that are used permanently for at least five years (ibid.). No comprehensive information is available for non-permanent grazing, which is typical for transhumant societies. Furthermore, data quality varies greatly among the individual nations as many ambiguities related to definitions exist (Asner et al. 2004; Ramankutty et al. 2008).

Reliable, comprehensive and consistent global maps for grazing are required for many purposes, including assessing land-use effects on the global carbon or nutrient cycles, developing approaches and options for achieving global food security, assessing the role of livestock in the global Earth system, such as its impacts on climate (Steinfeld et al. 2006) and delineating the existing option space associated with, for example, increases in livestock production (Erb et al. 2012; Ripple et al. 2013). Addressing all types of grazing is vital, for instance, for assessing the scale of human activities compared with natural processes (Vitousek 1997), such as by quantifying the human appropriation of net primary production (HANPP; see Method Précis on Human Appropriation of Net Primary Production and Chap. 4 in this volume).

Due to the ambiguities and restrictions of remote-sensing-derived data as well as census data, Erb et al. (2007) compiled a grazing land map by approximating global grazing land by a method of elimination rather than using such datasets. Consequently, the grazing land layer was approximated in a subtractive approach, subtracting all known land-uses as well as those regions that are apparently devoid of land use, from a layer of the terrestrial surface. Area not used for infrastructure, forestry or cropping and area that was not untouched or unproductive was assumed to be used for grazing. As input data, thematic maps for cropland and forestry, simple models for rural infrastructure and remote sensing data for urban areas were used. Methodological details can be found in Erb et al. (2007), but the definition of untouched areas is important and should be briefly mentioned here: the map excludes wilderness areas as identified by Sanderson et al. (2002) as well as areas with an aboveground productivity below 20 g dry matter/m^2/year, delineated with a global vegetation model (the Lund-Potsdam-Jena Dynamic Global Vegetation Model, LPJ-DGVM).

This subtractive approach generally leads to a much larger global extent of grazing land than the data on permanent pasture from FAO. According to FAO, the global permanent pastures in the year 2000 extended over 34 million km^2. In contrast, Erb et al. (2007) find 47 million km^2 of grazing land, which is still well in line with other estimates (see also White et al. 2000; Fig. 13.1 displays a country-level comparison of the two accounts). This discrepancy can, beyond questions of data uncertainty, be attributed to the inclusion of non-permanent grazing lands. Non-permanent grazing is, for instance, particularly to be found in Northern countries, such as in the Russian Federation, Norway and Finland. For other countries, such as India and Egypt, the subtractive approach also results in much larger grazing land than the permanent pastures figure given by FAO (Fig. 13.1). This allows us to illustrate—all caveats of data uncertainty and error propagation of the subtractive approach warranted—the underestimation of grazing land related to the permanent pasture category by FAO.

Surprisingly, however, FAO also overestimates the extent of permanent pastures, and, for a number of countries, the subtractive approach results in a downward correction of FAO figures (Fig. 13.1). Most noteworthy is the example of Saudi Arabia, but other examples include Yemen and Western Sahara. The case of Saudi Arabia allows us to highlight the definitorial heterogeneity of the FAO

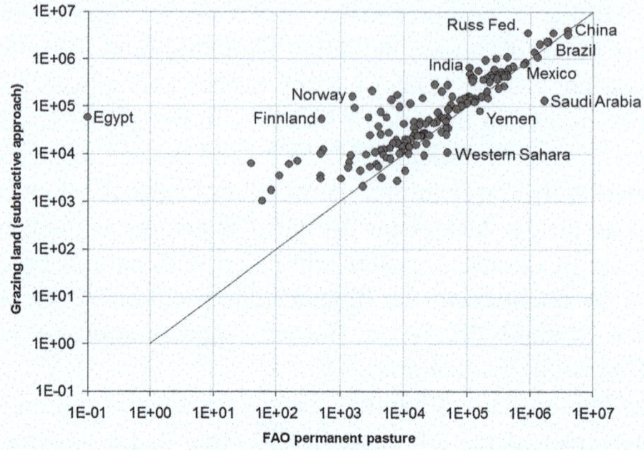

Fig. 13.1 National-scale comparison of grazing land according to Erb et al. (2007) with FAOstat data (FAO 2011) on permanent pasture. Each *dot* represents one country. The *solid* 45° line indicates a 100 % fit

account. Although permanent pastures are defined as being under use for at least five years, Saudi Arabia reports that this land-use class extends over 75 % of its territory despite climatic conditions that do not allow for permanent grazing in most of Saudi Arabia (Ramankutty et al. 2008). For other regions, especially for large countries such as Brazil, China and Mexico, the subtractive approach yields almost identical results as the census data reported by FAO (Fig. 13.1). Agreement at the national scale, however, is insufficient to draw conclusions about the quality of grazing land maps, as will be revealed in the subsequent sections.

In the last few years, four global maps of the extent of grazing lands have been developed. Figure 13.2 presents the spatial pattern of grazing land according to these four global maps. All maps have a spatial resolution of five arc minutes, or approximately 50 × 50 km at the equator, but they differ in their input data, their definition of grazing land and their downscaling procedures. Figure 13.2a displays the result from Erb et al. (2007). Figure 13.2b displays the pasture map provided by Ramankutty et al. (2008; hereafter, Ramankutty map), and Fig. 13.2c shows the map by Klein Goldewijk et al. (2007) developed for use in the IMAGE integrated assessment model (MNP 2006, hereafter, HYDE map, using data from the Hundred Year Database for Integrated Environmental Assessments). Figure 13.2d shows the grazing land map that was developed by the International Institute for Applied Systems Analysis (IIASA) in corporation with FAO (IIASA and FAO 2012) in the course of the Global Agro-Ecological Zones mapping exercise (hereafter, GAEZ map). The HYDE and Ramankutty maps share the approach of reconciling spatial, remote-sensing-derived patterns of land cover with national statistics for permanent pastures, but they differ in spatial allocation rules, input data and definitorial decisions, such as the authoritativeness of the national data on permanent pastures by FAO. The Ramankutty map, for instance, is based on census information for approximately 16,000 units, reconciled with satellite information, and it excludes all areas for grazing that are situated above 50°North. Obvious overestimations of the FAO permanent pastures dataset are corrected downward in this map. The HYDE map is based on national data (approximately

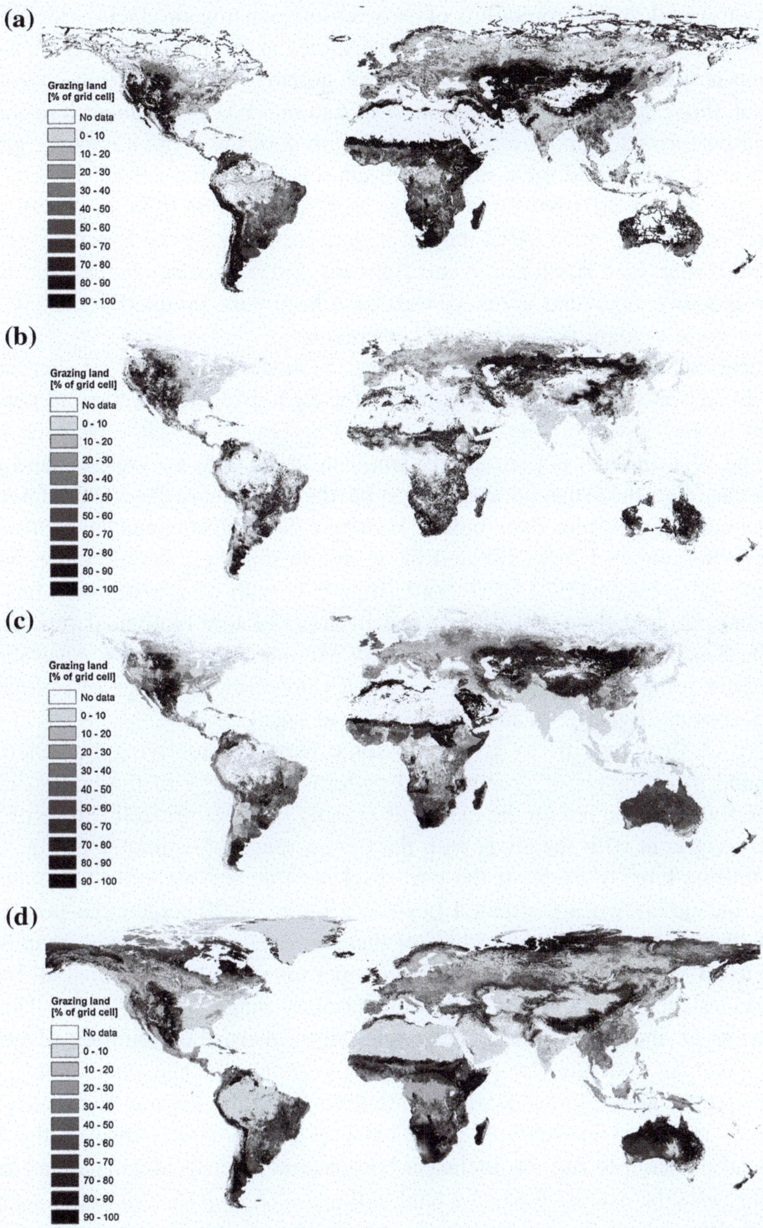

Fig. 13.2 Pattern and extent of global grazing lands in percent-per-gridcell representation according to different sources: **a** Erb et al. (2007), **b** Ramankutty et al. (2008), **c** Klein Goldewijk et al. (2007), **d** IIASA and FAO (2012)

200 nations) and an interpretation of two remote sensing products with the aim of reproducing the FAO permanent pasture values at the national level. A wilderness map is used to identify areas devoid of grazing, and no grazing is assumed to prevail above 50°North. No correction of national FAO permanent pasture data has been performed. In contrast, the GAEZ map does not rely on data for grazing land; instead, it is based on a subtractive approach. It differs from the Erb et al. map by the use of different input data as well as the definitions of grazing constraints. For instance, the GAEZ map uses data from the Global Land Cover 2000 (GLC2000) map (see Bartholomé and Belward 2005) on water bodies and barren or very sparsely vegetated areas as well as a minimum productivity of 10 g dry matter/m^2/year to identify areas devoid of grazing.

In general terms, all four maps have a moderate degree of agreement. Figure 13.3a shows the range of estimates for each gridcell and thus allows conclusions to be drawn on the spatial uncertainty involved with the four maps. The range of estimates is particularly small in areas with no grazing land (dark green areas, e.g., the Amazon and Congo basins, the Sahara desert, the taiga belt in Siberia and large areas over India). However, large discrepancies are found in many regions, most of them situated in or around relatively hostile environments (see dark red areas in Fig. 13.3a), bordering areas with no grazing land (e.g., the Sahel zone, around the Gobi desert) and in areas of low biological productivity (e.g., the Kazakh Steppe, areas in the Rocky Mountains, the Andes, Australia and the northern tundra zones). For many gridcells, however, the range of estimates is low to moderate (light green and yellow areas in Fig. 13.3a).

Figure 13.3b allows us to scrutinize some aspects underlying the uncertainties related to estimates of grazing land patterns. For 17 % of the terrestrial surface, the four maps agree on the extent of grazing land per gridcell (range of 20 % points), congruent with the areas with the lowest range of estimates in Fig. 13.3a. On 46 million km^2, or 36 % of the terrestrial ice-free surface, only two maps are similar; the other two are situated beyond a range of 20 percentage points (see Fig. 13.3b, c). On approximately half of these gridcells, the Erb et al. map is dissimilar to all others. This type of (dis-)agreement is spread over almost all biomes and includes the areas with the largest uncertainty ranges in Fig. 13.3a. This can be taken as an indication that discrepancies in terms of the definition of grazing land, not just data uncertainties, are exerting a certain influence. An example is the Kazakh steppe, where transhumance is dominant and the permanent pasture definition only relates to a subsample of the existing grazing land. However, the methodological decision to use the inclusive barren-area layer to identify areas devoid of grazing sets the extent of grazing land in the GAEZ map to zero in this region.

A situation where three maps show agreement and only one diverges is found on 30 % of the terrestrial surface (the '3 similar' group in Fig. 13.3b, c). This type of agreement is found for large areas in the boreal zones and tundra areas, where the map by GAEZ does not assume any restrictions to grazing in their subtractive approach, but all three others do by using a wilderness layer or by ruling out grazing per definition. On the Arabian Peninsula, in contrast, the encompassing definition by the HYDE database results in an outlier because it follows the permanent

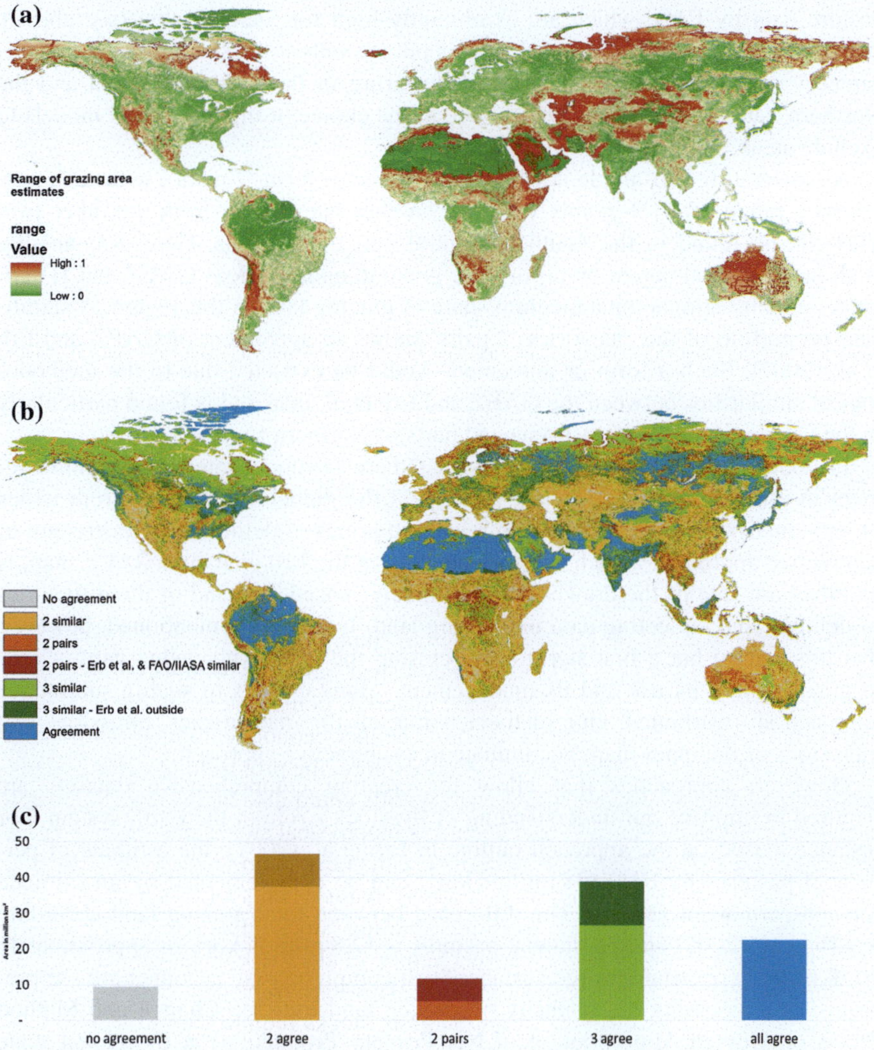

Fig. 13.3 Comparison of four grazing land maps. **a** Ranges of grazing land area estimates in each gridcell, **b** spatial pattern and level of agreement of the four maps, **c** histogram of global agreement levels in million km^2 related to map (**b**). In **b** and **c**, the level of agreement is set at ± 10 % points. 'Agreement' thus refers to a situation where all maps display values within a percentage range of 20 %. '2 similar' refers to a situation where two maps are within 20 % percentage points, and the other two are outside this range and also not close to each other. '2 pairs' relates to a situation where two maps are within a range of 20 % and two others are outside this range but close to each other (within 20 % points). Similar conditions apply to the other classes. For the situation '2 similar', '2 pairs' and '3 similar', the positions of the Erb et al. map are flagged individually: the *dark shaded* areas in (**c**) indicate agreement between Erb et al. and GAEZ

pasture data by FAO, which are exorbitantly high for Saudi Arabia (see above). Areas where the map by Erb et al. sheers out are only a smaller subsample of this category and are scattered with some clustering in India, Mesopotamia and the Northern Sahel fringe. This is partly due to the choice of input data and models to exclude areas.

An interesting subsample is where the four maps form '2 pairs', with each map within a range of 20 % points with one another but distinct from the other two. This can be found in the Southern-Central part of Australia. Here, Ramankutty et al. and Erb et al. agree on no or little grazing land, whereas GAEZ and HYDE agree in their estimate on a medium share of grazing land in this region. A significant proportion of the categories '2 pairs' shows an agreement of GAEZ and Erb et al. (2007). Such a form or agreement would be expected due to the methodological similarities between the GAEZ and Erb et al. map and is found particularly in the Congo basin, Southeast Asia and parts of Northern Europe.

In conclusion, Fig. 13.3 illustrates that there is much potential for improvement in estimates of grazing land patterns at the global scale. Aside from issues of data quality, this potential is also related to issues of definition. The strength of subtractive approaches, such as those pursued in the Erb et al. and GAEZ map, is that they are free of the drawbacks of remotely sensed data and of the difficulties of defining and collecting data on grazing land. It should be mentioned, however, that this can be but a first step toward closing the many knowledge gaps related to grazing land, its use and its management. Many caveats to such a subtractive approach are warranted. One of them is that all the inaccuracies, omissions and inclusions of the input maps accumulate in a subtractive approach.

However, approaches that allow for creating comprehensive datasets are required to improve our understanding of livestock's role in the Earth system. For instance, based on the approach outline in Erb et al. (2007), the estimate of permanent pastures by FAO (34 million km^2) can be complemented by an estimate for non-permanent grazing. The difference between total grazing land according to Erb et al. (2007) and permanent pastures is 12.8 million km^2, or approximately 10 % of the terrestrial ice-free surface. Such comprehensive accounts are the prerequisite to assessing the intensity of grazing land use (see Chap. 4 and Method Précis on Human Appropriation of Net Primary Production) at the global scale and in a wall-to-wall representation. Before we can explore this further, however, we must close another prominent data gap: the amount of biomass grazed by livestock. This will be discussed in the following section.

13.3 Estimating Biomass Grazed by Livestock

International databases such as those published by the FAO do not contain estimates of biomass grazed by livestock or mowed for livestock sustenance. In most census statistics, only data for the supply of market feed and fodder crops, including some grass-type crops, are available. National databases sometimes contain data on hay or clover harvest, but they often lack information on the amount of biomass grazed by roaming livestock. The resulting severe knowledge gap can be attributed to the fact that roughage and grasses are not of high economic value and are usually not traded.

However, given the importance of livestock in the Earth system as well as for many societies around the globe, improving the empirical database is a prerequisite to addressing issues such as food security or the manifold impacts grazing has on biodiversity and carbon. The only methods that have been proposed to estimate the amount of globally grazed biomass rely on estimates of grazing land and on subtractive approaches (Bouwman et al. 2005; Krausmann et al. 2008; Wirsenius 2000). These methods are based on a two-step process. First, the feed demand of a given livestock population (e.g., the livestock of a nation) is calculated, for example, by a livestock model that extrapolates feed demand from the amount of produce or daily feed requirements per livestock unit. In the second step, all known feedstuffs (usually market feed) from census statistics as well as (usually model-derived) amounts of cropland residues, such as straw used for feed, are subtracted from the total feed requirement. The remainder is the so-called 'grazing gap', or the amount of feed that is required to sustain the livestock population but not included in harvest statistics. This procedure is endorsed in the standard protocols of national material flow accountings as followed by, for example, the statistical office of the European Union and the Organisation for Economic Co-operation and Development (EUROSTAT 2012; OECD 2008).

Figure 13.4 displays a result of this approach, namely, the feed intake of ruminants in the year 2000 according to world regions, market feed and roughage. Roughage dominates the biomass inputs of livestock systems, but market feed plays a signification role in regions such as Eastern Europe, Central Asia and the Russian Federation, Northern America and Western Europe. Market feed supply is relatively high in regions with a large monogastric population of livestock; these regions are often also densely populated. In contrast, the share of grazed biomass is particularly high in Sub-Saharan Africa, Latin America and Oceania and Australia (all near 75 %), regions characterized by a large share of extensive livestock systems (Kruska et al. 2003; Seré et al. 1996).

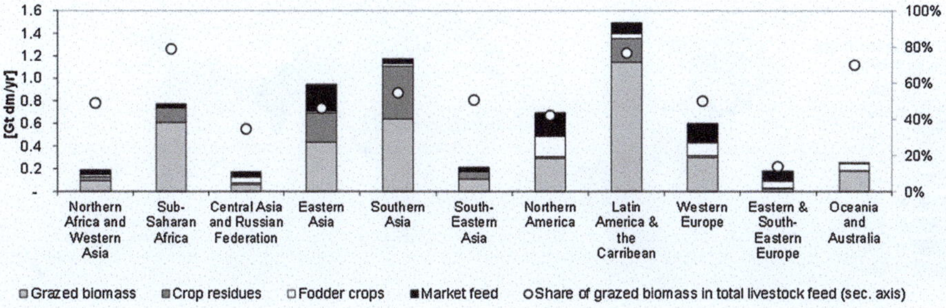

Fig. 13.4 Total feed supply of ruminants in the year 2000, breakdown by region. (Source: Haberl et al. 2007; Krausmann et al. 2008)

13.4 Putting the Pieces Together: Toward a Map of Global Grazing Intensity

The methods and databases described above allow us to advance our understanding of global land use related to grazing. In particular, integrating the information of grazing land patterns with the amount of grazing, both derived with subtractive approaches, allows us to (and is a precondition to) depict patterns of grazing intensity. This is a serious knowledge gap in current Land-System Science (Kuemmerle et al. 2013; see also Chap. 4 in this volume).

Figure 13.5 shows two maps of livestock grazing intensity. The first map in Fig. 13.5a is generated by intersecting the map of grazing extent with a downscaled map of grazed biomass (in this case, following the downscaling procedure from the national scale to the grid level described in Haberl et al. 2007). Such an approach would yield a simple map of grazing pressure. It reveals the regions where grazing plays a particularly strong role, such as Latin America, the Eastern part of North America, all of Europe, China, India and the Northern part of Sub-Saharan Africa. It also illustrates that from a global perspective, biomass flows due

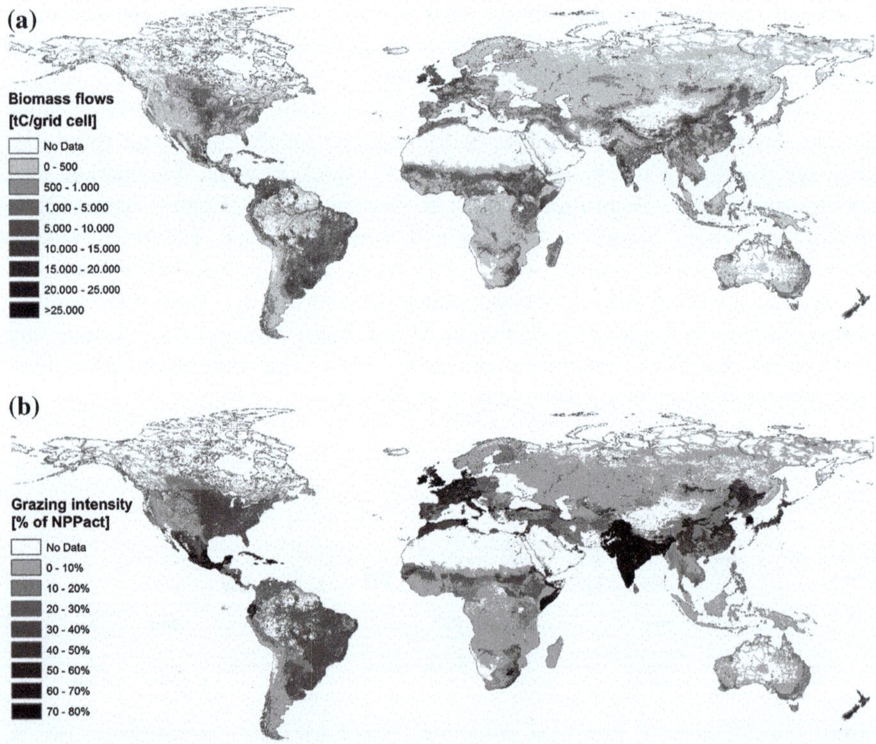

Fig. 13.5 Global map of grazing intensity. **a** Biomass grazed in tC/year per gridcell, **b** grazing intensity expressed as grazed biomass in percent of actual NPP on grazing land. (Source: derived from Haberl et al. 2007)

to grazing are rather small in the Northern and Central parts of Asia or the tropical basins in the new and old world.

However, the map presented in Fig. 13.5a does not really represent a map of grazing intensity or pressure; it only shows where large biomass flows occur due to grazing, whether because of a large share of grazing land in a region, a high grazing intensity or the combination of both. The construction of a true map for grazing intensity on grazing lands also requires taking into account the large productivity differences among grazing areas. As described above, grazing occurs across a wide range of habitats. These habitats spread over wide ranges of biological productivity, from intensively managed, highly productive pastures in Northern Europe to vast but moderately productive steppes and semi-deserts. Figure 13.5b displays a map that expresses grazing intensity as the amount of grazed biomass as a percentage of actual aboveground net primary production (NPP; see also Method Précis on Human Appropriation of Net Primary Production and Chap. 4 in this volume). It reveals that areas with particularly high grazing pressure are located in Southern Asia (particularly India), parts of China, Somalia, large parts of Western Europe, Northern Africa, Japan, Mexico and New Zealand. In these regions, grazing consumes a high share of the annually available NPP on grazing land.

In these regions of high livestock grazing intensity, the impact of grazing on ecological patterns and processes, such as biogeochemical cycling and carbon storage, is particularly high. There is also a high spatial concentration of many place-based ecological detriments associated with livestock (e.g., pollution of soil and water). It also shows where domestic livestock appropriates large shares of NPP and, consequently, where little remains for other heterotrophs, such as wild herbivores (Fritz and Duncan 1994). Thus, grazing intensity information such as that presented here allows us to empirically analyze, for instance, pressures on biodiversity (Chap. 18). Grazing intensity is also causally linked to livestock-induced soil degradation, a ubiquitous phenomenon but one that is difficult to quantify (Asner et al. 2004; Zika and Erb 2009). Finally, maps of grazing intensity contribute valuable information to the options space for future land-use intensification (see Chap. 14). In areas where grazing pressure is high, increasing stocking densities might not be suitable, and alternatives must be found, such as increasing cropland-based feedstuff supply.

From a methodological stance, this is the strength of the sociometabolic approach: it provides empirically robust, consistent and comparable information on the pressure related to many socioeconomic activities across different land-use types, and it allows researchers to systematically link this information to ongoing ecological processes.

13.5 Conclusions

The development of sustainable forms of land use, a central mandate in light of the ongoing global environmental changes and socioeconomic megatrends, needs to be based on robust and consistent information and monitoring systems.

For grazing, considerable knowledge gaps prevail. There are large uncertainties in the extent, patterns and associated biomass flows of grazing and the intensity with which different ecosystems are grazed. Consistent databases of sufficient quality and quantity that would allow for meaningful assessments of grazing, its social drivers and its socioecological impacts are lacking.

The socioecological metabolism concept provides an analytical framework that allows comprehensive and consistent accounts to be generated for livestock systems. It aims at comprehensive accounts that integrate permanent forms of livestock grazing as well as non-permanent forms, such as those prevailing in much of subsistence agriculture and in transhumance. The result is databases that overcome the limitations of narrow or ambiguous definitions and allow the data gaps to be closed, such as by applying subtractive approaches. A socioecological metabolism approach can provide a valuable contribution to initiatives dedicated to the observation and monitoring of the Earth system, such as the Global Earth Observing System of Systems (GEOSS), that currently struggle with the many requirements related to the establishment of an integrated, comprehensive and sustained earth-observing system (Grainger 2009; Turner 2011). The integrated perspective paired with the stringent application of a logic system boundary is well suited to inform and guide data collection and analysis on critical dimensions of land-use transitions, particularly for the most neglected land use in terms of scientific attention, livestock grazing.

Acknowledgements Funding from the European Research Council for the Starting Independent Researcher Grant ERC-263522 'LUISE' and the Austrian Science Fund (FWF) project P20812-G11 is gratefully acknowledged. This book chapter contributes to the Global Land Project (http://www.globallandproject.org).

References

Asner, G. P., Elmore, A. J., Olander, L. P., Martin, R. E., & Harris, A. T. (2004). Grazing systems, ecosystem responses, and global change. *Annual Review of Environment and Resources, 29*, 261–299.

Bartholomé, E., & Belward, A. S. (2005). GLC2000: A new approach to global land cover mapping from Earth observation data. *International Journal of Remote Sensing, 26*, 1959–1977. doi:10.1080/01431160412331291297.

Bouwman, A. F., Van der Hoek, K. W., Eickhout, B., & Soenario, I. (2005). Exploring changes in world ruminant production systems. *Agricultural Systems, 84*, 121–153.

Erb, K.-H., Gaube, V., Krausmann, F., Plutzar, C., Bondeau, A., & Haberl, H. (2007). A comprehensive global 5 min resolution land-use data set for the year 2000 consistent with national census data. *J. Land Use Sci., 2*, 191–224.

Erb, K.-H., Mayer, A., Kastner, T., Sallet, K. E., & Haberl, H. (2012). The impact of industrial grain fed livestock production on food security: An extended literature review. Commissioned by compassion in world farming, The Tubney Charitable Trust and World society for the protection of animals. Social ecology working paper. Institute of Social Ecology, Vienna.

Erb, K.-H., Kastner, T., Luyssaert, S., Houghton, R. A., Kuemmerle, T., Olofsson, P., et al. (2013). Bias in the attribution of forest carbon sinks. *Nature Climate Change, 3*, 854–856.

EUROSTAT. (2012). *Economy-wide material flow accounts compilation guide*. Brussels, Belgium: EUROSTAT.

FAO. (2011). FAOSTAT statistical database (WWW Document). URL http://faostat.fao.org/

Fritz, H., & Duncan, P. (1994). On the carrying capacity for large ungulates of African savanna ecosystems. *Proceedings of the Royal Society of London. Series B: Biological Sciences, 256*, 77–82.

Glatzel, G. (1999). Historic forest use and its possible implications to recently accelerated tree growth in central Europe. In T. Karjalainen., H. Spiecker., O. Laroussinie. (Eds.), *Causes and consequences of accelerating tree growth in Europe. Proceedings of the international seminar held in nancy* (pp. 65–74), France 14–16 May 1998, EFI Proceedings. European Forest Institute, Joensuu.

Grainger, A. (2009). Measuring the planet to fill terrestrial data gaps. *Proceedings of the National Academy of Sciences, 106*, 20557–20558. doi:10.1073/pnas.0912298107.

Haberl, H., Erb, K.-H., Krausmann, F., Gaube, V., Bondeau, A., Plutzar, C., et al. (2007). Quantifying and mapping the human appropriation of net primary production in earth's terrestrial ecosystems. *Proceedings of National Academy of Sciences, 104*, 12942–12947.

Herrero, M., Thornton, P. K., Gerber, P., & Reid, R. S. (2009). Livestock, livelihoods and the environment: Understanding the trade-offs. *Current Opinion Environment Sustainable, 1*, 111–120.

IIASA, FAO. (2012). *Global agro-ecological zones (GAEZ v3.0)*. Laxenburg, Rome: IIASA and Food & Agriculture Organization.

IPCC. (2007). Climate change 2007: *Synthesis report. Contribution of working groups I, II and III to the fourth assessment report of the intergovernmental panel on climate change, fourth assessment report of the intergovernmental panel on climate change*. Cambridge, UK: Cambridge University Press.

Kaimowitz, D., & Angelsen, A. (2008). Will livestock intensification help save Latin America's forests? *Journal of Sustainable Forestry, 27*, 6–24.

Klein Goldewijk, K., Van Drecht, G., & Bouwman, A. F. (2007). Mapping contemporary global cropland and grassland distributions on a 5 × 5 minute resolution. *Journal Land Use Science, 2*, 167–190.

Krausmann, F., Erb, K.-H., Gingrich, S., Lauk, C., & Haberl, H. (2008). Global patterns of socioeconomic biomass flows in the year 2000: A comprehensive assessment of supply, consumption and constraints. *Ecological Economics, 65*, 471–487.

Kruska, R. L., Reid, R. S., Thornton, P. K., Henninger, N., & Kristjanson, P. M. (2003). Mapping livestock-oriented agricultural production systems for the developing world. *Agricultural Systems, 77*, 39–63.

Kuemmerle, T., Erb, K.-H., Meyfroidt, P., Müller, D., Verburg, P. H., Estel, S., et al. (2013). Challenges and opportunities in mapping land use intensity globally. *Curr. Opin. Environ. Sustain., 5*, 484–493.

Millennium Ecosystem Assessment. (2005). *Ecosystems and human well-being: Current state and trends* (Vol. 1). Washington, D.C.: Island Press.

MNP. (2006). *Integrated modelling of global environmental change. An overview of IMAGE 2.4*. Bilthoven, The Netherlands: Netherlands Environmental Assessment Agency (MNP).

OECD. (2008). *Measuring material flows and resource productivity*. Synthesis report. Organisation for Economic Co-operation and Development, Paris.

Ramankutty, N., Evan, A. T., Monfreda, C., & Foley, J. A. (2008). Farming the planet: 1. Geographic distribution of global agricultural lands in the year 2000. *Global Biogeochemical Cycles 22*, GB1003.

Ripple, W., Smith, P., Haberl, H., Montzka, S., McAlpine, C., & Boucher, D. (2013). Ruminants, climate change, and climate policy. Nature Climate Change #NCLIM-13091080A, forthcoming.

Sanderson, E. W., Jaiteh, M., Levy, M. A., Redford, K. H., Wannebo, A. V., & Woolmer, G. (2002). The human footprint and the last of the wild. *BioScience, 52*, 891–904.

Seré, C., Steinfeld, H., & Groenewold, J. (1996). *World livestock production systems: Current status, issues and trends*. Rome: FAO.

Souttie, J. M., Reynolds, S. G., & Batello, C. (2005). *Grasslands of the world*. Rome: Food & Agriculture Organization.

Steinfeld, H., Gerber, P., Wassenaar, T., Castel, V., Rosales, M., & De Haan, C. (2006). *Livestock's long shadow: Environmental issues and options*. Rome, Italy: FAO/LEAD.

Szott, L., Ibrahim, M., & Beer, J. (2000). The hamburger connection hangover: Cattle, pasture land degradation and alternative land use in Central America. CATIE (Tropical Agricultural Research and Higher Education Center), Turrialba, Costa Rica.

Thornton, P., Kruska, R. L., Henninger, N., Reid, R. S., Atieno, F., Odero, A. N., et al. (2002). *Mapping poverty and livestock in the developing world*. Nairobi Kenya: International Livestock Research Institute.

Turner, D. P. (2011). Global vegetation monitoring: toward a sustainable technobiosphere. *Frontiers in Ecology and the Environment, 9*, 111–116. doi:10.1890/090171.

Vitousek, P. M. (1997). Human domination of Earth's ecosystems. *Science, 277*, 494–499.

White, R. P., Murray, S., Rohweder, M., Prince, S. D., Thompson, K. M. J., & Institute, W. R. (2000). Grassland ecosystems. World Resources Institute.

Wirsenius, S. (2000). Human use of land and organic materials: Modeling the turnover of biomass in the global food system (PhD Dissertation).

Wirsenius, S. (2003). Efficiencies and biomass appropriation of food commodities on global and regional levels. *Agricultural Systems, 77*, 219–255.

Zika, M., & Erb, K.-H. (2009). The global loss of net primary production resulting from human-induced soil degradation in drylands. *Ecological Economics, 69*, 310–318.

Method Précis: Using Geographic Information Systems in Social Ecology

Christoph Plutzar, Christine Kroisleitner, Tamara Fetzel and Karl-Heinz Erb

The Special About Spatial

The First Law of Geography, formulated by Tobler (1970), holds, 'All things are related, but nearby things are more related than distant things'. Most socioecological information has a spatial reference; hence, understanding spatial aspects and their implications is essential for socioecological research. Various model approaches, such as gravity models, distance decay models and the related spatial autocorrelation, are increasingly used in modeling environments. As Goodchild (2008) summarizes, 'A geographic world without this law would be impossible to learn about or describe, since every point would be independent of its most immediate surroundings' (p. 12).

Whereas non-spatial data can be addressed by asking 'what?', spatial (geographical) data have a spatial key that is based on (at least) two continuous dimensions that describe a geographical location on the curved surface of the Earth. This allows 'where?' questions. Geo-data combine attributes and their accompanying geo-information.

Geographic Information Systems

A tool to handle spatially explicit information of objects is called a geographic information system (GIS), an 'information system that is designed to work with data referenced by spatial or geographic coordinates. In other words, a GIS is both a system with specific capabilities for spatially-referenced data, as well as a set of operations for working [analysis] with the data' (Star and Estes 1990, p. 2). Hence, a GIS is not just a software system; it also consists of capable hardware, suitable data and people able to operate such a system and its tools.

Vector Versus Raster

There are (at least) two 'real geographical world' model views in a GIS:

1. The field view conceptualizes space as covered by surfaces with attributes that vary continuously across space.
2. The object view conceptualizes space as populated by well-defined indivisible objects, such as points, lines and polygons.

Human perception generally tends to group experiences of space into discrete objects such as buildings, trees and rivers. These objects are usually represented as vector data (points, lines or polygons) in a GIS. The discrete object view is deeply rooted in the history of geography and spatial analysis, where space is regarded as empty and filled with discrete objects (Smith et al. 2009).

In contrast to discrete objects, which are characterized by sharp boundaries and exact spatial locations, natural features such as climate, soil quality, atmospheric pollution, land-use intensity, agricultural yields or noise are characterized by fuzzy transitions that change with geographic scale and appear to vary continuously (Goodchild et al. 2007). Therefore, in contrast to the object-based view, the field-based view identifies the continuous surface (Smith et al. 2009) of a variable and is represented as raster data. Raster data consist of a matrix of cells, similar to picture elements (pixels) of photographic images.

Despite fundamental differences, both concepts can coexist and are, to a certain extent, interchangeable (Goodchild et al. 2007). The choice of perspective strongly depends on the type of information that is analyzed and displayed and the underlying research question. Combining vector- and raster-based information presents several difficulties. For instance, the scales of vector- and raster-based input data are different in most cases. Census or survey data are usually aggregated to administrative units (e.g., countries, regions), whereas raster data are defined by equal-sized grid cells according to their resolution. A high-resolution raster will yield a greater class diversity because it accurately reflects the actual extent of the classes. At lower spatial resolution, the grid cell will reflect the most dominant class.

Social Ecology and GIS

Applying GIS techniques to socioecological research questions entails several challenges. Whereas biophysical information, in most cases represented as raster data, can be incorporated in a GIS environment quite easily, 'the core questions of the social sciences are seen as difficult (even impossible) to address through these imaging techniques' (Geoghegan et al. 1998, p. 51).

Of course, several characteristics of human societies, such as demography and trade, can be handled by and displayed with spatially explicit tools. The majority of this information is related to distinct entities, such as countries and households, that can be analyzed based on vector-orientated data structures. A good example is harvest. Whereas land-use activities occur continuously over space and agricultural yields usually follow such continua, harvest data are usually reported at (sub-)national levels. Such is the case with harvest census data available for administrative units.

Census data, which are typical for research oriented in the Social Sciences, can be reconciled (disaggregated) to finer scales by using downscaling techniques incorporating environmental raster data sets with higher resolutions. For this process, a plausible allocation algorithm must address suitable local evidence. Such evidence can be provided by remote-sensing data and by suitability surface studies using climate, soil or biological data.

The aggregation of spatial data, the opposite of downscaling, is needed in environmental analyses that have moved from the local scale to larger regions (e.g., to match census data at administrative units). The aggregation of vector data denotes the process of collecting a set of similar, usually adjacent, polygons (with their associated attributes) to form a single, larger entity. In raster grids, the aggregated output cell contains the sum, mean, minimum, maximum or median of the input cells that are covered by the extent of that cell.

A recent development that links social behavior with environmental information is the combination of GIS and agent-based models (ABMs). This approach allows the interaction of social(-economic) agents such as persons or institutions (see Method Précis on Agent-Based Modeling) with spatially explicit biophysical units.

Social and natural processes are both highly diverse in space, which calls for appropriate tools and approaches. These tools need to combine the strengths and advantages of both perspectives, along with raster and polygon data and the field and object views, to link social and ecological patterns and processes appropriately.

References

Geoghegan, J., Pritchard, L. J., Ogneva-Himmelberger, Y., et al. (1998). Socializing the pixel and pixelizing the social in land-use/cover change. In J. Geoghegan., L. J. Pritchard., Y. Ogneva-Himmelberger., et al. (Eds.), *People and pixels* (pp. 51–69). Washington, DC, National Research Council.

Goodchild, M. F. (2008) Geographic information science: The grand challenges. In J. P. Wilson., & A. S Fotheringham (Eds.), *The handbook of geographic information science*. Blackwell.

Goodchild, Michael F., Yuan, May, & Cova, Thomas J. (2007). Towards a general theory of geographic representation in GIS. *International Journal of Geographical Information Science, 21*(3), 239–260.

Smith, M. J., Goodchild, M. F., & Longley, P. (2009). *Geospatial analysis. A comprehensive guide to principles, techniques and software tools* (3rd ed.). Leicester: Matador.

Star, J., & Estes, J. (1990). *Geographic information systems: An introduction*. Englewood Cliffs, NJ: Prentice-Hall.

Tobler, W. (1970). A computer movie simulating urban growth in the Detroit region. *Economic Geography, 46*(2), 234–240.

Chapter 14
Systemic Feedbacks in Global Land Use

Helmut Haberl, Karl-Heinz Erb, Thomas Kastner, Christian Lauk and Andreas Mayer

Abstract Land is a key resource, not only for human societies but also for all organisms—animals, plants and microorganisms—that inhabit terrestrial ecosystems worldwide. Humans use land for at least three purposes: resource supply, waste repository and living space (i.e., the area required for production, consumption, transport, recreation and many other activities). Land use involves the 'colonization of ecosystems', that is, purposive interventions into terrestrial ecosystems that aim to support these functions, usually by transforming natural into managed ecosystems (e.g., agro-ecosystems, managed forests, urban systems). Increasingly, land use also aims at other services, such as the conservation of habitats, species or ecosystems or increased carbon sequestration. Maximization of one function, such as biomass supply, often affects other functions, such as carbon sequestration or conservation. Along with the growth of the world population and its per-capita consumption, trade-offs among different functions are becoming more important. A particularly relevant example is the trade-off between food and fuel that has become apparent in the last few years as policies promoting bioenergy on agricultural lands have gained momentum. Although some of these trade-offs can only be mitigated but not completely avoided (e.g., biomass production requires

H. Haberl (✉) · K.-H. Erb · T. Kastner · C. Lauk · A. Mayer
Institute of Social Ecology (SEC), Alpen-Adria University,
Schottenfeldgasse 29, 1070 Vienna, Austria
e-mail: helmut.haberl@aau.at

K.-H. Erb
e-mail: karlheinz.erb@aau.at

T. Kastner
e-mail: thomas.kastner@aau.at

C. Lauk
e-mail: christian.lauk@aau.at

A. Mayer
e-mail: andreas.mayer@aau.at

H. Haberl et al. (eds.), *Social Ecology*, Human-Environment Interactions 5,
DOI 10.1007/978-3-319-33326-7_14

limited resources such as productive area and water), a sociometabolic approach can help identify potential synergies. For example, the use of wastes, by-products and residues ('cascade utilization') may help to increase biomass use efficiency and generate several outputs without resulting in resource competition. This chapter discusses such trade-offs and synergies in global land use with a view toward issues of resource supply (mainly food and energy) as well as various ecological conservation aspects (e.g., biodiversity conservation, carbon sequestration and environmentally less-demanding agricultural technologies).

Keywords Socioeconomic metabolism · Biomass flows · Land-use competition · Cascade utilization of biomass · Carbon sequestration

14.1 Introduction

In Ecology, the word 'colonization' is used when a species succeeds in extending its range into previously uninhabited terrain. In that sense, Homo sapiens is one of the world's most successful species, inhabiting almost all the planet's lands. Humans now use approximately three-quarters of the global land area,[1] more or less intensively, for living space, cropland, grazing and forestry. Only one-quarter of the earth's land is classified as (almost) natural (Ellis et al. 2010). Most of this land area may be highly valuable ecologically, but its biological productivity is low because it is dry or cold (e.g., deserts and arctic or alpine tundra). The remnants of natural forests, covering perhaps 5–7 % of the global land area, are among the few biologically productive but still largely pristine ecosystems (Erb et al. 2007).

Hence, on the vast majority of the earth's terrestrial surface, patterns, dynamics and functions of land systems emerge through intensive, recursive and complex interactions between natural and socioeconomic processes (Turner et al. 2007). Cultural landscapes are shaped by interactions between natural factors (such as geomorphology, landforms, climate and biotic communities) and socioeconomic activities (such as agriculture, forestry, settlement and infrastructure development and energy use). The analysis of cultural landscapes is a genuinely interdisciplinary endeavor and is a core research area of Social Ecology (Fischer-Kowalski et al. 1997).

With few exceptions, human societies organize land according to their needs and wants, namely, for the delivery of provisioning, regulating or cultural ecosystem services (Braat and de Groot 2012). The ecological notion of colonization is insufficient to capture this process. The colonization concept developed in Social Ecology therefore also encompasses purposive human interventions into natural

[1]In this chapter, we discuss a land surface of approximately 130 million km²; that is, all of the earth's land outside Greenland and Antarctica.

systems—in this case, terrestrial ecosystems—to optimize them in terms of their utility for human society. Colonization may transform ecosystems, which happens when forests are converted to croplands, meadows or pastures. However, colonization may also affect populations and organisms, such as through the breeding of crops and animals, and it may affect genomes directly through genetic engineering (Fischer-Kowalski et al. 1997). Hence, humans not only profit from ecosystem services delivered spontaneously by ecosystems but also colonize these ecosystems to increase or alter their service delivery—a process usually denoted as 'land use'. Land may be used with very different intensities, ranging from small interventions to strong modifications of ecosystems. Land-use intensity has three dimensions: socioeconomic inputs to the land, outputs from the land to human society and changes in the integrated socioecological system (see Chap. 4), as measured using the 'human appropriation of net primary production' (HANPP) approach (see Method Précis on Human Appropriation of Net Primary Production and Chap. 17 in this volume).

Because global land is finite and predominantly human-used (particularly the naturally fertile regions), almost any extension of area use for one purpose implies a reduction in the available area for other functions or services. To some extent, land can serve more than one function at a time ('multifunctionality') . For example, extensively used farmland can deliver food while also supporting valuable biotic communities, hence contributing to the conservation of biodiversity. Similarly, the use of by-product flows may result in synergies; increases in food crop production may raise by-product flows that can be used to feed livestock or for energy production. In many cases, however, maximization of one function (e.g., crop production) entails a reduction of other functions, such as biodiversity conservation, water retention capacity or carbon sequestration. In most cases, the maximization of one ecosystem service reduces others, resulting in trade-offs (Braat and de Groot 2012).

When different social groups profit from different services or suffer from adverse effects that result from the maximization of one specific product, land-use competition or even conflicts may arise. The extension of area required for food production may reduce the area available for carbon sequestration, biodiversity conservation or bioenergy production. A switch to organic agriculture has many positive ecological effects, including reduced pressure from pesticides and chemical fertilizers, improved soil quality and higher on-site biodiversity (IAASTD 2009), but it also reduces yields (Seufert et al. 2012). If demand remains the same, this implies increased demand for cropland and grazing areas, which may result in increased pressure on other ecologically valuable natural or semi-natural areas (Burney et al. 2010). The results of increases in land-use intensity resulting from mechanization, irrigation, fertilization and pesticides and from high-yield crop varieties and livestock breeds are also ambivalent; although they contribute to environmental pressures and problems such as inhumane animal husbandry systems, nutrient leaching, soil erosion, biodiversity loss and the toxic effects of pesticides (IAASTD 2009), they may reduce the area demand of agriculture and perhaps even reduce greenhouse gas (GHG) emissions due to land-use change

(Burney et al. 2010). In many industrialized countries, the emergence of substantial carbon sinks in biota and soils was made possible through massive increases in agricultural productivity per unit area (see Chap. 20). However, increased land-use intensity is no panacea, not only because of its potential ecological costs but also because it may induce socioeconomic feedbacks, the so-called 'rebound' effect: increased efficiency in production may result in rising consumption. Increased land-use intensity may even be a precondition for the adoption of resource-intensive consumption patterns such as increased consumption of meat and other animal products (Lambin and Meyfroidt 2011).

In this chapter, we discuss several important trade-offs and synergies related to global land use from a socioecological perspective to show how the concepts of metabolism and colonization can help us better understand systemic feedbacks—trade-offs as well as synergies—between different possible future changes in food consumption, cropland yields and livestock feeding efficiency.

14.2 Agriculture and Food Scenarios for 2050

The provision of sufficient amounts of nutritionally adequate food is one of the most important functions of global land use. Although biomass is used for additional purposes (fiber, energy), food supply is thought to occupy nearly half the earth's land area, that is, most of the area used as cropland and grazing land (Erb et al. 2007). The continuing growth of the human population and economic output (gross domestic product, GDP) are generally expected to result in a massive growth of food consumption, in terms of both total calories and the fraction of its most resource-demanding component: animal products (meat, milk, eggs). The UN Food and Agricultural Organization (FAO 2006) foresees a 70 % growth in the demand for agricultural products by 2050. Other studies even expect a doubling of global food demand (Tilman et al. 2011).

14.2.1 Dietary Change

In the last few decades, increased wealth was almost inextricably linked with increased consumption of animal products (meat, milk, eggs), sugar and oils and with reduced consumption of cereals, potatoes, rice and other staples (Erb et al. 2009). Future deviations from this trend are conceivable, but their likelihood is difficult to assess. There might be a reduced consumption of animal products in rich countries due to health concerns or greater environmental consciousness, but it is just as likely that currently undernourished regions might adopt European or US-American food consumption habits faster than foreseen by the FAO. To estimate the range of possible future food demand, we define several variants of possible future diets, the adoption of which would have massive consequences

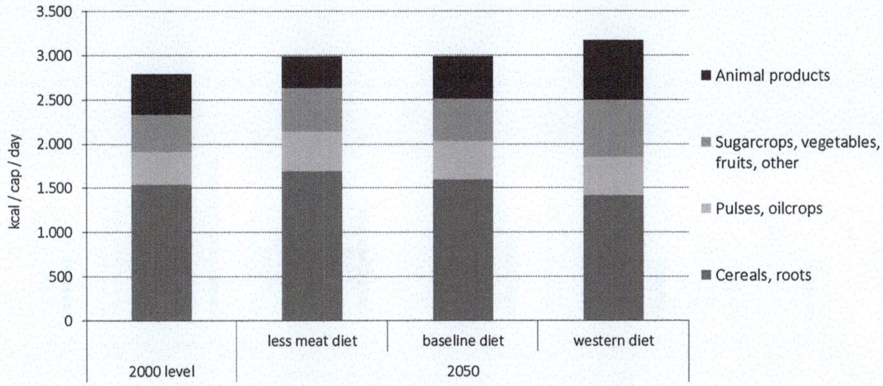

Fig. 14.1 Global calorific intake per capita for main food categories in 2000 and variants for 2050. (Source: Erb et al. 2012)

for future land requirements for food supply. In particular, the share of animal products in diets has enormous implications for land demand. In the year 2000, approximately 60 % of all biomass harvested and used by humans was required to feed livestock (Krausmann et al. 2008).

Figure 14.1 compares three possible future diet variants with the level and composition of global average food intake in the year 2000 (Erb et al. 2012). The 'baseline diet' is an extrapolation of current trends closely resembling FAO forecasts (FAO 2006). The 'less meat diet' assumes the same global average per capita calorie supply but a reduced share of animal products. The 'western diet' represents a scenario that assumes that poorer regions will catch up with Western European and US-American food habits faster than assumed in the baseline diet.

All these diet variants are nutritionally sufficient; they supply the world population in the year 2050 with sufficient calories and protein. In the 'less meat diet', protein deficiency is avoided through an assumed increase in the intake of protein-rich plant foods such as beans, lentils, peas and soybeans.[2]

14.2.2 Crop Yields

In an aggregate view, two parameters are most important on the supply side: (1) yields per unit area per year and (2) the conversion efficiency with which livestock converts feedstuff into meat, milk and eggs. Both parameters vary enormously

[2]We also analyzed variants of the 'baseline diet' by tweaking the production of animal products (a) toward pigs and poultry (+50 %, milk and ruminant meat reduced accordingly) and (b) toward ruminants by reducing pig and poultry products by 50 % and increasing ruminants accordingly. In both cases, the total consumption of animal products was assumed to remain the same as in the baseline.

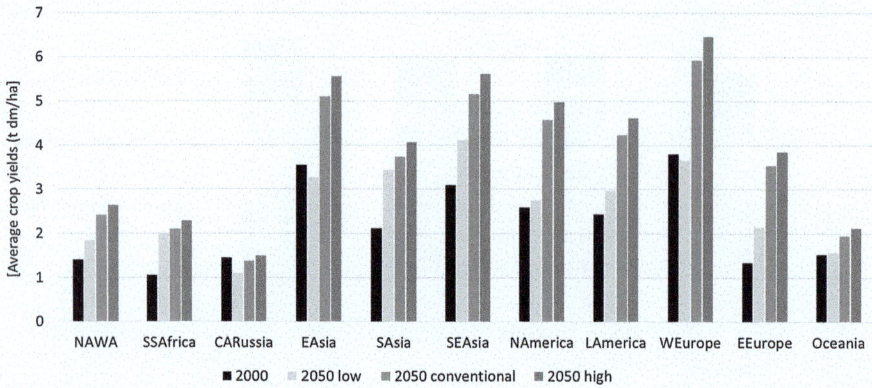

Fig. 14.2 Average crop yields in 2000 and 2050 for three different variants (see text) in a regional breakdown. Values are metric tons dry matter per hectare and per year. (Source: Erb et al. 2009)

among regions, crops and livestock species, and they strongly depend on agricultural technologies and management such as fertilization, soil management, breeding and livestock rearing conditions and herd management.

The FAO expects cropland yields to grow by more than 50 % globally by 2050. Approximately four-fifths of the projected growth in agricultural production is assumed to be the result of increased yields on existing croplands (FAO 2006). The 'Global Orchestration' scenario (Millennium Ecosystem Assessment 2005) assumes an even stronger growth of yields. Switching to organic agriculture reduces yields significantly. Although the yield of an organic wheat field may be nearly as high as that of a conventional one (Seufert et al. 2012), organic agriculture relies on intercrops and fallow to regenerate the soil nutrient pools and maintain soil fertility; over the entire crop rotation cycle, intercrops and fallows reduce yields considerably. According to a literature review (Erb et al. 2009), yields of organic agriculture are approximately 40 % lower over the entire crop rotation period than those of conventional, highly intensive agriculture. Based on these considerations, we derived three yield variants for the year 2050, displayed in Fig. 14.2, and compared them with the yield level achieved in the year 2000 in the respective world region.

The 'conventional' yield variant is based on FAO forecasts (FAO 2006). The 'high' variant was derived from the highest scenario within the Millennium Ecosystem Assessment (2005), which assumes 9 % higher yields than FAO in 2050. Both result in considerable yield increases in almost all world regions (Fig. 14.2). The 'low' variant assumes yields that could be achieved by fully switching to organic agriculture (Erb et al. 2009). The low-yield variant allows for modest increases in crop yields in some regions where yields are currently very poor (e.g., Sub-Saharan Africa and Asian regions, see Fig. 14.2): even organic agriculture can surpass yields achieved with traditional technologies prevailing in these regions. Yield reductions were only assumed for the fraction of cropland cultivated using intensive high-input methods, which is low in many regions.

14.2.3 Animal Husbandry: Feeding Efficiency

Measuring the efficiency with which livestock converts feedstuff into products is a complex endeavor. Conventionally, efficiency is input divided by output, but neither is easy to capture. The question of how 'products' or 'outputs' of livestock should be defined or what is even considered a product is less trivial than it sounds. In most industrial societies, the provision of animal-derived food such as meat, milk and eggs is seen as the dominant output of animal husbandry. Of course, dogs or cats serving as companions and horses serving for leisure riding also play a role.[3] In preindustrial settings, other outputs or even services may be very important, such as the work force of animals (draft animals), their importance as an economic buffer in bad times, their relevance as a status symbol and, not least, their contribution to the nutrient cycle, which can be harnessed by feeding animals with feed from grasslands or even forests and fertilizing the most intensively cropped areas with their manure. In industrial societies, the internal combustion engine and electric motors have replaced draft animals; cars and electronic gadgets have replaced livestock as symbols of wealth and power; and synthetic fertilizers have greatly reduced the importance of livestock in the (short-term) maintenance of soil fertility, at least in many intensively cropped regions. The services pets provide are difficult to name and quantify.

Measuring input is also not straightforward. Of course, it is possible to calculate the energy equivalent of feed (or the amount of protein it contains), but the resulting numbers are not always easy to interpret. For example, ruminants (cattle, sheep and goats) can digest roughage unsuitable for consumption by other species. The low digestibility of roughage results in a low conversion efficiency of feed to products, but the ability to digest roughage also means that these animals can dwell on resources not accessible to other species, especially humans. In contrast to species with a much higher 'feeding efficiency' in terms of pure energy input-output ratios (e.g., pigs and poultry), these animals do not compete with humans for food, and they allow the use of land that cannot (or at least not easily) be used for food production through cropping.

Moreover, the feeding efficiency of ruminants, measured as a calorie input/output ratio, improves if they are fed more easily digestible, high-calorie (high-protein) feeds. Hence, improved feeding efficiency depends, at least to some extent, on the increased use of high-quality feeds. These feeds, however, increase the competition between humans and livestock for food and area, except if livestock is fed on wastes not deemed acceptable for human consumption. The optimization of herd dynamics is another important determinant of feeding efficiency. If animal populations are optimized for meat output, animals are slaughtered when their

[3]Pets could not be modeled explicitly due to a lack of data. For the year 2000, their feed intake is included in the animal production/consumption data. Implicitly, this means they are scaled up/downward with changes assumed in animal product consumption in the different diet variants.

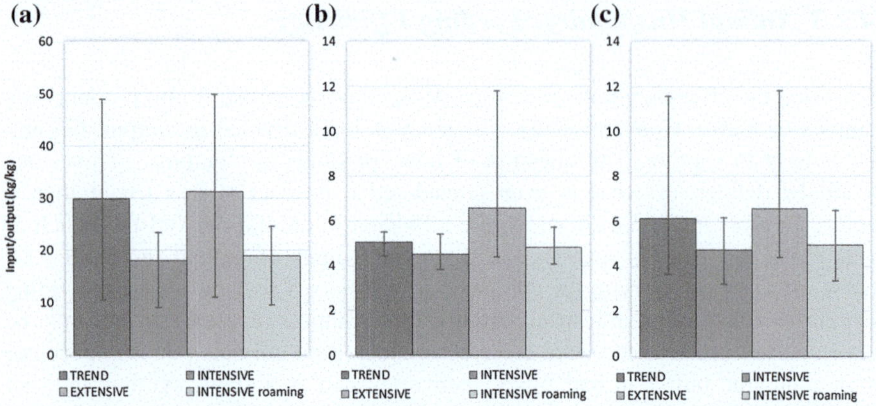

Fig. 14.3 Variants of global livestock feeding efficiencies in the year 2050: **a** ruminant meat, **b** monogastric species and **c** milk, butter and other dairy products. Values are unweighted arithmetic mean of regions. Whiskers indicate the maxima and minima of regional factors. Unit: dimensionless (kg dry matter/kg dry matter). Note that the scale of **a** differs from that of **b** and **c**. (Data Source: Erb et al. 2012)

body weight growth plateaus; continuing to feed them once the growth of their muscle mass declines would be a waste of feed. If, however, animals serve more purposes than just food production or if they are kept in regions with abundant grazing area, such 'optimization' may not be a primary target of the farmer, and feeding efficiencies in terms of input/output ratios can thus be surprisingly low (Thornton 2010).

The variants of feeding efficiency underlying our calculations are described in Fig. 14.3. They are based on the analysis of feed input-output ratios in 11 world regions in the last few decades (Bouwman et al. 2005; Erb et al. 2012; Krausmann et al. 2008). The feeding efficiencies reported in Fig. 14.3 were defined as the ratio of feed requirements of all livestock (including working animals and animals for reproduction) per unit of output of usable animal biomass, both measured as dry matter biomass. These variants reflect not only differences in management but also different strategies for feeding domesticated animals. The 'trend' variant is developed by extrapolating a forecast (Bouwman et al. 2005) for 2030–2050. The 'intensive' variant assumes intensive stable-kept rearing of animals using optimized herd management with a higher share of crop-based feed and accordingly reduced roughage demand. The 'intensive, roaming' variant is based on the same optimization of feeding and herd management but assumes that animals are allowed to roam according to standards of humane livestock rearing, which results in higher area demand and reduced feeding efficiency. The 'extensive' variant assumes a reduced share of grain-based feeds and, accordingly, higher roughage demand.

Figure 14.3 shows that feeding efficiencies differ quite strongly among variants and regions by factors of up to 5 for ruminant meat and by greater than 3 for meat of pigs, chicken and other monogastric species and for milk and other dairy products. Feeding efficiencies differ strongly between intensive and extensive variants, especially for ruminant meat, whereas animals managed otherwise intensively but allowed to roam are only slightly less 'efficient' than animals kept in stables. In other words, adopting humane livestock rearing conditions results in only minor losses in feeding efficiency (Erb et al. 2009).

14.2.4 Biomass Flows and Land Use 2050

Based on the assumptions explained above, we estimated the area required for global food supply in 2050 using the biomass-balance model BioBaM (Erb et al. 2009, 2012). BioBaM allows global biomass flows to be traced from production to consumption. It also allows feedbacks among changes in food demand, production technologies (yields, feeding efficiency) and land requirements to be assessed. It calculates the effects of assumed changes (called 'variants') in diets, yields and livestock feeding efficiency on the demand of cropland area and on the intensity with which grazing areas are exploited. Each specific combination of variants is denoted as a 'scenario'.

BioBaM is a purely biophysical biomass-balance model built on a global database that consistently integrates the global land-use pattern in the year 2000 (Erb et al. 2007) with a biomass balance for the same year (Krausmann et al. 2008). It contains neither dynamic simulation nor optimization algorithms, and it allows the transparent implementation of different assumptions and assessments of their consequences (Erb et al. 2009, 2012).[4] To derive estimates for the year 2050, land-use and biomass flow data for the year 2000 are modified to reflect 2000/2050 changes based on exogenous assumptions, the most important of which are discussed in the preceding sections. BioBaM determines whether any specific scenario (i.e., combination of variants) is 'feasible'. A scenario is classified as feasible if sufficient cropland and grazing land is available to produce the required volume of food products given the assumed yields and feeding efficiencies (within an uncertainty range of ±5 %).[5]

Cropland availability, which is used to decide whether the demand for cropland area can be met in each scenario, is derived from information on the developments of yield, cropping index and harvested area from FAO (2006). Because we are interested in a large option space, we doubled the cropland area increase estimated

[4]BioBaM distinguishes 11 world regions, seven crop aggregates and two different animal production systems (ruminants, monogastrics). The results can be disaggregated in geographic information system (GIS) grids with a five-minute geographic resolution (ca. 10 km at the equator) based on data by Erb et al. (2007).

[5]In all scenarios, urban and infrastructure areas are assumed to grow by +24 % until 2050. Cropland area demand is calculated from food demand according to the variants of yields and feeding efficiencies. The world population in 2050 is assumed to be nine billion.

by FAO (2006) using consistency checks as described in Erb et al. (2009). The expansion of cropland between 2000 and 2050 was thus +19 % on global average (in contrast to +9 % estimated by FAO 2006), with huge regional variations (e.g., +42 % in Latin America, +56 % in Sub-Saharan Africa). The cropland expansion is assumed to occur only on grazing land of high quality. For our scenario analysis, we assumed that cropland expansion does not result in deforestation; in other words, the analysis does not include trade-offs between forest protection and agricultural developments. Rather, it is assumed that cropland expansion reduces the area of grazing land and would—if all parameters were kept constant (e.g., feed demand)—result in an increase of grazing intensity (the same harvest volumes on smaller areas). The feasibility of scenarios related to grazing intensity is evaluated below.

Grazing land, which occupies a much larger area than cropland globally, is a complex issue (Chap. 13). Grazing land is an umbrella term for many different land categories that can potentially be used for grazing or mowing (hay, silage) with very different intensities. Some of this land is used intensively and is even irrigated and fertilized. This category also includes shrubland, tundra, mountains and other semi-natural lands used extensively or even intermittently for grazing. Different qualities of grazing land are reflected in a subdivision into four grazing land quality classes, where 1 is the most and 4 the least suitable quality class (Erb et al. 2007). For each of these quality classes, an upper limit of the percentage of aboveground net primary production (NPP) that can be grazed can be specified, thus allowing sustainability limits to be estimated for grazing pressure. To evaluate the feasibility of the scenarios, we assumed maximum grazing intensities of 75, 55, 40 and 20 % of NPP for grazing land of quality classes 1–4, respectively. It can be assumed that grazing land class 1 is potentially suitable for cropping (Erb et al. 2009), especially less-demanding energy crops such as herbaceous (e.g., Miscanthus sinensis or switchgrass) or woody (e.g., short-rotation coppice) second-generation energy plants (Haberl et al. 2011).

To compare the different scenarios, we calculated the area potentially available for purposes other than food production as follows. For all scenarios classified as feasible (see above), we calculated the area of cropland required to meet food/feed demand according to the yield level assumed in the scenario and grazing land in quality class 1 required to meet forage demand under the respective assumptions. We assumed that grazing and mowing were intensified up to the maximum sustainable use of the respective land area; that is, we assumed that the required roughage is extracted in a manner that minimizes the area required for grazing or mowing. In other words, we adopted a food-first approach to calculate the smallest possible area in each region's grazing land of quality class 1 that would suffice to cover roughage demand allocated to that quality class in each scenario. The area of cropland and grazing land (class 1) required for food supply was subtracted from the area of cropland and class 1 grazing land, revealing an upper limit for the area of cropland or grazing land that might be available for other purposes, such as energy crop production, the planting of new forests to sequester carbon or the establishment of nature conservation areas. Only grazing land quality class 1 is assumed to be potentially suitable for cropping at a reasonable level of investment (Coelho et al. 2012). The results are shown in Fig. 14.4. Note that 'baseline' means that diets, yields and feeding efficiencies develop according to FAO forecasts, not that the grazing land intensification assumed in calculating area demand will necessarily occur. In the absence of targeted policies or economic incentives, it seems unlikely that such an intensification of grazing areas would happen, and it is an open question whether such an intensification would be desirable or what it would cost to achieve it. The results may also be interpreted as an indication of grazing pressure; low area availability indicates high grazing pressure and vice versa.

Potentially available land is inversely related to the food system's land demand: the smaller the area of potentially available land, the higher the land demand of food production (assuming a standardized, high level of grazing pressure) or the pressure on grazing land. Figure 14.4 can hence be interpreted in terms of food-related land demand. It shows systemic interrelations and trade-offs among

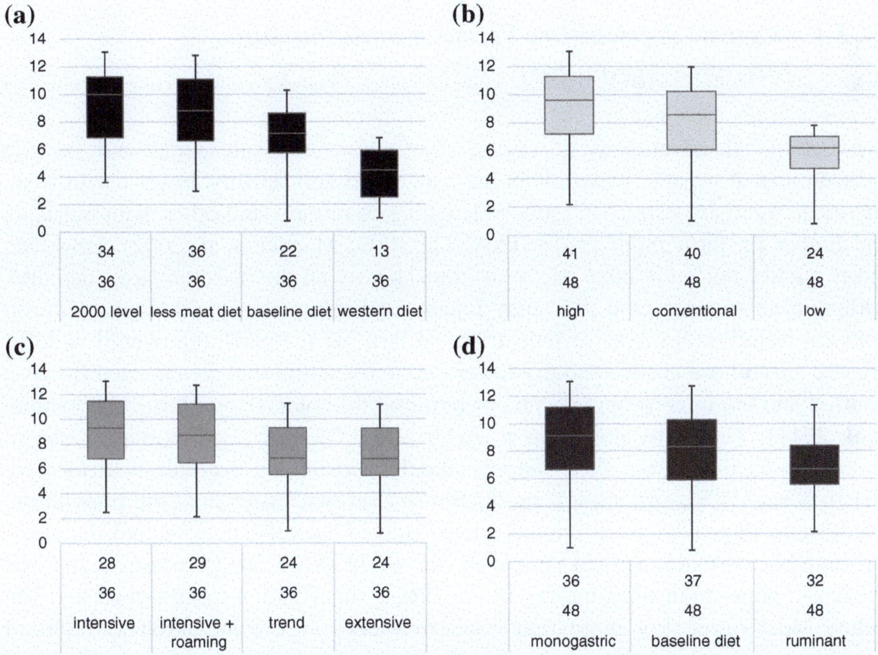

Fig. 14.4 Box plots showing the area potentially available in the year 2050 for purposes other than food production on good-quality land (i.e., cropland and high-quality grassland) in all scenarios classified as feasible; y-axis unit: million km^2. **a** Diet variants, **b** crop yield variants, **c** variants of livestock feeding efficiencies, **d** baseline diet tweaked toward higher shares of products from monogastrics or from ruminants. *Whiskers* show the full ranges of all feasible scenarios, *boxes* show quartiles (Q), i.e., the range from Q1 to Q3, and the *horizontal line* in each box shows the median (Q2). The numbers below the *box plots* indicate the number of all scenarios in the respective variants (*below*) and the number of 'feasible' scenarios (*above*)

different assumed future changes in the food and agriculture system. As expected, higher food demand, lower yields, lower feeding efficiencies and a higher share of ruminant products result in higher pressures of food production on the earth's land surface.

The choice of ruminant-based vs. monogastric-based animal products substantially affects area requirements of food production. The reason is the lower energy efficiency of ruminants compared to monogastrics, which is related to the much lower feed value of their diet (roughage). However, a higher fraction of monogastric products also results in more cropland, that is, more intensive land use. This can lead to the abandonment of ecologically valuable, high-biodiversity extensive grazing systems that store much carbon in the soil. Moreover, not all areas grazed by ruminants are suitable for cropping. Hence, these numbers need to be interpreted cautiously to avoid flawed conclusions or flawed policy recommendations.

14.3 Trade-Offs and Synergies

14.3.1 Organic Agriculture Versus Land-Sparing Intensive Agriculture

The ecological advantages of organic agriculture are substantial and include reduced loss of organic materials in the soil, better soil fertility, lower use of toxic chemicals such as pesticides, reduced leaching of nitrates and other plant nutrients and higher on-farm biodiversity (IAASTD 2009; Maeder et al. 2002). However, lower yields per unit area per year translate to an increase in area demand. Although an organic crop field may reach nearly the same yield level as a crop field cultivated with intensive industrialized farming methods, the overall yield of organic agriculture is substantially lower due to the additional area needed for crop rotation and the intercropping schemes required to replenish soil fertility (Guzman et al. 2011). This is by no means a trivial issue. Globally, approximately one in seven persons is malnourished (approximately as many are overfed; Godfray et al. 2010). Since 1950, agricultural production has grown faster than the population, resulting in improved nutrition and a reduction in both the fraction and the absolute number of malnourished people in the world (FAO 2013); whereas 26 % of the world population went hungry in the 1960s, this fraction has dropped to 12 % today. Most—probably more than three-quarters—of the increased agricultural production was achieved through intensification; the expansion of farmland played a much smaller role. The prolongation of these trends until 2050 may be impossible to achieve with organic agriculture.

Could organic agriculture feed the world? Our scenario calculations suggest it might be able to, but only if the 'less meat' diet is adopted, and even then, not for all variants of feeding efficiency. Moreover, at any given level of product supply, the area required for global food supply is larger when only organic agriculture is used compared to stronger yields increases based on industrialized farming methods. Larger area demand bears substantial ecological costs: it increases the pressure to extend farmland beyond its current boundaries, which usually results in deforestation and loss of other natural or semi-natural areas and leaves less area for biodiversity conservation, carbon sequestration and bioenergy production (Smith et al. 2013).

There are additional systemic feedbacks, such as the following:

- The biodiversity of areas farmed according to standards of organic agriculture is higher than that of conventional industrial farming. However, if organic farming requires more area, it may leave less area for biodiversity conservation and, hence, result in stronger pressures on biodiversity outside food farming areas. Whether 'land sharing' or 'land sparing' puts less pressure on biodiversity is a complex and unresolved question (Butsic et al. 2012).
- Organically grown food is usually more expensive than products from conventional farming. On the one hand, this would hamper the supply of sufficient food to poorer parts of the population as more widespread adoption of organic agriculture would result in an upward trend of food prices. This could lead to more hunger and malnutrition. On the other hand, the fact that products of conventional agriculture are cheaper may result in additional demand related to rebound effects that stimulate ecologically more demanding consumption patterns with a higher land demand (Lambin and Meyfroidt 2011). Ceteris paribus assumptions are, hence, of questionable value.

The nagging question remains whether the yield forecasts in the 'trend' variant, and even more so, the 'high' variant, are realistic. In some world regions, the growth of yields is already slowing. Soil degradation may hamper further yield growth in intensively used regions (Winiwarter and Gerzabeck 2012). Some strategies for yield improvements, such as improved harvest indices, may be about to reach physiological limits. In modern wheat cultivars, for example, the fraction of total aboveground biomass allocated to the commercial crop has already reached 40–60 % (Krausmann et al. 2008). It does not seem likely that plants could grow and remain productive and resilient with an even lower share of plant tissue allocated to leaves and stems.

14.3.2 Bioenergy, Carbon Sinks and Conservation Areas

The differences in potential area availability shown in Fig. 14.4 are by no means trivial. For example, the 4 million km^2 difference between the less meat diet and that of the western diet (Fig. 14.4a) is more than one-quarter of current global cropland. If planted with energy crops with a relatively modest assumed primary energy yield of 20 MJ/m^2/year (megajoules per square meter per year; 1 MJ = 10^6 J), this area would deliver 80 EJ/year (exajoules per year; 1 EJ = 10^{18} J) of primary bioenergy, or approximately 15 % of current global technical primary energy consumption, which would be a substantial contribution to the global energy supply. Hence, the magnitude of the bioenergy potential strongly depends on the choice of diet, the food crop yields achieved and the livestock feeding efficiencies, among many other factors (Coelho et al. 2012).

Even if productive land areas could be made available, other systemic feedbacks need to be considered before concluding that bioenergy is the best option. First, the full GHG costs of implementing such large-scale bioenergy schemes are unknown. For the area to become available, the intensification of livestock grazing is required, as described above. Such intensification would likely affect the amount of carbon stored in these areas because extensive grazing land stores far more carbon in soils than intensively used grasslands. Moreover, cultivating bioenergy plants would entail ploughing up substantial land areas, which could also result in carbon loss, although this effect depends on the energy crop chosen. Short-rotation coppice and perennial grasses can provide bioenergy while sequestering carbon when they are grown on soils that had been used for cropping, particularly when grown on degraded lands (Coelho et al. 2012), but whether this also applies when they are planted on lands that had previously been extensively grazed is unclear.

Second, using the land for bioenergy means it is unavailable for alternative options, such as carbon sequestration (apart from carbon sequestered by bioenergy plants, if such is the case) and biodiversity conservation. Afforesting available productive land may result in considerable carbon sequestration over long (decadal to centennial) periods. Even if no GHG costs of land conversion for bioenergy are factored in, it is not a priori clear whether use of the land for bioenergy production or for carbon sequestration is the superior option in terms of total GHG mitigation, at least over decadal time frames. Indeed, in many cases, carbon sequestration may be more beneficial for the climate than bioenergy (Smith et al. 2013). In many instances, carbon sequestration helps build up biologically more diverse biotic communities (Essl and Rabitsch 2013), so there are probably synergies between carbon sequestration and biodiversity protection.

14.4 Conclusions

Land is a unique resource for humans and for all other living beings on earth. Managing this limited resource in a manner that provides critical resources for humans while minimizing adverse effects for biodiversity or degrading critical ecosystem functions and services is a complex endeavor. On a planet where most of the terrestrial ecosystems are colonized to an extent that patterns and processes must be understood as coupled socioecological systems, changes in land-use practices and resource use create systemic feedbacks affecting ecosystems and the services they provide to human societies. Land-use changes are thus likely to affect the interests of many stakeholders, thereby raising issues of land-use competition or even conflicts.

Systemic feedbacks between demand and supply, among different land characteristics (e.g., biological diversity, landscape values, carbon sequestration and suitability for infrastructure, food provision or as living space for humans) and between different technologies and practices abound, and they carry the potential for enormous unforeseen or unintended consequences. The doubling or tripling of the price of many agricultural commodities in 2007, which most likely resulted from the coincidence of biofuel policies in the US and Europe with changes in demand for food products and a poor harvest (Coelho et al. 2012), is a good example of the possible magnitude of such systemic feedbacks. Another example is the GHG emissions that may result (indirectly through market-mediated effects) from the expansion of bioenergy production, the so-called 'indirect land-use change', or iLUC, effects. At present, these effects are poorly understood, but it has become clear that they are large enough to cast doubt on the potential positive outcomes of policies requiring a large fraction of the land surface of the planet—at least as long as these feedbacks have not been thoroughly studied.

What the sociometabolic approach shows is that demand reductions have positive synergistic effects. They reduce area demand and hence allow a reduction in the intensity of the colonization of ecosystems by reducing the need to boost yields and intensify grazing, or they allow land to be spared for uses other than food production, be it bioenergy, carbon sequestration and/or the conservation of biodiversity. The mixture of these options that is most beneficial is a difficult question for which there are no sweeping answers. Most likely, locally and regionally adapted solutions will help to increase benefits and reduce risks. As shown above, demand reductions can come from changes in diets. These could, in many parts of the world, also be beneficial in terms of health co-benefits. Another option is to reduce food waste, which has been estimated to exceed one-quarter of all food produced globally (Smith et al. 2013).

A largely complementary option suggested by the sociometabolic approach is to increase the efficiency with which biomass is used to generate a variety of products, including feed, food, fiber and energy. This strategy has been denoted the 'cascade utilization of biomass' (Haberl and Geissler 2000). It relies on using by-product and residue flows as well as reuse and recycling of biomass-based products whenever possible. Such optimization may help to generate more products and services from the same amount of primary biomass harvested. However, biomass residue backflows to the soil need to be considered when planning such measures. Otherwise, adverse effects on soil fertility as well as on the soil's carbon balance may ensue (Blanco-Canqui and Lal 2009).

Finally, it is important to question the ceteris paribus conditions invoked in many scenario analyses, namely, the assumption that everything else would stay the same if one factor, such as yields or feeding efficiencies, changed. This assumption, which is often a methodological necessity in mechanistic models, is quite unlikely to prevail in reality. BioBaM partly overcomes this limitation by systematically combining variants of many decisive land-use factors, thereby allowing a multitude of possible future options to be explored. This strategic orientation comes at a cost, however: it is impossible to judge which of the scenarios is more or less probable than others given certain socioecological developments or policy interventions. Furthermore, BioBaM cannot depict meta-level feedbacks in the land systems, such as rebound effects between technological progress and consumption levels. Moreover, some assumptions, such as the 'food first' and 'no deforestation' approaches, are unlikely to be an accurate description of future trajectories. It seems rather likely that such effects may occur. Food-fuel competition is likely to happen and deforestation may continue, even if it has slowed in some countries in recent years. In our view, increases in efficiency likely played a role in regions that adopted more wasteful lifestyles and diets. Policies based on fostering yield growth and efficiency may thus be ineffective in terms of reducing environmental pressures if not combined with efforts on the demand side in the same way that policies focused on organic agriculture may be ineffective if they do not succeed in changing demand patterns along with production and supply. Coping with trade-offs and maximizing synergies whenever possible is a central challenge in managing the earth's lands sustainably and to the benefit of humans and all other species on earth.

Acknowledgements Funding by the Austrian Science Fund (FWF) within the project P20812-G11, by the European Research Council within ERC Starting Grant 263522 LUISE and by the EU-FP7 project VOLANTE is gratefully acknowledged. This chapter was written in parts during Helmut Haberl's research stay at the Integrative Research Institute on Transformation in Human-Environment Systems (IRI THESys) at Humboldt-Universität zu Berlin.

References

Blanco-Canqui, H., & Lal, R. (2009). Crop residue removal impacts on soil productivity and environmental quality. *Critical Reviews in Plant Sciences, 28*, 139–163.

Bouwman, A., Van der Hoek, K., Eickhout, B., & Soenario, I. (2005). Exploring changes in world ruminant production systems. *Agricultural Systems, 84*, 121–153.

Braat, L. C., & de Groot, R. (2012). The ecosystem services agenda: bridging the worlds of natural science and economics, conservation and development, and public and private policy. *Ecosystem Services, 1*, 4–15.

Burney, J. A., Davis, S. J., & Lobell, D. B. (2010). Greenhouse gas mitigation by agricultural intensification. *Proceedings of the National Academy of Sciences, 107*, 12052–12057.

Butsic, V., Radeloff, V. C., Kuemmerle, T., & Pidgeon, A. M. (2012). Analytical solutions to trade-offs between size of protected areas and land-use intensity. *Conservation Biology, 26*(5):883–893.

Coelho, S., Agbenyega, O., Agostini, A., Erb, K.-H., Haberl, H., Hoogwijk, M., et al. (2012). Land and water: Linkages to bioenergy. In: T. Johansson., A. Patwardhan., N. Nakicenonivc, Gomez-Echeverri (Eds.), *Global energy assessment. International institute of applied systems analysis (IIASA)* (pp. 1459–1525). Cambridge University Press, Cambridge, UK.

Ellis, E. C., Klein Goldewijk, K., Siebert, S., Lightman, D., & Ramankutty, N. (2010). Anthropogenic transformation of the biomes, 1700 to 2000. *Global Ecology and Biogeography, 19*, 589–606.

Erb, K.-H., Gaube, V., Krausmann, F., Plutzar, C., Bondeau, A., & Haberl, H. (2007). A comprehensive global 5 min resolution land-use data set for the year 2000 consistent with national census data. *Journal of Land Use Science, 2*, 191–224.

Erb, K.-H., Haberl, H., Krausmann, F., Lauk, C., Plutzar, C., & Steinberger, J. K. (2009). Eating the planet: Feeding and fuelling the world sustainably, fairly and humanely—a scoping study. Social ecology working paper no. 116, Vienna, Potsdam.

Erb, K.-H., Mayer, A., Kastner, T., Sallet, K. E., & Haberl, H. (2012). The impact of industrial grain fed livestock production on food security: An extended literature review. Institute of social ecology, working paper social ecology no. 136, Vienna.

Essl, F., & Rabitsch, W. (2013). *Biodiversität und Klimawandel.* Springer, Heidelberg: Auswirkungen und Handlungsoptionen für den Naturschutz in Mitteleuropa.

FAO. (2006). World agriculture: Towards 2030/2050—Interim report. Prospects for food, nutrition, agriculture and major commodity groups. Food and Agriculture Organization of the United Nations, Rome.

FAO. (2013). *The state of food insecurity in the world 2013; the multiple dimensions of food security*. Rome, Italy: Food and Agriculture Organisation of the United Nations (FAO).

Fischer-Kowalski, M., Haberl, H., Hüttler, W., Payer, H., Schandl, H., Winiwarter, V., et al. (1997). *Gesellschaftlicher Stoffwechsel und Kolonisierung von Natur.* Amsterdam: Ein Versuch in sozialer Ökologie. Gordon & Breach Fakultas.

Godfray, H. C. J., Beddington, J. R., Crute, I. R., Haddad, L., Lawrence, D., Muir, J. F., et al. (2010). Food security: The challenge of feeding 9 billion people. *Science, 327*, 812–818.

Guzman, G. I., Gonzalez de Molina, M., & Alonso, A. M. (2011). The land cost of agrarian sustainability. An assessment. *Land Use Policy, 28*, 825–835.

Haberl, H., Erb, K.-H., Krausmann, F., Bondeau, A., Lauk, C., Müller, C., et al. (2011). Global bioenergy potentials from agricultural land in 2050: Sensitivity to climate change, diets and yields. *Biomass and Bioenergy, 35*, 4753–4769.

Haberl, H., & Geissler, S. (2000). Cascade utilization of biomass: Strategies for a more efficient use of a scarce resource. *Ecological Engineering, 16*, 111–121.

IAASTD. (2009). *Agriculture at a crossroads. International assessment of agricultural knowledge, science and technology for development (IAASTD).* Global report. Washington, D.C: Island Press.

Krausmann, F., Erb, K.-H., Gingrich, S., Lauk, C., & Haberl, H. (2008). Global patterns of socioeconomic biomass flows in the year 2000: A comprehensive assessment of supply, consumption and constraints. *Ecological Economics, 65*, 471–487.

Lambin, E. F., & Meyfroidt, P. (2011). Global land use change, economic globalization, and the looming land scarcity. *Proceedings of the National Academy of Sciences, 108*, 3465–3472.

Maeder, P., Fliessbach, A., Dubois, D., Gunst, L., Fried, P., & Niggli, U. (2002). Soil fertility and biodiversity in organic farming. *Science, 296*, 1694–1697.

Millennium Ecosystem Assessment. (2005). *Ecosystems and human well-being: Synthesis.* Washington, DC: Island Press.

Seufert, V., Ramankutty, N., & Foley, J. A. (2012). Comparing the yields of organic and conventional agriculture. *Nature, 485*, 229–234.

Smith, P., Haberl, H., Popp, A., Erb, K.-H., Lauk, C., Harper, R., et al. (2013). How much land based greenhouse gas mitigation can be achieved without compromising food security and environmental goals? *Global Change Biology, 19*, 2285–2302.

Thornton, P. K. (2010). Livestock production: Recent trends, future prospects. *Philosophical Transactions of the Royal Society B: Biological Sciences, 365*, 2853–2867.

Tilman, D., Balzer, C., Hill, J., & Befort, B. L. (2011). Global food demand and the sustainable intensification of agriculture. *Proceedings of the National Academy of Sciences, 108*, 20260–20264.

Turner, B. L., Lambin, E. F., & Reenberg, A. (2007). The emergence of land change science for global environmental change and sustainability. *Proceedings of the National Academy of Sciences, 104*, 20666–20671.

Winiwarter, V., & Gerzabeck, M. H. (2012). The challenge of sustaining soils: Natural and social ramifications of biomass production in a changing world. Interdisciplinary perspectives no. 1, Austrian Academy of Sciences Press, Vienna.

Method Précis: Human Appropriation of Net Primary Production (HANPP)

Helmut Haberl

The human appropriation of net primary production(HANPP) is an indicator of the intensity of the colonization of ecosystems, namely, the intensity of land use. HANPP is based on the quantification of human interventions in energy flows in ecosystems or, more precisely, in net primary production and the availability of the products of net primary production (primarily biomass) in ecosystems.

Net primary production (NPP) is a measure of the quantity of organic material produced by plants through photosynthesis from inorganic materials. In energetic terms, photosynthesis involves the transformation of radiant energy from the sun into energy stored in chemical compounds. This energy is initially stored in the biomass of plants and then either accumulates in the ecosystem or serves as food energy for humans, animals, fungi and some microorganisms (so-called 'heterotrophic' organisms). During photosynthesis, CO_2 is absorbed from the atmosphere and stored in a variety of chemical compounds in biomass. If this energy is released, for example, through combustion or the metabolism of heterotrophic organisms ('respiration'), then carbon is released into the atmosphere in the form of CO_2. In the short term, ecosystems may represent a 'carbon sink' (that is, absorb more CO_2 through photosynthesis than flows back to the atmosphere due to respiration and combustion) or a 'carbon source' (CO_2 outflows exceed photosynthesis). In the long term and across larger areas, the average absorption and release of CO_2 from ecosystems is largely balanced;[6] that is, CO_2 inflows equal CO_2 outflows (Körner 2009). NPP is an important process in ecosystems; it supplies the entire food energy for humans and all other heterotrophic food webs and provides the basis for the creation of vegetation cover and soils and their associated carbon stocks. NPP is one of the most important indicators of ecosystem capacity and forms the basis for the existence of all biodiversity (Vitousek et al. 1986; Wright 1990).

Insofar as humans use land for their purposes, they intervene in these processes. First, they replace natural ecosystems, such as forests and grasslands, with ecosystems utilized by humans, such as settlement areas, agricultural ecosystems and managed forests (possibly causing soil degradation in the process). The NPP of the ecosystems thus utilized often differs significantly from that of natural ecosystems. The difference between the NPP of potential natural vegetation (NPP_{pot}, NPP of the ecosystem with no human influence) and the vegetation that is predominant due to the land use at a particular point in time (NPP_{act}, *actual* NPP) is defined as $HANPP_{luc}$ (HANPP resulting from land use). Added to this—and this is, in many instances, the actual purpose of land use—is the harvest of biomass for

[6]Exceptions include raised bogs, which are able to create long-term carbon sinks because of the exclusion of oxygen in the soil.

human use ($HANPP_{harv}$, HANPP through harvest). In the current definition, which underpins the research presented in this book (see Haberl et al. 2007, 2014; the notation used here was taken from Krausmann et al. 2013, yet the concept remains the same), $HANPP_{harv}$ is relatively broadly defined and includes those parts of plants that, although they are not themselves economically utilized and actually removed, die off during the harvest. These include, for example, the roots of cereal crops and trees (by contrast, the rootstocks of perennial grasses survive the harvest and are therefore not included in calculations) and the harvest of by-products that remain on the field. In contrast to $HANPP_{harv}$, which is always greater than or equal to zero, $HANPP_{luc}$ can also be less than zero. This is the case when land use increases NPP, which is a common occurrence where artificial irrigation is employed in agriculture. However, land use can also increase NPP in humid regions, for example, in very intensively used agricultural regions. Nonetheless, the NPP_{act} of agricultural ecosystems is often smaller than the NPP_{pot}. The primary purpose of agriculture is to favor the cultivation of plants that produce a greater quantity of plant matter that can be utilized for human food, livestock feed or other economic purposes than natural vegetation would. Examples of usable plant matter are cereal grains and hay rather than unusable leaves or roots. Agriculture is primarily interested in an increase in the economically valuable parts of plants. Whether the NPP of the system rises or falls in the process is not per se important in economic terms but only inasmuch as this produces an increase in the desired harvest. HANPP can therefore be defined as follows (Haberl et al. 2007, 2014):

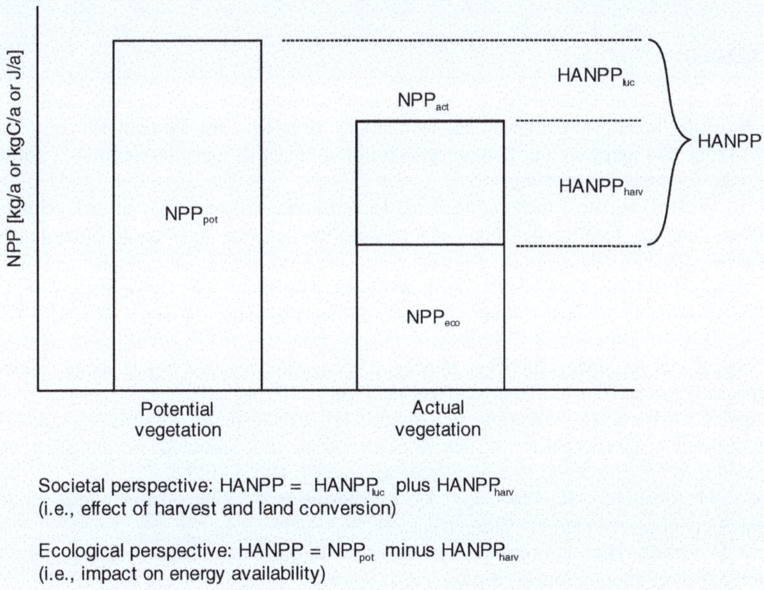

Fig. 14.5 The concept of the human appropriation of net primary production (HANPP)

$$HANPP := HANPP_{luc} + HANPP_{harv}. \quad (14.1)$$

If one subtracts the $HANPP_{harv}$ from the NPP_{act}, the result is the amount of NPP remaining in the ecosystem after harvest (and thus available to fulfill the ecosystem functions described above, i.e., the food required by heterotrophic organisms or the production/maintenance of carbon stocks). This is defined as NPP_{eco} (NPP remaining in the ecosystem). An equivalent definition of HANPP is, therefore (Fig. 14.5),

$$HANPP := NPP_{pot} - NPP_{eco}. \quad (14.2)$$

HANPP can be positive or negative, although a negative HANPP ($NPP_{eco} > NPP_{pot}$), as a rule, only occurs in arid areas with a low NPP_{pot}, which must be irrigated for agricultural purposes. In other words, HANPP is negative when $HANPP_{luc}$ is negative and the absolute value of $HANPP_{luc}$ is greater than $HANPP_{harv}$. This occurs in arid areas, not in humid regions where intensive agriculture is practiced.

In the literature, other definitions of HANPP are sometimes used, particularly the formulation of Vitousek et al. (1986). The definition used here is a further development of the definition produced by Wright (1990). The influential study by Imhoff et al. (2004) used a consumption-based approach similar to 'embodied HANPP' (Chap. 16). As shown by Haberl et al. (2007), the results of HANPP calculations vary significantly according to the definition used. It is thus of decisive importance that the particular definition used be taken into consideration when interpreting HANPP data.

References

Haberl, H., Erb, K.-H., Krausmann, F., Gaube, V., Bondeau, A., Plutzar, C., et al. (2007). Quantifying and mapping the human appropriation of net primary production in earth's terrestrial ecosystems. *Proceedings of the National Academy of Sciences, 104*, 12942–12947.

Haberl, H., Erb, K.-H., & Krausmann, F. (2014). Human appropriation of net primary production: Patterns, trends, and planetary boundaries. *Annual Review of Environment and Resources, 39*, 363–391.

Imhoff, M. L., Bounoua, L., Ricketts, T., Loucks, C., Harriss, R., & Lawrence, W. T. (2004). Global patterns in human consumption of net primary production. *Nature, 429*, 870–873.

Körner, C. (2009). Biologische Kohlenstoffsenken: Umsatz und Kapital nicht verwechseln (Biological carbon sinks: Turnover must not be confused with capital). *Gaia—Ecological Perspectives for Science and Society, 18*, 288–293.

Krausmann, F., Erb, K.-H., Gingrich, S., Haberl, H., Bondeau, A., Gaube, V., et al. (2013). Global human appropriation of net primary production doubled in the 20th century. *Proceedings of the National Academy of Sciences, 110*, 10324–10329.

Vitousek, P. M., Ehrlich, P. R., Ehrlich, A. H., & Matson, P. A. (1986). Human appropriation of the products of photosynthesis. *BioScience, 36*, 363–373.

Wright, D. H. (1990). Human impacts on energy flow through natural ecosystems, and implications for species endangerment. *Ambio, 19*, 189–194.

Chapter 15
A Burning Issue: Anthropogenic Vegetation Fires

Christian Lauk and Karl-Heinz Erb

Abstract Human-induced vegetation fires play a central role in past and present nature-society interactions. Tens of thousands of years ago, hunter-gatherers presumably employed fires as a hunting technique. Today, vegetation fires continue to be an integral part of shifting cultivation and traditional pastoralism, and they are a crucial tool for the clearing of forests. In industrial regions, however, vegetation fires are increasingly seen as a risk that threatens valuable infrastructures and contributes to climate change and air pollution. This chapter considers human-induced vegetation fires from a socioecological perspective. It begins with a quantitative estimate of the global relevance of human-induced vegetation fires and continues with a discussion of how these fires can be integrated into basic socio-ecological concepts. In a further section, we develop a global ideal typology of vegetation fires, which can serve as a basis for discussing their complex variety. We conclude with the question of to what extent and under which circumstances human-induced vegetation fires are sustainable. Overall, this chapter shows that human-induced vegetation fires continue to play a crucial role in society-nature interactions, and it demonstrates that Social Ecology provides important tools to analyze and conceptualize human-induced fires at different scales.

Keywords Vegetation fires · Biomass burning · Fire regimes · Shifting cultivation · Energy flows · Energy efficiency

---

C. Lauk (✉) · K.-H. Erb
Institute of Social Ecology (SEC), Alpen-Adria University,
Schottenfeldgasse 29, 1070 Vienna, Austria
e-mail: christian.lauk@aau.at

K.-H. Erb
e-mail: karlheinz.erb@aau.at

H. Haberl et al. (eds.), *Social Ecology*, Human-Environment Interactions 5,
DOI 10.1007/978-3-319-33326-7_15

15.1 'Fire as the First Great Force Employed by Man'

The controlled burning of hydrocarbons (i.e., of biomass and later of fossil fuels) might be considered the energetic basis of human civilization. For hunter-gatherers, fire rendered the food consumed more easily digestible and made it possible to settle in boreal and even polar regions. Accordingly, the anthropologist Stewart (1956) entitled his classic text on anthropogenic biomass burning *Fire as the First Great Force Employed by Man*. With the transition to agriculture and the parallel development of towns and cities, fire provided the energy required for the refining of those metals after which, in recognition of their social importance, archaeology has named entire eras. The central importance of fire also finds expression in mythology. Prometheus, for example, appears not only as the creator of culture and teacher of humans but also as the figure who brought humans the gift of fire, which had formerly been the sole privilege of the gods. The fact that the propagation of both cultural-technical skills and fire were united in a single figure is surely no coincidence. The industrial revolution, which marks the beginning of the contemporary age, is also based on the power of fire. The steam engine (which made it possible to overcome the natural limits of wind and water mills), the steam turbine and the combustion engine were all driven by the power of fire. Although non-combustion-based forms of energy generation, including nuclear energy and hydro and wind power, are gaining in importance, the combustion of what are now primarily fossil hydrocarbons provides more than 85 % of all the primary energy required for industrial use. However, the debate concerning biofuels and the key significance of bioenergy in long-term energy scenarios shows how difficult it will be to extricate ourselves from an energetic reliance on fire.

The control of fire started long ago with the burning of natural vegetation. Evidence indicates that long before the Neolithic revolution, hunter-gatherers employed controlled biomass burning both to extend the area of pasture lands and as a hunting technique. The power exercised through burning vegetation may have been so great that numerous researchers have theorized that the targeted burning of grassland by humans was a significant cause of the extinction of the megafauna inhabiting North America and Australia at the end of the Pleistocene era and of the global spread of grasslands. Despite the gradual marginalization of hunter-gatherers by agrarian societies, vegetation fires did not lose their significance in the context of land use. Especially in early forms of agrarian systems, namely, shifting cultivation and pastoralism, the burning of vegetation played a central role in the colonization of natural systems. With the intensification of land use, fire gained importance in the conversion of forest into crop and grazing land.

The use of fire in shifting cultivation and traditional pastoral farming continues today, and vegetation fires remain important in the context of large-scale deforestation. However, whereas the traditional use of vegetation fires remains important, mainly in tropical regions, in other regions, especially densely populated developed countries, vegetation fires have increasingly been perceived as a risk to materially and economically valuable structures and as a problem in the context of climate change and air pollution.

15.2 The Global Relevance of Anthropogenic Vegetation Fires

In this context, Seiler and Crutzen (1980) undertook the first comprehensive attempt to quantify the annual amount of biomass burned in vegetation fires on a global scale. This estimate was based on assumptions drawn from the literature and statistical data regarding land areas affected by specific practices involving vegetation fires (e.g., shifting cultivation and deforestation). In combination with assumptions about region-specific biomass densities (biomass volumes per area) and combustion efficiencies (the share of biomass affected by fire that is actually burned), the total biomass burned was derived. Further estimates published over the following two decades largely adopted the basic approach of Seiler and Crutzen (1980), yet partly modified the assumptions regarding burned area, biomass density and combustion efficiency (e.g., Andreae 1991). The uncertainty of these assumptions (e.g., the area on which shifting cultivation is practiced or combustion efficiencies) translates into a corresponding uncertainty about the resulting estimate of global volume of burned biomass, which was estimated at approximately ±50 % by Andreae and Merlet (2001).

A new approach to quantify the biomass burned through vegetation fires opened up at the turn of the millennium with the help of new remote sensing data. With the help of special satellite sensors, it became possible to detect areas burned at a resolution of 1 km^2. This created the possibility to derive, in combination with spatially resolved data about biomass densities, the volumes of burned biomass by another method (e.g., Ito and Penner 2004). However, despite a potentially higher degree of accuracy, this method has drawbacks. (1) At a resolution of 1 km^2, many smaller vegetation fires, which fall below this detection threshold, fail to be recorded. (2) Parts of the tropics are often veiled in cloud cover (especially during the burning season), which prevents or at least hinders the satellite identification of burned areas. Because of these first two drawbacks, small but presumably numerous fires are not detected, although they are allegedly highly relevant, particularly in shifting cultivation and deforestation. Finally, (3) remote-sensing-derived data are unable to provide information about the social context in which changes in land cover take place; that is, land cover is recorded, but not land use (or its transformation). This means that based on remote sensing data, no distinction can be made between natural and anthropogenic vegetation fires, let alone a more specific distinction between different socioecological contexts in which vegetation fires occur. From a socioecological perspective, however, these differentiations are crucial.

In this context, the work of Lauk and Erb (2009) combined recent remote sensing data on vegetation fires with literature data about the causes of vegetation fires to develop an improved estimate of the global volume of biomass burned as a result of anthropogenic vegetation fires. For this purpose, published data on biomass burned in vegetation fires (Ito and Penner 2004) were combined with data derived from numerous literature sources on country-specific shares of vegetation fires caused by humans. To take into account smaller fires, which were presumed to be missing in remote sensing data, the volume of biomass burned as a result of shifting cultivation, to which a large share of these smaller fires may be attributed, was also estimated. The latter estimate is based on a combination of the area used for shifting cultivation with the length of the fallow period, the biomass accumulated by the end of the fallow period per area unit and combustion efficiencies. Thus, the literature data and a range of statistical sources formed the basis for this research.[1]

The results produced by Lauk and Erb (2009) show that the biomass burned in anthropogenic vegetation fires represents a highly relevant energy and material flow. On a global scale, between 3.5 and 3.9 Gt (billion metric tons; weight in dry matter[2]) of biomass is burned each year through anthropogenic vegetation fires, which equals approximately one-third of all biomass used as food, feed, fibre and bioenergy. Each year, these fires produce 2 Gt of CO_2 emissions, representing nearly one-quarter of all CO_2 emissions produced by the use of fossil energy sources. It is important to note, however, that carbon emissions from the burning of biomass only lead to an increase in atmospheric CO_2 concentrations and thus contribute to global warming if they are linked to a reduction of carbon stocks in the biosphere or lithosphere, as in the case of permanent deforestation. In many cases, however, vegetation fires only accelerate the conversion of biomass to CO_2, such as when dry grass is burned on savannah lands that would otherwise have decayed over a longer period.

A literature review in Lauk and Erb (2009) revealed that the vast majority of all vegetation fires are caused by humans. Only in areas such as the boreal forests of Canada, Russia and the US do natural fires (i.e., those caused by lightning) play an important role (accounting for 85, 49 and 68 % of the vegetation fires occurring in Canada, Russia and the US, respectively). Based on these data, it is possible to further differentiate between two large classes of anthropogenic vegetation fires. Roughly two-thirds (2.5 Gt/year) of the biomass burned in anthropogenic vegetation fires can be attributed to large-scale fires, which are included in the relevant remote sensing data. The overwhelming majority of these large-scale fires can be observed in less-densely populated areas with savannahs, primarily the savannahs of Africa and, of less importance globally, the savannahs of Australia, the Cerrado of South America and the grasslands of Central Asia. In contrast, approximately

[1]See Lauk and Erb (2009) for a detailed description of the methodology and sources.

[2]Unless otherwise indicated, all subsequent references to biomass in mass units refer to dry matter.

one-third of the burned biomass (1.1–1.4 Gt/year) can be attributed to small-scale fires, primarily due to shifting cultivation. The practice of shifting cultivation and small-scale fire clearance presently takes place primarily in the tropical forested regions of Latin America, Southeast Asia and West and Central Africa.

One option to tentatively compare the societal relevance of anthropogenic vegetation fires in different world regions is to put the material flows induced by vegetation fires in relation to the total biomass extracted to be used as food, feed, fibre and bioenergy (Table 15.1). Such a viewpoint illustrates the extraordinary position of Sub-Saharan Africa in terms of vegetation fires. In this region, ca. 50 % more materials are converted by vegetation fires than by the entire societal utilization of biomass (as food and feed as well as industrial use). In Central Asia, Oceania and

Table 15.1 Comparison by region of burned area/biomass with biomass extracted for food, feed, fibre, and bioenergy

Region	Total area (1000 km^2)	Burned area (1000 km^2/year)	Burned biomass[a] (Mio. t dm/year)	Used biomass extraction[b] (Mio. t dm/year)	Relationship of burned to used biomass (%)
North Africa and West Asia	10,433	7	10	339	3
Sub-Saharan Africa	24,399	2342	2202	1454	151
Central Asia	26,288	190	157	388	41
East Asia	11,762	86	71	1703	4
South Asia	6787	61	149	1841	8
Southeast Asia	4957	86	336	841	40
North America	19,600	12	19	1639	1
Latin America & Caribbean	20,546	299	795	2186	36
Western Europe	3711	7	10	985	1
East & Southeast Europe	2283	32	49	441	11
Oceania	8012	370	140	350	40
Global	138,777	3492	3938	12,138	32
Global incl. bif[c]			4346	12,138	36

[a]Maximum estimate

[b]According to Krausmann et al. (2008). Includes harvested primary biomass (incl. crop residues), grazed biomass and wood removals

[c]bif: burning of crop residues in the field according to Yevich and Logan (2003); only developing countries are included

Latin America, the biomass flows induced by anthropogenic vegetation fires represent ca. 40 % of the used biomass extraction and are thus also remarkably high. From a purely material perspective, this comparison suggests that anthropogenic vegetation fires play barely any role in the regions of North America, Western Europe, North Africa and West and East Asia. However, even in this case, fires can be highly relevant in symbolic or economic terms, as evidenced by the news footage of large forest fires in Europe and North America that regularly garner attention in the media.

The methodology developed by Lauk and Erb (2009) thus enables to specify whether vegetation fires are caused by humans or by natural factors such as lightning. It also allows a quantitative distinction between, on the one hand, the biomass burned as part of shifting cultivation and, on the other hand, the biomass burned in large-scale forest, grassland or savannah fires. Nonetheless, due to the absence of better data, uncertainties of this estimate remain high. This approach is also very limited in accounting for the multiplicity of socioecological contexts in which vegetation fires are embedded. Therefore, in the following section, which takes a qualitative approach, vegetation fires are studied from the perspective of the socioecological model of interaction, which forms the basis for a suggested classification of socioecological fire regimes in Sect. 15.4. From this discussion, it becomes clear that the question of the boundary between natural and societal material and energy flows is of particular importance regarding the issue of anthropogenic vegetation fires.

15.3 Anthropogenic Vegetation Fires Within the Socioecological Model of Interaction

Anthropogenic vegetation fires are intimately linked to the ways in which humans transform and enter into material exchange with their natural environment. The basic conceptual tools of Social Ecology presented in detail in Chap. 2 provide a possible basis for a more differentiated view of anthropogenic vegetation fires, allowing us to understand them as a part of the interaction between society and nature. Social Ecology operationalizes society-nature interactions by means of two basic concepts: (1) social metabolism and (2) the colonization of nature. In the following section, anthropogenic vegetation fires will be discussed and located within this general conceptual framework. This discussion highlights the central importance of one particular aspect when studying material interactions between society and nature: the question of system boundaries, notably, the boundary between the material structure of societies and their natural environment.

Are anthropogenic vegetation fires a constituent part of social metabolism? The prerequisite for them to be categorized as such, according to the definition of social metabolism set out in Chap. 2, would be that such fires directly contribute to the production, maintenance and operation of the human population or its social artifacts. In this context, one might argue that in contrast to, for example, fuel that drives a combustion engine, vegetation fires do not directly serve the reproduction or operation of societal artifacts but rather affect colonized systems without an artifact transmitting this impact, as would be the case for a tractor powered by fossil fuels. For this reason, fires occur outside the socioeconomic system and might not be considered a part of social metabolism, at least in a strict sense. In contrast, the energy and material flows induced by biomass burned by power stations, heating facilities and (in the form of agricultural fuels) combustion engines are clearly a part of the metabolism of societies as they are used to reproduce or operate artifacts of society such as combustion engines.

What relevance does this discussion have? The question of whether vegetation fires form a part of the material and energy flows of human societies is closely linked to questions related to their energy consumption. If anthropogenic vegetation fires were to be counted as societal energy flows, then the burning of huge stocks of accumulated biomass would form part of societal energy use if these fires were intentionally and functionally induced, as in the case of shifting cultivation. This would turn those traditional societies dependent on the use of vegetation fires, such as swidden cultivators and pastoralists, into consumers of large amounts of energy.

This question concerning the extent to which burned fallow vegetation in shifting cultivation is a societal flow inspired a minor (yet interesting) debate in the field of Anthropology during the 1980s. In the publication *Pigs for the Ancestors*, Rappaport (1968), a US pioneer of Ecological Anthropology, calculated the energy efficiency of the shifting cultivation system practiced by the Tsembaga, a tribe living in the highlands of New Guinea. He arrived at the conclusion that for every unit of energy flowing into the shifting cultivation form of agriculture (consisting of the food required for work), the Tsembaga obtain 17 energy units in the form of food, in other words, that shifting cultivation is a highly efficient agricultural system. One decade later, anthropologist Terry Rambo (1980) fundamentally challenged this view, claiming that Rappaport failed to account for the energy released through the burning of fallow vegetation as part of the energy consumption of the analyzed shifting cultivation system. Rambo argued that such fires were not only intentionally induced but also had a specific function within the agricultural system. Therefore, the energy released through these fires, according to Rambo, must be accounted as a part of the energy consumption of the shifting cultivation system. Based on this principal argument, McGrath (1987) later recalculated the energy efficiency of the Tsembaga based on Rappaport's data and concluded that they used nine units of energy, particularly in the form of fallow vegetation, to produce one unit of energy in the form of food. This new calculation thus reversed the picture and transformed a highly efficient agricultural system into an extremely inefficient one in terms of the relation between energy inputs and outputs.

Which side is ultimately right in this debate? Possibly both. The difference in numbers can be ascribed to a different setting of the system boundary between natural and societal energy flows. The setting of this boundary cannot, however, be determined objectively but is rather dependent on the research question. This yields different narratives that, in their own way, can each lay claim to being correct. In the case of the example given here, Rappaport would surely not deny that the studied shifting cultivation system relies on the very large flows of energy induced by the burning of vegetation, yet he does not regard these flows as societal. In fact, McGrath also draws a boundary between societal and natural energy flows, if he counts as societal the energy released by the burning of wood but not the solar energy that was required to cultivate the wood. Thus, a third position might argue that Rambo and McGrath both made the mistake of neglecting the solar energy required for the growth of the vegetation that was subsequently burned by the swidden cultivators. This idea is not as incongruous as it might first appear; the concept of 'emergy', introduced by US systems ecologist Howard T. Odum as an abbreviated form for 'embodied energy', is defined as 'the available solar energy used up directly and indirectly to make a service or product' (Odum 1996) and is the basic category in many energy flow analyses that draw upon this concept. Based on Odum's emergy concept, one could argue that food produced by shifting cultivation has a very high emergy, corresponding to the solar energy flowing into the growth of the biomass burned as part of the shifting cultivation cycle. A debate such as this shows the extent to which statements about energy inputs and outputs and the energy efficiency of systems are dependent upon the way in which relevant system boundaries, above all those between natural and societal flows, are defined.

How should intentionally induced vegetation fires, such as those in shifting cultivation, be categorized within the framework of the underlying concept of social metabolism? If one regards the intentional use of vegetation fires in land-use systems such as shifting cultivation as a rudimentary tool for the colonization of natural systems, then it may also be argued that, at least in a broader sense, the energy released through intentionally induced vegetation fires is a part of social metabolism. In material and energy flow analysis (MEFA) as a standardized operationalization of social metabolism, this is taken into account insofar as a distinction is made between used and unused extraction of materials, including biomass (cf. Krausmann et al. 2008). The energy released through anthropogenic vegetation fires and the biomass that is burned in the process is accounted for in this framework as unused extraction. The terminology might provide some scope for debate because this biomass is not strictly unused. Beyond the question of terminology, the decisive issue here is the analytical differentiation that allows for anthropogenic vegetation fires to be included within the analysis of societal material and energy flows but at the same time allows for an analytical distinction between (a) energy released by combustion strongly mediated by technologies such as combustion engines and (b) the technologically relatively unmediated use of anthropogenic vegetation fires in shifting cultivation and pastoralism.

15.4 A Global Typology of Socioecological Fire Regimes

Analogous to sociometabolic regimes (see Chap. 3), in the following section, we introduce the term *socioecological fire regime* to describe the socioecological characteristics and roles of vegetation fires in different contexts. The idea of socioecological fire regimes can be viewed as an extension of the concept of ecological fire regimes, which has become established within the field of Fire Ecology and refers to the spatiotemporal patterns of vegetation fires specific to a particular ecosystem. In Fire Ecology, these patterns include characteristics such as the scale of the fire (dependent upon, e.g., whether a forest fire only burns vegetation near the ground or also burns the tree canopy) and the frequency with which fires occur as well as factors such as the intensity and seasonality of the fires. In nearly all ecosystems, fires also occur—albeit at very different intervals and scales—without any human inducement, and fires are thus an inextricable part of many ecosystems.

However, Lauk and Erb's (2009) overview of human-induced fires shows that today, vegetation fires, like ecosystems in general, are unaffected by human interventions only in a minority of cases. In most cases, ecological fire regimes have become—whether intentionally or as a side effect of other activities—socioecological fire regimes, distinguished by characteristic temporal and spatial patterns. For these patterns, natural interrelationships continue to play an important role. Societies need to consider these interrelationships, but they are largely co-determined by human activities. Socioecological fire regimes do not necessarily involve a more extensive use of vegetation fires compared with natural fire regimes; they can equally be characterized by a particular form of containment of naturally occurring fires. On this basis, the following section attempts to provide a broad global typology of socioecological fire regimes.

1. **Fires by Hunter-Gatherers and Pastoralists**: Hunter-gatherers probably used vegetation fires as early as several hundred thousand years ago, with a considerable impact on local vegetation. Omer C. Stewart reports in his seminal article on the use of vegetation fires that in the western US, indigenous people from more than 50 tribes record the use of vegetation fires in some form. Fire was used to create additional pastureland for game, and the burning of dried grasses at the end of the dry period triggered the growth of new vegetation with a higher nutritional value. Vegetation fires were also used to drive game into an unburned area, and fires simplified food gathering by the burning of litter, making the gathered food, such as nuts and acorns, more easily visible. In general, it is often assumed that human-induced vegetation fires have played a major role in the spread of grasslands. Today, because only a few hunter-gatherer societies remain, the use of fire during hunting can only be observed in a few cases, for example, among aboriginal Australians. However, the use of fire to assist in the gathering of 'non-wood forest products' is still quite common, particularly in areas of Southeast Asia. In addition, many fires are used today with the specific aim of promoting the growth of fire-resistant species, such as *Dipterocarpus tuberculatus*, which has large leaves that are used for packaging and as roofing material. The vegetation fires that continue to occur regularly in many grassland and savannah regions (typically every one to five years) are primarily caused by pastoralists. It is believed that the main function of such fires is to improve the productivity of extensively used grasslands by triggering the growth of fresh vegetation after the end of the dry season.

2. **Vegetation Fires in Shifting Cultivation**: Another type of colonization through fire continues to be employed today in forested areas through shifting cultivation, which involves the alternation of fallow periods lasting several years with two- or three-year cultivation periods. Most vegetation is cut down at the end of the fallow period and burned. In the central regions of the Amazon rainforest, particularly along the Amazon River, charcoal deposits can be found together with ceramic fragments and so-called 'Terra Preta', whose origins lie some 3000–6000 years in the past. These findings provide evidence that this socioecological fire regime was already present in the rainforest several thousand years ago. However, there are also signs in the case of temperate and boreal regions that shifting cultivation was practiced over a long period as long as the population density was still sufficiently low. The fact that shifting cultivation was the preferred form of agriculture in forested areas for a long period and still exists today in many regions is in no way indicative of backwardness. Rather, it may be explained by its labor productivity; shifting cultivation typically has a significantly higher rate of labor productivity than more-intensive forms of preindustrial agriculture, an important observation for the agronomist Ester Boserup's theory about the historical development of land use (Boserup 1965). In particular, the use of fire reduces the otherwise extremely labor-intensive practice of hand weeding as the heat of the fire kills the seeds living in the upper soil layers. However, increasing population density and the expansion of industrial forms of agriculture put area-intensive shifting cultivation increasingly under pressure even in tropical regions.
3. **Vegetation Fires as a Weapon**: Perhaps the first description of the use of vegetation fires as a weapon appears in a text written by the Greek historical chronicler Herodotus in ca. 447 BC. Herodotus reports how the nomadic Scythians in Central Asia burned the vegetation and the feed for the cattle of the local population as they retreated before the advancing troops of the Persian king Darius. In almost every era and in all regions of the world, it is likely that fire has been used as part of such a practice, particularly by those groups that possess knowledge of vegetation fires in another context, such as hunter-gatherers and later pastoralists and swidden cultivators. The use of vegetation fires within conflicts is still a living practice. A detailed analysis of the social catalyst for large-scale forest fires during the dry period of 1997/98 in Indonesia reached the rather surprising conclusion that they were partly started deliberately in the conflict between small farmers and the operators of large-scale plantations. However, just how widespread this form of fire use is in the global context has barely been examined to date.
4. **Vegetation Fires as Part of Fire Clearance Practices**: When fire is used for deforestation and shifting cultivation, trees are felled and subsequently burned. Whereas shifting cultivation is characterized by periodically alternating cultivation and fallow periods during which the periodically burned secondary woodland grows, deforestation is characterized by the permanent and large-scale transformation of woodland into cropland and pastures. There is an important difference in terms of the impact of these two practices on atmospheric CO_2 and, thus, on climate change. In the case of deforestation, carbon stocks in vegetation and soils

are reduced and therefore permanently released as CO_2 into the atmosphere, whereas carbon stocks within the system of shifting cultivation remain constant as part of a dynamic equilibrium as long as the area used for shifting cultivation and its fallow period remains above some threshold. Historically, there have been several waves of deforestation by fire. In North America, for example, radiocarbon dating of charcoal in soils has revealed that there was a significant increase in fire activity around 1750 and that this trend only reversed around 1870, a period that reflects the wave of deforestation carried out by European settlers in the region. Since the 1970s, deforestation rates have increased mainly in tropical regions with the deforestation of tropical rainforests, a trend that has become one of the most important drivers of global CO_2 emissions.

5. **Combatting and Preventing Vegetation Fires**: Whereas the socioecological fire regimes described thus far are relevant particularly in preindustrial systems and—in the case of deforestation—during the transition to industrial systems, vegetation fires largely lost their material function in later industrial societies. This led to a fire regime that was no longer defined by a particular use of vegetation fires but rather by the attempt to avoid vegetation fires altogether or to alleviate their impacts on the material structures of society. Essentially, two (not necessarily mutually exclusive) options are available in this respect. On the one hand, vegetation fires might be contained and extinguished as quickly as possible. This strategy is still considered the preferred option in many regions of the world today, such as in the Mediterranean region of Europe. However, fighting fires in this way leads to an increasing accumulation of biomass. Thus, fire fighting can result in an increased risk of particularly intensive vegetation fires. For this reason, many regions began to switch to the adoption of a particular fire prevention strategy of 'fight fire with fire', with the US playing a pioneering role in this regard. Through the controlled burning of vegetation, the volume of potentially combustible biomass is reduced before vegetation fires break out. In other words, the transition to a fire regime in which fire becomes a risk by no means leads to the complete disappearance of vegetation fires.

Considering anthropogenic vegetation fires in different regions from the perspective of the typology outlined above, it becomes clear that in many cases, these socioecological fire regimes cannot be clearly delimited from one another but are in fact interrelated and regularly merge with one another. For instance, although shifting cultivation in tropical regions is still widely practiced, it is often related to deforestation, with so-called pioneering swiddeners practicing shifting cultivation for some years, after which the land is permanently cleared and subsequently sold to pastoralists. Moreover, conflicts often arise between pastoralists, who use fire in a traditional way, and arable farmers who live in the same regions and who perceive the same fires as a danger in some circumstances. Furthermore, it is not always possible to make a clear distinction between the socioecological fire regime related to the intentional use of fire and fires that are caused unintentionally. For example, a significant part of large-scale forest fires in Indonesia during the 1997/98 season was caused by fires used for shifting cultivation that subsequently grew out of control. Countless reports on hunter-gatherers in the 19th century also describe how these groups generally left their campfires without extinguishing them, and it may be assumed that these often caused the outbreak of large-scale vegetation fires in savannah regions. Socioecological fire regimes should thus be understood as an ideal-typical concept that can help describe the actual complex diversity of anthropogenic vegetation fires.

15.5 The Role of Anthropogenic Vegetation Fires in the Climate System

In many respects, vegetation fires play a key role within the global carbon cycle and the dynamics of climate change. First, fires are an important tool in the deforestation process, making it possible to clear large forest areas with relatively little effort, eventually leading to the emission of carbon previously stored in soils and vegetation. An estimated 19 % of the additional radiative forcing that has occurred since 1750 can be ascribed to deforestation through fire (Bowman et al. 2009), with hot spots for deforestation activities and related CO_2 emissions presently located in regions with tropical rainforest.

Second, vegetation fires can act as catalysts within the transformation of forest into savannah and other non-forested biomes. Two factors play a role. The increasing dryness of forests due to climate change results in an increased frequency of vegetation fires. Global vegetation models predict that climate change will result in a significant increase in vegetation fires. For the boreal forests of Canada and Alaska, for example, a range of vegetation and climate models predicts an increase in burned areas by 50–100 % (Lavorel et al. 2007), and radiocarbon dating has shown a marked increase in fire activity during warmer climatic periods in these regions.

There is another, precarious relation among deforestation, climate change and fires: forest thinning related to the harvesting of valuable tropical wood resources leads, as a result of the opening of the tree canopy, to the desiccation of soils and vegetation and thus significantly increases the danger of forest fires in tropical forests. Without this effect, tropical forests, because of their high degree of humidity, are not particularly vulnerable to fire hazards (cf. Cochrane 2003). The first fires within such dried-up rainforests often only affect the vegetation close to the ground, thus sparing the large trees. However, they cause a further desiccation and inflammability of the forest, establishing a vicious circle among desiccation, forest fires, deforestation and climate change. In both the Amazon and in Southeast Asia, studies have shown that during the El Nino period of 1997/98, forest fires particularly affected forest areas that had previously been thinned (Bowen et al. 1999).[3]

[3]Due to these fires, approximately 0.8–2.6 Gt of carbon, corresponding to 13–40 % of all anthropogenic carbon emissions during the same period, were released from the drained peat soils of Indonesia (Page et al. 2002).

Vegetation fires also impact the climate system in more complex—and less well understood—ways than increased CO_2 emissions alone. For example, the regularly occurring fires in savannahs are linked with the South Atlantic and the Indian Ocean in complex ways. During a savannah fire, roughly two-thirds of the nitrogen contained in the burned biomass is released in the form of different nitrogenous emissions. Although a large part of this nitrogen remains on the continent because it falls back as rain, one-third of it finds its way into the oceans and is believed to form an important source of nutrition for phytoplankton, particularly in the Indian Ocean and the South Atlantic. This net primary production (NPP) of phytoplankton drives the so-called oceanic carbon pump, which has enabled the oceans to absorb part of the CO_2 emissions. Thus, reducing vegetation fires in savannahs could contribute to a weakening of the oceanic carbon pump, although it is largely unknown how large this effect might be.

15.6 Sustainable Vegetation Fires?

If sustainability is understood as a state of society in which natural resources are used in such a way that their current use does not restrict the options for the development of future generations, it becomes clear from the discussion above that the sustainability of anthropogenic vegetation fires depends on the very specific context in which these fires take place. Anthropogenic vegetation fires have been an integral part of certain practices such as shifting cultivation, pastoralism and even hunting and gathering for several thousands of years without placing the development of future generations at risk. Even if, by current standards, the burning of huge volumes of biomass may be seen in some respects as a very wasteful use of resources, it is a wastefulness that ‘affluent societies’ (Sahlins 1972) with a sufficiently low population density can afford.

The sustainability of the traditional use of fires in shifting cultivation and pastoralism is threatened, however, if the intensification of land use causes natural stocks to be reduced to a level at which the according practice cannot be continued. In the case of shifting cultivation, this takes the form of ever shorter fallow periods as the population density increases, until this system of agriculture can no longer be maintained (Boserup 1965). Many regions, particularly in the tropics, are currently in the middle of a transition from more extensive forms of land use involving vegetation fires to more intensive forms of land use in which such practices only play a marginal role. There are at least two reasons for the inertness of such a transition. On the one hand, there is often a lack of access to material resources that are necessary for such a transition, such as fertilizers required for more-intensive forms of agriculture or knowledge about agro-ecological practices that help maintain soil fertility. On the other hand, routines that have been practiced over centuries are often deeply embedded within cultural patterns and beliefs. Thus, considering the complex interrelation of material and cultural patterns is of paramount importance for an understanding of global anthropogenic vegetation fires and their change over time.

References

Andreae, M. O. (1991). Biomass burning: Its history, use, and distribution and its impact on environmental quality and global climate. *Global biomass burning—Atmospheric, climatic, and biospheric implications* (pp. 3–21). Cambridge, MA: MIT Press.

Andreae, M. O., & Merlet, P. (2001). Emission of trace gases and aerosols from biomass burning. *Global Biogeochemical Cycles, 15*, 955–966.

Boserup, E. (1965). *The conditions of agricultural growth: The economics of agrarian change under population pressure*. London, UK: Allen & Unwin.

Bowen, M. R., Bompar, J. M., Anderson, I. P., Guizol, P., & Gouyon, A. (1999). *Anthropogenic fires in Indonesia: A view from Sumatra*. Palembang: Government of Indonesia/European Union.

Bowman, D. M. J., Balch, J. K., Artaxo, P., Bond, W. J., Carlson, J. M., Cochrane, M. A., et al. (2009). Fire in the Earth system. *Science, 324*, 481–484.

Cochrane, M. A. (2003). Fire science for rainforests. *Nature, 421*, 913–919.

Ito, A., Penner, J. E., (2004). Global estimates of biomass burning emissions based on satellite imagery for the year 2000. *Journal of Geophysical Research 109*, D14S05.

Krausmann, F., Erb, K.-H., Gingrich, S., Lauk, C., & Haberl, H. (2008). Global patterns of socioeconomic biomass flows in the year 2000: A comprehensive assessment of supply, consumption and constraints. *Ecological Economics, 65*, 471–487.

Lauk, C., & Erb, K.-H. (2009). Biomass consumed in anthropogenic vegetation fires: Global patterns and processes. *Ecological Economics, 69*, 301–309.

Lavorel, S., Flannigan, M. D., Lambin, E. F., & Scholes, M. C. (2007). Vulnerability of land systems to fire: Interactions among humans, climate, the atmosphere, and ecosystems. *Mitigation and Adaptation Strategies for Global Change, 12*, 33–53.

McGrath, D. G. (1987). The role of biomass in shifting cultivation. *Human Ecology, 15*, 221–242.

Odum, H. T. (1996). *Environmental accounting: EMERGY and environmental decision making*. New York: Wiley.

Page, S. E., Siegert, F., Rieley, J. O., Boehm, H. D. V., Jaya, A., & Limin, S. (2002). The amount of carbon released from peat and forest fires in Indonesia during 1997. *Nature, 420*, 61–65.

Rambo, A. T. (1980). Fire and the energy efficiency of swidden agriculture. *Asian Perspectives, 23*, 309–316.

Rappaport, R. A. (1968). *Pigs for the Ancestors*. New Haven, CN: Yale University Press.

Sahlins, M. (1972). *Stone age economics*. New York: Aldine de Gruyter.

Seiler, W., & Crutzen, P. J. (1980). Estimates of gross and net fluxes of carbon between the biosphere and the atmosphere from biomass burning. *Climatic Change, 2*, 207–247.

Stewart, O. C. (1956). Fire as the First Great Force Employed by Man. In W. L. Thomas (Ed.), *Man's role in changing the face of the earth* (pp. 115–133). Chicago: The University of Chicago Press.

Yevich, R., & Logan, J. A. (2003). An assessment of biofuel use and burning of agricultural waste in the developing world. *Global Biogeochemical Cycles, 17*, 1095.

Chapter 16
How Far Does the European Union Reach? Analyzing Embodied HANPP

Helmut Haberl, Thomas Kastner, Anke Schaffartzik and Karl-Heinz Erb

Abstract International trade plays an increasingly important role in supplying societies with biophysical resources and products. In terms of land-based products, trade plays an ever-greater role in meeting the resource demand of densely populated industrialized regions such as Europe—not only with relatively small volumes of luxury products such as coffee, cocoa and tropical fruits but increasingly also with large volumes of resources such as staple crops and protein feed. Meanwhile, the land resources required to produce the products consumed in Europe are global, raising issues about consumer responsibility and the accounting and regulation of environmental impacts. This chapter discusses how global land demand related to Europe's consumption can be traced using the 'human appropriation of net primary production' approach (eHANPP). This approach aims to quantify and map the total HANPP ensuing in the supply chains of the products consumed in Europe. We discuss how eHANPP can be estimated using bilateral trade matrices of biomass-based products and how this approach can help us better understand trade-related global 'teleconnections' in the land system. We show that the EU27 increasingly depends on lands outside its territory, and we discuss the implications in terms of the European Union's land-related policies.

Keywords Colonization · Land use · Human appropriation of net primary production (HANPP) · Trade · Ecological effects of consumption

H. Haberl (✉) · T. Kastner · A. Schaffartzik · K.-H. Erb
Institute of Social Ecology (SEC), Alpen-Adria University, Schottenfeldgasse 29, 1070 Vienna, Austria
e-mail: helmut.haberl@aau.at

T. Kastner
e-mail: thomas.kastner@aau.at

A. Schaffartzik
e-mail: anke.schaffartzik@aau.at

K.-H. Erb
e-mail: karlheinz.erb@aau.at

H. Haberl et al. (eds.), *Social Ecology*, Human-Environment Interactions 5,
DOI 10.1007/978-3-319-33326-7_16

16.1 Introduction

Global trade is growing exponentially. This trade includes fossil energy carriers, minerals and biomass and biomass-based products. The growth of internationally traded biomass products surpasses both the growth of global biomass harvest and the consumption of biomass-based products. From 1961 to 2009, global gross biomass trade grew exponentially at a rate of 4 % per year, considerably faster than global biomass production, which grew at 2 % (FAO 2012).

The burgeoning biomass trade makes land use a prominent example of the growing disconnect between the places of production and consumption, exemplifying the increasing difficulty in relating cause and effect, such as in terms of environmental effects related to the consumption of products with increasingly global production chains. This growing challenge to land-related sustainability efforts has motivated the search for appropriate indicators. The notion of 'teleconnections' (or 'telecouplings') has been proposed to denote the interrelations among processes in different locations. These terms were borrowed from the Atmospheric Sciences, where they denote causal relations among different weather systems or, more generally, among earth system processes in distant, seemingly unconnected regions (Steffen 2006). Related to land use, teleconnections have been discussed in connection with trade (Erb et al. 2009; Haberl et al. 2009) and with city-hinterland relations (Liu et al. 2013; Seto et al. 2012).

Data from remote sensing are used to analyze and map many aspects of land cover and, to a certain extent, of land use. Aerial photography and satellite imagery provide a range of opportunities to derive reliable and highly resolved data to characterize land surfaces, such as in terms of climate (e.g., temperature, precipitation), vegetation and land cover (e.g., forests, cropland, infrastructure areas, tundra, glaciers), plant growth, net primary production (NPP) and many other important parameters. An increasing number of important aspects of land use can be described using data from remote sensing, including sowing dates, clear-cutting of forests, number of harvests per year, land abandonment and field sizes (Kuemmerle et al. 2013). Digital maps of land cover and, to some extent, land use are becoming increasingly reliable and spatially more highly resolved.

These data sources are important because they help researchers map and hence localize the impacts of cropping, livestock husbandry, infrastructure development and forestry. They can be used to depict the intensity of the socioeconomic colonization of ecosystems using indicators of land-use intensity such as the 'human appropriation of net primary production', or HANPP (see Method Précis on Human Appropriation of Net Primary Production Production as well as Chaps. 4, 14 and 17 in this volume; Erb et al. 2013), in a spatially explicit manner, for example, as maps. This information can be used to analyze how and where society changes the land, how strongly these activities affect ecosystems, how efficient they are in terms of the desired aims (e.g., provision of food or energy) and so on. Agriculture and forestry aim to procure biomass, energy-rich materials derived from plants and ultimately photosynthesis. They harness primary production, namely, the transformation of solar (radiant) energy into energy-rich organic materials built from inorganic substances such as carbon dioxide (CO_2), water and so-called plant nutrients, such as nitrogen, phosphorous and other elements needed by plants for their growth. In short, HANPP measures the aggregate effect of (1) changes in primary production resulting from land use and (2) the harvest of biomass for food, feed, fiber, bioenergy or any other socioeconomic use on the amount of NPP remaining in ecosystems each year as an input for all wild-living organisms. HANPP maps help localize impacts of agriculture and forestry, thereby providing a basis for all interventions aiming to influence the sustainability of production systems in relation to a defined territory (Sikor et al. 2013).

Although this type of information is highly relevant to certain types of governance, such as those related to a nation's own territory, it is incapable of accounting for environmental effects related to consumption let alone localizing such effects on a map. Ever-increasing volumes of trade mean that land-system governance must increasingly take product flows into account in addition to the traditional concerns focusing on a nation's territory (Sikor et al. 2013). The famous cup of strawberry yoghurt, which includes ingredients that have travelled several thousands of kilometers before being ready for consumption on one's breakfast table, is a case in point. The cow that produced the milk may have been fed with soy from Brazil or Argentina and may have grazed in the Po Valley or in Schleswig-Holstein. The yoghurt may even have been produced from a milk mixture of wide-ranging provenance in which many other regions were involved. The production chains of strawberries, sugar, cups and lids are usually similarly complex—one might go into any depth of detail here.

In this chapter, we will discuss how one (in our view, particularly important) aspect of this conundrum can be tackled. We present an indicator that has been denoted the 'embodied human appropriation of net primary production', abbreviated as eHANPP (Erb et al. 2009; Haberl et al. 2009). eHANPP is an account of the total amount of HANPP that accrued from the provision of a defined volume of products throughout the chain of production. To use the example of the strawberry yoghurt once more, we could ask how much NPP was appropriated when producing the market and non-market feed required to sustain the cows, how much was appropriated during the cows' grazing and how much was appropriated during the production of the strawberries. One might even go further, asking how much HANPP resulted from the infrastructures required to transport all the yoghurt's ingredients and to sustain the production processes in the food industry or in fertilizer production. We will present work that covers biomass flows involved in the production process (e.g. grazing, fodder and strawberry production in the yoghurt example), but infrastructure and other indirect effects could not be considered, even though such effects could be relevant. The methods we discuss are able to account for the HANPP related to a nation's consumption in a spatially explicit manner. Using bilateral trade matrices provided by the FAO, we established a database that can show, using approximately 500 products and 200 countries, how much HANPP occurred in which countries for the production of any product consumed in any one of the countries. Similarly, these accounts show us where the products stemming from any national territory were consumed. Using data on land use or population density, the information could even be disaggregated to a grid level, although with certain simplifying assumptions (see Erb et al. 2009).

16.2 The Embodied HANPP Concept

Embodied HANPP is an extension of the HANPP concept explained elsewhere in this volume (see Method Précis on Human Appropriation of Net Primary Production as well as Chaps. 4, 14 and 17 in this volume). Whereas HANPP is related to a defined territory, eHANPP is related to consumption, thus highlighting a neglected aspect of the HANPP approach. eHANPP can help account for a central driver of land-use change, namely, the consumption of products requiring land-based production, such as agriculture and forestry. Of course, eHANPP is complementary to HANPP, not an alternative. Measures of land-use intensity are important because they allow land use to be linked to its ecological impacts, e.g., in terms of biodiversity loss (Haberl et al. 2005). Whereas HANPP is useful in accounting for a nation's exploitation of the land area that makes up its territory, eHANPP can inform us about the claims on global land resources resulting from a nation's consumption. The same distinction can be made regarding any other spatially delineated land area. We can distinguish the socioeconomic colonization of ecosystems on that land from the claims on global land resources related to the consumption of products of the people living on that land. The difference is due to trade and will usually be larger the smaller the land unit under consideration. Of course, it will also depend on the degree of market-integration versus subsistence of the population living on the land.

It is the aim of the eHANPP concept to better link land use with consumption to be able to quantify environmental demands resulting from consumption. Although this can be done at the level of products, usually within life cycle assessment approaches (LCA, see Method Précis on Life Cycle Assessment), we are more interested in an aggregate concept of consumption, or the total consumption at the level of a national economy. At this level of aggregation, eHANPP accounts provide metrics for two factors: (1) the HANPP related to the production chains of all products that are part of a nation's 'apparent consumption' (defined as domestic production plus import minus export) and (2) the destination of the products resulting from the HANPP within the territory of that nation, that is, showing where the national HANPP 'embodied' in exported products ends up. Because the analysis we discuss in this chapter was derived from bilateral trade matrices, it is possible in both cases to localize both the origin of eHANPP imports and the destination of eHANPP in exported products at the country level. The eHANPP accounts discussed here refer to a large number (500) of biomass-based products, including food (e.g., cereals, milk, meat, eggs, vegetables), fiber (e.g., cotton), feed and energy carriers (e.g., fuel wood, pellets, agro-fuels, biogas).

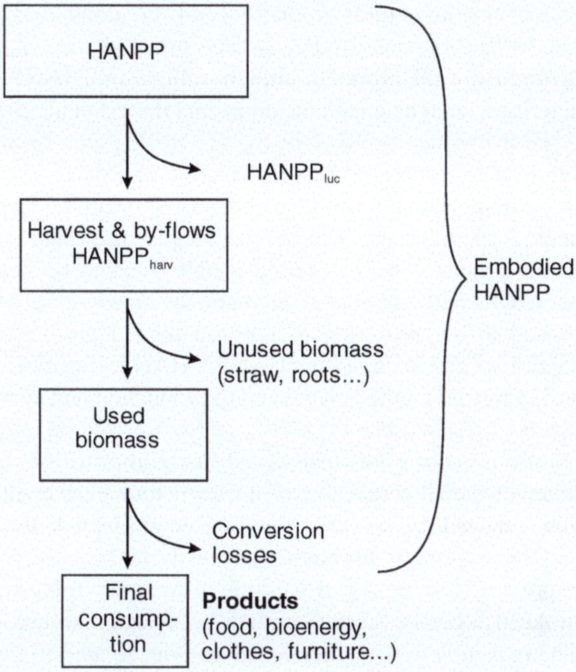

Fig. 16.1 Concept of the calculation of the embodied human appropriation of net primary production (eHANPP). *Redrawn* after (Haberl et al. 2009)

The calculation of eHANPP follows the product chain shown in Fig. 16.1: per unit of product, the calculation considers losses during the conversion of raw materials into final products, biomass flows included in harvested HANPP ($HANPP_{harv}$) but not recovered as raw materials (e.g., roots of felled trees, straw and other harvest residues left on the field) and land-use-related changes in NPP ($HANPP_{luc}$).

The way these flows are calculated depends on the aim of the study. Previous work (Erb et al. 2009; Haberl et al. 2009, 2012b) focused on net transfers between regions or nations but did not trace bilateral flows. Such studies can help researchers distinguish net-importing from net-exporting regions and map the spatial disconnect between production and consumption, but they are insufficient to depict telecouplings between individual countries or analyze them on a disaggregated level, such as in terms of individual products or product categories (e.g., biofuels). Doing so requires bilateral trade data that allow trade to be traced among individual countries at the product level.

Bilateral trade matrices such as those published by the Food and Agricultural Organization (FAO 2012) report trade flows among individual countries at a high level of disaggregation in terms of products. Such databases are a prerequisite for representing relations among regions and countries and allow the use of country-specific factors for important aspects such as area demand per unit of product (based on country-specific yields) and country-specific loss factors as well as factors for the natural productivity of the areas used (e.g., in terms of their potential NPP, abbreviated NPP_{pot}). The bilateral trade matrices of the FAO used here distinguish 500 products, thereby allowing eHANPP analysis regarding specific activities, such as meat consumption (Haberl et al. 2012a).

Globally, nearly 90 % of the HANPP results from agriculture and forestry and can hence be captured by the eHANPP approach. The remainder is HANPP related to settlements and infrastructures (4 %) and human-induced fires (7 %; Haberl et al. 2007; Lauk and Erb 2009). Because fires cannot be attributed directly to individual products, the associated HANPP cannot be 'reallocated' within an eHANPP approach. Settlement and recreational areas serve a purpose that is not directly related to the provision of biomass-related products; therefore, they cannot (and should not) be included in eHANPP. HANPP resulting from transport infrastructures and industrial facilities is related to production and consumption and could, in principle, be included in an eHANPP approach based on transport volumes, but this would require a major effort that would likely improve the results only slightly, given the relatively small percentage of infrastructure areas compared to croplands, grasslands and managed forests. Therefore, we did not include this in the present account.

One important problem that needs to be solved in such a calculation is re-exports. Consider soy. If we look at Austria's trade statistics, most of the soy used in Austria comes from Germany—a country known to produce little, if any, soy. The reason is that any economic activity related to the soy product, even if it is as insignificant as repackaging, results in Germany being considered the 'country of origin' in terms of trade statistics regardless of where the raw material was originally produced. For an eHANPP calculation, however, we need to know where the soy was originally grown. Trade statistics alone are insufficient to solve that question. They need to be combined with agricultural production statistics, which, in the case of Germany's soy exports, quickly reveal that most of the soy processed in Germany is imported from elsewhere, thereby allowing seemingly 'German' soy to be traced back to its origin, although often in several steps. The algorithms used to perform these calculations assume 'homogeneity' of the national supply, that is, that the average German soy supply structure can be applied to domestic consumption and export alike (see Kastner et al. 2011a, b). Although this assumption does not seem problematic when applied to market economies such as Germany, it might result in distortions in countries where a strong difference exists between export-oriented cash crop production and the supply of the same crops to domestic consumption based on subsistence or local markets. Although such intricacies are worth further study, the accounts discussed in this chapter do not incorporate such differentiations.

Another important problem is related to products stemming from longer chains of production. The most prominent examples are animal products that need to be traced back to the plant materials fed to the animals. Such problems were solved using the same methods as explained above for traded animal products as well as for traded feed, whereas non-market feed and grazing were allocated to the country in which the respective animal product was generated. More details on methods can be found in previous publications (Haberl et al. 2012a).

16.3 The European Union: Domestic Supply and Foreign Trade in the Last Two Decades

To show what can be done with the eHANPP approach, we discuss some results for the European Union (referring to the EU27 countries) for the 1986–2007 period. These results are derived from FAO trade matrices and FAO data on agriculture and forestry, combined with global HANPP data (Haberl et al. 2007; Krausmann et al. 2013), global land-use data (Erb et al. 2007; Monfreda et al. 2008) and data on global biomass flows (Krausmann et al. 2008). The methods are described in previous publications (Haberl et al. 2012a; Kastner et al. 2011a, b). The data required for this calculation are only available from 1986 onward for agricultural products and from 1997 onward for forestry. Statistics are available for more than 200 countries and more than 500 products, including 157 agricultural primary products and six animal primary products. The database we have established quantifies bilateral trade flows between any two countries down to the product level in terms of biomass flow as well as the land requirement and the eHANPP related to each product and country.

Figure 16.2a shows that the importance of the eHANPP related to net imports is rising slowly, reaching up to 20 % of the HANPP related to consumption in the EU27 in the later parts of the period. The inclusion of forest products (not included before 1997 due to lack of data) results in a small jump in the level of eHANPP but does not visibly affect the eHANPP related to net trade. Figure 16.2b shows the trade balance of the EU27 with the rest of the world, revealing a continuous increase in import volumes, whereas exports remain stable at an eHANPP value of roughly 200 Mt dry matter biomass per year (1 Mt = 1 million metric tons = 10^6 t).

As Fig. 16.2b shows, import from North America is losing importance, whereas import from South America plays an increasingly important role in Europe's supply with biomass-based products. Import from Asia fluctuates somewhat throughout the period, whereas Africa and the rest of the world (which includes European countries outside the EU27) play a relatively small and more or less constant role throughout the period.

The stagnation of the eHANPP related to imports and exports, which can be observed from 2001 onward in Fig. 16.2b, points to the increasing relevance of intra-European trade in the period around the 2004 enlargement of the EU. The trade among the EU27 member countries is not visible in these aggregate figures. In addition to this analysis of the EU27's supply and trade in biomass and biomass-based products at the supranational level, eHANPP data allow us to examine the role played by individual European countries in the region's overall HANPP. Figure 16.3 illustrates the level of self-sufficiency of the EU27 member countries in terms of the ratio of each nation's HANPP on its own territory to the eHANPP related to its consumption of biomass-based products. A glance at this map reveals that the 2004 enlargement of the EU almost exclusively added countries to the Union that were either balanced in terms of their eHANPP trade (Estonia, Malta, Poland, the Czech Republic, Slovakia) or were net exporters of eHANPP (Hungary, Latvia, Lithuania).

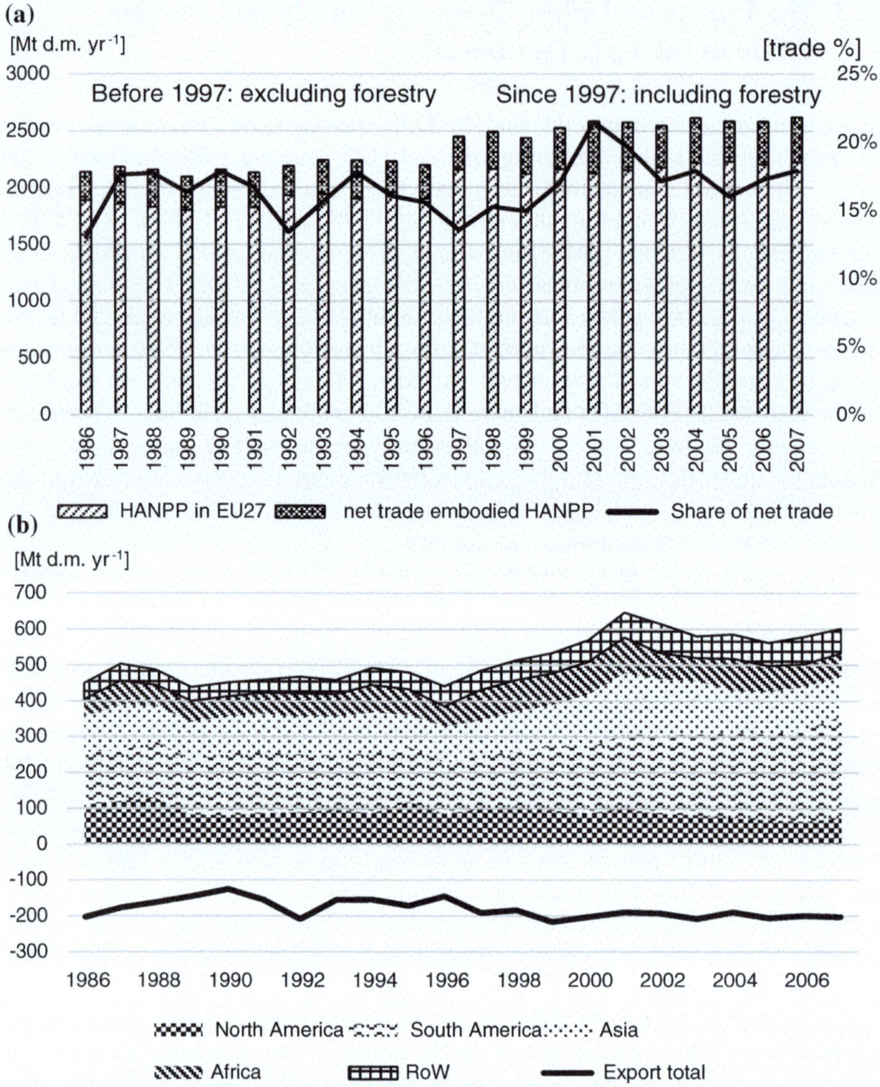

Fig. 16.2 Human appropriation of net primary production (HANPP) on national territory and embodied HANPP (eHANPP) related to consumption in the EU27. **a** HANPP on national territory and eHANPP related to net trade; **b** eHANPP related to imported products by region of origin (*positive* values) and to products exported from the EU27 (*negative* values)

Figure 16.3 shows that most Mediterranean countries (e.g., Italy, Spain, Greece) and most western European countries (France, Belgium, the Netherlands, the UK) are import-dependent, whereas central, northern and eastern European countries tend to be balanced (Germany, Poland, Czech Republic) or exporters (e.g., Hungary, the Ukraine, Finland). When interpreting the results in Figs. 16.2

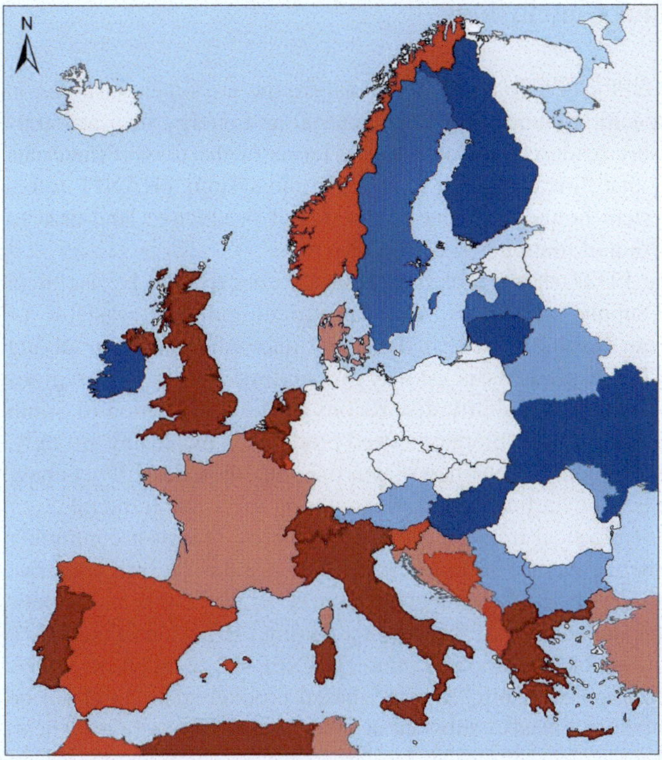

Fig. 16.3 National 'self-sufficiency' of European countries in the year 2007 in terms of biomass-based products, measured by the ratio of HANPP/eHANPP. *Warm* colors indicate import dependency. *Blue*-colored countries are net exporters in terms of eHANPP. For countries in *gray*, trade in terms of eHANPP is balanced. *Data* own calculations as described in the text

and 16.3, it is important to remember that the trade balance can differ strongly (and can even have the opposite sign) from the monetary trade balance when it is measured in physical units, such as tons, hectares or in appropriated NPP. In general, the reason is that in chains of production, the value of (intermediate) products usually rises faster than the amount of 'embodied' physical resources. For example, if a ton of timber is converted into products such as furniture, paper and pellets, losses are small, and almost all the material is used for making such products. The eHANPP of all the products resulting from one ton of timber is more or less the same as that of the ton of timber, but the monetary value of the products is, of course, much larger. Different physical indicators (e.g., the trade balance measured in tons of products or measured as eHANPP) can also differ quite substantially. For example, the eHANPP of one kilogram of beef consumed in Austria is approximately 33 kg dry matter, whereas that of bread or noodles is only approximately 2.5 kg eHANPP per kg of product (Haberl et al. 2012a).

16.4 Discussion and Conclusions

As shown above, eHANPP can help researchers analyze the resource consumption of nations in a very specific manner. Whereas material and energy flow accounts (MFA and EFA) measure resource consumption in terms of the tons of materials or the joules of energy that flow through a socioeconomic system, eHANPP refers to a quite different system boundary, namely, the use of productive land area to provide one specific group of materials (i.e., biomass).

The example of the EU27 shows that the eHANPP concept can be useful in detecting changes in supply structures. Whereas agricultural area is stable or declining in the EU, our results show that the importance of trade in supplying the EU with biomass-based products is (slowly) increasing. The role of regions with important land-related sustainability discussions (e.g., deforestation in South America) in supplying the EU with biomass-based products is increasing strongly. This suggests that displacement effects may play a role, and the eHANPP database presented here is an excellent starting point for examining this issue further.

In recent analyses of physical trade in mineral resources, the most commonly cited explanation for the increasing import-dependency of industrialized countries in meeting their demand, particularly in terms of fossil energy carriers and metals, is the depletion of domestic sources of these minerals (e.g., Bruckner et al. 2012). For biomass and biomass-based products, the eHANPP-based analysis shows that the EU27 not only imports ‘exotic’ biomass-based products that could not be harvested within its territory but also substantial amounts of cereals (e.g., wheat, maize and barley), oilseed (e.g., canola) and potatoes, all of which can be grown within the EU27. Therefore, in the analysis of trade-related eHANPP, factors that drive trade other than mere availability need to be considered, such as prices, trade agreements and government incentives and restrictions.

eHANPP accounts for the productive capacity of ecosystems ‘appropriated’ during the production of the products consumed in a national economy or region such as the EU27. It considers all land-demanding activities within the product chains considered, regardless of whether the products resulted from cropping, livestock husbandry or forestry. Land-use intensity is considered in terms of the impacts of land use on trophic energy (biomass) availability in ecosystems, thereby allowing the results to be aggregated across different land-use types in a meaningful way.

In contrast to approaches such as the ecological footprint, eHANPP considers only land use. It includes neither ‘hypothetical’ areas, such as C-sequestration areas related to fossil fuels, nor weighting factors related to different land qualities (e.g., cropland vs. grazing land or forest) or deviations between national or global yields (Haberl et al. 2001). Yield differences among different regions and points in time are considered in eHANPP accounts, as are differences in land-use intensity. In this sense, eHANPP differs from accounts of ‘actual land demand’ (Erb 2004) that sum all land required to meet national biomass demand measured in square meters (m^2) but do not differentiate differences in the intensity with which the land is used. Of course, eHANPP does not capture all aspects of land-use intensity, only those that relate to trophic energy flows in ecosystems. Other aspects of land-use intensity (see Erb et al. 2013) could be considered through appropriate extensions of the eHANPP approach.

A previous study (Haberl et al. 2012b) has shown that material flows, energy flows and the ecological footprint are highly correlated despite the different units underlying their measurement and although they attach different weights to different aspects of socioeconomic metabolism. Even emission indicators, such as fossil-energy-related CO_2 emissions, are strongly correlated with measures of total physical socioeconomic throughput and with economic activity, measured as gross domestic product (GDP). However, eHANPP behaves quite differently from indicators that mainly represent different concepts for tracing the total resource throughput of economies. Biomass use and eHANPP depend on different drivers, with population density and resource endowment among the most important, and are weakly (if at all) correlated with GDP (Haberl et al. 2012b; see Chap. 8).

This suggests that the aspect of biophysical resource use covered by eHANPP is complementary to throughput indicators and adds aspects not well covered in MFA, EFA or ecological footprint studies. In particular, it points toward issues such as the potential for efficiency increases related to closing yield gaps and for increasing the efficiency of biomass use. It also points toward the issue of substitutability (or non-substitutability) of land-based products with products based on mineral resources or other sources (Haberl et al. 2012b; Krausmann et al. 2009). In contrast, measures aiming at a substitution of products based on mineral resources with biomass-based products can have quite important feedback on land systems, including import-export relations among world regions, that could be traced using concepts such as eHANPP (Coelho et al. 2012). Hence, eHANPP can play an important role within a larger set of indicators of biophysical resource use in the sustainability context.

Acknowledgements This research was funded by the EU-FP7 project VOLANTE (grant agreement no. 265104), by the Austrian Federal Ministry of Agriculture, Forestry, Environment and Water Management (BMLFUW, contract UW.1.4.18/0081-V/10/2010), by the Austrian Science Fund (FWF), project P20812-G11 and by the Research Council (‘*Forschungsrat*’) of the Alpen-Adria Universitaet Klagenfurt, Wien, Graz. It contributes to the Global Land Project (http://www.globallandproject.org).

References

Bruckner, M., Giljum, S., Lutz, C., & Wiebe, K. S. (2012). Materials embodied in international trade—Global material extraction and consumption between 1995 and 2005. *Global Environmental Change, 22*, 568–576.

Coelho, S., Agbenyega, O., Agostini, A., Erb, K.-H., Haberl, H., Hoogwijk, M., et al. (2012). Land and water: Linkages to bioenergy. In: T. Johansson A. Patwardhan N. Nakicenonivc & Gomez-Echeverri (Eds.), *Global Energy Assessment* (pp. 1459–1525). International Institute of Applied Systems Analysis (IIASA), Cambridge University Press, Cambridge, UK.

Erb, K.-H. (2004). Actual land demand of Austria 1926–2000: A variation on ecological footprint assessments. *Land Use Policy, 21*, 247–259.

Erb, K.-H., Gaube, V., Krausmann, F., Plutzar, C., Bondeau, A., & Haberl, H. (2007). A comprehensive global 5 min resolution land-use data set for the year 2000 consistent with national census data. *Journal of Land Use Science, 2*, 191–224.

Erb, K.-H., Krausmann, F., Lucht, W., & Haberl, H. (2009). Embodied HANPP: Mapping the spatial disconnect between global biomass production and consumption. *Ecological Economics, 69*, 328–334.

Erb, K.-H., Haberl, H., Jepsen, M. R., Kuemmerle, T., Lindner, M., Müller, D., et al. (2013). A conceptual framework for analysing and measuring land-use intensity. *Current Opinion in Environmental Sustainability, 5*, 464–470.

FAO (FAOSTAT statistical database). (2012). Food and agriculture organisation of the United Nations (online) www.fao.org Accessed July 19, 2012 Rome.

Haberl, H., Erb, K.-H., & Krausmann, F. (2001). How to calculate and interpret ecological footprints for long periods of time: The case of Austria 1926–1995. *Ecological Economics, 38*, 25–45.

Haberl, H., Plutzar, C., Erb, K.-H., Gaube, V., Pollheimer, M., & Schulz, N. B. (2005). Human appropriation of net primary production as determinant of avifauna diversity in Austria. *Agriculture, Ecosystems & Environment, 110*, 119–131.

Haberl, H., Erb, K.-H., Krausmann, F., Gaube, V., Bondeau, A., Plutzar, C., et al. (2007). Quantifying and mapping the human appropriation of net primary production in earth's terrestrial ecosystems. *PNAS, 104*, 12942–12947.

Haberl, H., Erb, K.-H., Krausmann, F., Berecz, S., Ludwiczek, N., Martinez-Alier, J., et al. (2009). Using embodied HANPP to analyze teleconnections in the global land system: Conceptual considerations. *Geografisk Tidsskrift-Danish Journal of Geography, 109*, 119–130.

Haberl, H., Kastner, T., Schaffartzik, A., Ludwiczek, N., & Erb, K.-H. (2012a). Global effects of national biomass production and consumption: Austria's embodied HANPP related to agricultural biomass in the year 2000. *Ecological Economics, 84*, 66–73.

Haberl, H., Steinberger, J. K., Plutzar, C., Erb, K.-H., Gaube, V., Gingrich, S., & Krausmann, F. (2012b). Natural and socioeconomic determinants of the embodied human appropriation of net primary production and its relation to other resource use indicators. *Ecological Indicators, 23*, 222–231.

Kastner, T., Erb, K.-H., & Nonhebel, S. (2011a). International wood trade and forest change: A global analysis. *Global Environmental Change, 21*, 947–956.

Kastner, T., Kastner, M., & Nonhebel, S. (2011b). Tracing distant environmental impacts of agricultural products from a consumer perspective. *Ecological Economics, 70*, 1032–1040.

Krausmann, F., Erb, K.-H., Gingrich, S., Lauk, C., & Haberl, H. (2008). Global patterns of socioeconomic biomass flows in the year 2000: A comprehensive assessment of supply, consumption and constraints. *Ecological Economics, 65*, 471–487.

Krausmann, F., Haberl, H., Erb, K.-H., Wiesinger, M., Gaube, V., & Gingrich, S. (2009). What determines geographical patterns of the global human appropriation of net primary production? *Journal of Land Use Science, 4*, 15–33.

Krausmann, F., Erb, K.-H., Gingrich, S., Haberl, H., Bondeau, A., Gaube, V., et al. (2013). Global human appropriation of net primary production doubled in the 20th century. *Proceedings of the National Academy of Sciences, 110*, 10324–10329.

Kuemmerle, T., Erb, K.-H., Meyfroidt, P., Müller, D., Verburg, P. H., Estel, S., et al. (2013). Challenges and opportunities in mapping land use intensity globally. *Current Opinion in Environmental Sustainability, 5*, 484–493.

Lauk, C., & Erb, K.-H. (2009). Biomass consumed in anthropogenic vegetation fires: Global patterns and processes. *Ecological Economics, 69*, 301–309.

Liu, J., Hull, V., Batistella, M., DeFries, R., Dietz, T., Fu, F., et al. (2013). Framing sustainability in a telecoupled world. Ecology and Society 18, 26 http://dx.doi.org/10.5751/ES–05873–180226.

Monfreda, C., Ramankutty, N., & Foley, J.A., (2008). Farming the planet: 2 Geographic distribution of crop areas, yields, physiological types, and net primary production in the year 2000. Global Biogeochemical Cycles 22, doi: 10.1029/2007GB002947.

Seto, K. C., Reenberg, A., Boone, C. G., Fragkias, M., Haase, D., Langanke, T., et al. (2012). Urban land teleconnections and sustainability. *Proceedings of the National Academy of Sciences, 109*, 7687–7692.

Sikor, T., Auld, G., Bebbington, A. J., Benjaminsen, T. A., Gentry, B. S., Hunsberger, C., et al. (2013). Global land governance: From territory to flow? *Current Opinion in Environmental Sustainability, 5*, 522–527.

Steffen, W. (2006). The arctic in an earth system context: From brake to accelerator of change. *Ambio, 35*, 153–159.

Chapter 17
Africa's Land System Trajectories 1980–2005

Maria Niedertscheider, Tamara Fetzel, Helmut Haberl, Fridolin Krausmann, Veronika Gaube, Simone Gingrich, Christian Lauk, Christoph Plutzar and Karl-Heinz Erb

Abstract Global surges in biomass demand driven by an increasing world population, nutrition transitions in the developing world and increasing consumption of bioenergy are pressuring African land-use systems and threatening the sustainability of yet-intact ecosystems. Africa is not exempt from this land rush. In fact, much hope is placed in African lands for future increases in biomass production. However, it is still unclear whether and how African socioeconomic systems are able to meet these expectations without posing a further challenge to societal and natural integrity. In this chapter, we present an integrated analysis of land-use trajectories in Africa between 1980 and 2005

M. Niedertscheider (✉) · T. Fetzel · H. Haberl · F. Krausmann · V. Gaube · S. Gingrich · C. Lauk · C. Plutzar · K.-H. Erb
Institute of Social Ecology (SEC), Alpen-Adria University, Schottenfeldgasse 29, 1070 Vienna, Austria
e-mail: maria.niedertscheider@aau.at

T. Fetzel
e-mail: tamara.fetzel@aau.at

H. Haberl
e-mail: helmut.haberl@aau.at

F. Krausmann
e-mail: fridolin.krausmann@aau.at

V. Gaube
e-mail: veronika.gaube@aau.at

S. Gingrich
e-mail: simone.gingrich@aau.at

C. Lauk
e-mail: christian.lauk@aau.at

C. Plutzar
e-mail: christoph.plutzar@aau.at

K.-H. Erb
e-mail: karlheinz.erb@aau.at

H. Haberl et al. (eds.), *Social Ecology*, Human-Environment Interactions 5,
DOI 10.1007/978-3-319-33326-7_17

through the framework 'human appropriation of net primary production (HANPP)'. Based on our analysis, we will discuss the current stage and main determinants of African land systems and their change over the past 30 years. Since 1980, biomass harvest has increased markedly in Africa. However, this increase was chiefly driven by cropland expansion rather than yield increases. Consequently, the increase of biomass harvest was associated with a considerable HANPP that, for some regions, was more than twice the amount of biomass harvest. Exceptions are Northern and Southern Africa, where levels of land-use intensification were higher and, consequently, productivity losses were lower and decreasing. In the rest of Africa, productivity losses increased unhalted, in contrast to the land-use trajectories in most other parts of the world.

Keywords Africa · Human appropriation of net primary production (HANPP) · Land-use change · Land-use intensification · Agricultural expansion

17.1 The Role of Biomass in the African Context

At present, many African countries still heavily rely on agriculture and biomass production as central features of their economies. Although nonrenewable resources have become increasingly important during the last few decades, biomass still constitutes more than 50 % of Africa's domestic material extraction (EDAR 2012). Particularly in Sub-Saharan Africa (SSA), the share of the population employed in agriculture as well as the share contributed by agriculture to the total GDP are both far above the global average, and many households depend on subsistence farming for their livelihood and as their primary source of income. The situation is somewhat different in the Northern African oil-exporting countries as well as in the Republic of South Africa. Their economies are based to a much larger extent on nonrenewable resources, and a smaller fraction of their populations are employed in agriculture (FAO 2013).

In many regions such as Europe, North America and large parts of Southeast Asia, crop yields strongly increased throughout the second half of the 20th century. These increases were based on higher inputs of synthetic fertilizers and pesticides in connection with the use of high-yielding crop varieties and, in many regions, increased irrigation, all of which helped to support larger populations and/or richer diets. In Africa, yield increases played a much smaller role. Crop yields increased only slightly during the 20th century, especially in Central, Western and Eastern Africa. In terms of technological development, lower yield growth mainly resulted from the following: a much slower diffusion of high-yielding crop varieties in SSA; only marginal use of industrial inputs such as synthetic fertilizers (Evenson and Gollin 2003); low use of agro-ecological methods to replenish soil fertility (Pretty et al. 2006, 2011), as well as low levels of irrigation. Furthermore, slow agricultural productivity growth in many African countries has been associated with weak agricultural policies and institutions as well as with political turmoil and wars, which displaced labor force and livestock and prevented the emergence of an efficient system of agricultural production and distribution (FAO 2013).

Analyzing the causalities of Africa's agricultural performance on a more general level is a complex endeavor because it requires a profound understanding of the multitude of socioeconomic and natural factors that (co-) determine the trajectories of land use and associated biomass flows, particularly within the food system. Environmental constraints to agriculture include the particularly uneven rainfall distribution, reoccurring droughts and unfavorable soil properties in many African regions. These conditions reduce the suitability of marginal land for crop production (Showers 2012). In areas heavily affected by such natural limitations, high crop yields require considerable investments, such as in terms of external nutrient and water inputs, which are often difficult to secure given the particular socioeconomic conditions of agriculture in many African regions. High levels of poverty prevail in many rural areas, and a substantial share of farm families must subsist on less than 1 US dollar per capita per day. In addition, limited market access in combination with weak state institutions strongly reduces the capability of farmers to increase monetary income, to purchase fertilizers and machinery and to implement irrigation techniques. Consequently, meeting an increasing demand for biomass often forces farmers to abandon traditional fallow cycles and overstock their land, which leads to nutrient depletion and soil degradation, particularly in dryland areas (Sánchez 2010). Success stories of increasing crop yields in SSA, such as maize production in Malawi, have shown that state support schemes for agriculture and knowledge transfer are central for increasing agricultural outputs per unit of land (ibid.).

Nevertheless, in the last decade, SSA has become increasingly attractive to international investors due to increasing land demand for biomass production. This increasing demand is particularly driven by the booming biofuel industry and rising food security concerns (Cotula et al. 2009; Showers 2012). Foreign investments into African agriculture are often defended on the grounds that vast swaths of African land are currently unused or marginal and are thus available for crop production. In addition, foreign investment is promoted as an opportunity for technology and knowledge transfer. However, the extent to which these lands are really 'unused,' especially in terms of subsistence farmers or herders not accounted for in official economic statistics, is controversial (Showers 2012; Young 1999). At present, Africa heavily depends on food imports (Rakotoarisoa et al. 2011), and its food requirements are expected to grow strongly because it has the strongest population growth of all world regions—Africa's population is expected to more than double, from 1.1 billion people in 2013 to 2.4 billion in 2050 (UNDESA 2013). Given that diets in many African regions are among the poorest worldwide, food consumption per capita per day will also hopefully increase in terms of both quantity and quality. These considerations cast doubt on the notion that Africa has a high potential to become a major exporter of food or fuel in the next decades. Moreover, land deals are often characterized by an imbalance of power between selling and purchasing stakeholders. Thus, these land deals often threaten the land-use rights of local communities.

17.2 The HANPP Framework as a Tool to Analyze Land System Dynamics

Historical analyses of land-use change are particularly useful to better understand the patterns and trajectories of land-use change in Africa. They allow the tracing of possible drivers and legacies of land-use dynamics. This chapter discusses land-system change in Africa based on the human appropriation of net primary production (HANPP) framework for the period 1985 to 2005. HANPP is a systemic indicator framework that explicitly allows changes in the extent of land-use types as well as land-use intensity to be addressed (see Chaps. 4 and 14). The framework is built on the concept of net primary production (NPP). NPP is the amount of biomass produced by photosynthetic organisms (e.g., green plants), net of their own metabolic needs. Annual NPP is the energetic basis of almost all heterotrophic organisms on earth (animals, fungi, many microorganisms). It is an important measure of the productivity of ecosystems. One important measure required in calculating HANPP is potential natural productivity (NPP_{pot}), that is, the natural (hypothetical) NPP of the vegetation assumed to prevail in the absence of any human land use. In the HANPP system, NPP_{pot} is used as a reference with which current NPP or harvest is compared. HANPP is then defined as the difference between NPP_{pot} and the NPP that remains in ecosystems after harvest under current conditions (the latter is denoted NPP_{eco}). This difference is the sum total of two distinct processes: (1) alterations of NPP resulting from the conversion of natural vegetation to the currently prevailing vegetation ($HANPP_{luc}$) and (2) harvest of biomass, including biomass that is destroyed or burned during harvest ($HANPP_{harv}$). HANPP can be calculated consistently across time and space and has played a large role in helping us understand long-term trajectories of society-nature interactions in Long-Term Socioecological Research (LTSER) approaches (Singh et al. 2013). Details on the definition and quantification of HANPP are given in the Method Précis on Human Appropriation of Net Primary Production.

Our analysis of HANPP patterns and trends in Africa involves two interrelated steps. First, we present HANPP patterns on a country level based on medium-resolution spatially explicit data (10 × 10 km at the equator) originating from a global study on HANPP for the year 2000 (Haberl et al. 2007). In this part, we will also highlight the importance of dryland degradation for the African HANPP patterns based on results from a global study by Zika and Erb (2009). Second, we present HANPP trajectories from 1980 to 2005 for the five African macro-regions (Northern-, Southern-, Western-, Eastern-, Central-Africa) based on the results of a global study by Krausmann et al. (2013). All values are three-year averages around the given year and refer to total (above- and belowground) HANPP.

17.3 Patterns of HANPP in the Year 2000

Figure 17.1 presents HANPP values in tons carbon per hectare per year (tC/ha/yr). Compared to the global average (1.1 tC/ha/yr), the average African HANPP was

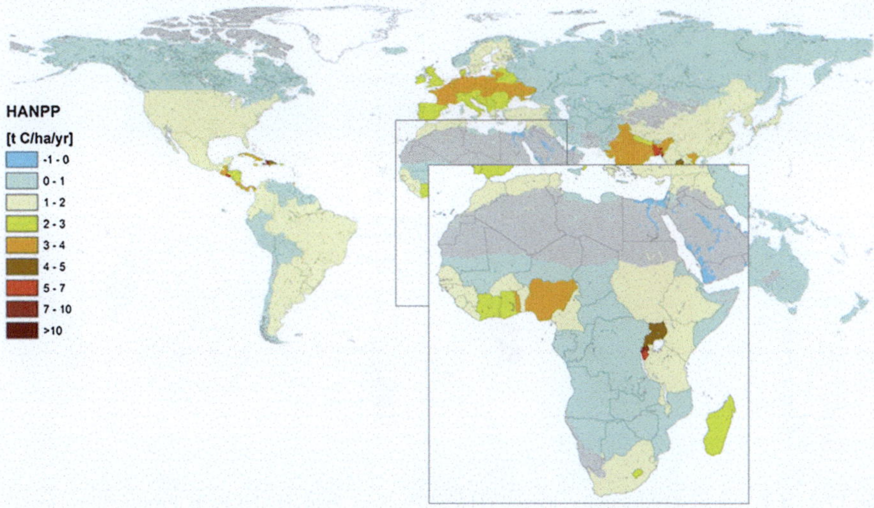

Fig. 17.1 HANPP per unit area (tC/ha/yr) in the year 2000, excluding unproductive areas

at an intermediate level (0.7 tC/ha/yr) in the year 2000. Non-productive areas are excluded from the calculation, such as areas covered with hyper-arid, vegetation-free deserts (gray color in Fig. 17.1). The highest HANPP levels worldwide (4–7 tC/ha/yr) can be observed in some Eastern African countries, basically covering the territories of Rwanda, Uganda and Burundi. Relatively high HANPP values (between 2 and 4 tC/ha/yr) can also be found along the Western Atlantic coast (Nigeria, one of the most densely populated countries on the African continent, Togo, Ghana, Cote D'Ivoire) and in Madagascar. These countries are hotspots of high HANPP. In all other countries, HANPP is rather low, ranging from 0 to 2 tC/ha/yr. These values are far below the average HANPP values in highly industrialized world regions such as Europe, Southeast Asia and Central America, where average HANPP levels were at or above 2.2 tC/ha/yr in the year 2000 (Haberl et al. 2007).

HANPP levels per unit area that are above average usually go hand in hand with high population density and a high fraction of total land area covered with croplands and settlement areas. Such regions are often characterized by a very high land-use intensity. On the African continent, this is particularly the case in Nigeria, as well as in Uganda and Rwanda. Similarly high population densities and HANPP levels can be found in countries outside Africa, such as in India, most European countries and Central America.

Disaggregating HANPP into its two components, $HANPP_{harv}$ and $HANPP_{luc}$, provides an indication of how much of the biomass appropriated is potentially useful for society ($HANPP_{harv}$) in comparison to the NPP lost through land conversion or land degradation ($HANPP_{luc}$). Figure 17.2a shows the ratio of productivity losses to total HANPP ($HANPP_{luc}$/HANPP, numbers are given as $HANPP_{luc}$ as a percentage of HANPP), which is an indicator of land-use inefficiency because

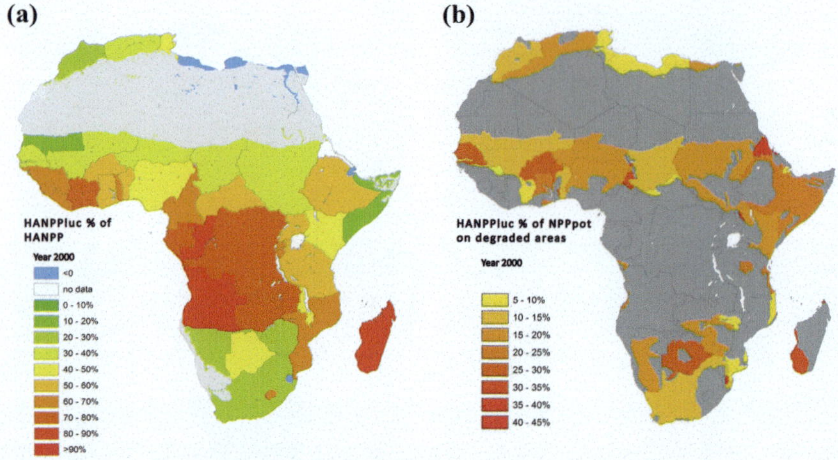

Fig. 17.2 **a** Ratio of productivity losses ($HANPP_{luc}$) to total HANPP; **b** Productivity losses ($HANPP_{luc}$) as percentage of the potential productivity (NPP_{pot}) resulting from human-induced dryland degradation. *Dark gray* areas indicate no prevalence of dryland areas

it yields high values when NPP losses represent a large fraction of the overall HANPP. Most of Africa is characterized by comparatively high $HANPP_{luc}$/HANPP ratios, with the African average at approximately 52 %, whereas globally, the fraction of $HANPP_{luc}$ to HANPP was approximately 45 % in 2000. In Western Europe, the ratio was only 18 %, indicating that due to the combined effects of natural and socioeconomic conditions, such as favorable climatic preconditions (abundance of water and fertile soils) and high degrees of land management (e.g., high levels of land-use intensity), the lion's share of HANPP consists of harvested NPP. It also indicates that productivity losses were minimized in the past.

In the Democratic Republic of Congo, Angola, Mozambique, Cote d'Ivoire and the Congo, $HANPP_{luc}$ was found to exceed 75 % of the total HANPP. Apparently, countries with a high level of HANPP per unit area (Fig. 17.1) also show high ratios of $HANPP_{luc}$/HANPP, which is particularly true for the densely populated areas of Nigeria and the Lake Victoria region (Uganda, Burundi and Rwanda).

As indicated in Fig. 17.2b, human-induced land degradation is one important factor that contributes to high $HANPP_{luc}$ values in many African dryland areas. Human-induced dryland degrada-tion is responsible for productivity losses of up to 40 % of the potential productivity in countries such as Uganda, Eritrea, Swaziland, Madagascar, Cameroon, Togo, Burkina Faso and Senegal. These losses occur in landscapes where water is chronically scarce and soils are fragile and are mainly used for subsistence agriculture and livestock grazing. Such is the case in the Sahel region, where nutrient mining and animal overstocking have been identified as crucial factors contributing to high degradation levels (Sissoko et al. 2011). The restoration of degraded soils depends on the degree of productivity losses, the investment of capital, technology and know-how (Daily 1995) and scarce resources, especially where poverty levels are high.

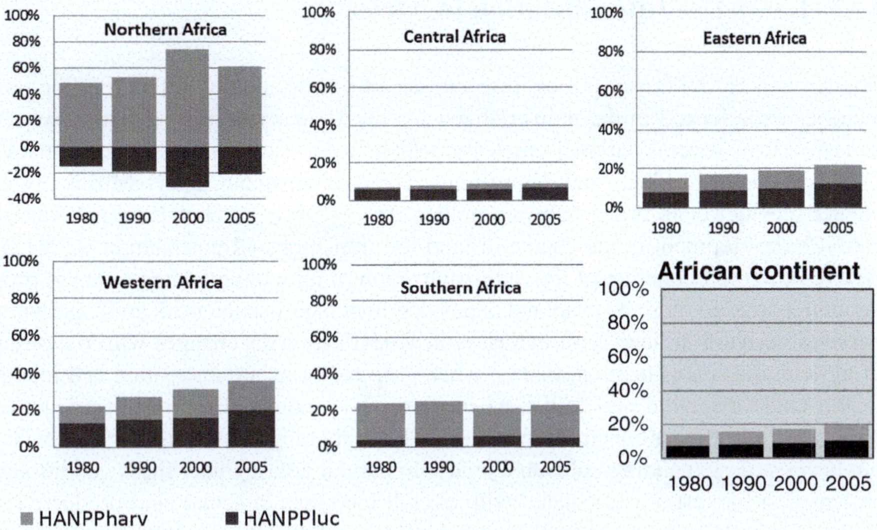

Fig. 17.3 The trend in HANPP and its components ($HANPP_{harv}$ = harvest; $HANPP_{luc}$ = productivity losses) as percentage of NPP_{pot} from 1980 to 2005

17.4 Past Trajectories of Biomass Flows

Although the African HANPP was at medium levels compared to other world regions in the year 2000, it has increased strongly since 1980 (Fig. 17.3). HANPP rose continuously over the past 25 years in almost all regions, with the exception of Northern Africa and Southern Africa. It increased most rapidly in West Africa (+ 64 %) and East Africa (+ 47 %), whereas the increase was lower in Central Africa (+ 29 %). However, rising HANPP was accompanied by rising harvest in all regions, predominantly harvest on croplands (not shown here). The strongest increases were found in Central Africa, where harvest nearly doubled between 1980 and 2005, followed by Western Africa, Eastern Africa and Northern Africa, where harvest increased by 78, 43 and 24 %, respectively. In Southern Africa, harvest declined slightly, by approximately 10 % between 1980 and 2005. Although increasing harvest was important for sustaining a growing African population, harvest increases did not keep pace with population growth. Consequently, all regions showed declining levels of $HANPP_{harv}$ per capita over the past 25 years (not shown here) except Western Africa, where the level remained constant.

Similar to harvest, $HANPP_{luc}$ increased along with rising HANPP levels in all regions except Southern and Northern Africa over the past 25 years (Fig. 17.3). This indicates that harvest increases were not derived from increased outputs per unit of land but from the expansion of cultivated land, without significant improvements in crop management. In Central Africa, Eastern Africa and the entire continent, these losses were even larger than the harvested fraction of $HANPP_{harv}$. High $HANPP_{luc}$ combined with low fractions of $HANPP_{harv}$ indicate low levels of HANPP efficiency.

17.5 Land-Use Intensification in Africa

The African HANPP trajectories in recent decades substantially differ from the trajectories experienced in the industrialized world. An analysis of long-term changes in HANPP in several national cases for which historic time series were available (Kastner 2009; Kohlheb and Krausmann 2009; Krausmann 2001; Musel 2009; Niedertscheider et al. 2012; Schwarzlmüller 2009) revealed archetypical patterns of HANPP development in the course of land-use transitions (Krausmann et al. 2012). Before or at the beginning of the industrialization of agriculture, when biomass production increased more by cropland expansion than land-use intensification, growing harvests go hand in hand with increases in $HANPP_{luc}$. This changes with the onset of agricultural intensification, that is, when crop rotations are intensified and inputs to the land (irrigation and fertilizer) increase. The industrialization of agriculture allows for a higher productivity per unit of land, thereby increasing harvests without necessarily expanding cultivated land into natural areas. Under these conditions, increases in harvest are associated with overall increases in actual land productivity (NPP_{act}), which leads to a reduction of $HANPP_{luc}$. For this reason, land use in industrialized systems is often characterized by a stabilization of HANPP at a high level (sometimes even a slight reduction), even when $HANPP_{harv}$ continues to increase. A concentration of land use to the most productive agricultural areas combined with high rates of land-use intensity can even lead to decreasing HANPP levels, which is particularly witnessed in areas where lands of marginal productivity are taken out of agricultural production and natural re-succession of forests occurs.

As discussed above, African HANPP trajectories in most countries show a tight coupling of increases in biomass harvest and $HANPP_{luc}$, which indicates that the industrialization of land use in most African regions is still in its infancy. This notion is supported by the trends of input intensification, namely, by the trajectories of fertilizer and machinery use. Figure 17.4a illustrates that African levels of fertilizer use are far below global average levels since 1980 and have not significantly increased over time.

South Africa is somewhat unique in this case because both its fertilizer consumption per hectare and its level of tractor use were at or slightly below the world average in 1980 but showed a strong decline thereafter. This reflects the economic crisis of the Apartheid regime in the Republic of South Africa (Niedertscheider et al. 2012). The share of people employed in agriculture declined in all regions but still was above the world average (with the exception of South and Northern Africa), indicating a strong dependency of African economies on agriculture as a sector of employment (Diao et al. 2007). This also indicates that apparently low levels of fertilizer and machinery use were offset by high inputs of labor. Cereal yields are closely linked to the intensity of agricultural management, particularly to agricultural inputs. Thus, it is not surprising that, similar to fertilizer consumption and machinery use, African cereal yields were below the world average and remained largely constant from 1980 to 2005 (Fig. 17.4d). Again, Southern and Northern Africa, which are characterized by a more industrialized pattern than the rest of Africa, show higher yields and faster yield growth rates than the rest of Africa.

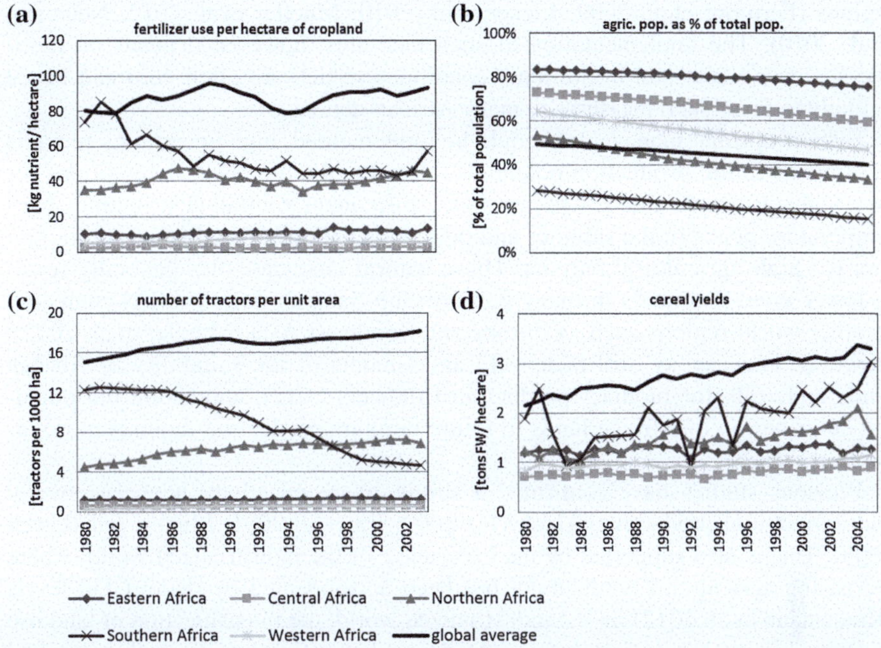

Fig. 17.4 Dynamics of indicators of land-use intensity in Africa compared to the world average. **a** Fertilizer use per hectare of arable land and permanent crops; **b** agricultural population as percentage of total population; **c** tractors per 1,000 ha of land; **d** cereals yields in tons fresh weight (FW) per hectare of arable land and permanent cropland. (Sources: (FAO 2013; own calculations))

17.6 Perspectives on African Land System Change

Three main aspects characterize the levels and developments of HANPP in Africa since 1980. First, HANPP in Africa is lower than the average of the industrialized world. Second, HANPP increased steadily in all African regions except Northern and Southern Africa from 1980 to 2005. Third, rises in $HANPP_{harv}$ have been associated with increasing $HANPP_{luc}$, indicating that the growth in harvest was the result of an expansion of agricultural land rather than of increasing productivity on existing croplands, which also resulted in rising losses of NPP. One major driver that resulted in increases of $HANPP_{luc}$ in many areas was human-induced dryland degradation. Yield improvements, by contrast, were moderate in most countries compared to other world regions.

Rising demand for biomass and shrinking land resources for further agricultural expansion have motivated research to identify potentials for future yield increases, globally as well as in Africa. Several studies indicate that yield gaps, defined as the difference between actual yields and region-specific potential yields, could be reduced through improved nutrient management, better market access, improved access to labor and the introduction of new crop genotypes in many African regions (Fermont et al. 2009; Licker et al. 2010; Mueller et al. 2012; Neumann et al. 2010). The implementation of such measures, however, depends on farmers' socioeconomic and institutional conditions, which are often poor and hence unlikely to favor such measures in many African regions.

Even if optimal management could be implemented, high uncertainty remains in terms of future production potentials related to the biophysical conditions for crop cultivation. In many African areas, infrequent rainfall distribution, high temperature peaks in the summer and poor soils are considered key limiting factors for high agricultural outputs. These natural characteristics generally result in lower yield potentials in many parts of Sub-Saharan Africa (SSA) compared to other world regions such as Europe and Southeast Asia (Mueller et al. 2012). However, because we still lack valid assessments of the suitability of African lands and soils for biomass production (Showers 2012), appraising the potential contribution of African lands to future biomass production requires more in-depth research.

Previous studies have confirmed a strong interconnectivity between poverty and agricultural productivity (IAASTD 2009; Irz et al. 2001; Minten and Barrett 2008). This is also supported by the experience of the industrialized world, where increasing agricultural productivity has been accompanied by rising GDP levels (Krausmann et al. 2012). In the industrialized world, the intensification of land use facilitated rising biomass outputs per land unit, but at the same time, they resulted in substantial environmental trade-offs and detriments, such as biodiversity losses, freshwater depletion, soil degradation and high levels of greenhouse gas (GHG) emissions (Fischer-Kowalski and Haberl 2007; Krausmann et al. 2008). Thus, from an environmental point of view, it is questionable whether African socioeconomic systems will benefit from a repetition of land-use trajectories, which in the past led to severe environmental externalities in other world regions. Moreover, according to Fischer-Kowalski and Haberl (2007), historic transitions from agrarian-based toward industrialized societies have always been marked by an amplification of per capita metabolic rates and by accelerating material consumption, factors that were mainly powered by the abundance of cheap and readily available fossil fuel energy carriers. However, because these energy sources are expected to become scarcer and gradually more expensive in the future, the preconditions for an industrial take-off in any region of the world, particularly in Africa, are likely to change in the near to mid-term future.

Introducing a second Green Revolution by means of sustainable intensification of agriculture has been suggested as a preferable option for increasing the productivity of African croplands while supporting smallholder livelihoods and keeping environmental impacts low (Pretty et al. 2006, 2011). However, no novel approaches to 'sustainable intensification' have been established yet, certainly not for implementation in structurally poor regions with a high risk of governance failure. Thus, evaluating the African potential for a second Green Revolution requires further research ranging from the field level (Lynch 2007) to the system level.

17.7 Concluding Remarks

Drawing conclusions on the sustainability of African land-use trajectories is extremely challenging. The tension underlying much of Sustainability Science, namely, balancing societal needs with ecological functions, is also a central conundrum. From an ecological perspective, Africa's low to average HANPP levels indicate a relatively low degree of human domination of ecosystems. Lower levels of HANPP imply that high fractions of productivity remain in ecosystems, with potential positive feedbacks on a set of ecosystem functions, such as biodiversity and carbon storage.

In contrast, the analysis of African HANPP shows that productivity losses are high but that harvested NPP is low in most regions of SSA. This is unfortunate from a socioeconomic point of view because a substantial share of the rural population depends on biomass harvest as the primary source of income and for sustaining their livelihoods. This situation is likely to be exacerbated in the near future considering that population numbers in these regions are expected to continue rising. Many African countries, particularly in Northern Africa, already rely to a high degree on biomass imports to sustain their food demand, which renders them vulnerable to price fluctuations on the world market. Because the prices of agricultural goods are expected to rise and because international trade is forecasted to surge, boosting national agricultural productivity is a prerequisite for guaranteeing food security for the African population, particularly in economically weak nations.

Hence, the fate of African HANPP trajectories will depend on the trends of regional and global biomass demand, on the one hand, and on the developments of HANPP efficiency in the coming decades, on the other hand. Rising African HANPP efficiency would have the potential to decouple harvest from HANPP trends, facilitating rising biomass production without boosting HANPP levels. Nevertheless, increasing HANPP efficiency while keeping environmental trade-offs low requires integrated measures to tackle the entire land-use system, which includes socioeconomic and institutional conditions and the technical conditions of agricultural production. However, investing in further in-depth research is mandatory to close the many knowledge gaps regarding the potentials, limits and complex dynamics of future African land system change.

Acknowledgements The results of this chapter are based on the UNCTAD's 2012 report *Economic Development in Africa: Structural Transformation and Sustainable Development in Africa* (available online: http://unctad.org/en/pages/PublicationWebflyer.aspx?publicationid=131). Furthermore, we gratefully acknowledge co-funding from the projects FWF P20812-G11 'Analysing Global HANPP' and EU-FP7 Volante 265104 as well as the ERC Starting Grant No. 263522 LUISE.

References

Cotula, L., et al. (2009). Land grab or development opportunity?: Agricultural investment and international land deals in Africa, Iied.

Daily, G. C. (1995). Restoring value to the world's degraded lands. *Science, 269*(5222), 350–354.

Diao, X., et al. (2007). The role of agriculture in development: Implications for sub-saharan africa, International Food Policy Research Institute (IFPRI).

EDAR, (2012). Economic development in africa report 2012: Structural transformation and sustainable development in africa, United Nations Conference on Trade and Development.

Evenson, R. E., & Gollin, D. (2003). Assessing the impact of the Green Revolution, 1960 to 2000. *Science, 300*(5620), 758–762.

FAO, (2013). FAOSTAT: Statistical database of the united nations food and agricultural organization. Retrieved from http://faostat.fao.org/.

Fermont, A. M., et al. (2009). Closing the cassava yield gap: An analysis from smallholder farms in East Africa. *Field Crops Research, 112*(1), 24–36.

Fischer-Kowalski, M., & Haberl, H., (2007). Socioecological transitions and global change: Trajectories of social metabolism and land use, Edward Elgar Publishing.

Haberl, H., et al. (2007). Quantifying and mapping the human appropriation of net primary production in earth's terrestrial ecosystems. *Proceedings of the National Academy of Sciences, 104*(31), 12942.

IAASTD, (2009). Agriculture at a crossroads. International assessment of agricultural knowledge, science and technology for development. The Global Report, Washington DC: Island Press. Retrieved from http://www.agassessment.org/.

Irz, X., et al. (2001). Agricultural productivity growth and poverty alleviation. *Development policy review, 19*(4), 449–466.

Kastner, T. (2009). Trajectories in human domination of ecosystems: Human appropriation of net primary production in the Philippines during the 20th century. *Ecological Economics, 69*(2), 260–269.

Kohlheb, N., & Krausmann, F. (2009). Land use change, biomass production and HANPP: The case of Hungary 1961–2005. *Ecological Economics, 69*(2), 292–300.

Krausmann, F. (2001). Land use and industrial modernization: An empirical analysis of human influence on the functioning of ecosystems in Austria 1830–1995. *Land Use Policy, 18*(1), 17–26.

Krausmann, F., et al. (2008). The global sociometabolic transition. *Journal of Industrial Ecology, 12*(5–6), 637–656.

Krausmann, F., et al. (2012). Long-term trajectories of the human appropriation of net primary production: Lessons from six national case studies. *Ecological Economics, 77*, 129–138.

Krausmann, F., et al., (2013). Global human appropriation of net primary production doubled in the 20th century. *Proceedings of the National Academy of Sciences, 21*.

Licker, R., et al. (2010). Mind the gap: How do climate and agricultural management explain the yield gap of croplands around the world? *Global Ecology and Biogeography, 19*(6), 769–782.

Lynch, J. P. (2007). Turner review no. 14. Roots of the second green revolution. *Australian Journal of Botany, 55*(5), 493–512.

Minten, B., & Barrett, C. B. (2008). Agricultural technology, productivity, and poverty in madagascar. *World Development, 36*(5), 797–822.

Mueller, N.D., et al. (2012). Closing yield gaps through nutrient and water management. Nature. Retrieved from http://www.nature.com/nature/journal/vaop/ncurrent/full/nature11420.html Accessed January 16, 2013.

Musel, A. (2009). Human appropriation of net primary production in the United Kingdom, 1800–2000: Changes in society's impact on ecological energy flows during the agrarian-industrial transition. *Ecological Economics, 69*(2), 270–281.

Neumann, K., et al. (2010). The yield gap of global grain production: A spatial analysis. *Agricultural Systems, 103*(5), 316–326.

Niedertscheider, M., Gingrich, S., & Erb, K.-H. (2012). Changes in land use in South Africa between 1961 and 2006: An integrated socio-ecological analysis based on the human appropriation of net primary production framework. Regional Environmental Change.

Pretty, J. N., et al. (2006). Resource-conserving agriculture increases yields in developing countries. *Environmental Science and Technology, 40*(4), 1114–1119.

Pretty, J. N., Toulmin, C., & Williams, S. (2011). Sustainable intensification in African agriculture. *International journal of agricultural sustainability, 9*(1), 5–24.

Rakotoarisoa, M. A., Iafrate, M., & Paschali, M. (2011). *Why has Africa become a net food importer*. Food and Agriculture Organization of the United Nations: Trade and Markets Division.

Sánchez, P. A. (2010). Tripling crop yields in tropical Africa. *Nature Geoscience, 3*(5), 299–300.

Schwarzlmüller, E. (2009). Human appropriation of aboveground net primary production in Spain, 1955-2003: An empirical analysis of the industrialization of land use. *Ecological Economics, 69*(2), 282–291.

Showers, K. (2012). Land use from below: Biofuels, urbanization and sustainable soil management in Europe and Africa. Retrieved from http://sro.sussex.ac.uk/11769/ Accessed October 11, 2013.

Singh, S.J., et al. (2013). Long term socio-ecological Research: Studies in society: Nature Interactions Across Spatial and Temporal Scales, Springer.

Sissoko, K., et al. (2011). Agriculture, livelihoods and climate change in the West African Sahel. *Regional environmental change, 11*(1), 119–125.

UNDESA. (2013). *World population prospects: The 2012 re vision*. United Nations, Department of Economic and Social Affairs, Population Division: Highlights and Advance Tables.

Young, A. (1999). Is there really spare land? A critique of estimates of available cultivable land in developing countries. *Environment, Development and Sustainability, 1*(1), 3–18.

Zika, M., & Erb, K.-H. (2009). The global loss of net primary production resulting from human-induced soil degradation in drylands. *Ecological Economics, 69*(2), 310–318.

Chapter 18
Of Birds and Bees: Biodiversity and the Colonization of Ecosystems

Christoph Plutzar, Karl-Heinz Erb, Veronika Gaube, Helmut Haberl and Fridolin Krausmann

Abstract Biodiversity is an important aspect of Earth's ecosystems, and is closely related to the provision of essential goods and services for human society. In this chapter, we discuss concepts and causes of biological diversity and, based on the socioecological interaction model, the relationship between human society and biodiversity. Furthermore, we present two empirical studies relating an energy-based pressure indicator to species richness numbers. This indicator, the human appropriation of net primary production (HANPP), can be linked to one of the major theories explaining biodiversity, the species-energy hypothesis, as well as to socioecological processes such as the consumption of biomass. Both studies are located in Austria. The first focuses on a transect covering cultural landscapes in the East of Austria and investigates several autotrophic and heterotrophic taxonomical groups. The second compares spatial patterns of avian diversity and HANPP for the whole Austrian territory. Both studies find clear evidence of a positive relationship between energy availability and species richness numbers, supporting the species-energy hypothesis. Moreover, these results suggest the suitability of applying an energy-related pressure indicator such as the HANPP to biodiversity.

Keywords Biodiversity · Colonization · Human appropriation of net primary production (HANPP) · Pressure indicator · Species-energy hypothesis

C. Plutzar (✉) · K.-H. Erb · V. Gaube · H. Haberl · F. Krausmann
Institute of Social Ecology (SEC), Alpen-Adria University,
Schottenfeldgasse 29, 1070 Vienna, Austria
e-mail: christoph.plutzar@aau.at

K.-H. Erb
e-mail: karlheinz.erb@aau.at

V. Gaube
e-mail: veronika.gaube@aau.at

H. Haberl
e-mail: helmut.haberl@aau.at

F. Krausmann
e-mail: fridolin.krausmann@aau.at

H. Haberl et al. (eds.), *Social Ecology*, Human-Environment Interactions 5,
DOI 10.1007/978-3-319-33326-7_18

18.1 Introduction

In the 1980s, recognition that plant and animal species were disappearing at an accelerating rate began to spread, resulting in increasing scientific and public attention to the decline of biodiversity and to issues of biodiversity in general. This has particularly been the case since 1992, when the United Nations Conference on Environment and Development in Rio de Janeiro concluded with the signing of the *Convention on Biological Diversity*, one of the conference's three framework conventions. The year 2010 was declared the International Year of Biodiversity.

This turned biodiversity into a key concept, the importance of which has increased in recent years. Thus, biodiversity has been granted an intrinsic value, and it is increasingly viewed as an indispensable foundation of society-nature interaction. The importance of biodiversity for ecosystem services, among others, has also been assigned monetary value.

In this ongoing discussion about biodiversity, its importance and value for society as well as its causes and foundations occupy a central place. There are numerous competing attempts to explain the emergence of biodiversity and the factors that determine current patterns of biodiversity. Monocausal explanations are probably inadequate in this respect. Empirical investigations and the models derived from them suggest that the significance of explanatory variables is strongly dependent on scale and that (geo) historical developments, geomorphology, climate and biological-ecological interactions all exert a powerful influence on observable patterns of biodiversity.

In this chapter, we discuss a socioecological perspective regarding biodiversity. We define the way in which human society purposively intervenes in natural processes—a phenomenon denoted 'colonization' by social ecologists (see Fischer-Kowalski and Haberl 1997; Chaps. 2 and 4 in this volume)—to utilize them for its own purposes. We discuss various consequences of colonizing interventions regarding biodiversity and its spatial distribution. We then present an empirical approach to the analysis of colonization and biodiversity that is based on a widely used and widely recognized explanatory approach: the species-energy hypothesis. Because this hypothesis can be linked with societal processes, we propose that this approach can serve as a starting point for investigations into human-nature interactions relevant to biodiversity. By this means, it is possible to address the problem of safeguarding biodiversity (at least partly) with the help of a broad understanding of social processes. This approach could make a valuable contribution to our understanding of potential and current conflicts, such as the use of the limited resource of 'land' for societal needs (and the restriction of its use to safeguard biodiversity). First, however, we wish to briefly discuss the ecological concepts of 'biodiversity'.

18.2 Ecological Concepts of Biodiversity

Despite the intensive engagement with this theme during recent decades, there is neither a universally accepted definition nor a comprehensive understanding of biodiversity. Originally, species richness, the number of biological species present in a particular area at a particular point in time, was used as the primary measure of biodiversity. Despite certain inadequacies, such as reliance on the concept of species, common difficulties in describing and delineating species and the laborious enterprise of preparing relevant data and information, species richness is still often cited as an indicator for biodiversity. Building upon species richness, derived indices such as the Shannon Diversity Index (Shannon and Weaver 1949) and Evenness (Pielou 1966) were developed. These indicators also take into account abundance, that is, the frequency with which a particular species is observed within the area being investigated. Whittaker (1972, 1977) draws a distinction among α-, β- and γ-diversity, the number of species within a small, relatively homogenous habitat (α), the diversity of a habitat within a larger area (β) and the overall diversity of species within a larger area (γ; $\gamma = \alpha \times \beta$). This concept made possible the integration of different scales and went beyond the pure recording of numbers of species and their respective abundances (Beierkuhnlein 2003).

Starting with the Rio *Convention on Biodiversity* (Johnson 1993), the following were integrated within the concept of biodiversity: the variability of higher integration levels (such as living communities, landscapes and ecosystems); the multiplicity of the functional-systemic processes occurring in these levels; and the diversity within species (such as genetic diversity). To be able to describe and register changes to this multiplicity of criteria, a range of appropriate and comparable sets of variables are required, beginning with the genotypes of selected groups of species (e.g., allelic diversity) and extending to the quantitative survey of ecosystem functions (e.g., nutrient retention) (Pereira et al. 2013).

The definition provided by Beierkuhnlein (2003), which touches on many aspects of biodiversity, is representative: 'Biodiversity is a measure for the qualitative, quantitative or functional diversity of biotic objects at all organizational levels in a concrete or abstract, spatial or temporal framework.' Particular attention is given today to species that are rare and/or endemic (i.e., only present in a very narrowly defined area) because they are disproportionately threatened by the risk of extinction (Dullinger et al. 2012).

Biodiversity develops through evolutionary processes in the interplay between speciation and extinction, in which organisms adopt different strategies to assert themselves in a constantly changing environment. The factors determining the patterns of biodiversity that are observable in the landscape or in different regions have been the subject of controversy and debate in recent decades. Attempts at explaining patterns of biodiversity regarding historical processes primarily consider dynamic geological developments and the speciation, migration and extinction processes related to them (Ricklefs and Schluter 1993). Evolutionary processes such as speciation play a major role in this context. Thus, many of the recent biodiversity patterns have been determined by geological occurrences such as continental drift and ice-age glaciation.

In contrast to these historical explanations, functional approaches place recent factors (in geological terms) in the foreground. Fundamentally, two large groups of functional approaches may be distinguished: (1) those based on structural heterogeneity and (2) those based on energy availability. The first group includes all those explanatory approaches that center upon the structural heterogeneity of a specific spatial segment. This can be described as the diversity of environmental conditions (Connell and Orias 1964), as habitat heterogeneity (Kerr et al. 2001; Rosenzweig 1995) or as topographical heterogeneity (Rahbek and Graves 2001; Richerson and Lum 1980). These concepts assume that an increase in structural diversity leads to a subsequent increase in biodiversity. In contrast, the intermediate disturbance hypothesis (Connell 1978) surmises that biodiversity is greatest when the frequency and magnitude of disturbances is intermediate in scale. That is, when disturbances are frequent and strong, only a few species can withstand the resulting harsh conditions. If, however, the disturbances are rare and weak, then a few successfully competing species will become dominant, and biodiversity decreases because less successful species disappear. However, the intermediate disturbance hypothesis has been strongly criticized (Fox 2013), and it remains to be seen how this discussion will develop in the future.

The approaches included in the second group look to energy as a source of explanation and have become known under the term species-energy hypothesis (Brown 1981; Wright 1983; Hutchinson 1959). This concept focuses on energy availability for the metabolic processes of organisms, particularly in the case of heterotrophic organisms (animals, many microorganisms, fungi), the trophic energy deriving from net primary production (NPP). Essentially, this hypothesis states that the number of species within an ecosystem is determined by the availability of biologically utilizable energy (Wright et al. 1993). This pattern can be observed, for example, in the increase in biodiversity from the poles toward the equator. The hypothesis also includes the consideration that other energy forms, such as the environmental temperature, exert an influence on physiological processes and metabolic rates and, thus, biodiversity (Allen et al. 2002; Turner et al. 1987, 1988).

Fundamentally, however, there is widespread consensus that biodiversity patterns cannot be explained with a monocausal approach and that different factors hold relevance depending on the ecological and historical contexts as well as spatial scales (Gaston 2000). In this respect, climatic conditions (which determine the availability of energy) are presumed to dominate from global to continental scales. At the local or landscape scale, there are other factors, such as climatic and geomorphological conditions and the scale and frequency of small-scale disturbances, that can have a significant impact on diversity (ibid.).

Biodiversity provides a basis for ecosystem functions and is therefore essential for ecosystem services that are useful for human society (Cardinale et al. 2012). Although there is evidence that diversity increases the stability of ecosystems, it is also clear that it is not biodiversity per se that is responsible for this relationship but rather the possibility of species communities to accommodate other species (groups) or functional groups, which can react to environmental changes in different ways (McCann 2000).

18.3 Colonizing Interventions and Biodiversity

Society intervenes in and alters natural structures and processes by means of different activities (e.g., land use, flood control). Over time, these interventions have become so significant that roughly three-quarters of the earth's land area and the habitats found there are affected to some extent by humans (Ellis et al. 2013; Erb et al. 2007; Sanderson et al. 2002). This situation has inspired the introduction of the term 'Anthropocene' to denote a new geological epoch (Crutzen 2002). Achieving a better understanding of the relationship between society and nature has thus become more important than ever. This relationship also affects biodiversity because a considerable proportion of terrestrial biodiversity worldwide is found in ecosystems that are more or less intensively managed by humans. The socioecological concept of colonization (Chap. 19) provides a conceptual framework and starting point for research into societal interventions in biodiversity because it explicitly links biodiversity with societal activities.

Figure 18.1 depicts the interactions between society's colonizing activities and biodiversity. The term 'colonization' refers to intentional changes made by society in natural processes so that these processes can be utilized more easily by human societies (Fischer-Kowalski and Haberl 1997). Colonizing interventions can directly aim at influencing biodiversity. Agriculture, for example, may aim for the greatest possible dominance of a particular plant species (and of this species, a highly specific, possibly even genetically modified variety); all other plants are consequently regarded as 'weeds' and are thus combatted by means of soil cultivation, weeding or herbicides. The restriction of biodiversity is—at least in the case of plants—an explicit target of crop production.

As a rule, however, colonizing interventions also have unforeseen side effects, often with severe impacts on biodiversity (Chap. 2). Colonizing interventions can

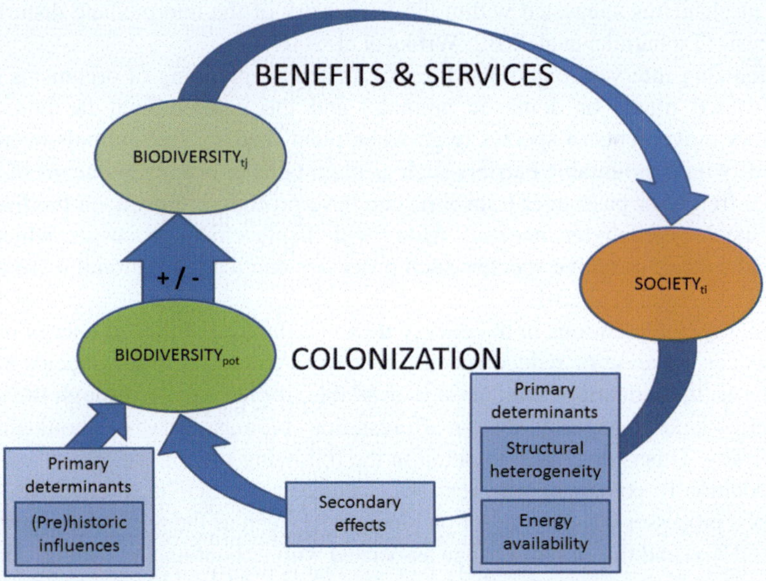

Fig. 18.1 Colonization and the relationship between society and biodiversity, based on the socioecological research paradigm

impact and modify important primary determinants of biodiversity, such as the structural heterogeneity of an ecosystem, the water balance, the availability of nutrients and energy and many other key parameters. Colonizing interventions impact an existing biodiversity that has typically already been affected by socioeconomic activities. Biodiversity patterns free of human impact ('potential biodiversity', $\text{BIODIVERSITY}_{pot}$) are rarely found today.

'Colonization' includes numerous activities undertaken within the context of land use, such as crop cultivation, livestock farming and forestry, as well as soil-sealing for the provision of infrastructure. The use of toxic chemicals (see the classic study on this issue, *Silent Spring*, Carson 1962) plays an important role here. These changes benefit some species but worsen the living conditions for others, the impact of which may lead to their extinction (at least locally).

Colonization can, however, also increase the diversity of habitats, such as through the creation of a pattern involving other forms of land cover in habitats that were previously relatively uniform in character. An example is cultivated landscapes in which natural woodland has been replaced by a mix of meadows, pastures, crop fields, gardens and other land cover forms utilized by humans. The small-scale mosaic structure of a cultivated landscape can lead to a localized increase in biodiversity in contrast to the potential biodiversity. In contrast, uniform, intensively used agricultural ecosystems such as cropland monocultures have a low level of biodiversity, not least because of the deliberate reduction in the diversity of plant species. Empirical analyses suggest that on the spatial scale of landscapes, such mechanisms can have the effect that species diversity is at its highest under conditions where interventions are of intermediate intensity, probably due to effects that are similar to the mechanisms suggested within the framework of the intermediate disturbance hypothesis (Tscharntke et al. 2012; Wrbka et al. 2004).

Colonizing interventions in the form of an active spreading of organisms (such as cultivated plants or domestic animals) and the unintentional facilitation of migratory movements of species (e.g., when plant seeds or live animals overcome previously insurmountable barriers such as mountains or oceans by means of international freight or passenger transport) can have profound impacts on biodiversity (alien species: neophytes, neozoa). Along with displacement processes, which can reduce the range of native species, such processes can lead to regional increases in biodiversity (Ellis et al. 2012).

Similarly, interventions in the energy flow and the availability of energy in ecosystems can increase or reduce species diversity. According to the species-energy hypothesis, the reduction or increase in available trophic energy through the modification of land cover (e.g., de- or afforestation, biomass harvest) should impact biodiversity. This is discussed in detail in the following section.

In addition to colonizing interventions, socioeconomic activities related to sociometabolic processes have an impact on biodiversity, such as the release of greenhouse gases (GHGs) and the climate change associated with it; hunting and fishing; and pollution due to toxins, heavy metals and acid emissions. The consequences of climate change for biodiversity have been intensively researched (Thomas et al. 2004), but such 'secondary' effects (Fig. 18.1) are not the primary focus of this contribution.

Activities with direct relevance for biodiversity ('pressures', e.g., land use), their underlying societal processes ('drivers', e.g., demand for particular products and services), and the resulting impacts on biodiversity, may be both spatial and temporal discontinuities. Thus, goods that serve to fulfill human needs (e.g., biomass) are often not produced at the location where they are finally consumed. This leads to a spatial decoupling between the interventions in ecosystems (and the related consequences for the biodiversity) and the consumption for which those interventions were undertaken (Erb et al. 2009; Chap. 16). Thus, for example, the demand for consumer products in one country may have impacts on biodiversity in other countries (Lenzen et al. 2012).

Extinction processes often appear with a time delay, so that a time lag, which is considerable at times, may lie between the causal influences and the actual extinction of a group of organisms in the affected region. This phenomenon, described as 'extinction debt' (Tilman et al. 1994), leads to a situation in which, due to current human interventions, the biodiversity that is currently observable is not necessarily indicative of the biodiversity that may be anticipated in the medium to long term. For this reason, the full scope of current negative impacts on the biodiversity that exists today will only become apparent in several decades. Empirical analyses show that the current pattern of species shown to be at medium to high risk of extinction in 22 European countries is more strongly determined by historically dominant 'pressures on biodiversity' (from 50 to 100 years ago) than those that are currently dominant. This is a clear sign that the pressures operating today will only manifest themselves to their full extent several decades from now (Dullinger et al. 2013).

18.4 Human Appropriation of Net Primary Production and Biodiversity

One indicator suitable for conducting empirical research on the impacts of colonizing interventions on biodiversity is the human appropriation of net primary production (HANPP; see Method Précis on Human Appropriation of Net Primary Production; Haberl et al. 2004, 2005, 2009). HANPP is an aggregated indicator that measures the impact of land use on energy availability (net primary production; NPP) in ecosystems (Haberl et al. 2007, 2014). It relates human activities to ecosystem processes and can be linked to explanatory theories regarding biodiversity that are based on the species-energy hypothesis (Vitousek et al. 1986; Wright 1983, 1990).

According to this hypothesis, a reduction in the energy flow that is available in the ecosystem for the ecological food web should lead to a reduction in species richness and, thus, in biodiversity (see Fig. 18.2). In this way, HANPP can serve as a connecting link between socioeconomic processes and biodiversity. The species-energy hypothesis supports the assumption that an increase in HANPP leads to a reduction in biodiversity, or at least exerts a pressure in this direction. The actual change in biodiversity may occur after a significant time delay—cue the 'extinction debt' (Dullinger et al. 2013).

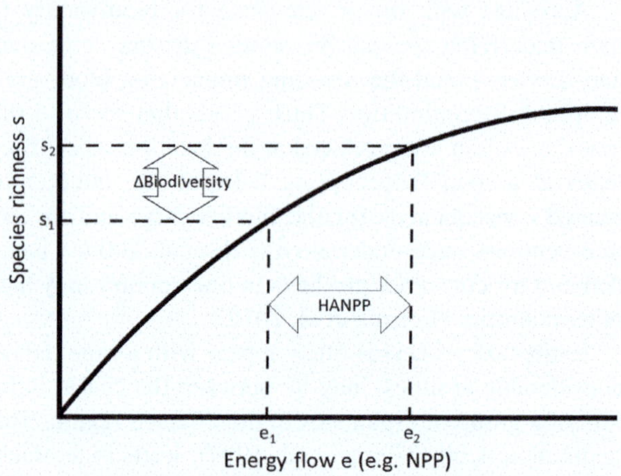

Fig. 18.2 Link between energy flow and species richness using the species-energy hypothesis: human-induced changes in the energy flow (HANPP) lead to changes in species richness (ΔBiodiversity). *Modified* from (Wright 1990)

It must be emphasized that HANPP is always related to a potential natural state, NPP_{pot} (see Method Précis on Human Appropriation of Net Primary Production), whereas data on species richness in an assumed natural state ($BIODIVERSITY_{pot}$) are either scarce or nonexistent and thus the change in biodiversity (ΔBiodiversity) is not known. Unfortunately, for both values to be compared in a meaningful way, quantitative data for diversity change (ΔBiodiversity) must be available. This problem must be taken into account when analyzing the relation between HANPP and biodiversity.

To test these hypothetical connections, several studies (Haberl et al. 2004, 2005) have compared the spatial patterns of NPP-relevant variables with the species richness of different taxonomic groups.

The starting points for the first such study (Haberl et al. 2004) were biodiversity patterns for several groups of terrestrial organisms that were surveyed in intensively used cultural landscapes of Eastern Austria (Sauberer et al. 2004). These data originated from 38 survey plots with an extent of 600 × 600 m. The organisms analyzed were mosses, vascular plants, snails, spiders, grasshoppers, ground beetles and ants. For these case studies, HANPP was calculated based on habitat mapping supported by orthophotography and on statistical harvest data. Linear regressions between the biomass remaining in the ecosystem after harvest (NPP_{eco}) and the data on species richness showed a significant increase in species richness together with increasing energy availability after the harvest. The strongest correlation was found for mosses ($r^2 = 0.758$; r^2 = coefficient of determination) and the weakest for ground beetles ($r^2 = 0.132$). Heterotrophic organisms had a weaker correlation ($r^2 = 0.549$) than autotrophic organisms ($r^2 = 0.723$). A comparison between HANPP—measured as a percentage of potential productivity (NPP_{pot})—and the biodiversity data produced consistently worse yet still significant results. As expected, species richness decreased with increasing HANPP.

These results support the species-energy hypothesis, which assumes a positive interrelation between available energy and species richness. The linear progression corresponds to the meta-analysis of Cusens and colleagues (2012), who found a dominance of positive-linear fits using 115 studies. Conclusive proof that HANPP leads to a reduction in species richness could not be found given the lack of data on biodiversity without anthropogenic influence. This study was limited to cultivated landscapes, namely, agricultural land, and did not contain any data on urban sites or forest ecosystems.

The second study (Haberl et al. 2005) compared the Austria-wide pattern of breeding bird richness with HANPP and other production ecology variables. The breeding bird inventory was estimated using the *Avifauna Archive of BirdLife Österreich*, a database with mapping data for a total of 213 breeding bird species (as of 1996). Of the 213 species, 94 were selected. These species were either only present locally and therefore had a range and distribution that was fairly well known and sufficiently well described, or they had a range and distribution that could not easily be modeled because of their particular way of life. For this range of species, the original distribution data of the Avifauna Archive were adopted. For the remaining 119 bird species, it had to be assumed that their distribution range had not been completely recorded. For these species, a geographic information system (GIS) was used to identify suitable potential breeding areas based on the living conditions in the known breeding areas. By superimposing these 119 model results with the 94 original mapping surveys, a map of the diversity patterns of Austrian breeding birds was produced (Plutzar et al. 2008). The spatial resolution of these data was 250 × 250 m, and a separate data set was compiled for endangered bird species.

The HANPP data were taken from an earlier work (Haberl et al. 2001) and were adapted for this study and aligned with the resolution of the bird data set. To investigate possible effects of scale, spatial resolutions of 1 × 1, 4 × 4 and 16 × 16 km were used in addition to the original resolution of 250 × 250 m, and both the NPP data and the data relating to breeding birds were processed for this mesh size.

The regression analyses between NPP_{eco} and breeding bird species richness showed a clear correlation between energy availability and species richness, whereby a nonlinear (quadratic), monotonically increasing relation for the entire observation area for the variation of NPP_{eco} was found between the NPP remaining after societal interventions in the ecosystem (NPP_{eco}) and species richness. The best coefficients of determination were provided by grid sizes 1 × 1 and 4 × 4 km, whereas grid sizes 250 × 250 m and 16 × 16 km showed worse results. Where the smallest scale is concerned, these poor results are probably due to data uncertainties; in the case of the largest scale, they may be due to the already considerable heterogeneity of such grid cells (a grid cell in this case has a size of 256 km^2, which is more than half the area of the federal county of Vienna).

Because one must assume that both the production ecology variables and the distribution of bird species are affected by climatic influences, this effect was excluded in a further step by a residual analysis. This was done to investigate the extent to which production ecology parameters have an additional explanatory value

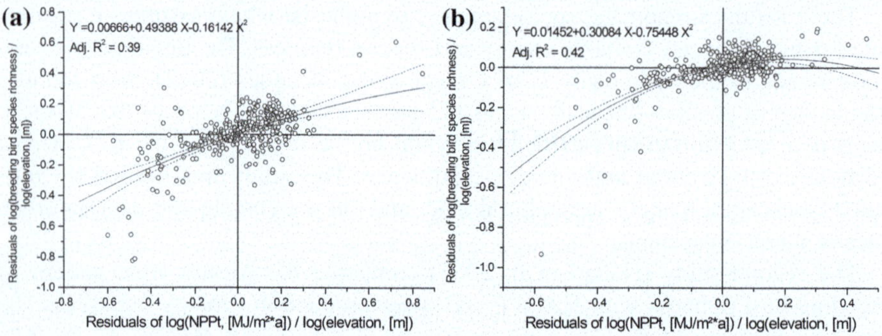

Fig. 18.3 Regression analysis between the NPP remaining after harvest ($NPP_t = NPP_{eco}$) and the values for breeding bird species richness in Austria for grid sizes **a** 1 × 1 km and **b** 4 × 4 km. Because both variables are highly dependent upon climatic conditions, the residuals of elevation (representing climatic conditions) were used

compared to climate, in which elevation is used to represent climate as a variable. In this way, the deviations regarding the productivity to be expected based on sea level could be compared with the deviations regarding species richness for each sampling area. After this step, there was still a significant, though weaker, correlation between the remaining energy flow after the harvest (NPP_{eco}) and the breeding bird species richness. As in the previous study, the statistical test (the Akaike Information Criterion, AIC) used favored quadratic regressions. As predicted by the species-energy hypothesis, a monotonically increasing relation between the residuals of NPP_{eco} and the residuals (i.e., of the variation not already explained by climate) of the species richness was found generally across the entire value range (Fig. 18.3).

The findings of this research support the species-energy hypothesis. It has to be noted, however, that the relevance of the HANPP for the state of and threat to biodiversity could only be shown indirectly because no data on biodiversity change (ΔBiodiversity) were available. On the one hand, the actual productivity deviating from the expected due to elevation can be interpreted as anthropogenic influence upon the energy flow. On the other hand, the number of endangered species rises with increasing HANPP. Findings from other studies (e.g., for the Long-Term Socioeconomic and Ecosystem Research (LTSER) region of Eisenwurzen, see Haberl et al. 2009) point in a similar direction.

18.5 Conclusions

Despite the high level of scientific and political attention to the issue of biodiversity, the systematic investigation of the processes by which human societies exert pressure upon biodiversity is still in its infancy. There is much evidence that the current strategy of placing species or habitats 'under protection' to either prevent their utilization or at least restrict certain uses (and possibly even encourage

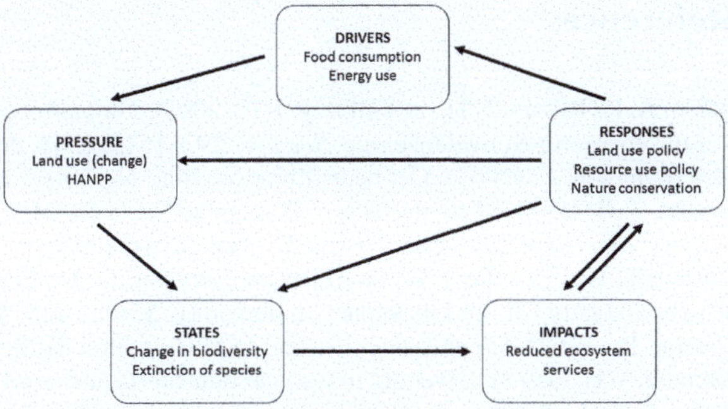

Fig. 18.4 DPSIR (Driving forces, Pressures, States, Impacts and Responses) model related to biodiversity. *Modified* from (Haberl et al. 2009)

some uses that are no longer economically viable) falls far short of achieving the declared goal of reducing the loss of biodiversity. It seems far more advisable to accompany this approach with a second strategy that integrates biodiversity protection into major areas of socioeconomic and political strategies (Spangenberg 2007). The interrelations between biodiversity and socioeconomic 'drivers' and 'pressures' also need to be understood (Fig. 18.4; see also Chap. 1) so that economic trajectories can be steered in a direction that is generally more biodiversity friendly instead of merely repairing the damage and placing individual, particularly valuable, elements of biodiversity under protection.

The manifold and strong interrelations depicted in Fig. 18.1 make the analysis of the relationship between colonization and biodiversity particularly promising. The analysis of the impacts of land-use intensity, measured using HANPP, on biodiversity is only a first step in this direction. The studies discussed have shown that the species richness of different groups of organisms in central European cultivated landscapes is significantly correlated with the size of ecological energy flows. This provides empirical confirmation for the species-energy hypothesis. However, due to a lack of data on the changes in biodiversity, only indirect evidence could be found that a change in energy availability arising from land use actually leads to a reduction in biodiversity. Further studies on this issue are required to test the relation between HANPP and biodiversity. Along with a broader empirical basis that includes other biogeographical regions, it would be important to investigate how the relation between HANPP and biodiversity varies in connection with different spatial and temporal scales. Based on the results obtained thus far, HANPP appears to be a relevant, highly aggregated pressure indicator for biodiversity losses. It is worthwhile to pursue this approach because it provides the opportunity to link biodiversity directly with socioeconomic activities and processes and, thus, to create the conditions for safeguarding measures to protect biodiversity.

Acknowledgements Funding from the European Research Council for the Starting Independent Researcher Grant ERC-263522 'LUISE', the Austrian Ministry of Education, Science and Culture (BMBWK) in the research program 'Cultural Landscapes Research (Kulturlandschaftsforschung)' and the Austrian Science Funds (FWF) within project P16692-G05 is gratefully acknowledged. This book chapter contributes to the Global Land Project (http://www.globallandproject.org).

References

Allen, A. P., Brown, J. H., & Gillooly, J. F. (2002). Global biodiversity, biochemical kinetics, and the energetic-equivalence rule. *Science, 297*, 1545–1548. doi:10.1126/science.1072380.

Beierkuhnlein, C. (2003). Der begriff biodiversität. *Nova Acta Leopoldina, 87*(328), 51–71.

Brown, J. H. (1981). Two decades of homage to Santa Rosalia: Toward a general theory of diversity. *American Zoologist, 21*(4), 877–888. doi:10.1093/icb/21.4.877.

Cardinale, B. J., Duffy, J. E., Gonzalez, A., Hooper, D. U., Perrings, C., Venail, P., et al. (2012). Biodiversity loss and its impact on humanity. *Nature, 486*, 59–67. doi:10.1038/nature11148.

Carson, R. (1962). *Silent spring*. Boston, MA: Houghton Mifflin.

Connell, J. H. (1978). Diversity in tropical rainforests and coral reefs. *Science, 199*, 1302–1310.

Connell, J. H., & Orias, E. (1964). The ecological regulation of species diversity. *The American Naturalist, 98*, 399–414.

Crutzen, P. J. (2002). Geology of mankind. *Nature, 415*, 23–24. doi:10.1038/415023a.

Cusens, J., Wright, S. D., McBride, P. D., & Gillman, L. N. (2012). What is the form of the productivity–animal-species-richness relationship? A critical review and meta-analysis. *Ecology, 93*, 2241–2252. doi:10.1890/11-1861.1.

Dullinger, S., Gattringer, A., Thuiller, W., Moser, D., Zimmermann, N. E., Guisan, A., et al. (2012). Extinction debt of high-mountain plants under twenty-first-century climate change. *Nature Climate Change, 2*, 619–622. doi:10.1038/nclimate1514.

Dullinger, S., Essl, F., Rabitsch, W., Erb, K.-H., Gingrich, S., Haberl, H., et al. (2013). Europe's other debt crisis caused by the long legacy of future extinctions. *Proceedings of the National Academy of Sciences, 110*, 7342–7347. doi:10.1073/pnas.1216303110.

Ellis, E. C., Antill, E. C., & Kreft, H. (2012). All is not loss: Plant biodiversity in the anthropocene. *PLoS ONE, 7*(1), e30535. doi:10.1371/journal.pone.0030535.

Ellis, E. C., Kaplan, J. O., Fuller, D. Q., Vavrus, S., Goldeweijk, K. K., & Verburg, P. H. (2013). Used planet: A global history. *Proceedings of the National Academy of Sciences, 110*, 7979–7985.

Erb, K.-H., Gaube, V., Krausmann, F., Plutzar, C., Bondeau, A., & Haberl, H. (2007). A comprehensive global 5 min resolution land-use data set for the year 2000 consistent with national census data. *Journal of Land Use Science, 2*, 191–224.

Erb, K.-H., Krausmann, F., Lucht, W., & Haberl, H. (2009). Embodied HANPP: Mapping the spatial disconnect between global biomass production and consumption. *Ecological Economics, 69*, 328–334. doi:10.1016/j.ecolecon.2009.06.025.

Fischer-Kowalski, M., & Haberl, H. (1997). Stoffwechsel und Kolonisierung: Konzepte zur Beschreibung des Verhältnisses von Gesellschaft und Natur. In M. Fischer-Kowalski, H. Haberl, W. Hüttler, H. Payer, H. Schandl, V. Winiwarter, & H. Zangerl-Weisz (Eds.), *Gesellschaftlicher Stoffwechsel und Kolonisierung von Natur* (pp. 3–12). Amsterdam: Gordon & Breach Fakultas.

Fox, Jeremy W. (2013). The intermediate disturbance hypothesis should be abandoned. *Trends in Ecology & Evolution, 28*, 86–92. doi:10.1016/j.tree.2012.08.014.

Gaston, K. J. (2000). Global patterns in biodiversity. *Nature, 405*, 220–227.

Haberl, H., Erb, K.-H., Krausmann, F., Loibl, W., Schulz, N., & Weisz, H. (2001). Changes in ecosystem processes induced by land use: Human appropriation of aboveground npp and its influence on standing crop in Austria. *Global Biogeochemical Cycles, 15*(4), 929–942.

Haberl, H., Schulz, N. B., Plutzar, C., Erb, K.-H., Krausmann, F., Loibl, W., et al. (2004). Human appropriation of net primary production and species diversity in agricultural landscapes. *Agriculture, Ecosystems & Environment, 102*, 213–218. doi:10.1016/j.agee.2003.07.004.

Haberl, H., Plutzar, C., Erb, K.-H., Gaube, V., Pollheimer, M., & Schulz, N. B. (2005). Human appropriation of net primary production as determinant of avifauna diversity in Austria. *Agriculture, Ecosystems & Environment, 110*, 119–131. doi:10.1016/j.agee.2005.03.009.

Haberl, H., Erb, K.-H., Krausmann, F., Gaube, V., Bondeau, A., Plutzar, C., et al. (2007). Quantifying and mapping the human appropriation of net primary production in earth's terrestrial ecosystems. *Proceedings of the National Academy of Sciences, 104*, 12942–12947. doi:10.1073/pnas.0704243104.

Haberl, H., Gaube, V., Díaz-Delgado, R., Krauze, K., Neuner, A., Peterseil, J., et al. (2009). Towards an integrated model of socioeconomic biodiversity drivers, pressures and impacts. A feasibility study based on three european long-term socio-ecological research platforms. *Ecological Economics, 68*, 1797–1812. doi:10.1016/j.ecolecon.2008.11.013.

Haberl, H., Erb, K.-H., & Krausmann, F. (2014). Human appropriation of net primary production: Patterns, trends, and planetary boundaries. *Annual Review of Environment and Resources, 39*(1), 363–391. doi:10.1146/annurev-environ-121912-094620.

Hutchinson, G. E. (1959). Homage to Santa Rosalia or Why are there so many kinds of animals? *The American Naturalist, 93*, 45–159.

Johnson, S.P. (1993). The earth summit: The united nations conference on environment and development (UNCED). Graham & Trotman, London.

Kerr, Jeremy T., Southwood, T. R. E., & Cihlar, Josef. (2001). Remotely sensed habitat diversity predicts butterfly species richness and community similarity in canada. *Proceedings of the National Academy of Sciences, 98*, 11365–11370.

Lenzen, M., Moran, D., Kanemoto, K., Foran, B., Lobefaro, L., & Geschke, A. (2012). International trade drives biodiversity threats in developing nations. *Nature, 486*, 109–112. doi:10.1038/nature11145.

McCann, K. S. (2000). The diversity-stability debate. *Nature, 405*, 228–233.

Pereira, H. M., Ferrier, S., Walters, M., Geller, G. N., Jongman, R. H. G., Scholes, R. J., et al. (2013). Essential biodiversity variables. *Science, 339*, 277–278.

Pielou, E. C. (1966). The measurement of diversity in different types of biological collections. *Journal of Theoretical Biology, 13*, 131–144.

Plutzar, C., Pollheimer, M., & Sauberer, N. (2008). Diversitätsmuster der Brutvögel Österreichs. In: N, Sauberer D, Moser & G, Grabherr (Eds.), Biodiversität in Österreich. Räumliche Muster und Indikatoren der Arten- und Lebensraumvielfalt. Zürich, Bristol-Stiftung; Bern, Stuttgart, Wien, Haupt.

Rahbek, C., & Graves, G. R. (2001). Multiscale assessment of patterns of avian species richness. *Proceedings of the National Academy of Sciences, 98*, 4534–4539.

Richerson, P. J., & Lum, K. (1980). Patterns of plant species diversity in California: Relation to weather and topography. *The American Naturalist, 116*, 504–536.

Ricklefs, R. E., & Schluter, D. (1993). *Species diversity in ecological communities: Historical and geographical perspectives*. Chicago: University of Chicago Press.

Rosenzweig, M. L. (1995). *Species diversity in space and time*. Cambridge: Cambridge University Press.

Sanderson, E. W., Jaiteh, M., Levy, M. A., Redford, K. H., Wannebo, A. V., & Woolmer, G. (2002). The human footprint and the last of the wild. *BioScience, 52*, 891–904.

Sauberer, N., Zulka, K. P., Abensperg-Traun, M., Berg, H.-M., Bieringer, G., Milasowszky, N., et al. (2004). Surrogate taxa for biodiversity in agricultural landscapes of Eastern Austria. *Biological Conservation, 117*, 181–190. doi:10.1016/S0006-3207(03)00291-X.

Shannon, C. E., & Weaver, W. (1949). *The mathematical theory of communication*. Urbana: University Illinois Press.

Spangenberg, J. H. (2007). Biodiversity pressure and the driving forces behind. *Ecological Economics, 61*, 146–158. doi:10.1016/j.ecolecon.2006.02.021.

Thomas, C. D., Cameron, A., Green, R. E., Bakkenes, M., Beaumont, L. J., Collingham, Y. C., et al. (2004). Extinction risk from climate change. *Nature, 427*, 145–148. doi:10.1038/nature02121.

Tilman, D., May, R. M., Lehman, C. L., & Nowak, C. A. (1994). Habitat destruction and the extinction debt. *Nature, 371*, 65–66.

Tscharntke, T., Tylianakis, J. M., Rand, T. A., Didham, R. K., Fahrig, L., Batáry, P., et al. (2012). Landscape moderation of biodiversity patterns and processes—eight hypotheses. *Biological Reviews, 87*(3), 661–685. doi:10.1111/j.1469-185X.2011.00216.x.

Turner, J. R. G., Gatehouse, C. M., & Corey, C. A. (1987). Does solar energy control organic diversity? Butterflies moths and the british climate. *Oikos, 48*, 195–205.

Turner, J. R. G., Lennon, J. J., & Lawrenson, J. A. (1988). British bird species distributions and the energy theory. *Nature, 335*, 539–541.

Vitousek, P. M., Ehrlich, P. R., Ehrlich, A. H., & Matson, P. A. (1986). Human appropriation of the products of photosynthesis. *BioScience, 36*, 368–373. doi:10.2307/1310258.

Whittaker, R. H. (1972). Evolution and measurement of species diversity. *Taxon, 21*, 213–251.

Whittaker, R. H. (1977). Evolution of species diversity in land communities. *Journal of Evolutionary Biology, 10*, 1–67.

Wrbka, T., Erb, K.-H., Schulz, N. B., Peterseil, J., Hahn, C., & Haberl, H. (2004). Linking pattern and process in cultural landscapes. an empirical study based on spatially explicit indicators. *Land Use and Sustainability Indicators, 21*, 289–306. doi:10.1016/j.landusepol.2003.10.012.

Wright, D. H. (1983). Species-energy theory: An extension of species-area theory. *Oikos, 41*, 496–506. doi:10.2307/3544109.

Wright, D. H. (1990). Human impacts on energy flow through natural ecosystems, and implications for species endangerment. *Ambio, 19*, 189–194.

Wright, D. H., Currie, D. J., & Maurer, B. A. (1993). Energy supply and patterns of species richness on local and regional scales. In R. E. Ricklefs & D. Schluter (Eds.), *Species diversity in ecological communities: Historical and geographical perspectives* (pp. 66–77). Chicago, London: University of Chicago Press.

Part IV
Empirical Approaches to Long-Term Socioecological Research

Chapter 19
Long-Term Risks of Colonization: The Bavarian 'Donaumoos'

Martin Schmid

Abstract This chapter presents a study in Environmental History that uses one of the key concepts of Social Ecology, 'colonization of natural systems', to reconstruct the dramatic transformation of a peculiar landscape. This landscape is the 'Donaumoos', a wetland along the left bank of the Upper Danube that was drained systematically from the 1770s onward. This colonization started during the first phase of the transition from an agrarian to an industrialized sociometabolic regime. Rich historical sources allow us to reconstruct the political, economic and cultural circumstances of such a large-scale intervention in a landscape, and they reveal that the environmental and social consequences of this project were heavily contested already among contemporaries. With a long-term perspective covering more than 250 years, this environmental history of the 'Donaumoos' exemplifies how societies are trapped in a 'risk spiral', where solving older problems of sustainability always results in new risks. Experts in the 18th century discussed major interventions into 'natural' systems with great passion. By revisiting a discourse of experts in the age of enlightenment, this chapter also contributes to a historical reflection on the term and the idea of 'colonization of nature'.

Keywords Long-term socioecological research (LTSER) · Industrialization · Transitions · Wetlands · Agroecological history

19.1 Introduction

'Colonizing natural systems', like 'social metabolism', is a key concept of Viennese Social Ecology. It essentially refers to controlling, targeted and sustained interventions by societies in natural systems (see Chap. 2). This contribution

M. Schmid (✉)
Institute of Social Ecology (SEC), Alpen-Adria University,
Schottenfeldgasse 29, 1070 Vienna, Austria
e-mail: martin.schmid@aau.at

H. Haberl et al. (eds.), *Social Ecology*, Human-Environment Interactions 5,
DOI 10.1007/978-3-319-33326-7_19

illustrates and discusses this admittedly abstract socioecological core concept within the context of the environmental history of the Danube. Particularly from the late Middle Ages onward, the transformation of Europe's second-longest river presents countless examples, spread across time and space, of colonizing interventions and their long-term impacts—for both nature and society. I have selected one such example as the focus of this essay.

The environmental history of the Bavarian Donaumoos, a lowland bog situated to the south of the Upper Danube between Neuburg (2477 river km) and Ingolstadt (2459 river km), is marked by a decisive transformation during the last decade of the 18th century. Its colonization—contemporaries termed this 'culture' ('*Kultur*' in German)—had dramatic, long-term regional consequences, the impacts of which are still felt today. Much of what we now characterize in retrospect as undesired side effects of a colonization taking place more than 200 years ago was in fact the recorded subject of an extremely heated debate by the actors involved at the time. The source material from this period—an early phase of the socioecological transition towards an industrialized regime—allows for a detailed reconstruction of the political context; the social, cultural and ecological preconditions; the economic expectations; and the social and economic consequences of a large-scale and radical colonizing intervention. However, we shall return to this story later. Let us begin with a question: why is 'colonization' such an important element in the formation of socioecological theory?

19.2 Societal Commitments Due to Colonization

From a Social Science perspective, the most important insight from the development and application of the 'colonization' concept is as follows: to the extent with which societies colonize 'natural systems', they also enter into a long-term commitment; they not only colonize 'nature', but, in a sense, they also colonize themselves. How might one imagine these different forms of commitment, and what causes of these forms can be identified? In the following section, three causes are discussed in detail.

19.2.1 Colonized Systems Are Socio-Natural Hybrids

Colonized systems maintain a form of life that is in opposition to societal requirements, and they keep a degree of resistance. The formulation 'renaturation tendency' is sometimes used in an attempt to conceptualize this characteristic (Fischer-Kowalski and Erb 2006, p. 47). Thus, if a mill weir on a small Alpine tributary of the Danube is not continuously maintained, it will eventually collapse, allowing water to flow unhindered once again and enabling species that thrive in flowing water to inhabit the area, while making the stretch of waterway once more passable for migratory fish. Ontologically, colonized systems should be conceptualized as hybrids (Fischer-Kowalski and Weisz 1999). Natural processes are always at work within them—before any human interventions take place, during these interventions and after human interventions have ceased. Colonization describes the targeted influencing of natural processes. The conditions of life for specific species (e.g., edible fish) are improved, while those of others (e.g., those competing for food with humans and species useful to them) are intentionally made worse. 'Colonies' in this sense are comparable to socio-natural arrangements (see Winiwarter and Schmid 2008; Winiwarter et al. 2013; see Chap. 23 in this volume), in which natural and social processes are inextricably interconnected. Their hybrid character should also be highlighted for another reason: colonization (as with the larger theoretical design of the socioecological model of interaction) is based on the analytical differentiation between society and nature. 'Society' versus 'nature' is the basic differentiation that underpins the socioecological concept of interaction.[1] Talking about the colonization of 'natural systems' only makes sense in the context of this intentional analytical differentiation (Schmid 2006). However, the risk of ontological misconception is always at hand (Schatzki 2003, p. 87). A pure form of nature cannot be found anywhere, and certainly not in European environmental history. When work began in 1870 on the excavation of the river landscape to allow the 'breakthrough' of a completely new riverbed extending several kilometers along the Danube near Vienna (Mohilla and Michlmayr 1996), it did not involve the colonization of a 'natural' system in an ontological sense. Intervention took place in an admittedly radical and ground-breaking fashion in a system that had been manipulated for at least half a millennium (with records dating back to the 14th century) by hydraulic and engineering measures (Hohensinner et al. 2013). In most areas of the world, humans do not colonize an untouched 'nature' but instead undertake colonizing interventions in hybrid systems they have inherited from earlier generations. Colonies always carry the traces of earlier utilization. Colonizing interventions are often reactions to earlier interventions and their consequences.

19.2.2 Colonization Has Side Effects

The second reason for societal commitment is that interventions have more than intended consequences. Sooner or later, they also result in unintended and often undesirable side effects to which society must react. Sometimes the colonizing actors consciously accept the trade-off with these undesirable side effects. Here too, an example may be given. Until the erection of barrage constructions in the Upper Danube and its tributaries, the Austrian Danube transported approximately

[1]Compare this with the differentiation between 'arrangements' and 'practices' in the concept of 'socio-natural sites' (SNSs).

500,000 m^3 of gravel and coarse sand (known as bed load) per year. This was accompanied by 5.5–7 Mt (1 Mt = 10^6 t = 1 million tons) of fine particulate matter as suspended sediment, in German known as *Letten*. Today, a large part of this material is intercepted by barrage constructions on the Alpine tributaries of the Danube. In the Danube itself, the numerous power stations hinder such transport further along the river (Hohensinner 2010, p. 39). The undesired yet consciously accepted side effect of this is that humans now have to take on the work of bed load transport themselves. Thus, approximately 200,000 m^3 of gravel is placed in the riverbed each year downstream from the Freudenau power station in Vienna to at least stem the further deepening of the riverbed and thus hinder, among other things, the decline in groundwater levels (and with this, the drying-out of the *Donauauen* National Park) and the formation of obstacles along river navigation channels (e.g., scouring). Formerly, the river's kinetic energy transported the sediment. We use this kinetic energy today to produce electricity. Excavating the sediment from one part of the river and placing it at another is part of the price we pay for this.

Other undesirable side effects only become apparent after a long time has passed. At the point of intervention, the colonizing actors are unaware of the impact related to this, which is often far-reaching. Evidence of engineering measures employed to prevent erosion on slopes and soils, particularly resulting from heavy precipitation events, can be traced in the Austrian Alps back as far as the 16th century. Success in solving a problem in one place (in the form of control structures for mountain streams, avalanches and landslides in the high Alpine region) leads to a new problem at another location. A lack of sediment results in an excess of kinetic energy in watercourses, and the power of the water begins to erode the streambed and banks (Wohl 2011, pp. 109–111).

19.2.3 Colonization Has Social Consequences

Above all, the social system has to organize itself in such a way that it has a sustainable capacity to perform the intervention services (particularly work) and monitoring that the colonization itself renders necessary. For an agrarian society based on solar energy, one of the most important functions of the Danube was the provision of long-distance transport for bulky goods such as fuelwood and timber from the hinterland to urban centers. This urban hinterland could be hundreds of kilometers away, and rivers ameliorated the transport limitations of the agrarian regime. Key resources that were particularly important in a qualitative sense, such as salt, were transported in barges hauled upstream, an endeavor that was logistically very intensive, entailing the work of dozens of men and horses (see Winiwarter and Schmid 2010, pp. 112, 184). In social terms, an agrarian utilization regime such as this requires the performance of certain preliminary services (e.g., the maintenance of towpaths and the monitoring and marking of upriver transport channels, which were subject to constant change due to the largely unregulated river), and it results in certain internal forms of social organization (e.g., highly organized barge transport). Before the industrial transition took hold (see Chap. 3) and before steam-powered rail transport became established, enlightened absolutist states, such as the Habsburg Monarchy during the second half of the 18th century, relied on waterways as a means to achieve the political, economic and military integration of their territories. In the 1770s, Maria Theresia installed 'navigation agencies' ('*Navigationsdirectionen*') to improve and maintain waterways within the Habsburg Monarchy. This also included the Danube, from the border with Bavaria right up to the city gates of Belgrade. Visionaries such as the engineer François-Joseph Maire from Lorraine dreamed of connecting European watersheds with canals, thereby linking the capital and court residence of Vienna with all European seaways (Maire 1786). How different is the colonization regime of a society that uses the river primarily for the production of electricity. Currently, roughly 66 % of Austria's electricity generation comes from hydropower (Wagner et al. 2015). In the 20th century, energy from the Danube and its tributaries profoundly changed social life. Light at the touch of a button at night and household appliances such as refrigerators and many others have profoundly altered the way we live and communicate, including our eating habits and the temporal rhythms of everyday life. This example makes it clear that colonizing interventions such as power stations have radically transformed not only the Danube but also social metabolism and ways of life.

19.3 The Bavarian Donaumoos: A Colonization Example from the End of the 18th Century and Its Consequences

In 1795, a book was published (Aretin 1795) that attracted great interest from the reading public of the age (Anonymous 1796). In this volume, Georg Freiherr von Aretin gave an account over some 250 pages of a colonizing intervention that was both profound in character and large in scale. The author was the son of one of the key figures of this intervention from 1790 onward (Hoser 2012, p. 214). The cultivation of the Donaumoos, a lowland bog directly bordering the Danube to the south in what is now Bavaria, was part of a series of both large- and small-scale projects to reclaim land and optimize agricultural practice under conditions of the agrarian regime, which took place across wide swaths of central and eastern Europe from around 1680. The drainage of the Oderbruch in the middle of the 18th century is just one prominent example among many (Reith 2011, pp. 25–28). Aretin's book was an adept attempt to discredit the opponents of cultivation and to garner public support for those in favor of intervention, which was the subject of heated debate.

19.3.1 Colonization and Public Opinion at the End of the 18th Century

The process of draining the Donaumoos was accompanied throughout by public debate between proponents and opponents. The arguments generated by this highly controversial subject saw the two camps implacably counterposed. Officials fulminated in the spirit of the late Enlightenment in the fiercest of terms against their opponents, and insinuation and personal accusations on both sides were not uncommon. However, it was not the sense of such an undertaking in itself that was contested but rather the advisability and feasibility of concrete interventions and the costs involved. Allegations of fraud, corruption (Aretin 1795, pp. 127–130, 1796, pp. 189–197) and sabotage (Aretin 1795, p. 72) were made. Primarily, however—and this is what makes this particular example of colonization so interesting for us—the potential short- and long-term consequences of the drainage proposal for nature and society were raised by both sides.

Those in favor of the project undertaken from 1790 saw the colonization or '*Kultur*'[2] of the bog as an undertaking that would be beneficial and bring blessings in all respects. To present the colonizing intervention in a good light, they depicted the bog's status before cultivation as unremittingly desolate. Rivers and streams '*left to themselves*' would have continually caused large-scale flooding (Aretin 1795, pp. 10f., 28, 35). The damming of badly planned millstreams, ponds for hemp retting[3] and watering places for livestock had prevented water from draining properly (ibid., p. 52). The vegetation in the bog largely served no agricultural purpose; wherever crops could have grown, grazing prevented them from doing so (ibid., pp. 1–13). The meadows could typically be cut only once a year and thus produced low yields for winter feed, in addition to being laid waste by the grazing livestock. Stagnant '*swamp water […] acidified*' the soils and through their '*evaporation*' threatened to cause '*the greatest damage*' to humans and livestock (ibid., pp. 24, 55). They created an unfavorable microclimate that would cause wildfires and grasshopper infestations in summer (ibid., pp. 29, 42) and hailstorms, thunderstorms, heavy rain (ibid., p. 40) and frost the damage to fruit trees in the spring and autumn (ibid., pp. 39, 43f.). It was also said that they led to the regular outbreak of disease among livestock (ibid., pp. 37f., 43, 55) and to epidemics among local inhabitants (ibid., pp. 42, 55) and were responsible for a lack of available wood (ibid., p. 38) and transport routes (ibid., pp. 40f.). The borders in the bog were unclear, and thus the ownership and control arrangements were said to be entirely unclear (ibid., pp. 16–24). As common land, the managed pastureland had often hampered the effort of the local people (ibid., p. 24). Those in favor of the bog's cultivation promoted private ownership of the land as the driver of modernization, and they saw grazing (particularly common grazing) as one of the major causes of the bog's desolate condition. Where they envisioned a future landscape well-ordered in every respect—agriculturally, morally and politically—with settlements, crop cultivation, orchards and flourishing trade, they instead saw nothing but wretched pastures into which the cattle were prone to sink,

[2]'*Kultur […] heißt […] die Oberfläche des Bodens in Rücksicht seiner künftigen Erträgnis verbessern*' (Aretin 1795, p. 145).

[3]Fibers from the hemp plant were used to produce, e.g., sailcloth and rope. Hemp retting was the process by which the hemp was rotted in order to separate the fibers from the woody inner core. To do this, the hemp plants were laid down for several days or weeks in stagnant water (also causing water pollution).

by which they were made sick by poor-quality feed and through which they were plagued by insects (ibid., pp. 25f.).

In the depiction by the proponents of cultivation, the Donaumoos before 1790 was an unproductive, entirely neglected and desolate agrarian ecosystem (for this term, see Chap. 21 in this volume). In the course of this public controversy, which was very much a propaganda exercise, writers depicted the area as a kind of hell on earth for both humans and animals. The endeavor, therefore, was to transform this hell into a '*paradise*' (Aretin 1795; p. 36; Stengel et al. 1791, pp. 21f.). In the spirit of the late Enlightenment, they saw themselves as having to fight against ignorance and 'prejudice'. In reality, the cultivation of the Donaumoos was an attempt to completely redesign an agroecosystem afresh within a period of only five years.

19.3.2 Colonization Has Political Prerequisites

The Donaumoos covers approximately 200 km^2, '*almost four square miles, and thus more than many a German principality*', as Aretin (1795, pp. 4, 22) noted in reference to the scale of the undertaking, not without some pride and alluding to the territorial fragmentation of the Old Empire before the Napoleonic Wars. Until 1777, a territorial border (dividing Bavaria and Palatinate-Neuburg) had actually run through the middle of the bog. In the Donaumoos in the 18th century, it was still a case of dynastic inheritance that created the political conditions that could facilitate intervention on such a large scale. With the unification of the Electoral Palatinate and Bavaria to create the Palatine-Bavaria Electorate under Elector Karl Theodore, the border across the Moos fell in 1777. In other regions, similar political and administrative prerequisites for such large-scale environmental interventions were not created until the beginning of the 19th century (in western and central Europe, often as a consequence of the Napoleonic Wars).[4] Large-scale and radical colonizing interventions not only have energetic and technological prerequisites (e.g., fossil-energy-powered machines) but also, and by no means least, political and administrative requirements. In the German-speaking world in particular, these two often coincided only from the early 19th century onward. Thus, how did the actors proceed in the Donaumoos, and what specific measures were implemented there?

19.3.3 The Intervention

When the political preconditions came into being in 1777, plans for the drainage were immediately drawn up, followed soon afterward by intervention measures,[5]

[4]See, for example, Blackbourn (2006, pp. 86–91) on the regulation of the Oberrhein by Tulla.

[5]Thus, the Catholic priest Johann Jakob Lanz worked on a project from 1778, which, although it was not realized in the form he envisaged, already saw some measures put into practice from 1790 and was regularly referred to by both proponents and opponents in the later arguments that took place. A description of Lanz' project may be found in Hoser (2012, pp. 206–208).

including the laying of a channel more than 17 m wide where the bog met the Danube (Aretin 1795, p. 59; Pechmann 1832, p. 32). This and earlier interventions in the bog were later qualified by Aretin largely as '*unsystematic*'. Commonly, these interventions had created more problems than benefits (Aretin 1795, pp. 53–55). In 1787, a '*culture commission*' was set up that focused on the '*preparation of a new culture system*' (ibid., p. 36). 'System' was one of the key words with which these late 18th-century experts chose to underline the considered nature of their approach in contrast to others. One of the most urgent tasks facing the newly founded commission was to clarify the nebulous ownership arrangements relating to the bog. In the writings of both proponents and opponents, countless pages are devoted to the most pedantic legal argumentation. Essentially, this related to the question of whether and to what degree the state was justified in intervening by cultivating privately owned land (Pechmann 1832, p. 33). Following from this question was the question of what responsibilities and rights might be derived from earlier legal arrangements both during and after the drainage of the bog. One must realize that a significant tract of land was being completely redesigned within only a few years and its usage had to be completely reorganized anew. Traditional usage rights, such as grazing rights, became obsolete because the pastureland in question ceased to exist. New usage opportunities had to be distributed among established and newly arrived population groups ('colonists'). Envy and resentment already took hold during the planning phase, and many felt themselves to have been cheated, whereas others wanted to use the opportunity to settle old scores (Pechmann 1832, p. 52). A specially established court, the 'Donaumoos Court', appears to have mediated in these numerous conflicts, at least temporarily (Aretin 1795, pp. 73–87).

From the spring of 1790, the cultivation project went full steam ahead. Clearing, leveling and diverting of streams was carried out, and drainage ditches, channels and dams traversed the bog. The main channel reached a length of more than 33 km, and the lower course had a width of more than 29 m (Pechmann 1832, pp. 38f.). The network of channels was to reach a combined length of more than 350 km within just a few years (Kling 1806, p. 30). At the same time, the bordering section of the Danube between Neuburg and Ingolstadt was straightened, for which five wide meanders were broken through and the course of the Danube rendered straight for a length of some 18 km (Aretin 1795, p. 68).[6] The altered outflow regime was intended to have an impact on river transport arrangements, to improve the protection of settlements and cultivated land bordering the Danube from erosion and flooding and to facilitate the drainage of the Donaumoos during ice jams (Pechmann 1832, pp. 46f.). In the Moos and at its perimeter, new highways and paths were laid down. In the very same year,[7] work began on the erection of the first 'colonies' (planned villages for new settlers). The bricks for these settlements came from a dedicated brickworks, where they were fired using the

[6]The naturalist Schrank made a detailed study (1795, pp. 21, 29f.) of the influence of the Danube on the bog and, in this context, referred to Luigi Ferdinando Marsigli's famous works on the Danube, written some two generations earlier (Marsigli 1726).

[7]Hoser (2012, p. 212) records that the first houses were constructed in 1792.

peat from the bog (Aretin 1795, p. 113).[8] The area was extensively cleared (ibid., p. XVI), hundreds of thousands of trees were planted (Pechmann 1832, p. 47), and uneven patches of land were leveled (Aretin 1796, p. 207).

Until 1795, the total costs of the work ran to '*not more than 530,000 Gulden*'. However, opponents of the project spoke of millions being senselessly wasted (Pechmann 1832, pp. 154f.). In any case, the financial resources involved were certainly significant. To raise the sums required, a relatively new institution in Bavaria was employed. In 1790, a shareholding company was founded. The large majority of the shares, however, were not owned by any private shareholders but belonged to the Elector himself and to his offices, officials and commissaries (Aretin 1795, pp. 61f.; Pechmann 1832, pp. 36–38).[9] On a daily basis, more than 1000 people, '*often 1500 to 2000*' (Aretin 1796, p. 160), were at work in the bog. The initial deployment of soldiers was soon replaced by paid laborers. In the case of less-heavy work, several hundred children from the villages in the surrounding area were 'rented' at little cost. In the view of the proponents of the scheme, this was, to put it in today's language, an opportunity to create local added value because the payment went to the parents of these child workers and thus to the local population. Heavy labor was undertaken—as was then common in businesses such as this (Maire 1786, p. 117)—by the homeless and petty criminals (Aretin 1795, pp. XVf.; Pechmann 1832, pp. 48f.). As is often the case, the colonization of nature and social control went hand in hand.

19.3.4 The Opponents and Their Arguments

Who were the scheme's opponents? Among their number were senior officials from Neuburg with a legal education who perhaps saw the entire undertaking as a Bavarian incursion into their territory. What were their major concerns?[10] Apart from the accusations concerning legal infringement already mentioned above, there was concern over fraud, land theft (including an alleged threat of violence), corruption, personal enrichment and the wasting of financial resources. We can describe a good portion of these concerns as agroecological arguments on the one hand and as social concerns on the other. In the following text, I wish to focus on these agroecological and social arguments. To summarize briefly, the opponents of

[8]The brickworks kiln using turf fuel was later abandoned, as the brick clay extraction sites were not productive enough and had become partially waterlogged (Pechmann 1832, p. 69).

[9]For more details on the financial arrangements, see Hoser (2012, pp. 208f.).

[10]It must have been the abovementioned Johann Georg Aretin himself who in 1794 put together a volume published anonymously two years later in which essays from influential opponents of the '*Donaumoos-Kultur*' appear alongside texts by his father Karl Albrecht von Aretin and others in favor of the scheme (Aretin 1796). This compilation presented the arguments of the opponents, only to demolish them point by point. The volume as a whole is a cleverly conceived apology for cultivation, upon which I have drawn in the following account to reconstruct the arguments of the project's opponents. For a summary, see also Hoser (2012, pp. 218f.).

the scheme did not believe that the new agroecosystem in the process of being created in the bog would be capable of functioning, and they warned of the social upheavals they feared would be a consequence of this colonizing intervention. They believed that the drainage plan would remove too much water from the soils in the heart of the moorland, as there were too many channels and they were too deep (Delagera 1794, p. 28). This would eventually lead to soil that was deprived of water, which would be eroded by the wind, taking any seed that was sown with it.[11] To render the bog soil suitable for cultivation, care needed to be taken not only to ensure an adequate supply of water but also to mix the soils of the bog with better soils from elsewhere (Schatte [n.d.], p. 59). It was also argued that the necessary transfer of cattle from the many pastures that would vanish due to the scheme to an indoor feeding regime would never be successful (Delagera 1794, pp. 1f.) because the regulated bog would not be able to provide the straw and meadow plants that would be required. Because there were not enough additional workers to work in the cowsheds, the livestock would suffer from being kept indoors (ibid., pp. 29f.). Fewer animals would be able to be kept after the drainage of the bog, and there would not be enough fertilizer for the pastureland as this would be needed for the cultivated land (Aretin 1795, pp. 146f.; Aretin 1796, p. 168). The newly established colonies would experience a lack of wood and forest litter, as wood would become scarcer and more expensive as the days went by (Delagera 1794, pp. 26f.).[12] The colonists would steal from the established inhabitants (ibid., p. 27; Schatte [n.d.], p. 86) and were altogether described as 'lumpen' and 'riff raff', as degenerates of the serving class and as ill-disciplined children (Delagera 1794, p. 25; Schatte [n.d.], p. 72).

It goes without saying that those in favor of the scheme had much to say in response to this and all other concerns by 1795, at which point contemporaries saw the cultivation as essentially signed and sealed (Anonymous 1796; Aretin 1795, 1796). Their counterarguments (e.g., Aretin 1795, pp. 149f.) drew upon scientific expertise, for example, from the renowned naturalist Franz von Paula Schrank, who had entered the bitterly contested debate with his published work *Naturhistorische und ökonomische Briefe über das Donaumoor* (Schrank 1795).[13] Overall, the proponents of the drainage plans still seem to have had the better arguments on their side in 1795. There was polemic on each side of the divide, but those in favor were well informed, smart and rational in their arguments. They were able to refute the often underhanded and personal accusations and concerns of their opponents point by point with well-founded arguments that were based on the expert knowledge of their age. Were it the job of any historian to come down on one side or the other of this contest, I would put myself on the side of Aretin

[11]'*Peat and sandy soil, which the first wind draws up like dust and steals away together with what little loosened fruit is there.*' (Delagera 1794, p. 22; Schatte [n.d.], p. 59).

[12]See the debate to which Radkau (1983) contributes on the alleged or real 'wood famine' in the 18th century, which the opponents also use in their arguments here.

[13]On Schrank as a fundamental proponent of cultivation, see also Hoser (2012, pp. 217f.).

and his allies—although only from the perspective of the sources available until 1795. The arguments brought forward to that date by opponents of the scheme were often formulated on the basis of comparatively little knowledge. Their warnings often appear pedantic, pessimistic and defeatist in comparison to the plans and arguments of those in favor. In the longer term, however, they were right on many points. With each passing year, and thus growing temporal distance from the far-reaching interventions undertaken in the bog area, it became clearer that the cultivation of the Donaumoos was leading to a disaster that was in equal measure ecological and social.

19.3.5 The Donaumoos by 1830—Taking Stock of a Socioecological Disaster

A little more than a generation later, poverty, begging and demoralization were widespread in the Donaumoos, and the region had a terrible reputation. By this time, approximately 3000 people were living in the region '*in physical and moral wretchedness*'. In 1829, a 'Donaumoos Association' (*Donaumoos-Verein*) was founded with the aim of finally turning the inhabitants into '*useful citizens*' (Donaumoos-Verein 1831, pp. 1, 12, 16f.).[14] The newly created agroecosystem was not profitable enough, and the social system threatened to collapse as a result. From all the expectations that had been so eloquently and knowledgeably expressed in the 1790s, very few had been fulfilled.

It should be noted that the microclimate had clearly improved through the drainage measures; there was less hail, mist and frost (Kling 1806, pp. 36–38); and mortality rates in and around the bog had apparently decreased significantly (Donaumoos-Verein 1831, p. 8; Pechmann 1832, p. 72). Some of the wounds scarring the landscape as a result of the intervention at the end of the 18th century had clearly healed with the passage of time. The opponents of the scheme had prophesied in the 1790 s that wind erosion would remove the desiccated peatland soils,

[14]A brief note regarding sources: in this 1831 publication, the *Donaumoos-Verein* had to pull off a rhetorical balancing act. Fundamentally, they wished to attract charitable patrons who were ready to provide loans to the Association from which those donating could not expect to receive any financial profit (Donaumoos-Verein 1831, pp. 18f.). The situation of the bog must therefore not appear to be without hope, yet the conditions still had to be described in such a way that would awaken sympathy and a desire to help. Above all, it had to be convincingly shown that the colonists currently living in the bog were suffering in desperate conditions largely through no fault of their own. The central statements made by the Association are, however, confirmed by contemporary publications (particularly Pechmann 1832). That a critical view of the Donaumoos was not a unique opinion is also shown by the following judgment reached between 1827 and 1837 by the Bavarian encyclopaedist J.A. Schmeller: '*The sums used for the cultivation of the [Donaumoos] appear to have been wasted, where new efforts do not use up what is available [...] Nothing comes of nothing [...] beggars remain [...] even as colonists, generally beggars.*' (Schmeller 1872 [1827–1837], p. 1673).

and they were proven right.[15] There had been heated debate surrounding indoor feeding regimes for livestock. Because it had been assumed that the colonists would thus require less pasture land, they had been allocated too little land altogether. Both cropland and livestock became scarce, as livestock had been unequally distributed among the colonists. The settlers should have purchased fertilizer, yet most did not have enough money to do so (Donaumoos-Verein 1831, pp. 14f.). At the drawing board stage, the 'Culture Commission' had planned a closed agroecosystem, but it remained reliant on external input. Forty years after the area's intensively discussed, fully considered and precisely planned cultivation, the Donaumoos had become an extremely vulnerable landscape. In the summer of 1800, a wildfire raged across the desiccated bog for six consecutive days. Over a large area (more than 20 km^2 affected), the fire ate its way through the peat, even to a depth of 70 cm in some places (Kling 1806, pp. 6f.). At the end of June 1831, a '*hurricane*' blew through the bogland, destroying dozens of colonists' homes, razing cultivated areas and leaving an '*extraordinary flood*' in its wake, as a result of which the potato harvest rotted (Donaumoos-Verein 1831, pp. 22f.).

Was this the paradise promised by Commission Director Stengel (1791) and his commissioners? How could things have reached this point?

19.3.6 Social Responses to the Disaster

Contemporaries searched for answers and came up with the following explanations. Aretin and his like had simply promised too much.[16] The first colonies had been too hastily established. The planners should have waited longer after the drainage measures until the peat soils had recovered and developed greater resistance to wind erosion. The terrible reputation of the Donaumoos went much further back than 1830 and had thus attracted the least capable rather than the most suitable colonists (Donaumoos-Verein 1831, pp. 9, 11f.). From 1802, as a result of the War of the First Coalition, families from the Rhineland had settled in the Donaumoos, but they had no experience working with land of this type (Kling 1806, p. 21). Others living in the Moos included Württemberg and Frankish natives and people from various regions of Bavaria. Unscrupulous investors had also taken beggars off the streets and turned them into colonists (Pechmann 1832, pp. 75f.).

All contemporaries seem to have been united on one issue: the most significant reason for the desolate condition of the bog by 1830 was its '*neglect*' (Pechmann 1832, p. 73). The writings of Aretins and other supporters had led people to

[15]'*Fifteen to 20 years ago [i.e., until ca. 1815], it was not uncommon for the wind to carry away the light peat soil as it does the snow, yet such occurrences are certainly not observed in the present time.*' (Donaumoos-Verein 1831, p. 9).

[16]This observation, although correct in itself, did not prevent the Association 40 years later from itself promising '*one of the most beautiful and flourishing provinces in Bavaria within 15 years*' (Donaumoos-Verein 1831, p. 19).

believe that after the massive intervention between 1790 and 1795, this unparalleled tour de force would basically be complete. In the language of the socioecological concept of colonization, the supporters had underestimated to what extent the social system was entering into a long-term commitment imposed by this colonizing intervention. It might be that this group of actors deliberately underplayed the possible long-term effects to increase public acceptance for their project. The fact is that (apart from all the social upheavals already noted) the new agroecosystem after 1795 was not stable in any respect. Instead, a more profound decline gradually set in, particularly through the 1820s.

Heinrich Pechmann, who, as a water engineer, became well-known particularly for his later canal project linking the Danube and the Main rivers, focused on the Donaumoos from 1820 onward. He saw the decline of the Donaumoos as an opportunity for '*instructive and useful experiences*' (Pechmann 1832, p. III). In his book published in 1832, he devoted several chapters to the question of how and why events had taken the course the way they did. According to him, the Commission itself had made some minor errors. For example, the main channel was too wide. As a result, the water was unable to transport silt, leading to an overall rise in the level of the river bed (ibid., p. 42). The project planners should not have removed all mills from the bogland (ibid., p. 43).[17] Instead, they should have prevented new ones from being built (ibid., p. 73). Still, Pechmann was an ardent supporter of the '*Kultur*' of the bog.[18] For him, the fault lay not primarily with the planners of the 18th century but with unscrupulous investors who had brought in unsuitable colonists and had proceeded to treat them badly.[19] This was compounded by the incorrect management of the desiccated peat soils. Instead of bringing these under the plough, cattle grazing should have been continued in the region (ibid., pp. 75–78, 102). The fact was that the channels and drainage ditches were not adequately dredged and maintained, so they fell into disrepair. Flooding episodes returned. The trees that had been planted in the 18th century were felled, paths through the bogland were not maintained, and most of the bridges had collapsed. Around 1818, many of the drained areas began to revert to bogland (ibid., pp. 99–102; Donaumoos-Verein 1831, p. 12).

The social system was clearly unable to cope with maintaining the drainage infrastructures over the long term. The desired benefits never came to pass, and inhabitants returned to practices such as cattle grazing, which had dominated before the colonizing intervention took place. However, the landscape was no longer the same as it had been before the intervention. The arrangements had undergone a complete transformation in the intervening period. The grazing livestock completely destroyed the newly created channels. The vegetation that had thrived in the waterlogged peat bog had vanished from the desiccated ground, while the plants that were adapted to the new conditions were unable to grow because they were immediately eaten by the grazing animals

[17]Both Father Lanz and Aretin had already insisted upon this, believing that the mills were one of the major obstacles to the 'culture' (Aretin 1795, p. 52; on Lanz, see Hoser 2012, p. 207).

[18]Pechmann begins his book with a chapter '*On the Harmfulness of the Swamps and the Usefulness of Draining the same*'.

[19]For more details, see Hoser 2012, p. 220.

(Pechmann 1832, p. 77). We can see how in such a precarious transitional situation, old practices collided with new arrangements. This too made the problems worse and hastened the bog's decline at the beginning of the 19th century.

How did society react to such a social and ecological disaster? Colonizing interventions are irreversible—there is no way back to the time before the intervention. Too much money and effort had already been invested to leave the colony and the people who spent their lives there to themselves. For some forms of social response to the failure, the word experimentation seems appropriate. People began to conscientiously observe the colony (and themselves). Further targeted interventions were undertaken, and the way in which the colonized system reacted to these interventions was precisely noted. Fundamentally, an attempt was made to understand what had actually been created and how and for what it could be used.

Johann Peter Kling did precisely this. As a senior official in the service of the Bavarian authorities, he focused on questions of agricultural development. In 1801, he began an open-air experiment in the Donaumoos after several years of preparation. Kling believed it was time '*to ask Nature herself, whether and under which conditions the culture of this significant area of land would be advisable*' (Kling 1806, p. 8, 41). He leased 34 ha of land in the Donaumoos and began to run an experimental farming project there (Hoser 2012, p. 214). For a period of five years, Kling cultivated 15 different crop plants there and observed them as they flourished and decayed, noting the produce achieved under different weather conditions and the impact of pests. He experimented with various patterns of crop rotation and fertilization and had the quality of dye plants, for example, tested by experts (Kling 1806, p. 41–101). Kling was convinced that it would be possible to improve the fertility of the desiccated peat soils by burning the upper layer of peat and then mixing it into the lower layers (ibid., p. 5). He viewed the large-scale fire of 1800 as a providential stroke of luck, albeit one that had not been taken advantage of because the ashes were blown away before they could be ploughed into the soil (ibid., p. 7). Kling was not the only person who wished to use a model farm experiment in the first half of the 19th century to demonstrate that agricultural activities had a future in the Donaumoos (Hoser 2012, p. 215).

A businessman from Boston proposed the introduction of millions of sugar maple trees (*Acer saccharum*) from North America to transform the Donaumoos into a region producing sugar and hardwood. This suggestion, however, earned nothing but derisive comments (Kling 1806, p. 6; Pechmann 1832, p. 79). Others had other plans regarding the peat. After the crop and fruit cultivation and indoor livestock-keeping plans of the 18th century had clearly failed, peat was rediscovered as a resource around the turn of the century. Thus, a Kommerzienrat Bresselau planned to use the entire bogland from 1798 as a peat-digging area (Hoser 2012, p. 212) to supply Bavaria and Austria, as far as Vienna via the Danube, with this fuel (Kling 1806, p. 5; Pechmann 1832, p. 79).[20] Around 1830, the Donaumoos Association also wanted to promote the removal of the

[20]In Hoser's account (2012, pp. 212f., 234), the primary responsibility for the decline of the Donaumoos after 1800 is attributed to Bresselau. Because he had placed all his faith in peat extraction, he had shown no interest in the maintenance of the drainage infrastructure and had allowed them to fall into disrepair.

peat, which was seen as a resource that could not be exhausted even over centuries (Donaumoos-Verein 1831, p. 20). For this purpose, a peat storehouse was to be erected by the Danube and the drainage channels adapted to allow for the transportation of peat by barge. Making these channels navigable was a project for which Pechmann, the water engineer and designer of the Danube-Main canal, was the most prominent advocate (Pechmann 1832, pp. 127–129, 152). These projects seem to be symptomatic of the early phase of the industrial transition. To fulfill the growing need for energy, people wished to rely again on peat, and at that time—a few years before the establishment of steam-powered rail transport—it was envisaged that the large-scale transport of the fuel should still take place in accordance with agrarian models, by water transport. None of this was actually carried out. The key question remained unanswered: what have we actually created, and how could it be useful?

19.3.7 The Donaumoos up to the Present Day: Long-Term Consequences of Colonization[21]

Agrarian usage of the Donaumoos remained the dominant social interest in further colonizing interventions until well into the 20th century despite all the associated problems. Yet, the question of how this goal was to be achieved received a number of different answers. In the language of Social Ecology, social systems always intervene in colonized systems based on a particular mode of perception and specific expectations. These culturally driven perceptions—how the colony functions and, therefore, which interventions are believed to ensure the desired outcome—are extremely variable in historical terms (Schmid 2006, pp. 67f.). This is evident, for example, in the question of whether the Donaumoos required irrigation or drainage to finally become more productive in an agricultural sense. Until 1870, irrigation was largely seen as the answer. This view peaked in 1866/67, when several Donaumoos communities called for drainage activities to cease completely (Hoser 2012, p. 223). Starting in the 1870s, at least two decades were spent alternating between '*regulated*' irrigation and drainage. In 1873, some suggested 'water reservoirs' to control the in- and outflow of water, whereas others placed their faith in the installation of additional floodgates in the drainage channels. From the 1890s onward, there was a return to the 18th-century paradigm and a reliance on thorough drainage measures. This reached a peak in 1990, with the demand by one village community that the bog be completely drained of water. Regardless of whether current thinking favored drainage or irrigation, measures were always aimed at improving the conditions for agricultural production in the Donaumoos. The intensity of colonization was never reduced; in fact, it escalated

[21]The following abbreviated account of the history of the Donaumoos from the second half of the 19th century until the present day is based on Hoser (2012, pp. 221–232).

beginning in 1906, when the 'new regulation' of the Donaumoos was initiated. Both World Wars delayed this major project, which dragged on for more than four decades and was only regarded as concluded in 1947. A few years later, however, from the 1950s until the 1970s, more drainage measures were needed. The reason for this appears to have been a paradoxical effect of the drainage project. The peat soils subsided, reducing the depth to groundwater level, and the land became well and truly saturated. In the 1970s, it was found that a peat layer 3 m in depth and covering some 6000 ha of the bog (roughly one-third of the entire Donaumoos) had been lost since the 18th century (Hoser 2012, p. 231). In the mid-1980s, the latest paradigm shift came into play. An ecological expert opinion ('*Ökologisches Gesamtgutachten*') recommended localized irrigation to combat the problem of peat depletion. This led to protests among the local population, some of whom saw this as a regression to the time before the great cultivation project (ibid., p. 232).

Flooding represents one of the major challenges facing the Donaumoos today. Heavy precipitation events have led and continue to lead to large-scale flooding in the Donaumoos, for example, in 1923 and 1936 (Hoser 2012, p. 229) and recently in January 2011. Because the soils remain subject to progressive subsidence due to oxidative decomposition as a result of drainage, making it even more difficult for water to flow unhindered into the Danube, the bog is becoming waterlogged (Donaumoos Zweckverband 2010). The further extraction of peat has also exacerbated this trend. As early as the second half of the 19th century, it was recognized that peat extraction, although an important source of income, also made the continual deepening of drainage channels necessary (Hoser 2012, p. 224). Agricultural production was the principal victim of these floods, and yet it was the intensification of agricultural activity from the end of the 19th century and into the 20th century that had contributed to the subsidence of the bogland. From the 1870s onward, the chronic lack of fertilizer in the bogland could be counteracted by the increased application of artificial fertilizers, thus achieving success in the short term. However, the artificial fertilizer accelerated the mineralization of the peat soils that was already occurring due to the drainage measures and thus also increased the rate at which subsidence was taking place (ibid., p. 227). We may thus see how the solution to one problem creates a new one or, as here, compounds an existing problem. The history of the Donaumoos provides us with an abundance of examples of this relationship, which environmental historians term the 'risk spiral' (Müller-Herold and Sieferle 1997; see also Chap. 6 in this volume).

In 1991, several local authorities joined to form an association named '*Donaumoos-Zweckverband*'. The association's *Development Concept*, drafted in 2000, aims to harmonize the interests of flood protection, agriculture, species and habitat protection and peatland conservation by 2030. By then, it will have been 240 years since the radical and large-scale colonizing intervention took place, whose direct and indirect, desired and undesired consequences society will still have to address. That much is certain.

19.4 Conclusions

The environmental history of the Donaumoos is, in many respects, a particularly dramatic example of the 'colonization of natural systems' and its long-term consequences for both nature and society. However, more general recommendations for a better understanding of 'colonization' may be drawn from this.

The cultivation of this lowland bog from 1790 onward was a radical project in the sense that it promised to create an agroecosystem from scratch on a relatively large tract of land within a short space of time. This required specific political conditions and, at the end of the 18th century, public acceptance, which had to be deliberately manufactured. The result is a rich legacy of historical sources. If we did not have such a wealth of sources to draw upon, we would be inclined to characterize many of the later consequences of this intervention as 'unintended side effects'. In the case of the Donaumoos, we know that contemporaries of the intervention had intensively discussed problems such as wind erosion, a lack of fertilizer, social upheavals and many other issues while still seeing the intervention through in practice, having given it much thought based on the expert knowledge of the time. A few decades later, the result was a social and ecological disaster. Although failure in the case of the Donaumoos may be particularly conspicuous, colonization reveals itself through this example to be a social activity that is high-risk in principle. This is partly because society never has a full understanding of the system in which it intervenes; the intervention is always based on historically specific and highly variable (cultural) perceptions (of nature). The experiences society undergoes in the process are, however, not solely cultural. They are absolutely tangible: people suffer starvation; livestock perish; and the land burns in wildfires, subsides and becomes subject to flooding. It is precisely in precarious transitional situations—when new arrangements encounter traditional practices—that the level of uncertainty is particularly high. To understand what they have actually created through colonization, people begin new interventions and invest in the 'monitoring' of the colony. New experiences must be made, where possible, under controlled conditions. How does the colony react to this or other interventions? How could it be useful?

There is much in this environmental history of the Donaumoos that falls within the scope of the term 'risk spiral'. Long-term historical observation of an example of colonization reveals more than the new risks that result from the solution to earlier problems. The risk spiral conceals not only the fact that colonies continually create new problems through their resistance to stresses and independent existence but also the fact that over time, societal interests, utilization requirements and technological possibilities undergo fundamental change. New contradictions in terms of utilization develop both within a particular period and throughout time. Colonies are inherited socio-natural hybrids that always carry the traces of earlier uses. A colony that was once—more or less successfully—created for crop farming will later on demonstrate resistance to other forms of utilization. This is also true in the case of so-called 'renaturation projects' in recent times. The environmental history of the Bavarian Donaumoos confirms what Historical Ecology has also found upon examination of similar (former) wetlands in Switzerland: the transformation of these special landscapes is not merely a story of habitat destruction. Forms of utilization and conservation were and continue to be defined by the geomorphological characteristics of these landscapes. Today, former bogs can be protected and rehydrated through localized measures because, despite (or perhaps because of) centuries of invested effort by society, they are in principle less well-suited for agricultural production than other landscapes (Bürgi et al. 2010; Gimmi et al. 2011).

Social Ecology interprets 'colonization' as targeted, regulatory and sustained interventions in natural systems. The colonization concept of the 18th and 19th centuries focuses primarily on the incoming settlement of people—in other words, it focuses precisely on the founding of 'colonies' (Krünitz 1776/1855). Yet what the experts of the 18th century understood by the term '*culture*' was actually very similar to what Social Ecology today terms 'colonization'. The experts of the later Enlightenment period were convinced that it was possible to intervene in the Donaumoos in a targeted, regulatory and sustained way. They were certain that their '*systematic*' approach would significantly improve the utility of this tract of land. They did not anticipate that succeeding generations might be overwhelmed by the effort of maintaining the infrastructures they had created, that aims are subject to change and that complete regulation and control turn out to be an illusion. The history of the Donaumoos and the example it sets may also be read as a warning against the hubris of experts in any form.

Acknowledgements This contribution is based on research I conducted while at the Rachel Carson Center for Environment and Society (RCC) at the Ludwig-Maximilians-Universität München in 2011. I wish to thank the team and the Fellows of the RCC for their professional support and inspiring discussions. My thanks go to Martin Knoll for his reference to J.A. Schmeller and his wise comments on an earlier version of this text. A longer German-language version of this text is planned for publication in 2015 in a volume that brings together contributions to the annual conference of the Institut für Donauschwäbische Geschichte und Landeskunde, organized by Márta Fata in Tübingen in November 2013, on the subject of 'The colonization of bogs and wetlands: Land reclamation and migration between 1700 and 1850—a comparative study of the Habsburg Monarchy, Prussia and other German states'. Thanks to Ursula Lindenberg for translating this text from German into English.

Sources and References

Printed Source Materials

Anonymous. (1796). Anonymous [signed „Dr."]: [Review by] Georg Freyh. v. Aretin, Actenmäßige Donaumoos-Cultursgeschichte. In: Neue allgemein deutsche Bibliothek. Vol. 25/1. Kiel, pp. 89–92. (Online: http://www.ub.uni-bielefeld.de/diglib/aufklaerung/)

Aretin. (1795). Aktenmäßige Donaumoos-Kulturs-Geschichte. Mannheim. Bayerische Staatsbibliothek, Signatur: 4. Bavar. 135 a. (Online: http://www.bsb-muenchen.de/)

Aretin (1796). Anonymous [probably Johann Georg von Aretin]: Vier wichtige Aktenstücke zur Kultursgeschichte des Donaumooses in Bayern. Ein Beitrag zu einer allgemeinen Kultursgeschichte des Landes. Germanien [i.e. Augsburg]. Bayerische Staatsbibliothek, Signatur: Bibl. Mont. 3132. (Online: http://www.bsb-muenchen.de/)

Delagera, N. (1794). Wahrhaft patriotische, ganz uneingennützige [...] Gedanken [...] über die famose Donau-Moos-Kulturs-Geschichte [...]. In: Aretin 1796, pp. 1–30.
Donaumoos-Verein (1831). Beschreibung des Donaumooses in Beziehung seines Einflusses auf die Umgegend vor und nach der Austrocknung dann, seiner Produktionsfähigkeit und der Kolonisierungsweise mit einem Actien-Plane. Augsburg. Bayerische Staatsbibliothek, Signatur: Bibl. Mont. 3218-78#Suppl.13 (Online: http://www.bsb-muenchen.de/)
Kling, J. P. (1806). Beschreibung eines Kulturversuches im Donaumoos [...] München. Bayerische Staatsbibliothek, Signatur: Bavar. 1463 (Online: http://www.bsb-muenchen.de/)
Krünitz, J. G. (1855). Oekonomische Enzyklopädie [...], vol. 8, Lemma „Colonie, Pflanzstadt". (Original work published 1776) (Online: http://www.kruenitz1.uni-trier.de/)
Maire, F. J. (1786) Bemerkung über den inneren Kreislauf der Handlung [...] oder Hauptentwurf der zu eröfnenden [sic!] schiffbaren Wasserstraßen von allen Meeren Europens bis nach Wien. Straßburg and Leipzig.
Marsigli, L. F. (1726). Danubius Pannonico-Mysicus, Observationibus Geographicis, Astronomicis, Hydrographicis, Historicis, Physicis Perlustraturs [...], 6 vols. Den Haag and Amsterdam.
Pechmann, H. F. V. (1832). Geschichte der Austrocknung und der Cultur des Donaumooses in Bayern. München. Bayerische Staatsbibliothek, Signatur: Bavar. 2068 f (Online: http://www.bsb-muenchen.de/)
Schatte, F. V. [n.d.]: Pro-Memoria. In: Aretin 1796, pp. 31–103.
Schmeller, J. A. (1872). Bayerisches Wörterbuch, Lemma „Mos", p. 1673. (Online: http://www.bayerische-landesbibliothek-online.de/schmeller) (First editions published between 1827 and 1837).
Schrank, F. V. P. (1795). Naturhistorische und ökonomische Briefe über das Donaumoor. Mannheim. Bayerische Staatsbibliothek, Signatur: 4 Bavar. 1675 (Online: http://www.bsb-muenchen.de/)
Stengel, S. V., Mettenleiter, J. M., & Riedl, A. V. (1791). Die Austrocknung des Donaumooses. An dem Jahrtage der Akademischen Stiftung in einer öffentlichen Versammlung. München. Bayerische Staatsbibliothek, Signatur: 4 Bavar. 1762. (Online: http://www.bsb-muenchen.de/)

References

Blackbourn, D. (2006). The conquest of nature: Water, landscape and the making of modern Germany. London: Random House.
Bürgi, M., Straub, A., & Gimmi, U. (2010). The recent landscape history of Limpach valley, Switzerland: Considering three empirical hypotheses on driving forces of landscape change. *Landscape Ecology, 25*, 287–297.
Donaumoos Zweckverband. (2010). Kernprobleme: Torfschwund und Moorsackung – Folgen der Entwässerung. (Online: http://www.donaumoos-zweckverband.de/das-donaumoos/kernprobleme.html)
Fischer-Kowalski, M., & Erb, K.-H. (2006). Epistemische und konzeptuelle Grundlagen der Sozialen Ökologie. In *Mitteilungen der österreichischen Geographischen Gesellschaft* (pp. 148, 33–56).
Fischer-Kowalski, M., & Weisz, H. (1999). Society as a hybrid between material and symbolic realms. Toward a theoretical framework of society-nature interaction. *Advances in Human Ecology, 8*, 215–251.
Gimmi, U., Lachat, T., & Bürgi, M. (2011). Reconstructing the collapse of wetland networks in the Swiss lowlands 1850–2000. *Landscape Ecology, 26*, 1071–1083.
Hohensinner, S. (2010). „Sobald jedoch der Strom einen anderen Lauf nimmt ...": Der Wandel der Donau vom 18. Zum 20. Jahrhundert. In: Verena Winiwarter, Martin Schmid (eds.), Umwelt Donau: Eine andere Geschichte. St. Pölten, pp. 39–55.
Hohensinner, S., Lager, B., Sonnlechner, C., Haidvogl, G., Gierlinger, S., Schmid, M., Krausmann, F., Winiwarter, V. (2013). Changes in Water and Land. The Reconstructed Viennese Riverscape 1500 to the Present, in: Water History, 5, pp. 145–172.
Hoser, P. (2012). Die Donaumooskultivierung und ihre Folgen. In: Rolf Kießling, Wolfgang Scheffknecht (eds.): Umweltgeschichte in der Region (=Forum Suevicum. Beiträge zur Geschichte Ostschwabens und der benachbarten Regionen 9), 205–235.

Mohilla, P., & Michlmayr, F. (1996). Donauatlas Wien. Geschichte der Donauregulierung auf Karten und Plänen aus vier Jahrhunderten, Wien.

Müller-Herold, U., & Sieferle, R. P. (1997). Surplus and survival: Risk, ruin, and luxury in the evolution of early forms of subsistence. *Advances in Human Ecology, 6*, 201–220.

Radkau, J. (1983). Holzverknappung und Krisenbewußtsein im 18. Jahrhundert. *Geschichte und Gesellschaft, 9*, 513–543.

Reith, R. (2011). Umweltgeschichte der Frühen Neuzeit (=Enzyklopädie deutscher Geschichte 89), München.

Schatzki, T. R. (2003). Nature and technology in history. *History and Theory, 42*, 82–93.

Schmid, M. (2006). Herrschaft und Kolonisierung von Natur: ein umwelthistorischer Versuch zur Integration von Materiellem und Symbolischem. *Mitteilungen der Österreichischen Geographischen Gesellschaft, 148*, 57–74.

Wagner, B., Hauer, C., Schoder, A., & Habersack, H. (2015). A review of hydropower in Austria: Past, present and future development. *Renewable and Sustainable Energy Reviews, 50*, 304–314.

Winiwarter, V., & Schmid, M. (2008). Umweltgeschichte als Untersuchung sozionaturaler Schauplätze? Ein Versuch, Johannes Colers „Oeconomia" umwelthistorisch zu interpretieren. In T. Knopf (Ed.), Umweltverhalten in Geschichte und Gegenwart: Vergleichende Ansätze. Tübingen (pp. 158–173).

Winiwarter, V., & Schmid, M. (Eds.). (2010). Umwelt Donau: Eine andere Geschichte. St. Pölten.

Winiwarter, V., Schmid, M., Hohensinner, S., & Haidvogl, G. (2013). The environmental history of the Danube River basin as an issue of long-term socio-ecological research. In S. J. Singh, H. Haberl, M. Chertow, M. Schmid & M. Mirtl (Eds.), *Long-term socio-ecological research. Studies in society—Nature interactions across spatial and temporal scales* (=Human—Environment Interactions 2, pp. 103–122).

Wohl, E. (2011). A world of rivers: environmental change of ten of the world's Great Rivers. Chicago.

Method Précis: Working with Historical Material

Verena Winiwarter

Long-term socioecological studies and environmental histories, like most other forms of interdisciplinary approaches, rest on a variety of input data, with written material dating back hundreds or thousands of years among the most important. This kind of material is the mainstay for historians. Historians call any material handed down from the past a 'source' (for an introduction, see Howell and Prevenier 2001). Written sources have dominated the work of historians and still play a major role, but maps, plans and images are becoming increasingly important. Historians aspire to a claim of truth, and they apply methodical rigor to reach it. Whenever they use sources produced by humans, they act as observers of observations—that is, as second-order observers of incomplete evidence—which makes conjecture necessary. To piece together a story, historians select from the available extant material, which, in all likelihood, is a very incomplete witness to a historical event.

However, incompleteness of the extant material is not the only problem for second-order observers. Walter Benjamin already noted in 1940, 'History is the object of a construction whose place is formed not in homogenous and empty time, but in that which is fulfilled by the here-and-now [*Jetztzeit*].' (Benjamin 1974, para. XIV) If this is true, then historical writing—unlike the work of scientists—is not aimed at an abstract truth but is written with a purpose arising from the present. This is where the insistence on methods comes into play. Whatever their way of dealing with sources, historians want their work to be discernible from literary storytelling: they claim to write non-fiction. Carlo Ginzburg has shown the similarities between a detective's work and that of a historian in piecing together evidence and making conjectures, thereby elucidating a general methodical principle of historians (Ginzburg 1984).

In historical theory, the primacy of narrative ('emplotment') has been discussed widely, with Hayden White leading the way (e.g., White 1987/1990). The main argument is that historians write literature, which means that the narrative has primacy. Historians cannot deliver transcendent reflections of reality. Historical writing is a cultural artifact, a complex form of literature, but it is constructed under the constraint of a protocol that will evoke in the reader the label 'non-fiction'. The textual intention of the historical narrative is a claim of truth (Partner 1998). This construction of a truth-claiming narrative consists of two operations. One is the evaluation of the quality of input data (commonly called 'sources'), and the other is the construction of the narrative itself.

To evaluate the feasibility of a particular piece of historical source becoming part of a truth-claiming narrative, several steps are performed. Reconstructing the context of production is often the first step. Who produced the material we are looking at, and when, where and why? Who commissioned the work, if it is not the person who produced the work? The context of use is also important. How and why was the source used, and by whom, when and where? The third contextual question is related to the reason for keeping the source. Why was the source kept and not discarded? (Clanchy 2012) One must take into account that most of the historical sources are lost. We are dealing with incomplete material. Fires in libraries and willful discarding of material that is no longer necessary have made historical work both difficult and possible (we would be hopelessly swamped if we had everything that was ever written at hand). Not all extant material falls into the category of 'kept material'. Historians can also find discarded materials particularly interesting as they are often of a different character than the materials purposely kept. A case in point is the ancient rubbish dump near Oxyrhynchus in Egypt discovered in the 19th century. The manuscripts from the dump date from the 1st to the 6th century CE. They include thousands of Greek and Latin documents, among them shopping lists and letters. But they also contain fragments of otherwise lost works by Greek playwrights and the Roman historian Livy.

Interpreting the content of the source is another necessary step. Usually, both steps are performed in parallel as an understanding of content helps one evaluate context and vice versa. Historical ancillary sciences such as Diplomatics (the study of charters or, more generally, documents) or Metrology (the study of historical measures) are needed to interpret sources, particularly those from the pre-print period, that is, anything before the 16th century CE. Paleographic skills are needed to decipher manuscripts. The form of each letter in a text can also be used as a dating device, as writing styles changed over time (e.g., 'Carolingian minuscule'). If text is written on stone, the result is called an 'inscription'. Epigraphy deals with this type of evidence. The study of inscriptions has changed the way we understand ancient history a fair bit. The identification of new sources (such as inscriptions) is commonly called heuristics in the historical profession.

What has been said above about the interpretation of these materials is true for any material artifact, a well-known problem in Archaeology. 'Artefacts mean nothing. It is only when they are interpreted through practice that they become invested with meanings […] Our knowledge is not grounded upon the material evidence itself, but arises from the interpretive strategies which we are prepared to bring to bear upon that evidence' (Barett 1994, pp. 168, 171).

In Long-Term Socioecological Research (LTSER) and Environmental History, learning about past environmental conditions is indispensable. Obviously, any written matter is written from someone's viewpoint, that of the author(s), turning us into second-order observers. We observe the information contained in the written matter as informing us about the perception of the world (or the environment) the authors had and not directly about the world they perceived. For biological and geological archives such as tree-rings, sediments and pollen, we are first-order observers as these archives are a direct way to reconstruct past environments. We have to pay attention to the methodical limits of each method to decipher these archives and not make overstated or unfounded claims. Combining multiple sources of different origin with different methodical limits greatly enhances the validity of the narratives we construct.

Let us now turn to the study of past society-nature interactions. A variety of source types can be used. Anything pertinent to land use, such as tithe registers or books kept by landlords about peasant holdings, are useful. Any type of yield-related information, most often contained in lists of taxes and tariffs, is particularly welcome. Over the course of history, societies have tended to write down increasingly more in increasingly greater detail. The new science of statistics gave rise to particularly detailed inventories of natural resource availability and extraction from the early 19th century onward. Because quantitative methods are more easily applied using such sources, LTSER and historical studies in the realm of Social Ecology have focused on this period. Innovative combinations of quantitative and qualitative evidence are currently under way that will expand the time frame for such studies. Historians of climate and historical climatologists have together perfected the art of combining different source materials and statistically testing the validity of these materials.

Whereas administrative documentation, such as a 15th-century land register, would in essence be written as an account of reality (forgeries notwithstanding), sources such as diaries, travelogues and other subjective narrations are also important. We need to analyze the context of such subjective sources and evaluate the truth claims we are making based on them, but without them, we would be unable to study the past. In fact, an author's perception of a particular environment can be the very basis of our analysis of the material ramifications it produced. All sources teach us about human practices. These practices are the basis of arrangements created from the natural world that, in turn, influence future practices (see Chap. 19).

In accordance with the great questions of Environmental History, material on infrastructures is of particular relevance in addition to information on land use. Alpine paths, channels, railway lines and accounts of bridge repairs can all be used to work on questions related to infrastructural connections.

The more people in the past disagreed on the claims they made (particularly about possessions and rights to use resources), the more written sources we tend to find. Legal material (so-called normative sources) such as laws and court records greatly enhance our potential to reconstruct the management of natural resources in the past. Public media became an increasingly widespread and important source from the 17th century onward, and the learned literature of any given time is always an important source to reconstruct the prevalent understanding of natural phenomena. Information on technology is important, too. Here, written and pictorial types of matter are best combined with the study of artifacts. Pictorial matter, particularly maps, is a key resource for LTSER and Environmental History. This material must be viewed as critically as written material, because it is based on the perceptions and interests of its makers. Photography also cannot be taken at face value. Decisions about the motif, the angle and the viewpoint influence the pictorial representation in important ways. In addition to studying the intentions behind and functions of sources, we must always ensure that the matter at hand is what it claims to be and is not forged. Oral history, the production of narrative sources by historians through interviews and film, is not exempt. Oral history involves a painstaking evaluation of the informational content of oral sources, as can be seen when making transcriptions that include every sigh, pause, or cough by the interviewer and interviewee. Even a sharp intake of air is an informative part of the material.

Questions of land-use and infrastructure, of production and consumption, of trade and technological advances have been studied by economic and social historians and historians of technology for a long time. Statistical analysis of serial sources, such as long time series of prices or historical demography, is performed using the same software used for data from the recent past and present. Such analyses have greatly enhanced our ability to extract meaningful information out of vast arrays of data. However, such statistical data need to be interpreted with a pinch of salt because they are often inconsistent, the areas covered can change, the data are never error-free, and fragmented evidence calls for particularly careful analysis.

These briefly sketched tools for addressing historical sources are often taken for granted by historians, and books on historical methods are still rare.

Let me briefly pick up the thread on the construction of narratives that we discussed at the beginning of this reflection on methods, but from a slightly different angle. Historians have pioneered the study of the analytical constraints arising from the fact that histories are stories told as stories, and they cannot be told otherwise. Any chronological listing of events needs to be turned into a narrative sequence or a network of causally related and structurally coherent histories (for the difference between chronicle and story, see, e.g., White 1987/1990). This construction is not a post hoc operation on a fixed, neutral chronicle but rather the central task of the historian. Historians are observers. They differentiate and denominate, performing the two main operations of observation (compare, e.g., Schmidt 1997). The related discussion in the profession has been called 'the narrative turn' and can be dated to the 1970s and 1980s, but the methodical issue has concerned historians much longer. The construction of meaningful narratives is both unavoidable and problematic. Historians of future generations will study our own historical analyses as sources, and they will see the limitations of our interpretations.

According to sociologist Niklas Luhmann, society is constituted through recursive communication (e.g., Brandhoff 2009, p. 309). This communication is driven by the need for the construction of identity. Social memory, according to Schmidt (1997), functions like the immune system of a body by distinguishing between foreign and innate. Societies, like individuals, do not possess a repository of memory. Knowledge is embodied in the possibilities for perception, cognition and action of the members of a society. They organize interactions according to their social differentiation. Societies recollect by the activation of components of their organized connectivity, which leads to a process of heterarchical self-regulation. Thus, by offering the 'emplotment' of LTSER or Environmental History narratives, we are part of the larger effort of society to regulate itself with both its limitations and its potential to intervene in interaction patterns via historical reflection.

References

Barrett, J. C. (1994). *Fragments from antiquity. An archaeology of social life in Britain, 2900–1200 BC*. Oxford: Blackwell.

Benjamin, W. (Trans. D. Redmond) (1974). On the Concept of History. In R. Tiedemann & H. Schweppenhäuser (Eds.), Gesammelte Schriften: Vol. I,2. Abhandlungen. Frankfurt am Main: Suhrkamp. Retrieved from http://members.efn.org/~dredmond/ThesesonHistory.html

Brandhoff, B. (2009). Autopoietic systems, not corporate actors: A sketch of Niklas Luhmann's theory of organisations. *European Business Organization Law Review, 10*(2), 307–322. doi:10.1017/S1566752909003073.

Clanchy, M. T. (2012). From memory to written record: England 1066–1307 (3rd ed.). New York: Wiley (1st ed. 1979).

Ginzburg, C. (1984). Morelli, Freud, and Sherlock Holmes: Clues and scientific method. In U. Eco &. T. A. Sebeok (Eds.), *The sign of three: Dulpin, Holmes, Peirce* (pp. 81–118), Bloomington: Indiana University Press.

Howell, M. C., & Prevenier, W. (2001). *From reliable sources. An introduction to historical methods*. New York: Cornell University Press.

Partner, N. (1998). Hayden White: The form of the content. *History and Theory, 37*(2), 162–172.

Schmidt, S. J. (1997). Geschichte beobachten. Geschichte und Geschichtswissenschaft aus konstruktivistischer Sicht. *Österreichische Zeitschrift für Geschichtswissenschaft ÖZG, 8*(1), 19–55.

White, H. (1990). Die Bedeutung der Form, Frankfurt am Main: Fischer. [Original title: The Content of the Form: Narrative Discourse and Historical Representation. Published 1987]

Chapter 20
A Forest Transition: Austrian Carbon Budgets 1830–2010

Simone Gingrich, Christian Lauk, Thomas Kastner, Fridolin Krausmann, Helmut Haberl and Karl-Heinz Erb

Abstract The concept of forest transitions was introduced by geographers in the 1990s to describe the observation that forests regrow with industrialization in many parts of the world. We use the case of Austria to discuss the forest transition in the context of Social Ecology based on empirical evidence on Austria's carbon budget in the period 1830–2010. In this period, Austria's forests grew not only in area but also in wood density, resulting in a carbon sink of 23 %, or ca. 240 MtC (megatons carbon). This process was accompanied by increasing societal use of carbon, due in part to the surge in fossil fuel use and a fivefold increase in societal carbon stocks, or a sink of ca. 110 MtC, in 2010. As in ecosystems, (construction) wood was the main component driving rising carbon stocks in society. Although somewhat significant in extent, annual carbon sink rates are well below fossil fuel emissions to the atmosphere. We argue that the carbon sink in Austria's ecosystems and society was a by-product of increasing societal carbon throughput in the course of industrialization, fuelled by the use of fossil energy, and that carbon sequestration is therefore an unsuitable strategy to mitigate carbon emissions.

Keywords Forest transition · Carbon budget · Industrialization · Austria

S. Gingrich (✉) · C. Lauk · T. Kastner · F. Krausmann · H. Haberl · K.-H. Erb
Institute of Social Ecology (SEC), Alpen-Adria University,
Schottenfeldgasse 29, 1070 Vienna, Austria
e-mail: simone.gingrich@aau.at

C. Lauk
e-mail: christian.lauk@aau.at

T. Kastner
e-mail: thomas.kastner@aau.at

F. Krausmann
e-mail: fridolin.krausmann@aau.at

H. Haberl
e-mail: helmut.haberl@aau.at

K.-H. Erb
e-mail: karlheinz.erb@aau.at

H. Haberl et al. (eds.), *Social Ecology*, Human-Environment Interactions 5,
DOI 10.1007/978-3-319-33326-7_20

20.1 Introduction

Forests are an important issue in environmental discourse. From the mid-1970s to the late 1980s, environmentalists were concerned about the forest dieback syndrome in Central Europe that mainly resulted from air pollution from heavy industries. In the early 1990s, the global destruction of tropical rain forests became an issue, threatening, among others, biodiversity and local livelihoods. A different perspective on forest development emerged in the 1990s, when geographers introduced the concept of a 'forest transition' (Grainger 1995; Mather 1992) to capture the observation that after centuries and even millennia of agricultural expansion during which forest areas declined, forests actually expanded in the 19th and 20th centuries in many industrialized regions of the world.

Recently, the concept of forest transitions has been picked up in Land-System Science and Environmental History, and empirical evidence has been published on forest transitions in various regions of the world (e.g., Kauppi et al. 2006; Rudel et al. 2005), including non-industrialized countries and the globe as a whole. Some effort has also been made to understand the socioeconomic drivers of forest transitions. Two main explanatory narratives of forest transitions have been established, though they may not equally apply to all regions: (1) economic development pulling farmers off of their land and allowing for forest regrowth and (2) forest scarcity leading to legal regulations aiming at forest conservation. In addition, several contributing factors have been put forward, such as agricultural adjustment to more favorable land or the perception of an ecological crisis fostering legislative action (for a more detailed account of the discourse on the forest transition, see Meyfroidt and Lambin 2011).

The issue of forest regrowth has recently entered the debate on anthropogenic climate change and its accounting methods (particularly in carbon accounting; see also Method Précis on Carbon Accounting in this volume). Climate change mitigation strategies foster reforestation to mitigate the increase in atmospheric carbon dioxide (CO_2) concentration. Carbon (C) storage due to increasing forest stocks is also considered a positive contribution to national greenhouse gas accounts. In light of the forest transition, this accounting rule has been criticized because the C stored now was emitted during deforestation processes in historic periods. Thus, the accounting rule only settles old scores.

In this contribution, we take a socioecological perspective on the forest transition. To this end, we take an equally detailed look at societal and ecological change. In line with recent work (e.g., Kuemmerle et al. 2011), we base our analysis on stocks and flows of C rather than forest area. The empirical basis for our analysis is a data set on socioeconomic and ecological C stocks and flows in Austria in the time period 1830–2010. Forests are among the land-use categories with the highest C densities (i.e., C stock per unit of area). Moreover, C densities are directly linked to forest management intensity. Therefore, changes in ecosystem C stocks and flows are closely connected to the dynamics of forests, both in terms of forest management and in terms of forest area. However, C is also an important element in social metabolism. Assessment of the societal C budget allows us to understand how harvested wood is used. It also contextualizes the socioeconomic use of wood with that of other important materials processed in a society, particularly food, feed and fossil energy carriers.

The analysis of Austria's C budget thus allows us to consistently link changes in ecosystems (the C budget in biota and soils) with the (biophysical) activities of societies (the societal C metabolism). In addition, because C plays a crucial role in global climate change, an understanding of Austria's C budget sheds light on Austria's contribution to global climate change. Our study thus also provides background information for the development of climate change mitigation policies and the evaluation of the corresponding measures.

20.2 Methods and Data

Austria is a good example for tracing the forest transition. It is a small, highly industrialized Central European country with a medium population density (area 83,000 km^2, current population approximately 8 million, 96 inhabitants per km^2). More than 60 % of the country's area is covered by the Alps. The high share of mountainous and hilly regions is related to Austria's large forest cover, currently 47 %, which is distinctly higher than the EU-average of 40 %. Austria was chosen as a case study because of the significance of its woodlands and because of the typical pattern of forest regrowth after centuries of deforestation, a pattern that can be observed in all highly industrialized European countries. But Austria was also chosen for methodological reasons. The forest transition is part of a long-term historic process of land-use change. To gain insights into the dynamics and drivers of this process, a centennial perspective is required. Reconstructions of historical land-use patterns and material flows are built on archival sources. Analyzing such sources is a time-consuming process that requires specific skills, such as the ability to read and understand handwritten topographic descriptions and the critical evaluation of these sources.

Obviously, the geographical proximity of such sources in Austrian archives, as well as their common language, fostered the choice of Austria as a case study. Another advantage is rich data availability, particularly in the 19th century. The *Franciscean Cadaster* provides an extensive historic land account for the first half of the 19th century, before large-scale industrialization took off in Austria. The cadaster provides plot-level information on land use by municipalities for the largest part of the Austro-Hungarian Empire. It contains data on forest cover, rotation period and wood harvest for different forest categories (e.g., high forest, coppice). In addition to the cadaster, a detailed inventory of forests in governmental tenure ('Staats- und Fondsforste'), covering 10 % of Austria's total forest area (Wessely 1882), is available for the late 19th century. This allows us to investigate Austria's forests at a second point in the 19th century. For the 20th century, the investigation of forest dynamics can build on a 50-year time series of detailed forest inventories (Weiss et al. 2000). We present a comprehensive data set of Austria's C budget in the period 1830–2010, including data on (1) C stocks in ecosystems (i.e., vegetation and soils), (2) yearly ecosystem C flows, (3) yearly socioeconomic C flows and (4) socioeconomic C stocks. As indicated above, data from a variety of sources were used and consistently linked to generate this database. Many of the data presented here have been published elsewhere (Erb et al. 2008; Gingrich et al. 2007; Krausmann 2001), but they were updated to include the year 2010 by applying the same methodology to recently published data sources.

Let us start by explaining how we addressed Austria's ecosystem C budget. The main pillar for this data set was an assessment of ecosystem C stocks at ten points in time between 1830 and 2010. C stocks were assessed separately for six different land-use categories as products of (1) area and (2) C density (i.e., amount of C per unit of area). Both factors were derived, where possible, directly from published statistical data (from the late 19th century onward) or from the evaluation of archival material (early 19th century). Where necessary, calculations and estimation procedures allowed us to derive the average amount of C per (average) unit of forest area from original source information on wood harvest, turnover period and tree species composition. Belowground C stored in roots and other belowground biomass ('belowground standing crop') was calculated as a land-use-specific share of the respective aboveground standing crop. Soil organic C was assessed by using land-use-specific coefficients. A detailed description of the land-use data set was first published by Krausmann (2001), whereas the C stock calculation was presented by Gingrich et al. (2007). Yearly C exchange rates with the atmosphere (gross primary production, GPP) were calculated based on net primary production (NPP) values from a previous study (Krausmann 2001), assuming a general GPP/NPP ratio of 2:1 (see Erb et al. 2008). The source or sink function of ecosystems was assessed from the difference between C stocks at different points in time.

The socioeconomic C budget, including inputs, stocks and outputs of C, was compiled based on data on yearly flows of the socioeconomic extraction, trade and use of C-rich materials derived from a national energy flow account for Austria (Krausmann and Haberl 2007). These energy flow data were converted to units of C by applying the respective C content of the different energy carriers. The methods and factors used are described by Erb et al. (2008). In addition to the previously published data described above, we present a new estimate of socioeconomic C stocks for the period 1830–2010. These results were assessed based on data on yearly inflows of industrial roundwood, paper, plastic and bitumen since 1700, which were combined with product specific C densities to account for product-specific C inflows. By combining these inflows with product-specific half-lives and biophysical depreciation rates of the stocks derived thereof, we simulated the annual accumulation of C stocks, starting with a hypothetical C stock of 0 in the year 1700 (see Lauk et al. 2012 for a detailed description of this method). This assumption can be made because most of the C stocks of 1700 were depreciated by 1830, the starting year of our results. According to the boundary of the socioeconomic system, we did not include stocks in landfills. The amount of C stored in human bodies, livestock and food stored for later use was neglected. An assessment of these stocks on the global level reveals that their extent is of minor importance (ibid.).

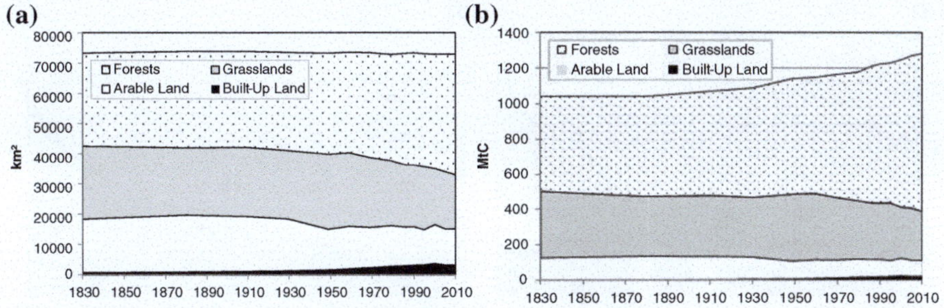

Fig. 20.1 The forest transition in Austria, 1830–2010. **a** Land use in Austria, 1830–2010. **b** Carbon stocks in Austria's terrestrial ecosystems, 1830–2010

20.3 The Forest Transition in Austria—Ecosystem C Stock Changes 1830–2010

After centuries of gradual deforestation, Austrian forest areas started to regrow some time in the 18th or 19th century. The exact turning point of the forest transition is difficult to assess because reliable quantitative information on land use and the extent of forest areas is only available from the early 19th century onward. However, empirical evidence shows that since 1830, when forests covered approximately 41 % of the total land area, Austria's forest areas have increased at the expense of agricultural land to the present value of 47 % forest cover (Fig. 20.1a). The forest transition also affected the density of timber stocks, that is, the amount of C stored per unit of forest area (C density). The C density grew even faster than the forest area.

Two rather distinct stages can be observed in the long time series. In the 19th and early 20th centuries, change occurred gradually. Between 1830 and 1949, forest area increased by less than 10 %, resulting from annual growth rates below 0.1 %. The C density in forests also grew rather slowly in this period, at similar annual growth rates. The total C stock increase between 1830 and 1949 in Austria's terrestrial ecosystems in this roughly 120-year period was 9 %. The period after World War II is characterized by a much more rapid increase in both forest area and forest C density. C stocks in forests grew by another 13 % in the 50-year period from 1949 to 2010. This can again be attributed to both growing forest area and growing C densities on forest land, that is, a recovery or aging of wood stocks. Whereas forest area grew by an average of 0.3 % per year, forest C density displays annual growth rates of 0.6 % per year throughout this period.

The expansion of forest areas and the increase in C density in forests created a C sink in Austria's ecosystems of 243 Mt (megatons; 1 Mt $= 10^6$ t), or 23 % of total ecosystem C stocks, between 1830 and 2010 (Fig. 20.1b), which overcompensated the slight decrease in C stocks in all the other land-use categories. The share of C stored in forests grew from 52 % (1830) to 70 % (2010) of total ecosystem C stocks. This underlines the importance of the forest transition for Austria's ecosystem C sink.

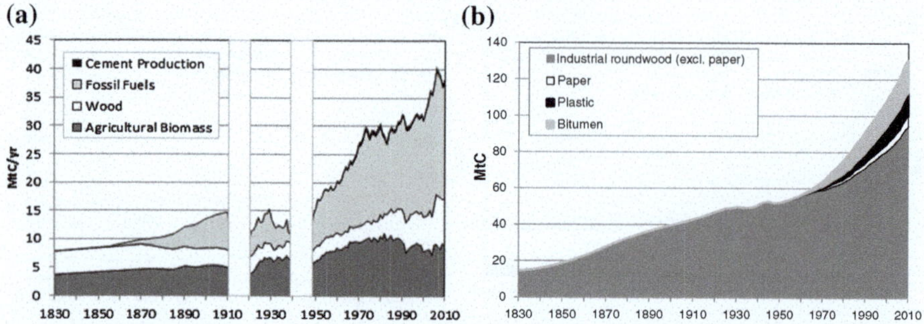

Fig. 20.2 Austria's socioeconomic carbon budget 1830–2010. **a** Domestic carbon consumption (DCC; annual carbon flows). **b** Socioeconomic carbon stocks

20.4 Socioeconomic Change: Austria's C Metabolism and Socioeconomic C Stocks

What were the drivers of the above-described C sink in Austria's forests? Although we cannot—within the scope of this contribution—assess all possible social, economic and ecological drivers of this development, we do consider the C budget in Austria's 'socioeconomic system', the C contained in the biomass and fossil energy carriers extracted and used within the national economy as well as the foreign trade of C-rich products. To obtain a comprehensive picture of the changes in the Austrian C metabolism, we also assess the socioeconomic stocks of C, that is, the amount of C accumulated in built structures and durable goods.

Austria's C metabolism underwent fundamental transformations in the 180-year period from 1830 to 2010. Domestic carbon consumption (DCC; i.e., the annual C turnover in harvest of biomass, fossil fuel combustion and net imports) rose nearly fivefold, from approximately 8 MtC/year (megatons carbon per year) in 1830 to nearly 40 MtC/year in 2010 (Fig. 20.2a). The origin of the C used in society also changed. In the early 19th century, C was consumed almost exclusively in the form of biomass, distributed equally between agricultural biomass and wood. Wood was the main provider of technical energy. From the late 19th century onward, increasing amounts of fossil fuels (first coal, then, particularly after World War II, crude oil and natural gas) added to the DCC. By the early 21st century, the consumption of fossil fuels had overtaken the use of biomass at 17 MtC/year. Austria is poorly endowed with deposits of fossil energy carriers, and the extraction of coal, oil and natural gas was comparatively low. Therefore, the shift to the use of fossils went along with increasing import dependency.

By far, the largest part of all C-rich materials that enter the social metabolism is consumed within one year. Most of the biomass is digested by humans and their livestock, and most of the fossil materials are combusted for the generation of energy. Therefore, the majority of DCC leaves the economy as gaseous emissions in the form of carbon dioxide (CO_2) or methylene (CH_2) within one year, and only a small fraction of C enters socioeconomic stocks. The share of C emitted within the year of use was high throughout the entire period, decreasing from 95 % in 1830 to 81 % in 2010.

With the growing annual C turnover in Austria's economy, C stocks accumulated in the socioeconomic system in built structures such as roads, buildings, furniture and paper. Total stocks grew by a factor of eight, from 15 MtC to roughly 130 MtC, between 1830 and 2010 (Fig. 20.2b). Throughout the entire period, industrial roundwood, wood stored in buildings, infrastructure and other construction works, was the dominant category. It was complemented only after World War II by stocks of paper, plastic and bitumen that grew rapidly and contributed 28 % of the total socioeconomic C stocks in 2010.

Interestingly, the increase in the C stocks of Austria's society followed an almost linear trend. To a certain degree, this can be explained by the method of calculation and the fact that stocks (as opposed to flows) accumulate year by year: a decline in stocks occurs only when net inflows are significantly lower than outflows for a noticeable period. Constant inflows of wood at a higher level than the depreciation rate result in what appears to be a linear growth of stocks. One important reason for the strong increase in C stored in Austria's society is that the use (i.e., annual inflow) of construction wood went up in the past two centuries: while only 10 % of wood was used for construction in 1830, this share exceeded 40 % in the early 21st century. Therefore, the amount of wood that went into societal C stocks each year grew more than the amount of wood used each year.

20.5 Linking Societal and Ecosystem C Budgets

The socioecological perspective applied in this study allows us to consistently link the data compiled for Austria's ecosystems to those describing its national economy. This enables us to understand the coevolution of Austria's natural and social systems over time and to tackle the question of how the forest transition fits into changes in the interrelations between the two. Let us now look at Austria's socioecological C budget since the onset of industrialization. How have C stocks and flows in ecosystems and society compared since the early 19th century, and how have their dynamics differed?

Under agrarian conditions (ca. 1850) and after industrialization (ca. 1990), the ecosystem C stocks were much larger than the societal C stocks (they differed by more than a factor of ten in both periods, Fig. 20.3). Furthermore, the exchange flows between atmosphere and territory are much larger (just below a factor of ten) in ecosystems than in society. Despite these differences, societal activities already substantially influenced ecosystem C dynamics in the 19th century: the C outtake of ecosystems at 8.5 MtC/year amounted to roughly 10 % of GPP and 20 % of NPP. Also, ecosystem C stocks were significantly lower at 1043 Mt than in potential ecosystems (i.e., ecosystems that would prevail without any societal

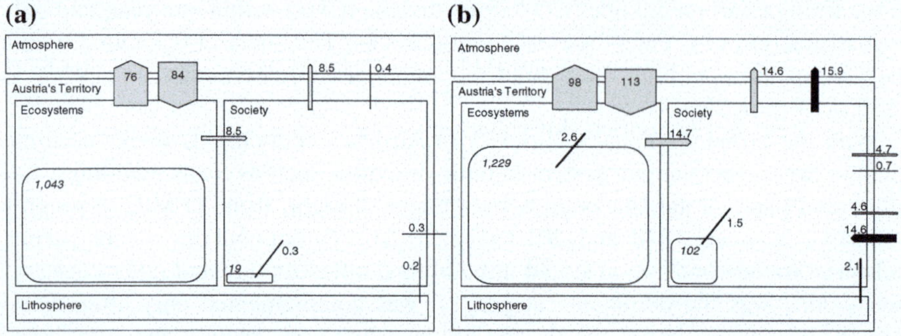

Fig. 20.3 Carbon (C) flows and stocks in Austria's ecosystems and society in MtC/year and Mt (*italic numbers*), **a** ca. 1850, **b** ca. 1990. The areas of the *arrows* (C flows) *and boxes with round* corners (C stocks) correspond to the extent of the respective flow or stock. *Grey arrows* signify C flows exchanged by or in the form of biomass, and *black arrows* refer to fossil C flows. Arrows to stocks reflect yearly net stock increases

interference). According to Erb (2004) and Gingrich et al. (2007), Austria's potential ecosystems would store over 2000 MtC, nearly twice the stock of 1850. This means that over the centuries and millennia of deforestation and agricultural activity in Austria until the 19th century, humans removed roughly half of all the C stored in ecosystems. This trend was reversed in the 19th century.

By the end of the 20th century, all of Austria's C stocks and flows, both in ecosystems and society, had increased. Although Austria's ecosystems still exceed society in terms of C stocks and flows, the difference between the two has significantly decreased. The C stocks in Austria's ecosystems grew by 20 % (i.e., one-fifth) in the 150-year period, whereas they increased fivefold in society. The total C sink in Austria's ecosystems in the period 1830–2010 was 243 MtC, only twice that of the socioeconomic C sink (116 MtC). C flows through society also increased significantly, but the more remarkable change in societal C flows was qualitative: with industrialization, biomass ceased to be society's main C flow. Fossil fuels, stemming mostly from imports and, to a small extent, from domestic extraction, added to society's C metabolism and reached the same order of magnitude as biomass flows in the late 20th century.

Let us now focus on the net addition to C stocks in ecosystems and society. Ecosystem C stocks did not change significantly between our first two data points in 1830 and 1880. Therefore, we assume no annual C uptake in this period. This could indicate that the forest transition (i.e., the change from deforestation to afforestation) took place only between these two points in time or slightly before. The C uptake in society, at 0.3 MtC/year, represented a minor flow, comparable in size to the C contained in the use of fossil fuels at that time. By the end of the 20th century, net addition to C stocks in ecosystems and society had increased to 2.6 and 1.5 MtC/year, respectively.

These numbers are striking for two reasons. First, it is noteworthy that Austria's society acts as a C sink comparable to that of its (highly forested) ecosystems. The C stored in society each year accounts for 10 % of the yearly C emissions from fossil fuels, which is far above the global average of approximately 3 % (Lauk et al. 2012). The second important aspect in Austria's strongly increasing C stocks is the ratio between stocks and flows (the turnover). In the late 20th century, annual net addition to stocks made up 0.2 % of the total stocks in ecosystems, whereas the ratio was 1.5 % in society. The question of how the relation between stocks and flows influences system dynamics and the potential for future development makes for an interesting future research topic.

20.6 The Forest Transition: A Side Effect of the Socioecological Transition

How is it possible that forests regrew and wood stocks in society increased when the harvest of wood did not substantially change in the past two centuries? To understand these seemingly paradoxical occurrences, we adopt the idea of a socioecological transition (SET; Fischer-Kowalski and Haberl 2007). An SET describes the relatively rapid change from one more or less stable socioecological regime to another, that is, from one pattern of resource use to another. Industrialization can be conceptualized as one such SET. Before or during the onset of industrialization, the functioning of Austria's national economy was, from a biophysical perspective, based largely on the use of biomass as an energy carrier. This changed with industrialization, when fossil fuels added to and increasingly dominated Austria's energy provision. The shift of Austria's energetic base had effects not only on economic development and social structure but also, as we have demonstrated in this chapter, on ecosystems.

Before the onset of industrialization, Austrian forests were used differently than they are today. The removal of wood for energetic and material use was only one of several extractive activities. Forests also played an important role in livestock management. Forest grazing, the extraction of litter used as bedding material in stables and even the cutting of live leaves from trees to supply feed (pollarding) were among the most common practices of forest use in Austria until the mid-20th century, affecting the shape of forests and the amount of C stored in forest ecosystems. This changed with industrialization. Erb et al. (2007) argue that the increase in Austria's ecosystem C stocks was possible only because of the rising use of fossil fuels. They describe the phenomenon as a 'fossil-fuel powered carbon sink', linking the use of fossil fuels to the advancement of land-saving technologies in agriculture. The energy-intensive production of mineral fertilizer and the introduction of fossil-fuel-based agricultural machinery such as tractors were key factors in boosting agricultural yields (see also Chap. 21). These technological innovations allowed for Austrian agriculture to retreat to the more favorable areas as agricultural output increased. The areas freed from agricultural use were often reforested.[1] At the same time, forests lost their importance in locally integrated agricultural production, and material extraction from forests focused increasingly more on wood. This led to increasing C densities in forests.

Our study also shows that the use of wood changed during industrialization from being the most important source of technical energy to becoming a raw material used mainly for construction purposes (or as paper). It is often argued—with good reason—that the increasing use of fossil fuels did not replace the societal use of biomass but merely added to it or even allowed for increasing biomass use. In Austria, however, fossil energy did, to some extent, substitute for the use of wood as a technical energy carrier. Our data suggest that the share of wood fuel in total wood use fell from 90 % in 1830 to only 20 % in 2000, resulting in an absolute decline in wood fuel use of roughly 70 %. At the same time, increasingly more wood was used as construction material and paper, adding to increasing socioeconomic C stocks.

Wood thus played a crucial role in Austria's changing C budget during industrialization. C stock increases in both ecosystems and the socioeconomic system resulted from increasing stocks of wood. During the SET, the socioeconomic function of wood changed dramatically. From being a prime provider of technical energy, it became a resource used mainly for material purposes. This change is directly connected to a strong increase in the half-life of wood within the socioeconomic system: whereas fuelwood is usually burned within one year, timber remains in society for several decades and contributes to increasing stocks.

The forest transition, therefore, can be linked to a shift in the socioeconomic use of wood. This has far-reaching consequences in terms of sustainability. In a sustainable society that (largely) foregoes the use of fossil fuels, fuelwood could become more important again, and the dynamics in C budgets of ecosystems and societies are likely to reverse. Therefore, policies for sustainable development should take a broad perspective in terms of C accounting (i.e., consider all C flows in their environmental indicators) and in terms of scenario building (i.e., consider the long-term effects of an energy shift toward biomass).

References

Erb, K.-H. (2004). Land-use related changes in aboveground carbon stocks of Austria's terrestrial ecosystems. *Ecosystems, 7*(5), 563–572.

Erb, K.-H., Haberl, H., & Krausmann, F. (2007). The fossil-fuel powered carbon sink. Carbon flows and Austria's energetic metabolism in a long-term perspective. In M. Fischer-Kowalski & H. Haberl (Eds.), *Socioecological transitions and global change: Trajectories of social metabolism and land use* (pp. 60–82). Cheltenham, UK, Northampton, USA: Edward Elgar.

[1]Although somewhat different in focus, this argument is quite in line with 'old' arguments brought forward in the first discussions of the forest transition. Mather and Needle (1998) also discuss the fact that forest areas increase while agricultural areas retreat to the most favorable sites; however, they attribute this process to learning by farmers rather than technological change or energy availability.

Erb, K.-H., Gingrich, S., Krausmann, F., & Haberl, H. (2008a). Industrialization, fossil fuels and the transformation of land use: An integrated analysis of carbon flows in Austria 1830-2000. *Journal of Industrial Ecology, 12*(5–6), 686–703.

Fischer-Kowalski, M., & Haberl, H. (2007). Conceptualizing, observing and comparing socioecological transitions. In M. Fischer-Kowalski & H. Haberl (Eds.), *Socioecological transitions and global change: Trajectories of social metabolism and land use* (pp. 1–30). Cheltenham, UK, Northampton, USA: Edward Elgar.

Gingrich, S., Erb, K.-H., Krausmann, F., Gaube, V., & Haberl, H. (2007). Long-term dynamics of terrestrial carbon stocks in Austria. A comprehensive assessment of the time period from 1830 to 2000. *Regional Environmental Change, 7*(1), 37–47.

Grainger, A. (1995). The forest transition: An alternative approach. *Area, 27*, 242–251.

Kauppi, P. E., Ausubel, J. H., Fang, J., Mather, A. S., Sedjo, R. A., & Waggoner, P. E. (2006). Returning forests analyzed with the forest identity. *Proceedings of the National Academy of Sciences of the United States of America, 103*(46), 17574–17579.

Krausmann, F. (2001). Land use and industrial modernization: An empirical analysis of human influence on the functioning of ecosystems in Austria 1830–1995. *Land Use Policy, 18*(1), 17–26.

Krausmann, F., & Haberl, H. (2007). Land-use change and socio-economic metabolism. A macro view of Austria 1830–2000. In M. Fischer-Kowalski & H. Haberl (Eds.), *Socioecological transitions and global change: Trajectories of social metabolism and land use* (pp. 31–59). Cheltenham, UK, Northampton, USA: Edward Elgar.

Kuemmerle, T., Olofsson, P., Chaskovskyy, O., Baumann, M., Ostapowicz, K., Woodcock, C. E., et al. (2011). Post-Soviet farmland abandonment, forest recovery, and carbon sequestration in western Ukraine. *Global Change Biology, 17*(3), 1335–1349.

Lauk, C., Haberl, H., Erb, K.-H., Gingrich, S., & Krausmann, F. (2012). Global socioeconomic carbon stocks and carbon sequestration in long-lived products 1900–2008. *Environmental Research Letters, 7*, 034023.

Mather, A. (1992). The forest transition. *Area, 24*(4), 367–379.

Mather, A. S., & Needle, C. L. (1998). The forest transition: A theoretical basis. *Area, 30*(2), 117–124.

Meyfroidt, P., & Lambin, E. F. (2011). Global forest transition: Prospects for an end to deforestation. *Annual Review of Environment and Resources, 36*, 343–371.

Rudel, T. K., Coomes, O. T., Moran, E., Achard, F., Angelsen, A., Xu, J., & Lambin, E. (2005). Forest transitions: Towards a global understanding of land use change. *Global Environmental Change Part A, 15*(1), 23–31.

Weiss, P., Schieler, K., Schadauer, K., Radunsky, K., & Englisch, M. (2000). *Die Kohlenstoffbilanz des österreichischen Waldes und Betrachtungen zum Kyoto-Protokoll*. Wien: Umweltbundesamt.

Wessely, J. (1882). Forstliches Jahrbuch für Oesterreich - Ungarn. Oesterreichs Donauländer. II. Theil: Spezial-Gemälde der Donauländer. Carl Fromme, Wien.

Method Précis: Carbon Accounting

Karl-Heinz Erb

Carbon (C) is the fourth most abundant element in the universe and is a key component of socioecological systems. The processes of life are inextricably linked to carbon: long-chained carbon compounds, formed mainly through the process of photosynthesis, represent the energetic basis of practically all life on Earth and, thus, of the endosomatic metabolism of human society. Additionally, carbon-rich materials, such as fossil fuels and wood, dominate society's exosomatic energy metabolism (Chaps. 2, 3 and 8). Furthermore, carbon compounds such as carbon dioxide (CO_2) and methane (CH_4) are powerful greenhouse gases (GHGs) that are central components of the global climate system.

Society intervenes in the global carbon cycle, the flow of carbon in various chemical forms between carbon pools in the atmosphere, lithosphere, oceans and terrestrial vegetation. Two main processes leading to net carbon emissions can be discerned: (1) By mobilizing an inert carbon pool, the combustion of fossil fuels and cement production currently adds 8.7 ± 0.5 PgC/year (petagrams carbon per year; $1\ \text{Pg} = 10^{15}\ \text{g} = 10^{9}\ \text{t} = 1\ \text{Gt}$) to the atmosphere (in 2008; see Le Quéré et al. 2009). (2) Land-use change, most prominently deforestation, reduces the carbon stored in ecosystems, resulting in the emission of large amounts of carbon to the atmosphere (a flow of approximately 10–20 % of the volume of global fossil fuel emissions; Le Quéré et al. 2009). Accounting for, monitoring and modeling the human interventions in the global carbon cycle is a central focus of Sustainability Science.

Several carbon accounting schemes exist. A systematics of carbon flows and carbon accounting approaches is provided in Fig. 20.4 (for a discussion of system boundaries, see Chap. 2 in this volume). Socioeconomic carbon flows (flow C in Fig. 20.4) include carbon emissions due to fossil fuel combustion, cement manufacture and the socioeconomic use of biomass. These flows can be quantified in a substance flow analysis (SFA), an approach that is closely related to the material and energy flow accounting (MEFA) framework (see Method Précis on Material Flow Analysis). In SFA, MEFA data are combined with material-specific information about carbon content. The carbon content of materials varies widely, ranging from ca. 85 % for crude oil and 75 % for coal to 0.3 % for steel. Biomass dry matter contains 45–55 % carbon; an accepted average value is 50 % (Eggleston et al. 2006). In the SFA approach, national C-emissions are assessed according to apparent consumption, defined as the sum of domestic extraction (DE) and imports, minus exports.

C-flows between the atmosphere and ecosystems (soil and biota) are characterized by large gross and small net flows (Houghton 2003). For instance, ecosystems absorb approximately 120 PgC from the atmosphere each year (gross primary production; GPP)—the largest single annual flow of the global carbon cycle. Approximately half the GPP is immediately returned to the atmosphere due to autotrophic (plant) respiration. The other half (net primary production; NPP)

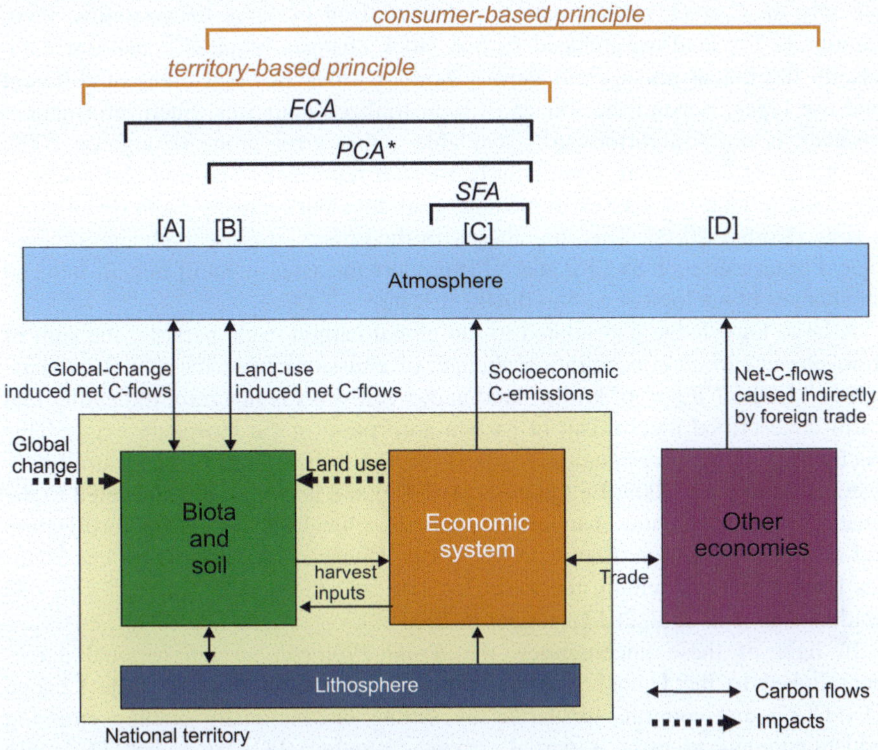

Fig. 20.4 Systematics of carbon flows and carbon accounting systems. Flows between marine ecosystems and the atmosphere are not included. SFA = substance flow analysis. PCA = partial carbon accounting (* only selected ecological C-flows such as flows related to reforestation, afforestation and deforestation are accounted for in a PCA). FCA = full carbon accounting

enters heterotrophic food chains, and much of it returns to the atmosphere within the same year. The net flows, e.g., the global balance of deforestation and reforestation (1–2 PgC/year; see Le Quéré et al. 2009) are considerably smaller than the gross flows. However, only those flows stemming from changes in carbon stocks in biota and soils are relevant in the context of climate change. Calculating net flows by balancing gross flows has proved impractical. Instead, the assessment of human-induced natural carbon flows (due to land use, land-use change and forestry—LULUCF; flow B in Fig. 20.4) often relies on an assessment of the difference between C-stocks measured or modeled at two points in time (Erb et al. 2008; Gingrich et al. 2007). Such approaches can make use of forest inventories, databases on the state of forests that are commonly available at the national level for many industrialized nations and cover mainly the temperate and boreal zones (Houghton 2003). Forests play a crucial role in the carbon cycle because they store between five and 50 times more carbon than any other vegetation type. Thus, ecological C-flows are largely dominated by forest dynamics. However, approaches that rely on C-stock assessments are characterized by large uncertainties. First, accounting for land-use-induced carbon stock changes requires consistent datasets on the extent and carbon density (carbon stock per unit area) of different land-use types, beyond the forests of industrialized countries. Such information, however, is not straightforwardly available and is error prone (Houghton 2003. Second, most ecosystems (particularly forests) represent 'rapid-out-slow-in' systems, where land-use causes immediate emissions but recovery from disturbance is slow (Körner 2003). This, together with the effects of climate change on ecological processes such as GPP and NPP, renders the assessment of human-induced net changes of ecological carbon flows intricate.

A large knowledge gap related to the global carbon cycle persists: the sum of atmospheric carbon concentration change, of emissions from fossil fuel combustion and LULUCF and of known land and ocean sinks is not zero, indicating that a still-undescribed mechanism of carbon absorption in the biosphere exists. This mechanism, termed 'residual sink', is suspected to be situated in terrestrial ecosystems (Houghton 2003; Le Quéré et al. 2009) and is commonly attributed to the effects of environmental change (for example, climate change, CO_2 fertilization and nitrogen deposition, flow A in Fig. 20.4). However, there are huge uncertainties related to the strength, the exact location and the underlying causes of this residual sink (Erb et al. 2013; Houghton 2003).

In light of these uncertainties, the *Kyoto Protocol*, an international agreement linked to the United Nations Framework Convention on Climate Change (UNFCCC) that commits its Parties by setting internationally binding emission reduction targets, proposes a 'partial carbon accounting' (PCA) approach. PCA only accounts for flows related to predefined activities, such as specific land-use activities causing forest area change (e.g., afforestation, reforestation and deforestation). In this vein, the Intergovernmental Panel on Climate Change (IPCC, under the auspices of the UN) provides guidelines for national carbon flow accounting (e.g., related to LULUCF: Eggleston et al. 2006). The IPCC has proposed a tier approach that combines different accounting methods, depending on the availability of data.

A major critique of the PCA approach is that 'leakage' effects, such as carbon effects induced by shifts of emission-intensive activities to geographic regions that are not committed to the *Kyoto Protocol* or substituting accounted-for with non-accounted-for activities, remain unrecognized. In response to the leakage and uncertainty challenges, a full carbon accounting (FCA) approach has been proposed (Nilsson et al. 2007). FCA encompasses all carbon flows between the atmosphere and the various other compartments of the Earth system, including global-change-induced and land-use-induced changes between ecosystems and the atmosphere (i.e., flows A, B and C in Fig. 20.4). The challenges of this approach are related to the large data gaps, but because FCA allows one to combine and mutually constrain top-down and bottom up assessments and databases, knowledge and data uncertainty can be explicitly quantified (and reduced), which is critical for the verification and validation of carbon mitigation strategies. Furthermore, full carbon accounts are robust against problem-shift because all flows between all compartments are covered.

The SFA, PCA and FCA accounting schemes follow a territory-based principle: they account for carbon flows between a specific territory and the atmosphere. Consumer-based principles, in contrast, have been proposed to attribute direct and indirect (also denoted 'upstream') carbon flows to the consumption of final products. These accounts, encompassing flows B, C and D in Fig. 20.4, deliver complementary information to territory-based accounts that allow the discussion of questions of responsibility and leakage as well as the analysis of the driving forces behind carbon emissions (Hertwich and Peters 2009). Relatively robust methodologies following the consumer-based principle have been proposed for the assessment of fossil-fuel-related C-flows, based either on LCA approaches (Method Précis on Life Cycle Assessment) or multi-regional input-output models and commonly denoted as 'carbon footprint' (e.g., Hertwich and Peters 2009). Less straightforward, however, is the assessment of indirect land-use-related C-flows associated with foreign trade. This is mainly due to the difficulties of systematically linking biomass harvest flows to human-induced carbon stock changes in vegetation. The response function of vegetation to harvest depends not only on harvest intensity (i.e., the amount of biomass harvested per unit area) but also on the type and location of the ecosystem, global change and the history of land-use (Erb et al.2008; Kastner et al. 2011). Like territory-based accounting schemes, consumer-based accounting schemes have to be constructed in such a way that they yield double-counting free accounts.

References

Eggleston, S., Buendia, L., Miwa, K., Ngara, T. & Tanabe, K. (Eds.), (2006). *2006 IPCC guidelines for national greenhouse gas inventories* (Vol. 4: Agriculture, forestry and other land use). Hayama, Japan: Institute for Global Environmental Strategies (IGES).

Erb, K.-H., Gingrich, S., Krausmann, F., & Haberl, H. (2008). Industrialization, fossil fuels, and the transformation of land use. *Journal of Industrial Ecology, 12*, 686–703.

Erb, K.-H., Kastner, T., Luyssaert, S., Houghton, R. A., Kuemmerle, T., Olofsson, P., & Haberl, H. (2013). Bias in the attribution of forest carbon sinks. *Nature Climate Change, 3*, 854–856.

Gingrich, S., Erb, K.-H., Krausmann, F., Gaube, V., & Haberl, H. (2007b). Long-term dynamics of terrestrial carbon stocks in Austria: A comprehensive assessment of the time period from 1830 to 2000. *Regional Environmental Change, 7*, 37–47.

Hertwich, E. G., & Peters, G. P. (2009). Carbon footprint of nations: A global, trade-linked analysis. *Environmental Science & Technology, 43*, 6414–6420.

Houghton, R. A. (2003). Why are estimates of the terrestrial carbon balance so different? *Global Change Biology, 9*, 500–509.

Kastner, T., Erb, K.-H., & Nonhebel, S. (2011). International wood trade and forest change: A global analysis. *Global Environmental Change, 21*, 947–956.

Körner, C. (2003). Slow in, rapid out–carbon flux studies and Kyoto targets. *Science, 300*, 1242–1243.

Le Quéré, C., Raupach, M. R., Canadell, J. G., Marland, G., Bopp, L., Ciais, P., et al. (2009). Trends in the sources and sinks of carbon dioxide. *Nature Geosci, 2*, 831–836.

Nilsson, S., Shvidenko, A., Jonas, M., McCallum, I., Thomson, A., & Balzter, H. (2007). Uncertainties of a regional terrestrial biota full carbon account: A systems analysis. In D. Lieberman, M. Jonas, Z. Nahorski, & S. Nilsson (Eds.), *Accounting for climate change* (pp. 5–21). Netherlands: Springer.

Chapter 21
From Energy Source to Sink: Transformations of Austrian Agriculture

Fridolin Krausmann

Abstract Traditional low-input agriculture had to organize local land, labor and livestock resources in a way that maintained soil fertility and stable yields, albeit at a low level. Industrialization transformed the socioecological functioning of agriculture and its role in social metabolism. Agriculture turned into a high input/high output system that obtains high yields but consumes more energy than it produces. By formalizing the functional interrelations of agricultural systems into a sociometabolic model, we are able to reconstruct this process of transformation for the case of Austria.

Keywords Land use • Material and energy flow accounting (MEFA) • Social metabolism • Colonization • Rural systems • Sustainable agriculture • Long-term socioecological research (LTSER) • Livestock • Soil fertility • Land-use intensity

21.1 Introduction

For 10,000 years, agriculture has been the most significant interaction between society and its natural environment. During this period, agriculture has driven far-reaching regional and global environmental changes. Agricultural activities have altered the earth's vegetation cover, reduced species diversity, changed soil conditions and water regimes and contributed to climate change (Simmons 2008). Consequently, agriculture has been and remains a major theme in sustainable development. On the one hand, agriculture, along with its intensification, has drastically increased the earth's carrying capacity for humans, driven population growth and facilitated the provision of a continuously growing global population

F. Krausmann (✉)
Institute of Social Ecology (SEC), Alpen-Adria University,
Schottenfeldgasse 29, 1070 Vienna, Austria
e-mail: fridolin.krausmann@aau.at

H. Haberl et al. (eds.), *Social Ecology*, Human-Environment Interactions 5,
DOI 10.1007/978-3-319-33326-7_21

with food and raw materials. On the other hand, agriculture has changed the biosphere and still causes far-reaching ecological damage that threatens sustainability at all scales. Although the concept of sustainability was formulated much later, it was key to the functioning of historic agrarian societies: there was an immediate necessity to farm sustainably; to avoid overexploiting local renewable resources; and to make use of nonrenewable resources in a way that would maintain soil fertility and allow cultivation without a decline in productivity and food output. Maintaining soil fertility was the main sustainability challenge in the agrarian metabolic regime, and it was a difficult and often labor-intensive task that required the careful organization and management of locally available resources (Mazoyer et al. 2006; McNeill and Winiwarter 2006).

Industrialization changed the metabolic functioning of agriculture and its role in socioeconomic metabolism. Novel fossil-fuel-based technologies facilitated the emancipation of agriculture from traditional environmental constraints. Within a few decades, large-scale irrigation and fertilization, tractors and machinery, chemical pest control and the development of new cultivars adapted to these industrial technologies changed the character of farming and increased the intensity of society-nature interactions. This development led to unprecedented changes in productivity, production and the intensity of land use. It changed the scale of human domination of terrestrial ecosystems and the quality and quantity of sustainability problems related to agriculture (Vitousek et al. 1997).

In recent years, the investigation of the sociometabolic transformation of agriculture and of the links between social metabolism and land use has gained attention in Long-Term Socioecological Research (LTSER) and Environmental History (Haberl et al. 2006). Concepts and methods to study agriculture and its evolution as a coupled socioecological system have been advanced and applied in local case studies (Bayliss-Smith 1982; Cunfer and Krausmann 2009; Cusso et al. 2006). We follow this line of research and develop a socioecological perspective on the transformation of agriculture and farming systems during European industrialization. How did farmers maintain soil fertility as they cultivated the same land over decades and centuries? How did they transfer energy and plant nutrients across the landscape to fertilize crops? How did they structure landscapes (field, pasture, woodland, water) to sustain communities, ensure long-term productivity, produce profits over time and, in the process, create valuable, highly diverse cultural landscapes? How did industrialization change agriculture as a socioecological system, the material and energy flows in agricultural production systems and the quality of society-nature interactions? How did the role of agriculture in social metabolism change?

This text draws from research in a number of different projects that investigated the socioecological characteristics of preindustrial farming systems in Austria and their change during industrialization in regional and national case studies and that analyzed farming systems and their metabolism in the early 19th and late 20th century (Krausmann 2004, 2008; Umweltgeschichte 2000; Sieferle et al. 2006). These studies investigated local case studies in different agroecological zones using with detailed historical sources (an important criterion for long-term socioecological studies) that provide sufficient information for quantitative socioecological research. The local information from regional cases was complemented with results from the national scale. At this level, data are available in annual resolution, which allows for better insights into the temporal dynamics of change. It also places the results from the local level in a larger spatial context. Because space is limited, this text focuses on one of the case studies, the village of Theyern, to illustrate the more general picture of the socioecological transformation of agriculture we derived from the whole multi-scale sample.

21.2 Methods and Approach

Agriculture has been the main economic activity in the preindustrial economy and has been observed meticulously by statistical institutions since the early 19th century. Hence, a wealth of quantitative and qualitative information on agriculture, land use and production is available in historical sources. We use different sources to reconstruct farming systems and land-use patterns. The most important source for the first half of the 19th century is records from the *Franciscean Cadaster*, a tax survey conducted between 1829 and 1854 (Sandgruber 1979). It provides detailed qualitative, quantitative and spatial information concerning population, livestock, land use, crop yields, manuring, labor input and so forth for individual farmsteads and cadastral municipalities. We complement this information with data from statistical records, regional topographic descriptions and historical maps. For the late 20th century, information from field surveys and from regional statistics is used to quantify land-use patterns and material and energy flows in the case study regions. The main units of analysis are cadastral units (19th century), municipalities (20th century) and the national scale. The information from sources is used to reconstruct a quantitative picture of material and energy flows in farming systems and to calculate a number of key socioecological indicators (Singh et al. 2010).

We use knowledge about functional interrelations among stocks (e.g., livestock, population) and the respective energy and material flows required to sustain them to construct a formal model (Krausmann 2008). A major strength of such a model-based sociometabolic approach is that it combines quantitative and qualitative information from historical sources with biophysical models. Data gaps can be filled, and a more complete picture of historical production systems and their functioning can be achieved. For example, historical sources have very limited information on livestock grazing and manure production, information that is of crucial importance for the analysis of land-use intensity and the establishment of nutrient balances. The metabolic modeling approach combines historical records on livestock populations, milk yields and livestock management with knowledge about species-specific feed demand and the physiological characteristics of livestock species. This allows the establishment of local feed and nutrient balances. Temporal patterns of human and animal labor input can be estimated by using information on land use and labor requirements for specific agricultural tasks and spatial information on the location of fields to calculate transport efforts (Schaschl 2007).

21.3 Agriculture in the Early 19th Century

Agricultural communities in the early 19th century in Austria typically combined different types of land use and livestock husbandry at the farm or village level (mixed farming). The local combination of cropland, meadows, pastures and woodland and the significance of the different livestock species varied according to the local climate, terrain and soils.

This article focuses on a rural community in the fertile crop production region in the lowlands of Lower Austria to illustrate the specific characteristics of pre-industrial agriculture in Austria in the first half of the 19th century. Theyern is a small village approximately 60 km west of Vienna at 250–370 m a.s.l. It was part of the nearby Göttweig monastery until the land reform of 1848. In 1830, it was inhabited by 102 people living on 17 farmsteads and a total area of 2.3 km^2. More than half of the available land was used to grow crops; most of the remainder was woodland. Meadows and pastures only covered a small fraction of the land. On average, each farm cultivated 7.2 ha of cropland, but differences between the few larger holdings and the many smallholders were considerable.

In 1830, the land was still farmed in traditional three-course rotation, with one-third of the arable land sown in fall with winter cereal, one-third sown in spring with summer cereal and one-third kept fallow. The fallow field was tilled several times and used for grazing during summer and fall. The other fields were grazed after harvest in the fall. The farmers in Theyern were already beginning to grow new crops and to adopt new farming systems when the survey officers made their observations for the new cadaster around 1830. According to their accounts, one-third of the fallow field was planted with potato and clover, more than in any other regions we have investigated. Whereas the focus of the production system was on cereal production (rye), livestock was an integral element of farming systems. The farms in Theyern kept cows, oxen, heifers, horses, pigs and sheep, for a total of approximately 54 large animal units (LAUs),[1] of which 72 % were cattle. This corresponded to a comparatively high livestock density of 24 LAU/km^2. These animals were of key importance for the organization and functioning of the agricultural production system. Horses and oxen provided draught power for working the croplands and carrying carts with wood, harvested crops or manure from field to farm or market. The farm animals also played a key role in plant nutrient management by providing manure and transferring nutrients from the grassland and

[1]The large animal unit (LAU) is a standardized measure for livestock; all livestock is converted into animal units of 500 kg live-weight; i.e., a cow weighing 250 kg equals 0.5 LAU.

forests where they grazed to croplands and gardens. Finally, farm animals were a source of food and raw materials. In Theyern, animal products contributed roughly one-quarter to the overall food output.

Despite the high number of roughage-consuming animals (cows, oxen, horses and sheep), only a few percent of the land was dedicated to pasture. Instead, the animals were grazed on the fallow field, on stubble fields after harvest and in the woodlands, and they were fed straw and other byproducts and wastes. Furthermore, clover, still a new crop at that time, was increasingly grown in the fallow field and contributed rich winter feed for the animals. Only a very small fraction (9 %) of all feed was edible crops, such as oats, which were mostly fed to the horses. Rather than competing with humans for valuable crop products, animals were a way for the farmer to make use of crop residues and byproducts and of land that was unsuitable for crop production to provide power, manure, food and raw materials. Even in crop-producing regions, the largest part of all agricultural biomass harvested (more than 95 % in Theyern) was used in the livestock subsystem either as feed or bedding material. It was exactly this multifunctional use of livestock—combining the provision of labor, soil fertility management, biomass conversion and the production of food and raw materials—that made farm animals an integral part of traditional farming systems. Animals were a means to use local resources in an efficient and optimized way—even if this meant that the conversion efficiency of feed to food was low by modern standards. This also underpins that narrow efficiency measures focusing on product output are insufficient for comparing multifunctional farming systems; more-encompassing assessments are required.

Maintaining soil fertility and stabilizing yields was a major challenge in pre-industrial farming. To gain insights into soil fertility management, we quantified nitrogen flows in the production system (Fig. 21.1). Nitrogen was extracted from agricultural land at a rate of 26 kg/ha/year (kilograms per hectare per year) by harvesting biomass or grazing. Stable yields required that this outflow of nitrogen from agricultural soils be replaced. Natural inputs such as wet and dry deposition or microbial fixation contributed to this replacement, as did nitrogen fixation in leguminous crops grown on the fallow field. In Theyern, these two nitrogen sources added up to an annual input of approximately 14 kg/ha. The nitrogen input from animal manure (taking losses during storage and application into account) added another 10 kg/ha/year. Even if the uncertainties involved in such calculations are taken into account, this indicates that nitrogen was roughly balanced on agricultural areas, although this was probably not equally true for all land-use types or plots of land. In general, the better plots of land received most of the manure. Grazing of farm animals and the extraction of bedding material from woodlands resulted in a transfer of plant nutrients from woodlands (and pastures, in other regions) to fields, whereas the replacement of nutrients on this extensively used land was left to natural processes only. This placed considerable pressure on the soils of woodlands and pastures and reduced wood and grazing yields. Overall, the land was used quite intensively in the 19th century.

(a)

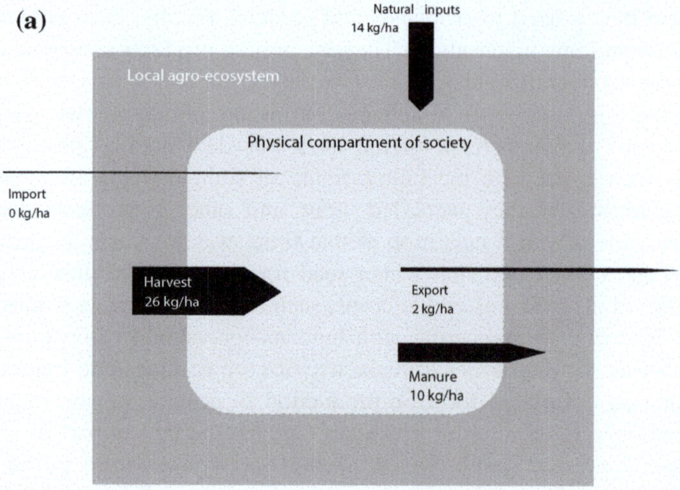

[kgN/ha]	Agro-ecosystem	Physical compartment of society
Inputs	24	26
Outputs	26	12
Balance	-2 (N-deficit)	14 (losses)

(b)

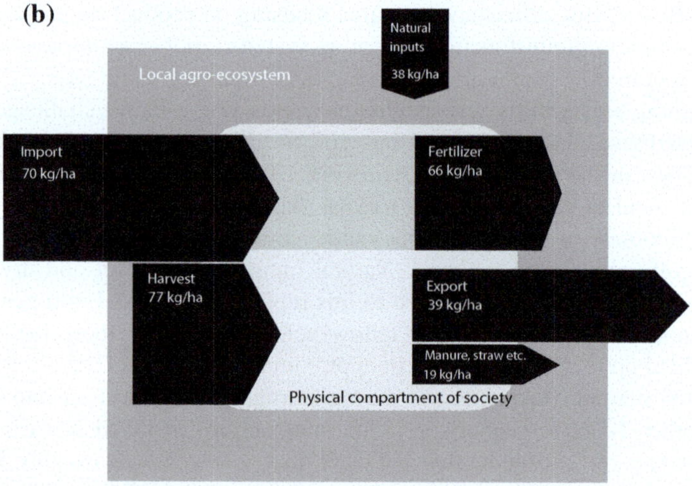

[kgN/ha]	Agro-ecosystem	Physical compartment of society
Inputs	123	144
Outputs	77	124
Balance	46 (losses)	20 (losses)

Fig. 21.1 Nitrogen flows in the agricultural production system of Theyern in 1830 (**a**), and 1999 (**b**) in kg per hectare and year

In Theyern, this type of low-input farming helped keep yields stable but at a rather low level, at least compared to modern standards. The cereal yield in 1830 was only 0.8 t/ha/year (tons per hectare per year), of which farmers had to reuse a considerable share as seed for the next year. This reduced the net yield to 0.64 t/ha/year. Aggregate food output in the village was 4.4 GJ/ha/year (gigajoules per hectare per year) of agricultural land per year. Every person active in agriculture produced 9.4 GJ/year. Cereals made up two-thirds of the total output, milk another 20 % and meat only 7 %. One hectare of agricultural land in Theyern could provide enough food for approximately one person. Overall, this was more than was required to nourish the local population. According to these calculations, a surplus of 25–30 % of the total food production was available. Part of this surplus had to be delivered to the landlord as tithe; the reminder could be sold in markets. From a sociometabolic perspective, another productivity or efficiency measure is important to consider. Under the conditions of the controlled solar energy system, agriculture is society's main energy-providing activity (Sieferle et al. 2006). An important measure of agricultural efficiency is, therefore, the so-called energy return on investment (EROI), which measures the ratio of socioeconomic energy expenditure in farming (e.g., labor or fuel) to the amount of energy contained in the food (or other outputs) produced in agriculture. Our rough quantification of the energy efficiency of the farming system in Theyern reveals that approximately 6 J of energy contained in food could be gained per joule of energy invested in the form of human labor, the only energy expenditure at that time (Fig. 21.2). In the other cases, this value was at a similar level.

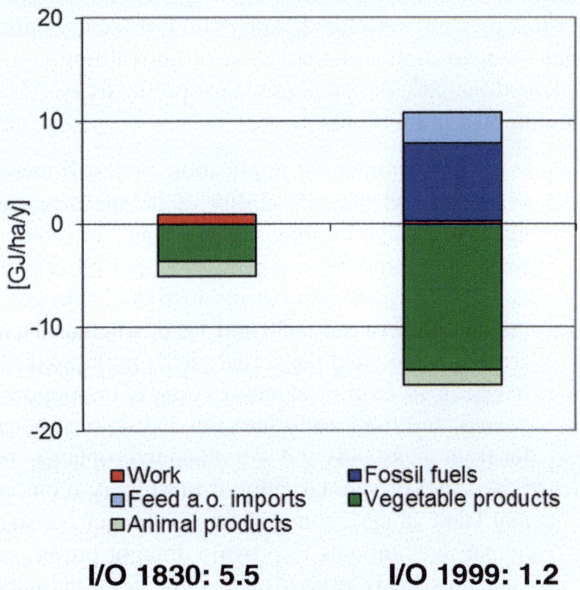

Fig. 21.2 Agricultural energy inputs and outputs in Theyern in 1830 (*left bar*) and 1999 (*right bar*). I/O denotes the ratio of energy input to energy output. Positive values designate energy inputs, negative values are energy outputs. An I/O of 5.5 means that for every joule of energy input, an output of 5.5 J was produced

21.4 The Socioecological Transformation of Agriculture

Unfortunately, we do not have any records that allow us to follow the development of the farming system in Theyern during the 19th and 20th centuries. At the national scale, we do know that the productivity of farming increased considerably during the 19th century. These productivity increases were mostly based on biological innovations, which were already beginning to gain significance in Theyern in 1830. New crops drove a shift from three-field to permanent crop rotation, increased the availability of nitrogen in cropland soils and provided more forage for livestock. This helped improve livestock production and increased the availability of manure. Hand in hand with institutional changes (e.g., the land reform and the abolishment of serfdom in 1848) and industrial tools (e.g., the iron plough), this doubled both land and labor productivity in the 19th century (Table 21.1) and further improved the agricultural energy efficiency of agriculture. Although these were far-reaching changes in farming, the socioecological principles of agriculture as we have outlined them for Theyern in the 1830s did not fundamentally change. External inputs remained low, and the maintenance of soil fertility continued to be largely based on natural inputs, manure and mixed farming, thus perpetuating the domination of the multifunctional use of livestock. This began to change at a moderate pace in the late 19th century with the introduction of the first industrial fertilizers, machinery and the use of steam power in farming.

However, radical change only began around World War II, when fossil-fuel-powered agricultural technologies became available at a large scale and at economically feasible terms. This triggered a rapid transformation process that, within only a few decades, resulted in a fundamental restructuring of the agricultural production system at all scales (cf. Grigg 1992; Langthaler 2003). The consequences of this process were the disintegration of locally optimized land-use systems; spatial specialization and separation of agricultural land uses; and steep and unprecedented increases in area and labor productivity. Above all, three aspects must be highlighted in this context.

(a) With the commercial application of the Haber-Bosch process for ammonia synthesis and the availability of inexpensive energy, nitrogen, probably the most important limiting soil nutrient in preindustrial agriculture, went from scarce to abundant in a very short period. This eliminated the nutrient limitation for growth in area productivity. Nitrogen inputs grew by an order of magnitude between 1950 and 1970, whereas the area planted with leguminous crops declined and the cereal yields multiplied (Table 21.1). The need to keep livestock in crop production systems to support nutrient management disappeared, and the locally integrated system of nutrient concentration and transfer from grasslands and woodland to croplands became obsolete.
(b) Fossil-fuel-driven agricultural machinery replaced the use of human and animal labor in agriculture. It was no longer necessary to keep large numbers of 'expensive' animals to provide draught power, and a considerable amount of land thus far reserved for feeding these animals became available for other

Table 21.1 The industrialization of Austrian agriculture in the 20th century: selected indicators. (Sources: Own calculations based on data and sources presented in Krausmann (2004); Krausmann et al. (2003); Sieferle et al. (2006))

		1830	1930	1950	1960	1970	1980	1990	2000	2010
Agricultural labor force	[1000]	1520*	1260	1092	776	432	290	214	185	159
Agricultural area[a]	[%]	44	40	39	37	36	34	32	30	29
Draft animals[b]	[1000]	616	666	633	363	50	0	0	0	0
Tractors	[1000]	0	0.7	14	118	268	324	351	335	330
Installed power[c]	[1000 kW]	521	536	504	1880	5583	8872	12,075	15,281	16,592
Industrial fertilizer[d]	[1000 t]	0	25	54	209	438	402	312	263	170
Nitrogen fertilizer	[kg/ha cropland/y]	0	2.5	11.8	28.4	81.8	102.3	93.3	86.6	76.0
Leguminous crops	[1000 ha]	109	240	280	220	120	73	54	70	90
Meat production[e]	[% of NV]	5	11	13	14	16	17	17	18	17
Cereal yield	[t/ha]	0.91	1.46	1.55	2.51	3.20	4.54	5.61	5.46	5.48
Area productivity[f]	[GJ_{NV}/ha]	3	6.1	5.0	8.9	10.5	13.5	16.2	16.6	19.5
Labor productivity[f]	[GJ_{NV}/cap]	9	21	23	46	95	166	266	303	384

*The value refers to the year 1869
[a]Agricultural area (excl. alpine pastures) as % of total
[b]Horses and oxen used to provide draft power
[c]Installed power refers to the potentially available power and includes human labor (0.1 kW per person), draft animals (horses 0.7 kW; oxen and cows 0.5 kW) and agricultural machinery (ca. 10–100 kW)
[d]Nutrient content of artificial fertilizer
[e]Meat production as % of total food production (measured in Joule nutritional value)
[f]Productivity: net output of plant- and animal based food (measured in GJ nutritional value) per ha of agricultural area (excl. alpine pastures) and per agricultural labor force

purposes. Although the number of draught animals declined from 630,000 in 1950 to only 50,000 in 1970, a rising number of tractors multiplied installed power (Table 21.1).

(c) Finally, the possibility of low-cost transportation of large quantities of bulk material drove the spatial separation between cropland and grassland and a specialization of farming and land use. Because of this transportation, it was no longer necessary to produce grain for food and feed locally under unfavorable climatic or topographic conditions, and the transportation facilitated large livestock operations that exceeded the local potential of feed supply.

From a sociometabolic perspective, these were the major drivers[2] of the transformation of the socioecological characteristics of agricultural production. It took only a few decades for traditional preindustrial land use and the corresponding agricultural landscapes to vanish and the current form of industrialized agriculture that is now practiced across Central Europe to emerge. The once beneficial local combination of cropland farming and multifunctional livestock keeping largely disappeared, and the function of livestock was reduced to the production of animal protein. This has led to specialization, specifically, grassland-based cattle farming in the alpine regions and cropland farming in the most fertile lowlands. It has also led to a large-scale spatial separation of livestock production and (market) feed production, as illustrated by the following numbers. Although the ratio of cropland to grassland has been stable at 1:1.4 in Austria on a national level, significant regional segregation processes have been observable since the 1950s. In 35 of 87 agricultural production regions, the acreage of cropland increased. In these regions, the ratio of cropland to grassland grew from 2:1 in 1950 to 4:1 in 1995. In the remaining 52 production regions, cropland decreased and the relation of cropland to grassland fell from 1:3 in 1950 to 1:6 in 1995 (Krausmann et al. 2003). Agricultural area as a whole, however, decreased by 25 % in the 1950–2010 period (Table 21.1).

What happened to agriculture in the village of Theyern? We have calculated the same set of socioecological indicators for the farming system in Theyern for the late 20th century.[3] In accordance with the national trend, the share of land used for agriculture decreased by 16 %, mostly because areas of marginal productivity were taken out of production and converted to woodlands. Although the population did not change much, only three of the households are still engaged in farming. Cropland is still the major land-use type, but agriculture has shifted toward either fruit cultivation or the production of feed grains combined with the fattening of pigs, which now account for half the livestock measured in LAUs.

Overall, livestock, particularly cattle and horses, has been reduced. The livestock density is now only 8 LAU/km^2, or 25 % of the 1830 value. Farm machines took over the role of humans and farm animals and have increased the installed power by two orders of magnitude, from 0.2 kW/ha (kilowatts per hectare) agricultural area in 1830 to nearly 10 kW/ha in 1995. The function of livestock as a converter of roughage and wastes into food has largely been abandoned, and the share of cereals and protein feed in total feed has grown from 9 % to more than 40 %. This has increased the production of animal products. Despite the massive

[2]These basic innovations were closely related, directly or indirectly, to a series of other fossil-fuel-based technologies, e.g., breeding technologies, biotechnology, pesticides and herbicides, irrigation systems, industrial processing of raw materials, and refrigeration and conservation systems.

[3]The spatial entities for which statistical data are recorded have changed between 1830 and 1999. The results for 1999 cover a considerably larger area. Averages and relative values can, however, be compared with reasonable accuracy over time.

reduction in livestock, nearly twice the amount of animal-based calories is produced per unit of land.

In contrast, livestock's role in nutrient management has declined. Only 10 % of the available nitrogen for plants originates from manure, whereas the largest part comes from industrial fertilizer (Fig. 21.1b). The amounts of both extracted and applied nitrogen have also changed. Agricultural nitrogen extraction increased threefold in Theyern, and roughly half of this nitrogen is exported in agricultural products. Similarly, nitrogen imports have multiplied, amounting to 70 kg/ha in 1999. All these changes have helped drastically raise area and labor productivity. Net cereal yield has multiplied sevenfold since 1830, and average food output per unit of agricultural area has grown from 4.4 to 31.2 GJ/ha. One hectare of agricultural area now produces enough food for seven persons, and one person active in agriculture produces enough for 23 people. The cost of these tremendous productivity gains was a loss in energy efficiency. Fuel and electricity for agricultural machinery and energy requirements to produce the large amounts of agrochemical inputs have grown much more than the energy contained in agricultural produce. Thus, the energy efficiency of farming in Theyern, and in all other case studies, has fallen to an input-output ratio close to 1 (Fig. 21.2).

21.5 Conclusions

From a socioecological perspective, agricultural modernization resulted in the transformation of locally integrated production systems and low levels of material flows into throughput systems, increasing biomass, energy and nutrient flows by one to two orders of magnitude. Imports and exports of energy, feed and plant nutrients to and from agricultural production systems have multiplied. Moreover, the extraction of biomass per unit of land and the internal turnover of biomass have increased.

Consequently, the intensity of land use has changed in quality and quantity. The human impact on the environment shifted in terms of scale from the local to the global level. Furthermore, socioeconomic interference with global biogeochemical cycles, which was more or less negligible in the agrarian regime (with the important exception of considerable amounts of CO_2 released to the atmosphere due to the expansion of agricultural areas, see Chap. 20), reached an unprecedented level (Vitousek et al. 1997).

Direct and indirect inputs of fossil fuels into agriculture transformed the role of agriculture within social metabolism. Due to a novel abundance of energy, the ultimate prerequisite of preindustrial agriculture to produce a net gain of socioeconomically useful energy in the long run became obsolete. Fossil fuels allowed for tremendous increases in output.

Consequently, both labor and area efficiency increased (Table 21.1), but they did so at the cost of energy efficiency. Agriculture changed from being the central activity for providing society with primary energy to an energy-consuming provider of food and raw material (cf. Leach 1976; Pimentel and Pimentel 1996). In other words, the example of the industrialization of Austrian agriculture shows us that the industrial metabolic transition changed the significance of ecosystem colonization for social metabolism. In traditional farming systems, agriculture, that is, the colonization of terrestrial ecosystems, secures nearly all resource inputs into society's metabolism. In industrial systems, this relation has turned upside down, and society's metabolism provides large amounts of mineral and fossil materials to support high agricultural labor and area productivity.

References

Bayliss-Smith, T. P. (1982). *The ecology of agricultural systems*. Cambridge: Cambridge University Press.

Cunfer, G., & Krausmann, F. (2009). Sustaining soil fertility. Agricultural practice in the old and new worlds. *Global Environment, 4*, 9–43.

Cusso, X., Garrabou, R., & Tello, E. (2006). Social metabolism in an agrarian region of Catalonia (Spain) in 1860 to 1870: Flows, energy balance and land use. *Ecological Economics, 58*(1), 49–65.

Grigg, D. B. (1992). *The transformation of agriculture in the west*. Oxford: Blackwell.

Haberl, H., Winiwarter, V., Andersson, K., Ayres, R. U., Boone, C. G., Castillio, A., et al. (2006). From LTER to LTSER: Conceptualizing the socio-economic dimension of long-term socio-ecological research. *Ecology and Society, 11*(2), 13. http://www.ecologyandsociety.org/vol11/iss2/art13/

Krausmann, F. (2004). Milk, manure and muscular power. Livestock and the industrialization of agriculture. *Human Ecology, 32*(6), 735–773.

Krausmann, F. (2008). *Land use and socio-economic metabolism in pre-industrial agricultural systems: Four nineteenth-century Austrain villages in comparison*. Social Ecology Working 72. Vienna: Inst. of Social Ecology.

Krausmann, F., Haberl, H., Schulz, N. B., Erb, K.-H., Darge, E., & Gaube, V. (2003). Land-use change and socio-economic metabolism in Austria. Part I: Driving forces of land-use change: 1950–1995. *Land Use Policy, 20*(1), 1–20.

Langthaler, E. (2003). Agrarwende in der Ebene. Eine region im niederösterreichischen Flach- und Hügelland (1880–2000). In E. Bruckmüller, E. Hanisch & R. Sandgruber (Eds.), *Geschichte der österreichischen Land- und Forstwirtschaft im 20* (pp. 651–726). Jahrhundert. Regionen, Betriebe, Menschen. Ueberreuter, Wien.

Leach, G. (1976). *Energy and food production*. Guildford: IPC Science and Technology Press.

Mazoyer, M., Roudart, L., & Membrez, J. H. (2006). *A history of world agriculture: From the neolithic age to the current crisis*. London: Earth Scan.

McNeill, J. R., & Winiwarter, V. (Eds.). (2006). *Soils and societies: Perspectives from environmental history*. Isle of Harris: The White Horse Press.

Pimentel, D., & Pimentel, M. (1996). *Food, energy and society*. Niwot: University Press of Colorado.

Sandgruber, R. (1979). Der Franziszeische Kataster und die dazugehörigen Steuerschätzungsoperate als wirtschafts- und sozialhistorische Quellen. *Mitteilungen aus dem niederösterreichischen Landesarchiv, 3*, 16–28.

Schaschl, E. (2007). *Rekonstruktion der Arbeitszeit in der Landwirtschaft im 19. Jahrhundert am Beispiel von Theyern in Niederösterreich*. Social Ecology Working Paper 96. Inst. of Social Ecology, Vienna.

Sieferle, R. P., Krausmann, F., Schandl, H., & Winiwarter, V. (2006). *Das Ende der Fläche. Zum gesellschaftlichen Stoffwechsel der Industrialisierung*. Köln: Böhlau.
Simmons, I. G. (2008). *Global environmental history 1000 BC to AD 2000*. Edinburgh: Edinburgh University Press.
Singh, S. J., Ringhofer, L., Haas, W., Krausmann, F., & Fischer-Kowalski, M. (2010). *Local studies manual: A researcher's guide for investigating the social metabolism of rural systems*. Social Ecology Working Paper 120. Vienna: Inst. of Social Ecology.
Umweltgeschichte, P. (2000). Historische Entwicklung von Wechselwirkungen zwischen Gesellschaft *und Natur. Forschungsschwerpunkt Kulturlandschaft, CD-ROM, 7*.
Vitousek, P. M., Mooney, H. A., Lubchenco, J., & Melillo, J. M. (1997). Human domination of earth's ecosystems. *Science, 277*(5325), 494–499.

Chapter 22
The Philippines 1910–2003: A Century of Transitions

Thomas Kastner, Karl-Heinz Erb and Fridolin Krausmann

Abstract The Philippines experienced a rapid land-use transition during the 20th century. Forest cover decreased from approximately 70 % in 1900 to less than 25 % in 2000, whereas cropland areas and grasslands expanded. At the turn of the millennium, however, the land cover patterns appear more stable. We analyze this trajectory by linking it to the transition from an agrarian sociometabolic regime to an industrial regime. During the 20th century, the Philippines changed from a sparsely populated to a densely populated nation, with population numbers increasing more than tenfold. This rapid population growth went hand in hand with fundamental changes in the nation's agricultural system. The challenge of maintaining food supply levels for the ever-growing number people was mainly met by the expansion of cultivated areas during the first part of the century. Later, intensification (i.e., higher output per unit land) became the dominant mode of increasing agricultural output. Here, the availability of fossil resources, offering non-land-based energy subsidies for agriculture, played a crucial role. In this way, land cover could be stabilized, albeit under massive changes in land-use intensity associated with negative environmental impacts and increased dependency on fossil fuels and mineral resources. We conclude with an outlook on the near future: the Philippines is still in the midst of the transition toward an industrial society, and the option space for future development is markedly smaller than for past trajectories of now-industrialized nations, mainly due to global resource competition.

T. Kastner (✉) · K.-H. Erb · F. Krausmann
Institute of Social Ecology (SEC), Alpen-Adria University,
Schottenfeldgasse 29, 1070 Vienna, Austria
e-mail: thomas.kastner@aau.at

K.-H. Erb
e-mail: karlheinz.erb@aau.at

F. Krausmann
e-mail: fridolin.krausmann@aau.at

H. Haberl et al. (eds.), *Social Ecology*, Human-Environment Interactions 5,
DOI 10.1007/978-3-319-33326-7_22

Keywords Land-use transition · Sociometabolic transition · The Philippines · Agricultural change · 20th century

22.1 Introduction

Around the globe, pressures on natural resources are increasing due to growing population numbers and rising consumption levels. Land that can provide humans with food, fiber and fuel is clearly a limited resource. Overall, human pressure on land resources has increased throughout the past century, albeit at a slower pace than population numbers, indicating that human societies have found ways to sustain more people per unit of land (Krausmann et al. 2013). These efficiency gains came at the cost of increasing dependency on fossil resources and negative environmental impacts of agricultural intensification (e.g., eutrophication, pesticide contamination). In this chapter, we explore the rapid changes the Philippines' land system experienced in the 20th century through concepts that are central in Land-Change Science and in socioecological research.

Land-Change Science seeks to understand how human-environment relations shape land cover and land use (Turner et al. 2007). The notion that land-use transitions follow a typical pattern (Boserup 1993; DeFries et al. 2004) of increasing human domination of terrestrial ecosystems provides a valuable starting point for analyzing processes of land-use change. This concept describes the replacement of pristine ecosystems with increasingly human-dominated ones as agricultural and forestry activities spread. With increasing population numbers, agriculture is intensified and the market integration of food systems increases, pushing subsistence communities back into niches. At a certain point of (population) growth, intensification (more output per unit of land) becomes more important for increasing food supply than extensification (more land under cultivation), and the land cover picture stabilizes.

Analogously, the concept of sociometabolic regimes and transitions, which is related to the interaction of natural and social systems, has been used to describe fundamental changes in human resource use (Fischer-Kowalski and Haberl 2007; Singh et al. 2012). Sociometabolic regimes are characterized by archetypical modes of subsistence (hunter-gatherers, agrarian societies and industrial societies), and sociometabolic transitions describe changes from one mode to the other along gradients of increasing population density and energy demand. Within the concept of sociometabolic transitions, land use plays a prominent role. Transitions from agrarian to industrial societies, starting in the 18th century and continuing in many parts of the world today, are above all characterized by decreasing dependence on land-based energy sources and increasing dependence on fossil fuels and mineral resources in quantitative terms (see Chap. 3).

The 20th-century Philippines provides a valuable case study to explore the strong links between the concepts of land-use transitions and sociometabolic transitions, as the country experienced high dynamics in virtually all relevant variables. Population numbers grew more than tenfold throughout the last century. At the same time, rapid land-use changes occurred throughout much of the century, implying massive clearing of originally forested lands. With the International Rice Research Institute established approximately 60 km south of Manila in 1960, the Philippines became one of the centers of the Green Revolution, promoting agricultural intensification through the large-scale introduction of high-yielding varieties of rice and other crops that are highly responsive to increased agrochemical and water inputs (Hayami and Kikuchi 1999). Fossil energy started to play a significant role in the nation's energy systems only during the final decades of the 20th century, and its use has experienced steep increases since then. The country is not endowed with significant deposits of fossil energy carriers and almost fully depends on imports of these resources. Furthermore, average per capita income levels are low, though they are increasing. Thus, the country can be considered representative of a large share of the global population: presently at low income levels and low levels of per capita resource consumption but aspiring to increased affluence.

22.2 A Land-Use Transition and the Intensification of Agriculture

The most striking development when looking at land cover change in the Philippines during the 20th century (Fig. 22.1a) is the rapid decline of forest land, from approximately 65 % of the total land area in 1910 to less than 25 % by 1986.

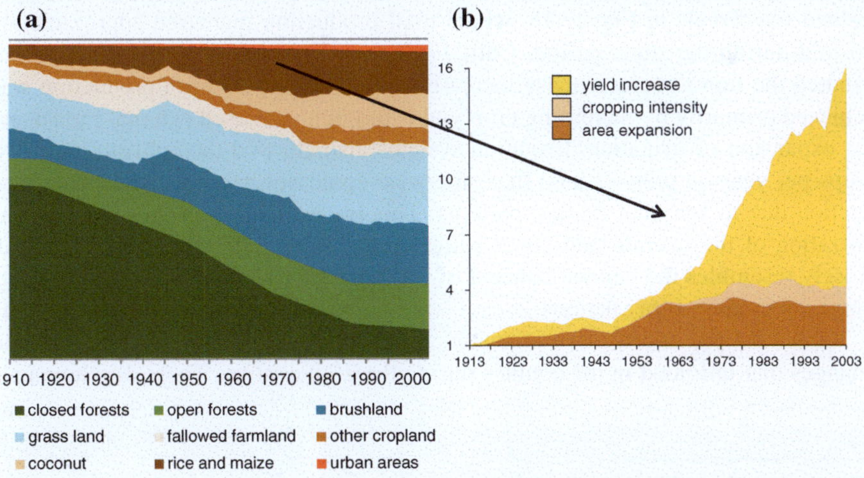

Fig. 22.1 **a** Land use/cover in the Philippines 1910–2003 (total area = 30 Mha); **b** indexed (1913 = 1) development of cereal (rice and maize) production according to increases in area, cropping frequency and yields (Sources: Kastner 2009; Kastner and Nonhebel 2010)

The loss of closed forests was particularly drastic. By the year 2000, their extent had dropped to less than one-fifth of the 1910 value. This decrease in forest land occurred at the expense of increases in cultivated area and in secondary vegetation types of grass- and bush land. Deforestation came to a halt in the mid-1980s. Since then, the general land cover pattern has appeared relatively stable. A comparison of the recent patterns of land cover change in the Philippines to the developments in other nations (Lambin and Meyfroidt 2010) indicates that the country might be in the midst of a forest transition, that is, a change from net deforestation toward a net increase in forest area.

Changes in agricultural practices and modes of human subsistence can explain much of the land-use transition depicted in Fig. 22.1a. Before the beginning of the 20th century, much of the existing primary vegetation was already converted by human use. Historical accounts mention that only 20 % of the Philippine forests were considered to constitute truly primary vegetation in the early 20th century, with the remaining parts recovering from past shifting cultivation (Kolb 1942). Throughout the 20th century, the cultivation frequency increased as the population grew, and more permanent forms of agriculture became widespread. The cropping frequency expressed as an, the cropped area divided by the total land useable (and previously used) for cropping times 100 (Ruthenberg 1971),[1] shows the following trend at the national level: the value doubled from 17 in 1920 to 34 in 1960 and further increased to 49 by 1980, and it has not changed much since then. This trajectory indicates the rapid increase in cropped land relative to the available land base. The national totals mask large regional differences between intensive lowland agriculture (with hardly any fallow period and multiple harvests per year) and the shifting cultivation systems prevailing in the uplands.

During the second half of the century, increases in biomass output were achieved by the further intensification of annually cropped lands. This is highlighted for cereals in Fig. 22.1b, where total production increased approximately 16-fold during the study period. Until 1960, the expansion of cropped areas contributed the lion's share of these increases. After 1960, growth in production was achieved primarily by increasing crop yields through increased external inputs and the expansion of irrigation. Irrigation systems also allowed the cultivation of two crops per year on plots of land that otherwise could not have sustained such production due to seasonal rainfall patterns. This intensification, along with the stabilization of the overall land cover pattern toward the end of the studied period, closely resembles the general concept of the land-use transition and developments observed in many other nations.

Table 22.1 presents a set of indicators that characterize the massive agricultural changes that unfolded in the country during the 20th century. These developments

[1]The presented values include the following lands: cropland area, bush land, open forest and 60 % of closed forests. Grassland was excluded because in the Philippines they are mostly Imperata grasslands, which, once established, require considerable investment to be taken back into crop cultivation.

Table 22.1 Development of population, gross domestic product (GDP) and indicators of agricultural intensification in the Philippines during the 20th century (Sources: FAO 2013; Kastner 2009)

		1920	1940	1960	1980	2000
Population						
[million capita]	Population	10.4	16.4	27.4	48.3	76.5
Share agricultural population in total population		N.d.	N.d.	63 %	52 %	39 %
capita/km^2	Population density	35	55	92	162	257
Economic development						
['000 1990 pesos/ capita]	GDP	13	15	15	24	24
Share agriculture, fishing and forestry in total GDP		37 %	37 %	30 %	23 %	20 %
Agricultural intensification						
[t dm/ha/year]	Overall crop yields	1.4	2.0	1.9	2.9	3.7
[t dm/ha/year]	Rice yields	1.1	1.2	1.2	2.9	4.4
[kg/ha cropland]	Artificial nitrogen fertilizer use	N.d.	N.d.	11	38	85
[number/1000 capita]	Use of tractors	N.d.	N.d.	0.15	0.22	0.77
[million capita]	Population economically active in agriculture	N.d.	N.d.	6.4	9.6	12.4
[million head]	Water buffaloes (draught animals)	1.5	3.0	3.7	2.9	3.0
Metabolic profile						
[GJ/capita/year]	Energy use per capita	37	38	34	37	36
[GJ/ha/year]	Energy use density	14	20	31	60	93
[%]	Biomass share in energy use	93 %	95 %	86 %	65 %	48 %
Human Appropriation of Net Primary Production (HANPP)						
[t dm/capita/year]	HANPP per capita	11.0	8.0	6.1	3.9	2.8
[t dm/ha/year]	HANPP density	3.9	4.4	5.6	6.3	7.1
[t dm/t dm]	HANPP per unit biomass harvest	1.6	1.6	1.7	1.5	1.6
[t dm/t dm]	HANPP per unit used biomass extraction	6.1	4.2	2.7	2.9	2.8

N.d.: No data; t dm: metric tons of dry matter biomass

must be seen within the context of the exponential increase in population numbers. The high (albeit declining) relevance agriculture still holds in the country is reflected in the high shares of the primary sector in total GDP (20 % in 2000) and of agricultural population in total population (just below 40 % in 2000). Average crop yields, in tons of dry matter per hectare per year, increased by a factor 2.6 from 1920 to 2000 (Table 22.1). For rice, the nation's crucial staple food, this increase was fourfold, with yield growth taking off only in the 1960s. These yield improvements were achieved by means of the so-called *Green Revolution* (Hayami and Kikuchi 1999), the provision of varieties more responsive to artificial fertilizers and pesticides. The use of these inputs rose rapidly, as reflected by a nearly eightfold increase in mineral fertilizer application per unit cropland in only four decades from 1960 to 2000, reaching values similar to that observed in fully industrialized nations in recent decades (FAO 2013). The picture looks different with regard to the mechanization of Philippine agriculture. The number of tractors per capita grew during the second half of the century (from 0.15 tractors per 1000 inhabitants in 1960 to 0.77 in 2000). By 2000, however, the number of tractors was only 2 and 5 % of the levels for the Western industrial countries Austria and the US, respectively. The number of combined harvesters per capita in the Philippines was just 1 % of US and Austrian levels (FAO 2013). Human labor input into agriculture grew throughout the century; the number of people economically active in agriculture nearly doubled from 1960 to 2000. This trend should be placed within the context of continued population growth at high rates. Additionally, the water buffalo stock indicates that working animals still play an important role in Philippine agriculture; the numbers were fairly constant throughout the century, with no marked decline in recent decades.

The Philippines' labor intensification trajectory differs markedly from past developments in Western countries, where industrialization was linked to a strong decline in agricultural labor force and the vanishing of working animals, often preceding the rise of inorganic fertilizer use (Chap. 3). This divergence can be explained by differences between temperate crop farming and paddy rice farming and by the Philippines slow economic development in other sectors, which means there are no great incentives spurring a migration of labor out of agriculture, as experienced in other nations (Bautista 1995).

22.3 On the Way from an Agrarian to an Industrial Metabolic Regime?

The framework of sociometabolic regimes has been widely used to scrutinize long-term trends in society-nature interactions (Fischer-Kowalski and Haberl 2007). Sociometabolic regimes are above all characterized by their energy system, which is the overall and per capita magnitude of energy flows required to sustain a society and the sources and technologies that are used to provide this energy. The changing relation of land use and energy provision is of central importance (Chap. 3): preindustrial societies (hunter-gatherer as well as agrarian societies) rely primarily on land-based energy inputs, i.e., various forms of biomass (such as food, feed, fiber and fuel). In contrast, the industrial sociometabolic regime is characterized by the massive impact of the availability of fossil fuels, i.e., energy sources not needing relevant amounts of land per unit energy provided (Smil 2008). By emphasizing the role of land use in energy regimes, the concept of sociometabolic transitions can also provide a strong and intuitive framework to explain land-use transitions such as those discussed for the Philippines in the previous section (Fig. 22.1a). This concept allows us to understand that along the metabolic transition, the role of agriculture in the overall social metabolism changes fundamentally. Agriculture transitions from a system that largely has to provide for its energy supply internally into a system that receives large energy and material subsidies (in the form of inorganic fertilizers, pesticides, machinery) from outside to maximize the output of food production. The case of massive cereal yield increases in the Philippines (Fig. 22.1b) subsidized by rapidly rising fertilizer inputs per unit of cropland (Table 22.1) is one example.

The human appropriation of net primary production (HANPP) has been developed as an indicator to comprehensively link human land-use pressure to the social metabolism concept (for a concise overview of the HANPP method, refer to Method Précis on Human Appropriation of Net Primary Production). HANPP provides a measure of how much of the potentially available (i.e., without human presence) annual net primary production (NPP) is appropriated by humans through two main processes: conversion of vegetation through land use ($HANPP_{luc}$) and biomass harvest ($HANPP_{harv}$). Following the terminology common in material flow accounting (MFA), the latter can be further broken down into used biomass extraction (biomass harvested for further socioeconomic use) and unused biomass extraction (biomass killed in the process of harvest but not entering the socioeconomic system; see Method Précis on Material Flow Analysis for details). Following these definitions, Fig. 22.2 presents a comprehensive breakdown of societal energy use in the Philippines throughout the 20th century, differentiating used biomass extraction (Fig. 22.2a), unused biomass extraction (Fig. 22.2b), a decrease in NPP due to land conversion (Fig. 22.2c) and energy use from non-biomass sources (Fig. 22.2d).

The development of used biomass extraction (Fig. 22.2a) is further broken down into crop harvest, industrial wood harvest, wood fuel harvest and grazed biomass. All subcategories grew throughout time, with the exception of industrial wood harvest. Timber harvest in the Philippines was largely export driven and was associated with large-scale deforestation. This exploitation of timber stocks far exceeded natural regeneration limits and could not be sustained in the long run. Industrial timber harvest peaked in the late 1960s and has declined strongly since then, with the levels in 2000 at approximately 20 % of the peak levels. This trend can be explained by the largely export-driven exhaustion of Philippine forest stocks (Kastner 2009). Of the other subcategories, crop production grew fastest, representing the only subcategory with growth rates that kept up with population development.

Unused biomass extraction (Fig. 22.2b) associated with the used extraction shown in Fig. 22.2a is broken down into biomass destroyed through slash-and-burn practices and other unused extraction such as harvest residues or felling losses from forestry operations. Overall, unused extraction remained rather constant throughout time, with the relevance of slash-and-burn fires declining.

Figure 22.2c shows the reduction in NPP levels due to land-use practices, as denoted by the main land cover classes ($HANPP_{luc}$). Overall ecosystem productivity was considerably and increasingly lowered through human activities throughout the century. In the first half of the century, this trend was driven by declining

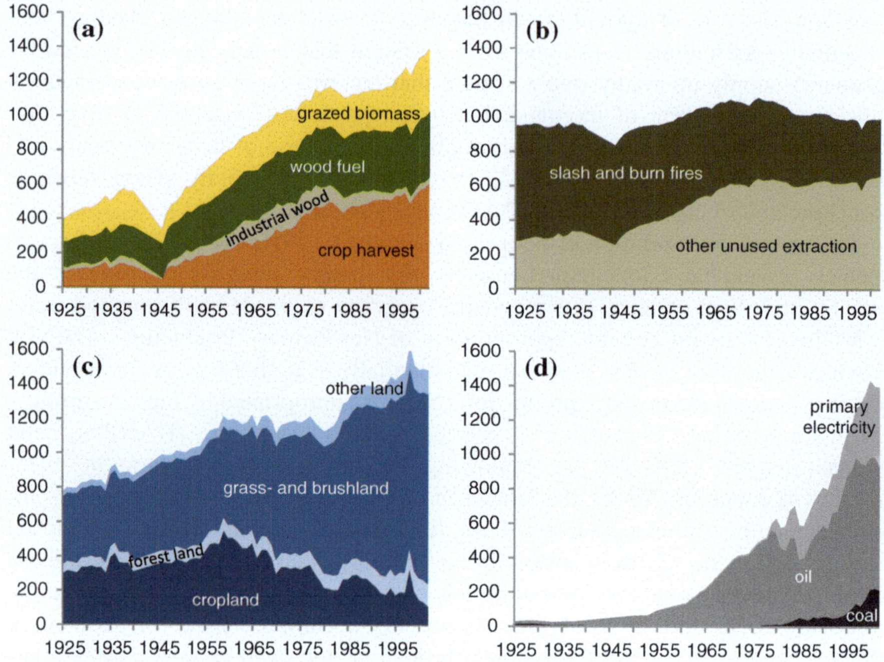

Fig. 22.2 Energy use in the Philippines throughout the 20th century, with a focus on the nation's biomass metabolism: **a** used biomass extraction, **b** unused biomass extraction, **c** reduction of NPP resulting from land conversion ($HANPP_{luc}$) and **d** energy use from non-biomass sources. All values are in petajoules (1 PJ = 10^{15} J) per year. (Source: Kastner 2009)

forest land at the expense of cropland above all. Until the 1960s, these lands received little external inputs and exhibited NPP levels much lower than those of the natural forests that would have prevailed within the archipelago without human influence. Along with the intensification of agriculture, the NPP per unit of cropland increased greatly, as reflected in the decline of $HANPP_{luc}$ on cropland throughout the last decades of the 20th century. This trend—increasing biomass harvest with decreasing land-use-induced reduction of NPP levels—is a common feature of agricultural intensification and has been found in various settings around the world (Krausmann et al. 2012, 2013). The fact that the overall $HANPP_{luc}$ also increased during the second part of the century makes the Philippine case different from other long-term HANPP studies (Krausmann et al. 2012). This increase was caused by continued deforestation until the mid-1980s at the expense of secondary grassland and bush land. These lands exhibit lower productivity levels than natural forests and are prone to degradation, especially grasslands. Based on local case studies, it was assumed that the effect of this degradation on the lowering of NPP levels would have increased throughout the century. However, the high uncertainty regarding this trend of increasing degradation should be noted (for details see Kastner 2009), and it is unclear how representative such a trend is for similar regions around the globe.

Per capita HANPP values (Table 22.1; overall HANPP can also be derived from adding the values in Fig. 22.2a–c) declined rapidly throughout the century (from 11 t dm/cap/year; metric tons of dry matter biomass per capita per year; in 1920 to 2.8 t dm/cap/year in 2000). However, the rate at which the values declined was less than the rate at which the population rose, leading to a steady increase in overall HANPP pressure on the nation's territory (from 3.9 t dm/ha/year in 1920 to 7.1 t dm/ha/year in 2000). Examining the relations between the individual subcomponents of HANPP provides insight into how the increases in land-use efficiency were achieved; used extraction grew at fast rates throughout the century, whereas the other two subcomponents combined saw slower increases, resulting in 2.8 t dm of HANPP per t dm used biomass in 2000 (a decline of more than 50 % from the 1920 value of 6.1 t). This view on the changes in the biomass metabolism of the Philippine economy complements the land-use transition narrative very well. The gains in efficiency can be explained by shifting cultivation with long fallow periods increasingly being replaced by systems with shorter cropping cycles, implying a declining relevance of unused extraction in the form of slash-and-burn fires. Through the use of fire, shifting cultivation typically destroys large amounts biomass per unit used extraction. Additionally, strong yield increases on cropped land (Fig. 22.1b) contributed to massive efficiency gains in terms of useful output per unit of land. In addition to the increased input of human labor, these efficiency gains came at the cost of increased dependency on non-land-based resources, particularly fossil fuels (Fig. 22.2d) and inorganic fertilizers (Table 22.1). These means of increasing land-use efficiency in terms of output per unit of land were, in turn, associated with negative environmental impacts on land systems (e.g., eutrophication, pesticide contamination).

Figure 22.2d presents the development of Philippine energy consumption from sources other than biomass, split up into coal, oil and primary electricity (hydropower and geothermal energy). The graph reveals very low levels of energy use from these carriers before 1960 and a rapid increase since then, mainly driven by oil consumption. Overall, the per capita energy use in the Philippines (including used extraction only, Fig. 22.2a and d) did not change greatly throughout the century (37 GJ/cap/year; gigajoules per capita per year; in 1920, and 36 GJ/cap in 2000; see Table 22.1[2]). However, the energy use density in terms of used energy per unit land increased more than sixfold from 1920 to 2000 due to the rapid population growth. A major trend was the decline in the share of biomass in the energy mix; this share was at 95 % in 1940 and decreased to approximately half this value (48 %) in 2000 (Table 22.1). This development can be seen as evidence that the country is the midst of an energy transition from an agrarian regime to an industrial metabolic regime. Krausmann et al. (Chap. 3) put typical shares for

[2]The inclusion of unused biomass extraction and NPP decreases due to human land (Fig. 22.2b and c) changes this trend toward a pronounced decline in per capita energy use from 185 GJ/cap/year in 1920 to 69 GJ/cap in 2000. See Chap. 15 for a methodological discussion of this issue of defining system boundaries.

energy use from non-biomass sources at 5 % for agrarian societies (well in line with values for the Philippines until mid-20th century) and 70–90 % for industrial societies. In absolute values, Western nations showed a per capita use of fossil fuels more than ten times the Philippine value (Steinberger et al. 2010) in the year 2000. These numbers imply that the country is still far from the saturation phase observed for established industrial societies. Increasing shares of fossil energy carriers in overall energy use are typically not achieved by a substitution of fossil fuels for biomass but by a much faster growth in fossil-fuel consumption compared to biomass and a fast overall growth of the level of energy use; globally, absolute quantities of biomass used as fuel more than doubled in the period 1850–2000 (Fernandes et al. 2007). This situation highlights that the rapid rise in fossil fuel consumption is not expected to lower pressure on the Philippines' biomass resources and terrestrial ecosystems.

22.4 Conclusion

This chapter studied the rapid land-use transition of the Philippines during the 20th century through the lens of the sociometabolic transition concept. The comprehensive focus on sociometabolic energy regimes provides a clear framework that complements prevailing theories and narratives of land-use transition processes with agriculture taking a central role. The availability of fossil resources offers non-land-based energy subsidies and enables massive increases in output per unit of land, shifting the principal mode of production increases from extensification toward intensification. In this way, land cover can be stabilized, albeit under massive changes in land-use intensity associated with negative environmental impacts and increased dependency on fossil fuels and mineral resources. With this focus on societal energy use, the comprehensive framework of sociometabolic transitions is complementary to and compatible with studies on social drivers and the consequences of land-use transition processes (e.g., Lambin and Meyfroidt 2010) and in-depth accounts of agricultural change (e.g., Evans 1998; Hayami and Kikuchi 1999; Mazoyer and Roudart 2006).

Our study opens room for a speculative outlook on the Philippines' future. If the country were to follow past trajectories of industrial nations, there would likely be a large additional demand for non-biomass energy without a substantial reduction in per capita biomass demand. Although further efficiency gains in terms of overall biomass appropriation per unit used biomass appear possible (the country is still above the global average in terms of these indicators), ways to limit the negative environmental impacts of further intensification will have to be found. Population projections estimate the country's population at 155 million by 2050 (FAO 2013), more than twice the value in 2000. With similar population dynamics and increasing per capita resource use in many developing nations along with global competition for limited resources, it is difficult to imagine that the country will reach values of per capita resource use typical for established industrial regimes. In this context, it can be argued that the option space for ongoing socio-metabolic transitions towards an industrial regime is markedly smaller than it was for past transitions.

Acknowledgements Funding by the Austrian Science Fund (FWF) within the project P21012, by the European Research Council within ERC Starting Grant 263522 LUISE and by the EU-FP7 project VOLANTE is gratefully acknowledged.

References

Bautista, R. M. (1995). Rapid agricultural growth is not enough: The Philippines, 1965–80. In J. W. Mellor (Ed.), *Agriculture on the road to industrialization* (pp. 113–149). London: Johns Hopkins University Press.

Boserup, E. (1993). The conditions of agricultural growth: The economics of agrarian change under population pressure.

DeFries, R. S., Foley, J. A., & Asner, G. P. (2004). Land-use choices: Balancing human needs and ecosystem function. *Frontiers in Ecology and the Environment, 2*, 249–257.

Evans, L. T. (1998). *Feeding the ten Billion: Plants and population growth*. Cambridge: Cambridge University Press.

FAO. (2013). FAOSTAT statistical database. Last accessed: August 13, 2013 [WWW Document]. URL http://faostat.fao.org/

Fernandes, S. D., Trautmann, N. M., Streets, D. G., Roden, C. A., Bond, T. C. (2007). Global biofuel use, 1850–2000. *Global Biogeochemical Cycles 21*, GB2019.

Fischer-Kowalski, M., & Haberl, H. (Eds.). (2007). *Socioecological transitions and global change: Trajectories of social metabolism and land use*. Cheltenham, Northampton: Edward Elgar.

Hayami, Y., & Kikuchi, M. (1999). *A rice village saga: Three decades of green revolution in the Philippines*. Houndmills: Macmillan.

Kastner, T. (2009). Trajectories in human domination of ecosystems: Human appropriation of net primary production in the Philippines during the 20th century. *Ecological Economics, 69*, 260–269.

Kastner, T., & Nonhebel, S. (2010). Changes in land requirements for food in the Philippines: A historical analysis. *Land Use Policy, 27*, 853–863.

Kolb, A. (1942). Die Philippinen, Geographische Handbücher. K.F. Koehler Verlag, Leipzig.

Krausmann, F., Erb, K.-H., Gingrich, S., Haberl, H., Bondeau, A., Gaube, V., et al. (2013). Global human appropriation of net primary production doubled in the 20th century. *PNAS, 110*, 10324–10329.

Krausmann, F., Gingrich, S., Haberl, H., Erb, K.-H., Musel, A., Kastner, T., et al. (2012). Long-term trajectories of the human appropriation of net primary production: Lessons from six national case studies. *Ecological Economics, 77*, 129–138.

Lambin, E. F., & Meyfroidt, P. (2010). Land use transitions: Socio-ecological feedback versus socio-economic change. *Land Use Policy, 27*, 108–118.

Mazoyer, M., & Roudart, L. (2006). *A history of world agriculture: From the neolithic age to the current crisis*. London: Earthscan.

Ruthenberg, H. (1971). *Farming systems in the tropics*. Oxford, UK: Clarendon Press.

Singh, S. J., Haberl, H., Chertow, M., Mirtl, M., & Schmid, M. (Eds.). (2012). *Long Term socio-ecological research: Studies in society-nature interactions across spatial and temporal scales*. Dodrecht, The Netherlands: Springer.

Smil, V. (2008). *Energy in nature and society: General energetics of complex systems*. Cambridge, MA: MIT Press.

Steinberger, J. K., Krausmann, F., & Eisenmenger, N. (2010). Global patterns of materials use: A socioeconomic and geophysical analysis. *Ecological Economics, 69*, 1148–1158.

Turner, B. L., Lambin, E. F., & Reenberg, A. (2007). The emergence of land change science for global environmental change and sustainability. *Proceedings of the National Academy of Sciences, 104*, 20666–20671.

Chapter 23
How Tourism Transformed an Alpine Valley

Robert Groß

Abstract Tourism moves global flows of capital, people and knowledge and thereby fundamentally transforms materiality, social relations, communities and life-worlds. This paper examines the environmental history of an alpine community (Damüls/Austria) under the influence of tourism in the 20th century. Environmental History seeks to understand historical society-nature relations and people's perceptions of nature in the past. Such an endeavor poses a twofold challenge to environmental historians, who conceptualize 'nature' as an independent parameter of history without reducing it to a social construct but at the same time address the social construction of 'beautiful landscapes' as an integral part of the tourism industry. These two viewpoints can only be bridged dialectically. In this article, the concept of socio-natural sites (SNSs) is used to bridge that gap and to analyze the long-term impact of ski lifts on materiality and the cultural representation of winter sport landscapes. SNSs are constituted by the nexus of social practices and material arrangements. Damüls is a telling example of the restless transformation of an Alpine sport arena built for skiers.

Keywords Environmental history · Regional history · Long-Term socioecological research (LTSER) · Socio-natural sites (SNSs) · Spiral of risk

---

This essay was written within the framework of the FWF-funded project 'How skiers' sensations shaped Alpine valleys during the 20th century' (P24278-G18).

R. Groß (✉)
Institute of Social Ecology (SEC), Alpen-Adria University,
Schottenfeldgasse 29, 1070 Vienna, Austria
e-mail: robert.gross@aau.at

H. Haberl et al. (eds.), *Social Ecology*, Human-Environment Interactions 5,
DOI 10.1007/978-3-319-33326-7_23

23.1 Introduction

'Peripheries' are often the focus of highly ambitious schemes. Peripheral rural spaces are well suited as settings for socially desirable utopias because these spaces are seen in terms of their 'empty', 'primordial' and entirely unexploited potential. The real impact of industrialization on rural spaces often has little in common with these utopias, but it does have important consequences for sustainable development. If one follows the fortunes over time of a small municipality in Bregenzerwald/Austria, now known as a tourist destination, the interlinkage of industrial centers and rural peripheries becomes apparent.

This text investigates the socioecological transformation of a tourist resort in the Alps. The socio-natural site (SNS) of Damüls has undergone a process of transformation since the late 19th century from an Alpine agrarian regime to a modern industrial winter sports location, involving far-reaching changes in the type and intensity of the colonization of the Alpine ecosystem and the local metabolism. As a result, a new SNS is being created. Although solutions to some specific traditional sustainability problems can emerge during a socioecological transition (SET), new problems arise, and a spiral of risk begins to turn. According to the paradigm of Social Ecology, industrialization is understood as a transition from a biomass-based to a fossil-energy-based socioecological regime. Two concepts may be used to describe a socioecological regime: social metabolism (see Chap. 3) and the colonization of natural systems (see Chap. 18). These two concepts have a dialectical relationship with one another and are interdependent.[1] How should this be understood? Colonization presupposes work, or the investment of energy by humans, working animals or machines. This work must be performed in a sustained way for a colonized system to be maintained in a particular condition; thus, colonization implies societal commitment. Therefore, when a society embarks upon an SET, it is most often accompanied by the differentiation of new colonizing strategies. Long-term social consequences are described as 'societal commitment' (see Chap. 19) and produce long-term 'legacies' (see Chap. 6).

In the research example discussed here, land uses at two points in time are empirically surveyed. The first survey concerns 1857, when the Damüls municipality was still largely agrarian in character. The second survey examines Damüls in 2010. An initial hypothesis arising from the land-use comparison suggests that the SET toward an industrialized society in Damüls did not take place as a result of a transformation of agriculture. The transition was fostered by a completely new form of socioeconomic structure and its infrastructures being superimposed upon an agrarian system. In simplified terms, Damüls was industrialized through tourism. Thus, if we wish to understand the SET in Damüls, we must focus on the

[1]Colonization is very often carried out to provide a specific social metabolism. Industrialization alters the relationship between social metabolism and colonization. One might almost say that it reverses the relationship.

mass tourism of the 20th century. Winter sports and holidays were and are a social institution for travelers that is coupled with the industrialized livelihood. For hosts, however, the provision of accommodation and hospitality for guests was initially a practice of economic risk minimization in an agrarian system that, because of the economic growth in the valleys, was progressively sliding into a state of relative impoverishment.

23.2 The Consequences of Industrialization in a Mountain Village

The political community of Damüls is located at the center of the Austrian Federal Province of Vorarlberg. The municipal area of 20.9 km^2 has an altitude ranging from 1150 m.a.s.l. at Argenbach to 2095 m.a.s.l. at the Mittagspitze. The major water courses are the Krumbach, Brägatzbach and Argenbach. The region is built from complex geology composed of marls, limestones, flysch and sandstones. The geology forms a landscape of south-facing, gently inclining slopes and north-facing, steeply sloping ravines (Staudinger 2009, p. 9). Because of its exposed position on the northern edge of the Alps, the municipality has a climate characterized by frequent and heavy precipitation (1894 mm long-term yearly average), a snow cover of 2–3 m that remains for 5–6 months of the year and low summer temperatures (with a mean July temperature of 13 °C). This results in a short growing season. Hay meadows are mostly one-cut, with two-cut meadows only found in the lower-lying parts of the municipality. In other words, this is a parcel of land that is only conditionally suited to year-round habitation (Frei 1970, p. 137).

In the 14th century, 'Walsers' from Switzerland settled the exposed higher altitude parts of Damüls, which were already being used as *Alpe*[2] [mountain pasture, transl. by R.G.] at that time. The settlers cleared a significant part of the woodland over the course of time and transformed the cleared land into mountain pastures, meadows and fields (Feuerstein 1929, p. 16). The 'Walsers' colonized the high Alpine ecosystems to make them produce higher yields. Officials of the Bavarian land survey of 1810 identified a 14.6 km^2 area of fields and pastures in Damüls (VLA Bayerischer Steuerkataster 1810). Approximately half of these areas were categorized as *Räuhe* [less-productive grazing pasture, transl. by R.G.]. The other half was divided into the categories one-cut and two-cut *Fettgut* [fat meadows, transl. by R.G.]. At this time, the majority of grasslands were either used communally as *Alpe* or as *Vorsäß* [preliminary stage to the alp, being used in spring and autumn, transl. by R.G.]. These areas were and still are communally owned, with their use regulated through a complex system of rights passed down through the generations.

[2]'Alp' is a regional term for seasonally used mountain pasture land (a German equivalent term here is *Alm*), not to be confused with 'the Alps' as a term for the central European mountain range.

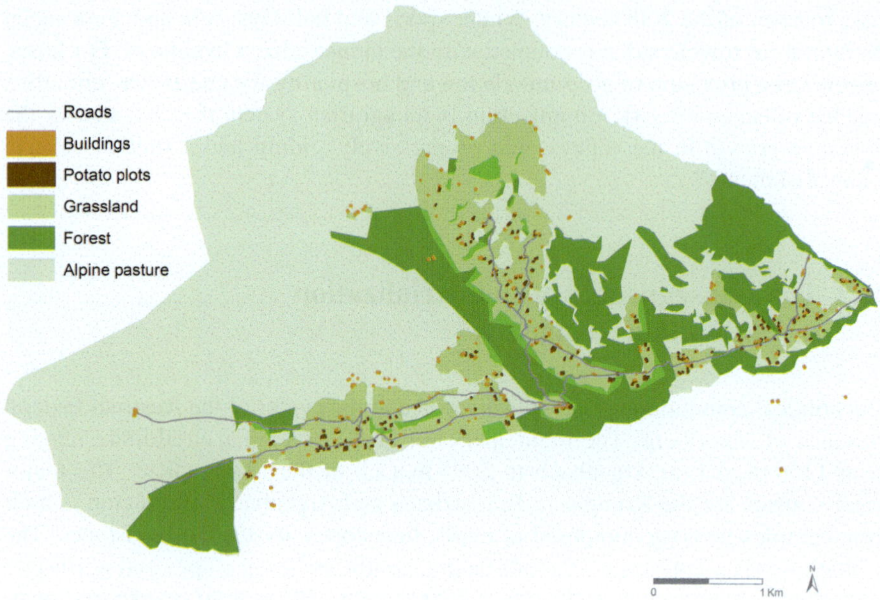

Fig. 23.1 Author's depiction of land cover in Damüls in 1857 based on the Austrian Land Register. (Source: VLA 1857; mapped by T. Fetzel)

The same was true for woodland. Therefore, there are no records pertaining to woodland in the Bavarian land survey of 1810 (VLA Bayerischer Steuerkataster 1810). According to the land survey, there was no cropland in Damüls, although cereal cultivation had previously been widespread in this region, as evidenced by the presence of threshing floors, grain stores and millstones (Feuerstein 1929, p. 17). In similarly exposed villages, the cultivation of cereal crops ceased in the early part of the 19th century and was replaced by potato cultivation (Bilgeri 1947, p. 33).

In Damüls, cropland first appears again in the Austrian cadastral survey of 1857, as Fig. 23.1 shows.

The potato plots are marked in brown in Fig. 23.1.[3] In total, 226 plots with an area of 5.8 ha can be identified in the land survey. These were in the immediate proximity of houses (marked in yellow). Practically every household in Damüls possessed a field with an area of 0.01–0.28 ha. The average size of these plots was 0.09 ha. According to estimates produced by Robert McC. Netting for Törbel in

[3]According to Benedikt Bilgeri, from the mid-19th century, potatoes were the main crop in the rainy north Alpine regions of Vorarlberg. The situation was different in the municipalities of southern Vorarlberg, which were characterized by the drier climate of the central Alps. Here, the cultivation of rye and barley survived the introduction of potato crops.

Switzerland,[4] a field of 0.09 ha can produce between 1250 and 1620 kg of potatoes. Using these numbers, between 87 and 104 t of potatoes could be produced on the entire cropland of the village (Netting 1981, p. 41). Apart from those used as seed potatoes (20 %), 0.5–0.6 kg of potato was available daily for each inhabitant of Damüls. Although these figures are estimates at best, a closer examination of local sources provides evidence that farmers in Damüls produced a surplus sold to other villages as 'Damülser Vieläugler' (Feuerstein 1929, p. 22 f.). According to Robert McC. Netting, potatoes were responsible for a decline in child mortality and for population growth in Törbel/Switzerland, and such is the case in Damüls.[5] Between 1820 and 1830, 63 children under six years of age died. In the decade between 1840 and 1850, this figure fell to 36, and child mortality rates stagnated at this level until the 1880s. Simultaneously, the birth rate rose slightly from the middle of the 19th century (VLA Hackspiel 1947, p. 32). From the mid-19th century, the population of Damüls began to grow slowly. In 1869, the population census records 383 inhabitants. The discussed factors of land use and demography thus suggest that the middle third of the 19th century was a period of relative prosperity.

With the progressive industrialization of the valleys at the end of the 19th century, life also changed for the inhabitants of Damüls. The railway via the Arlberg, which opened in 1884, brought cheap agricultural imports (field crops, cotton, cereals) to the region. The mountain farmers could no longer compete with those of the lowlands, and the *Saumwege* [mule tracks, transl. by R.G.] fell into disuse.[6] At the same time, the labor-intensive textile industry in the Rhine Valley was booming and attracted ever more people to work. Between 1880 and 1924, 27 families emigrated from Damüls. This had a particular impact on those parts of the municipality that were farther away from the road, which connected the village to the valley below from 1892 onward. Between 1869 and 1924, the population fell from 383 to 204 inhabitants (Groß 2012, p. 37). Because it was especially young people who emigrated, the available workforce declined, which had an impact on land use, that is, upon the intensity of colonization in the Alpine ecosystems. Potato cultivation disappeared almost entirely, the scale of the pastures was reduced, livestock decreased, and marginal production areas were abandoned.

[4]Robert McC. Netting calculated 150–180 kg potatoes per 100 acres for Törbel (see Netting 1981, p. 39).

[5]Robert McC. Netting traces this back to the fact that children's food could be easily produced from potatoes and dairy products. Little effort was thus required for children to receive better quality nutrition, and they survived longer as a result (see Netting 1981, p. 159). From a contemporary perspective, the introduction of the potato has been seen as an important factor in the population growth that began with industrialization (see Sandgruber 1982, p. 49).

[6]*Saumwege* constituted a transport system in agrarian society that used horses to transport goods across Alpine passes. These packhorse trails provided an important source of additional income for the farming communities in the mountain regions.

The houses, barns, granaries and livestock stalls were either sold, in the best instances, or fell into ruin (Feuerstein 1929, p. 25). In the course of industrialization and economic growth in the valleys, Damüls became a problem case.

23.3 The Tourist Socio-Natural Site (SNS) in Damüls

When the first skiers arrived in Damüls in the mid-1920s, they required accommodation, catering and skiing lessons. In meeting this need, economic conditions began to improve for the inhabitants. Within a decade, novel tourism-related practices emerged in Damüls, leaving significant marks on the agricultural arrangements of the village. Uninhabited farmhouses were turned into guesthouses, and alpine huts were converted into ski lodges.

What is the issue here? From an Environmental History perspective, winter sports destinations can be described as socio-natural sites (SNSs). An SNS is constituted by the nexus of practices (e.g., skiing, providing accommodation, mobility) and arrangements (e.g., hotels, ski lifts, ski slopes). Practices and arrangements are abstract terms for observable human activities and their material consequences in the world. The term ‘practice’ also emphasizes that human activities are always guided by pre-reflexive elements (interests and traditions). The arrangements constitute the material matrix, which makes certain practices possible while rendering others impossible (Winiwarter and Schmid 2008, p. 162). Through perception and representation, tourism SNSs are open to stereotyping through visual and textual discourses. These representations become guidelines for action. Actions, in turn, have an impact on the structure of tourist practices (Winiwarter 2000, pp. 48–53).

The newly created arrangements of tourism in Damüls were clearly distinct from farms because of their long-term consequences, or ‘legacies’, and they altered the metabolism of the communities in the following three ways. (1) They structured biomass flows (milk, eggs, cheese, meat, potatoes, wood) in the municipality. (2) The energy system of the municipality was altered by the pioneers of tourism in 1932, when a small electric power station was built. Until the 1940s, electricity for lighting was only available for the tourism arrangements; the farmers sat in darkness as they had always done. (3) The construction of new hotels and the conversion of buildings into hotels required financial means that would allow their owners to obtain loans and make regular repayments on those loans (Groß 2012, p. 40). Thus, the tourist SNS in Damüls was, from its very inception, coupled with a business logic that was clearly distinct from the subsistence-oriented logic of the local farmers.

The economic ‘legacies’ of the arrangements generated competition among hotel operators. This competition became more pronounced when visitors stayed away for economic (global economic crisis), political (the 1000-Reichsmark-Sperre[7]) or climatic (lack of snow) reasons. Given the economic difficulties of the Austrian national economy and its dependence on foreign currencies, the Austrian *Ständestaat* [fascist, corporative state, transl. by R.G.] initiated a hotel renovation program in 1933. This program envisaged low-interest loans with a long-term borrowing scheme to protect indebted hoteliers from bankruptcy. National Socialist hoteliers—of which there were also some in Damüls—were clearly excluded from participation in this scheme. Moreover, for the first time, a planned destination management program and coordinated promotional campaign abroad were developed with the aim of establishing Austria as an alternative winter sports destination that was cheaper than Switzerland. In Vorarlberg, only Lech am Arlberg profited from these activities, together with the 'Kleinwalsertal'. Damüls, by contrast, felt the impact of the global economic crisis and the 1000-Reichsmark-Sperre very strongly (Groß 2012, p. 43). Here, any farmer who provided guest accommodation or rented alpine huts to visitors as a secondary income was in a better position than the full-time hoteliers, who were already greatly in debt.

Damüls differed little in terms of its arrangements and practices through the 1940s from the period between the wars. The major development took place at the end of the 1920s and the early 1930s. The tourism boom, which resulted from the *Anschluss* of Austria to Germany, proved to be short lived. Similarly, the modernization efforts of the National Socialists in Damüls remained very narrow in scope. As a result of the global economic crisis and the 1000-Reichsmark-Sperre, the tourist transformation of the SNS ground almost entirely to a halt.

23.4 The Technological Transformation of the Tourist SNS After 1950

In 1952, the road to Damüls was improved sufficiently by the provincial government of Vorarlberg as part of a reconstruction program. In so doing, the old mobility practices of horse-drawn transport were replaced by diesel-powered post buses and automobiles. This removed a significant bottleneck in terms of access to the village and fostered the transformation of the SNS (Groß 2012, p. 46). The actual postwar boom in tourism began in Damüls when the first ski lifts were constructed. Ski lifts and cable cars had already been successfully introduced in the noble ski resorts of Arlberg, Switzerland and Salzburg and had begun to transform skiing practices. As early as the 1930s, it was seen as a winter version of Alpinism. Positive connotations were attached to the ascent, which promised to cleanse skiers both mentally and physically of the ballast of an industrialized

[7]The '1000-Reichsmark Ban' was established by Adolf Hitler between 1933 and 1936. It obliged each citizen of the German Reich wishing to cross the German border into Austria to pay a total of 1000 Reichsmarks (equal to approximately €4000 in 2012).

Fig. 23.2 T-bar lift '1800', looking toward Faschina. Picture postcard of the '1800 T-bar Lift'. (Photo: Risch-Lau; Vorarlberger Landesbibliothek, Bregenz)

society's lifestyle in the mist-shrouded valleys. Through the introduction of ski lifts, skiing lost its alpinist pathos. The technologization of the arrangements transformed cross-country skiing [*skilaufen*] into alpine skiing [*skifahren*], in which the emphasis no longer lay on the ascent but on the descent. Downhill skiers were now able to complete several downhill runs in succession, and they learned to ski far quicker than before. Initially, the ski lifts were added to the skiing lessons. Soon thereafter, the newly introduced arrangements were used to make undeveloped, high altitude terrain accessible for people with a lesser degree of personal fitness by means of technical energy. Moreover, it became apparent that the ski lifts brought considerable profits. The expansion of ski resorts drew a greater number of skiers. When there was too little snowfall, these skiers could move by ski lift to the higher-altitude regions (Groß 2012, pp. 97–100).

Damüls' first large-capacity T-bar lift was built in 1957. The arrangement is still mentioned by older inhabitants of Damüls. In interviews, it became apparent that people use the date of construction to orient their historical reminiscences, giving structure to their personal impressions of the world around them. They talk about the ski lift in a positive manner, demonstrating that the large-capacity T-bar lift, known as the '1800',[8] came to represent the tourist boom in the area. Even the critical farmers were prone to believe that Damüls had a future as a ski resort. Immediately following its construction, a range of new businesses were founded, and for the first time since 1880, the population again showed significant growth (Groß 2012, p. 53). The arrangement appeared frequently as a motif in a variety of tourist representations of the ski resort through advertising and on picture postcards.

Figure 23.2 shows the '1800' T-bar lift from a typical bird's eye perspective that in classical fine art served to represent aristocrats' claims to power over the lands

[8]The name '1800' referred to the altitude above sea level to which it brought access. The ski lift operators in Damüls used the topographical characteristics of the terrain to represent the arrangements in the discourse of tourism. This form of representation was and is still very commonly employed in Alpine tourism arrangements, such as when guesthouses are named after Alpine flowers such as the alpine rose *(Rhododendron hirsutum)*.

they owned. The view across the panoramic landscape communicates to the viewer a feeling of grandeur and of dominance over nature. Already by the early 20th century, cable cars were the most often used symbols for progress in the Alpine region because they made the domestication of nature visible and comprehensible for all (Tschofen 1994, p. 126). The arrangement in photography served a dual purpose. To inhabitants of the municipality, the arrangement inspired a sense of identity and helped anchor their self-image as a winter tourism resort. It represented the departure from a rural past, which was associated with farming practices and existential threats. To the outside world, the arrangement and its visual representation functioned as user guidelines for a modern alpine skiing practice that no longer had anything in common with the strenuous scaling of mountain peaks. With the T-bar lift, one glided uphill in a standing position, apparently effortlessly, saving one's energies for the pleasure of the downhill run.

The outward appearance of Damüls as a winter sports region in media representation was perpetuated by the cableway company entering a technological dynamic of expansion. Because the construction of ski lifts initially required large amounts of capital investment, the inhabitants of Damüls sought a loan in 1961 via the Ministry of Trade and Reconstruction from ERP funds[9] of 1.7 million Austrian Schilling for the construction of a chairlift on the Uga-Alpe (AdR 10). The chairlift was intended to provide access to a ski area that had been very popular with ski tourists since the 1920s. Because of its exposed position, the Uga-Alpe offered very challenging snow and wind conditions, and the ski region was thus only suitable for practiced skiers. Thus, the chairlift was barely able to produce the kind of income that was necessary to fulfill the loan repayments during the first few years. To avoid bankruptcy, it was necessary to improve conditions on the pistes sufficiently to increase the profitability of the chairlift. This required the acquisition of a snow groomer in the early 1960s, which not only prevented the cableway company from going bankrupt but also raised the popularity of the region to such an extent that it became possible to construct another chair lift in 1969 (Groß 2012, p. 53). In the winter resort brochure of 1968, the snow groomer was elevated to the status of a tourist attraction in itself, as the photo in Fig. 23.3 shown below.

The snow groomer in Fig. 23.3 represents the capacity of Damüls' inhabitants to gain control over the unyielding nature of the terrain and the snow conditions by means of diesel and combustion engines. The order between technology and nature is represented in the image through the proportions given to the snow groomer on one hand and Damüls' 'own' mountain, the Mittagspitze, seen in the background, on the other. The proportional relationship between the snow groomer and the mountain give those who view the picture a sense of the

[9]Within the framework of the European Recovery Programme (ERP), ca. 305 million Austrian Schilling was made available for the development of the tourist industry in Austria. These funds were provided by the USA between 1948 and 1955 and given to businesses in the form of low-interest, long-term loans. New loans are being provided from the repayments of the original loans even today. The ERP represented the most important national assistance instrument for the development of tourism arrangements until the late 1980s.

Fig. 23.3 Snow groomer. (In: tourism brochure *Damüls, the Sun Terrace in Vorarlberg…* (ca. 1968), published by Damüls Tourist Association; in the archives of Damüls—Faschina Tourismus, Kirchdorf 138, Damüls)

effortlessness with which the snow groomer would be able to push the mountain aside. Snow groomers transform snowy terrain into ski pistes by conferring a uniform quality to the varied snow conditions on the individual sections of the piste. Thereafter, the clearly delineated piste areas are graded according to their level of difficulty (red, blue and black pistes). The information about different pistes is communicated to the skiers through posters, maps and signs. This practice of piste management by the ski lift operating companies increased the predictability of the terrain experience and enhanced the physical sensation during the downhill run. In this way, the practice of piste management has contributed to the rationalization and increased efficiency of skiing enjoyment. The picture fails to show the 'legacies' the ski pistes eventually had, which were difficult to predict for contemporary actors. The heavy piste machine compacted the snow, delaying the thawing process. Moreover, where there was too little snow, the practice of piste management using snow groomers caused damage to the alpine grassland, leading to erosion and harvest losses among the landowners, who were largely farmers, whereas the operators of the ski lifts remained unaffected (Groß 2012, p. 56).

Ski piste arrangements were managed during the summer months by traditional agriculture. Thus, unique agricultural practices emerged over time in relation to the delayed snowmelt. One of these was the so-called *Schwärzen* [blackening, transl. by R.G.]. This involved the use of 'Thomasmehl', a fertilizer rich in phosphates that was spread across the frozen terrain. The darkened surface of the snow absorbed more energy from solar radiation and thus melted more quickly. This practice was in use until 1995 in Damüls and contributed to the eutrophication of the soils there. Thereafter, 'Thomasmehl' was replaced by stone powder (Groß 2012, p. 58). In the ski resorts of Vorarlberg, with very high numbers of visitors, farmers used other means to keep the snow groomers away from their land, such as erecting fences or planting lines of trees. This practice was soon brought to a halt by the provincial government, which enacted a Law on Sport. From 1969, this law provided for the interests of ski lift operators to be protected in the first instance. The ski lift operators argued their case on the basis of an 'immemorial right of way' arising from decades of use of a portion of terrain as a ski piste. In 1971, the law was amended after pressure from the provincial Chamber of Commerce, the Landowners Association and a petition for which a private interest party had collected 300 signatures. Since then, ski lift operating companies have been obliged to pay compensation to the landowner for any damage and repairs to the land. The law simultaneously obliges the landowner to allow the use of snow groomers and any construction works deemed necessary to maintain the ski operations on their land. In addition, the Law on Sport regulates the farming practice of using fertilizer on the ski pistes (VLA Sportgesetz).

In 1980, there were five ski lifts in operation in Damüls that, together, were capable of transporting a total of 5000 people per hour uphill. In 1987, two further lifts were constructed. The entire facility then had a transportation length of 7323 m and was able to transport a total of ca. 7000 people uphill each hour. The ski resort had a prepared piste area of ca. 50 ha. From 1984 to 1997, the company operating the ski lifts aimed to progressively extend this transportation capacity through so-called 'qualitative expansion'. The construction of the first four-seater chairlift in 1996 marked the beginning of a new dynamic of exploiting further territory. In 2000, together with the neighboring municipality of Mellau, the Damüls cable car company began to develop the long-discussed 'Skischaukel' Mellau-Damüls (an integrated ski region), for which purpose two further chairlifts were built in 2007. The connection with the Mellau ski region required the blasting of a section of rock, the erection of a 'ski tunnel' and extremely complex groundwork. The integration work was completed in December 2009 (Groß 2012, pp. 57–64). Since then, the ski lift arrangements in Damüls have been able to transport 20,000 people per hour. In 2010, the prepared pistes covered a total area of 54.4 ha. Between 1970 and 2010, the hourly transportation rate had thus become 9.2 times greater. Meanwhile, the piste area had become 2.8 times larger (ibid., p. 63).

The decoupling of the piste areas from the transportation capacity took place in Damüls in the mid-1990s and was synchronous with the installation of a snowmaking system. From the end of the 1980s onward, snowmaking systems were installed throughout the Alpine region because the duration of snowcover in the Alps was becoming noticeably shorter. At the same time, Alpine tourism was experiencing a crisis. Market liberalization and globalization were increasing competition among the ski resorts and between these and other holiday destinations beyond the Alps. The many investments that had been made during the previous decades to support the development of ski lifts and hotel facilities made every winter with too little snow a nightmare for business operators (Groß 2012, p. 142). In the 1989/90 season, Damüls suffered a 66 % decline in the number of visitors using the ski lifts compared with the 1988/89 season. This cost the ski lift operating company alone a total of 14 million Austrian Schilling (VLA Beschneiungskonzept). The indebtedness and a lack of snow were factors that prepared the way for the new practice of snowmaking.

The first snowmaking facility was installed in Damüls in 1991, whereupon the next risk spiral already began to turn. The water required for the snow cannons was initially drawn from the Argenbach stream. From there, it was pumped via a system of pipes laid underground for several hundred meters up the mountain to the ski pistes to provide snow cover for 6.4 ha (11 % of the entire piste area). The artificial snow was very expensive to produce, so to reduce the cost of production, the cable car company began to remove any knolls or crests in exposed piste terrain, upon which snow melted more quickly. The soil was used to fill in any hollows in which the costly artificial snow might accumulate. In 1996, the use of artificial snow was extended to a total area of 9 ha (16 % of the entire piste area). With this expansion, the cable car company soon realized that the ecological sustainability of taking water from the Argenbach had its limits. First, recourse was taken from smaller water reserves, although these soon proved insufficient. According to the founder of the cable car company, in 2010, 80 % of the piste areas were at least temporarily covered with artificial snow cover. Because both the small capacity of the water reservoir and the strict regulations applied to water removal from the Argenbach limited the use of the snowmaking equipment, in 2007, the company created a snowmaking reservoir at the foot of the Mittagspitze to ensure that the practice of snowmaking could continue (Groß 2012, p. 61). The ski lift company thus managed to overcome the economic risks entailed in the event of too little snowfall. At the same time, a further portion of high alpine landscape was irreversibly damaged.

Figure 23.4 depicts the Damüls SNS in 2010, transformed through tourism.

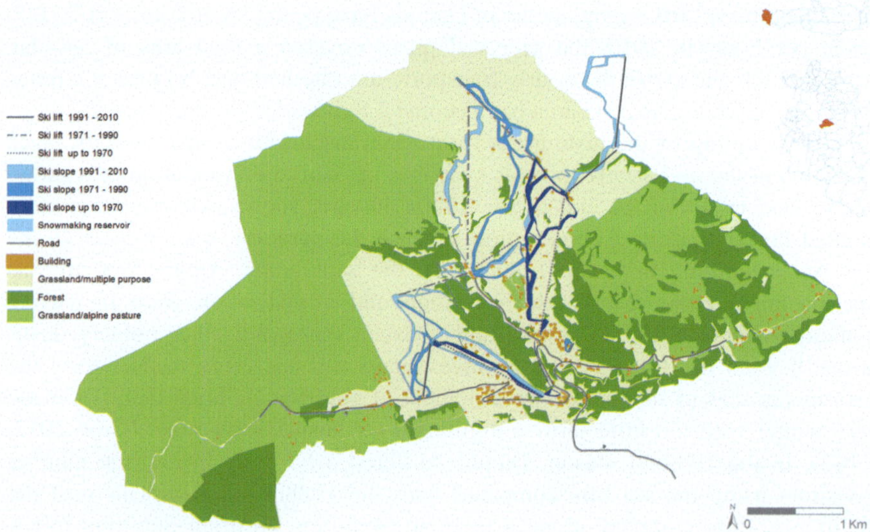

Fig. 23.4 Development of the piste areas and ski lift facilities in the Damüls ski region. Terrain map of the Damüls ski region, produced by the company Doppelmayr. (*Mapping source* http://www.vorarlberg.at/atlas, BEV DKM 30.9.08, Office of the Environmental Ombudsman Vorarlberg. Mapped by T. Fetzel)

By 2010, the parts of the Oberdamüls and Uga district that were used for agriculture were subsumed by a network of ski pistes and ski lifts, where the practices described above for altering terrain, managing pistes, snowmaking, ski sports and ski lift use were taking place. In Damüls in 2010, 83 % of the municipality's population belonged to the tertiary (service) sector and only 7 % to the primary sector, meaning that the majority of the population dedicated their available working power, knowledge and attention to maintaining the tourist SNS in Damüls. This has produced a wide range of opportunities in terms of jobs and incomes, such as working on the ski lifts, serving as ski instructors and working in the catering and hotel industries. Significant parts of local practices are structured by the maintenance requirements of the tourist SNS. Ski pistes, which would have fallen into disuse because of their low productivity and the decline of the agricultural sector, continue to be managed in the interests of ski tourism. All available natural resources in Damüls, such as soil, water, biomass, biodiversity and air, are integrated into the tourist SNS and have been affected by technological transformation since the 1950s. In 2010, the entire infrastructure of public services (sewerage services, electricity supply, water supply, waste management) were designed to sustain the tourist SNS, in which the ski lifts function as the central arrangements. Even the settlement structure had concentrated itself around the valley and mountain stations of the ski lifts and pistes. The scattered farming settlement is now a thing of the past. Instead, the settlement centers are clogged with hotels, guest houses, après-ski bars and discotheques, which make a truly bleak impression during the summer months.

23.5 Discussion

As Bill McKibben argues, the beginning of Environmental History is marked by 'The End of Nature'. There are no longer any natural places, in the sense of places that are not in some way altered, on the entire planet. What we find at the beginning of the 21st century is tracts of land that have been technologically altered, shaped by cultural or political processes, integrated through economic activities into all manner of economic utilization processes or penetrated by the globalized side effects of human economics (McKibben 1989, as cited in Jorgensen et al. 2013, p. 10). The socioecological model of interaction between society and nature places great emphasis on the systemic character of both coupled systems. According to this conceptualization, nature, understood as the material world, and social existence interact through the 'colonization of natural systems' and 'social metabolism'. This approach has proven its value in the analysis of environmental and sustainability processes and in interdisciplinary communications. However, the dichotomous conceptualization of 'natural' and 'social' systems and their overlapping areas also raises questions for environmental historians. Has the degree of intersection between 'natural' and 'social' systems been altered through the course of industrialization, as the quote from Bill McKibben exemplifies?

The research example presented here discusses tourism as a practice that has a transformative impact on SNSs. So how do practices relate to the socioecological concept of colonization? Both terms can be understood in relation to 'work', which is invested in ecosystems and which brings material consequences with it. Further commonalities include the focus on long-term effects and societal commitments, or legacies, which structure societal interventions in the material world. However, with both concepts, different questions can be addressed in each case. Abstract areas of overlap between 'natural' and 'social' systems can be conceptualized as SNS, as networks of functionally different practice-arrangement nexuses. From this perspective, the question of the degree to which 'natural' and 'social' systems overlap recedes into the background, whereas the differentiation between practice-arrangement nexuses and forms of interaction between functionally different SNSs now stands in the foreground. In this research example, the tourist SNS—which has been symbolically, materially and economically very strongly interconnected with the valley since the early 20th century—is superimposed upon a collapsing agrarian colony.

Several central issues in the environmental history of Damüls may be expressed using the concept of the risk spiral developed by Rolf Peter Sieferle and Ulrich Müller-Herold. This concept is coupled with the colonization concept, and it identifies the permanent pressure to innovate as the driving force behind the transformation of societies (Sieferle and Müller-Herold 1996, p. 141). The most significant sustainability problem of the socioecological system in Damüls at the beginning of the 20th century was the municipality's progressive depopulation. Although tourism in Damüls can be understood as a practice of risk minimization in a contemporary perspective, tourism was and is vulnerable to climatic, economic and political disturbances.

The representation of the SNS as a tourist destination in tourism promotion represents a very important practice of risk minimization. Promotion aims to convince potential customers of the suitability of a destination for winter sports practices. Many of the technological innovations in an SNS are communicated through visual and textual representations in the public sphere. In this way, a second form of risk spiral becomes established, where individual destinations enter into competition with one another for potential customers. In this competition, only those destinations that have the means to permanently enrich their advertising with new attractions and places of interest stay in the running (Urry 1995, p. 175). The Austrian Prime Minister Julius Raab said at the opening of the Valluga-Bahn in 1953 in Arlberg: 'We must build, build, build…we must keep on tirelessly, building, building […]' (Anonymous in Vorarlberger Nachrichten 1953, p. 5). These words have lost none of their relevance today.

Embedded in a network of ski lift infrastructure, ecologically degraded ski pistes and snowmaking facilities, villages such as Damüls lie 'frozen' on the winter tourist development path. However, new risks and insecurities induced through global climate change call this development path into question in many cases.

Even making increased investments in snowmaking equipment, ski lifts and piste management systems cannot overcome these uncertainties—indeed, they create new vulnerabilities. Environmental History is challenged in this context to review past decisions and patterns of thought and action and thereby to undertake a critical examination of the present.

Sources and References

Unprinted Material

Archiv der Republik (AdR). 10 Verkehr BMfV/Präs. Sch. 667 ERP Kredite. Listen nach Bundesländer 1964–66.

Vorarlberger Landesarchiv (VLA). Bayerischer Steuerkataster 1810, 1/18. LG Sonnenberg: Damüls.

VLA 1857, VLA Österr. Grundkataster, Bauparzellen Protokoll 1857, 91006 KG Damüls, Hs 1.

VLA Beschneiungskonzept, VLA, AVLR, Abt. VIIa 422.05. Sch. 268 Damüls Beschneiungskonzept.

VLA Sportgesetz, VLA AVLR, Abt. PrsG 306/1972. Sch. 19 Sportgesetz.

VLA Hackspiel, K. (1947). Heimatkundliche Stoffsammlung für die Volksschule Damüls.

References

Anonymous (1953). Eröffnung der Valluga-Gipfelbahn. Vorarlberger Nachrichten (January 22), p. 5.

Bilgeri, B. (1947). Der Getreideanbau im Lande Vorarlberg. Ein Beitrag zur Wirtschafts-. Siedlungs- und Stammesgeschichte. Sonderdruck aus der Zeitschrift Montfort 7/12.

Feuerstein, A. (1929). Damüls—die höchste ständige Siedlung im Bregenzerwald. Geographischer Jahresbericht aus Österreich XIV/XV, pp. 1–28.

Frei, Hans. (1970). Der Fremdenverkehr in seiner Bedeutung für die Bergbevölkerung am Beispiel von Damüls im Bregenzer Wald. *Mitteilungen der Geographischen Gesellschaft München, 55*, 135–154.

Groß, Robert. (2012). *Wie das 1950er Syndrom in die Täler kam*. Regensburg: Umwelthistorische Überlegungen zur Konstruktion von Wintersportlandschaften am Beispiel Damüls in Vorarlberg.

Jorgensen, Finn Arne, et al. (2013). Entangled environments: historians and nature in the Nordic countries. *Historisk tidsskrift, 92*(1), 9–34.

Netting, R. M. (1981). *Balancing on an Alp: Ecological change and continuity in a Swiss mountain community*. Cambridge.

Sandgruber, R. (1982). Die Anfänge der Konsumgesellschaft. Konsumgüterverbrauch, Lebensstandard und Alltagskultur in Österreich im 18. und 19. Jahrhundert. Wien.

Sieferle, Rolf Peter, & Müller-Herold, Ulrich. (1996). Überfluß und Überleben—Risiko, Ruin und Luxus in primitiven Gesellschaften. *GAIA, 5*(3–4), 135–143.

Staudinger, M. (2009). Biotop. Aktualisierung des Biotopinventars Vorarlberg. Gemeinde Damüls. Vorarlberger Landesregierung.

Tschofen, B. (1994). Die Seilbahnfahrt. Gebirgswahrnehmung zwischen klassischer Alpenbesteisterung und moderner Ästhetik. In: Tourimus und Regionalkultur, hg. von Burkhard Pöttler (Buchreihe der Österrichischen Zeitschrift für Volkskunde 12). Wien. pp. 107–128.

Urry, J. (1995). Consuming places. London, New York.
Winiwarter, V. (2000). Landschaft auf Vierfarbkarton: Betrachtungen zur kulturellen Konstruktion des Blickens. In ZOLLtexte 35/4, pp. 48–53.
Winiwarter, V., Schmid, M. (2008). Umweltgeschichte als Untersuchung sozionaturaler Schauplätze? Ein Versuch, Johannes Colers „Oeconomia“ umwelthistorisch zu interpretieren. In: Thomas Knopf (Hg.), Umweltverhalten in Geschichte und Gegenwart: Vergleichende Ansätze. Tübingen. pp. 158–173.

Chapter 24
Cleaning a Metropolis: The History of Vienna's Sewage System

Sylvia Gierlinger and Michael Neundlinger

Abstract In this chapter, we take a close look at 19th century Vienna and discuss the sociometabolic transition from an agrarian to an industrial sociometabolic regime in a local urban context. We identify the challenges this rapid transformation of the city posed for the existing disposal infrastructure, and we investigate how city officials responded to these challenges and what type of legacies for present river-city-relations were created. In the second half of the 19th century, Vienna was transforming to an industrial city. Population numbers were rising rapidly, and widespread urbanization took place. With the increase in urban metabolism, output flows multiplied. The older disposal system was incapable of coping with this increased amount of waste and wastewater, which placed enormous pressure on the urban waterscape. After years of intense debate about the optimal disposal system, urban authorities decided to construct a water-borne sewage system. The many small streams intersecting the urban area were vaulted and integrated into the sewage system. A disposal system was created that constantly needed a certain amount of water input and regular maintenance to fulfill its functions. The disposal system created structures of river-city relations that persist to the present day. The aim of this chapter is to shed new light on how large-scale processes such as changes in urban metabolism, urbanization, industrialization and the transition of the energy system are realized at the local level and in an urban context.

Keywords Long-Term socioecological research (LTSER) · Environmental history · Vienna · Danube · Urban metabolism · Sociometabolic transition

S. Gierlinger (✉) · M. Neundlinger
Institute of Social Ecology (SEC), Alpen-Adria University, Schottenfeldgasse 29, 1070 Vienna, Austria
e-mail: sylvia.gierlinger@aau.at

M. Neundlinger
e-mail: michael.neundlinger@aau.at

H. Haberl et al. (eds.), *Social Ecology*, Human-Environment Interactions 5,
DOI 10.1007/978-3-319-33326-7_24

24.1 Introduction

The transition from an agrarian to an industrial sociometabolic regime implies major changes in society-nature interactions. Urbanization plays a key role in this transformative process. When cities grow, the organization of rural metabolic systems is completely restructured. Flows become integrated at a much larger scale. This development challenges existing urban infrastructure, such as supply and disposal systems. In this chapter, we consider the case of Vienna and its sewage system from a long-term socioecological perspective (Singh et al. 2013). We discuss the sociometabolic transition in Vienna and ask what this meant for the disposal regime. The chapter starts with an introduction of the specific geographical situation of the city and its waterscape in the 19th century and today. The second section depicts the sociometabolic transition taking place in Vienna in the second half of the 19th century and subsequent challenges for the existing urban sanitation regime. Thereafter, we discuss societal responses to new sanitary challenges and nuisances. In the last section, we conclude with a discussion of socioecological legacies created by the construction of a water-borne sewage system.

24.2 Vienna and Its Waterscape

Vienna, the former capital of the Austro-Hungarian Empire and the capital of Austria today, is located at one of the major European waterways and water transport routes, the Danube. The city's large river and its tributaries played an important role in urban development, water supply and the discharge of urban waste (Winiwarter et al. 2013). The specific geographic situation of Vienna and its waterscape was crucial for the common history of the city and its disposal system. The River Danube intersects the urban area from northwest to southeast (see Fig. 24.1), flowing at the relatively high speed of 1–3.5 m/s. The annual mean discharge can range from 910 m^3/s or less (low water) to 1915 m^3/s (mean water) to 5700 m^3/s or more (high water) (Schratt-Ehrendorfer 2011). With its discharge regime and swift velocity, the urban portion of the river provided ideal conditions for a comprehensive sanitation system that could efficiently dissolve and wash away urban waste. Before the river was brought into its present day course in the 1870s, the highly dynamic riparian landscape included small side arms, river branches and islands of different sizes. Within the so-called 'Great Danube Regulation' from 1869 to 1875, most side arms of the Danube were cut off, and the main branch of the river was brought into a straight riverbed. Only the southernmost side arm of the Danube (*Donaukanal*) and a stretch in the north, the so-called *Alte Donau* (old Danube), remained in place. Between 1972 and 1988, a 21 km long flood bypass, the so-called *Neue Donau* (new Danube), was built next to the main arm of the Danube (Hohensinner et al. 2013).

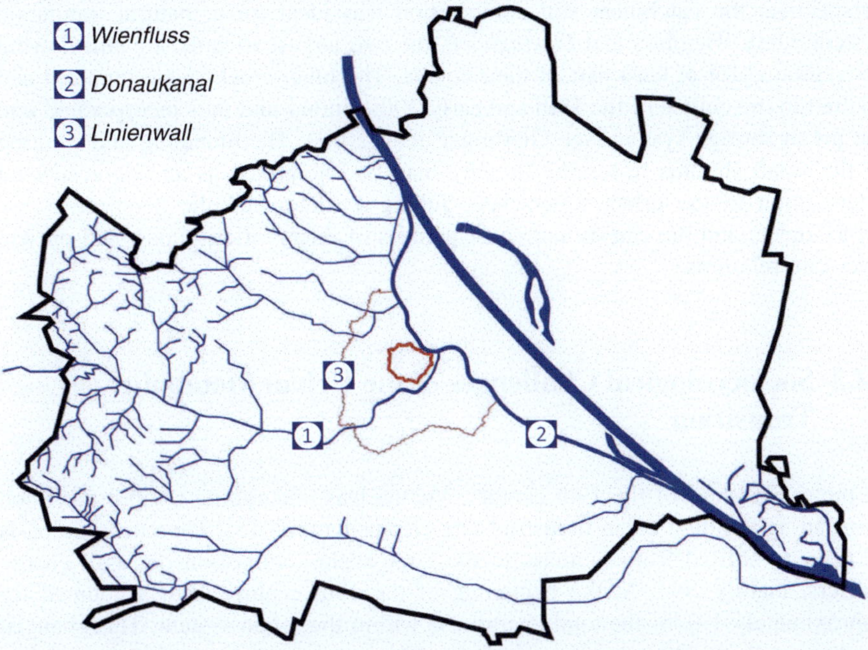

Fig. 24.1 Vienna and its waterscape in its current boundaries. *Note* Most of the Danube's tributaries depicted have been integrated into the urban sanitation system. The course of the Danube is shown after the comprehensive regulation in the 1870s. Present-day Vienna extends over an area of 414 km^2. Between 1706 and 1890/92, the area within the city tax boundary, the so-called *Linienwall*, had an expansion of approximately 60 km^2. (Source: Neundlinger et al. 2014, p. 326)

Even though the main arm of the Danube forms a dominant landscape element in Vienna today, the *Donaukanal* had been the main side arm for centuries. It directly connected the city center with the river; it played an important role in the transport of food, feed, wood and other materials; and it served as a runoff ditch for discharge and waste (Gierlinger et al. 2013).

Additionally, a complex network of watercourses intersects catchment areas between Viennese hills in the west and the *Donaukanal* (see Fig. 24.1). These small streams originate in the so-called *Wienerwald* (Viennese Woods), flow through the urban area from west to east and finally discharge into the Danube. The terrain elevation ranges from 542 m.a.s.l. in the west of the city, in the Viennese Woods, to 151 m.a.s.l. at the lowest point, a floodplain area along the Viennese Danube. In medieval and early modern times, these creeks were highly variegated. Broader sections narrowed down to small rivulets, and steep banks alternated with flat edges. Riverbeds ranged from course-grained pebble stones to sandy sediments (Neundlinger et al. 2014). The catchment area soils were mainly composed of marl lime, marl clay and sandstone. The high clay mineral content in the soil led to impermeability (Summesberger 2011). From a pedological perspective, the catchment soil composition was ideal for a 'natural' sanitation system. Only *Wienfluss* and *Liesingbach*, the two largest streams, are still flowing aboveground for at least part of their course. The other creeks were tunneled and vaulted in the course of the 19th and early 20th century and thus incorporated into the urban sewage system (see Gierlinger et al. 2013). The tunneling and vaulting of the small streams traversing the city may be interpreted as an unprecedented intervention in the urban waterscape. Taking a sociometabolic perspective, we try to understand the causes and consequences of intervention in past and present river-city-relations.

24.3 Socioecological Challenges of the Urban Metabolic Transition

In the second half of the 19th century, Vienna was undergoing a rapid transition from an agricultural to an industrial city (Krausmann 2013). The energetic basis of the city shifted from biomass to the large-scale combustion of fossil energy carriers, mainly coal. At the beginning of the 19th century, coal accounted for approximately 1 % of the total energy use within the urban system. This changed fundamentally in the second half of the 19th century. Around 1900, more than 75 % of the primary energy used was derived from coal. In the same period, Vienna faced a massive increase in population. The population rose from approximately 200,000 inhabitants at the beginning of the 19th century to more than 2 million inhabitants in 1910 (Fig. 24.2a). Industrialization and population growth entailed widespread urbanization processes, leading to a growing demand for additional settlement areas. With the Great Danube Regulation, new areas suitable

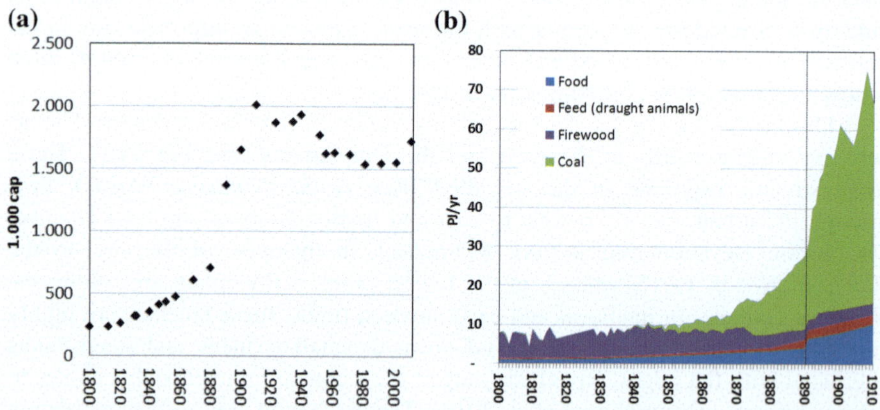

Fig. 24.2 **a** Population of Vienna 1800–2011. (Data from Magistrat der Stadt Wien, MagSW 1883–2010) **b** Domestic energy consumption in Vienna 1800–1910. (Data from Krausmann 2013)

for building houses were created. In the following decades, settlements along the Danube showed the highest growth rates of all Viennese districts (Haidvogl et al. 2013). In close connection with the sociometabolic transition, accelerated urbanization and rising population, the total consumption of material and energy (see Fig. 24.2b) increased markedly. Consequently, the increasing amount of urban sewage and waste posed new challenges for the existing disposal infrastructure.

Vienna's geomorphologic situation proved ideal, as many streams traversing the city already served as a water-borne sewage system. By 1830, more than 80 % of all houses within the city's tax boundary were connected to sewers, open drains or open streams (Kohl 1905). Even part of the night soil from cesspools was discharged into the streams or directly into the Danube downstream of the residential area. However, this disposal system created socioecological pressures, especially as the amount of waste and sewage grew. The dynamically changing water table was perceived as a major threat to urban hygiene (D' Avigdor 1873). In case of a low water table, waste could not be completely flushed away but remained in the riverbeds close to settlement areas. In case of a high water table, floods could occur, as the soils in the catchment area are nearly impermeable (Seebacher et al. 2011). In case of a flood, waste and sewage were flushed back into the streets, cellars and wells and created sanitary nuisances. Another problem was that the many small sewers were not connected to each other. Waste and excrement could remain in the sewers for weeks because there was not enough water from the houses to flush them away. Numerous contemporary accounts vividly depict sanitary nuisances of that period (Wiener Stadtphysikat, WSTP, various years). With the increasing development of industry, manufacture and other commercial infrastructure and the growing population along the small streams, pollution of the urban waterscape peaked (Pollack 2012). The Viennese city physician Nikolaus Theodor Mühlibach considered the open sewers the main reason for growing epidemiological threats in the city: 'Two flowing waters, the *Wienfluss* and the *Alserbach*, in those suburbs that they flow through, contribute much to the contamination of the air. Most children of those people living near these two watercourses suffer from adenopathy, almost all look pale and bloated. The odor coming from these impure springs is unbearable for those not used to it and it spreads far' (Mühlibach 1815, as cited in Payer 1997, p. 53). After the great cholera epidemics of the 1830s, urban authorities mandated the construction of intercepting sewers[1] along the *Wienfluss*, the main tributary of the Danube. Within the *Linienwall*, some of the streams were tunneled and vaulted. These early improvements in the sanitary situation of Vienna were unconnected to each other and were local in scale (Neundlinger et al. 2014). The transformation process to an industrial sociometabolic regime had just begun. In the second half of the 19th century, growth in material and energy use and in population accelerated. The as-yet unchanged disposal system had to handle a massive increase in waste and discharge. Estimates

[1]An intercepting sewer collects the wastewater from the smaller sewers that are connected to the houses and transports the collected sewage to the place where it is discharged.

of the amount of excrement disposed of daily by the Viennese suggest 420 t/day in 1830 compared to 2600 t/day in 1910 (see Gierlinger et al. 2013). Within a relatively short period, urban outputs increased drastically, so the capacity of the old system was reached, causing sanitary challenges and nuisances.

24.4 Societal Response

In the subsequent decades, Viennese city authorities responded to these new challenges. In the 1860s, the public health sector was reorganized (Meissl 2001; Senfelder 1908). The urban health authority, the so-called *Stadtphysikat*, received more financial resources and responsibilities than before the sanitary crisis. According to the new sanitary law (*Reichssanitätsgesetz*) of 1870, the physicians of the urban health authority became public officers directly subordinated to the urban administration. In the following decades, numerous health-related parameters were comprehensively assessed on a yearly basis. Various reports and statements on sanitary issues were published by *Stadtphysikat*. Technicians and engineers alike consulted *Stadtphysikat* for expertise when planning houses, water supply and disposal systems as well as public buildings. Accordingly, *Stadtphysikat* played an important role in the discourse on hygiene and sanitation. They perceived the existing sewage system as one of the main reasons for the high death rates of the Viennese population, so a citywide and long-lasting solution to the sanitary question was needed (WSTP 1872, p. 9). Different options for a new disposal system were debated among urban health authorities, engineers, economists and urban authorities. Similar debates were held all over Europe (Barles 2007; Goddard 1996; Winiwarter 2001). Experience from other cities such as London, Hamburg and Amsterdam showed that an urban-wide, comprehensive disposal system could be introduced and maintained efficiently. Members of the Viennese *Stadtphysikat* argued for a system that would dispose excrement as fast and as undecomposed as possible, moving it out of the urban area quickly to avoid infiltration of air and ground (WSTP 1872, p. 9). Several options for this massive transformation of the urban infrastructure were discussed. These options can be subsumed under two main alternative models for a new disposal system. One option was to systematically collect large quantities of urban excreta in bins or buckets to make use of them as fertilizer. This model was favored by economists such as the *Volkswirtschaftlicher Verein* (see WSTP 1872, p. 9). The other model was to improve the existing sewage system and convert it to a water-borne sewage system similar to London and Hamburg. This option was favored by urban health authorities. Initially, the potentials of a water-borne system were widely debated as huge quantities of water would be needed to guarantee efficient transport of excreta and waste. Only after the installation of the alpine water pipeline (*Hochquellenwasserleitung*) in 1873, which brought drinking water in sufficient quantities from alpine regions 80 km southwest of Vienna into the city center, did

Fig. 24.3 Constructing the main intercepting sewer: construction site on Marxergasse. (Source: Kohl 1905, p. 205)

the availability of water no longer seem to be a problem. In the subsequent decades, an urban-wide[2] water-borne sewage system was constructed (see Fig. 24.3).

The new system differed from the old in several important aspects. One of the differences was the materials used for its construction. Technological innovation in cement and pipe production allowed the building of sewers that were more waterproof than bricks and mortar. Infiltration of the ground with sewage was prevented rather efficiently—which was one of the prime objectives for urban health authorities. The installation of intercepting and interconnected sewers was another major difference. They collected the domestic wastewater from the small sewers connected to the houses and channeled it to the next-largest branch of the sewage network. Most of the former streams were incorporated into the sewage system as intercepting sewers, inside and outside the *Linienwall*. Basically, all sewers were connected to each other, which ultimately led to two main sewers along the *Donaukanal*. Almost all sewage of the entire city was discharged at a single point downward of Vienna's center in the *Donaukanal*. At the beginning of the 20th century, this was not perceived as problematic because the outflow was situated in an industrial area outside the residential zone (see Generalstadtplan 1912[3]).

[2]Including the suburbs incorporated in 1890/92.

[3]Retrieved from http://www.wien.gv.at/kultur/kulturgut/plaene/generalstadtplan.html [Accessed: April 6, 2013].

One of the most striking differences from the old sewage system was the increase in water throughput. After the opening of the *Hochquellenwasserleitung* in 1873, large parts of Vienna's urban infrastructure were quickly connected to the new water pipeline. Six years after the opening, more than 70 % of the houses had been connected to the new water pipeline (MagSW, various years). Accordingly, household water consumption increased enormously, growing from 12 million m^3 in 1876 to 43 million m^3 in 1910[4] (ibid.). After use, the water was discharged into the sewers—either in the houses or through the newly built gutters on the street. Consequently, sewers could be flushed on a daily basis, which had not been the case before. In addition to the increased water input from households, sewers were flushed regularly with water from the *Hochquellenleitung* (Kohl 1905). Excrement and other wastes were removed from the residential area faster and more regularly than before. In 1910, the urban health authorities announced that the newly introduced disposal system satisfactorily fulfilled all tasks (WSTP 1872, p. 153).

24.5 Socioecological Legacies

In the last decades of the 19th century, the water-borne sewage system was mainly constructed for sanitary purposes. The basic structure of this sewage system remains in place today. This intervention in Vienna's urban waterscape created long-lasting legacies for the urban disposal infrastructure and still influences the structure of the whole city. In the following paragraphs, we will briefly elaborate on different long-term socioecological legacies created.

First, the disposal system in place needs a constant amount of water throughput to maintain its vital functions. Sufficient quantities of water are needed to flush waste and excrement away. For this reason a so-called combined sewage system[5] was built. To dimension a combined sewage system adequately, urban authorities and engineers considered certain socioeconomic aspects, such as water use and the number of inhabitants connected to the sewage system; but they also considered meteorological aspects, such as rainfall patterns. At the turn of the 19th century, two main intercepting sewers were built along the *Donaukanal* under the assumption of linear population growth rates, which the engineers and city authorities predicted to be more than 4 million inhabitants by the 1950s (Kohl 1905). This increase in population numbers never occurred. The number of inhabitants peaked before World War I, with slightly more than 2 million inhabitants (see Fig. 24.2a). In the year 2010, Vienna had approximately 1.7 million inhabitants

[4]Numbers refer to the city of Vienna in its changing boundaries.

[5]In a combined sewage system, rainwater and wastewater from household and industry are discharged together. In a separated sewage system, rainwater is discharged separately from wastewater.

(MagSW 2010). In the aftermath of the collapse of the Austro-Hungarian Monarchy after World War I, immigration from the former provinces of the Monarchy ceased, and out-migration to the succession states took place (Weigl 2000). The projections of population growth and its consequences for the sewage system were counterbalanced by the assumption of engineers that water consumption per capita would remain constant at approximately 90.5 l per capita per day (Kohl 1905). Contrary to this projection, water consumption levels rose markedly. In the year 2007, the per capita water consumption (including industry and leakage) accounted for approximately 200 l per day (MagSW 2007).

Second, rainfall patterns needed to be considered for the construction of a combined sewage system. Due to recurring heavy rainfall, overflow facilities were installed along the *Wienfluss* and *Donaukanal* and their sewage facilities. As soon as the capacity of the sewers was reached, part of the urban wastewater was discharged into the rivers via the overflow facilities. At the end of the 19th century, urban construction authorities expected them to be in use four to five times a year (Kohl 1893), which was not perceived as problematic. A century later, environmental consciousness and actual overflow had changed significantly (Nowak 2005). As the sewage system had expanded spatially the amount of rainwater drained increased, so the overflow facilities were used more often. In 1997, in the context of adaptations to EU legislation on environmental conservation, urban authorities (*Gemeinderat*) enacted measures concerning sewage management and water pollution control (ibid.). Absorption channels and basins along the sewage system were constructed.

Finally, the implementation of a water-borne sewage system may be read as a long-term intervention in local nutrient cycles (Winiwarter 2001). Instead of using excrement and fractions of organic waste as fertilizer, they were discharged into the Danube. Night soil and sewage mainly contain nutrients, nitrogen and phosphorus. The amount of nitrogen discharged into the urban waterscape rose from approximately 900 t in 1830 to approximately 6500 t in 1910 (Gierlinger et al. 2013). In the second half of the 19th century, several European cities experimented with different kinds of disposal systems, directly using or converting sewage and/or night soil into fertilizer. Sewage farms in Paris (Barles 2007), the pneumatic Liernur system[6] in Amsterdam and systems based on collection in bins and conversion into fertilizer, as in Zürich and Heidelberg (WSTP 1872), are some examples. Similar projects were discussed but not realized in Vienna. High transportation costs, little demand for fertilizer in the surrounding agricultural systems and the advantageous geomorphologic situation allowing for an early proto-water-borne sewage system may explain why large-scale fertilizer collection was of minor importance in Vienna. Additionally, the Danube, with its high discharge rate (mean discharge 1915 m^3/s), absorbed larger amounts of organic waste than

[6]In the Liernur system, named after a Dutch military engineer, rainwater and wastewater were disposed of separately. The wastewater was drained pneumatically via pipes and was collected for agricultural reuse (Buiter 2006).

most other urban rivers in European cities could (Gierlinger et al. 2013). Nevertheless, discharge from the city increased socioecological pressures on the waterscape. Especially after World War II, pollution was increasingly perceived as a threat to urban aquatic ecosystems (Bihl 2006). In 1969, the urban administration decided to install a wastewater treatment plant. The first city-wide wastewater treatment facility was opened in 1980. A combined sewage system in which rainwater and wastewater are discharged together posed a challenge for the treatment of wastewater (Nowak 2005). A combined sewage system is heavily dependent on regional rainfall patterns. With a combined sewage system, the capacity of the wastewater treatment plant (WWTP) needs to be higher than with a separated sewage network. In Vienna, the amount of rainwater drained greatly fluctuates. To warrant a constant amount of water input to the wastewater treatment plant, absorption basins and channels along the sewage system had to be constructed, allowing collected water to be stored and released to the WWTP flexibly (Nowak 2005).

24.6 Conclusions

In the 19th century, Vienna found itself in the middle of a socioecological transition with many implications for the urban disposal infrastructure. The new coal-based sociometabolic regime, urbanization processes and a massive increase in population created new requirements for the existing disposal infrastructure. City officials responded to structurally new problems by introducing a water-borne sewage system. Its construction may be interpreted as a large-scale intervention in the urban waterscape and local nutrient cycles. The prime objective of urban authorities was to remove excrement and wastewater from industry and households in residential areas as fast as possible. Local biophysical conditions and arrangements and specific actor relations and decisions were crucial for the specific ways the urban waterscape was transformed during the metabolic transition. Contemporaries' perceptions, needs and demands were decisive for the implementation of the new system. Urban actors debated several options for a new disposal system, the positive and negative effects of each system were evaluated, and projections about future developments that would influence the functioning of the sewage system were made. The specific discharge regime of the Danube, the dense network of small streams traversing the city and the introduction of the *Hochquellenleitung*, which brought a considerable amount of water to the urban households, provided ideal conditions for the construction of a water-borne disposal system. Socioecological pressure, urban development and changing attitudes toward environmental problems required further interventions in the system, such as the installation of a WWTP and the construction of absorption channels and basins. The basic layout of the sewage system built in the 19th century is still intact.

Taking a long-term socioecological research perspective may shed new light on how large-scale processes such as changes in urban metabolism, urbanization, industrialization and the transition of the energy system are realized at the local level and in an urban context. During Vienna's sociometabolic transition, metabolic output flows had to be rearranged spatially. Consequently, the biophysical disposal infrastructure of the time was adapted to changing metabolic realities. The disposal system introduced was intensely debated among urban actors. The long-term sociometabolic view of the city as sketched here shows that input and output flows of urban metabolism are closely linked and that interventions on the input side may have impacts on the output side. With the increase in urban metabolism, output flows multiplied. The older disposal system was incapable of coping with this increased amount of waste and wastewater, which placed enormous pressure on the urban waterscape. Interestingly enough, an innovation on the input side—the introduction of the *Hochquellenwasserleitung*—allowed for the installation of an efficient new disposal system.

References

Barles, S. (2007). Feeding the city: Food consumption and flow of nitrogen, Paris, 1801–1914. *Science of the Total Environment, 375*, 48–58. doi:10.1016/j.scitotenv.2006.12.003.

Bihl, G. (2006). Wiederaufbau und Ausbau der städtischen Infrastruktur. In P. Csendes & F. Opll (Eds.), *Geschichte einer Stadt von 1790 bis zur Gegenwart* (pp. 564–567). Wien: Böhlau.

Buiter, H. (2006). Transforming water infrastructures in Amsterdam and Utrecht, 1860–2000: Power relations, social functions and urban identities. International Summer Academy on Technology Studies—Urban Infrastructure in Transition.

D'Avigdor, E. H. (1873). Der Wienfluss und die Wohnungsnoth. Ein Vorschlag. Wien.

Generalstadtplan (1912): WStLA, Kartographische Sammlung, Sign. 1701a.

Gierlinger, S., Haidvogl, G., Gingrich, S. & Krausmann, F. (2013). Feeding and cleaning the city: The role of the urban waterscape in provision and disposal in Vienna during the industrial transformation. *Water History*, *5*(2), 219–239.

Goddard, N. (1996). "A mine of wealth"? The Victorians and the agricultural value of sewage. *Journal of Historical Geography, 22*, 274–290.

Haidvogl, G., Guthyne-Horvath, M., Gierlinger, S., Hohensinner, S. & Sonnlechner, C. (2013) Urban land for a growing city at the banks of a moving river: Vienna's spread into the Danube island Unterer Werd from the late 17th to the beginning of the 20th century. *Water History*, *5*(2), 195–217.

Hohensinner, S., Lager, B., Sonnlechner, C. H., Haidvogl, G., Gierlinger, S., Schmid, M., Krausmann, F., Winiwarter, V. et al. (2013). Changes in water and land: The reconstructed Viennese riverscape from 1500 to the present. *Water History*, *5*(2), 145–172.

Kohl, J. (1893). Über das Project für den Haupt-Sammelcanal am linken Donaucanal—Ufer in Wien. *Zeitschrift des oesterreichischen Ingenieur- und Architekten-Vereins, 18*, 266–270.

Kohl, J. (1905). Die Entwässerung. In P. Kortz, Österreichischer Ingenieur- und Architektenverein (Eds.), *Wien am Anfang des XX. Jahrhunderts. Ein Führer in technischer und künstlerischer Richtung*. Wien: Gerlach und Wiedling.

Krausmann, F. (2013). A city and its Hinterland: Vienna's energy metabolism 1800–2006. In S. J. Singh, H. Haberl, M. Schmid, M. Mirtl, & M. Chertow (Eds.), *Long term socio-ecological research*. New York: Springer.

MagSW = Magistrat der Stadt Wien (= Vienna Municipal Administration), 1883–2010. Statistisches Jahrbuch der Stadt Wien.

Meissl, G. (2001). Hochquellenleitungen und Unratsschiffe. Zur Geschichte der Wiener Wasserver- und Entsorgung vor 1914. In S. Hahn & R. Reith (Eds.), Umwelt-Geschichte. Arbeitsfelder—Forschungsansätze—Perspektiven. (pp. 157–179). Wien: Verlag für Geschichte und Politik.

Neundlinger, M., Gierlinger, S., Pollack, G. & Krausmann, F. (2014). An environmental history of the Viennese sanitation system—From Roman to modern times. In T. Tvedt & T. Oestigaard (Eds.), *A history of water* (Series 3, Vol. 1, Water and urbanization). London: I.B. Tauris.

Nowak, R. (2005). RTC—Die zukunftsweisende Kanalnetzsteuerung in Wien. Österreichische Wasser—und Abfallwirtschaft Heft, 11–12.

Payer, P. (1997). Der Gestank von Wien. Über Kanalgase, Totendünste und andere üble Geruchskulissen. Wien: Döcker Verlag.

Pollack, G. (2012). Verschmutzt—verbaut—vergessen: eine Umweltgschichte des Wienflusses von 1780 bis 1910. Thesis, University of Klagenfurt.

Schratt-Ehrendorfer, L. (2011). Donau und Auenlandschaft. Ein Lebensraum voller Gegensätze. In R. Berger & F. Ehrendorfer (Eds.), *Ökosystem Wien: Die Naturgeschichte einer Stadt* (pp. 328–391). Wiener Umweltstudien 2. Wien, Köln, Weimar: Böhlau.

Seebacher, F., Mrkvicka, A. & Kroiss, H. (2011). Flüsse und Bäche: eine Herausforderung für die Stadt. Schutz vor dem Hochwasser und Schutz der Natur. In: R. Berger & F. Ehrendorfer (Eds.), *Ökosystem Wien. Die Naturgeschichte einer Stadt* (pp. 117–122). Wiener Umweltstudien 2. Wien, Köln, Weimar: Böhlau.

Senfelder, L. (1908). *Geschichte des Wiener Stadtphysikates*. Wien: Im Selbstverlag des Verfassers.

Singh, S. J., Haberl, H., Schmid, M., Mirtl, M., & Chertow, M. (2013). *Long term socio-ecological research*. New York: Springer.

Summesberger, H. (2011). Vom Tropenmeer zur Eiszeittundra. 250 Millionen Jahre Wiener Erdgeschichte. In R. Berger & F. Ehrendorfer (Eds.), *Ökosystem Wien. Die Naturgeschichte einer Stadt* (pp. 58–87). Wiener Umweltstudien 2. Wien, Köln, Weimar: Böhlau.

Weigl, A. (2000). *Demographischer Wandel und Modernisierung in Wien*. Wien: Pichler Verlag.

Winiwarter, V. (2001).Where did all the waters go? The introduction of sewage systems in urban settlements. In C. Bernhardt (Ed.), *Environmental problems in European cities in the 19th and 20th century* (pp. 105–119). München: Waxmann.

Winiwarter, V., Schmid, M. & Dressel, G. (2013). Looking at half a millennium of co-existence: The Danube in Vienna as a socio-natural site. *Water History*, *5*(2), 101–119.

WSTP = Wiener Stadtphysikat (= Vienna health authority), 1864–1913. Berichte des Wiener Stadtphysikates über seine Amtsthätigkeit. Braumüller, Wien.

Part V
Empirical Approaches to Working with Stakeholders

Chapter 25
Planning, Residential Decisions and Energy Use in Vienna

Veronika Gaube, Alexander Remesch and Barbara Smetschka

Abstract Urban planning must address both a changing urban population size and increasing sustainability issues in terms of providing good socioeconomic and environmental living conditions. Households play a major role in that they are affected by urban planning decisions but are partly responsible for the environmental performance (e.g., energy use) of a city. Here, we present an agent-based decision model of the city of Vienna, the capital of Austria, with a population of approximately 1.7 million. The model results are used to assess the spatial patterns of energy use caused by different household types. The outcomes show that changes in households' preferences regarding the presence of nearby green areas have the most important impact on the distribution of households across the small-scaled city area. Additionally, the results demonstrate the importance of the distribution of different household types regarding spatial patterns of energy use.

Keywords Sustainable urban planning · Energy demand · Vienna · Agent-based model (ABM) · Residential location choice

V. Gaube (✉) · A. Remesch · B. Smetschka
Institute of Social Ecology (SEC), Alpen-Adria University,
Schottenfeldgasse 29, 1070 Vienna, Austria
e-mail: veronika.gaube@aau.at

A. Remesch
e-mail: alexander.remesch@aau.at

B. Smetschka
e-mail: barbara.smetschka@aau.at

H. Haberl et al. (eds.), *Social Ecology*, Human-Environment Interactions 5,
DOI 10.1007/978-3-319-33326-7_25

25.1 Linking Urban Planning, Households' Decision-Making and Energy Demand

Given the United Nations' prediction that 60 % of the world's population will live in cities by the year 2030, the importance of cities as living space has received increasing attention in the context of sustainability debates. The role of cities in reducing socioeconomic material and energy flows is increasingly recognized (Weisz and Steinberger 2010). A better understanding of cities as socioecological systems—particularly regarding energy use and the relationship between urban planning and the socioeconomic structure of a city—is required. Urban areas face increasing sustainability problems when it comes to providing good socio-economic and environmental living conditions. Urban planning must address these challenges by considering processes of growth in new areas, decay and abandonment in others and restructuring and rehabilitation. Modern cities are the result of both planned and spontaneous development. In both cases, the emerging patterns of urban structures are influenced by factors such as available building technologies, urban planning regulations, real estate markets, investment strategies of public and private institutions, public policies (related to, for example, housing, transport, environment and taxation) and institutional traditions. Other influencing factors are environmental and social factors and individual lifestyle choices and behavior. Consequently, we understand urban planning as involving restrictions to human freedom due to various constraints related to infrastructure, demography, social requirements and the economy.

The decisions made by households regarding residential preferences may have an impact on the spatial distribution of energy use due to the spatial allocation of different household types. These decisions also have an impact on total urban energy use resulting from changes in transport distances and modes, population and dwelling density and building age and insulation. In addition, the response of individual households to urban planning strategies is an important issue in designing a sustainable city if we assume that sustainable urban development is among other parameters characterized by a balanced distribution of households with different socioeconomic structures across the city area.

A large majority of people live in cities, which means that their energy demand matters significantly. Urban planners need an in-depth understanding of the energy demand of their city to efficiently lay out the urban infrastructure. Urban energy use can be understood from a demand perspective, not only for final energy forms such as electricity and transportation fuels but also for energy services (Jochem 2000; Lovins 1977). Each household and economic activity in urban areas has a need for energy services, such as mobility (physical access to certain destinations and certain goods), ambient temperature (hotter or colder than the local climate) and working appliances (e.g., for households, office and industry, communication). These urban energy services are common to most urban areas, but the energy needed to provide them varies greatly. Household demand for energy services depends on several factors, which can be categorized as economic, demographic and behavioral. The factors that determine urban energy consumption do not work in isolation; they are linked and exhibit feedback behavior, precluding simple linear relations. The interaction among these factors may change from city to city. Moreover, many of these factors are dynamic and path dependent.

Urban populations may have significantly smaller household sizes than rural populations due to smaller family sizes and a larger generation gap as well as smaller dwellings. The evidence of a connection between age and urbanity level is mixed. The most important impacts of age may be apparent in changing household sizes and changing income levels. In many European cities, demographic growth is rather moderate or even negative and is mainly driven by migration. The composition of household types within many European cities varies from a mixture of one-person to more than five-person households to a dominance of single and couple households within the city and an allocation of family households in suburban areas. This process is based on the residential location decisions of individual households. Dieleman (2001), Coulombel (2010) and Knox and Pinch (2010) each provide comprehensive literature overviews on the residential decision-making of urban households.

This chapter draws from research from a series of completed and ongoing research projects. This research allows for the development of an agent-based model (ABM) for Vienna to estimate its energy consumption as a function of individual household decision-making based on quality of life preferences. First, we develop an ABM to simulate the residential choices of different household types (singles, couples, families). The spatial distribution of household types serves as a basis to calculate the spatial distribution of household energy consumption. This ABM simulates the spatial distribution of household types under different urban planning and household preference scenarios. We also use the ABM to test assumptions on time-use preferences, which are a key factor in household decision-making.

25.2 Building an Agent-Based Model (ABM) for the City of Vienna

The pattern of household locations in a city results from the everyday decision-making of single households. In many cases, these decisions also affect the distribution and the amount of energy consumption in the city. In this chapter, we describe an ABM analyzing the effect of households' residential location decisions on the spatial pattern of urban energy use for the city of Vienna.

The capital of Austria, Vienna, currently has a population of approximately 1.7 million. After decades of decline, the population of Vienna grew by approximately 120,000 inhabitants between 1987 and 1994. The reasons for this rapid increase may lie in the new geopolitical status of Vienna after the fall of the Iron Curtain. Austria's accession to the European Union at the beginning of 1995 continued the population increase in Vienna. This growth in population also led to a growing demand for housing and jobs. By the end of the 1980s, the construction of subsidized apartments had dropped to an annual rate of approximately 4000. The rising demands on the quality of accommodation, increased housing demand in general and the growing number of (single-person) households are the main factors in the great need for new subsidized flats Vienna has experienced since the beginning of the 1990s. Given these new framework conditions for Vienna, the city government decided at the beginning of the 1990s to increase the building rate of subsidized housing to 10,000 new units annually.

Demand for new housing does not depend solely on the quantitative development of the resident population. It also depends on changing expectations regarding the quality of housing in terms of living space per inhabitant, the quality of the infrastructure and the quality of the environment (e.g., private and public green spaces). Therefore, even if the population were to stagnate, there would still be demand for new housing.

25.2.1 Modeling Residential Choice

Urban models are becoming increasingly detailed in terms of the number of household categories, agent types beyond households and fine-grained spatial units (on a grid cell level). Many of these models consist of several modules combining different sectors, activities and preference structures. The need to disaggregate households, primarily on socioeconomic criteria, from eight (Putman 2010) to more than 100 categories (Simmonds 2010) is a common feature of residential location models. In contrast to this trend, we chose to limit our model to categories regarding household types and spatial units that are relevant but simple enough to address socioecological research issues.

To keep the model output and processes comprehensible, we decided to focus on household members' age, household size and income because these are the most important factors in energy consumption. Age and size taken together define the 'household type'. There are seven household types in the model: single young, single old, couple young, couple old, single parent, small family and large family. Each of the household types has its own preference profile regarding residential mobility. Each individual household is assigned to one specific household type by virtue of its household properties. During the model run, households change their family structure and thus their household type through demographic dynamics. The income is set to change only by rearrangement of the household members, that is, when a household merges with another household or splits.

Due to a lack of detailed micro-level data, the model population of approximately 770,000 households and 1.5 million household members was synthesized in a Monte Carlo sampling process as described by Wilson and Pownall (1976). As input data, the population and household numbers per cell were taken from the 2001 census (Statistik Austria 2004, pp. 195 ff.), and the distributions of other sociodemographic parameters, such as age and sex, depending on household size and type were estimated using the combined public available samples of the micro-census datasets from 2006 to 2008 (Statistik Austria 2008). The process of generating the synthetic population is described in full detail in Gaube and Remesch (2013). Micro-level data on approximately 900,000 dwellings from the 2001 census are available from Statistik Austria (data retrieved from http://www.statistik.at).

The number of persons in a household can change due to biographical events, which in turn can influence space needs, household income and residential preferences. Many authors (Bauer-Wolf et al. 2003, p. 18; Fontaine and Rounsevell 2009, p. 1240; IZT Institut für Zukunftsstudien und Technologiebewertung 2003, p. 112; Schneider and Spellerberg 1999, p. 126) attribute strong effects on residential mobility to biographical events. We consider the following life-course events in our demographic sub-model: (1) birth of a household member; (2) death of a household member; (3) leaving the parental home; and (4) moving in together or marriage and the founding of a household. The demographic sub-model is strictly based on the probability of a certain biographic event for a certain household based on its household or member properties.

The resolution of the model is based on 59 spatial units, the so-called 'small-scaled city areas' defined by the Viennese spatial planning administration (Stadt Wien 2007, p. 64 ff.) using parameters such as density and dominant type and age of buildings. The small-scaled city areas are characterized by the share of green area and access to infrastructure such as public transport and services. Each small-scaled city area contains a number of dwellings. Each dwelling has a certain size and a specific price per m^2.

Every year, some households are affected by demographic events, but many are not. However, all households evaluate their current living situation and decide whether to start looking for a new dwelling. This process can be examined through a stress/resistance-model (Benenson 2004, p. 6; Knox and Pinch 2010). The model starts with roughly 770,000 household agents and runs in yearly time steps. The main interaction in the model takes place between households and spatial units. Each household agent in the model has a preference profile regarding residential location defined by its assigned household type.

The attractiveness of living in a certain area in the city can be described by several factors. The only available empirical study on factors influencing residential satisfaction in Vienna is from Zucha et al. (2005). These authors use data from interviews of 8300 persons in Vienna conducted between May and October 2003 by the Viennese urban planning department (MA18) to conduct a structural equation analysis of attachment to the neighborhood, with residential satisfaction as a main component. We used these results to determine the most important influencing factors for the residential satisfaction/stress-to-move calculation in our model and to derive the relative weights of these factors.

Based on the study of Zucha et al. (2005), we decided on characteristics such as environmental amenities (e.g., green areas), services through infrastructure, centrality, satisfaction with the dwelling in terms of cost and size and, finally, the social prestige of an area. We define social prestige as the relation between the average income in a small-scaled city area and the income of a specific household. It adds to a household's stress-to-move if the average income in the neighborhood is lower than the household's own, but if income is the same or above, the household is not worried about social prestige.

When a household searches for a new place to live, it compares the potential residential satisfaction with its current situation.

Figure 25.1 shows a calculation example. On the right-hand side down the columns, each household defines its preferences using importance scores for a set of selected criteria. On the left-hand side along the rows, each spatial unit offers fulfillment scores for a certain set of attributes. Each attribute reflects one criterion (criteria 1 and attribute 1 are the same: for example, share of green area). The compliance between preferences and attributes is calculated each year for those households that look for a new residence. Those spatial units that offer the highest residential satisfaction to the particular household will be the regions aspired to in the subsequent search for a dwelling.

If a household cannot find a dwelling that promises a higher residential satisfaction, it will remain at its current living place. Transaction costs for moving are included using a stochastic residential satisfaction threshold modifier value on the individual household level. To reflect the incomplete knowledge of the household, the household may consider only a certain number of dwellings per year (Semboloni 2007, p. 61). If a more suitable dwelling can be found, the household will relocate, and the old dwelling will be free on the market for other households.

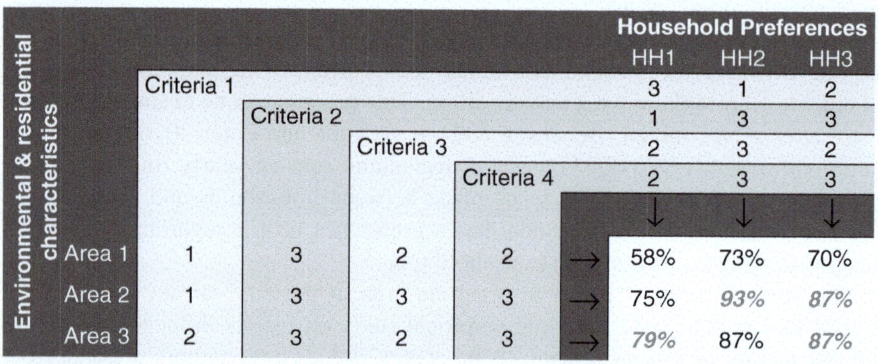

Fig. 25.1 Example of an implemented procedure for residential location decision-making using weights ranging from 1 to 3: Households have different preferences for criteria 1–4 (depending on household type, income, etc.); areas fulfill these criteria with properties 1–4 to different degrees (which changes through urban planning actions). This results in specific preferences of each household for each area

25.2.2 Scenarios: Sustainable Mix and Green Areas

We use the residential mobility model for Vienna as a tool to simulate future scenarios depending on changes in (1) urban development planning, (2) economic framework conditions in terms of dwelling prices and (3) changes in a household's preferences in terms of environmental amenities. We focus on two scenarios: we contrast the no-green-area-preference scenario with the conventional urban planning scenario, which carries forward the initial values as inputs. It might seem counterintuitive to choose this scenario as it describes the opposite of a sustainable city. Environmental amenities seem to be a key criterion in the decision-making process of Viennese households (see Zucha et al. 2005, p. 50). Given the assumption that the sustainable development of a city is best secured with a thoroughly mixed population structure, analysis of the effects of an extreme behavior of a household regarding the key criterion 'environmental amenities' becomes important. Fortunately, models allow for testing such experiments.

The **conventional urban planning scenario** assumes that the conditions of the current planning policies remain constant over the 50 years of the simulation period, supporting past spatial development trends (densities and configurations of the urban fabric). For future city planning, those projects that are part of the *Urban Development Plan of Vienna* are assumed to be realized by 2025 or by 2050. The assumptions are based on the available planning documents of the city. This scenario demonstrates how the future could unfold if neither external nor internal factors were to change over the next few decades.

In the **no-green-area-preference scenario**, we assume that the preference for green areas in the desired spatial unit is equal for all households, namely, zero. The assumption implies that the residential choice of households is independent of the share of green areas in the neighborhood and is thus exclusively focused on the centrality, infrastructure and price level of a dwelling. Even for family households, the share of green areas no longer has relevance in the decision-making process. This change in household preference structures is the only difference between our two scenarios, meaning that urban planning and residential building projects are the same as in the conventional urban planning scenario.

The density per small-scaled city area stays almost the same because under the condition of strong population growth (through inbound migration), the number of available dwellings in each spatial unit mainly defines it. Vienna, with its densely populated inner city emanating from the historical old city (1st district), still has a rather monocentric city structure. Important subordinated centers (the 2nd to 9th districts) are grouped around the original center and are densely built-up zones containing many buildings designated as cultural heritages. Experts from the ÖIR (Österreichisches Institut für Raumplanung; Austrian Institute for Regional Studies and Spatial Planning) assume that substantial densification of these areas is unrealistic.

Changes in political, economic and planning strategies affect the socio-economic structure of urban areas, that is, the share of households of different

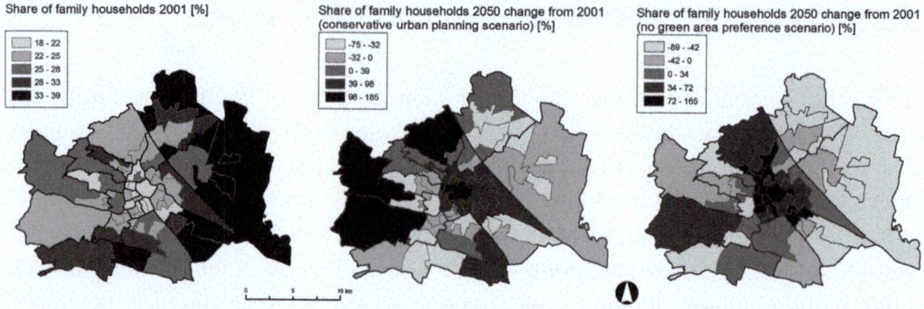

Fig. 25.2 Distribution of family households in 2001 (census data from Statistik Austria 2004, p. 195, Table 22) and 2050 in the conservative urban planning scenario and in the no-green-area-preference scenario

income, age and family structure in each urban area rather than the overall density. The **no-green-area-preference scenario** shows that this preference makes a substantial difference. Comparing the distribution of households by household types (families, couples, singles and others), this scenario shows that the allocation of families who would otherwise favor environmental amenities as one of the strongest residential decision criteria differs clearly from the results of other scenarios. In this scenario, families are more or less evenly distributed across the small-scaled city areas (Fig. 25.2). Families, as the largest group representing middle income households, have a very broad set of satisfactory residence locations as soon as the factor of green areas becomes unimportant. In turn, this behavior of family households allows young, mostly single households with low incomes to move to the outer districts of Vienna, taking the more affordable space made available by the lower concentration of family households there.

In general, the overall picture of this scenario shows a more even distribution of the different household types among the districts compared with other scenarios. If the removal of a preference for a high share of green area allows a mixed socioeconomic structure across the whole city, one can conclude that urban planning strategies supporting the establishment of green spaces even in dense city areas might allow for a sustainably mixed socioeconomic structure in the city.

25.2.3 Energy Consumption Patterns of the City

The model outputs indicating scenario-dependent patterns of the city's socioeconomic structure serve as a basis for assessing changes in the energy consumption patterns. We use existing data on average yearly consumption of energy from electricity and natural gas in 2007 in Austria from Wegscheider-Pichler (2009). For transport, we use data on yearly energy use for public and individual transport in Austria from Endl (2010). By combining these datasets per household

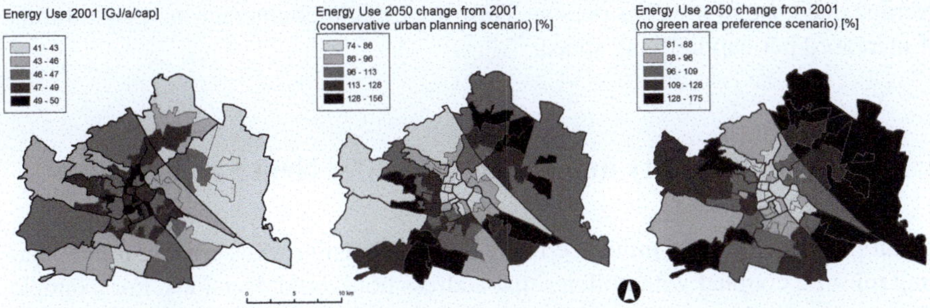

Fig. 25.3 Distribution of households' energy consumption for heating, electricity and transport

energy consumption parameters with the distribution of household types over the small-scaled city areas, the yearly consumption of energy of private households in Vienna per small-scaled city area can be calculated (Fig. 25.3).

Based on the spatial distribution of household types and data on average energy use per household size, we calculated the energy use for heating, electricity and transportation in the 59 small-scaled city areas for all scenarios. Household energy demand depends on many factors. Space heating, for example, one of the most important energy use categories of private households, depends on technical factors such as type of dwelling and on the income and family structure of a household. For example, in the Northeast of Vienna, the average energy use strongly increases (see Fig. 25.3) due to the increasing proportion of single households in these areas, as shown in Fig. 25.2.

In the **no-green-area-preference scenario**, both the population and the total yearly energy use in Vienna grow by 38 % between 2001 and 2050 (Table 25.1). Other studies (e.g., Druckman and Jackson 2008), in contrast, find an increase in energy use up to three times the increase in population growth. They interpret this finding as an increase in the proportion of small households (single, couples and small families). Druckman's results show the effect of the socioeconomic trend of an increasing number of single households living in cities. Single households have the largest amount of energy use per capita and therefore determine the spatial pattern of energy use. However, the increase in population in Vienna, which is also characterized by an increase in small households of 30 % by 2050, might be based on a densification in terms of family structure and living space per person. The large share of migration-driven population increase in Vienna could cause an

Table 25.1 Total population, number of households and estimated energy use for Vienna in 2001 and for two scenarios in 2050

	Population	Households	Energy use (GJ/yr)
2001 (census data)	1,531,440	771,710	71,311,645
2050 trend	2,062,670	1,072,970	98,372,481
2050 no-green-area	2,043,060	1,057,050	97,430,478

average decrease in area per person. This trend might counteract the general trend of increased per capita living space.

25.3 Sustainable City and Quality of Life: Next Steps

Urban planning is a key influence in residential decisions. Sustainable urban planning must be coupled with policies that address household decisions. For example, if land-use policies promoting high-density development succeed, the preferences of users for lower-density development must be addressed. If we try to influence household decisions concerning residential choice, modal split and energy use, we must take questions of quality of life into consideration.

Time use can serve as a key indicator for quality of life. Lack of time often results in everyday decisions that translate into spending more energy (as well as more money). Certainly, the time-use structure of both individuals and households is essential for their physical and emotional well-being. The issue of time use has a strong influence on household decisions concerning choice of living space, consumption patterns and means of transportation. All of these activities are energy consuming, so we expect a significant link between time use and energy use. Consequently, phenomena such as ‘time squeeze’ and problems with the synchronization of activities of different members of a household could provide a missing link for reconciling household decisions, quality of life and sustainable city planning.

Changing gender relations can be another link between household decision-making and the specific situation of its household members in the past century in industrialized countries. Gender relations have an impact on family size as well as trends concerning marriage, separation and living arrangements of singles, couples and families. We assume that women’s employment changed the demand of energy services for household and family work.

It is along these lines that current research at the Institute of Social Ecology tries to gain knowledge on the link between urban spatial planning and time-use patterns. Beyond what has been shown so far, the project ‘UTE—Urban Time and Energy’ works on the development of a model linking energy use with time use in Vienna. We owe much knowledge on this to Mikko Jalas (Jalas 2005) and Angela Druckman et al. (2012), who performed similar research in Finland and the UK. We are developing a computer model on time and energy use of households in Vienna. We invite citizens to participate in discussing and finding solutions regarding their problems with time use. Moreover, if we know more about how time-use patterns are linked to energy and material use, we can ask the next questions: How does urban infrastructure shape the patterns of how citizens spend their time, for instance, by influencing housing and transport infrastructure? How can these patterns be changed in a way that allows for both a higher quality of life and lower material and energy use?

Contemporary models of urban architecture and spatial planning aim to align themselves again with the historical city. Concepts of the 'compact city' or the 'city of short distances' are favored in contrast to the inflexible, divided and homogenous use of spaces. In the term 'slow city', a notion of time is already included. The interplay between the time-use structure of household members, the energy demand of households and the spatial organization of cities seems obvious and is part of everyday common knowledge. Everything else being equal, reaching the goal of an activity faster requires more energy for the same achievement. Similarly, larger distances between locations of activities require either more time or more energy or both. In contrast, 'time sovereignty', which indicates some freedom of choice over one's time use, is considered a key feature of quality of life. Available time governs everyday decision-making just as much as available money. Lack of time often translates into spending more money and more energy, severely constraining individual and household choices. Could an awareness of these interrelations help improve spatial settings and infrastructures in cities as a win-win connection between quality of life and energy savings?

In UTE, we use the residential choice model for Vienna described above and expand it by including time-use and energy-use data to find the links between these components. We hope this can help researchers understand paths toward the sustainable city of the future from a socioecological perspective. Ongoing and future research at the Institute aims at gaining better insight into the drivers of time-use patterns and energy use. Time-policy measures (Mairhuber and Atzmüller 2008) that consequently impact the energy and material consumption activities of urban households could help narrow the gap between current lifestyle preferences and the goals of sustainable city development.

Acknowledgements We acknowledge financial support by the European Union for the project 'Sustainable Urban Metabolism for Europe' (SUME) within its 7th Framework Programme and the Austrian Science Fund project 'Urban Time and Energy' (UTE) (project TRP214).

References

Bauer-Wolf, S., Roth, M. C. M., Baumfeld, L., & Riesenfelder, A. (2003). *Stadt-Umland migration Wien*. Wien: Erforschung zielgruppenspezifischer Interventionspotentiale.

Benenson, I. (2004). Agent-based modelling. From individual residential choice to urban residential dynamics. In: M. F. Goodchild, & D. G. Janelle (Eds.), *Spatially integrated social science*. Examples in best practice (pp. 67–95). Oxford: Oxford University Press.

Coulombel, N. (2010). *Residential choice and household behavior: state of the art*. Sustaincity working paper 2.2a. Cachan. Retrieved from http://www.sustaincity.org/publications/WP_2.2a_-_Residential_Choice_and_Household_Behavior.pdf

Dieleman, F. M. (2001). Modelling residential mobility. A review of recent trends in research. *Journal of Housing and the Built Environment, 16*, 249–265.

Druckman, A., & Jackson, T. (2008). Household energy consumption in the UK: A highly geographically and socio-economically disaggregated model. *Energy Policy, 36*, 3177–3192.

Druckman, A., Buck, I., Hayward, B., & Jackson, T. (2012). Time, gender and carbon: A study of the carbon implications of British adults' use of time. *Ecological Economics, 84*, 153–163. doi:10.1016/j.ecolecon.2012.09.008.

Endl, A. (2010). *Energy consumption patterns for mobility in Austrian households. A socio-economic analysis*. Thesis, Vienna.

Fontaine, C. M., & Rounsevell, M. D. A. (2009). An agent-based approach to model future residential pressure on a regional landscape. *Landscape Ecology, 24*(9), 1237–1254.

Gaube, V., Remesch, A. (2013). *Impact of urban planning on household's residential decisions: An agent-based simulation model for Vienna*. Environmental Modelling & Software.

IZT Institut für Zukunftsstudien und Technologiebewertung. (2003). Mobilität und Wohnen. Werkstattbericht Nr. 61. IZT Institut für Zukunftsstudien und Technologiebewertung, Berlin.

Jalas, M. (2005). The everyday life context of increasing energy demands: Time use survey data in a decomposition analysis. *Journal of Industrial Ecology, 9*, 129–145. doi:10.1162/1088198054084644.

Jochem, E. (2000). Energy end-use efficiency. In J. Goldemberg (Ed.), *World energy assessment: Energy and the challenge of sustainability* (pp. 173–217). New York: United Nations Development Programme (UNDP), United Nations Department of Economic and Social Affairs, World Energy Council (WEC).

Knox, P., & Pinch, S. (2010). *Urban social geography. An introduction*. Harlow: Prentice Hall, Pearson Education.

Lovins, A. B. (1977). *Soft energy paths: Toward a durable peace*. Cambridge: Ballinger.

Mairhuber, I., & Atzmüller, R. (2008). *Machbarkeitsstudie: Chancen und Möglichkeiten kommunaler Zeitpolitik*. Wien: FORBA.

Putman, S. H. (2010). DRAM residential location and land use model: 40 years of development and application. In F. Pagliara, J. Preston, & D. Simmonds (Eds.), *Residential location choice* (pp. 61–76). Heidelberg: Springer.

Schneider, N., & Spellerberg, A. (1999). *Lebensstile, Wohnbedürfnisse und räumliche Mobilität*. Opladen: Leske & Budrich.

Semboloni, F. (2007). The management of urban complexity through a multi-agent participatory simulation. *disP, 170*(3), 57–70.

Simmonds, D. (2010). The DELTA residential location model. In F. Pagliara, J. Preston, & D. Simmonds (Eds.), *Residential location choice* (pp. 77–97). Heidelberg: Springer.

Stadt Wien. (2007). Leben und Lebensqualität in Wien. Kommentierte Ergebnisse und Sonderauswertungen der Großstudien "Leben in Wien" und "Leben und Lebensqualität in Wien". Stadt Wien - MA18, Wien.

Statistik Austria. (2004). Volkszählung. Hauptergebnisse II – Wien. Retrieved from http://www.statistik.at/web_de/dynamic/services/publikationen/2/publdetail?id=2&listid=2&detail=262

Statistik Austria. (2008). Mikrozensus Testdatensätze 2006–2008. Retrieved from http://www.statistik.at/web_de/services/mikrodaten_fuer_forschung_und_lehre/datenangebot/standardisierte_datensaetze_sds/index.html

Wegscheider-Pichler, A. (2009). *Strom- und Gastagebuch 2008*. Wien: Strom- und Gaseinsatz sowie Energieeffizienz österreichischer Haushalte - Projektbericht.

Weisz, H., & Steinberger, J. K. (2010). Reducing energy and materials flows in cities. *Current Opinion in Environmental Sustainability, 2*(3), 185–192.

Wilson, A. G., & Pownall, C. E. (1976). A new representation of the urban system for modelling and for the study of micro-level interdependence. *Area, 8*(4), 246–254.

Zucha, V., Rapa, S., Putz, I., et al. (2005). *Wohnzufriedenheit und Wohnqualität in Wien*. Wien: SORA-Institut for Social Research and Analysis.

Method Précis: Agent-Based Modeling

Alexander Remesch and Veronika Gaube

Why Model?

'Building a model is a well-recognized way of understanding the world: something we do all the time, but which science and social science has refined and formalized. A model is a simplification—smaller, less detailed, less complex, or all of these together—of some other structure or system' (Gilbert and Troitzsch 2005, p. 2).

Everyone with a theory about an unfolding social dynamic or with a projection is essentially a modeler; they are running some implicit model with hidden assumptions, untested internal consistency, unknown logical consequences and an unknown relation to empirical data. Hence, the choice is not whether to build models but whether to build explicit ones, where the assumptions are laid out in detail and others are able to replicate the results (Epstein 2008).

Apart from analyzing predictive scenarios, any model can be used to better explain processes for an improved understanding of its subject and to provide a 'laboratory' that allows experiments with the model dynamics, eventually leading to new research directions.

What Are Agent-Based Models?

An agent is a computational representation of a real-world actor. Formally, it is a small autonomous unit situated in time, and frequently also in space, that behaves according to rules. Agent-based modeling (ABM) is a computational method that allows researchers to create, analyze and experiment with models composed of many such agents that dynamically interact with each other and their environment. This approach produces bottom-up models that aim to show phenomena at aggregate levels that are not 'intended' and are often not even 'known' or 'perceived' by the individual agents but are the result of the interaction of a large number of independent agents.

This process, termed 'emergence', can be formally described as occurring 'when interactions among objects at one level give rise to different types of objects at another level' (Gilbert and Troitzsch 2005, p. 11). Examples include traffic jams unintentionally caused by the independent behavior of motorists, the phenomenon of temperature caused by the independent movement of atoms, patterns of a flock formed by independent birds or fish and a picture visually emerging from many tiles of a mosaic.

Why Use Agent-Based Models (ABMs) to Analyze Socioecological Systems?

In the field of Social Ecology, agents are usually implemented as distinct parts of a computer program that are used to represent social actors. Agents are located in a computational environment. This environment is an abstraction of the real environment—a 'virtual world'—in which the social actors operate and interact.

Within the context of socioecological systems (SESs), an ABM is composed of a population of agents and a landscape within which they act and interact.

Rounsevell et al. (2012, p. 260) describe the ABM process as an iterative cycle in which conceptual and theory-driven models are used to evaluate key research questions and to determine conditions under which spatial and temporal patterns of behavior can emerge. These results may provide new knowledge about how the SES under study may change in simple contexts.

To define the assumptions of an ABM, the following key aspects must be considered:

1. **Definition of agent attributes**: characteristics of an agent (e.g., age, income, education, marital status) influencing decision-making, identified through social research methods or from existing data. Agent attributes may act to enable or constrain agent behavior and may alter decisions made by agents. Furthermore, knowledge of these attributes by other agents may enable or prevent agent interactions.
2. **Definition of decision-making strategies**: (1) Heuristic (decision trees) (if… then…else), which closely correspond to conceptual decision-making models mixing qualitative and quantitative data (expert knowledge); (2) utility maximization and bounded rationality; (3) learning and adaptation as the ability of agents to retain knowledge and change their behavior.
3. **Condensation of agent typologies**: To simplify SES models: (1) inductive analysis (clustering, participative approaches, use of national social surveys, statistical databases); (2) deductive reasoning (based on theory or expert opinion, cultural theory, plural rationality, etc.).
4. **Changes in agent population**: (1) Population changes over time (by using census or housing survey data); (2) using closed systems with fixed agent population.

Integrated Socioecological Modeling

Integrated studies of coupled human and natural systems have offered unique interdisciplinary findings that cannot be gained from ecological or social research alone (Liu et al. 2007). Thus, there is a need for integrated socioecological models that incorporate knowledge from different disciplines to gain a better understanding of the interrelations among various drivers behind the trajectories of SESs and thereby capture the complexity of these systems as a whole (Milne et al. 2009).

In addition to being developed in a transdisciplinary environment by incorporating insights from disciplines such as Sociology, Ecology and Economics and sometimes aided by the participation of stakeholders, integrated socioecological models often combine different modeling techniques, such as agent-based, GIS (for details, see Method Précis on Using Geographic Information Systems in Social Ecology in this volume) and stock-flow models (see Method Précis on Carbon Accounting).

Integrated socioecological models have been shown to be particularly useful for simulating complex systems that are difficult to conceptualize. Constructing integrated models can be of great help in promoting interdisciplinary integration and mutual learning in interdisciplinary teams (Gaube et al. 2009).

References

Epstein, J. M. (2008). Why model? *Journal of Artificial Societies and Social Simulation*, *11*(4), 12. Retrieved from http://jasss.soc.surrey.ac.uk/11/4/12.html

Gaube, V., et al. (2009). Combining agent-based and stock-flow modelling approaches in a participative analysis of the integrated land system in Reichraming, Austria. *Landscape Ecology*, *24*(9), 1149–1165. Retrieved from 10.1007/s10980-009-9356-6

Gilbert, N., & Troitzsch, K. G. (2005). Simulation for the social scientist, McGraw-Hill International.

Liu, J., et al. (2007). Complexity of coupled human and natural systems. *Science*, *317*(5844), 1513–1516. Retrieved from http://www.sciencemag.org/content/317/5844/1513.abstract

Milne, E., Aspinall, R., & Veldkamp, T. (2009). Integrated modelling of natural and social systems in land change science. *Landscape Ecology*, *24*(9), 1145–1147. Retrieved from 10.1007/s10980-009-9392-2

Rounsevell, M. D. A., Robinson, D. T., & Murray-Rust, D. (2012). From actors to agents in socio-ecological systems models. *Philosophical Transactions of the Royal Society B: Biological Sciences*, *367*(1586), 259–269. Retrieved from http://rstb.royalsocietypublishing.org/content/367/1586/259.abstract

Chapter 26
Time Use, Gender and Sustainable Agriculture in Austria

Barbara Smetschka, Veronika Gaube and Juliana Lutz

Abstract Available time—as much as available money—governs the everyday decision-making of individuals concerning their living space, consumption patterns and means of transportation. Time-use research can serve as an integrative means to encompass social aspects in sustainability research, to integrate a gender perspective in sustainability research and to enable transdisciplinary work. We show how we worked toward these objectives in the project GenderGAP. Here, time use is a crucial factor in decisions concerning production strategies on Austrian farms. Farmers aim to avoid longer working hours and less income than employees from other sectors. Technological change can diminish the workload of farmers, mainly in regions favorable to large-scale industrialized agriculture. Sustainable agriculture with a focus on mixed production and maintenance of cultural landscapes in a lively region should not place a greater burden on the farmers. If organic and small-scale farming increases the workload on women in a traditionally gendered working environment, there are two options for addressing the issue. Either farmers opt for less sustainable methods of production or cease agricultural activity entirely, or farmers opt to adapt to socioeconomic changes and find ways of producing for the increasing market for sustainable products with a new work organization that is attractive to young people and does not place a greater burden on farm women than on men.

Keywords Time use · Participatory modeling · Transdisciplinary research · Sustainable agriculture · Gender studies · Sustainability triangle

B. Smetschka (✉) · V. Gaube · J. Lutz
Institute of Social Ecology (SEC), Alpen-Adria University,
Schottenfeldgasse 29, 1070 Vienna, Austria
e-mail: barbara.smetschka@aau.at

V. Gaube
e-mail: veronika.gaube@aau.at

J. Lutz
e-mail: julia.lutz@aau.at

H. Haberl et al. (eds.), *Social Ecology*, Human-Environment Interactions 5,
DOI 10.1007/978-3-319-33326-7_26

26.1 Time Use in Transdisciplinary Sustainability Research?

Time is abundant or scarce, flows slowly or rapidly and is spent according to individual preference or entirely according to external constraints. We all have different experiences and perceptions of time, and these vary widely over space and history. Nevertheless, no one would deny the importance of time in our everyday lives. From this point of departure, we take a closer look at the following statement: time seems to be an important factor in decision-making. It is a resource as well as a means that structures how we manage other resources. Consumption, production and reproduction patterns are influenced by our notion of time as well as by our time-use patterns and time management.

Sustainability research needs forms of knowledge production involving various actors to enhance the probability of implementation of innovative and sustainable solutions. This requires a common effort of stakeholders, experts and scientists from various disciplines in an inter- and transdisciplinary research design. Scientists can support this process by providing (1) data on different scales and issues; (2) knowledge on ecological, economic and social contexts; and (3) skills in transdisciplinary process-oriented methods.

The first step in any transdisciplinary research process is a joint effort to frame the problem to be solved (Dressel et al. 2014; Hirsch Hadorn et al. 2008; Pohl and Hirsch Hadorn 2007). Life-world problems and scientific questions have to be explained, translated and discussed among all involved parties until a decision on the focus of one concrete aspect of a complex problem to be tackled can be reached. Deliberation and decision-making can be much easier if made based on intermediary concepts or even terms (Vinck 1999). Time and how we use it serves as such a common term. Time does not belong to one of the different spheres, so it can be related to very different perspectives, be it individual everyday experience, professional expertise or perspectives from the Social or Natural Sciences or Humanities. 'Time', therefore, can be used as an intermediary concept that is both a means for translation and something to which other aspects of a complex problem can be related.

In GenderGAP,[1] the project that illustrates our approach, the time use of Austrian farmers was the focus of a series of in-depth interviews, field excursions and four workshops. We established a focus group consisting of eight farm women and four experts from the Agricultural Chamber. In these workshops (see upper line of yellow bubbles in Fig. 26.1), we addressed the problems of small-scale agriculture in Austria and discussed the computer model, future scenarios and recommendations on actions to be taken.

[1]'GenderGAP: A gender perspective on the impact of the reform of EU's Common Agricultural Policy' 2005–2009 was financed by the Austrian Ministry of Science in its program on transdisciplinary methods, TRAFO.

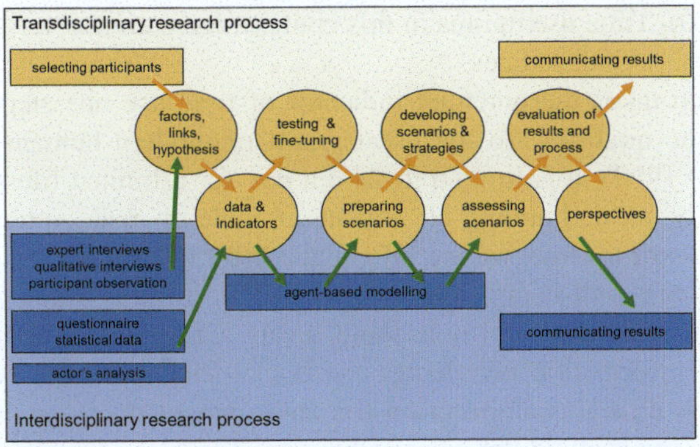

Fig. 26.1 Overview of the transdisciplinary research process. (Reproduced from Smetschka et al. 2014; reprinted with permission)

26.2 Time Use in Social Ecology?

Socioecological research focuses on the patterns of nature-society interactions and the possibility of changes within these patterns. Changes in consumption and production, mobility and transport and material, energy and land use are therefore central to socioecological research. The sustainability triangle helps draw attention to the interlinked character of the economic, ecological and social factors driving these interactions in their interrelations and individually (see Fig. 26.2 and Sect. 5.3.1).

Social factors include individual well-being and overall social welfare (Munasinghe 2013). Changes can be measured with demographic and economic data. However, the need for better measures of the quality of life (QOL) is acknowledged and addressed by international bodies (Costanza et al. 2007;

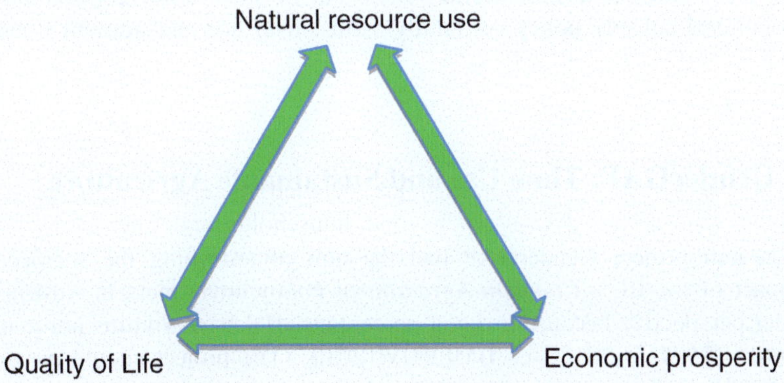

Fig. 26.2 The 'sustainability triangle'

Eurofound 2010). Time use figures in this context; recreation or leisure time is one of the indicators of QOL indices.

Our approach takes the potential indicator of time use one step further. Time has some unique qualities. To the great regret of modern humans (Grönemeyer 1993; Nowotny 1989), it is limited to 24 h a day for a limited lifespan. However, it is also equally divided among all humans, female or male, young or old, poor or rich. With regard to time, we are all equals; thus, we can easily talk about it as a dimension of our well-being. Available or lacking time is a key factor in many decisions at the individual and household levels. Therefore, time-use data could provide useful information on changes in social inequality and socioeconomic structure and on changing impacts on the environment due to society's material and energy use (see Method Précis on Functional Time Use Analysis).

How can time be linked to the core concepts of Social Ecology, social metabolism and colonization? In various research projects, we attempt to acquire data on how time use and material or energy use are linked via consumption and production processes and are therefore useful for studies on society's metabolism. The colonization of natural systems is the basis for this social metabolism. Colonizing activities can be regarded as interventions into natural systems set by individuals, households, the economy and society. Human activities in total can be classified as activities set to (re-)produce the individual, the family/household, the economy and society as a whole. These activities have an impact on natural systems, they imply a particular use of materials and energy, and they endure for a certain amount of time. The time used for certain activities, therefore, can provide information about changes in metabolism and colonization (see Method Précis on Functional Time Use Analysis). Moreover, as explained above, time use can serve as an intermediary concept among researchers from the Natural and Social Sciences as well as stakeholders and experts in transdisciplinary research. Thus, time-use analysis can play an important role in informing and supporting actors and social systems on actions and decisions toward more sustainable solutions.

The farmers in the project GenderGAP face the problem of how best to continue as a family-run farm. We developed a model that can link time-use-related scenarios for farms and agricultural stocks and flows to analyze the impacts of an agricultural subsidy policy on family farms, land use and nutrient flows at a regional level (Gaube et al. 2009a).

26.3 GenderGAP: Time Use and Sustainable Agriculture

The research project GenderGAP had the aim of analyzing the socioecological impact of the EU's *Common Agricultural Policy* on farmers in Austria from a gender perspective because a rising percentage of Austrian farms are managed by women (36 % in 2012; see BMLFUW 2013). The project should provide an example of both the integrative and the analytical assets of time use for transdisciplinary research and socioecological research.

Land use and agricultural activities are crucial for sustainable development. Although the economic importance and the extent of agricultural activities have decreased in Austria over the last century, as in all industrialized countries, they are still important for food production and for their impact on the Austrian landscape. The intensification of agriculture with an increased use of energy and materials in some areas and reforestation in less-favorable areas may be providing larger carbon sinks but less 'cultural landscape' to be used for recreation and tourism—but these are the two sides of the same development (see Chap. 21).

26.4 Keeping Small-Scale Farmers in Agricultural Production: A Case Study

Despite global trends to industrialize agriculture, small-scale family farms are still important in Austria. According to Statistics Austria, in 2010, 71 % of farms worked on less than 30 ha. This is mostly due to geography; it is hardly possible to industrialize agriculture in alpine regions with steep slopes and limited accessibility. It is also due to the fact that, in addition to a tradition of value and pride in nationally produced milk and meat, organic farms are on the rise, as are their products.

In GenderGAP, we studied how farmers make decisions concerning their agricultural production. We wanted to know how these decisions are influenced by factors internal to the farm (such as the social, economic and environmental situation) and by factors external to the farm (such as the EU and national systems of subsidies, prices, regional infrastructure and land available). The objective of the project was to build a computer model in a participatory way to gain better systems knowledge on decision-making by small-scale family farms and to support farmers and experts in developing more sustainable solutions for the problems encountered by these farms (Smetschka et al. 2014).

We used data on farming and domestic working time from two comprehensive studies from 1979 and 2002 (Blumauer et al. 2002; Wernisch 1979) and from agricultural statistics (Handler et al. 2006; Pöschl 2004) for Austria. This material was verified through several workshops and a series of qualitative interviews with men and women farmers and adapted for use in the model.

The study was conducted in a region of Lower Austria that represents the varieties of Austrian landscapes and the main agricultural production types. We chose two villages as case studies for the model:

- Hainfeld, in the southern mountainous part of the region, is a village with 4000 inhabitants. It lies at 439 m a.s.l. and has 45 % forest area. The number of farms and people working in agriculture has diminished over the last few decades. In 2005, the village had 100 farms, 60 % of them farmed full-time, with dairy, forest or mixed production and 30–100 ha area per farm.

- Nussdorf, in the north, has 1500 inhabitants. It lies at 249 m a.s.l. and has a milder climate, more fertile soils and less forest area. We started the model with 107 farms, 40 % full-time, with mainly cropland and permanent cultures—in this case, vineyards and 1–20 ha area per farm.

Although they represent the wide range of Austrian agriculture, the farms in both villages face similar problems. Structural agricultural changes toward intensification of farm production with a higher degree of technology and more energy input resulted in fewer but larger farms over the last few decades. Small-scale family farms have little chance of following this trend. They either give up production or opt for part-time farming. They need to find niches and diversify production. They depend considerably on national and EU agricultural subsidies. Beyond but related to these economic constraints, all farmers interviewed referred to the high demand on physical labor and the unwillingness of their heirs to continue with this kind of farm production as problems. A popular TV show, *Farmer Wants a Wife*, sums up the problem of small-scale family farms: the number of young people willing to work more and harder than the average employee for less income is in decline.

Farmers' associations such as the Women Farmers Association, our project partner, look for solutions to such problems. For them, sustainability primarily means finding ways to sustain the family farm over generations. Family farms do not primarily want to expand territory or profits. Their main goal is to have enough income to be able to work the land they inherited and pass it on to their children. They want to keep producing their own food. Some farmers actually told us that they want to keep their livestock, even knowing that it would mean less work and possibly more money to change production.

26.4.1 Time Use on the Farm

Individuals use their time for production and reproduction in four areas of life (Table 26.1) (adapted from Fischer-Kowalski et al. 2010; see also Method Précis on Functional Time Use Analysis). To reproduce our life daily, we spend time reproducing our own person/body and our family/household, and we support the systems of economy and society so that they can be reproduced as a whole. This concept helps broaden the definition of working time, which in the current

Table 26.1 The four areas of time use for production and reproduction

Reproduction of	Encompasses	Division of labor possible	Working hours in model
Person	Individual sphere	No	No
Household	Family	Yes	Yes
Economy	Farm business	Yes	Yes
Community	Society, politics & culture	Yes	No

model includes family and household work, all types of para-agricultural tasks and farm work per se. The model thus includes reproduction and subsistence activities, which are often the activities of women (see also Bennholdt-Thomsen 1999; Oedl-Wieser 1997), as working time.

We classify time as divided among the following:

- personal time
  - sleeping, eating, hygiene, regeneration
  - leisure time
  - education/training
- household and caring time
  - household
  - caring for household members
- working time
  - agricultural work
  - para-agricultural tasks (garden, food preservation, direct marketing, farm holidays, educational services)
  - extra-agricultural work (other employment)
- community time
  - community events
  - voluntary and honorary work
  - activities in associations
  - political engagement

Studies on the situation of women in farming show that the traditional gender-related division of labor is changing slowly because women farmers tend to have a higher workload due to their combination of various roles and responsibilities (Inhetveen and Schmitt 2004; Oedl-Wieser 2008).

26.4.2 How Time Use Is Conceptualized in the Model

We built an agent-based model (ABM) with family farms as agents. To address the issue of many working hours in farm production and its importance for decisions about changes in agricultural production, we introduced time use as data to represent the quality of life (QOL) of farmers. Each of the agents/farms was described along the sustainability triangle with data on income, land conditions and availability, livestock and family time budgets. Figure 26.3 shows the links between the three corners of the triangle and the ways they influence each other. These impacts can be read in a cycle: agricultural production is constrained by the amount of hours available in the family time budget, it produces income, and this income is an important factor in enabling or constraining QOL.

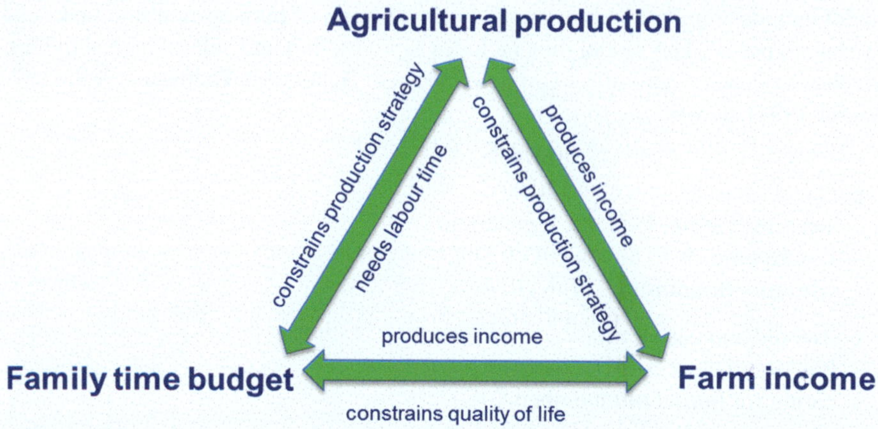

Fig. 26.3 The 'sustainability triangle' for the farm agent. (Reproduced from Smetschka et al. 2014; reprinted with permission)

The family time budget or time use on a farm varies depending on requirements of the type and extent of agricultural production and the size and composition of the household. To operationalize this for the farms in the model, we defined three categories of time:

- **Available time**: 24 h short of 10.5 h for biophysical reproduction for an adult
- **Free time**: personal reproduction other than biophysical reproduction; this is the residual between available and working time
- **Working time**: has three subcategories
  - Maximal possible working time
  - Necessary working time
  - Desired working time

Maximum possible working time could be equal to available time or 13.5 h per day. It would be perilous for a person's well-being if the maximum available time were used for working. Drawing from literature and statistical data, we could establish the necessary working time for different agricultural tasks in direct relation to the size of the production in terms of area and number of livestock. According to time-use data and interviews, we established values for household and caretaking time in relation to family size and composition. The default desired working time was set for a fully working adult at 2160 h/year (270 days × 8 h).

Drawing from the discussions in the focus groups, we constructed decision trees for the agents/farms. During a model run, the agents make annual decisions on changes in production and the utilization of working time. These decisions are influenced by the following:

- external conditions
  - different forms of subsidies
  - prices for conventional or organic products
  - costs for conventional or organic products
  - availability of extra land on a rental market
  - soil conditions
  - climate conditions
- internal structure
  - the number of adults, children and elderly people on the farm
  - the degree of agricultural training and open-mindedness to innovative behavior
  - special events occurring in family life: birth, death, marriage, separation
  - leisure time preferences
  - attractiveness of extra agricultural labor market
  - income preferences

The farmers and experts in the workshops discussed which of these conditional factors were important and interesting for further discussions. The 'controls' about which agreement was achieved were displayed on the interface of the computer model. They work as sliders and can be changed with each model run. It is therefore possible for a group of people to design a model run and discuss its outcomes (see Fig. 26.4).

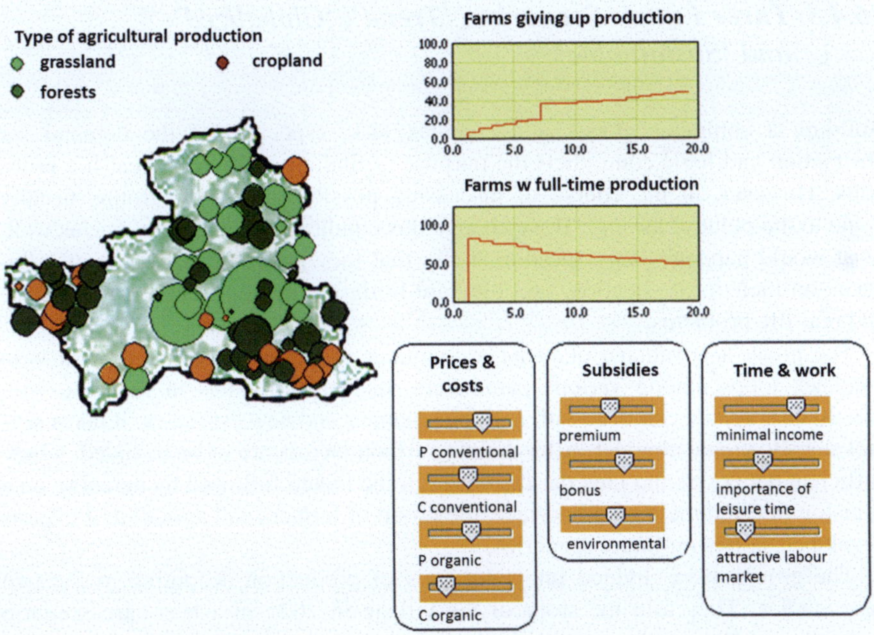

Fig. 26.4 Model interface: sustainability scenario in the village of Hainfeld. (Reproduced from Smetschka et al. 2014; reprinted with permission)

For example, a farm could decide to give up dairy products in favor of raising cattle. This decision would result in fewer working hours on the farm and the possibility of taking on external employment. Dairy production is mainly female work, so the family would be looking for employment for a female member of the family. If dairy farmers do not change their production, it is primarily men who look for outside work. One farmer described the situation poignantly: 'If the woman finds work, they will not go on with extensive dairy farming, having the cows out on the meadows, leaving the calves with the herd'.

Farmers' possible alternatives to solve the problem of too much work for too little income, as discussed in the focus groups and described in the literature, are intensification or extensification, expansion or contraction and farm abandonment or changing production. If intensification and expansion are not options for small-scale family farms because of time-use constraints and if the family does not want to abandon agricultural production altogether, the options are to move to diversification of production or to supplementary income activities, such as taking on external labor or direct marketing.

Our model results clearly show the limits of such an approach. Aiming for a mixed form of production is ecologically and economically sound but demands more labor. Diversifying and settling for niche products often means a higher workload for farm women.

26.4.3 *Three Future Scenarios: 'Trend', 'Globalized' and 'Sustainable'*

Building a computer model is a complex task, especially if the demand for abstraction and logic contradicts the purpose of finding solutions to real-life problems. However, in the course of discussing possible solutions, groups usually come to the point of asking, 'If we change this condition and the other one as well, what would happen?' This question shows that they grasp the relevance of influences on their local situation, and they are bridging the gap between abstraction and real-life problems.

The greatest advantage of using a computer model is its ability to deal with feedback loops among various parameters. Its versatility lies in the ease with which the software can deal with multiple changes and new estimates. It takes several steps to create a model. After defining which factors are to be included, where sufficient data exist and how the factors are to be linked followed by agreeing on a decision tree and the model interface, the group of farmers and agricultural experts developed stories on future scenarios.

The groups were divided into two types of production according to the two case studies. They told the story of the village in 2025 in a best-case scenario that attempted to encompass all the important factors agreed upon in the previous workshop. They came up with a rich picture of a thriving region with good

Table 26.2 Settings for scenarios

	Trend	Globalized	Sustainable
Hourly wages extra-agricultural	10	8	12
Basic subsidies	−25 %	−80 %	Constant
Subsidies for sustainable production	−15 %	−80 %	+10 %
Price of conventional product	Constant	−20 %	+10 %
Cost of conventional product	Constant	−20 %	+10 %
Price of organic product	Constant	−20 %	Constant
Cost of organic product	Constant	−20 %	Constant

infrastructure and optimal conditions for agricultural production in terms of product prices for organic and/or niche products. The opposite of this scenario was described as a future in which agricultural production was highly market oriented, with few subsidies and lower wages because of globalization. The stories were translated into sets of parameters and labeled TREND, GLOB and SUST Scenario as shown in Table 26.2.

The results of a model run within a chosen scenario can be seen on the model interface. The two yellow graphs on the screenshot below (Fig. 26.4) show how many farms leave agricultural production and how many change to part-time farming, as seen in this example for Hainfeld.

In all three scenarios, full-time farms decline by at least 40 %. Between 25 and 40 % of farms will abandon agricultural production altogether. In the GLOB scenario, most surviving farms cultivate forests. Only in the SUST scenario can land-use diversity with a mix of grassland, forest and cropland be maintained. The cultivated agricultural area diminishes by 50–80 %. Animal stock is reduced by 50–90 %.

The model runs showed that regional labor market, regional infrastructure, production costs and prices/subsidies are the sensitive parameters. Quality of life, operationalized in the parameters ‘minimum income’ and ‘leisure time’, was not set within the scenarios but can be chosen individually per model run. The TREND scenario, with a high preference for leisure time, has the same results as the GLOB scenario: 70 % of the farms abandon agricultural production. Most farms with mixed agricultural production, and hence, most cultivated area, are found in the SUST scenario, which also provides a higher amount of leisure time for the farmers.

Examining working hours more closely, it becomes clear that although the SUST scenario seems favorable for the environment and farmers alike, the work demand for women is higher in this scenario than the work demand for men. Women farmers are not likely to opt for this solution. In the discourse on the feminization of agriculture (Boserup 1970; Inhetveen and Schmitt 2004), the high labor burden on females is often linked to rural poverty and food insecurity. Sustainable solutions to cultivating landscapes and producing food will not succeed in Austria if a higher workload is demanded from women farmers.

The farmers in our working group agreed upon the necessity of rethinking and changing the sexual division of labor on a farm. Training and agricultural education offer a means to promote this change. Regional infrastructure to assist families with caring duties is another way to manage the unequal distribution of reproductive work. Working opportunities within the region foster decisions toward a sustainable form of agricultural production and allow for short-term changes according to personal needs and preferences. In short, a higher quality of life on a farm would require regional development, regional infrastructure and agricultural education that allows for the participation of women and men, young and old farmers and that might result in their finding solutions to halt the decline in small-scale family farms.

Inviting women to participate in finding solutions, making decisions and representing farmers is another measure to curb inequality and to identify inadequate policies to promote sustainable development (Oedl-Wieser 2008). Finding new ways of collaboration on farms, between farms (Gaube et al. 2009b) and among farms, retailers and civil society (consumers) would be essential for creating more sustainable methods of agricultural production and ways of spending time.

26.5 Time-Use and Sustainability Research: Conclusions and Further Cases

Sustainability transitions analyzed from the perspective of time-use research enhance our knowledge of the social aspects of the changes involved in the transition. Changes in time-use patterns are indicators of changes in society in terms of changes in inequality, inclusion and the participation of different groups in societal production and reproduction. They can also be used as indicators of changes in society-nature interactions via changing material and energy use.

In GenderGAP, we show that support for sustainable development requires a close look at the quality of life in terms of time pressure. The sustainability triangle can also be interpreted as a vicious circle: more income equals more quality of life and means more resource use. How could decoupling this inherent mechanism work? We need to find alternatives that combine a higher quality of life with less impact on the environment. There are many good ideas and approaches along these lines (Jackson 2009; Schaffer and Stahmer 2006). Working less is one solution promoted by a number of authors (Hartard et al. 2006; Haug 2008; Schor 2010).

Urban gardening can be viewed as a para-economic activity: producing organic food and spending time in a way that brings meaning and quality of life and reduces the need for money from employed work. It may be one example of new ways of collaboration. Community-supported agriculture may be another.

Enriching studies on sustainable consumption with socioecological thinking and time-use analyses guides our research in the next projects (see box below).

We hope that with a focus on time use, we can shed light on farmers' decisions for sustainable production and can use this approach to gain a better knowledge of daily decisions regarding material and energy use.

In the ongoing project 'UTE—Urban Time and Energy' (see also Chap. 25), we try to establish the link between time use and energy use in Vienna. We owe much knowledge on this topic to Mikko Jalas (Jalas 2008) and Angela Druckman (Druckman et al. 2013), who performed similar research in Finland and the UK. We are planning to build a computer model on the time use and energy use of households in Vienna. We invite citizens to participate in discussing and finding solutions regarding their problems with time use in their household and in their city. If we know more about how time-use patterns are linked to energy and material use, we can ask the next questions: how does urban infrastructure shape the patterns of how we spend our time through, for instance, urban planning, housing and transport infrastructure? How can these patterns be changed in a way that allows for both a higher quality of life and lower material and energy use?

References

Bennholdt-Thomsen, V. (1999). *Subsistenzkultur und bäuerliche Ökonomie*. Diskussionsbeiträge zur Subsistenz, Bielefeld: ITPS.

Blumauer, E., Handler, F., & Greimel, M. (2002). *Arbeitszeitbedarf in der österreichischen Landwirtschaft*. Irdning: Bundesanstalt für alpenländische Landwirtschaft Gumpenstein.

BMLFUW. (2013). *Grüner Bericht 2013: Bericht über die Situation der österreichischen Land- und Forstwirtschaft*. Wien: Bundesministerium für Land- und Forstwirtschaft, Umwelt und Wasserwirtschaft. Retrieved from http://www.gruenerbericht.at/cm3/download/finish/154-gb2013/852-s-326-1-ant-frauen-alter/0.html

Boserup, E. (1970). *Women's role in economic development*. New York: St Martin's Press.

Costanza, R., Fisher, B., Ali, S., Beer, C., Bond, L., Boumans, R., et al. (2007). Quality of life: An approach integrating opportunities, human needs, and subjective well-being. *Ecological Economics, 61*, 267–276.

Dressel, G., Berger, W., Heimerl, K., & Winiwarter, V. (2014). *Interdisziplinär und transdisziplinär forschen*. Praktiken und Methoden. Bielefeld: transcript.

Druckman, A., Buck, I., Hayward, B., & Jackson, T. (2013). Time, gender and carbon: A study of the carbon implications of British adults' use of time. *Ecological Economics, 84*, 153–163.

Eurofound. (2010). How are you? Quality of life in Europe. Dublin: Eurofound.

Fischer-Kowalski, M., Singh, S. J., Ringhofer, L., Grünbühel, C., Lauk, C., & Remesch, A. (2010). Sociometabolic regimes in indigenous communities and the crucial role of working time: A comparison of case studies. Social Ecology Working Paper 121.

Gaube, V., Kaiser, C., Wildenberg, M., Adensam, H., Fleissner, P., Kobler, J., et al. (2009a). Combining agent-based and stock-flow modelling approaches in a participative analysis of the integrated land system in Reichraming, Austria. *Landscape Ecology, 24*, 1149–1165.

Gaube, V., Reisinger, H., Adensam, H., Aigner, B., Colard, A., Haberl, H., et al. (2009b). Agentenbasierte Modellierung von Szenarien für Landwirtschaft und Landnutzung im Jahr

2020, Traisental, Niederösterreich. *Verhandlungen der Zoologisch-Botanischen Gesellschaft in Österreich, 146*, 79–101.

Grönemeyer, M. (1993). *Das Leben als letzte Gelegenheit. Über Sicherheitsbedürfnisse und Zeitknappheit*. Darmstadt: Wiss. Buchgesellschaft.

Handler, F., Stadler, M., & Blumauer, E. (2006). *Standardarbeitszeitbedarf in der österreichischen Landwirtschaft*. Ergebnis der Berechnung der einzelbetrieblichen Standardarbeitszeiten, (vol. 48). Wieselburg: Francisco Josephinum.

Hartard, S., Schaffer, A., & Stahmer, C. (2006). *Die Halbtagsgesellschaft. Konkrete Utopie für eine zukunftsfähige Gesellschaft*. Baden- Baden: Nomos Verlag.

Haug, F. (2008). *Die Vier-in-einem-Perspektive. Politik von Frauen für eine neue Linke*. Hamburg: Argument.

Hirsch Hadorn, G., Hoffmann-Riem, H., Biber-Klemm, S., Grossenbacher-Mansuy, W., Joye, D., Pohl, C., et al. (2008). *Handbook of transdisciplinary research*. Stuttgart, Berlin, New York: Springer.

Inhetveen, H., & Schmitt, M. (2004). Feminization trends in agriculture: Theoretical remarks and empirical findings from Germany. In H. Buller, & K. Hoggart (Eds.), *Women in the European Countryside. Perspectives on Rural Policy and Planning* (pp. 83–102). Hampshire: Ashgate Publishing.

Jackson, T. (2009). *Prosperity without growth. Economics for a finite planet*. London: Earth Scan.

Jalas, M. (2008). The everyday life context of increasing energy demands. Time use survey data in a decomposition analysis. *Journal of Industrial Ecology, 9*, 129–145.

Munasinghe, M. (2013). Sustainability triangle. In *Encyclopedia of the Earth*. Retrieved from http://www.eoearth.org/view/article/156365/

Nowotny, H. (1989). *Eigenzeit. Entstehung und Strukturierung von Zeitgefühl*. Frankfurt: Suhrkamp.

Oedl-Wieser, T. (1997). *Emanzipation der Frauen am Land. Eine explorative Studie über Ambivalenzen und Lebenszusammenhänge*. Bundesanstalt für Bergbauernfragen (Rep. No. 40). Wien: Bundesanstalt für Bergbauernfragen.

Oedl-Wieser, T. (2008). The rural gender regime: The Austrian case. In I. Asztalos Morell, & B. Bock (Eds.), *Gender regimes, citizen participation and rural restructuring* (pp. 283–297). Oxford: Elsevier.

Pohl, C., & Hirsch Hadorn, G. (2007). Systems, targets and transformation knowledge. In G. Hirsch Hadorn, H. Hoffmann-Riem, S. Biber-Klemm, W. Grossenbacher-Mansuy, D. Joye, C. Pohl, U. Wiesmann, & E. Zemp (Eds.), *Principles for designing transdisciplinary research* (pp. 36–40). Munich: Oekom-Verlag.

Pöschl, H. (2004). Frauen in der Landwirtschaft. Ein nachrangiges Thema in den Agrarstatistiken. *Wirtschaft und Statistik, 9*, 1017–1027.

Schaffer, A., & Stahmer, C. (2006). Women's GDP – A time-based Input-output Analysis. *Schweizerische Zeitschrift für Volkswirtschaft und Statistik, 142*, 367–394.

Schor, J. B. (2010). *Plenitude: The new economics of true wealth*. New York: Penguin Press.

Smetschka, B., Gaube, V., & Lutz, J. (2014). Working time of farm women and small-scale sustainable farming in Austria. In M. Fischer-Kowalski, A. Reenberg, A. Mayer, & A. Schaffartzik (Eds.), *Ester Boserup's legacy on sustainability: Orientations for contemporary research* (pp. 221–238). Dordrecht: Springer.

Vinck, D. (1999). Les objets intermédiaires dans les réseaux de cooperation scientifique. Contribution à la prise en compte des objets dans les dynamiques sociales. *Revue française de sociologie, 40*(2), 385–414.

Wernisch, A. (1979). Wieviel arbeitet die bäuerliche Familie? *Der Förderungsdienst, 2*, 44–51.

Method Précis: Functional Time Use Analysis

Lisa Ringhofer and Marina Fischer-Kowalski

With functional time use (FTU), we follow the same systemic logic as with material and energy metabolism while largely complying with the generally established classifications used by survey-based time use studies (e.g., Robinson and Godbey 1997). We treat human time as a key resource of social systems. Although each individual has 24 h per day at his or her disposal, the *stock* of available time in a social system (e.g., a village) is 'created' by population size, reproduction (the birth of new individuals) and the average number of years its members stay in the community. Concerning the *flows* of human time, one fraction of daily time use is expended on certain metabolic functions (such as sleeping or eating) necessary for an individual's basic reproduction, whereas the remainder is used according to sociocultural norms, economic necessities or simply individual preferences. We distinguish among flows serving four functional subsystems, each of which requires time for reproduction: the *person system,* the *household system*, the *economic system* and the *community system* (for a discussion on methodology, see Singh et al. 2010).

Such a systemic analysis offers a perspective on how much human time is available and what it is used for in the *whole* social system, thereby helping us to understand the specific *opportunities* and *constraints* a society faces in its interaction with the natural environment. At the same time, because the lifetime/labor time ratio is calculated for all the age/sex groups in this system, FTU sheds light on the 'labor burden' or 'time poverty' some of these groups bear with regard to important aspects of social inequality.

Classifications and Methods

The *person system* functionally serves personal reproduction and includes all those activities that cannot be delegated or 'outsourced' to others. It holds all the physiologically necessary functions for a person's self-reproduction, such as sleeping and eating, and it encompasses functions for extended reproduction, such as studying, leisure activities and idling. Breaking it down into single activities, the *person system* comprises sleeping, eating, hygiene, rest and idleness, leisure activities and study and education.

The *household system* involves those personal reproduction functions of its members that need or allow for collaboration and a division of labor The household system is typically organized as an exchange of unpaid labor according to the social norms regulating age and gender roles in the local system. Time use for the household system contains the following sub-activities: care for dependents, food preparation, house building, repair/maintenance work and domestic chores.

The time invested in reproducing the *economic system* is what we usually refer to as 'labor time'. The economic system implies and relies upon a social division of labor and, at least in market-based societies, usually involves monetary transactions. In subsistence societies, economic activities may simply be an additional function of households or communities. The economic system entails the following activities: agriculture, hunting, fishing, gathering, trading, wage work, kitchen work, gardening, manufacture of handicraft and animal husbandry.

The *community system* is the reference system for activities contributing to the reproduction of services on the community level, reciprocal relationships, social cohesion, politics and religion. It subsumes public sports and games, visiting friends and relatives, ceremonies and festivals, communal work and political participation.

FTU data collection in the field may use various methods. In some case studies, 'time-frame' analysis was applied, focusing only on certain activities (such as building a house or repairing a boat) of functional importance. The duration of these activities and their participants (in terms of gender and age) were recorded and subsequently weighted according to their annual frequency. This method is practical for understanding the importance of specific focal activities, but it hardly allows for constructing a comprehensive time budget on the system's level. In more recent empirical studies, time use data were collected more systematically, with samples of people and households observed for days during waking hours. In addition to these samples, spot checks by direct observation (who does what at a certain time) allow the creation of independent statistical estimates that can—in combination with household interviews—be used for cross-checking. Optimally, people have to be 'shadowed' at different times of the year, thus covering seasonal differences. The reliability of these measurements as well as the annual estimates that need to take into account the variability of the number of people present in the system (because of some leaving for seasonal work, for example) and the variability of seasonal tasks (such as sowing or harvesting) still requires some improvement. However, repeated application and refinement of these methods across various case studies seems to produce increasingly robust results, which we illustrate for one of our latest case studies below.

Empirical findings from Campo Bello (see Ringhofer 2010, 2013), an indigenous subsistence community in the Bolivian Amazon, provide insights into the share of labor of each age/sex group from a system-level perspective (Table 26.3). The economic system predominantly draws on male labor; however, 38 % of all working hours are supplied by women. For household work, the share of female labor amounts to three-quarters of overall time investment in this system. Taken together, the female share in labor hours exceeds the share of the female population (55 % of labor hours relative to a 46 % share in population).

Children of both sexes (6–15 years) contribute quite significantly to labor activities within the economic system, accounting for nearly one-third. The same is true for the 'household economy'. One-third of all the labor invested in the upkeep of the household system is provided by children. Clearly, children perform lighter work and may perform it less efficiently than adults, but these results lend support to the argument that in agricultural communities, children have high use value in terms of labor contribution.

Table 26.3 Daily labor time invested by age/sex groups in Campo Bello (Bolivia) as observed in 2004 and 2006

	Boys 6–15	Male adults 16–60	Male adults > 60	Girls 6–15	Female adults 16–60	Female adults > 60	Total hours per day	Share of children	Share of females
Population Numbers (n)	37	41	6	27	38	6	3720	41.3 %	45.8 %
Household System (hours/day) %	66.6	45.1	13.2	78.3	258.4	30	491.6		
	13.5 %	9.2 %	2.7 %	15.9 %	52.6 %	6.1 %	13.2 %	29.5 %	74.6 %
Economic System (hours/day) %	96.2	237.8	15	70.2	133	13.8	566		
	17.0 %	42.0 %	2.7 %	12.4 %	23.5 %	2.4 %	15.2 %	29.4 %	38.3 %
Total Daily Working Time %	**162.8**	**282.9**	**28.2**	**148.5**	**391.4**	**43.8**	**1057.6**		
	15.4 %	26.7 %	2.7 %	14.0 %	37.0 %	4.1 %	28.4 %	29.4 %	55.2 %

In effect, the labor demand on women of all age groups competes with the time to spend on personal reproduction activities, such as sleeping, eating, studying, leisure and idleness (Fig. 26.5).

Thus, FTU also serves as a tool for assessing the gender impact of development innovations. In the case of Campo Bello, the use of labor-efficient rice seeders, a technology initially introduced as part of a larger development program (and that has since turned into a highly solicited agricultural tool), does not particularly benefit women. Rice seeders only increase the efficiency of rice planting, an activity that is largely undertaken by men. Other labor-intensive crop management tasks such as weeding and harvesting largely remain in the hands of women—with no particular labor-saving devices introduced. A similar conclusion applies to the innovation of nylon fishing nets. These labor-saving devices are only used by men; women still use traditional fishing armory, such as hooks and lines and machetes. However, the mending of the fishing nets is assigned to women, thereby adding to their labor burden rather than reducing it.

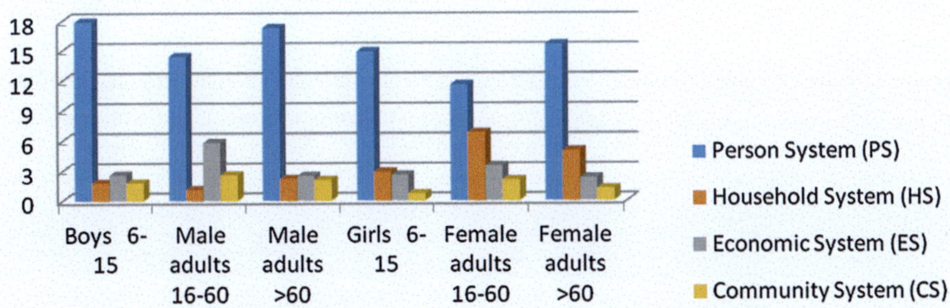

Fig. 26.5 Age/sex group segregation in time use, Campo Bello 2004 and 2006

Other uses of this methodology allow for structural comparisons between different social systems and sociometabolic regimes. Examples that continue in the tradition of Ester Boserup's work include the analysis of the energy intensity of working hours, labor investment per unit area or per unit harvest and the impact of increasing fossil fuel use in changing these relationships (see, for example, Fischer-Kowalski et al. 2010), thus enhancing our understanding of the impact of development trajectories in both social and natural systems.

References

Fischer-Kowalski, M., Singh, S. J., Ringhofer, L., Grünbühel, C. M., Lauk, C., Remesch, A. (2010) *Sociometabolic regimes in indigenous communities and the crucial role of working time: A comparison of case studies*. Social Ecology Working Paper 121

Ringhofer, L. (2010). *Fishing, foraging and farming in the Bolivian Amazon: On a local society in transition*. Dordrecht, Heidelberg, London, New York: Springer Science + Business Media B.V.

Ringhofer, L. (2013) *Introducing functional time use (FTU) analysis: A gender-sensitive approach to labor time in Gendered Perspectives on International Development*, GPID Working Paper 302. (http://gencen.isp.msu.edu/documents/Working_Papers/WP302.pdf)

Robinson, J. P., & Godbey, G. (1997). *Time for life: The surprising ways Americans use their time*. Pennsylvania: Pennsylvania State University Press.

Singh, S. J., Ringhofer, L., Haas, W., Krausmann, F., Fischer-Kowalski, M. (2010). *Local studies manual: a researcher's guide for investigating the social metabolism of local rural systems*. Social Ecology Working Paper 120 (http://www.uni-klu.ac.at/socec/downloads/WP120_Web.pdf)

Chapter 27
Complex Disasters on the Nicobar Islands

Simron J. Singh and Willi Haas

Abstract This chapter is a case study of a local rural system affected by the 2004 Indian Ocean tsunami. The Asian tsunami clearly revealed the vulnerability of coastal communities with respect to dealing with ecological hazards. An area that was greatly affected was the Nicobar Islands, an archipelago belonging to India and located in the Bay of Bengal. Critiquing disaster management and humanitarian aid structures, the chapter considers how an indigenous, subsistence, island community of hunter-gatherers was transformed into an aid-dependent monetary economy embedded in the regional market. Drawing on the concept of social metabolism and transitions, the chapter presents various scenarios of consumption and the consequences these will have on future material and energy demand, land use and time use for the local population. The case reveals the inherent metabolic traps in terms of the islands' sustainable future, both ecologically and socially, and the role of disaster response in driving them to their biophysical limits as islands in the aftermath.

Keywords Nicobar Islands · Local studies · Social metabolism · Transitions · Tsunami · Disasters

S.J. Singh (✉)
Faculty of Environment, University of Waterloo,
200 University Avenue West, Waterloo, Ontario N2L 3G1, Canada
e-mail: simron.singh@uwaterloo.ca

W. Haas (✉)
Institute of Social Ecology (SEC), Alpen-Adria University,
Schottenfeldgasse 29, 1070 Vienna, Austria
e-mail: willi.haas@aau.at

H. Haberl et al. (eds.), *Social Ecology*, Human-Environment Interactions 5,
DOI 10.1007/978-3-319-33326-7_27

27.1 Introduction

In this chapter, we examine the dynamics of a local community to understand sociometabolic transitions. We describe the environmental relations of an indigenous community of hunters, gatherers and coconut growers inhabiting the Nicobar Islands (India) in the Bay of Bengal. In December 2004, the islands were subject to massive destruction by the Asian tsunami. Aid organizations flocked to the scene, and the overwhelming aid that followed catapulted the islands into a new sociometabolic regime with severe consequences for sustainability. This chapter compares the metabolic profile of the archipelago at two points in time—before and in the aftermath of the tsunami. The aim of this chapter is to illustrate how a local community organizes material and energy flows with its environment under conditions of an intervention from higher system levels that prompts accelerated industrial transformation.

There is a long tradition of studying local communities within the Social Sciences—Anthropology, Rural Sociology, Development Studies and Human Geography are but a few prominent disciplines. However, the study of local systems where socioecological systems are a unit of analysis and with sustainability as a point of departure is relatively new (Grünbühel et al. 2003; Mehta and Winiwarter 1997; Ringhofer 2010; Schandl et al. 2006). Local systems are the basis of national and global economies. They provide critical ecosystem services, such as access to materials and energy (provisioning services) and the regulation of water, diseases and climate (regulating services), and they allow for spiritual and recreational benefits (cultural services). Thus, they are also vulnerable to socioecological change arising from extraction, production and waste deposition and from conservation efforts and the appropriation of landscapes through tourism. In other words, micro-level studies combine a local with a global perspective by paying special attention to scale interactions. They seek to understand the ways by which the 'local' is altered by global processes through interventions such as subsidies, markets, legal frameworks, creation of infrastructure and the introduction of services such as health and education. Analysis at local scales is gaining importance because it also provides insights into local actions and decisions that lead to a cumulative effect on the global environment.[1]

'Local' refers to the scale at which direct empirical observation and primary data collection take place. It is the scale at which secondary data either do not exist or are not readily available in the form required, therefore involving a certain amount of fieldwork. It also refers to the level of engagement between researcher and the researched. Local studies rely on the use of Social Science methods such as participant observation, interviews, focus groups and surveys to generate the data; therefore, social and trust-building skills are essential. Often, the research encounters challenging situations due to close proximity to local actors and stakeholders and, in some instances, engagement in local power dynamics.

[1]The theme of the 2014 *Global Land Project Open Science Meeting* was 'Land Transformations: Between Global Challenges and Local Realities', and the theme for the 2014 *Ecosystem Services Partnership* conference was 'Local Action for the Common Good'.

This chapter builds on fieldwork totaling 2½ person-years in the Central Nicobar archipelago[2] by the first author between 1999 and 2010. Prior to the tsunami, research aimed to gain insight into the biophysical exchanges between society and nature (in terms of material, energy, time and land use) and the role of sociocultural, political and economic institutions in organizing these flows. Physical accounting methods applied to local scales are described in Singh et al. (2010) and in the Method Précis on A Methodological Guide to Local Studies in this volume. Trinket Island, with 399 inhabitants, was the core sample for in-depth enquiries related to biophysical flows, and sociocultural, economic and political systems were investigated across the archipelago. The methods used included participant observation and in-depth semi-structured interviews with dozens of inhabitants (such as village elders, elected representatives, priests and heads of households).

In the post-tsunami period, research aimed at searching for viable future options for the islands. Official data as well as direct observations formed the basis of biophysical analysis in the post-tsunami period. Once again, participant observation and structured and semi-structured interviews were conducted with more than 300 Nicobarese. In addition, interviews with government officials, aid workers and local journalists provided insights into the effects of tsunami aid on society-nature interactions and future sustainability.

We begin with a brief description of the Nicobar Islands and then present several indicators to show an economic portfolio of a community that combines horticulture, hunting and gathering activities with elements of industrialization and a market economy. The second part considers the aftermath of the tsunami and changes in environmental relations as a consequence of aid.

27.2 The Nicobar Islands: Geographical and Cultural Context

Located some 1200 km off the east coast of India, the Nicobar Islands are part of the larger Andaman and Nicobar archipelago that runs from north to south like an arched chain in the Bay of Bengal. The 24 tropical islands spread over an area of 1841 km^2 are administered as a union territory of India. They are home to an outstanding terrestrial and marine biodiversity, one-tenth of which is endemic. Most of this area is protected forests and mangroves. These islands are also home to an indigenous community—commonly referred to as the *Nicobarese*—with a current population of approximately 23,500 (Census of India 2011).[3] The Nicobarese have remained relatively isolated for a long time. Nevertheless, one may find traces of a Southeast Asian cultural complex reflected in layered cosmologies, secondary burial of the dead, spirit mediumship and carved figures to attract or ward off spirits (Singh 2003).

[2]The Central Nicobars, also known as the Nancowrie group of islands, comprise six islands: Kamorta, Nancowrie, Trinket, Katchal, Chowra and Teressa.

[3]Prior to the tsunami, the Nicobarese population in 2001 was 26,565 (Census of India 2001).

Due to their location on a historically important sea route to Southeast Asia, these scattered islands between mainland India and Indonesia offered an attractive resting harbor for traders and sailors for a very long time. Trade was used as a pretext for the use of the natives' safe harbors. Later, from the 15th century, the islands drew the attention of several European powers (mainly the Dutch, Portuguese, Danes, Austrians and British), who saw them as a strategic military location for maintaining supremacy over trade and territory in Southeast Asia (Chakravarti 1994; Gupta 1994).

The Nicobar Islands have remained a sensitive military area, and entry into them has been highly regulated ever since 1869, the year when the islands were officially handed over by the Danes to the British. Even under independent India, the Indian government has followed the same restrictive policy under the *Andaman and Nicobar Protection of Aboriginal Tribes Regulation* (ANPATR) of 1956. Thus, contacts with the Nicobarese have been limited only to specific forms of interactions, namely with government employees and their contractors, traders from the neighboring Andaman Islands (who arrive and stay illegally) and occasional researchers with special permits from the government. Despite such protectionism, the local population has been subject to several welfare programs, such as primary health and education, an inter-island ferry service, construction of wells and (partial) electrification of villages.

Nicobari villages are (or 'were', prior to the tsunami) located along the coast, usually sheltered behind mangroves or within a bay. A typical Nicobari dwelling is perched on stilts facing the sea with coconut palms in the background. Outrigger-canoes provide for easy access to villages along the coast or to ferry across to nearby islands. Largely subsistent, the Nicobarese exhibit an economic portfolio comprising hunting-and-gathering, fishing, pig-rearing and selling copra in lieu of rice, sugar, cloth, kerosene and other necessities. Although their dependency on the market has increased considerably, capital accumulation is still largely absent. *Copra*[4] is produced when there is a requirement of food or other commodities from the market. Some Nicobarese maintain food gardens where they grow an assortment of crops such as bananas, pineapples, yam, sugarcane, oranges, lemons, papaya and jackfruit. Additionally, the Nicobarese select from a wide-ranging abundance of edible leaves, tubers and fruits from the forest and (protein-rich) seafood from the surrounding mangroves and coral reefs (Singh 2003).

In many senses, the Nicobarese are a traditional society. *Traditional* society refers to the political, social and economic characteristics of Trinket. Although administered by India, the elected headmen of the villages (called 'captains') wield social and political influence. The extended family and village solidarity dominate the Nicobari social structure. Despite the existence of a few specialized roles in the society, such as the doctor-priest, midwife and teacher, social stratification is nearly egalitarian on the village level. This is not so on the regional level, where a traditional system of tribal leaders and an elected 'Tribal Council' is in place.

As with most indigenous cultures across the world, the various segments of the socioecological system in the Nicobars are inextricably linked to each other.

[4]Copra is desiccated coconut flesh that is dried over fire for several hours. It serves as raw material for the extraction of coconut oil.

Elaborate festivals, rituals and ceremonies, some lasting months, reproduce society in terms of power relations, established hierarchies and access to and regulation of resources. To give one example, the shift in the wind direction (during October and November) marks the observance of the *Oliov* festival.[5] With the organization of *Oliov*, restrictions on the harvest and on the consumption of certain varieties of food are imposed until the next season, whereas restrictions that were previously in place are lifted. Such regulation through cultural expressions and social institutions ensures the availability of resources year round and prevents the overuse and eventual extinction of a particular food when it is scarce. Thus, intervening in any one aspect of the Nicobarese life and culture will affect other aspects of their life.

27.3 The Environmental Relations of the Nicobarese (2000–2004)

A society's environmental relations may be expressed on two basic levels: first, at the level of cultural representation (e.g., rituals, nature worship, taboos); second, at a biophysical level. In this section, we will focus on the latter, drawing on the sociometabolic approach. The sociometabolic approach rests on the premise that a society organizes material and energy exchanges with its environment to satisfy its requirements for reproduction and maintenance. The quantity and structure of matter and energy a society draws from its environment largely depends on people's mode of subsistence and lifestyle, which are related to technology (Fischer-Kowalski and Haberl 1997).

Trinket Island, with an area of 36 km^2 and a population of 399 in the year 2000, was our core sample for an in-depth analysis of social metabolism.[6] The indicators used to describe the material and energy flows (MEFA) for Trinket are:

Direct Material Input (DMI): Domestic (material) Extraction (DE) + Material Import
Domestic Material Consumption (DMC): DMI—Material Export
Direct Energy Input (DEI): Domestic (energy) Extraction (DE) + Energy Import
Domestic Energy Consumption (DEC): DEI—Energy Export.

Looking at Fig. 27.1, the materials extracted from within Trinket's domestic environment consisted (in 2000) mainly of biomass (wild catch from sea and land, forest produce, and fuelwood) and minerals (sand, gravel). The imports consisted of biomass (rice and sugar), minerals (cement, steel), fossil fuels and consumer goods (such as clothes and soaps). On the output side, we accounted only for exported materials. Accounting for waste and emissions was problematic for a society without a system of waste collection and treatment. Exports comprised

[5]Singh (2006) provides a detailed account of some of the most common Nicobarese festivals and observances.

[6]For a more in depth analysis of the social metabolism of Trinket Island, see Singh and Grünbühel (2003) and Singh et al. (2001).

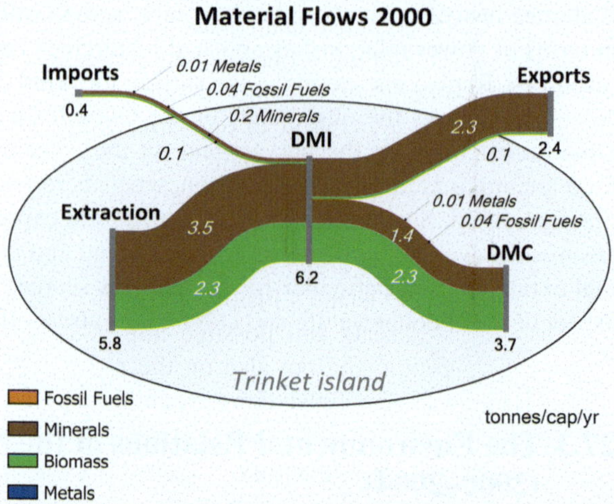

Fig. 27.1 Material flows on Trinket Island, 2000–2001

sand (for the construction of buildings by and for government establishments) and copra (for industrial use). Although sand greatly exceeded copra by mass, the economic gain from copra was much higher than that from sand.

The DMI in 2000 for Trinket was 6.2 t/cap/year (metric tons per capita per year). Of this, minerals accounted for 3.7 t/cap/year, biomass 2.4 t/cap/year, fossil fuels 0.04 t/cap/year and other products (e.g., soaps, cloth) 0.01 t/cap/year. As these figures indicate, the bulk of the DMI is due to the movement of minerals on the island. Mineral inputs consisted of imported cement and steel and domestically extracted sand and gravel. Of this, only one-third was used for local construction activities; the other two-thirds are exported to the neighboring island to be used in the construction of government headquarters.

Most of the biomass (2.3 t/cap/year) is harvested domestically (coconuts, fish, tubers, timber, fuelwood, grass, etc.). Only a small amount is imported in the form of rice, flour and sugar. Although small in volume, the dependency on imported biomass in the form of rice and sugar is rather high and is strongly indicative of Trinket's dependency on the industrialized world and its transition from a subsistence to a non-subsistence economy. Except for biomass, the majority of the materials (fossil fuels, minerals and some products) were introduced only in the early 1990s. Exports are mainly minerals, sand (2.3 t/cap/year) and copra (0.1 t/cap/year).

The domestic material consumption (DMC = DE + imports − exports) for Trinket was 3.7 t/cap[7] in 2000, with biomass being 2.3 t/cap, minerals 1.4 t/cap and fossil fuels and other products being the same as in DMI. This difference between DMI and DMC suggests how much a society is willing or is able to

[7]This compares to 1.6 t/cap/year in Campo Bello (a swidden village in Bolivia), 2.6 t/cap/year in Nalang (a subsistence rice cultivating village in Laos) and 3.6 t/cap/year in Sang Saeng (an intensive rice cultivating village in Thailand) (Fischer-Kowalski et al. 2011).

Fig. 27.2 Energy flows on Trinket Island, 2000–2001

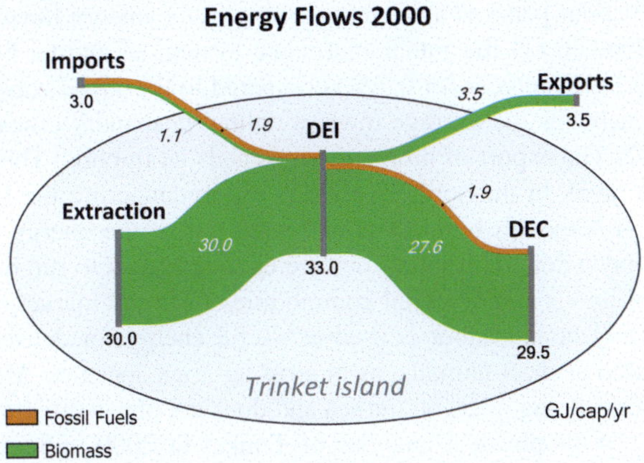

produce or harvest in addition to its own domestic consumption to enter into a trade relation with other societies for its own sustenance and reproduction. Typically, in the case of a subsistence society, DE would more or less satisfy DMC. For Trinket, this difference between DE and DMC is an indicator of its transition to a non-subsistence society dependent on trade and outside relations.[8]

Following methodological guidelines analogous to material flows (Fig. 27.2), we calculated the direct energy input (DEI) and domestic energy consumption (DEC) for Trinket. The DEI was 33 GJ/cap/year (gigajoules per capita per year) in 2000. Approximately 9 % (or 3 GJ/cap/year) of this DEI is imported, of which biomass is 1.1 GJ/cap/year (e.g., rice, sugar, flour) and fossil fuels 1.9 GJ/cap/year. The remaining 91 % (or 30 GJ/cap/year) is domestically extracted. Almost all of the locally harvested energy is biomass, two-thirds of which is for the bio-metabolism of humans (3.7 GJ/cap/year) and for livestock (17 GJ/cap/year) on the island.[9] The remaining one-third of harvested biomass is coconuts for copra production (6.2 GJ/cap/year), fuelwood used in copra production (1.6 GJ/cap/year) and firewood for domestic cooking (1.4 GJ/cap/year). A solar energy plant was introduced on the island by the local administration in 1990 and currently delivers 0.0009 GJ/cap/year for household lighting. Subtracting exports (all of which is copra) from DEI, we obtain a DEC of 29.5 GJ/cap/year.[10] Fossil fuels accounts for only 6.4 % of the total DEC on the island. The total efficiency of energy use is 7 % for Trinket Island.[11]

---

[8]The difference between DE and DMC is not the only indication of dependency on trade with other societies. One may find cases where DMI equals DMC if the volume of imports equals that of exports (e.g., exporting cash crops to import fertilizers, machinery and other industrial goods).

[9]The total energetic intake of pigs was calculated to be 9.2 GJ (including energy from scavenging), chickens 1.3 GJ, cows 5.4 GJ and goats 0.7 GJ.

[10]This compares to 20.6 GJ/cap in Campo Bello, 26.3 GJ/cap in Nalang and 40.5 GJ/cap in Sang Saeng (Fischer-Kowalski et al. 2011).

[11]Energetic efficiency is the percentage of useful energy (2.32 GJ) in relation to the DEI (33 GJ).

Four patterns are rather striking in the energy flows and conversion processes on Trinket: (1) the rather inefficient system of animal husbandry. The output is only 0.1 GJ/cap/year (or 0.7 %) compared to the 17 GJ/cap/year input of biomass energy, far below the average modern animal husbandry efficiency, which is roughly 10 %. (2) The export of biomass far exceeds its imports. This one-way flow of nutrients is a break in the soil's nutrient cycle, endangering the future of the local ecosystem. (3) Although biomass comprises most of the energy input on Trinket, fossil fuels play a dominant role. They are used primarily to run out-boats for the transportation of goods and essential commodities from the market on the neighboring island. (4) Only human labor is used as useful energy for delivering work. Livestock are not used as draft animals, as in most agrarian societies. Mechanical energy is used only for running boats on the sea and does not play a direct role in altering land resources.

A biophysical analysis of Trinket in 2000 indicates it is a society that is still rather traditional and largely subsistent, although it is also rapidly moving toward a market economy. Economically, fishing, pig rearing and food gathering from the forest are primarily undertaken to meet people's daily nutritional requirements. Of the total nutritional intake, 71 % is still from domestic sources, whereas the rest is imported. Copra production is undertaken solely to satisfy immediate needs for food and for products unavailable on the island, not for accumulating capital. Household interviews indicated that almost all coconuts available on the island were harvested, of which 48 % were used for copra, 32 % were fed to pigs, 11 % were consumed by chickens, and only 9 % were domestically consumed within the household.

Although pig-rearing is highly inefficient,[12] one-third of the coconuts produced on the island are fed to pigs. Trinket society could thus produce at least 32 % more copra if they wanted to. However, pigs have a strong cultural meaning, they add to the social status, and they are significant during festivals and ceremonies as sacrifice.[13] In this way, the significance of culture still outweighs the western concept of efficiency in Trinket society.[14]

27.4 The 2004 Tsunami and the Aftermath

For the Nicobarese, the 2004 tsunami underlines a deep incision in their memory that separates the 'then' from the 'now' because life will never resemble the past in any of its sociocultural and economic complexities. The tsunami of December 26,

[12]Pigs consume 12.6 times more energy (total feed across their lifetime) than their output to the social system in terms of pork meat.

[13]Wildenberg (2005) ran a computer simulation to show that pig festivals on Trinket have a positive effect on the resilience of the local socioecological system as a whole. Removal of the pigs would drastically alter the system.

[14]The inefficiency of pig husbandry, along similar methodological concepts, has also been studied by Rappaport (1971) among the Tsembaga population of New Guinea. In addition to the importance of pigs in rituals and in regulating relations between local groups, he discusses ecological reasons for their importance, such as being part of a food chain and converting vegetable carbohydrates into high-quality protein.

2004, was triggered by the Sumatra-Andaman earthquake that occurred at a magnitude of 9.3 (Thakkar and Goyal 2006). Being close to the epicenter, large-scale destruction in the Nicobar Islands was inevitable. The earthquake rocked the islands for several minutes, followed by eight consecutive tsunami waves that caused most of the damage. Spread out in the open sea and topographically flat, the islands were easy victims of the high waves. With settlements invariably along the coast, the Nicobarese had little chance of escape. In some cases, the gigantic waves recklessly washed the islands from one end to the other, taking with them thousands of human lives and destroying both the built and natural environment. When the tsunami subsided, some of the islands had sunk nearly two meters from their original level, and hardly any sign of human settlement could be seen (Thakkar and Goyal 2006). Trinket Island was broken into three parts. The sea around was clogged with uprooted trees, and the coast was littered with smashed corals and debris from the houses and other infrastructure.

In a matter of minutes, some 4000 Nicobarese were washed away by the gigantic waves, and the villages were either completely destroyed or were affected beyond recognition, along with their cultural artifacts (some of them hundreds of years old) and livestock. Every coconut tree (the main basis of the local economy) standing within a kilometer of the sea was uprooted or killed as the ocean water passed over. The destruction of anchored boats and vessels rendered the survivors immobile, preventing commute to the government headquarters or the ability to help those still floating in the sea. The earthquake and the consequent sinking of the islands resulted not only in the destruction of the mangrove and coral ecosystems that had been the main source of (protein-rich) seafood but also in large areas of land being lost to the ocean, creating a new coastline and making navigation difficult. In short, life had changed entirely for the Nicobarese in a few minutes.

The national and international response to the tsunami was overwhelming. The aid sector was joined by governments, corporations, academic institutions and hundreds of thousands of individuals who involved themselves in some way or another in bringing relief and rehabilitation to the victims. Approximately 14 billion US dollars was donated or pledged to address the enormity of the disaster in what is known as the world's largest fund-raising exercise. Special arrangements were made to transport volumes of goods from across the country to the affected populations. In a matter of weeks, the Nicobarese were swamped with a variety of goods, some of which they could use and others they had never seen before or had no use for. In addition to the wide distribution of 'relief' material, the government ensured a constant food supply to the several relief camps on each island.

However, good intentions and the enormous amount of resources made available to the tsunami victims were not sufficient to meet the challenges at hand. Confronted for the first time with the idea of aid and development, the Nicobarese found it difficult to grasp their dynamics. Because the islands had been protected under the *Protection of Aboriginal Tribes Regulation* of 1956 and entry to them was highly regulated, the interaction of the Nicobarese with the outside world had been very limited. Now, for the first time, they were approached by large donor organizations, each of which gave the impression of fulfilling a large part of relief and rehabilitation needs single-handedly. This obviously did not impress the Nicobarese. Unable to work and rebuild their lives, they were extremely agitated and suffocated in the relief camps that were set up for them. 'Leave us alone. We can manage on our own. We don't need biscuits and chips. We need to make our homes and plant our gardens. Give us tools, if you wish to help us', is what some had begun to say. Some even believed that outside interference and non-indigenous settlers had caused the tsunami. 'This is our land. Please leave us alone. Otherwise we are sure to die', was the remark of a leader from Katchal.

The Nicobarese were not left to themselves. The event was much too large and the international media attention too high to leave the Nicobarese on their own. Thus, the government took it upon itself to launch a series of temporary and permanent rehabilitation programs for the affected population. This included cash compensation for losses, intermediate shelters, permanent housing, intensification of welfare programs, free food supply, measures to revive the local economy and development of infrastructure. The question of the location of intermediate and permanent shelters became crucial, as did the question of design. Following coastal security regulations, the new settlements were built on higher hinterlands, in several cases, 2–3 km from the shore. This meant extensive discussion with those who owned the land and discontent over being located far from the shore. The sea had always been an integral part of the Nicobarese world in all its social, cultural and economic complexity, but the tsunami was no small event. The law took its course, and on grounds of security against future calamities, land surveys and construction programs were initiated accordingly.

Soon after the tsunami, the government announced an immediate relief of 2000 Rupees[15] per family. When local officials realized that Nicobarese live in large extended families, they suggested (in good faith) splitting up families into nuclear units so that each could be entitled to the sanctioned amount. It took quite a while to educate the Nicobarese on the concept of a nuclear family, quite alien to them so far. The list, when finally ready, became the blueprint for all compensation packages that followed. Bank accounts were opened for all heads of nuclear families for the issue of cheques related to several forms of compensation. At the same time, construction contracts for 7000 single-family homes were given to large companies. This was the beginning of the disintegration of the extended family system and of future conflicts.

In compliance with the national policy, the government announced a package of cash compensation to the next of kin for each person who was missing or had died in the tsunami. Another package offered compensated land and crop loss per hectare. Most families received amounts up to several hundred thousands of rupees. Conflicts arose in both forms of compensation because traditional rules did not match the Indian legal framework. For example, in the Central Nicobars, it is the norm that the husband must go and live with his wife as an *ungrung* (slave). Thus, he has no right to the wealth of his wife or her family. According to Indian law, the next of kin in case of the death of the wife is the husband. Without due consideration of

[15]One Euro equals approximately 70 Indian Rupees.

the traditional system, cheques were issued in the husbands' names, which changed power dynamics and induced conflicts. Indeed, the possibility of receiving large sums of money further spawned greed and jealousy, which were visible in the conflicts over who was next of kin for those dead. Moreover, compensation for land and crop loss (also payable to nuclear families alone) caused the splitting of land that was previously jointly held, thus leading to conflicts in several households.

Finally, the generosity of the government in compensating losses with cash was a predicament in itself. Never before had the Nicobarese possessed so much money at one time. Pre-tsunami, copra was bartered immediately for rice, sugar, cloth or other necessities. The concept of investment and saving for the future was alien to a common Nicobari. Now, with large amounts of cash at hand, demand surged for consumer goods such as motorbikes, televisions, DVD players, cell phones, stereo systems and junk food. Substantial amounts of money were spent on cheap 'red alcohol' at exorbitant prices because these bottles had to be brought to the islands illegally. Aside from damage to health, the money had burned a hole in the tribal pocket, as the Nicobarese ended up paying two to three times the going price to immigrant traders for all commodities.

Aid organizations in the Nicobars mainly engaged in the distribution of relief materials (household goods, tools, clothes, boats, etc.) and in organizing a few training/capacity-building workshops in the first few months after the tsunami. Although the effectiveness and usefulness of most of these training programs are questionable, some of the relief materials and necessities were urgently needed. These included a generous distribution of household goods, utensils, tools, clothes and a large number of motor-driven boats for fishing and transport. However, there were several products for which there was little use and that only created conflicts for want of possession of these limited exotics. Examples are radios that run on mechanical power, woolen blankets (unfit for a tropical climate), saris (which the women did not know how to wear), ceiling fans (when there was no provision for electricity), bicycles (without pathways to ride them) and a variety of other consumer goods and food products alien to the local culture and conditions. Some of the interventions were even absurd. For example, an international organization imported several galvanized water tanks from Australia, each with a capacity of a million liters. These were installed at the highest points of the five islands. The effort took nearly a year, but no one questioned where the water would come from. Thus, in effect, they remained empty.

In most instances, it was clear that projects were 'supply driven' rather than 'need driven'. The large volumes of money that had been collected had to be spent, regardless of how, and tangible results had to be reported back to the donors. This 'one-size-fits-all' approach has also been criticized in Abid's (2006) report: 'These organizations are driven by their own agendas and they have heedlessly introduced new concepts, ideas, schemes and projects without taking into account the socio-cultural milieu of the district'.

Humanitarian aid and interventions in general had a detrimental effect on the instruments of social containment. The death of a large number of elders in the community was in itself a great social loss. The institutions, mechanisms and individuals that survived were not only rendered ineffective but at times even contributed to further instability. Discontented families who received less than others began to stir up turbulence in society. Village captains were under pressure to contain this unrest, but some of these captains were young and incompetent. In some villages, old and experienced captains who had died in the tsunami were replaced by temporary captains (cynically referred to as 'tsunami captains') who were ill informed of the history of land use and family structures. The situation became even more complicated when the Tribal Councils confronted differences in opinions among the various indigenous leaders on future options. Whereas the elders preferred options close to the traditional lifestyles, the younger leaders opted for new alternatives. The discussions were endless and included topics such as whether they should give up sea transport altogether and take to roads, use concrete housing or traditional huts, depend on a land-based economy versus off-land labor and employment and whether to continue as traditional joint family units or nuclear families.

The stakes were high and directed at the very foundation of their sociocultural sense of being. Whereas modern options were a sign of development, the economic sustainability of these options was in doubt. Aid money would not flow incessantly—eventually the local economy had to sustain it all. With most coconut trees destroyed, and with them the economy, the question of their subsistence was still open. At the same time, rapid cultural transformation was seen as hazardous. The disintegration of the traditional joint family system would have implications for the entire socioeconomic structure as it was directly linked to land and resource distribution and, thus, social control.[16]

27.5 The Sustainability of Aid

Increasing dependency on aid and the new affluent lifestyle has had consequences for the society and the environment. Instead of supporting the Nicobarese to construct their traditional dwellings as they always have, the government took it upon itself to provide 7000 houses, or 'permanent shelters', for nuclear families using construction materials not available on the islands. The estimated 200,000 metric tons of construction materials brought in from the neighboring Andamans or from the Indian mainland resulted in an eightfold increase in the built stocks per capita compared to pre-tsunami figures. Consequently, the maintenance of these houses will require a ceaseless flow of materials in the future—at the expense of the Nicobarese. Because the new homes were electrified, per capita demand for water and use of energy rose substantially. According to the Andaman Public Works Department (APWD), water consumption increased to 70–100 L per capita per day (compared to the Indian average of 40 L). Calculations indicate a 20-fold increase

[16]A more detailed analysis of institutional change due to tsunami aid in the Nicobar Islands is described in Ramanujam et al. (2012).

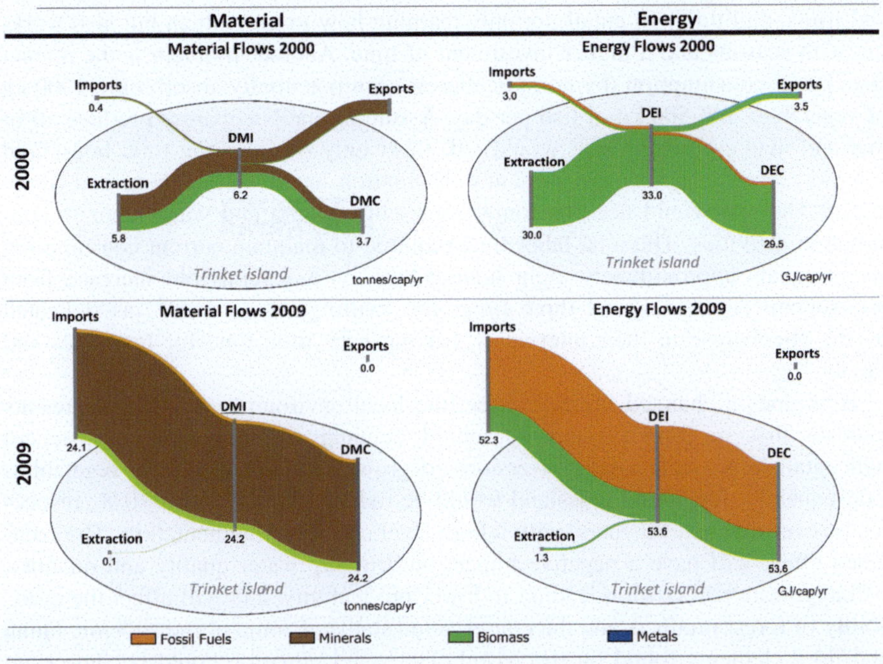

Fig. 27.3 Comparing metabolic throughput of materials and energy—pre- and post-tsunami

in the consumption of fossil fuels compared to pre-tsunami figures, much of it used to produce electricity, with the remainder used for motor-bikes, cooking gas and boats. Consequently, in the two periods, the DMC increased sixfold, from 3.7 to 24 t/cap/year, whereas DEC nearly doubled, from 30 to 54 GJ/cap/year (Fig. 27.3).

Five years of incessant aid flow led the Nicobarese to adopt a new way of living based on a consumption much higher in quantity and that fundamentally changed in quality compared to what it was before. The issue is that hardly any of this is locally produced but must be obtained from outside as aid, subsidy or trade financed by compensation money. Change in lifestyle is inevitably accompanied by an increase in material and energy flows and dependency on resources from outside the islands. This concern is further aggravated when considering what their new economy will be. As a hunter-gatherer society, producing copra for the market when there is a need for commodities did not require a disciplined investment of working time. Coconut trees, once planted, provided fruit for nearly 100 years all year round without much maintenance and without having to worry about seasons. The pigs scavenged the forest for three-quarters of their diet, and hunting and fishing were combined with leisure.

How will the Nicobarese sustain a newly adopted lifestyle once they have spent the compensation money? Presumably, the only means of livelihood that is readily accessible to the Nicobarese is growing and selling a variety of vegetables, fruits and fish to the local market. Unfortunately, very few know how to grow vegetables and fruits, and this will entail not only learning how to grow them but also working with seasons and a higher investment of time. Another problem is the market size. Local consumption (by non-Nicobarese) can potentially absorb only 1000 kg of vegetables and 500 kg of fish per day. Assuming the Nicobarese produce all of that and manage to sell it, it would still meet only 40 % of the total household income required to maintain present consumption levels (Wildenberg and Singh 2012). The rest would need to come from selling copra and wage labor in construction activities. The total labor time required to maintain current consumption levels equals approximately eight hours/adult/day, a several fold increase from pre-tsunami conditions and three times the 'willingness to work', as indicated by the Nicobarese in their interviews, leaving very little time for festivities and rituals.

How does a changed lifestyle affect the local environment? Modeling results indicate that meeting present household demand (of cash and subsistence) will entail large scale land conversions to grow more coconuts and vegetables. Consequently, forest and grassland would be reduced by 15 % and 10 %, respectively, over the next 30 years, with a high level of forest fragmentation. The combined effect will have a negative impact on drinking water quality and quantity, will aggravate soil erosion leading to lower productivity, and will affect the availability of forest products and the conservation status of some of the endemic fauna and flora elements found on the island. The water situation could become even more critical if we consider the decline in water availability with a scenario where water demand is likely to increase due to population growth and agriculture would have to move from being rain-fed to irrigated due to climate change predictions. Finally, the negative impact of top soil erosion might not be limited to land degradation—it may also have an undesired effect on the coral reefs surrounding the islands (Wildenberg and Singh 2012).

27.6 Conclusions

'Local studies' provide us with valuable insights at the case level, and they help us gain a deeper understanding of the phenomenon of global environment change. One crucial insight is the systemic character of society-nature interactions. Even well-intentioned interventions can elicit a chain of consequences in the sociometabolic system that can be detrimental to the social or ecological system or both. Humanitarian aid and traditional development policies tend to overlook the intricate ties among demography, sociocultural arrangements, labor time, subsistence patterns and regulation of natural resources. Seeking viable pathways requires a thorough consideration of each option in terms of benefits in the long term because the outcomes may be very different for different socioecological systems. Projects often follow a sectorial approach when designing interventions. The focus is to boost the efficiency of one variable alone, such as to increase land productivity or enhance dairy output, without an awareness that doing so will lead to an overall system change, part of which may not be desirable or sustainable in the end. Development policies need to become sensitive to these systemic interrelations and the trade-offs involved therein.

Another issue is the complexity of scale interaction. We often hear the phrase, 'Think globally, act locally'. Although this message is attractive and catchy, policies and interventions from higher scales greatly limit the option space for local self-determination. As in the case of Trinket, the islanders seek pathways within given opportunities and constraints, and they orient themselves accordingly. A comparison of culturally diverse local systems across the world (Bolivia, Thailand, Laos and India) has clearly revealed patterns in material and energy use consistent with their production systems in response to interventions from higher scales (Fischer-Kowalski et al. 2011). In each of these cases, the trajectory observed is unidirectional: growing consumption, increasing dependency on the outside, a move away from subsistence to integration in the market, a shift from renewables to fossil fuels and minerals and the eventual degradation of the local resource base with concomitant changes in demography, social structures from simple to complex, social inequalities and the increasing differentiation of roles. A number of microeconomic studies have documented similar outcomes. For example, there is evidence that the introduction of railways and roads in rural India resulted in increased trade, increases in real income levels and consumption, higher use of fertilizers, adoption of new technologies and increased dropouts from school (Aggarwal 2014; Donaldson 2010). It is no surprise that the cumulative effect of such local trends mirrors the current global environmental crisis. Hence, a focus on the local is a crucial link to understanding this global challenge. Sustainable development requires a broader search for pathways where short- and long-term benefits for the people come at the lowest possible environmental cost while avoiding increasing the burden and stress on the people in terms of working time. The concepts of social metabolism and sociometabolic transitions offer powerful tools to analyze some of these dynamics across scale.

References

Abid, M. (2006). Relief and rehabilitation. Role of civil society organizations. Report submitted to the Andaman and Nicobar Administration.

Aggrawal, S. (2014). Do rural roads create pathways out of poverty? Evidence from India. http://people.ucsc.edu/~saggarw1/

Census of India. (2001/2011). Retrieved from office of the registrar general & census commissioner website, Ministry of Home Affairs, Government of India: http://www.censusindia.gov.in/

Chakravarti, A. (1994). Andaman and Nicobar: Early history. In: V. Suryanarayan & V. Sudarsen (Eds.), Andaman and Nicobar Islands challenges of development, Vol. III (pp. 20–36). New Delhi: Konark Publishers Pvt. Ltd.

Donaldson, D. (2010). Railroads of the Raj: Estimating the impact of transportation infrastructure. National Bureau of Economic Research, Working paper 16487. http://www.nber.org/papers/w16487.pdf?new_window=1

Fischer-Kowalski, M.,& Haberl, H. (1997). Tons joules, and money: Modes of production and their sustainability problems. *Society and Natural Resources, 10*(1997), 61–85.

Fischer-Kowalski, M., Singh, S. J., Ringhofer, L., Grünbühel, C. M., Lauk, C., & Remesch, A. (2011). Socio-metabolic transitions in subsistence communities. Boserup Revisted. In: Human Ecology Review. Vol. 18 (2).

Grünbühel, C. M., Haberl, H., Schandl, H., & Winiwarter, V. (2003). Socio-economic metabolism and colonisation of natural systems in a northeast Thai village: material and energy flows, land use, and cultural change in SangSaeng. *Human Ecology, 31*(1), 53–86.

Gupta, A. D. (1994). The Andaman-Nicobar Islands in the Trade and Navigation of the Bay of Bengal 1500–1850. In: V. Suryanarayan, & V. Sudarsen, (Eds.), Andaman and Nicobar Islands challenges of development, Vol. IV, (pp. 37–50). New Delhi: Konark Publishers Pvt. Ltd.

Mehta, L., & Winiwarter, V. (1997). Stoffwechsel in einem indischen Dorf: Fallstudie Merkar. Vienna: IFF Social Ecology (Social Ecology Working Paper 49).

Ramanujam, R. V., Singh, S. J., & Vatn, A. (2012). From the ashes into the fire: Institutional change in the post-tsunami central Nicobar Islands. *Society and Natural Resources, 25*(11), 1152–1166.

Rappaport, R. A. (1971). The flow of energy in agricultural society. *Scientific American, 224*(3), 116–132.

Ringhofer, L. (2010). Fishing, foraging and farming in the Bolivian Amazon. On a local society in transition. New York: Springer.

Schandl, H., Hobbes, M., & Kleijn, R. (2006). Local material flow analysis in social context in Tat Hamlet, Northern Mountain Region, Vietnam. IFF Social Ecology (Social Ecology Working Paper 90).

Singh, S. J. (2003a). *In the Sea of influence: A world system perspective of the Nicobar Islands, Lund studies in human ecology 6*. Lund: Lund University Press.

Singh, S. J. (2006). *The Nicobar Islands: Cultural choices in the aftermath of the Tsunami*. Vienna: Oliver Lehmann, Czernin Verlag.

Singh, S. J., & Grünbühel, C. M. (2003). Environmental relations and biophysical transitions: The case of Trinket Island. *Geografiska Annaler* (*Human Geography*), *85B*(4), 187–204.

Singh, S. J., Grünbühel, C. M., Schandl, H., & Schulz, N. (2001). Social metabolism and labour in a local context: Changing environmental relations on Trinket Island. *Population and Environment, 23*(1), 71–104.

Singh, S. J., Ringhofer, L., Haas, W., Krausmann, F., & Fischer-Kowalski, M. (2010). Local studies manual: A researcher's guide for investigating the social metabolism of rural systems. Vienna: IFF Social Ecology (Social Ecology Working Paper 120).

Thakkar, M. G., & Goyal, B. (2006). Historic submergence and tsunami destruction of Nancowrie, Kamorta, Katchall and Trinket Islands of Nicobar district: Consequences of 26 December 2004 Sumatra-Andaman earthquake. *Current Science, 90*(7), 989–994.

Wildenberg, M. (2005). Ecology, rituals and system-dynamics. An attempt to model the Socio-Ecological System of Trinket Island. Vienna: IFF Social Ecology (Social Ecology Working Paper 80).

Wildenberg, M., & Singh, S. J. (2012). Integrated modelling and scenario building for the Nicobar Islands in the aftermath of the Tsunami. In: Glaser, B. et al. (Eds.), Human/nature interaction in the anthropocene. Potentials for socio-ecological systems analysis. Routledge.

Method Précis: A Methodological Guide to Local Studies

Simron J. Singh and Willi Haas

Undertaking a local study can be a thrilling experience at both an intellectual and personal level. Because data are not readily available at this scale, researchers need to be innovative in generating data first-hand. It is useful to have conceptual and analytical skills, the ability to think in terms of systems and feedback loops, innovative and logical thinking in the field to generate reliable quantitative data and social and process skills to generate qualitative data on the sociocultural system. This chapter briefly outlines some of the methodological steps one needs to follow to undertake an analysis of a local rural system and is a highly truncated version of the local studies manual produced at the Institute of Social Ecology Vienna and available online (see Singh et al. 2010; for further reading, see also Fischer-Kowalski et al. 2011; Ringhofer 2010; Singh 2003).

Identifying an Appropriate Unit of Analysis

The first key step is to identify the 'focal system' one wishes to study, that is, an *appropriate unit of analysis*. To make a viable decision, it is important to determine which territory, which population and which other biophysical stocks (i.e., infrastructure, artifacts and livestock) 'belong' to the focal system. This helps establish the *system's boundaries* toward (a) other sociometabolic systems and (b) the (natural) environment. The two boundaries allow one to distinguish between the extracted and appropriated materials from the domestic environment, termed *domestic extraction* (*DE*), on the one hand, and *imported* materials from other social units, on the other. On the *output* side, wastes and emissions deposited into the domestic environment are distinguished from material *exports* to other social units.

A *social system's territory is a geographical area that is under legitimate control of this particular social system.* Legitimate control is meant both politically (governing body, institution with a right to sanction standards for social behavior on that territory) and economically (right to exploit the resources, land, soil, minerals, water, etc.).

Identifying the biophysical stocks of this society is the next step. In so doing, one must consider first the several material components or biophysical structures of the society that have been produced and maintained through the labor process. These are typically human population, domesticated livestock and durable artifacts such as buildings, roads, machines and anything that stays in the system for longer than a year (see Method Précis on Material Flow Analysis).

Generating Data on Society's Stocks

Human Population A society's *human population* is a set of people defined as having a legitimate right to that society. This right may be conferred through (a) *political membership* (e.g., voting rights, the right to participate in decision-making processes, the right to voice an opinion), (b) *economic membership* (e.g., some degree of entitlement to the use and governance of local resources, receiving economic support from the social system or an obligation to economically contribute to the social system) or (c) *factual presence*. Factual presence implies a certain minimum use of resources (space, air, freshwater) regardless of the legitimacy of this use. Factual absence, in contrast, can be associated with economic contribution to the community's resource base (e.g., through remittances) as well as the consumption of community resources during absence.

Understanding the system size—in terms of numbers, weight, composition by age and gender and flows in terms of births, deaths and migration—begins with a human census by gender and age groups. Flows for a human population mean the number of births and deaths that take place each year as well as migration patterns (seasonal, annual, permanent).

Livestock In the case of livestock, ask the following three key questions: are these animals culturally defined as 'belonging' to (members of) this society? Is the reproduction of these animals deliberately and to a great extent controlled by (members of) this society? Is human planning and labor invested in their feeding, health and breeding? It is important to be clear about which animals belong to the category 'livestock' and which do not because only their material and energy requirements (feedstuffs) are counted as flows attributable to the social system (even if the animals graze in the wild).

For a stock and flow account, we require the average number of *heads* for each of the main livestock species recorded during one year as well as their respective average *live-weight*. This can be collected by interviewing the household heads and through direct observation during feedings.

Artifacts The definition of artifacts is guided by key questions similar to those for livestock. Operationally, this may become quite a fuzzy endeavor because the timeframes of use and maintenance may be highly variable. On the one hand, artifacts that are typically used up or out of use within a year are considered 'flows', not 'stocks'. These short-lived artifacts should not be included in the stock account. It is important to identify the different building structures and their use. It is also important to identify artifacts the people own in their households that remain longer than a year and to figure out a reasonable typology (or categories) of the houses, buildings and infrastructures in the village. The classification may look something like this: residential houses, kitchen huts, out-houses for storage, animal stalls or enclaves, pathways, school, dispensary, wells.

Generating Data on Material and Energy Flows

Biomass Flows In rural local systems, biomass usually accounts for the most significant material. It is required for feeding the local human and livestock population and for cooking and heating dwellings. It is also used as building material and for generating income by selling biomass products on the market. Usually, rural local systems are characterized by a combination of subsistence and market economies. In addition to supply and use, there are two further distinctions: *primary* production (e.g., crops) and *secondary* production (e.g., livestock products).

To obtain estimates on biomass flows, try to first obtain an estimate of the area under crop production by crop type and calculate the yields for each crop. Estimate yields per unit land for each of the crops by observing actual harvests if possible or by interviewing households. Become familiar with local units such as bags and bundles and find out their actual weights. Estimating the ratio produced for the market and for household consumption might offer interesting insights.

Non-biomass Flows Accounting for flows for artifacts can be quite demanding and requires analytical effort in problem solving. The most striking problem is that many products are purchased rather infrequently, sometimes with long intervals between two purchases. This is obviously because many artifacts, such as durable goods and infrastructures, may have a life span of any number of years. Thus, depending on your sampling period, you may record large flows (e.g., when a new house is built) or none at all. Neither scenario, however, is representative of the general dynamics. To address this issue, we suggest two ways of approaching the estimation of annual flows for artifacts, both of which may be driven by different motivations (again, what is the research question?) and may lead to fairly different results: (1) *actual flows*: the actual flows occurring during a specific year; (2) *discounting approach*: here, you estimate the weight of the stock and the average lifespan. Finally, combine this information (dividing mass by lifetime in years) to yield an average annual flow.

Consumer Goods Consumer goods in rural contexts are certain varieties of food and beverages, toiletries, clothes, fertilizer, batteries and other goods. The easiest way to generate data is to look at the supply structure. You may be lucky to find just one shop where you can interview the owner or observe the buying frequency for a couple of days. Another, more time-consuming (and occasionally culturally inappropriate) method includes the production of consumer diaries by the members of a household. Because income levels and household size are usually influential factors for per capita consumption, it is useful to create homogenous household clusters. You will have to determine the composition and weight of these goods to estimate the flows for that period.

Fossil Energy Carriers Fossil energy carriers include a range of products, including diesel, petrol, kerosene, coal and gas. They are used for transport, agricultural equipment and other machinery, lighting, cooking and heating. Data on fossil energy carriers are meaningful for understanding sociometabolic transitions, the degree of dependency on imports and the ratio between renewable and nonrenewable energy sources.

As a first step, gain an overview of the uses of fossil energy carriers where residents buy them (gas stations, general dealers, etc.). Sometimes it is easier to obtain information on import flows rather than consumption. Often, consumption equals imports.

References

Fischer-Kowalski, M., Singh, SJ., Ringhofer, L., Grünbühel, CM., Lauk, Ch., & Remesch, A. (2011). Socio-metabolic transitions in subsistence communities. Boserup Revisted. In: Human Ecology Review. Vol. 18(2).

Ringhofer, L. (2010). Fishing, foraging and farming in the Bolivian Amazon. On a local society in transition. New York: Springer.

Singh, S. J. (2003b). *In the Sea of influence: A world system perspective of the Nicobar Islands*. Lund: Lund University Press.

Singh, S. J., Ringhofer, L., Haas, W., Krausmann, F., Lauk, C. & Fischer-Kowalski, M. (2010). Local studies manual: A researcher's guide for investigating the social metabolism of rural systems. Social Ecology Working Paper, Vol. 120. Vienna: IFF Social Ecology, 1–69. pp. 1726–3808. http://www.uni-klu.ac.at/socec/downloads/WD120_Web.pdf

Chapter 28
Island Sustainability: The Case of Samothraki

Panos Petridis and Marina Fischer-Kowalski

Abstract The very 'insularity' of islands makes them excellent focal points for sustainability studies that systematically analyze the interactions between human activities and the environment. In this chapter, we seek to explore the factors that cause island societies to prosper and sustain themselves and those that lead to collapse. A number of historical cases of collapse have occurred on the island we investigate (Samothraki, Greece) in the sense of a breakdown of social complexity and rapid population decline. At present, there is a fragile situation of slow population decline and ecological challenges that might be brought to a 'tipping point' by the impacts of the Greek economic and governance crisis and by climate change. The island community has decided to make an effort to turn the whole island into a Biosphere Reserve by UNESCO standards. Building upon a sociometabolic understanding of socioecological systems and using systems thinking (and, to a certain degree, modeling), we attempt to identify environmental and social 'tipping points' for Samothraki. Moreover, in line with the Long-Term Socioecological Research (LTSER) tradition, we argue that analyzing society-environment relations for different phases of the island's history and gaining insights from past collapses can help to identify threats and possible ailments. Finally, this chapter will reflect not only on the outcome but also on the process of performing transdisciplinary research, that is, research that aims to achieve a practical outcome.

Keywords Island sustainability · Transdisciplinary research · Socioecological transition · Tipping points · Collapse

P. Petridis (✉) · M. Fischer-Kowalski
Institute of Social Ecology (SEC), Alpen-Adria University, Schottenfeldgasse 29, 1070 Vienna, Austria
e-mail: panos.petridis@aau.at

M. Fischer-Kowalski
e-mail: marina.fischer-kowalski@aau.at

H. Haberl et al. (eds.), *Social Ecology*, Human-Environment Interactions 5,
DOI 10.1007/978-3-319-33326-7_28

28.1 Introduction: Managing a Process with Both Transdisciplinary and Scientific Goals

The present chapter is the outcome of several years of socioecological research and communication efforts on the Greek island of Samothraki, an ecologically rich and culturally significant place at the cross-roads of development pathways. Our research aim is to analyze society-environment relations on Samothraki for different phases of its history. We seek to explore the factors that cause societies to prosper and sustain themselves on islands and those that lead to collapse. On Samothraki, a number of historical cases of collapse have occurred in the sense of a breakdown of social complexity and rapid population decline (see Tainter 1988). At present, there is a fragile situation of slow population decline and ecological challenges that may be brought to a tipping point by the impacts of the Greek economic and governance crisis and climate change.

We are simultaneously supporting the local population in a process of placing the island on a path toward a sustainable future by 'transforming' it into a United Nations Educational, Scientific and Cultural Organization (UNESCO) Biosphere Reserve.[1] The challenge is how to firmly place such a remote and ecologically rich area—which is already at the crossroads of wider economic influence and potential destructive development paths (as well as impacts of climate change)—under an effective Biosphere Reserve management system as a model of sustainable development. The research team gives scientific support to this process by generating deeper insights from the past regarding collapse as a phase in the history of societies and by identifying the current threats and possible ailments. This has been described as 'an experience in transdisciplinarity' (Fischer-Kowalski et al. 2011).

This chapter is intended to serve a dual task. On the one hand, we show how socioecological thinking can be applied to the study and analysis of island sustainability. To this end, we will draw on the notion of socioecological collapse using population as a key indicator. On the other hand, beyond pure basic research, we reflect upon the transdisciplinary research approach employed, understood as an effort to achieve a practical outcome, namely, supporting the society living on this island. An interesting but challenging fact is that the two processes usually have different time frames and different control factors. Basic research is controlled by the research team and available funding, whereas the process on the ground is influenced, and even controlled by, a series of local and regional stakeholders as well as events such as the current Greek crisis.

[1] Following a feasibility study and comprehensive consultations with stakeholders from 2008 to 2011 (see Fischer-Kowalski et al. 2011), an application to UNESCO was unanimously supported by the municipal council and signed by the island's Mayor. The application was submitted by the Greek National Man and the Biosphere (MAB) committee to UNESCO and is currently under review. UNESCO's Seville Strategy aims to make Biosphere Reserves the principal internationally designated areas dedicated to sustainable development in the 21st century (UNESCO 1996).

28.2 The Island Context: The Case of Samothraki

The defining feature of islands—namely, their 'insularity' or, better, their 'islandness' (Baldacchino 2006)—provides most of them with fairly distinct features, broadly categorized as issues of scale and issues of isolation (Kerr 2005). For ecological science, islands are particularly useful model systems because they have clear physical boundaries, cover delimited geographic areas and are governed by driving forces that can be disaggregated and experimentally controlled. However, in a modern, interdependent world, these same properties confront island populations with the challenges of limited resource availability, tenuous resource security and limited natural carrying capacity (Deschenes and Chertow 2004). Thus, populated islands typically have fragile ecosystems and economies, are heavily dependent on imports for a broad range of goods and suffer from size constraints in the development of resilient water, sanitation, energy and waste management systems. This makes islands excellent focal points for studies that systematically analyze the interactions between human activities and the environment (but difficult settings for the people who live there) in an attempt to move toward systems and practices that are sustainable in the long term. For example, Industrial Ecology explicitly studies model flows of materials and energy at the island system level, using the analytical results to offer recommendations for sustainable practices.

The Greek island of Samothraki, located in the northern Aegean, has 2840 permanent inhabitants (Greek Census 2011). Approximately three-quarters of its total surface area of 178 km^2 is included in the Natura 2000 network.[2] Samothraki has a rich ecological and cultural heritage. In 2011, after several years of research and communication efforts, the mayor of Samothraki submitted a proposal to UNESCO that the island be included in the World Network of Biosphere Reserves (Fischer-Kowalski et al. 2011). The idea of transforming Samothraki into a Biosphere Reserve started as a bottom-up process: it was initiated by a regular visitor to the island and was gradually transmitted to local stakeholders. This was followed by several years of research investigating the socioeconomic feasibility of this transformation as well as the opinions of the different stakeholders regarding the future development of the island. The basic idea was to use the Biosphere Reserve concept as a tool to implement sustainable development on the island with reference to the two main areas of economic activity: agriculture and tourism (Petridis 2012). For this to be achieved, the research community will need to support the future Biosphere Reserve management by preparing detailed plans of key areas of intervention toward sustainable development, namely, the economy, natural resource management and infrastructure. Essentially, it will be necessary to identify future activities and to prepare the ground with structured information and

[2]'Natura 2000' is an EU-wide network of nature protection areas, established under the 1992 Habitats Directive, and is the centerpiece of the EU nature & biodiversity policy (See: http://ec.europa.eu/environment/nature/natura2000/index_en.htm).

convincing arguments to encourage acceptance by local authorities and the cooperation of local stakeholders. This approach closely follows the recent concept of Biosphere Reserves as 'Learning Laboratories', as outlined by Ishwaran and Persic (2008) and operationalized, among others, by Nguyen et al. (2011).

28.3 Research Approach

A model of interaction between cultural and natural spheres of causation is our guiding paradigm for understanding the self-reproduction of socioecological systems (Fischer-Kowalski and Weisz 1999). In the realm where these two spheres interact, namely, the hybrid (that is, both naturally and culturally governed) compartments of the system, we deal with a stocks and flows model. The system and its various compartments reproduce themselves as long as the flows required for maintaining the stocks can be organized. Figure 28.1 gives a comprehensive overview of this model. The center is occupied by the core (hybrid) compartments of the socioecological system on the island: the local population, the visitor population and the three key economic sectors, including the biophysical infrastructures (in the case of agriculture: the livestock) these sectors require and maintain. The local population invests labor in the economic sectors and receives income and

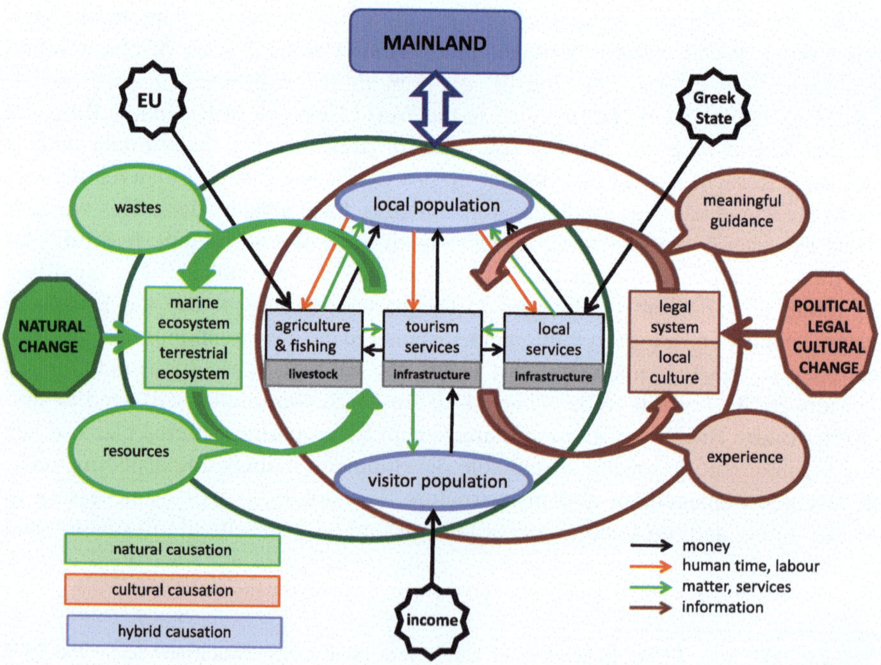

Fig. 28.1 Comprehensive model of the island's socioecological system

services in return. The visitor population brings money from outside and receives services in return. All economic sectors draw on certain resources from the marine and/or terrestrial environment and generate wastes in return. The behavior of all actors is guided by the island's legal and cultural system, and this system, in turn, may incorporate new experiences.

The socioecological system of the island strongly depends on the outside world. There is a shipping lifeline with the mainland that provides transportation for people's mobility (both the local population and visitors), for imports and exports, for services and for information (the latter today is less dependent on physical transportation than in the past). Marine and terrestrial ecosystems, as well as most hybrid structures, are influenced by natural changes (such as climate change), and the island's legal and cultural system is affected by political, legal, cultural (for example, religious) and economic changes in the outside world. Additionally, there are currently three regular major income flows originating from outside: the money visitors bring with them (as mentioned above), European Union (EU) subsidies for the agricultural sector and the payments made by the Greek state for local services (such as communal administration, schools, medical services, social security payments, energy services, public transport and public infrastructure).[3] Most elements pinpointed in this model can be assumed to be relevant for the contemporary situation as well as for the past.

The ability of any socioecological system to reproduce itself hinges on the interdependencies of its stocks and flows. When critical stocks cannot be reproduced, the system collapses. We assert that the societal metabolism needs to operate between certain margins. High sociometabolic demands and the subsequent path dependencies can lead to an unsustainable use of natural resources and a degradation of the resource base. Likewise, a standard of living that is too low (relative to the rest of the society of which the island is part) may lead to inhabitants leaving the island for more prosperous parts of the country, thus inducing a breakdown of certain basic institutions (e.g., family support systems, schools, health services) and eventually sending the local society into a downward spiral of unsustainability. Thus, on the one hand, there exist ecological/biophysical tipping points (such as the overuse of resources) that, if crossed, can lead to ecological collapse. Overfishing, deforestation and massive erosion following overgrazing serve as examples. On the other hand, there are social tipping points that, if crossed, can lead to societal disintegration and collapse and, finally, to the die-out of the local population. An example of the latter is the maintenance of a school. If a local secondary school cannot be maintained, young parents would be forced to withdraw from the island, which in turn would threaten the caretaking of the old, who would therefore also have to leave the island.

These issues, familiar from 'dying rural peripheries', are even more pronounced in island communities that face more stringent limitations or inefficiencies.

[3]There is also a return flow of taxes from the island, but here we assume there is a net balance of public monetary flows to the island.

Bussing children to school in a neighbor community may be an option, but shipping them regularly across the sea is not. Thus, tipping points become more critical and, if crossed, can lead to a mass abandonment of the island. In the words of Deschenes and Chertow (2004), '[T]he island context shortens the planning horizon over which sustainability concerns become important' (p. 204). Whereas the fertility of the land and protection from raids may have been critical for sustaining the local population in agrarian societies of the past, in modern society, the critical level of the sociometabolic standard and complexity is the relevant issue.[4]

Comparable dynamics, perhaps of a different qualitative nature, can be identified for different historical periods on the island. Identifying and categorizing historical 'tipping points' would help us better understand the system under study, its main transformative forces and its system imbalances. These may include a fluctuation in the number of island visitors (in history, as pilgrims), interrupted trade relations and connectivity with the mainland and singular events such as earthquakes and pirate attacks. Identifying the main dynamics of the system at different historical periods would help us to better understand future challenges and outline sustainable pathways.

28.4 Methods

Following a systemic approach, we focus on the socioeconomic system of Samothraki and the domestic environment where it is embedded (Fig. 28.1), breaking it down to its various subsystems and identifying the interrelations among them (and the potential links to other socioecological systems). Our intention is to operationalize the system compartments and flows as pictured in Fig. 28.1, focusing on key linkages (metabolism, labor and monetary flows), and to generate, as far as possible, quantitative estimates. We will then attempt to link the stocks and flows in a formal model simulating the system dynamics of the respective time period. With regard to the present, we will explore the option space for sustainability and use qualitative methods to determine whether cultural change toward collaboration and self-empowerment may occur and widen the range of possible solutions during a prolonged economic crisis.

The two main themes to be addressed are the roles of agriculture and tourism. Agriculture feeds the population but demands substantial labor and causes environmental problems/sustainability issues. A reconstruction of the agricultural system now and in the past reveals possible system imbalances that affect the

[4]A recent study (PLANISTAT 2002) identified a population of 4000–5000 people as a key threshold for the provision of an important part of services locally. This still does not include 'superior services' (e.g., hospitals, tertiary education, cinemas); for those, one must travel to a larger urban center (Spilanis et al. 2012).

vulnerability or robustness of the system under study. To account for the spatial dimension, two issues are important: harvest and land cover change. Moreover, tourism brings benefits and costs and can be seen as problem solving (providing opportunities) as well as problem creating (environmental degradation). Tourism on Samothraki, both in the ancient past (in the form of pilgrim visitors) and during the last few decades, has revitalized the island and saved it from population collapse. However, more visitors means more infrastructure and, thus, an increase in complexity.

A series of interdisciplinary methods are used to generate data. First, archival work is necessary. This includes scanning and reviewing official reports and scientific and popular documents to reconstruct the ecological, socioeconomic and legal conditions of Samothraki in the past few decades. It also includes data mining of statistical accounts on the island's demography, material and energy flows. To reconstruct Samothraki's current socioecological system, a series of methods are required, including ecological (land-use) methods, sociological methods and 'hybrid' socioecological methods. Hybrid methods bridge the ecological and sociological methods by estimating the current social metabolism of the island in terms of material/energy flows and by reconstructing the sectors of the island's economy in biophysical terms. Moreover, to learn from past crises, a reconstruction of Samothraki's socioecological system during periods of past collapses can provide important insights regarding current tipping points. This could be achieved by combining socioecological methods with archaeological data, such the reconstruction of land use, and with the socioeconomic and paleo-ecological system on the island.

Some of these methods have already been applied on Samothraki (see Petridis et al. 2013): After a thorough inquiry into existing statistical documents, the qualitative method of focus groups with various stakeholders on the island was used to determine initial viewpoints from the local inhabitants of Samothraki on their livelihoods, ways of life and the environment surrounding them and to explore current challenges and alternative visions for the future of the island. We chose to arrange seven focus groups, divided according to socioeconomic status, to obtain an array of perspectives: elderly people; parents of small and school children; craftsmen; farmers and livestock herders; small-scale fishermen; tourism-related professionals; and professionals in the tourism accommodation sector. Next, we used distance sampling, an established quantitative method involving line transects, to produce independent estimates of the livestock burden deemed responsible for many problems such as erosion and loss of biodiversity on the island. Our two methods were conducted and analyzed in parallel to each other. The results of the line transects, revealing an estimate of the population of small ruminants on the island, helped to confirm much of what was said in the focus groups and allowed cross-checking of the livestock population count in official documents.

28.5 Results and Discussion

28.5.1 Social and Ecological Challenges and Opportunities

Within the past 20 years, there has been exponential growth in the number of sheep and semi-wild goats on the island, largely due to the agricultural policies of the EU. According to the latest estimations based on slaughtering statistics, the number of domestic and free-roaming goats and sheep totals approximately 60,000–80,000 (Greek Ministry of Agriculture 2008). This range corresponds well with the livestock estimates arrived at by our counts. Overgrazing, coupled with the steepness of the terrain, has led to dramatic levels of soil erosion. This erosion also occurs within the Natura 2000 area and poses a major threat to its conservation goals. The erosion is also destroying roads. Illegal logging and collection of firewood from the forests has led to further erosion and deterioration of the forests and their regenerative capacity. Rising feed prices and market fluctuations may force farmers to become more independent from financial support and to change their current practices. This could be seen as a 'fresh start' toward a better utilization of livestock (i.e., reaching the same income with a substantially reduced number of animals) through marketing and innovations that would improve the value-chain of (organic) agricultural products (Petridis 2012).

The tourism sector faces the same problems as all seasonal vacation destinations. On the one hand, the sector is affected by a general decrease in visitation in the off-season. On the other hand, according to the shipping data analyzed, it experiences a highly concentrated tourist season of less than two months. Under current conditions, most infrastructures on the island are being overused for the peak months and then remain underutilized for the rest of the year. The challenge of development toward a more sustainable form of tourism should be met by efforts to reduce the environmental burden associated with tourism while seeking to increase the local income derived from it, including the generation of more highly qualified jobs that would allow young, educated people to stay on the island and sustain their lives there. The goal, then, should be to identify attractions, activities, information channels and target groups to populate Samothraki with visitors at other times of the year and to provide incentives for longer stay (e.g., family opportunities).

28.5.2 Future Prospects of the Transdisciplinary Process

Within the local population, two main tendencies can be identified. On the one hand, there is a large group of young and educated people working directly or indirectly in the public and tourism sectors who wish to protect the natural and cultural heritage of the island and who are motivated to look into innovative, collaborative ways of doing so. They are highly supportive of the idea of establishing a Biosphere Reserve on Samothraki, and they believe that they could individually and collectively benefit from such a scenario. On the other hand, there is a more conservative group of middle-aged males with low levels of education who are mainly occupied with farming and livestock herding. They appear to be slightly skeptical and even indifferent to the idea of a Biosphere Reserve, which is seen as an external enforcement that would limit their ability to keep large subsidized herds of free-roaming goats. Nevertheless, the current financial situation seems to be forcing them to look into more collaborative ways out of the crisis, and a new generation of farmers with a fresher attitude toward the utilization of agricultural produce seems to be in disagreement with those holding more traditional views.

Another aspect is the importance and the special position of the so-called 'outsiders' or 'newcomers', who are often the group of residents who are initiating and implementing new ideas. Perhaps this happens because of their ability to take a different, more distant view on the situation, or perhaps they have—compared to the average resident—a different background, coming from cities, the mainland and different countries. This may allow us, as the research group, to act as outsiders and initiate new development. Despite this impression of the importance of outsider initiators, it is also necessary to involve the locals so that the whole process of becoming a Biosphere Reserve is a process of real participation. We gained the impression of general mistrust in all things official, governmental, subsidized and other top-down arrangements, such as middle men and previous dealings with cooperatives. This makes it even more important to frame the process of establishing the Biosphere Reserve as a transparent and bottom-up process.

A common strategy on Samothraki is to reduce external dependency and the cost of living by practicing some degree of subsistence agriculture and trying to increase income by engaging in various activities such as tourism, handicraft and agriculture. Self-sufficiency has been of increasing importance, and most people on the island own a piece of land and a few olive trees, produce their own vegetables or wine or own some goats. Thus, despite the lack of opportunities, the island still provides the basic requirements for self-subsistence as well as a social network that is stronger than in any larger city of mainland Greece.

Nevertheless, there is a general awareness that things are currently unsustainable and that change is needed to be able to support the livelihoods on the island in the near future. The generalized financial crisis, coupled with the widespread mistrust in higher institutions, does not leave room for hope of external support. Rather, and despite negative experiences with earlier efforts with cooperatives, people seem increasingly convinced that the only way to combat the current crisis is by seeking a collaborative way out. Most focus group participants appeared highly motivated to change their traditional attitudes and cooperate for a common future on the island. Practically speaking, this would mean, first and foremost, resuming more organized communication among different interest groups, developing joint strategies and, at a later stage, setting up local cooperatives.

The tourism sector is considered important for a more sustainable Samothraki. Tourist-dependent professionals seemed well aware of this. They seemed especially eager to pursue a well-defined vision of sustainable tourism that would extend the season without adding an extra environmental burden. This burden would be avoided by attracting groups of visitors who would respect the unique cultural and natural assets of the island and potentially give something back. This would require the development of a joint strategic plan as well as clear marketing strategies. This attitude is very much in line with the Biosphere Reserve concept, which has generally been positively received by most focus group participants. The benefits of increased visitation were clearer to professionals working in the accommodation sector. However, the benefits would also be enjoyed by people working in the primary sector via an enlarged market for high-quality organic agriculture, an improved value chain of processed agricultural products and the promotion of agro-tourism.

28.5.3 Sustainability and Collapse

Tainter (1988, 2011) distinguishes collapse from other transitions by highlighting the abrupt reduction in complexity. *Sociopolitical collapse* is defined as a rapid simplification or the loss of an established level of social, political or economic complexity (Tainter 1988, 2006). What drives collapse? In Tainter's view, collapse occurs when the level of social complexity can no longer be sustained because the costs of additional complexity generation in the course of problem solving exceed the benefits. Following the sociometabolic approach, we conceptualize collapse as a situation where social metabolism ceases to function properly. In a socio-ecological system, or, more precisely, a socioeconomic system with ecological hinterlands, collapse occurs when one or more subsystems can no longer reproduce themselves. Complex systems may have several *tipping points*. Sometimes referred to as critical transitions or catastrophic thresholds, these are points at which a system shifts abruptly from one state to another (Scheffer et al. 2009). This may happen when the ecological resilience of a system is exceeded (Briske et al. 2010), or it may be triggered by social stressors, such as demographic, economic governance and environmental-perception factors (Dearing et al. 2010). The research challenge remains whether we can identify such ecological or social tipping points in advance and then link this concept of collapse to specific points in Samothraki's history.

Drawing from our data, we should be able to provide general answers to the following questions:

- What is the role of shared beliefs and collaboration? How is collaboration established, and what causes it to disintegrate? How does a functioning division of labor come into being, and how can it be governed? What role does a feeling of shared identity and 'specialness' play within the island population? In this context, the findings of Ostrom (2009) are highly relevant.
- What appear to be key conditions for the sustainability/resilience of island communities? What, on the other hand, triggers collapse? What are the implications for the future of Samothraki?

28.6 Conclusions

The Greek island of Samothraki is endowed with high cultural and unique natural assets and is in a crucial phase of development toward a more sustainable future path as it is soon to become the first 'new generation' UNESCO Biosphere Reserve in Greece. In line with the nature of the process and research goals, two distinct but complementary future directions can be identified. On the one hand, future efforts should deepen and systematize knowledge on island sustainability both now and in the past. To avoid collapse in the present, we need to generate deeper insights from past collapses as well as identify current threats and possible ailments. On the other hand, a practical focus remains on how to drive a process toward a solid local development using, for example, a collaborative attitude for it to succeed. We need to explore the option space for sustainability and use qualitative methods to determine whether in a situation of crisis, cultural change toward collaboration and self-empowerment may occur and widen the range of possible solutions.

There is special interest in the ways the two processes, distinct as they may be, are mutually supportive and reinforce each other. For example, focus group interviews were primarily used to generate data, but they were also instrumental in moving the process forward by providing a platform for local stakeholders to form networks and to explore future opportunities for their island. Forming a local association to act as a Biosphere Reserve management body has been instrumental in facilitating the local development process, but was also an opportunity to network with other scientists performing research work on the island and, in so doing, generate new scientific knowledge as a solid foundation for scientifically guided decision-making on the island.

Acknowledgements We are indebted to Lazaros Xenidis for his preparatory research from 2007 to 2009, to Peter Fleissner for advice on modeling strategies, to Sheba Schilk for inquiring into the possibilities of realizing these strategies with archaeological and historical data and to all the students who contributed to the focus group data analysis. We are at least as indebted to the mayors of Samothraki, Petroudas and Chanos for their continuous political support and to the non-governmental organization (NGO) 'Samothraki in Action' and its active female members for their initiative, local information and persevering networking. Last but not least, we would like to gratefully acknowledge financial support from the Austrian Academy of Sciences as well as support and encouragement from the Austrian UNESCO-MAB Committee.

References

Baldacchino, G. (2006). Islands, Island studies, island studies journal. *Island Studies Journal, 1*(1), 3–18.

Briske, D. D., Washington-Allen, R. A., Johnson, C. R., Lockwood, J. A., Lockwood, D. R., Stringham, T. K., & Shugart, H. H. (2010). Catastrophic thresholds: A synthesis of concepts, perspectives and applications. Ecology and Society *15*(3), 37. [online] url:http://www.ecologyandsociety.org/vol15/iss3/art37/

Dearing, J. A., Braimoh, A. K., Reenberg, A., Turner, B. L., & van der Leeuw, S. (2010). Complex land systems: the need for long time perspectives to assess their future. *Ecology and Society 15*(4), 21. [online] url:http://www.ecologyandsociety.org/vol15/iss4/art21/

Deschenes, P. J., & Chertow, M. (2004). An Island approach to industrial ecology: Towards sustainability in the island context. *Journal of Environmental Planning and Management, 47*(2), 201–217.

Fischer-Kowalski, M., & Weisz, H. (1999). Society as hybrid between material and symbolic realms: Toward a theoretical framework of society-nature interaction. *Advances in Human Ecology, 8*, 215–252.

Fischer-Kowalski, M., Xenidis, L., Singh, S. J., & Pallua, I. (2011). Transforming the Greek Island of Samothraki into a UNESCO biosphere reserve. An experience in transdisciplinarity. *GAIA, Ecological Perspectives for Science and Society, 20*(3), 181–190.

Greek Census. (2011). National Statistical Service of Greece (EL.STAT.). Retrieved from www.statistics.gr

Greek Ministry of Agriculture. (2008). Annual report on agriculture on Samothraki Island.

Ishwaran, N., & Persic, A. (2008). Concept and practice: The case of UNESCO biosphere reserve. *International Journal of Environment and Sustainable Development, 7*(2), 118–131.

Kerr, S. A. (2005). What is small island sustainable development about? *Ocean and Coastal Management, 48*, 503–524.

Nguyen, N. C., Bosch, O. J. H., & Maani, K. E. (2011). Creating "learning laboratories" for sustainable development in biospheres: A systems thinking approach. *Systems Research and Behavioral Science, 28*, 51–62.

Ostrom, E. (2009). A general framework for analyzing sustainability of social-ecological systems. *Science, 325*(5939), 419–422.

Petridis, P. (2012). Perceptions, attitudes and involvement of local residents in the establishment of a Samothraki Biosphere Reserve, Greece. *Journal on Protected Mountain Areas Research, 4*(1), 59–63.

Petridis, P., Hickisch, R., Klimek, M., Fischer, R., Fuchs, N., Kostakiotis, G. et al. (2013). Exploring local opportunities and barriers for a sustainability transition on a Greek island. Social Ecology Working Paper 142, Vienna, Austria. Available at: http://www.uni-klu.ac.at/socec/downloads/WP142_WEB.pdf

PLANISTAT. (2002). Analyse des régions insulaires de l'Union Européenne. Report prepared for the European Commission.

Scheffer, M., Bascompte, J., Brock, W. A., Brovkin, V., Carpenter, S. R., Dakos, V., et al. (2009). Early-warning signals for critical transitions. *Nature, 461*, 53–59.

Spilanis, I., Kizos, T., & Petsioti, P. (2012). Accessibility of peripheral regions: Evidence from Aegean Islands (Greece). *Island Studies Journal, 7*(2), 199–214.

Tainter, J. A. (1988). The collapse of complex societies. New studies in archaeology (reprinted in 2003), Cambridge University Press, Cambridge, England.

Tainter, J. A. (2006). Social complexity and sustainability. *Ecological Complexity, 3*(2), 91–103.

Tainter, J. A. (2011). Energy, complexity, and sustainability: A historical perspective. *Environmental Innovation and Societal Transitions, 1*(1), 89–95.

UNESCO. (1996). *Biosphere reserves: The Seville strategy and statutory framework of the world network*. Paris: UNESCO.

Method Précis: Transdisciplinary Research

Willi Haas

Within Sustainability Science, the recognition of the relevance of engaging social actors in addressing complex societal problems has been growing in recent years (Erb et al. 2013; Hirsch Hadorn et al. 2008; Lang et al. 2012). The involvement of actors from outside the scientific community is perceived as helpful in achieving more meaningful research results, particularly in terms of their adequacy and usefulness for stakeholders. In other words, with such an approach, scientists attempt to increase the effectiveness of their insights in non-science domains. The involvement of social actors has led to an approach called 'transdisciplinary research'. With transdisciplinary research, science goes beyond the idea of interdisciplinarity, a research approach that aims at transgressing scientific disciplines within academia (Dressel et al. 2014).

However, interpretations of transdisciplinarity differ among world regions and scientific fields. In the context of socioecological research, the definition proposed by Christian Pohl and Trude Hirsch Hadorn is widely used. They suggest that transdisciplinary research should deal with socially relevant problems in such a way that it can '(a) grasp the complexity of problems, (b) take into account the diversity of life-world and scientific perceptions of problems, (c) link abstract and case-specific knowledge, and (d) constitute knowledge and practices that promote what is perceived to be the common good. Participatory research and collaboration between disciplines are the means of meeting requirements (a)–(d) in the research process' (Pohl and Hirsch Hadorn 2007, p. 30).

This approach is not just a new label. Taken seriously, it has far-reaching consequences for the types of results that are sought and, consequently, how the research process should be organized. Thus, the involvement of so-called stakeholders entails the promotion of problem solutions relevant to the non-scientific social actors involved in addition to generating novel scientific insights. This enlarged scope is based on the premise that knowledge exists and is also produced in societal fields other than science (Hirsch Hadorn et al. 2008; Klein 2004). Consequently, the entire research process needs to be designed appropriately.

Various authors have proposed conceptual models for a transdisciplinary mode of research that enables a better and shared understanding of the endeavor among the various actors in terms of quality assurance, evaluation and self-reflection. Such schemes suggest the inclusion of the various actors right from the start, that is, within the process of planning and problem formulation (e.g., Smetschka et al. 2008, p. 6).

A scheme that has received wide recognition conceptualizes the transdisciplinary research process as the generation of integrated knowledge that can finally be incorporated into practical problem solving strategies as well as into new interdisciplinary knowledge (Jahn 2008, p. 31). To achieve such integrated knowledge, it is necessary to integrate a societal as well as a scientific understanding of the problem at hand to constitute the common research subject (Fig. 28.2).

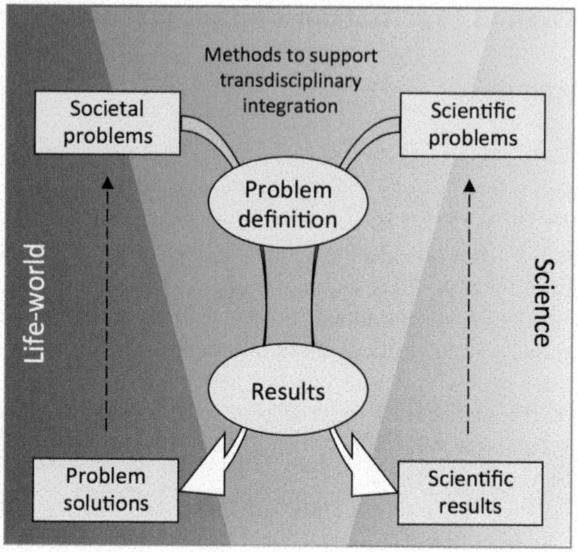

Fig. 28.2 Scheme for a transdisciplinary research process that aims at integrating life-world and science perspectives in generating integrated knowledge. (Based on Jahn 2008, p. 31)

To this end, several methods to include society in participatory processes—from problem framing to seeking solutions—are in use. Within the field of participatory technology assessment, the involvement of non-science actors has taken the most elaborate and reflected forms, including citizens' panels, scenario workshops, round tables and consensus conferences, 21st century town meetings, charrettes, citizens' juries, technology festivals, world cafés, expert panels, focus groups, planning cells, as well as deliberative polling and participatory assessment, monitoring and evaluation (PAME) (for an excellent overview of these approaches, see Elliott et al. 2005). In addition to these participation tools, within the post-normal science discourse the social multi-criteria evaluation (SMCE) has been developed as a formalized procedure for the entire research process aiming at a participatory, strategic and integrated assessment of a problem. The emphasis is explicitly on the quality of the 'decision process' rather than the 'final choice' (Munda 2004; see also Chap. 5 in this book).

Among the various methods used to engage stakeholders, scenario workshops have proven to be both practical and useful. Scenarios are best employed if large changes are anticipated and if the expected challenges appear to be complex. Such a situation requires thorough consideration of the many driving forces and alternative pathways. Scenario workshops are interactive and are perfectly suited to the development of shared storylines and visions of alternative futures. They promote cross-group communication on driving factors as well as the pros and cons of various scenarios, and, if the goal be such, they help stakeholders arrive at a consensus on a future scenario or perspective. This method of engaging stakeholders is best suited to developing an informed understanding of future local or regional sustainability.

A further approach that has been proven useful in the field of socioecological transdisciplinary research is agent-based modeling (ABM)—for details, see Method Précis on Agent-Based Modeling in this volume.

The limitations of transdisciplinary approaches that often entail the employment of time-consuming methods must be seriously considered. Seeking consensus among a variety of perspectives and interests can be both limiting in scientific outcome and frustrating. Of course, if such an inclusive science with an emphasis on integration across disciplines is successful, it holds enormous potential for effectively addressing complex sustainability challenges beyond science.

References

Dressel, G., Berger, W., Heimerl, K., & Winiwarter, V. (Eds.) (2014). Interdisziplinär und transdisziplinär forschen, transcript-Verlag Bielefeld.

Elliott, J., Heesterbeek, S., Lukensmeyer, C. J. & Slocum, N. (2005). Leitfaden partizipativer Verfahren. Ein Handbuch für die Praxis. Brüssel-Wien: kbs-ITA.

Erb, K.-H., Gaube, V., Fischer-Kowalski, M. (2013): Leaping over disciplinary shadows. Global Change Magazine, October 21; http://www.igbp.net/download/18.29d2af531409c09b24364c/1382360686522/NL81-interdisiplinary.pdf

Hirsch Hadorn, G., Hoffmann-Riem, H., Biber-Klemm, S., Grossenbacher-Mansuy, W., Joye, D., Pohl, C., et al. (Eds.). (2008). *Handbook of transdisciplinary research*. Stuttgart, Berlin, New York: Springer.

Jahn, T. (2008): Transdisziplinarität in der Forschungspraxis. In: Matthias Bergmann/Engelbert Schramm (eds.): Transdisziplinäre Forschung. Integrative Forschungsprozesse verstehen und bewerten. Frankfurt/New York: Campus Verlag, pp. 21–37.

Klein, J. T. (2004). Prospects for transdisciplinarity. *Futures, 36*, 515.

Lang, D. J., Wiek, A., Bergmann, M., Stauffacher, M., Martens, P., Moll, P., et al. (2012). Transdisciplinary research in sustainability science: Practice, principles, and challenges. *Sustainability Science, 7*(1), 25–43.

Munda, G. (2004). Social multi-criteria evaluation: Methodological foundations and operational consequences. *European Journal of Operational Research, 158*, 662–677.

Pohl, C., & Hirsch Hadorn, G. (2007). Systems, targets and transformation knowledge. In: G. Hirsch Hadorn, H. Hoffmann-Riem, S. Biber-Klemm, W. Grossenbacher-Mansuy, D. Joye, C. Pohl, U. Wiesmann, & E. Zemp, (Eds.), Principles for designing transdisciplinary research. (pp. 36–40). Munich: Oekom-Verlag.

Smetschka, B., Gaube, V., & Lutz, J. (2008). Gender als forschungsleitendes Prinzip in der transdisziplinären Nachhaltigkeitsforschung. In E. Reitinger (Ed.), *Transdisziplinäre Praxis* (pp. 23–34). Heidelberg: Carl-Auer Verlag.

Chapter 29
Health Through Socioecological Lenses—A Case for Sustainable Hospitals

Ulli Weisz and Willi Haas

Abstract We reflect on the interrelations of two important societal concerns: sustainable development and health. In this field of research, the key notion of 'health co-benefits' has been coined. Health co-benefits aim to utilize synergies between both respective strategies. This approach is mainly applied in climate change mitigation and is intended to inform policies. 'Cross-cutting' issues, such as energy, agro-food systems and transport, are receiving increasing international recognition. However, the significant case of the health care system has rarely been addressed. While responsible for the reproduction of human health, with its energy- and material-intensive forms of therapy, it contributes to environmental problems. Therefore, the health care system itself threatens human health. In a transdisciplinary series of hospital projects involving scientists and health care practitioners, we asked how sustainability can be conceptualized for hospitals in line with both a socioecological understanding of sustainable development and 'hospitals' reality'. Our approach aims to avoid unintended long-term and side effects of health care—hospitals' core business—by expanding quality criteria for decision-making to include sustainability and health gain improvement. The results of the testing phase convinced political actors in the health care system, and they demonstrate that health co-benefits are a valuable additional argument within the sustainability debate.

Keywords Sustainable development · Human health · Health promotion · Health co-benefits · Sustainable hospitals

---

The German version of this chapter is part of the lead author's dissertation, which contains more comprehensive literature references than is possible to include here (see Weisz 2015).

U. Weisz (✉) · W. Haas
Institute of Social Ecology (SEC), Alpen-Adria University,
Schottenfeldgasse 29, 1070 Vienna, Austria
e-mail: ulli.weisz@aau.at

W. Haas
e-mail: willi.haas@aau.at

H. Haberl et al. (eds.), *Social Ecology*, Human-Environment Interactions 5,
DOI 10.1007/978-3-319-33326-7_29

29.1 Two Societal Concerns

'[...] ultimately, the entire report is about health' (Brundtland 1989, p. 52). This summary, by no less than Gro Harlem Brundtland herself, refers to the report commonly named after her, officially titled *Our Common Future* (WCED 1987). As is widely known, in 1987, the year in which the *Brundtland Report* was published, she was the Chair of the World Commission on Environment and Development (WCED), Environment Minister and (for the second time) Prime Minister of Norway. What is probably less well known is that Harlem Brundtland, originally a physician with a public health specialty from Harvard, assumed the leadership of the World Health Organization (WHO) as General Director for five years in 1998. Thus, she personally embodies two central societal concerns that have lost none of their relevance today: sustainable development and health promotion. The prominent politician is not alone in this. Since the debate over sustainable development began, emphasis has been placed on references to both issues at the highest echelons of political power. Health is seen as an important prerequisite to and as an outcome of sustainable development (cf. Weisz et al. 2011). This is underpinned by the assumption that addressing these issues together makes synergies, known as co-benefits, possible.

If we look back at the beginnings of these movements, we find astounding parallels. Sustainable development and New Public Health[1] are interpreted as reactions to the unintended negative impacts of industrial modernization (Pelikan, personal communication; Wegleitner 2012), which have become increasingly evident through the second half of the 20th century. As both a special characteristic and a historical first, these problems acquired global dimensions (cf. Fischer-Kowalski 1997). This volume contains a wealth of examples showing the core biophysical problems of industrialization, such as the pattern of industrial metabolism, the impacts of which are particularly evident in climate change and resource scarcity. New health problems and diseases, which have developed in parallel to the achievements of this epoch, are accordingly described as diseases of civilization '[...] that emerge in a way that is difficult to control as the result of a way of life [...]' (Luhmann 1990, p. 182 f., author's own translation).

Thus, it is no coincidence that the founding documents for each movement appeared almost contemporaneously (the *Ottawa Charter*, WHO 1986; and the *Brundtland Report*, WCED 1987; cf. Weisz et al. 2011). Ultimately, these movements pursue the same goal: the wellbeing of all humans and a 'holistic human development', as it is formulated in the public health literature (Dooris 1999, p. 366). It is therefore all the more astounding that, despite these commonalities

[1]The origins of the New Public Health movement can be traced to the *Ottawa Charter for Health Promotion*, which takes a broad approach to health policy (WHO 1986).

and analogies, the development of both movements occurred largely in parallel yet without interconnection, as Dooris describes in detail (ibid.).[2]

For us, this is of twofold interest: first, in terms of a more detailed understanding of the relation between these two concerns as referred to in the political documents; second, in terms of the question of the potential co-benefits that may be realized if we engage with the theme of health through our socioecological perspective, using our socioecological understanding of sustainable development. We discuss this in our contribution by drawing upon a research example from the hospital setting.

Let us begin with the first of these points. In the international literature, there are many examples where both themes are addressed in combination. However, it appears that efforts to ensure that health considerations are given more prominence in the sustainability debate and the conception of common strategies for solutions are primarily instigated by the public health community.[3] It is stressed that the effects of sustainability strategies on health should be (more strongly) taken into account (as called for in politics). Finally, and very forcefully, this message was conveyed within the framework of the Rio+20 Conference (The Lancet 2012). Nonetheless, attempts to operationalize this interconnection and an interdisciplinary connection among the respective scientific communities remain largely absent.

29.2 Health Co-benefits—Looking at the International Literature

Within the public health community, the relationships between these two themes have attracted increased interest since the late 1990s, particularly within the context of climate change and its negative consequences for health. These relationships have also become the focus of political attention (e.g., WHO 2009) and research (see, for example, the series on 'Energy and Health' and 'Health and Climate Change' published in *The Lancet* in 2007 and 2009, respectively). So-called 'cross-cutting issues' have been addressed, such as energy, nutrition, food production, transport, mobility, urban planning and (occasionally) the role of health systems. Health also finds a place in the relevant assessment reports (e.g., Smith et al. 2014) and in cost analyses within research that examine the consequences of climate change (e.g., Ciscar et al. 2011; Stern 2006). This research is

[2]The attempt at linking the 'healthy cities' and 'sustainable cities' programs represents an exception in this context [e.g., Dooris 1999, see also www.who.int/healthy_settings/en (accessed on March 1, 2013)].

[3]Recent publications note that the issue is now attracting interest beyond the public health community (e.g., Shindell et al. 2012; Tilman and Clark 2014).

contextualized according to the notion of 'health co-benefits' (i.e., co-benefits between climate change mitigation and human health) as a contribution to a 'sustainable transition'.[4] It aims to propel the implementation of climate protection measures—which, depending on the inertia of the climate system, might only show results after several decades—through public health benefits that can often be generated more quickly and that have an impact at local scales.

A few examples may serve to illustrate this point.

A prominent example is the motorized personal transport that exists in industrialized countries. It is responsible for a considerable and growing share of global greenhouse gas (GHG) emissions. At the local level, such emissions (fine particles, ozone precursors, noise) place pressure on health and quality of life for large population groups. At the same time, motorized personal transport promotes a lack of physical exercise, which has been identified as partly responsible for a variety of diseases associated with civilization (e.g., cardiovascular diseases, obesity related health problems). In this context, the health co-benefits approach investigates altered mobility patterns in regions with a high population density: a reduction in motorized transport in favor of physical mobility (cycling, walking) reduces certain individual health risks. If this approach, based not only on individual initiatives but also on (urban) policy measures, leads to a tangible reduction in urban traffic volume, then air quality and noise pollution problems will be improved in these areas along with an effective contribution to climate protection (e.g., WBGU 2011).

A further example of health co-benefits is provided by the food chain (agriculture, food production, nutrition). Public health experts, nutritional scientists and physicians are addressing the problem of unhealthy eating habits among large sectors of the population in industrialized countries.[5] These poor habits are seen as the leading cause of various lifestyle diseases (e.g., cardiovascular diseases, a range of cancers, chronic degenerative diseases, excess bodyweight and obesity) with all their attendant problems. A transition to a healthy (sustainable) diet, which would include eating less meat, would go hand in hand with a reduction in livestock breeding and, consequently, a significant reduction in GHG emissions (e.g., Friel et al. 2009; Westhoek et al. 2014). McMichael and colleagues highlight another aspect: a fairer global distribution of meat consumption would have the dual benefit of reducing the impact of health problems caused by overeating in highly industrialized nations and reducing the malnutrition suffered in poor countries (McMichael et al. 2007).

---

[4]This research thus links to an earlier tradition that predominated at the beginning of industrialization, where health and environmental problems were addressed together. This integrated approach evidently came to a halt when environmental problems began to go far beyond the bounds of hygiene problems and acquired a global dimension, characterized by a spatial and temporal distance between cause and effect (cf. Smith and Ezatti 2005). Environmental medicine approaches and research into 'environmental (public) health' are an exception in this context, although they have what we would regard as a too 'narrow' epidemiological focus in most cases.

[5]A trend that is also observable in the developing world (cf. Tilman and Clark 2014).

Of particular interest for our concerns are the potential co-benefits within the health care systems of industrialized countries. Above all, as the lead organizations within these systems, hospitals pollute the natural environment through their services and preliminary services and thereby contribute to global problems of sustainability. This has a negative impact on human health and quality of life, which the health care systems then have to address. According to the calculations of the Stockholm Environment Institute (SEI), the total CO_2 emissions produced by the National Health Service (NHS) in England in 2004 amounted to 18.6 Mt/year (megatons per year).[6] This represents 2.6 % of all emissions produced through consumption by the UK (SDC-SEI 2008). According to our own estimates, Austria's hospitals emit ca. 2.4 Mt/year CO_2 and produce 4.5 % of the annual national CO_2 emissions.[7] A shift from the currently dominant material- and energy-intensive 'repair' medicine in favor of a prevention-oriented health-promoting medicine could help interrupt this dynamic. Alongside directly positive health impacts for the population, this co-benefits strategy would lead to a reduction in environmental risks to health and would simultaneously relieve the pressure (also in economic terms) on our health care systems. Meanwhile, health systems and hospitals will be challenged to increase their engagement with this issue (McMichael et al. 2009; WHO and HCWH 2009). Nonetheless, a systematic approach, let alone an implementation-oriented approach, is still largely missing.

At this juncture, we wish to introduce our research example from the hospital setting. It is taken from a longstanding inter- and transdisciplinary cooperation (see Table 29.1) that had the aim of identifying potential synergies or (to adopt the terminology of public health literature) co-benefits between sustainable development and health in an implementation-oriented approach. We were aware that we could not take the 'ivory tower' approach to research. This contributes to explaining our decision to choose a hospital and, thus, the organization level as the investigation and intervention level for our study—in other words, why we decided in favor of 'Social Ecology in hospitals'. With this example, we traverse 'socioecological frontiers' in a twofold sense: we address the theme of health and explore the organizational level, encountering the actors within hospital practice.

[6]Consumption-based survey including import-related emissions (for methodological details, see SDC-SEI 2008).

[7]Including preliminary services. The share of other emissions, such as NOx, or toxic waste, at 4–7 %, is thus on a similar scale (Weisz et al. 2011).

Table 29.1 'The Sustainable Hospital': Key features of the project series

Year	Sustainable hospital project series	Funded by	Project partners
2004–2005	**Feasibility study**	Austrian Federal Ministry for Transport, Innovation and Technology (BMVIT) within the transdisciplinary program *Factory of Tomorrow*	**Science partners** An interdisciplinary team of researchers from the • Institute of Social Ecology, Vienna/IFF (project coordinator) • Ludwig Boltzmann Institute Health Promotion Research (LBIHPR) • ARECon GmbH (testing phase) • IUW/Vienna University of Economics and Business (feasibility study) • WIHO/IFF (reflection phase) **Hospital partners** • Otto Wagner Hospital, Vienna (pilot hospital) • Vienna Hospital Association • Immanuel Diakonie Group, Berlin (consulting observers)
2006–2008	**Testing the sustainable hospital** key areas: • sustainable business management: creating orientation • sustainable provision planning: innovative planning • sustainable service provision: making sustainability work in day-to-day business	BMVIT/ program *Factory of Tomorrow* and Austrian Research Promotion Agency (FFG)	
2009–2010	**Dissemination**	BMVIT/ program *Factory of Tomorrow* Research Council of the Alpen-Adria University Klagenfurt, Vienna, Graz	
2011	**Reflecting the sustainable hospital**	No external funding	

Link to project description and downloads: http://www.fabrikderzukunft.at/highlights/ (retrieved August 24, 2013)

29.3 Social Ecology in the Hospital—Theoretical Prerequisites

Although sustainability problems are global in character, the causes of these problems and most of the actors who need to respond to them are anchored at the local level. Thus, organizations, as important decision-making and active operational bodies at local level, must be examined when implementing sustainability strategies.

Hospitals are special organizations (see Grossmann 1997; Mintzberg 1993). As the lead organizations of our health and patient treatment system, they are responsible for the reproduction of health—specifically, the maintenance and recovery of public health, an (existentially) important public good.[8] This places these organi-

[8]In this sense, these organizations also belong to the few social function systems that continue to concern themselves with aspects of social metabolism (interacting directly with nature) and to co-organize these. The nature with which the health system is directly confronted is human nature.

zations in both a special and (as already described above) paradoxical situation. In carrying out their core business, they contribute, through a material- and energy-intensive 'medicine of repair', to the intensification of environmental problems such as climate change. Conversely, this has a negative impact on public health. Thus, in a certain sense, hospitals are doubly affected by sustainability problems (cf. Weisz et al. 2011).[9] For this reason, according to our assumption, sustainability strategies must be particularly effective in this case, and hospitals must be particularly attuned to sustainability issues.

A 'Social Ecology in the hospital' necessitates several preliminary theoretical considerations. We begin with the crucial question of a definition of health that is compatible with our purpose. This search for a definition may also function as a brief introduction to the research theme for the 'sustainability researchers' among our readership who do not deal traditionally, at least professionally, with health.

29.3.1 Which Concept of Health Is Compatible with Social Ecology?

There are at least as many suggested definitions of health in the health debate as there are definitions of sustainable development in the sustainability debate. The most well known and most often cited definition of health is the World Health Organization's definition from 1946, which conceives of health in a holistic sense as the interplay of physical, psychological and social factors and no longer as the 'absence of disease': 'Health is a state of complete physical, mental and social well-being and not merely the absence of disease or infirmity' (WHO 1946, p. 100). This description of an optimal condition has been much criticized for representing an unattainable ideal. Thus, in health promotion (see WHO 1986), in a development that may be interpreted as a reaction to this definition, health is formulated as a process with an emphasis on its function. Health is not interpreted as a primary goal but much more as a resource for daily living that may be expressed in the degree of quality of life (see Pelikan 2007; Pelikan and Halbmayr 1999).[10]

Today, human health (the health of individuals as well as the health of the population) is conceptualized in terms of its determinants. Dahlgren and Whitehead, in their oft-cited 'conceptual model of the main determinants of health-layers of

[9]Furthermore—and here, too, hospitals play a special role—they act as 'social organs of perception', registering changes in public health (cf. Fischer-Kowalski 1997).

[10]Perhaps the most wonderful definition, which we wish to share with readers and which emphasizes the function of health, is generally credited to Sigmund Freud: 'Health is the capacity to be able to work and love'. In the original, however, Freud provides a more 'dry' definition, describing the aim of psychoanalytical treatment as: 'Nach Möglichkeit leistungs- und genussfähig zu machen' ('Rendering a person, as far as possible, capable of both work and enjoyment', author's own translation, Freud 1923, p. 226 as cited in Nedelmann 2009, p. 338).

influence' (Dahlgren and Whitehead 1991), distinguish four interacting levels of factors influencing health: (1) the individual (the genetic disposition, constitution, behavior and lifestyle of the individual), (2) the social environment, (3) living and working conditions, and (4) economic, cultural and physical environmental conditions. We are concerned that the factors that constitute the fourth level are named jointly and, thus, that natural conditions and social/cultural conditions, which follow such different 'logics', are in some sense rendered analogous. We cannot pursue the argument related to this concern here.[11] It should be sufficient at this point to note that health is determined both through natural and social factors, that is, through factors from both the material and the symbolic realms (see Fischer-Kowalski and Weisz 1999). Thus, if we attempt to find an approach to the concept of health from our socioecological perspective, the most significant aspects have already been addressed; we see them through the approach to health determinants and the influence of health upon quality of life.

We view medical practice as interesting and problematic in equal measure. The diagnosis 'results negative', which merely expresses that no pathological deviations have been detected, is an operational reference to the long-outdated health definition described above, which defines 'health as the absence of disease'. Why does this not discourage us from placing our research focus on the hospital setting? The health promotion strategies described at the beginning, which have led (among others) to the international Health Promoting Hospitals network (HPH, see http://www.hphnet.org/) of the WHO, help us here as they enable us to speak about health, at least in the context of Health Promoting Hospitals.

The next step would be to offer a sustainability concept for organizations—in our case, for hospitals—that would fulfill our requirements in two ways. First, it should be scientifically coherent (for us, this means it should be consistent with the socioecological conceptualization of sustainable development) while simultaneously being related to 'hospital reality'. Second, a sustainability concept for hospitals is only likely to be implemented when the effort demanded of hospital actors is at an acceptable level and the anticipated benefits seem attractive. Such a sustainability concept must therefore relate to the real challenges facing hospitals, must promise solutions and must expand in a meaningful way upon those concepts and strategies that have already been successfully introduced rather than competing with them (cf. Weisz et al. 2011).

[11]The socioecological approach (the model of interaction between society and nature, which positions the human population at the 'hybrid' interface) also does not focus any further on the 'special case of human nature', and it can be of no further help to us here.

29.3.2 Conceptualizing Sustainable Development in Health Promoting Hospitals

The literature on sustainability and environmental management does not, to our knowledge, offer any sustainability approaches for organizations that satisfy our criteria. In the contributions, often characterized as 'corporate social responsibility' (CSR) or 'corporate sustainability', the definitions are rather vague.[12] Moreover, the integrated sustainability approaches, which attempt to take equal account of the three dimensions of sustainable development (i.e., ecological, social and economic aspects) and their interaction with one another, are barely represented (cf. Stubbs and Cocklin 2008; Weisz et al. 2011). Looking at this issue in isolation allows for every short-term individual solution and any kind of improvement to be subsumed within the term 'sustainability'. In the process, potential negative side effects and the transfer of problems to other organizational areas or in environments (externalizations) and positive synergistic effects (namely, potential co-benefits) remain unaccounted for.

We also encounter this problem in hospital practice. Despite the diverse, important and quite successful efforts in ecological building design, energy efficiency, and environmental management systems that have been undertaken to improve hospitals' ecological performance (see the 'green hospitals' initiative) and to address social concerns (such as improvements in working conditions), these strategies are pursued in an isolated way, without reference to the actual challenges facing hospitals. These challenges include increasing quality demands and ever greater societal expectations regarding the 'feasibility of health' (Kickbusch 2006), as exemplified by political pressure in favor of efficiency and reducing costs. Even the relevant strategies (such as quality strategies[13]) remain unaffected by environmental and sustainability concerns. In the core business, when it is a matter of patient health, environmental concerns do not play a role at all—indeed, they are not allowed to play a role, according to general opinion. 'It takes what it takes'[14] is the claim made by doctors.

For this reason, we decided to develop a socioecological sustainability concept for hospitals that is capable of differentiating between integrated, which we term 'multi-dimensional', and 'one-dimensional' solutions (Weisz et al. 2011). In so doing, we built upon the theoretical considerations of the socioecological concept for global sustainable development, the key message of which can be briefly summarized as follows. In global terms, (world) society can only fulfill its function

---

[12]The 'Brundtland definition' (WCED 1987) is often used, although it is too vague to be used as a guide for action.

[13]These exist within the context of an evidence-based medicine, which has been introduced partly to control the cost efficiency of services (Weisz et al. 2011).

[14]From an interview with a chief hospital physician from a project on 'sustainability monitoring in hospitals' (project MOKA, cf. Weisz 2015).

in a sustained way if it does not destroy the natural conditions for existence upon which it depends. The development of society must therefore be compatible with the maintenance of natural systems (nature). Accordingly, from a socioecological perspective, (global) sustainable development means the sustained preservation of society-nature interaction. This addresses the physical interactions between multiple dynamic systems: that of the biosphere and its natural systems on the one hand and the anthroposphere and its social systems on the other (Fischer-Kowalski and Haberl 1997). The normative requirement involves behaving toward the environment (nature) in the present in such a way that it will remain available to sustain the existence of future generations (cf. the Brundtland definition WCED 1987). 'Sustainability, therefore, is an anthropocentric notion: it means that human-induced changes in ecosystems must not threaten the exchange processes between society and its natural environment in ways that affect society's survival or well-being' (Haberl et al. 2004, p. 200). The key distinctions made here concern the relationship between a system and its environment (society and nature) and the relationship between present and future generations (the present and the future). According to this understanding, the interactions between society and nature are at risk when societal problems are externalized in spatial, temporal or factual terms, with unintended side effects and long-term impacts as a result.

How can the principles of this concept, which have been developed for the global level, be applied to organizations—in our case, to hospitals? From a socio-ecological point of view, sustainable development concerns the long-term maintenance of the interrelationships between society and nature at the global level. We have produced an analogous formulation for the level of organizations: the capability of organizations to maintain their relationships with their (social and natural) environments over the long term is a prerequisite for the long-term functioning of organizations (Weisz et al. 2011). In contrast to the global perspective, at lower scales, the direct relationships with nature are reduced, whereas the relationships with relevant social environments increase. When formulated in terms of actors, these are the stakeholders. The relevant social environments of hospitals include patients and employees as internal environments and the political environment and civil society (the public sphere) as external environments (Pelikan and Halbmayr 1999). To continue with this analogy, a hospital may be deemed to be acting sustainably when it does not transfer or outsource (externalize) problems in spatial, temporal or factual terms to its social and natural environments. These considerations lead to the stipulation that a wider perspective should be applied when organizations evaluate the impacts of their own (systemic) activities.[15]

These system-theoretical considerations may aid our efforts to achieve a better theoretical understanding of how sustainability may be conceptualized at an organizational level. However, where 'the reality of hospitals' (i.e., the concrete cooperation with actors) is concerned, these considerations have shown

[15]This extension beyond the direct system boundaries also finds expression in the concept of corporate social responsibility (CSR).

themselves to be too 'abstract', not easily communicable and not helpful regarding the question of 'concrete implementation', which is always of interest for actors. Luhmann insists that the differentiation between a system and its environments has practically no significance in the exercise of a hospital's core function, the treatment of patients. 'No patient can be helped by medicine establishing itself as a system that is capable of differentiating between the internal and the external' (Luhmann 1990, p. 176, author's own translation), hence his conclusion that it is 'only of limited relevance for medical practitioners' (ibid.). How did we make the link to 'the reality of hospitals'?

29.3.3 Linking to Hospital Practice

It became evident at the start of our work with actors from the sphere of 'practice'[16] that the so-called 'sustainability triangle' in its socioecological extension (Fischer-Kowalski 1997) is well suited for use in transdisciplinary communication on sustainable development (cf. Weisz et al. 2009 and Sect. 5.3.1 in this volume). This simple systems model attempts '[…] to depict the dynamic of industrial society's development' (Fischer-Kowalski 1997, p. 201, author's own translation). It describes this dynamic as the self-reinforcing interactions via positive feedback mechanisms among the three components (or dimensions of sustainability) of quality of life (well-being), prosperity (economic activity) and material use (social metabolism). The dynamics of development within societies are linked to the natural world via negative feedback loops (at the 'ends' of social metabolism and quality of life)[17] (ibid.). This perspective provides a suitable basis for considering potential strategies for interrupting unsustainable dynamics (decoupling strategies). In cooperation with hospital actors, we can use this model to enable us to direct attention, via hospitals and the wider healthcare system, toward global relationships. In our experience, this works particularly well when we relate it to the determinants of health and accordingly depict health explicitly, as a key resource for quality of life (and the negative impact on health via the natural environment), within the 'sustainability triangle'.

Of particular interest in our case is that the sustainability triangle can be formulated not only for the level of global society but equally for social sub-systems, such as economic sectors (e.g., for agriculture; see Chap. 26), or for organizations.

[16]In the transdisciplinary research process, however, the differentiation between science and practice is shown to be unsuitable for medical practitioners who themselves participate in research (indeed, often with greater success than researchers in a transdisciplinary setting) (Weisz et al. 2014).

[17]This approach may be interpreted as an 'anti-progress model', as a critique of a now untenable idea of progress that depicts the prospect of economic growth, increased living standards and the simultaneous conservation of nature through the 'three-pillar model' (cf. Fischer-Kowalski 1997).

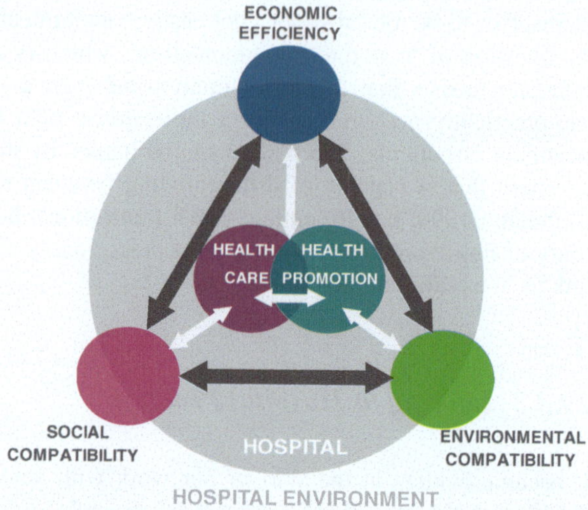

Fig. 29.1 The sustainability triangle for hospitals (Weisz et al. 2011, p. 195)

Figure 29.1 shows the 'sustainability triangle for hospitals', which we have formulated together with hospital actors (from management, medicine and nursing) and our partners in the sociology of health and medicine and health promotion research.

This triangle places health care and health promotion (as core services) at the center of the system and introduces additional quality criteria (sustainability criteria) for decision-making regarding the core services. Thus, we can formulate the following: sustainable development at the level of a hospital is an optimization activity in which the fulfillment of its core business, along with medical and health promotion quality criteria, is investigated with reference to economic, social and environmental sustainability (Weisz et al. 2011). In other words, we interpret sustainability in hospitals as the extension of quality criteria. Thus, this approach is applicable to the quality concepts and strategies employed by modern hospital organizations, for example, the European Foundation for Quality Management (EFQM) and the *Kooperation für Transparenz und Qualität im Gesundheitswesen* (KTQ, Cooperation for Transparency and Quality of Healthcare).

The application of the 'sustainability triangle for hospitals' thus offers several benefits at once: (1) with the help of this model, we are able to create interest and understanding among actors in the (global) issue of sustainable development; (2) health benefits (through patient care and health promotion) can be linked to savings in physical (material and energy) resource use and financial advantages and, thus, (two fold) to an increase in quality of life; (3) the model brings environmental considerations into the key decision-making process (in decisions regarding patient care and health promotion) of hospitals, such as decisions about service provisions, as we will illustrate in our example.

Having attempted to provide a suitable terminological and conceptual context for our area of concern, we now wish to illustrate what a socioecological approach in the hospital setting can mean in 'practical' terms, that is, in the sense

Table 29.2 Planning of care provision for long-term ventilated patients: Comparison of the 2-step model (ICUs and RCU) with the proposed 3-step model (Source: Weisz et al. 2011, p. 196)

Results		
	2-step model	**3-step model**
Wards	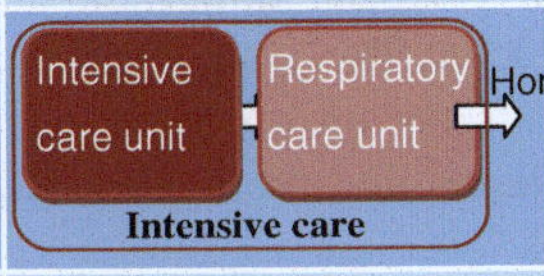	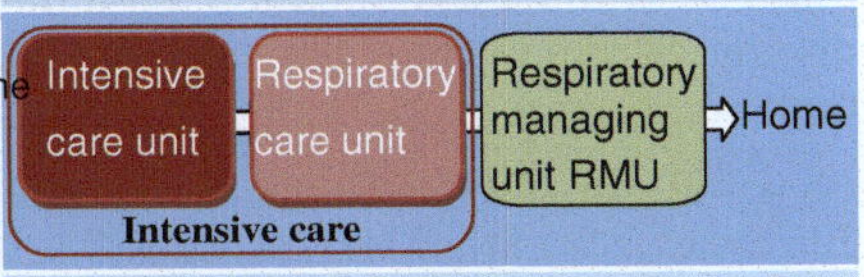
Length of stay in ward (patient days)	21,534 (ICUs)/2,522 (RCU)	18,628 (ICUs) / 2,389 (RCU) / 3,039 (RMU)
Costs for patient group (million €/yr)	46.3	42.6 i.e., a reduction of 3.7 million €/yr or 8 %
Cost-revenue relation	The 3-step model shows a marginal improvement (using conservative assumptions, e.g., not considering additional revenue through the reallocation of beds)	
Material use (t/yr) -gross weight (including packaging) -net weight	4,056 687	3,657 i.e., a reduction of 318 t/yr or 7.8 % 625 i.e., a reduction of 62 t/yr or 9 %
Health gain	Empowerment of patients (RMU); reduction of health risks such as infections (because in the 3-step model, less time is spent in intensive wards [ICUs and RCU] and more time is spent on training)	

of concrete implementation. To do this, we use an example from the 'test phase' of our project series (see Table 29.1) that refers to the question presented at the start of this chapter regarding possible synergies or co-benefits.

29.4 An Example of Co-benefits in the Hospital: Sustainable Provision Planning[18]

The example is taken from the area of respiratory medicine and was implemented at the 1st Department of Internal Medicine at the Otto Wagner Hospital in Vienna, our pilot hospital. It concerns the planning of the service palette for long-term

[18] This example is based on Weisz et al. 2009, 2011.

ventilator-dependent patients, that is, those with chronic critical illness. Medical provision planning is particularly well suited for our purposes because it concerns the core business of the hospital and determines the physical and financial resource use for several years in advance. In view of the ageing population, the future healthcare provision for persons with chronic illness is a highly relevant issue for health policy. Health promotion plays an important role here because only by 'strengthening the healthy characteristics' (see Pelikan and Halbmayr 1999) of these patients is it possible (with certain preconditions) to facilitate a 'good' life for those affected (i.e., the patients and their families), wherever possible away from in-patient facilities.

The background to this example is as follows. Ventilator-dependent patients with chronic lung diseases are cared for in the 1st Department of Internal Medicine at the Otto Wagner Hospital in two interconnected intensive care wards (2-step model). The intensive care unit (ICU) is in charge of acute care for critically ill patients, often through organ replacement and artificial respiration. The respiratory care unit (RCU),[19] which is unique in Austria, was set up to care for patients with prolonged ventilator dependency following acute illness. This unit specializes in weaning patients from artificial respiration.[20] Because patients with chronic respiratory problems are prepared for home ventilation, the further tasks of this unit involve training patients and their relatives and transmural case management[21] so that the patients can achieve the appropriate level of safety and quality in respiration while living at home. These patients are readmitted at regular intervals to the RCU for check-ups and further care. In cases involving acute problems, doctors from the ward continue to function as contact partners.

Experience over 15 years shows that training, remobilization and check-ups could take place under better conditions outside the RCU. The resource use of intensive care wards in terms of apparatus and staff is not only unnecessary but is even a hindrance; patients and relatives frequently find a direct transfer from the intensive care ward to the home overwhelming, which leads to unplanned readmission and frequent contact with the ward. Furthermore, the intensive care setting endangers patients by exposing them to nosocomial infections and other health risks, including psychological stresses. The RCU management therefore suggested that the 2-step model be extended to a 3-step model (a so-called 'Weaning Center') by assigning a ward outside the intensive care setting, the respiratory management unit (RMU), to address all tasks associated with preparing patients for ventilation at home.

We were interested in comparing the two model variants regarding the criteria of sustainable development and health promotion. This comparison included a

[19]Intensive care units fall within the highest category of intensive care (Class 3), and respiratory care units fall within Classes 1 and 2.

[20]This affects ca. 9 % of all ventilated patients and 30 % of those patients with underlying chronic obstructive pulmonary disease (COPD).

[21]'Transmural' refers to the interface between hospital and home care.

survey of potential savings in patient days spent on the intensive care wards and an estimation of potential savings thereby accrued in costs and material resource consumption.[22] A prospective needs survey at the ICUs within the Vienna Hospital Association conducted in 2007 showed that 13 % of the patient days ventilated patients spent in ICUs could have been transferred to the RCU; that is, the beds were misallocated for that time.[23] Moreover, 56 % of the patient days spent in the RCU at the pilot hospital could have been spent outside the intensive care setting at an RMU. This represents a total of more than 3000 days ventilated patients with long-term respiration needs, a relatively small group of patients, spend unnecessarily in intensive care wards within the Vienna Hospital Association.

The results of the model comparison (see Table 29.2) show that—according to conservative estimates—the 3-step model could result in savings of ca. 8 % in both costs (entire cost of in-patient care) and material use (gross material use). This represents a potential savings of 3.7 million Euro and ca. 300 t of material per year. In addition, health risks can be reduced and health gains increased by reducing the length of stay in intensive care wards and by establishing a ward outside the intensive care setting that specializes in training and instruction, thereby empowering patients. Thus, a 3-step care model for patients receiving long-term ventilation would lead to an improved care situation and to better conditions for health promotion measures for patients. The investments required to implement this model are justifiable in view of the anticipated cost savings thereby accrued.

This investigation allowed us to show that significant economic and social improvements can lead to simultaneous savings in physical resources if the criteria of sustainable development and health promotion—in addition to care quality criteria—are included during provision planning. These arguments convinced not only our project partners in the pilot hospital but also those responsible for health policy within the Vienna Hospital Association. In 2012, the head office of the Vienna Hospital Association commissioned the construction of the RMU, and this unit was opened in February 2013.

These results exceeded our original goal of conducting an ‘implementation-oriented pilot’ (also resulting in continued cooperation with ‘our’ pilot hospital). They are also remarkable in terms of the status of environmental strategies in hospitals. This example from hospital practice enabled us to demonstrate that in addition to the successes to date, significant ecological improvements are achievable when socioecological criteria are brought into central decision-making processes about the (planning of) core business in hospitals. In other words, it is precisely in the area of patient health that it makes sense to address the issues of environmental protection and social and economic questions of patient care together and not to allow ‘ecological sustainability’ to become an isolated marginal issue.

---

[22]To estimate material use, we surveyed 80 % of the most expensive consumer goods (excluding investment goods, pharmaceuticals, infusions and (blood) transfusions). For methodological details, see Weisz et al. 2009. Surveying energy use at the ward level was not possible.

[23]Misallocation of beds refers here to instances in which patient care does not take place in the optimal setting.

29.5 Health as a Socioecological Argument

For us, the decisive issue and central conclusion is as follows. Considering questions of the interrelationships between health and sustainable development creates the prerequisite conditions for identifying synergies and developing strategies for co-benefits. Those addressed in our example were ultimately the political actors within the health system. We argue that the criteria of sustainable development, that is, economic, social and ecological criteria and their interactions, should be observed and included on an equal basis in core decision-making regarding patient care and health promotion—in the key decisions on health.

If we turn our attention to 'the sustainability community', which traditionally does not address questions of health, or to 'sustainability policy', we can easily reverse this argumentation and formulate the following thesis: sustainability strategies prove themselves to be more effective and cost efficient when health impacts are included in considerations and in calculations. The literature on forecasts for the health costs of climate change emphasizes that the potential health gains that may be generated by climate protection measures have the advantage of being more quickly and locally effective—in contrast to the positive impacts on climate—and thus have a good chance of seeing political implementation.

Thus, we believe it is worthwhile for 'sustainability research' to engage with these issues at a more complex level and to conceive of health as an additional issue and argument in favor of successful sustainability strategies. In this sense, health also becomes a socioecological argument. We believe this understanding will open the way for fruitful areas of investigation beyond the health system. Public Health research on the co-benefits of climate change mitigation and human health provides important insights in this respect. Perhaps Harlem Brundtland's quote, with which we began this chapter, that '[...] ultimately, the entire report is about health', can be interpreted precisely in this way.

References

Brundtland, G. H. (1989). Statement to the forty-first world health assembly, 3 May 1988. *Health Promotion International*, *4*(1), 51–55.

Ciscar, J. C., Iglesias, A., Feyen, L., Szabó, L., van Regemorter, D., Amelung, B. et al. (2011). Physical and economic consequences of climate change in Europe. Edited by Hans-Joachim Schellnhuber. *PNAS*, *108*(7), 2678–2683.

Dahlgren, G., & Whitehead, M. (1991). *Policies and strategies to promote social equity in health*. Stockholm: Institute for Future Studies.

Dooris, M. (1999). Healthy cities and Local Agenda 21: The UK experience—challenges fort he New Millenium. *Health Promotion International, 14*(4), 365–375.

Fischer-Kowalski, M. (1997). Dynamik und Selbststeuerung industrieller Gesellschaften. In: M. Fischer-Kowalski, H. Haberl, W. Hüttler, H. Payer, H. Schandl, V. Winiwarter, & H. Zangerl-Weisz, (Eds.), Gesellschaftlicher Stoffwechsel und Kolonisierung von Natur. Ein Versuch in Sozialer Ökologie. Gordon & Breach Fakultas, Amsterdam, pp. 203–221.

Fischer-Kowalski, M., Haberl, H. 1997. Stoffwechsel und Kolonisierung. Konzepte zur Beschreibung des Verhältnisses von Gesellschaft und Natur. In: M. Fischer-Kowalski, H. Haberl, W. Hüttler, H. Payer, H. Schandl, V. Winiwarter, & H. Zangerl-Weisz, (Eds.), Gesellschaftlicher Stoffwechsel und Kolonisierung von Natur. Ein Versuch in Sozialer Ökologie. Gordon & Breach Fakultas, Amsterdam, pp. 3–12.

Fischer-Kowalski, M., & Weisz, H. (1999). Society as hybrid between material and symbolic realms. Toward a theoretical framework of society-nature interaction. *Advances in Human Ecology 8*, 215–251.

Friel, S., Dangour, A. D, Garnett, T., Lock, K. Chalabi, Z., Roberts, I., Butler, A., Butler, C. D., Waage, J., McMichael, A. J, Haines, A. 2009. Public health benefits of strategies to reduce greenhouse-gas emissions: food and agriculture. Series Health and Climate Change 4. *The Lancet 374* (9706), 2016–2025.

Grossmann, R. (Ed.). (1997). Besser Billiger Mehr. Zur Reform der Expertenorganisationen Krankenhaus, Schule, Universität. Iff-texte, vol. 2 (pp. 24–35). Springer, Wien/New York.

Haberl, H., Fischer-Kowalski, M., Krausmann, F., Weisz, H., & Winiwarter, V. (2004). Progress towards sustainability? What the conceptual framework of material and energy flow accounting (MEFA) can offer. *Land Use Policy, 21*(3), 199–213.

Kickbusch, I. (2006). Die Gesundheitsgesellschaft: Megatrends der Gesundheit und deren Konsequenzen für Politik und Gesellschaft: Verlag für Gesundheitsförderung. Gamburg.

Luhmann, N. (1990). Der medizinische Code. In: Luhmann, N.: Soziologische Auflärung 5. Konstruktivistische Perspektiven. 4. Auflage. VS Verlag für Sozialwissenschaften: 176–188.

McMichael, A. J., Powles J. W., Butler C. D., & Uauy, R. (2007). Food, livestock production, energy and health. Series Energy and Health 5. *The Lancet, 370,* 1253–1263.

McMichael, A. J., Neira, M., Bertollini, R., Campell-Lendrum, D., & Hales, S. (2009). Climate change: A time of need and opportunity for the health sector. *The Lancet, 374*, 2123–2125.

Mintzberg, H. (1993). *Structure in fives*. New Jersey: Designing Effective Organizations.

Nedelmann, C. (2009). Nach Möglichkeit leistungs- und genussfähig machen. *Betrachtungen zum psychoanalytischen Behandlungsziel. Forum Psychoanal, 25*, 337–347.

Pelikan, J. (2007). Gesundheitsförderung durch Organisationsentwicklung. *Prävention und Gesundheitsförderung., 2*(2), 74–81.

Pelikan, J. M., Halbmayr, E. (1999). Gesundheitswissenschaftliche Grundlagen zur Strategie des Gesundheitsfördernden Krankenhauses. In: Das gesundheitsfördernde Krankenhaus. Konzepte und Beispiele zur Entwicklung einer lernenden Organisation. Edited by J.M. Pelikan, S.Wolff. Weinheim: Juventus. 13–36.

SDC-SEI. (2008). NHS England carbon emissions: carbon footprinting study. London: Sustainable Development Commission. London.

Shindell, D. et al. (2012). Simultaneously mitigating near-term climate change and improving human health and food security. Science *335*, 183.

Smith, R. K., & Ezzati, M. (2005). How environmental health risks change with development: The epidemiologic and environmental risk transitions revisited. *Annual Review of Environment and Resources, 30*, 291–333.

Smith, K. R., Woodward, A., Campbell-Lendrum, D., Chadee, D. D., Honda, Y., Liu, Q., Olwoch, J. M., Revich, B. & Sauerborn, R. (2014). Human health: impacts, adaptation, and co-benefits. In *Climate Change 2014. Impacts, Adaptation, and Vulnerability. Part A. Global and Sectoral Aspects. Contribution of Working Group II to the Fifth Assessment Report of the Intergovernmental Panel on Climate Change*, pp. 709–754. Cambridge University Press, Cambridge, United Kingdom and New York, NY, USA.

Stern, N. (2006). *The economics of climate change. The stern review*. Cambridge: Cambridge University Press.

Stubbs, W., & Cocklin, C. (2008). Conceptualizing a "sustainability business model". *Organization and Environment, 21*(2), 103–127.

Lancet, The. (2012). Global health in 2012: Development to sustainability. *Editorial, The Lancet, 379*, 193.

Tilman, D., & Clark, M. (2014). Global diets link environmental sustainability and human health. *Nature, 515*(7528), 518–522.
WBGU. (2011). World in Transition. A Social Contract for Sustainability: Flagship Report. German Advisory Council on Global Change.
WCED World Commission on Environment and Development. (1987). *Our common future*. New York: Oxford University Press.
Wegleitner, K. (2012). Nachhaltige regionale Selbstentwicklung von Palliative Care in der flüchtigen Moderne. Partizipative, transdisziplinäre Entwicklung von kommunalen Solidaritätsnetzwerken: eine reflexive Form von Network - Governance in der Gesundheitspolitik etablieren. Dissertation, Universität Wien.
Weisz, U. (2015). Das nachhaltige Krankenhaus – sozialökologisch und transdisziplinär erforscht. Ein Beispiel für Synergien zwischen nachhaltiger Entwicklung und Gesundheit auf Organisationsebene, Dissertation in Human und Sozialökologie, Alpen-Adria Universität Klagenfurt, Wien, Graz.
Weisz, U., Haas, W., Pelikan, J. M., Schmied, H., Himpelmann, M., Purzner, K., Hartl, S., & David, H. (2009): Das nachhaltige Krankenhaus. Erprobungsphase. Social Ecology Working Paper 119; Social Ecology Vienna.
Weisz, U., Haas, W., Pelikan, J. M., & Schmied, H. (2011). Sustainable hospitals: A socio-ecological approach. *GAIA, 20*(3), 191–198.
Weisz, U., Karner, S., Grossmann, R., & Heintel, P. (2014). Zwischen Welten. Transdisziplinäre Forschungsprozesse realisieren. In: G. Dressel, W. Berger, K. Heimerl, & V. Winiwarter (Eds.), Interdisziplinär und transdisziplinär forschen. Methoden und Praktiken. transcript Verlag Bielefeld: 121–134.
Westhoek, H., Lesschen, J. P., Rood, T., et al. (2014). Food choices, health and environment: Effects of cutting Europe's meat and dairy intake. *Global Environmental Change, 26*, 196–205.
WHO World Health Organization. (1946). Preamble to the Constitution of the World Health Organization as adopted by the International Health Conference, New York, 19–22 June, 1946 (Official Records of the World Health Organization, no. 2).
WHO World Health Organization. (1986). Ottawa Charter for Health Promotion. WHO Geneva.
WHO World Health Organization. (2009). Protecting health from Climate Change. Connecting science, policy and people. WHO Geneva.
WHO and HCWH, World Health Organization and Health Care Without Harm. (2009). Healthy hospitals, healthy planet, healthy people: Addressing climate change in healthcare settings. Discussion draft. WHO Department of Public Health and Environment, HCWH. Geneva.

Perspectives on Social Ecology: Learning for a Sustainable Future

Verena Winiwarter

With contributions from Marina Fischer-Kowalski, Simone Gingrich, Robert Groß, Willi Haas, Daniel Hausknost, Panos Petridis, Anke Schaffartzik, Martin Schmid, Ulli Weisz and Dominik Wiedenhofer

Abstract Social Ecology embraces the Social and Natural Sciences and the Humanities. It can therefore offer an interdisciplinary long-term perspective on global and local relations between society and nature. In a micro-poll experiment, members of the Vienna School of Social Ecology offered their insights regarding promising avenues for moving Social Ecology forward into the 21st century. It became clear that learning for a sustainable future requires more than one type of scholarly approach because different types of knowledge and learning are necessary. Suggestions for a promising focus are varied. From a study of what a sustainable socioecological metabolism would look like to the setting up of diversity funds or funds for scholars from the Danube Area, a wide range of themes was suggested. Taken together, they show both the broad reach and the great potential of Social Ecology.

Keywords Organizational learning • Transformative learning • Wicked problems • Sustainable development • Mode-3-science

Social Ecology is a manifold and varied scholarly endeavor. It embraces the Social and Natural Sciences and the Humanities, offering an interdisciplinary long-term perspective on global and local relations between society and nature. Social Ecology's explicit conceptual foundation as an attempt to integrate perspectives from several scholarly realms is both a necessity for interdisciplinary cooperation and one of the outcomes of interdisciplinary communication. It is also a precondition for working on the questions Social Ecology is addressing. Social Ecology is empirically rich, methodologically sophisticated and theoretically outspoken. As this volume vividly testifies, over the three decades the Vienna School of Social Ecology has

H. Haberl et al. (eds.), *Social Ecology*, Human-Environment Interactions 5,
DOI 10.1007/978-3-319-33326-7

been in the making, practitioners of the craft have learned a great deal and continue to refine methods, apply the concept to new questions and carefully design processes at the science/policy interface. Social Ecology continues to be in the making.

When it came to conceptualizing a final, perspectival chapter for this book on 'Perspectives', a long discussion among the editors ensued. Summarizing the wealth of diverse material presented did not make much sense. In the preceding pages, Social Ecology has been put into context, applications have been shown, and methodical issues have been discussed. Any summary would be both repetitive and lacking. The editors decided for an open end and invited all authors to write a paragraph or two about how they would spend one million Euros to further Social Ecology. The answers were as wide ranging as we expected, but at second glance, a basic, if implicit, principle of Social Ecology became visible in the answers; they showed that learning for a sustainable future requires more than one type of scholarly approach because different types of knowledge and learning are necessary.

Wicked Problems and the Knowledge Needed for a Sustainable Future

The eminent German philosopher Jürgen Mittelstraß suggested distinguishing between 'orientational' and 'positive' knowledge. Science and Technology, so Mittelstraß said, lead to knowledge about causes, effects and means. He called this kind of knowledge 'positive knowledge'. His critique was that this kind of knowledge alone does not suffice to solve problems. Knowledge to guide action is needed, which he calls 'orientational knowledge'. This type of knowledge is ethical in the sense that it provides orientation about what we should, and not just about what we can do (Mittelstraß 1984, p. 64).

Many other researchers have discussed the types of knowledge necessary in the 'real world' or 'lifeworld', the world of messy problems that has become the main driver of the environmental sciences at large since the 1970s.[1] In 1972, Barbara Ward and Paul Dubos declared that 'laymen' should play a much more prominent role in investigations. They called scientists within their disciplinary boundaries 'parochial' and held that perceptive and informed nonprofessionals could contribute as much as technical experts to policies concerning the human environment. As Rolf-Peter Sieferle has noted, the Post-World War II world was different with respect to environmental problems. These problems originated not so much from polluted production sites, as had been the case during the Industrial Revolution, but were chiefly caused by the mass use of the products of the now much cleaner industrial operations (Sieferle 1995, p. 49). Dealing with the effects of mass consumption and the increasingly interconnected, systemic problems created by inventions such as DDT (*dichlorodiphenyltrichloroethane*) required both social judgment and specialized scientific knowledge (Ward and Dubos 1972).

[1] The following paragraphs are based on Winiwarter (2014).

Many of the problems Social Ecology addresses fall within the category of 'wicked'. In 1973, Horst Rittel and Melvin Webber suggested discerning between problems they called 'wicked' and those they called 'tame' in the context of planning. Wicked problems, they argued, are different, as can be seen by ten defining characteristics. 'There is no definitive formulation of a wicked problem', they state, because 'the information needed to *understand* the problem depends upon one's idea for *solving* it'. They further state, 'The formulation of a wicked problem is the problem! The process of formulating the problem and of conceiving a solution (or re-solution) are identical, since every specification of the problem is a specification of the direction in which a treatment is considered' (Rittel and Webber 1973, p. 161). The second specific quality of wicked problems is that they have no stopping rule. When dealing with societal problems, 'there are no ends to the causal chains that link [the] interacting open systems' (ibid., p. 162).

With more effort invested, a better solution is always reachable. The end of the research project (which is ultimately determined by money) and not the arrival at a solution terminates work on the issue. Wicked problems can only be resolved, never solved. This fits another of their defining characteristics: 'wicked problems do not have an enumerable (or an exhaustively describable) set of potential solutions, nor is there a well-described set of permissible operations that may be incorporated into the plan' (Rittel and Webber 1973, p. 164). Rather, agreement based on trust and credibility leads to common-sense 'realistic judgment' because ill-defined problems cannot have well-defined solutions. Rittel and Webber further propose that solutions to wicked problems are not true or false but good or bad. If stakeholders with different interests deal with a problem, it is highly unlikely that any (re-)solution will yield only winners. There are no objective truths. Therefore, '[…] none [of the parties involved] has the power to set formal decision rules to determine correctness. Their judgments are likely to differ widely to accord with their group or personal interests, their special value-sets, and their ideological predilections' (ibid., p. 163). To complicate things further, 'There is no immediate and no ultimate test of a solution to a wicked problem. […] With wicked problems, […] any solution, after being implemented, will generate waves of consequences over an extended—virtually an unbounded—period of time. Moreover, the next day's consequences of the solution may yield utterly undesirable repercussions which outweigh the intended advantages or the advantages accomplished hitherto' (ibid.). On a related note, 'Every solution to a wicked problem is a 'one-shot operation'; because there is no opportunity to learn by trial-and-error, every attempt counts significantly' (ibid.). Every implemented solution leaves traces. One cannot wipe the slate clean after an intervention in a social system, and because all wicked problems involve interventions into social systems, each solution leaves a unique legacy. Likewise, every problem is unique and is likely the symptom of another problem. Because it is a matter of choice how to frame the discrepancy between the desired and actual state of things, the choice of explanation determines the nature of the problem's resolution. If social inequality is chosen to be the reason for the deterioration or the problematic state of a natural system, the solution is different from that arrived at when the design of a technical system ('the polluter') is identified as a reason. As the authors put it, 'Planners are liable for the consequences of the actions they generate; the effects can matter a great deal to those people that are touched by those actions' (ibid., p. 167). This is not just true for planners but for every scholar involved in sustainability research. To name but one example, it does matter if the project increases or decreases the choice of options of the population involved. A planner, the authors formulate provocatively, 'has no right to be wrong' (ibid., p. 166). This is also true for sustainability professionals, who are often called in as 'experts' to deal with messy, wicked problems.

These insightful observations on the nature of wicked problems show that as early as 1973, some researchers were thinking about their roles and their associated limitations in the messy technological world of nature modified by humans. What later became denoted as 'Mode 2' research—that is, 'a new paradigm of knowledge production ('Mode 2'), which was socially distributed, application-oriented, trans-disciplinary, and subject to multiple accountabilities' (Nowotny et al. 2003, p. 179), nowadays known as a call for 'socially robust knowledge' (Nowotny 2003, p. 155)—had its origins, perhaps not coincidentally, in the recognition in the 1970s that there was a veritable environmental crisis.

Social Ecology provides expert knowledge. It is an endeavor to provide basic knowledge about the interfaces between society and nature, and it is concerned with the quality of its empirical database. However, it also seeks to provide orientational knowledge and works toward cooperative knowledge production, which is one of the ways to deal with wicked problems (Fischer-Kowalski et al. 2014).

Transformative Learning as Prerequisite for Sustainable Development

Many questions surround the types of knowledge needed for sustainable development and the processes deemed able to produce it. For this brief look into how Social Ecology can contribute to sustainable development, we shall focus on the question of learning, which has gained momentum in recent years, perhaps in conjunction with the UNESCO Decade of Education for Sustainable Development.[2] It is useful to ask what kind of learning is necessary to produce the kind of orientational, embedded and democratized expert knowledge so many authors have called for in the context of environmental problems.

Sterling (2011) suggested 'transformative learning' as key to a sustainable development. Sterling follows Paul Raskin's assessment that '[t]he path actually taken will rest with the reflexivity of human consciousness: our capacity to think

[2] http://www.unesco.org/new/en/education/themes/leading-the-international-agenda/education-for-sustainable-development/

critically about why we think what we do—and then to think and act differently' (Raskin 2008, p. 469). Sterling turns to learning theory and draws on Gregory Bateson's work on learning. First-order change, change within boundaries, is distinguished from second-order change, a significant change in thinking as a result of examining assumptions and values (Sterling 2011). Other researchers have tried to capture the same difference with different notions. Reflections on organizational learning were offered early on (Argyris 1976). 'Single-loop' and 'double-loop' learning, as the different types are called in this context, have become an influential notion.

> Argyris and Schön (1996) developed the theory of double loop learning as a way of creating more robust knowledge (Argyris and Schön 1996). Their concern was that most organisations only undertook single loop learning, which left the values and norms underpinning a strategy or action unchanged. This lack of change prevents organisations learning from their errors and, potentially, leads to failure. As a result they advocate double loop learning which will promote inquiry, challenging current assumptions and actions and leading to new theories-in-use. (Blackman et al. 2004, p. 17 f.)

However, Bateson actually distinguished three types, the most profound being epistemic learning, a learning that critiques the construction of learning itself as part of the learning process. After his evaluation of the literature, Sterling (2011) summarizes the three orders of change and learning. First-order learning/change, which seeks effectiveness or efficiency, is conformative and can be summarized as 'doing things better'. Second-order learning/change seeks to examine and change assumptions. It is reformative and can be described as 'doing better things'. The third type of learning/change, epistemic learning, leads to a paradigm shift and is transformative. It can be summarized as 'seeing things differently'. Uwe Schneidewind and Mandy Singer-Brodowski use this tripartite distinction as the basis of their discussion on the type of scholarship reflexive modernity needs. Table 1 is an English translation of their summary of the three levels of learning/change and transformation they distinguish.

Mode-3-science aims at producing highly contextualized knowledge on the system level and the level of goals and aims, and it seeks to generate transformative knowledge. For Mode-3-science, civil society is an important actor of knowledge production and is institutionalized in the organization of scholarship. The so-created transformative knowledge is heterodox, coming from real-laboratories and from actual processes of transformation. It is clear that such a mode of knowledge production needs cooperative structures, and hence, quality systems will have to coevolve with the evolving interaction between science and society (Schneidewind and Singer-Brodowski 2014, p. 123). Social Ecology is an important forerunner of Mode-3-science. To provide but one example, Social Ecology researchers experiment with Mode-3-science by producing highly and multiply contextualized knowledge gained from trying to foster a sustainability transition for hospitals and preparing products aimed at non-scientific stakeholders (hospital personnel) as an offer to enhance their self-reflexivity (see Chap. 29 in this volume).

Table 1 Three levels of learning/change (Reproduced from Schneidewind and Singer-Brodowski 2014, p. 81), translated by V. Winiwarter

Level	Individual	Organization	Academia
First-order learning/change	Ambition within given framework, learning as adaptation	Efficiency/operative controlling	Mode-1-science, 'efficiency gap' diagnosed, linear efficiency logic
Second-order learning/change	New ways of living aspired to, reflection of individual values and orientations	Changing goals, strategic controlling Reflection of cooperative/organizational strategies of action	Mode-2-science Science for sustainability
Transformative learning/ change	Wisdom, presencing consciousness about the embeddedness of the individual/ being in relations	New models of meaning, organizational change in the reflection of societal transformation	Transformative Mode-3-science

Perspectives for a Social Ecology of the 21st Century

When we asked the authors of this volume to suggest action to be taken with one million Euros at hand, we received exciting answers. A kaleidoscope of interesting future options for Social Ecology emerged. Many of the projects suggested could be characterized as transformative, but there were also suggestions that Social Ecology could be expanded with first-order learning, second-order learning and transformative learning or combinations thereof.

Over the past few decades, we have gathered huge amounts of data and now understand a great deal about the sustainability challenges faced by current and historic resource use in various socioeconomic, institutional and environmental settings. As the contributors have made clear, however, there is also room to 'make things better', providing a better basis for first-order learning.

Dominik Wiedenhofer, suggesting such first-order learning, argues that the systemic interactions and long-term dynamics between material stocks and flows of material, energy, waste and emissions are of paramount interest. Therefore, understanding and quantifying how the composition, spatial configuration, size and dynamics of the stock shapes resource use while providing important functions for society is an important research area. Interaction between stocks and flows, although at the beginning of Viennese Social Ecology, could be the target of monitoring and management, as Dominik Wiedenhofer suggests, to facilitate a sustainability transformation.

Simone Gingrich asks what a sustainable world would look like, and particularly what a sustainable socioecological metabolism would look like. How could the needs of society be met based on sustainable resource throughput? How would this compare to historic and present patterns of resource use in different parts of the world? On the ecological side, we ask which level of resource extraction can be considered sustainable in the long run, whereas on the socioeconomic side, we clarify which needs of society should be met and which technologies exist that make most efficient use of natural resources. She bases her suggestion on the assessment that a sustainable global socioecological metabolism will not only yield prosperity for the global population but will also create new option spaces.

Which institutional and geopolitical framework conditions need to be established to implement a sustainable metabolism? As a political scientist, **Daniel Hausknost** contributes to this question by identifying the challenge of governance. One of the great challenges of our century is how to establish stable forms of democratic governance that make do without continuous economic growth and energy input. It is therefore decisive to understand the underlying mechanisms of legitimation at work in the fossil-energy regime and to replace them with new mechanisms designed for a post-fossil world. He suggests analyzing and reconstructing the coevolution of the fossil energy regime and the symbolic order of modernity. In particular, he wants to investigate the extent to which modern political institutions (including those of parliamentary democracy) are structurally dependent on a high-energy metabolism and/or on a growth-based capitalist economy. The results of this investigation will have decisive implications for our ability to construct the institutional setup of a post-fossil, post-growth democratic society. This research will combine quantitative long-term socioecological analyses with historical institutional analysis and the history of political thought.

Like Daniel Hausknost, **Anke Schaffartzik** suggests we turn to historical data for the design of a more sustainable future. Socioecological research on the past developments of material consumption and trade of countries and regions clearly shows one thing: 'business as usual' is not an option for the future of global resource use, no matter how we define what 'usual' means. If the growth rates in material consumption of the past 50 years were to continue for the next 35 years, we would see a tripling of material use compared to the 2010 level by 2050. If we even try to imagine the environmental impact associated with the extraction and use of materials at this rate, the picture becomes apocalyptic. Global material use, as it is currently organized, is a heavy weight, and once heavy weights are in motion, they are characterized by inertia. Any transition toward a 'sustainable' future must build up enough momentum to redirect the inertia of our current trajectory.

Based on an in-depth analysis of the social, political and economic conditions that accompanied the major metabolic transitions of the past, hypotheses regarding different possible sets of conditions that would have to accompany a future sustainability transition will be formulated. The definition of these conditions takes into account the changes in our point of departure (in terms of material and energy use, land use, stocks and technology) as opposed to transitions of the past. Whereas Simone Gingrich's suggestion is concerned with what a sustainable future might look like, Anke Schaffartzik focuses on developing a 'map' that will show us how to get to that future or that will at least allow us to mark in bright red the areas to which we definitely do not want to go.

Willi Haas, starting from a very similar assessment of the present, arrives at a different conclusion with regard to the most promising research focus: there is ample evidence that a continuation of the present fossil-fuel-dominated world energy regime is not feasible. Consequently, smart scientists have been drafting alternative pathways over the last few decades. After all these efforts, a 'low-carbon' society seems to be technically feasible. Internationally, however, there is hardly any progress in the transition away from fossil fuels. Even in the European Union, a political entity that has committed itself to change, policies are vague, and implementation has become an issue of low priority with hardly visible results. It seems that mainstream political and economic trajectories continue to be unimpressed by the urgent need for change, as expressed by numerous scientists. The situation calls for relocating research topics from details of alternative pathways to the spheres of societal decision-making. Key research questions deserving attention are the following: What decisions are crucial for a 'low-carbon' society? What hinders political actors from asserting themselves in the political and economic sphere when it comes to major decisions about Europe's future? What changes in framework conditions are fore seeable that might alter the likelihood for 'low-carbon' decisions? What would be the most effective contributions science could make to finally persuade decision-makers to favor 'low-carbon' decisions—at least in Europe?

Panos Petridis' suggestion addresses one of the main conundrums of sustainability transitions: they might have, and indeed are likely to have, unwanted social effects. Although they might bring a new type of global prosperity in the long run, they may have adverse side effects on global justice. He seeks ways to support the emergence of more ecologically sustainable and just societies. Such research should (1) have a good understanding of the system, past and present; (2) have an effect in practice; and (3) leave a legacy for the future. He would include historical research on past egalitarian societies and less resource-intensive modes of social organization and combine it with a detailed analysis of examples of current societies at various scales, resource intensities and patterns of social organization. A novelty of his suggestion is the combination of this approach with on-the-ground experiments to identify and promote positive feedbacks between initiatives and institutional reform mixes that drive a system toward certain predefined normative goals. In effect, he plans to plant 'socioecological seeds', that is, the initiation of stakeholder processes in the study areas to create dynamics toward a consensus after the project finishes.

Whereas Panos Petridis would work at a specific locale to embed his research into practice, **Ulli Weisz**'s approach focuses on one sector: searching for co-benefits in the hope that they could be a transformative contribution. The identification and pursuit of 'health co-benefits' aims at utilizing synergies between two important societal concerns: sustainable development and public health. Health gains are used as an additional argument to support climate policy (e.g., WBGU 2011). Current research addresses human health and the scale of organizations. The project 'ClimbHealth', for example, investigates climate and health co-benefits from changes in urban mobility and diet for Austria in both economic terms and in quantifiable terms beyond economics. The results are intended to increase political acceptance for the Austrian transition toward a low-carbon society with respect to urban mobility and diet because climate and health policy concerns are addressed simultaneously. In the project 'Sustainable Care', nursing students are invited and guided to reflect on potentials for sustainable care in hospitals. The students address the question of how treatment and care could be organized in the future as we face demographic transitions and ever-scarcer resources. The result should be a greater awareness of future actors in the health care system working toward the possibilities and conditions of a 'caring society'. These projects are just two examples of what could be termed 'socioecological public health research'. Public health and sustainability might go together when it comes to nutrition, as there is a striking coincidence between the ever-increasing societal metabolism and rising obesity rates. The new field, 'socioecological public health research', should be established with the hope of providing a model for synergetic measures for the transformation toward sustainability. Substantial second-order learning is involved in this project proposal.

What about the messy world of conflict and prevailing military interventions? **Verena Winiwarter** calls for a Social Ecology of warfare and military spending, seeing a necessary link to the powerful subsystem of the military and sustainability. Wars and their effects on the environment—and, vice-versa, the question of resource wars—has been a neglected theme in Social Ecology, despite its affinity with Political Ecology. Never before has humanity conducted so many and such resource-intensive wars than in the 20th century. The nuclear legacy of test sites, production and mining areas will pose a threat to human health and safety and to ecosystems for hundreds of thousands of years. The prognosis of how long cleanup of the US nuclear production site Hanford will take has been revised upward over the past decade, currently standing at the 2060s. The environmental movement and the peace movement were essentially one movement at the beginning; 'no nukes' went beyond a call for peace by implicitly including the environment. This was not a coincidence, and perhaps Social Ecology can support the reemergence of a joint agenda of these movements by conducting research on the long-term socioecological investigation of war and the environment in a transdisciplinary fashion. Without peace, Verena Winiwarter is convinced, there is no sustainability, but there is also no sustainability without peace as wars are enormous gobblers of resources and bring with them environmental destruction.

A Social Ecology of the 21st century will have to develop more diverse ties to the Social Sciences, as **Marina Fischer-Kowalski** points out in a call for meaningful interdisciplinarity, which, after decades, remains a desideratum. Social Ecology lacks a Social Science, an Economics and historians willing to respect and learn from each other. The academic system is averse to this kind of endeavor, even if interdisciplinarity remains a buzzword. Interdisciplinary specialization, differentiated and incompatible reward systems and a lack of shared responsibility for solving problems prevail. Mutual ignorance and status fights between camps dominate the scene. Natural scientists are often not interested in contributing to a publication in journals with an impact factor of 1.5 or lower, which is a typical value among the Historical and Social Sciences.

Meanwhile, the International Social Science Council (ISSC) publishes an *Encyclopedia of the Social Sciences on Sustainability*—a laudable effort, although it has resulted in many incoherent chapters on many seemingly incoherent issues. Integrated assessment models (IAMs) link diverse elements ranging from climate to consumer behavior, but they are so complex that any new module can only be accommodated by the handful of people who know how the links work. Millions of working hours have been sunk into these highly sophisticated modeling tools, often the property of one or a few institutions that make their living on them. These are tools whose basic assumptions are 30 years old and will not change because doing so is unaffordable as it would ruin the economic base of their host institutions. However, scientists still overestimate their own capacity to judge how things work. To make a dire story short, the landscape is dominated by a long history of scientific specialization in theory and methods based uncritically on the premise that differences in the perception of social systems and processes, their development trajectories and options are a matter of values and opinions, not of scientific and rational analysis under conditions of uncertainty. A reform of the scholarly system, of teaching, of funding, of valuing contributions is a prerequisite for a Social Ecology of the 21st century. Marina Fischer-Kowalski makes a very clear call for transformative learning in her perspective. One million Euro could go a long way in this direction, but it would have to be spent differently than is common.

The remaining two contributions are very clear instances of calls for transformational learning. **Robert Groß** and Martin Schmid both propose setting up funds to involve new actors as researchers and contributors. Robert Groß's plan seems like a direct reaction to Marina Fischer-Kowalski's analysis, although they were developed independently. He suggests setting up a 'liminality fund', applying Victor Turner's notion of 'liminality' to Social Ecology and Environmental History. The one million Euro funding scheme will support experimental formats and bring together multidimensional scientific backgrounds, plurality in educational biographies and varieties in lifestyles. Liminality can increase the functionality and adaptability of any social system in transformative periods (Turner 1992). In contrast, any vital interdisciplinary scientific culture is confronted with internal and external forces narrowing diversity. Edgar Zilsel described the internal force he called 'induration' as early as 1930. He argued that constant cooperation of too like-minded scholars leads to constrictions of thinking (Zilsel 1930). Internal restrictions are economically enforced by modern scientific funding systems, which tend to consolidate individual career paths rather than foster their plasticity. Following Victor Turner, (scientific) culture can lose its innovative capacity and potential diversity if funding systems become too powerful. One million Euro would help establish a research program that fosters methodological, theoretical and sociocultural diversity within Social Ecology and Environmental History. The 'liminality fund' would sponsor projects on the current margins of Social Ecology and Environmental History. Researchers from a variety of bio-geographical regions would be sponsored and hence, become involved.

Martin Schmid's DUNA-Fellowships to support Long-Term Socioecological Research (LTSER) and Environmental Histories of Central and South-Eastern Europe are a regional rather than biographical and disciplinary application of third-order learning. The overarching aim of this fellowship program is to improve knowledge and understanding of the interaction of human societies with the environment in Central and South-Eastern Europe from the earliest to most recent times. Fellows would be expected to rewrite a peculiar aspect of the long and multifaceted history of this part of Europe. They would thus contribute to a larger history in which human societies and nature are irresolvable. In turn, this would provide the long-term perspective and socioecological insight required to assess the future sustainable development of the region. The fellowship program would support writing as well as fieldwork and archival research. It should be open for applicants from all disciplines pursuing a historical, long-term and interdisciplinary approach. The research program would help build research capacity in the Danube region, so applicants from these countries would be preferred.

Conclusions

All contributions to the one million Euro exercise show that a Social Ecology for the 21st century will build on its foundations in being global and local, practical and theoretical, inter- and transdisciplinary. It will entail learning to do things better, doing better things and transforming Social Ecology to be able to 'view things differently', to start a process of third-order learning. Although it might seem that metabolic approaches have dominated the examples above, land-use is inextricably linked to the metabolic profiles and requirements of societies; hence, the Social Ecologies of land-use will remain an important area of study with a distinct method development and a different community. To support the transformation into a sustainable society, Social Ecology will not only produce research but will also try to implement it in international bodies such as the Intergovernmental Panel on Climate Change (IPCC) and the United Nations Environment Programme (UNEP), as has been done in the past. Newly emerging institutions might become new targets for such science-policy dialogues, such as the Intergovernmental Platform on Biodiversity and Ecosystem Services (IPBES) and Future Earth. But the Vienna School of Social Ecology will have to remain experimental to remain productive, and it will continue to foster young scholars and students in their quest for a better, more sustainable world. It will thus remain an academic endeavor with strong ties to practical communities rather than become a new non-governmental organization (NGO) itself—unless transformative knowledge will someday suggest this move as beneficial.

References

Argyris, C. (1976). Single-loop and double-loop models in research on decision making. *Administrative Science Quarterly*, 363–375.

Argyris, C. S., & Schön, D. (1996). *Organizational learning II: Theory, method and practice*. Reading, PA: Addison-Wesley.

Blackman, D., Connelly, J., & Henderson, S. (2004). Does double loop learning create reliable knowledge? *The Learning Organization, 11*(1), 11–27.

Fischer-Kowalski, M., Pallua, I., Xenidis, L., & Singh, S. J. (2014). Samothraki. Die Geschichte einer griechischen Insel, die sich aufmachte, ein UNESCO-Biosphärenreservat zu werden. In G. Dressel, W. Berger, K. Heimerl, & V. Winiwarter (Eds.), *Interdisziplinär und transdisziplinär forschen: Praktiken und Methoden* (pp. 239–246). Bielefeld: Transcript Verlag.

Mittelstraß, J. (1984). Zur wissenschaftlichen Rationalität technischer Kulturen. *Physikalische Blätter, 40*(3), 64–68.

Nowotny, H. (2003). Democratising expertise and socially robust knowledge. *Science and Public Policy, 30*(3), 151–156.

Nowotny, H., Scott, P., & Gibbons, M. (2003). Introduction: "Mode 2" Revisited: The new production of knowledge. *Minerva, 41*(3), 179–194.

Raskin, P. D. (2008). World lines: A framework for exploring global pathways. *Ecological Economics, 65*(3), 461–470.

Rittel, H. W., & Webber, M. M. (1973). Dilemmas in a general theory of planning. *Policy Sciences, 4*(2), 155–169.

Schneidewind, U., & Singer-Brodowski, M. (2014). *Transformative Wissenschaft. Klimawandel im deutschen Wissenschafts- und Hochschulsystem* (2nd revised ed.). Marburg: Metropolis-Verlag.

Sieferle, R. P. (1995). Naturlandschaft, Kulturlandschaft, Industrielandschaft. *Comparativ, 4*, 40–56.

Sterling, S. (2011). Transformative learning and sustainability: Sketching the conceptual ground. *Learning and Teaching in Higher Education, 5*, 17–33.

Turner, V. (1992). Prozeß, System, Symbol: Eine neue anthropologische Synthese. In R. Habermas & N. Minkmar (Eds.), *Das Schwein des Häuptlings: Beiträge zur Historischen Anthropologie* (pp. 130–146). Berlin: Wagenbach.

Ward, B., & Dubos, R. (1972). *Only one earth: The care and maintenance of a small planet*. Harmondsworth: Penguin Books Ltd.

WBGU (Wissenschaftlicher Beirat Globale Umweltveränderungen). (2011). *Welt im Wandel: Gesellschaftsvertrag für eine Große Transformation. Hauptgutachten* (2nd edn.). Retrieved from http://www.wbgu.de/fileadmin/templates/dateien/veroeffentlichungen/hauptgutachten/jg2011/wbgu_jg2011.pdf

Winiwarter, V. (2014). Designing trans-disciplinary projects for sustainable development. In S. Brumat, & C. Macchia (Eds.), *DIAnet International School Proceedings*, 2014. The role of natural heritage for the sustainable development of the Danube Region, Gorizia 22nd–31st March 2014 (pp. 45–59). Trieste: EUT Edizioni Università di Trieste.

Zilsel, E. (1930). Soziologische Bemerkungen zur Philosophie der Gegenwart. Der Kampf, 23, 417 f.

Index

A

ABM, *see* Modeling > Agent-based model(ing)

Actor, l, 125, 501, 506, 555, 568. *See also* Agent / Interdisciplinarity / Scientific expertise / Stakeholders / Transdisciplinarity > Transdisciplinary research

  civil society, 581

Africa, 362

  Savannah, 338. *See also* Fire / Grassland

  Sub-Saharan Africa

    Sub-Saharan Africa, crop yield, 320, 324, 362, 370. *See also* Cropland / Yield

    Sub-Saharan Africa, energy and material use, 207. *See also* Energy > Energy consumption/use / Material and energy flow accounting/analysis (MEFA) / Material consumption/use

  Sub-Saharan Africa, feed intake of ruminants, 305. *See also* Feed and fodder / Grazing / Livestock > Ruminant

    Sub-Saharan Africa, material flows, 221. *See also* Material and energy flow accounting/analysis (MEFA) / Material flow accounting/analysis (MFA) / Material flows

    Sub-Saharan Africa, vegetation fires, 339. *See also* Fire / Vegetation fire

Agency, l, 127. *See also* Institution / Organization / Transdisciplinarity > Transdisciplinary research

  human agency, xlii. *See also* Agent

Agent, 127, 492, 501, 511. *See also* Actor / Agency > Human agency / Modeling > Agent-based model(ing) / Stakeholders

Agent Orange, Vietnam (herbicide/chemical weapon), 161. *See also* Legacy

Agrarian metabolism, 434. *See also* Agrarian mode of subsistence / Agrarian society / Metabolic regime > Agrarian metabolic regime / Sociometabolic regime

Agrarian mode of subsistence, 107. *See also* Agrarian metabolism / Agrarian society / Human subsistence / Industrial mode of subsistence / Metabolic regime > Agrarian metabolic regime / Sociometabolic regime

Agrarian society, 37, 70. *See also* Agrarian metabolism / Agrarian mode of subsistence / Metabolic regime > Agrarian metabolic regime / Sociometabolic regime

Agrarian-industrial transition, lv, lvi, 63, 82, 107, 200, 370, 391, 433, 452, 460, 476. *See also* Industrialization / Transition

Agricultural inputs, 96. *See also* Agricultural intensification > Fertilization / Agro-food system / Input intensification / Labor / Pesticide

Agricultural intensification, 449. *See also* Agriculture > Intensive agriculture / Food > Food demand / Grazing > Grazing intensity / High-input agriculture / Input intensification / Land use > Land-use intensification / Output intensification / Pesticide / Soil > Soil fertility / Unintended outcomes/side effects

  fertilization, 82, 106, 317, 368, 404, 406, 434, 440, 480, 483. *See also* Yield

  manure, 435. *See also* Nutrient management / Yield

H. Haberl et al. (eds.), *Social Ecology*, Human-Environment Interactions 5,
DOI 10.1007/978-3-319-33326-7

Thomasmehl (phosphate fertilizer), 468. *See also* Nutrient management / Yield
mechanization, 82, 107, 317, 434. *See also* Industrialization / Labor / Technology
tractors, 155, 368, 434. *See also* Agricultural modernization / Technology
sustainable intensification, 96, 370. *See also* Agriculture > Intensive agriculture / Food > Food demand / Grazing > Grazing intensity / Input intensification / Land use > Land-use intensification / Output intensification / Pesticide / Soil > Soil fertility / Sustainability / Unintended outcomes/side effects
Agricultural involution, 103. *See also* Agrarian-industrial transition / Food > Food supply / Industrialization / Labor > Human labor > Knowledge work / Land use > Land-use change
Agricultural modernization, 443. *See also* Agricultural intensification > Mechanization / Technology
Agricultural productivity, 362. *See also* Agricultural intensification / Food > Food production / Mass production / Productivity / Yield
Agriculture, 350, 548. *See also* Farming
high-input agriculture, 440. *See also* Agricultural intensification / Grazing > Grazing intensity / Land use > Land-use intensification / Pesticide
intensive agriculture, 326. *See also* Agricultural intensification / Grazing > Grazing intensity / Land use > Land-use intensification / Pesticide
organic agriculture, 320. *See also* Sustainability
sustainable agriculture, 505. *See also* Sustainability
Agro-food system, lviii. *See also* Agricultural inputs / Farming / Food > Food production / Knowledge / Transport
Air pollution, 157. *See also* Greenhouse gas (GHG) emissions / Legacy
fossil fuel emissions, 428. *See also* Greenhouse gas (GHG) emissions / Legacy / Material categories > Fossil fuel
silicosis, 157. *See also* Colonial mining / Health / Legacy
sulfur dioxide (SO_2), 157
Animal husbandry, 317, 436. *See also* Grazing > Grazing animals / Livestock / Pastoralism
cow, 351. *See also* Animal poduct / Farming / Livestock
domesticated animals, 322. *See also* Livestock
draught animals, 155, 436, 441. *See also* Agricultural intensification > Mechanization / Farming / Labor > Animal labor / Livestock
goats, 550
Animal product, 318. *See also* Diet / Food / Grazing / Livestock
meat, 40, 296, 318, 352, 439. *See also* Animal husbandry / Diet / Farming / Food
milk, 40, 296, 351, 352, 439. *See also* Animal husbandry / Diet / Farming / Food
Anthrome, 102. *See also* Ecosystem / Habitat / Land use / Socioecological system
Anthropocene, 150, 379. *See also* Colonization of natural processes and systems / Cultural Anthropology / Society-nature interaction / Socioecological system
Archaeology, 412
Archive, 412
Area productivity, 82, 440. *See also* Land use / Productivity / Yield
Arrangements, 132, 138, 162, 413. *See also* Hybrid / Practice / Socio-natural arrangements / Socio-natural practices / Socio-natural site (SNS)
Artifacts, 32, 42, 66, 106, 142, 279, 340, 412, 540. *See also* Biophysical stocks / Buildings / Infrastructure
Atmosphere, 428
Austria, *see* Europe
Autopoiesis, 127
Autopoietic system, 32
Autotrophic organisms, 53. *See also* Energy > Energy forms > Trophic energy / Heterotrophic organisms / Photosynthesis / Species

B

Biodiversity, xliii, 105, 109, 375. *See also* Conservation / Species > Species richness
abundance, 376. *See also* Species > Species richness

alien species, 380. *See also* Species > Species richness
α-, β- and γ-diversity, 377. *See also* Species > Species richness
biodiversity conservation, 317. *See also* Conservation / Species > Species richness
evenness, 377. *See also* Species > Species richness
extinction, 381. *See also* Conservation / Species > Species richness
extinction debt, 381. *See also* Conservation / Species > Species richness
genetic diversity, 377. *See also* Species > Species richness
migration of species, 163, 377. *See also* Species > Species richness
Shannon Diversity Index, 377. *See also* Species > Species richness
speciation, 377. *See also* Species > Species richness
Biohistory, 15. *See also* Society-nature coevolution
Biological productivity, 316. *See also* Agricultural productivity / Area productivity / Productivity
Biomass, *see* Material categories
Biophysical limits, 69
Biophysical stocks, 11, 234. *See also* Artifacts / Buildings / Infrastructure / Livestock / Population
Biophysical structures, 3, 9, 22, 23, 106
Biosphere Reserve, (UNESCO), 544. *See also* Man and the Biosphere (MAB) Reserve
Biotic communities, 316
BMR, *see* Metabolic rate > Basic metabolic rate
Boserup, Ester, 10, 37, 98, 180
Bourdieu, Pierre Félix, xlvii, 138. *See also* Praxeology
Boussingault, Jean-Baptiste, 97
Breeding, 320. *See also* Animal husbandry / Cropping / Livestock / Yield
Brundtland, Gro Harlem, 560. *See also Our Common Future*
Brundtland definition, 568
Brundtland Report, 560. *See also Our Common Future*
Buildings, 278. *See also* Artifacts / Infrastructure / Maintenance

C
Carbon (C), 418, 428
carbon budget, lv, 417. *See also* Europe > Austria, carbon budget
ecosystem carbon budget, 423. *See also* Ecosystem / Europe > Austria, carbon budget / Socioecological system
carbon cycle, 346
global carbon cycle, 101
carbon flows, 153. *See also* Europe > Austria, carbon stocks and flows / Greenhouse gas (GHG) emissions
socioeconomic carbon flows, 420
carbon footprint concept, 112, 431. *See also* Ecological footprint / Material footprint / Water footprint concept
carbon pool, 428
cement production, 428. *See also* Material categories > Nonmetallic minerals > Construction minerals
carbon sequestration, 315
carbon sink, li, 332. *See also* Greenhouse gas (GHG) emissions
terrestrial carbon sink, 153. *See also* Greenhouse gas (GHG) emissions
carbon stock, lv, 153, 423, 430. *See also* Europe > Austria, carbon stocks and flows / Greenhouse gas (GHG) emissions
carbon stock reduction in vegetation, 112. *See also* Greenhouse gas (GHG) emissions
Caring society, 585. *See also* Human health / Public health
Carrying capacity, 69. *See also* Area productivity / Cropping / Ecosystem / Material categories > Biomass / Output intensification / Yield
Cash crop, 354. *See also* Cereal / Cropping / Export / Soy / Trade / Yield
Cereal, 318. *See also* Cash crop / Cropping / Yield
cereal yield, 368. *See also* Cash crop / Cropping / Yield
Chayanov, Alexander V., 99
Christianization, 161
City-hinterland relations, 39, 72, 98, 350, 394. *See also* Local studies / Periphery / Rural community / Transport / Urban planning / von Thünen, Johann Heinrich

Climate change, 346, 561. *See also* Kyoto Protocol / Sustainability
Climate protection, 574. *See also* Sustainability
Climax vegetation, 109. *See also* Ecosystem / Forest / Grassland / Renaturation / Undisturbed
Co-benefits, 560. *See also* Health co-benefits
Coevolution, 6, 50, 128. *See also* Society-nature coevolution
Collapse, 114, 544, 552. *See also* Disaster / Risk spiral / Tipping point
Colonial mining, 156. *See also* Air pollution / Europe > Spain, as colonial power / Indigenous people / Labor > Human labor > Forced labor
  mercury mining, 157. *See also* Air pollution / Europe > Spain, as colonial power / Indigenous people / Labor > Human labor > Forced labor
    Peru, town of Huancavelica, 157. *See also* Air pollution / Europe > Spain, as colonial power / Indigenous people / Labor > Human labor > Forced labor / Material categories > Metals
    Slovenia, town of Idrija, 157. *See also* Air pollution / Material categories > Metals
    Spain, town of Almadén, 156. *See also* Air pollution / Material categories > Metals
  mining-related health issues
    mercury poisoning, 157, 159. *See also* Health / Legacy / Material categories > Metals
    silicosis, 157. *See also* Air pollution / Health / Legacy
  silver mining, 157. *See also* Europe > Spain, as colonial power / Indigenous people / Labor > Human labor > Forced labor / Material categories > Metals
    amalgamation, 157. *See also* Material categories > Metals
    Bolivia, town of Potosí, 159. *See also* Europe > Spain, as colonial power / Indigenous people / Labor > Human labor > Forced labor / Material categories > Metals
Colonization of natural processes and systems, 34, 109, 126, 133, 316, 375, 391, 460. *See also* Disturbance / Human appropriation of net primary production (HANPP) / Land use / Society-nature interaction / Socio-natural arrangements / Socio-natural practices / Unintended outcomes/side effects
  colonizing interventions, 34, 37, 46. *See also* Disturbance / Human appropriation of net primary production (HANPP) / Land use / Society-nature interaction / Socio-natural arrangements / Socio-natural practices / Unintended outcomes/side effects
Communication, 8, 129, 414. *See also* Information and communication technology (ICT)
  interdisciplinary communication, 138. *See also* Actor / Interdisciplinarity / Transdisciplinarity
Compact city, 499. *See also* City-hinterland relations / Periphery / Urban planning
Complex system, xliv, 503
Complexification, 48
Complexity, 544
Consciousness, 143
Conservation, 408. *See also* Environmental protection
Conservation of mass and energy, 106. *See also* Energy > Energy consumption/use / Energy > Energy flows / Material and energy flow accounting/analysis (MEFA) / Material consumption/use / Material flows / Thermodynamics
Constitutive operation, 131
Construction minerals, *see* Material categories > Nonmetallic minerals
Consumer-based principle, 431. *See also* Territory-based principle
Consumption, 83, 239, 241, 542. *See also* Domestic energy consumption (DEC) / Domestic material consumption (DMC) / Energy consumption/use / Material and energy flow accounting/analysis (MEFA) / Material consumption/use
  mass consumption, 86. *See also* Domestic energy consumption (DEC) / Domestic material consumption (DMC) / Energy consumption/use / Material and energy flow accounting/analysis (MEFA) / Material consumption/use
Corporate social responsibility (CSR), 567
Country of origin, 354. *See also* Re-export / Trade
Cow, 351. *See also* Animal husbandry / Animal poduct / Farming / Livestock

Cradle to grave, 253. *See also* Life cycle analysis/assessment (LCA)
Crop residue, 215. *See also* Cereal / Material categories > Biomass / Yield
Crop rotation, 320, 404. *See also* Agriculture / Cash crop / Cereal / Farming / Yield
Cropland, 436. *See also* Agriculture / Cash crop / Cereal / Farming / Yield
  expansion, 368
  yield, 318
Cropping, 324. *See also* Agriculture / Cash crop / Cereal / Farming / Yield
Cropping frequency, 450. *See also* Agriculture / Cash crop / Cereal / Farming / Yield
CSR, *see* Corporate social responsibility
Cultural Anthropology, 5
Cultural landscape, 316, 434
Culture, 128

D
de Saussure, Ferdinand, 97
DE, *see* Domestic extraction
DEC, *see* Energy > Energy consumption/use > Domestic energy consumption
Decision-making, 489, 505
Decoupling of economic growth and resource use, xlii, 12, 80, 205, 230, 381, 516, 569. *See also* Economy / Resource use / Sustainability
Deforestation, 79, 160, 297, 324, 337, 344, 429, 453, 461. *See also* Biodiversity / Forest / Fire / Greenhouse gas (GHG) emissions / Reforestation
Degradation, 79. *See also* Land degradation / Soil > Soil degradation / Yield
DEI, *see* Direct energy input
Demand, 329. *See also* Consumption / Final demand / Food > Food demand / Supply
Dematerialization, 205
Diet, 318. *See also* Animal product / Food / Malnourishment / Staple crop
Direct energy input (DEI), 214. *See also* Energy
Disaster, 401, 523. *See also* Collapse / Tipping point
  nuclear accident (Chernobyl, Ukraine; 1986 / Fukushima, Japan; 2011), 164. *See also* Radioactive contamination / Waste > Hazardous waste management / Waste > Nuclear waste
  tsunami (Nicobar Islands, India, 2004), 523. *See also* Collapse / Tipping point / Humanitarian aid
Discourse analysis, 138
Discrete classification systems, 100. *See also* Modeling / Land cover / Land use
Disturbance, 104. *See also* Colonization of natural processes and systems / Land use / Society-nature interaction / Undisturbed
DMC, *see* Domestic material consumption
Domestic extraction (DE), 12, 220, 234, 242. *See also* Extraction / Extraction of raw materials / Resource extraction / Unused extraction
Domestic material consumption (DMC), 67, 81, 201, 220, 236, 248, 282, 527. *See also* Domestic extraction (DE) / Material and energy flow accounting/ analysis (MEFA) / Material consumption/use / Metabolic profile
Downcycling, 267. *See also* Material flows / Recycling / Resource efficiency / Waste
DPSIR (Driving forces, Pressures, States, Impacts and Responses) model, 15, 385
Drivers, 381, 385. *See also* Biodiversity / Demand / DPSIR-model / Energy > Energy consumption/use / Food > Food demand / Material consumption/ use / Population > Population growth / Pressures / Resource use
Driving force, 219. *See also* Demand / DPSIR-model / Energy > Energy consumption/ use / Food > Food demand / Material consumption/use / Population > Population growth / Pressures / Resource use
Durkheim, David Émile, 7, 37, 150

E
Eco-design, 273. *See also* Buildings / Life cycle analysis/assessment (LCA) / Sustainability
Ecological footprint, 13, 358. *See also* Carbon > Carbon footprint concept / Lifestyle / Material footprint / Sustainability / Water footprint concept
Ecological modernization, 7
Ecological sustainability, 573. *See also* Sustainability
Economy, 399. *See also* Decoupling of economic growth and resource use / Gross domestic product (GDP) / Trade
  capital, 7. *See also* Marx, Karl
    growth-based capitalist economy, 399. *See also* Marx, Karl

circular economy, lii, 259. *See also* Loops / Material flows / Nonrenewable resources / Recycling / Resource efficiency / Resource use
economics
bioeconomics, 13
ecological economics, 11. *See also* Interdisciplinarity
oil price shock, 205. *See also* Material categories > Fossil fuel > Oil/Petroleum
subsistence economy, 6
Economy-wide material flow accounting (EW-MFA), 220, 234, 240, 282. *See also* Material flow accounting/analysis (MFA)
Ecosystem, 315. *See also* Colonization of natural processes and systems / Habitat / Human appropriation of net primary production (HANPP) / Socioecological system
managed ecosystem, 315
agro-ecosystem, 315, 438
terrestrial ecosystem, 434
Ecosystem services, 10, 104, 316, 378
cultural services, 104
provisioning services, 104
regulating services, 104
supporting services, 104
EFA, *see* Energy flow accounting/analysis
Efficiency, 205
eHANPP, *see* Human appropriation of net primary production > Embodied human appropriation of net primary production
Electrification, 83. *See also* Energy > Energy consumption / use / Industrialization / Technology
Emergence, 501
Empathy, 176, 182
Emplotment, 411
Endemic species, 377. *See also* Biodiversity / Conservation / Species
End-of-life waste, 267. *See also* Circular Economy / Loops / Recycling / Waste
Endosomatic metabolism, 66, 228. *See also* Exosomatic metabolism / Feed and fodder / Material categories > Biomass
Energetics, 5
Energy, 9, 212, 350, 422. *See also* Power
energy consumption/use, 63, 199, 455, 489
Domestic energy consumption (DEC), 67, 81, 201, 527. *See also* Metabolic profile
household energy consumption, 497
urban energy consumption, 491
energy efficiency, 341, 439, 443
energy flows, xliii, l, lvii, 107, 527, 541. *See also* Domestic extraction (DE) / Energy > Energy consumption/use > Domestic energy consumption (DEC) / Direct energy input (DEI) / Material and energy flow accounting/analysis (MEFA) / Resource flow
energy flow accounting/analysis (EFA), 212. *See also* Domestic extraction (DE) / Energy > Energy consumption/use > Domestic energy consumption (DEC) / Direct energy input (DEI) / Material and energy flow accounting/analysis (MEFA) / Resource flow
energy forms
bioenergy, 100, 327, 351. *See also* Biomass
chemical energy, 212
final energy, 213
fossil energy, 245
fossil energy, fossil fuel-based energy system, 16. *See also* Material categories > Fossil fuel / Nonrenewable resources / Point resources
geothermal energy, 214, 269. *See also* Energy > Energy forms > Thermal energy
hydropower, 214
hydropower, hydropower plant, 162. *See also* Biodiversity / Riverine landscape, interventions
hydropower, run-of-river hydropower plant, 163. *See also* Biodiversity / Riverine landscape, interventions
hydropower, watermill, 162. *See also* Riverine landscape, interventions > Millstream
kinetic energy, 212. *See also* Energy > Energy forms > Wind power / Kinetic efficiency
mechanical energy, 212, 296
nuclear power, 163, 212. *See also* Disaster > Nuclear accident / Waste > Nuclear waste
potential energy, 212
primary energy, Total primary energy supply (TPES), 213
radiant energy, 212
solar energy, 15, 65, 269
thermal energy, lii. *See also* Energy > Energy forms > Geothermal energy / Thermodynamics

trophic energy, 378. *See also* Biodiversity / Human appropriation of net primary production (HANPP) / Net primary production (NPP) / Species
useful energy, 213
wind power, 214, 269. *See also* Energy > Energy forms > Kinetic energy
Energy return on investment (EROI), 73, 103, 180, 439
energy service, 214. *See also* Transport
energy supply density, 68
energy system, 64
fossil fuel-based energy system, 16. *See also* Energy > Energy forms > Fossil energy / Material categories > Fossil fuel / Nonrenewable resources / Point resources
energy transition, xliv, 63, 80, 169, 204, 269, 273, 455. *See also* Industrialization / Transition
energy unit, 212
British thermal unit (Btu), 213
Calorie (Cal), 213
Joule (J), 212
Kilowatt-hour (kWh), 213. *See also* Power
Tons of coal equivalent (Tce), 213
Tons of oil equivalent (Toe), 213
Environmental History, 128, 149, 391, 411, 418, 434, 459, 475, 587. *See also* Long-Term Socioecological Research (LTSER)
Environmental impacts, 4, 253, 255. *See also* DPSIR-model / Ecosystem services
Environmental management systems, 567
Environmental protection, 573. *See also* Conservation / Sustainability
Environmental relations, 527
Epigraphy, 412
Epistemological, 126
EROI, *see* Energy > Energy return on investment
ERP, *see* European Recovery Programme
Eternity costs, 150, 162
EU, *see* Europe > European Union
Europe
Austria, liv, lvi
Austria, agrarian-industrial transition, 155. *See also* Agrarian-industrial transition
Austria, agricultural transformation in Theyern/Lower Austria, 433. *See also* Agrarian-industrial transition
Austria, carbon budget, 417. *See also* Carbon > Carbon budget
Austria, carbon stocks and flows, 153. *See also* Carbon > Carbon flows / Carbon > Carbon stock
Austria, HANPP and biodiversity, 382. *See also* Biodiversity / Human appropriation of net primary production (HANPP)
Austria, material flows and consumption, 239. *See also* Material and energy flow accounting/analysis (MEFA) / Material consumption/use / Material flow accounting/analysis (MFA) / Material flows
Austria, sewage system of Vienna, 475. *See also* Sewage system
Austria, skiing tourism in Damüls / Vorarlberg, 459. *See also* Skiing / Tourism
Austria, urban planning, residential decisions and energy use in Vienna, 489. *See also* Energy > Energy consumption/use / Residential choice / Urban planning
Austria, workload of farming, 505. *See also* Farming / Labor / Time use > Working time
Central and South-Eastern Europe, 587
European Union, 355
Europe, feed intake of ruminants, 305. *See also* Feed and fodder / Grazing / Livestock > Ruminant
Europe, material stocks, 278. *See also* Material stocks
Finland, Onkalo Nuclear Storage Site, 162. *See also* Waste > Nuclear waste
Germany, colonization of the Donaumoos region/Bavaria, lv, 391. *See also* Colonization of natural processes and systems / Riverine landscape, interventions
Greece, island of Samothraki, 543. *See also* Colonization of natural processes and systems / Riverine landscape, interventions
Spain, as colonial power, 157. *See also* Colonial mining
Western Europe, grazing pressure, 307. *See also* Grazing / Land use / Livestock / Pressures
European Recovery Programme (ERP), 467
Evolution, 377
Evolution of culture, 30

EW-MFA, *see* Economy-wide material flow accounting
Exosomatic metabolism, 64, 228. *See also* Endosomatic metabolism / Material categories
Expectations, 131
Experimental farming, 404. *See also* Agriculture / Farming
Explanation, 138
Export, 240. *See also* Domestic energy consumption (DEC) / Domestic material consumption (DMC) / Import / Physical trade balance (PTB) / Re-export / Trade
Extended phenotype, 31
Externalizations, 567
Extraction, 201. *See also* Domestic extraction (DE) / Resource extraction / Unused extraction
Extraction of raw materials, 11. *See also* Domestic extraction (DE) / Resource extraction / Unused extraction

F

Farmers, 505. *See also* Agriculture / Small-scale farmers / Smallholders
Farming, lvi. *See also* Agriculture / Animal husbandry / Cropping / Land use / Livestock
  experimental farming, 404. *See also* Agricultural intensification > Fertilization / Agriculture / Crop rotation / Yield
  mixed farming, 75. *See also* Agriculture / Animal husbandry / Cropping / Land use / Livestock
Farming system, 433. *See also* Agriculture / Farming / Land use
FCA, *see* Full carbon accounting
Feed and fodder, 297, 321, 351, 400. *See also* Animal husbandry / Cropping / Farming / Livestock / Market feed / Roughage / Yield
Feeding efficiency, 321
Fertilization, *see* Agricultural intensification
Fiber, 351
Field size, 350
Field view, 311. *See also* Geographic information system (GIS)
Final demand, 242, 257. *See also* Demand
Final treatment, 266
Fire, liii, 65. *See also* Deforestation / Hunting and gathering
  fire prevention, 345
  fire regime, 342. *See also* Hunting and gathering
  forest fire, 343
  vegetation fire, liii, 335
  wildfire, 402
First Law of Geography, 311
Food, 203, 350. *See also* Agriculture / Animal husbandry / Animal product / Cereal / Cropping / Farming / Livestock / Staple crop
  food demand, 318. *See also* Malnourishment
  food imports, 363. *See also* Trade
  food production, 317, 439. *See also* Agricultural productivity / Agriculture / Animal husbandry / Animal product / Cereal / Cropping / Farming / Livestock / Productivity / Staple crop
  food security, 116, 299, 371
  food supply, 324
  food web, 381
Forest, 418, 437. *See also* Carbon > Carbon sink / Deforestation / Ecosystem / Reforestation / Renaturation / Wood
Forest fires, 340, 346. *See also* Deforestation / Fire / Hunting and gathering
Forest management, 418
Forest transition, 114, 417, 450. *See also* Transition
Forestry, 350
Fossil energy carriers, *see* Material categories
Fossil fuel, *see* Material categories
FTU, *see* Functional time use
Full carbon accounting (FCA), 429. *See also* Partial carbon accounting (PCA)
Functional time use (FTU), 519. *See also* Time use
  community system, 520
  economic system, 520
  household system, 519
  person system, 519
Future Earth, 588

G

GDP, *see* Gross domestic product
Gender studies, 505
Geographic information system (GIS), 311, 383. *See also* Modeling
Geomorphology, 479
GHG emissions, *see* Greenhouse gas emissions

GIS, *see* Geographic information system
Global, xli
Globalization, 240
Grassland, 297, 321, 442, 461. *See also* Climax vegetation / Ecosystem / Grazing
Grazing, 295, 366, 403, 436, 461. *See also* Animal husbandry / Feed and Fodder / Grassland / Land use / Livestock / Roughage
  grazing animals, 321
  grazing gap, 305
  grazing intensity, 307, 324. *See also* Agricultural intensification / Agriculture > Intensive agriculture
  grazing land, 324. *See also* Grassland / Land use
  overgrazing, 161, 297, 550. *See also* Agriculture > Intensive agriculture / Animal husbandry / Livestock / Soil erosion / Sustainability
Great acceleration, lviii. *See also* Environmental impacts / Greenhouse gas (GHG) emissions / Society-nature interaction / Sociometabolic transition
Greenhouse gas (GHG) emissions, xliii, 12, 297, 370, 562. *See also* Air pollution
  carbon dioxide (CO_2), 153, 428
    CO_2-fertilization, 155
  methane (CH_4), 428
Gross calorific value, 214. *See also* Energy
Gross domestic product (GDP), 229, 247, 318. *See also* Economy / Money
Groundwater, 406

H

Habitat, 307, 315. *See also* Ecosystem
  Habitat, diversity of, 380. *See also* Biodiversity
  habitat heterogeneity, 378. *See also* Biodiversity / Species > Species richness
HANPP, *see* Human appropriation of net primary production
Health, 318. *See also* Human health / Hospital / Public health
Health care systems, 563
Health co-benefits, lviii, 561. *See also* Co-benefits / Human health
Health promotion, 560. *See also* Human health / Hospital
  health promoting hospitals, 566
Heat, 212. *See also* Energy / Thermodynamics
Herd management, 320. *See also* Animal husbandry / Farming / Livestock
Heresy, xlvii
Heterotrophic food chains, 110, 332, 429. *See also* Autotrophic organisms / Energy > Energy forms > Trophic energy / Species
Heterotrophic organisms, 332, 378
Heterotrophic species, 53
Historical Climatology, 413. *See also* Environmental history
Historical legacy, xliv. *See also* Legacy
Historiography, xlvii
Hospital, 563. *See also* Health / Human health / Public health
Household, lvi
Human appropriation of net primary production (HANPP), liv, 53, 110, 332, 364, 381, 453. *See also* Colonization of natural processes and systems / Ecosystem / Energy / Land use / Material categories > Biomass / Net primary production (NPP)
  embodied human appropriation of net primary production (eHANPP), 349. *See also* Land use / Net primary production (NPP)
  HANPP efficiency, 367
  $HANPP_{harv}$, 364
  $HANPP_{luc}$, 364
Human exceptionalism, 7
Human health, 559. *See also* Health / Hospital / Public health
Human population, *see* Population, 106
Human practices, 413. *See also* Arrangements / Practice / Socio-natural arrangements / Socio-natural practices
Human subsistence, 30. *See also* Agrarian mode of subsistence / Economy > Subsistence economy / Industrial mode of subsistence
Humanitarian aid, 531. *See also* Disaster / Subsidy
Hunting and gathering, 6, 64, 525. *See also* Fire
Hybrid, 128
Hybrid system, l, 32

I

IAM, *see* Modeling > Integrated assessment model
ICT, *see* Information and communication technology

iLUC, *see* Land use > Land-use change > Indirect land-use change
Import, 240. *See also* Domestic energy consumption (DEC) / Domestic material consumption (DMC) / Export / Physical trade balance (PTB) / Trade
Indigenous people, 160, 523. *See also* Hunting and gathering
  Guachichites (Mexico), 160. *See also* Hunting and gathering / Latin America > Mexico, colonial mining
Individual, 142. *See also* Actor / Agency / Agent / Stakeholders
Industrial Ecology, 11. *See also* Energy > Energy flows / Material flows
Industrial metabolism, 434. *See also* Industrial mode of subsistence / Industrialization / Metabolic regime > Industrial metabolic regime / Sociometabolic regime
Industrial mode of subsistence, 107. *See also* Agrarian mode of subsistence / Human subsistence / Industrial metabolism / Industrialization / Metabolic regime > Industrial metabolic regime
Industrial revolution, 65. *See also* Industrialization / Transition
Industrialization, li, 80, 200, 441, 461. *See also* Agrarian-industrial transition / Technology
Information and communication technology (ICT), 88, 192. *See also* Communication / Technology
Infrastructure, 42, 78, 84, 278, 403, 408, 460, 471, 476. *See also* Artifacts / Buildings / Maintenance / Transport
Inorganic material, 332
Input intensification, 103. *See also* Agricultural intensification / Labor / Output intensification / Pesticide / Trade / Yield
Input-output (IO), 243, 257. *See also* Trade
  environmentally extended input-output, 257
  input-output ratio, 98
Installed power, 176. *See also* Power / Watt
Institution, 134. *See also* Agency / Organization / Stakeholders
Interdisciplinarity, xlii, 502, 555. *See also* Actor / Communication / Institution / Organization / Scientific expertise / Stakeholder / Transdisciplinarity
  interdisciplinary communication, 138. *See also* Actor / Communication / Transdisciplinarity
Intergovernmental Platform on Biodiversity and Ecosystem Services (IPBES), 588
Intermediate disturbance hypothesis, 378. *See also* Biodiversity / Disturbance
IO, *see* Input-output
IPAT (Impact = Population x Affluence x Technology) equation, 13
IPBES, *see* Intergovernmental Platform on Biodiversity and Ecosystem Services
Irrigation, 405
Island studies, 543. *See also* Local studies

J

Jevons paradox, 97

K

Kinetic efficiency, 176. *See also* Energy > Energy forms > Kinetic energy / Power
Knowledge, 176. *See also* Learning / Labor > Human labor > Knowledge work
  knowledge production ('Mode 2'), 580
  mode-3-science, 581
  orientational knowledge, 578
  positive knowledge, 578
Kyoto protocol, 430. *See also* Climate change / Greenhouse gas (GHG) emissions / Partial carbon accounting (PCA)

L

Labor, xliii, 35, 69, 71, 169, 435. *See also* Functional time use / Time use
  animal labor, 181, 435, 440, 452. *See also* Agricultural intensification > Mechanization / Animal husbandry / Farming / Grazing > Grazing animals / Livestock
    draught animals, 155
  human labor, 169, 435, 440
    child labor, 399, 520
    family work, 177
    forced labor, 159. *See also* Colonial mining
      Mita (forced labor), 159
      serfdom, 178
      slavery, 177
    household-based family labor, 182. *See also* Household
    knowledge work, 188. *See also* Information and communication technology (ICT) / Knowledge
    wage labor, 178

labor, division of, 171. *See also* Functional time use > Household system
labor efficiency, 72
labor, institutional form of, 171, 178
labor productivity, 75, 440. *See also* Productivity
physical labor, 181. *See also* Physical power
working hours, 174, 505, 520. *See also* Functional time use / Time use > Working time
Land, 325, 434
land-related policy, 358
Land abandonment, 350
Land cover, liv, 99, 298, 350, 449
land-cover change, 95
Land degradation, 364. *See also* Degradation / Soil degradation / Yield
Land sparing, 326. *See also* Agricultural intensification / Intensive agriculture / Grazing intensity / Sustainable intensification
Land-sparing vs. land-sharing debate, 100
Land use, xliii, 295, 315, 433
land-use change, 419
indirect land-use change (iLUC), 329
land-use competition, 116, 315
land-use efficiency, 111
land-use intensification, 361. *See also* Agricultural intensification / Agriculture > Intensive agriculture / Grazing > Grazing intensity
land-use intensity, 93, 317, 352, 358, 448
land use, land-use change and forestry (LULUCF), 429. *See also* Forestry
land-use transition, 447. *See also* Transition
Land use, Social science-based analysis of, 101
Latin America, 201
Central America, 156
Latin America, crop yield, 324. *See also* Cropping / Yield
Latin America, energy and material use, 205. *See also* Energy > Energy consumption/use / Material consumption/use
Latin America, feed intake of ruminants, 305. *See also* Feed and fodder / Grazing / Livestock > Ruminant
Latin America, material flows, 224. *See also* Material flows
Latin America, vegetation fires, 339. *See also* Fire
Mexico, colonial mining, 156, 160. *See also* Colonial mining / Material categories > Metals
Mexico, grazing pressure, 307. *See also* Grazing
South America, 156. *See also* Colonial mining / Material categories > Metals
Bolivia, daily working time in the town of Campo Bello, 173, 521. *See also* Functional time use / Labor / Time use
Bolivia, silver mining in the town of Potosí, 159. *See also* Colonial mining / Material categories > Metals
Peru, mercury mining in the town of Huancavelica, 157
Law of diminishing returns, 98
LCA, *see* Life cycle analysis/assessment
Learning, 581. *See also* Knowledge
Double-loop learning, 581
First-order learning, 582
Second-order learning, 582
Single-loop learning, 581
Transformative learning, 582
Legacy, li, lvi, 149, 161, 579
benign, problematic and wicked legacies, 163
legacy effects, 155
Leguminous, 160
Leontief inverse, 258
Life cycle analysis/assessment (LCA), 106, 244, 253, 264, 352
Lifestyle, 66, 490, 527, 534. *See also* Quality of life
Liminality, 587
Limits to growth, xlii. *See also* Consumption / Population > Population growth / Quality of life / Sustainability
Lithosphere, 428. *See also* Soil > Soil formation
Livestock, 40, 106, 295, 321, 400, 402, 435, 540. *See also* Agriculture / Animal husbandry / Farming / Grazing / Labor > Animal labor / Pastoralism
livestock density, 436
livestock feeding efficiency, 318. *See also* Animal husbandry / Feed and fodder / Grazing / Pastoralism
livestock grazing, liii
ruminant, 296, 321. *See also* Animal husbandry / Farming / Grazing
cow, 358. *See also* Animal husbandry / Animal poduct / Farming / Grazing

goats, 550
Local, xli. *See also* Global
Local studies, 136, 523, 539. *See also* Island studies / Rural community / Socioecological system > Local socio-ecological system / Socio-natural site (SNS)
Long-Term Ecological Research (LTER), 152
Long-term effects, 162. *See also* Environmental History / Unintended outcomes/side effects
Long-Term Socioecological Research (LTSER), xli, liv, 16, 128, 149, 411, 434, 587. *See also* Environmental History
Longue durée, 151
Loops, 268. *See also* Circular Economy / Material flows / Recycling / Resource efficiency
Low-carbon society, 585. *See also* Greenhouse gas (GHG) emissions
LTER, *see* Long-term ecological research
LTSER, *see* Long-term socioecological research
Luhmann, Niklas, 129, 142, 414, 569
LULUCF, *see* Land use, land-use change and forestry
Lyell, Charles, 97

M

MAB, *see* Man and the Biosphere programme, United Nations
Maintenance, 280, 394, 403, 408. *See also* Buildings / Infrastructure
Malnourishment, 326. *See also* Diet / Food / Poverty / Staple crop
Malthus, Thomas Robert, 5, 98. *See also* Neo-Malthusian
  Malthusian trap, 10, 95
Man and the Biosphere programme, United Nations
  Man and the Biosphere (MAB) Reserve, vi. *See also* Biosphere Reserve, UNESCO
Management, 101
Man's Role in Changing the Face of the Earth (conference and book title), 5, 99
Maps, 413
Market feed, 305. *See also* Feed and fodder / Roughage
Marsh, George Perkins, 97
Marx, Karl, 5, 170. *See also* Economy > Capital / Metabolic rift / Social metabolism
Mass consumption, 86. *See also* Consumption / Domestic energy consumption (DEC) / Domestic material consumption (DMC) / Energy consumption/use / Material and energy flow accounting/analysis (MEFA) / Material consumption/use
Mass production, 86. *See also* Production / Productivity
Material and energy flow accounting/analysis (MEFA), 43, 105, 200, 215, 342, 428, 433, 527 *See also* Domestic extraction (DE) / Domestic material consumption (DMC) / Energy consumption/use / Material consumption/use / Material flow accounting/analysis (MFA) / Social metabolism / Substance flow analysis (SFA)
Material and energy interactions, 288
Material categories, 234, 268
  biomass, 63, 153, 201, 218, 234, 245, 271, 332, 420, 422. *See also* Energy > Energy forms > Bioenergy / Renewable resources / Ubiquitous resources / Wood
    biomass-based product, 350
    biomass flows, 363, 541
    biomass stocks, 153
    biomass supply, 315
    biomass use, 201
      cascade utilization, 316
    charcoal, 160. *See also* Carbon / Deforestation / Energy / Nonrenewable resources / Point resources / Wood
    peat, 404, 406. *See also* Carbon > Carbon sink / Energy > Energy forms > Bioenergy / Renewable resources / Resource extraction / Ubiquitous resources
  bulk materials, 225
  fossil energy carriers, 235, 245, 269, 542. *See also* Energy > Energy forms > Fossil energy / Nonrenewable resources / Point resources / Resource extraction
  fossil fuel, 80, 107, 225, 422, 584
    coal, 203
    fossil fuel emissions, 428. *See also* Air pollution / Greenhouse gas (GHG) emissions
    Oil / Petroleum, 203
      Oil price shock, 205. *See also* Economy

peat, 404, 406. *See also* Carbon > Carbon sink / Energy > Energy forms > Fossil energy / Nonrenewable resources / Point resources / Resource extraction
industrial minerals, 203, 226
metallic minerals, 235
metal ores, 221, 245. *See also* Nonrenewable resources / Point resources
metals, 84, 203, 269
mercury, 157. *See also* Colonial mining
silver, 157
nonmetallic minerals, 218, 235, 245, 270. *See also* Nonrenewable resources / Point resources
construction minerals, 203, 248, 278
cement production, 428. *See also* Carbon > Carbon pool
Material consumption/use, lii, 63, 199, 584. *See also* Consumption / Domestic material consumption (DMC) / Material and energy flow accounting/analysis (MEFA) / Material flow accounting/analysis (MFA) / Resource use
Material flow accounting/analysis (MFA), 234, 239. *See also* Economy-wide material flow accounting (EW-MFA) / Material and energy flow accounting/analysis (MEFA)
estimation procedures, 235
Material flows, lii, 217, 527, 541, 583. *See also* Domestic extraction (DE) / Domestic material consumption (DMC) / Economy-wide material flow accounting (EW-MFA) / Material and energy flow accounting/analysis (MEFA) / Material categories > Biomass > Biomass flows / Material flow accounting/analysis (MFA) / Resource flow
Material footprint, 236. *See also* Carbon > Carbon footprint concept / Ecological footprint / Water footprint concept
Material input, 268. *See also* Input-Output (IO)
Material intensity (MI), 206, 230. *See also* Decoupling of economic growth and resource use / Domestic material consumption (DMC) / Material consumption/use / Nonrenewable resources / Renewable resources / Resource use
Material requirements, 258
Material stocks, lii, 204, 266, 277, 583. *See also* Stock accounting methods
Meaning, 128
Mechanization, *see* Agricultural intensification
MEFA, *see* Material and energy flow accounting/analysis
Mesquite tree, 160. *See also* Deforestation / Feed and fodder / Material categories > Biomass > Charcoal
Metabolic profile, 64, 66, 73, 83, 221. *See also* Domestic energy consumption (DEC) / Domestic material consumption (DMC) / Energy consumption/use / Material consumption/use / Social metabolism
Metabolic rate, 12, 66, 204, 221. *See also* Social metabolism
basic metabolic rate (BMR), 176
Metabolic regime, 63. *See also* Agrarian-industrial transition / Social metabolism
agrarian metabolic regime, 71
industrial metabolic regime, 80
Metabolic rift, 11. *See also* Marx, Karl
Metabolic transition, 63, 218, 229, 447 *See also* Social metabolism / Transition
global metabolic transition, 231
urban metabolic transition, 478
Metabolism, 218
Metallic minerals / Metals, *see* Material categories
MFA, *see* Material flow accounting/analysis
MI, *see* Material intensity
Military conflicts, 178. *See also* War
Mind, 131
Minerals, *see* Material categories
Mining, *see* Colonial mining
Mistvieh (dung-animal), 296
Mixed farming, 75. *See also* Farming / Land use
Mobility, 66, 73, 78, 86. *See also* Lifestyle / Technology / Transport
Modeling, 128, 435, 491, 501. *See also* Scenario
agent-based model(ing) (ABM), xxv, 135, 491, 501, 511. *See also* Actor / Agent / Stakeholders
distance decay model, 311
global vegetation model, 299
gravity model, 311
integrated assessment model (IAM), 586
integrated socioecological model, 502

multi-regional input-output models (MRIO), 44. *See also* Input-output
participatory modeling, 134, 506. *See also* Actor / Participatory processes / Stakeholders
socioecological interaction model, liv. *See also* Colonization of natural processes and systems / Society-nature interaction
sociometabolic model, lv. *See also* Metabolism / Social metabolism / Socioeconomic metabolism / Sociometabolic regime
Money, 131, 244. *See also* Gross domestic product (GDP)
MRIO, *see* Modeling > Multi-regional input-output model
Multifunctionality, 317

N
Narratives, 414
National accounting, 43
Nature-society boundary, 214. *See also* System boundaries
Neolithic revolution, 65, 164
Neo-Malthusian concept, 6. *See also* Malthus, Thomas Robert
Net primary production (NPP), xlviii, 105, 110, 153, 209, 224, 364, 378. *See also* Ecosystem / Human appropriation of net primary production (HANPP) / Material categories > Biomass / Photosynthesis
NPP_{eco}, xxix
Netting, Robert McCorkle, 99, 462
New Public Health, 560. *See also* Health / Human health / Public health
Nicobar Islands, *see* Southeast Asia > India
Nitrogen (N), 437, 483. *See also* Agricultural intensification > Fertilization > Manure / Cropping
Nominal scales, 100
Nonmetallic minerals, *see* Material categories
Nonrenewable resources, 206, 223.*See also* Energy > Energy forms > Fossil energy / Material categories > Fossil fuel / Point resources / Resource competition
NPP, *see* Net primary production
Nutrient management, 369, 436. *See also* Agricultural intensification > Fertilization / Cropping / Plant nutrients / Yield

O
Object view, 311. *See also* Geographic information system (GIS)
Observation, 138, 414
second-order observation, 138, 414
self-observation, 133
Oceans, 428. *See also* Ecosystem
Ontology, 126
Operational closure, 48
Organization, 564. *See also* Agency / Institution
Ostrom, Elinor, 16
Ottawa Charter, 560
Our Common Future (report title), 560. *See also* Brundtland, Gro Harlem > *Brundtland Report*
Output intensification, 102. *See also* Agricultural intensification / Input intensification / Yield
Outsourcing, 240
Overgrazing, 161. *See also* Agriculture > Intensive agriculture / Animal husbandry / Grazing / Livestock / Soil erosion / Sustainability

P
Partial carbon accounting (PCA), 429. *See also* Full carbon accounting (FCA)
Participatory processes, 556
scenario workshop, 556
social multi-criteria evaluation, 134, 556. *See also* Actor / Modeling > Participatory modeling / Stakeholders / Transdisciplinarity
Pastoralism, 40, 76, 336. *See also* Animal husbandry / Land use / Livestock
PCA, *see* Partial carbon accounting
Pechmann, Heinrich, 403
Perception, 405. *See also* Actor
Periphery, 460. *See also* City-hinterland relations / Rural community
Permanent pastures, 304
Pesticide, 106, 434. *See also* Agricultural inputs / Yield
Philippines, *see* Southeast Asia
Photography, 413. *See also* Pictorial representation
Photosynthesis, 332. *See also* Autotrophic organisms / Net primary production (NPP)
Physical power, 176, 180. *See also* Labor > Physical labor / Power

Physical trade balance (PTB), 220, 248. *See also* Material and energy flow accounting/analysis (MEFA) / Material flows / Trade
Physiological combustion, 268
Pictorial representation, 413. *See also* Photography
Planning, 579. *See also* Sustainability
Plant nutrients, 437. *See also* Nutrient management
Plutonium (Pu), 162. *See also* Energy > Energy forms > Nuclear power / Radioactive contamination / Waste > Nuclear waste
Point resources, 223. *See also* Energy > Energy forms > Fossil energy / Material categories / Nonrenewable resources
Political Ecology, 586
Politics, 397
Pollen, 412
Population, 96, 204, 540. *See also* Statistical data > Census data
  population density, 37, 349, 433
  population growth, 10, 83, 205, 367, 447. *See also* Demand / Food > Food demand / Limits to growth / Resource use / Statistical data > Census data / Sustainability / Waste
  population projections, 456. *See also* Scenario / Limits to growth
Post-fossil world, 583. *See also* Energy / Material categories / Renewable resources
Potential vegetation, 110, 332. *See also* Human appropriation of net primary production (HANPP) / Net primary production (NPP)
Poverty, 370. *See also* Food / Malnourishment / Ragpicker
Power, 213. *See also* Energy / Watt
Practice, 132, 138. *See also* Arrangements / Human practices / Socio-natural arrangements / Socio-natural practices / Socio-natural site
Practices and arrangements, 464. *See also* Arrangements / Human practices / Socio-natural arrangements / Socio-natural practices / Socio-natural site
Praxeology, 138. *See also* Bourdieu, Pierre Félix
Pressure indicators, 109. *See also* DPSIR-model / Land use
Pressures, 381, 385. *See also* DPSIR-model / Land use
Primary production, 332, 350. *See also* Human appropriation of net primary production (HANPP) / Net primary production (NPP)
Processing, 245
Product life cycle, 45. *See also* Consumption / Domestic extraction (DE) / Extraction / Life cycle analysis/assessment (LCA) / Loops / Production / Recycling / Waste
Production, 241
  mass production, 73, 86. *See also* Supply
Productivity, 82, 109, 206, 440. *See also* Output intensification / Production / Supply / Yield
  agricultural productivity, 362. *See also* Agricultural intensification > Fertilization / Yield
  area productivity, 71, 75, 76, 440, 444. *See also* Land use / Yield
  labor productivity, 75, 82, 86, 440, 443. *See also* Labor
Propaganda, 395, 397. *See also* Military conflicts / Politics / War
Protein, 319. *See also* Diet / Food
Provision planning, 571. *See also* Planning
PTB, *see* Physical trade balance
Public debate, 395. *See also* Actor / Politics
Public health, 564. *See also* Health / New Public Health / Sanitation / Sewage system
Purposeful interventions, 109. *See also* Colonization of natural processes and systems / Society-nature interaction

Q

Quality management, 570
Quality of life, 498, 507, 511, 562. *See also* Lifestyle

R

Radioactive contamination, 162. *See also* Disaster > Nuclear accident / Energy > Energy forms > Nuclear power / Legacy / Plutonium / Waste > Nuclear waste
  Chernobyl, Ukraine, 164
  Fukushima, Japan, 164
  Hanford Site for nuclear production, United States, 586. *See also* Energy > Energy forms > Nuclear power / Legacy / Plutonium / Waste > Nuclear waste

Ragpicker, 261. *See also* Poverty
Raster data, 312. *See also* Geographic information system (GIS)
Raw material consumption (RMC), 236, 248. *See also* Material consumption/use / Material flows
Raw material equivalents (RMEs), 12, 44, 236, 242. *See also* Export / Import / Material consumption/use / Material flow accounting/analysis (MFA) / Material flows / Trade
Rebound effect, 97, 263
Recycling, 259, 262, 278. *See also* Downcycling / Reuse / Waste
Re-export, 354. *See also* Country of origin / Export / Trade
Reforestation, 10, 429. *See also* Carbon > Carbon stocks / Deforestation / Forest / Renaturation / Society-nature interaction
Regime stability, 141
Remote sensing, 95, 99, 298, 337, 350
Renaturation, 392. *See also* Reforestation / Riverine landscape, interventions
Renewable resources, 206, 218. *See also* Material categories > Biomass / Resource competition / Ubiquitous resources
Reproduction of the self, 172. *See also* Lifestyle / Quality of life / Self-reproduction
Residential choice, 490. *See also* Lifestyle / Quality of life
Residential satisfaction, 493
Residual sink, 430
Resilience, 552
Resource competition, 316. *See also* Extraction / Nonrenewable resources / Renewable resources
Resource efficiency, 12, 263. *See also* Circular Economy / Loops / Recycling
Resource endowment, 224. *See also* Nonrenewable resources
Resource extraction, 44, 583. *See also* Domestic extraction (DE) / Extraction / Extraction of raw materials
Resource flow, xliii. *See also* Energy > Energy flows / Material flows
Resource use, 203, 583. *See also* Decoupling of economic growth and resource use / Material consumption/use
Reuse, 260. *See also* Downcycling / Recycling
Ricardo, David, 5, 98
Risk, 7. *See also* Collapse / Tipping point
Risk spiral, 35, 48, 114, 165, 406, 407, 472. *See also* Collapse / Colonization of natural processes and systems / Society-nature interaction / Tipping point
Riverine landscape, interventions, *See also* Biodiversity / Colonization of natural processes and systems / Ecosystem / Soil erosion / Wind erosion
  canal, 398
  drainage of bogs/wetlands, 161, 391, 395
  flooding, 406
  hydropower plant, 163. *See also* Biodiversity / Energy > Energy forms > Hydropower
    run-of-river hydropower plant, 163
  millstream, 162
    watermill, 162
  river regulation (Danube), 398, 476
  waterscape, 476
RMC, *see* Raw material consumption
RMEs, *see* Raw material equivalents
Roughage, 297, 305, 321. *See also* Animal husbandry / Feed and fodder / Grazing / Livestock
Routines, 132
Rural community, 137, 436, 459, 523, 539, 543, 547. *See also* Agriculture / City-hinterland relations / Farming / Infrastructure / Local studies / Periphery / Transport

S

Samothraki, Greece, *see* Europe > Greece
Sanitation, 476. *See also* Sewage system / Wastewater
Scale interactions, 524, 537. *See also* Globalization / Local studies / Rural community
Scenario, xliv. *See also* Modeling
  Scenario, agricultural production in Hainfeld, Austria by 2025, 506, 514. *See also* Agriculture / Europe > Austria / Farming / Modeling > Agent-based model(ing)
  scenario, global food supply by 2050, 318. *See also* Food / Modeling
  scenario, household residential decisions and energy use in Vienna, Austria by 2050, 495. *See also* Europe > Austria / Household / Modeling > Agent-based model(ing) / Residential choice

scenario, recycling in the EU by 2020, 284. *See also* Europe > European Union / Mineral categories / Recycling / Waste
Scientific expertise, 400, 480, 502, 506, 578. *See also* Interdisciplinarity / Transdisciplinarity
Sediments, 412. *See also* Soil > Soil formation
Self-reproduction, 172, 519. *See also* Reproduction of the self
Serial sources, 414
Service lifetime, 267. *See also* Stock accounting methods
SES framework, *see* Social-Ecological Systems framework
SET, *see* Socioecological transition
Sewage system, lvi, 475. *See also* Infrastructure / Sanitation / Wastewater
SFA, *see* Substance flow analysis
Shifting cultivation, 75, 335. *See also* Swidden agriculture
Short-rotation coppice, 328. *See also* Crop rotation / Grassland
Sieferle, Rolf Peter, 9, 15, 30, 578
Skiing, 464. *See also* Colonization of natural processes and systems / Infrastructure / Tourism
  artificial snowmaking, 470
  Skischaukel, 469. *See also* Tourism
  ski lifts, 465. *See also* Colonization of natural processes and systems / Infrastructure / Tourism / Transport
  snow groomer, 467. *See also* Colonization of natural processes and systems / Tourism
Small-scale farmers, 509. *See also* Farmers / Smallholders
Smallholders, 436. *See also* Farmers / Small-scale farmers
Smith, Adam, 5, 170
SNS, *see* Socio-natural site
Social construction, 126
Social control, 399. *See also* Politics / Legacy
Social metabolism, ix, 11, 240, 435, 527, 569. *See also* Metabolism / Social metabolism / Socioeconomic metabolism
Social ontology, 130
Social-Ecological Systems (SES) framework, 17
Society-nature coevolution, 15, 22. *See also* Biohistory / Coevolution / Society-nature interaction / Socio-natural arrangements / Socio-natural practices / Socio-natural site (SNS)
Society-nature interaction, liii, 19, 126, 376. *See also* Colonization of natural processes and systems / Social metabolism / Society-nature coevolution / Socio-natural arrangements / Socio-natural practices / Socio-natural site (SNS)
Socio-natural arrangements, 54, 393, 403, 407. *See also* Arrangements / Society-nature interaction / Hybrid / SNS
Socio-natural practices, 54, 403, 407. *See also* Human practices / Hybrid / Practice / Society-nature interaction / Socio-natural site (SNS)
Socio-natural site (SNS), xx, 138, 149, 459. *See also* Local studies / Society-nature interaction / Socioecological system
Socioecological legacies, 476, 482. *See also* legacy
Socioecological system, lvii, 136, 502. *See also* Ecosystem / Local studies / Social-Ecological Systems (SES) framework / System
  local socioecological system, 524, 537, 546
Socioecological theory, 286
Socioecological transition (SET), xliii, 18, 155, 425, 460, 552. *See also* transition
Socioeconomic activity, 212. *See also* Energy
Socioeconomic metabolism, l, li, 105. *See also* Metabolism / Social metabolism
Sociometabolic regime, 19, 448, 452. *See also* Metabolism / Social metabolism
Soil, 75, 156, 406
  soil-sealing, 380
  soil degradation, 110, 327. *See also* Degradation / Land degradation / Yield
  soil erosion, 400, 401, 536, 550. *See also* Degradation / Land degradation / Overgrazing / Unintended outcomes/ side effects / Wind erosion
  soil fertility, 434. *See also* Agricultural intensification > Fertilization / Yield
  soil formation, 161. *See also* Lithosphere / Sediments
Somerville, Mary, 97
Source, 411
Southeast Asia
  India, 524
    India, grazing pressure, 307. *See also* Grazing
    India, Nicobar Islands, lvii, 523. *See also* Disaster > Tsunami / Humanitarian aid

Philippines, lv, 447
Southeast Asia, vegetation fires, 339. *See also* vegetation fires
Soy, 354. *See also* Cash crop / Country of origin / Trade
Spatial autocorrelation, 311. *See also* Modeling
Spatial disconnect, 353
Species, 315. *See also* Biodiversity / Conservation
autotrophic organisms, 32. *See also* Energy > Energy forms > Trophic energy / Heterotrophic organisms / Photosynthesis
endemic, 377, 525, 536. *See also* Biodiversity / Conservation
heterotrophic species, 53. *See also* Energy > Energy forms > Trophic energy / Heterotrophic food chains
species richness, 377. *See also* Biodiversity / Conservation
species-energy hypothesis, 376, 378. *See also* Biodiversity
Stakeholders, 127, 545, 555, 568. *See also* Actor / Institution / Organization / Transdisciplinary research
Ständestaat (corporative state), 465
Standing crops, 105. *See also* Cropping
Staple crop, 349. *See also* Cereal / Cropping / Diet / Food / Trade
Statistical analysis, 414
Statistical data, xliv
census data, 101, 298, 312, 493. *See also* Population
Franziscean Cadaster, 419, 435. *See also* Land use / Population / Yield
uncertainties, 302
Stock accounting methods, 280. *See also* Service lifetime
Stock dynamics, 281. *See also* Material stocks
Stocks and flows, 540, 546. *See also* Material flows / Material stocks
Stocks policy, 287. *See also* Material stocks
Structural coupling, 32, 129, 142
Structure, 125
Subsidy, 82, 453, 492, 510, 515, 535, 547. *See also* Economy / Money
Substance flow analysis (SFA), 428, 429
Supply, 329. *See also* Demand / Production / Productivity
Surface, 326. *See also* Land cover / Land use
Sustainability, 87, 232, 460. *See also* Agriculture > Sustainable agriculture / Decoupling of economic growth and resource use / Ecological sustainability / Recycling
sustainable development, xlii, 277, 560
sustainable development, education for, 580. *See also* Knowledge / Learning
sustainability problem, 212, 564
sustainability sciences, 20
sustainability transformation, 583
sustainability transition, xlii, xlix
sustainability triangle, 134, 507, 511, 516, 570
for hospitals, 570
Swidden agriculture, 113. *See also* Shifting cultivation
Symbolic, 128
Symbolic structures, 130
System, 125, 398. *See also* Socioecological system
System boundaries, 234, 241, 264, 340, 539. *See also* Nature-society boundary / Socioecological system
Systematic, 408
Systemic feedback, 315, 327
Systemic imperatives, 141
Systems theory, 126. *See also* Socioecological system

T

Tainter, Joseph A., 51, 165, 552
Technical combustion, 268
Technological change, 434. *See also* Industrialization
Technological development, 5
Technology, 55, 434, 465. *See also* Industrialization
Teleconnections, 350
Terrestrial vegetation, 428
Territory, 9
Territory-based principle, 431. *See also* Consumer-based principle
Textual intention, 411
Thermodynamics, 212. *See also* Conservation of mass and energy / Energy / Heat
Entropy Law, 212
Tier approach, 430
Time sovereignty, 499. *See also* Functional time use
Time use, lvi, 172, 498, 505. *See also* Functional time use
working time, 42, 521. *See also* Labor > Working hours

Tipping point, lvii, 547, 552. *See also* Collapse / Disaster / Risk spiral
Topographical heterogeneity, 378. *See also* Biodiversity
Total primary energy supply (TPES), 213. *See also* Energy > Energy forms > Primary energy
Tourism, 459, 548. *See also* Lifestyle / Quality of life / Rural community / Skiing
TPES, *see* Total primary energy supply
Trade, xliii, 44, 203, 224, 239, 349. *See also* Economy / Export / Import / Material flows / Physical trade balance (PTB) / Transport
  bilateral trade matrix, 349, 353
  international trade, liii
Trade
  Trade-off, 315
Transdisciplinarity, xliii, 134, 544. *See also* Actor / Communication / Institution / Interdisciplinarity / Organization / Participatory processes / Scientific expertise / Stakeholder
  transdisciplinary research, x, 506, 555, 563, 557
    transdisciplinary research, limitations of, 557
Transition
  agrarian-industrial transition, lv, lvi, 65, 82, 107, 200, 370, 391, 434, 452, 460, 476, 591
  energy transition, xliv, 80, 169, 204, 273, 597 *See also* Energy / Industrialization / Technology
  forest transition, 114, 417, 450 *See also* Forest
  land-use transition, 447. *See also* Land use
  metabolic transition, 63, 218, 229, 447. *See also* Metabolism / Social metabolism
    global metabolic transition, 231
    urban metabolic transition, 478. *See also* Metabolism / Social metabolism
  socioecological transition (SET), ix, xli, 18, 155, 425, 460, 552
  sustainability transition, vi, xli. *See also* Sustainability
Transitions management, 18
Transport, 39, 71, 73, 98, 271, 394, 405, 441, 469, 496, 547, 562. *See also* Energy service / Greenhouse gas (GHG) emissions / Infrastructure / Material categories > Fossil fuels / Trade / Urban planning
  limitation, 394
  transportation networks, 278
Tree-rings, 412
Truth, 411

U

Ubiquitous resources, 223. *See also* Biomass / Point resources / Renewable resources
Undisturbed, 152. *See also* Climax vegetation / Colonization of natural processes and systems / Disturbance / Ecosystem / Land use / Society-nature interaction
Unintended outcomes/side effects, 109, 393, 407. *See also* Air pollution / Colonization of natural processes and systems / Degradation / Land use / Risk spiral / Soil > Soil erosion / Wind erosion
Unit of Analysis, 539
Unused extraction, 236. *See also* Domestic extraction (DE) / Extraction / Extraction of raw materials / Material and energy flow accounting/analysis (MEFA) / Material flow accounting/analysis (MFA) / Resource extraction
Upstream, 246
Upstream flow, 215. *See also* Energy
Urban planning, 489. *See also* City-hinterland relations / Periphery
Urbanization, 476. *See also* City-hinterland relations / Periphery
Use value, 170

V

Vector data, 312. *See also* Geographic information system (GIS)
Vegetation fire, liii, 335. *See also* Deforestation / Fire / Hunting and gathering
von Humboldt, Alexander, 97
von Thünen, Johann Heinrich, 98
  von Thünen rings, 98. *See also* Farming / Land use / Transport

W

War, 397, 402, 406. *See also* Military conflicts / Politics / Propaganda
Waste, 261, 278. *See also* Downcycling / Recycling / Reuse
  hazardous waste management, 164

night soil, 479. *See also* Agricultural intensification > Fertilization / Sanitation / Sewage system / Wastewater
nuclear waste, 162. *See also* Energy > Energy forms > Nuclear power / Plutonium
Onkalo Nuclear Storage Site, Finland, 162. *See also* Energy > Energy forms > Nuclear power / Plutonium
waste, discharge of, 476. *See also* Recycling / Sewage system
waste, prevention of, 272. *See also* Downcycling / Recycling / Reuse
Wastewater, 475. *See also* Sewage system / Sanitation
Water footprint concept, 112. *See also* Carbon > Carbon footprint concept / Ecological footprint / Material footprint
Water pipeline, 480. *See also* Infrastructure / Sanitation / Sewage system
Water supply, 480. *See also* Food supply
Waterscape, 476. *See also* Riverine landscape, interventions
Watt (W), 213. *See also* Power
WCED, *see* World Commission on Environment and Development
WHO, *see* World Health Organization
Wicked problems, 578
Wildfire, 402. *See also* Fire / Vegetation fire
Wind erosion, 400, 401. *See also* Degradation / Land degradation / Soil > Soil degradation / Soil > Soil erosion / Unintended outcomes/side effects
Wind power, 214. *See also* Energy > Energy forms
Wood, 422. *See also* Energy > Energy forms > Bioenergy / Material categories > Biomass
Wood resources, 160. *See also* Energy > Energy forms > Bioenergy / Material categories > Biomass
Working time, 42, 521. *See also* Functional time use (FTU) / Labor / Time use
World Commission on Environment and Development (WCED), 560
World Health Organization (WHO), 560
World War II, lvi, xlv. *See also* Military conflicts / Politics / War

Y

Yield, 102, 358. *See also* Agriculture / Cereal / Cropping / Farming / Input intensification / Output intensification / Staple crop
cereal yield, 368. *See also* Agriculture / Cereal / Cropping / Farming / Staple crop
yield gap, 369. *See also* Agriculture / Cropping / Farming
yield gap analysis, 111

The manufacturer's authorised representative in the EU is Springer Nature Customer Service Centre GmbH, Europaplatz 3, 69115 Heidelberg, Germany. If you have any concerns regarding our products, please contact ProductSafety@springernature.com

Printed and bound by CPI Group (UK) Ltd, Croydon, CR0 4YY
05/10/2026
02228127-0018